Advanced Two-Dimensional Nanomaterials for Environmental and Sensing Applications

Advanced Two-Dimensional Nanomaterials for Environmental and Sensing Applications provides state-of-the-art progress developments in the design strategies of 2D-based nanomaterials. It covers specific focused applications in respective environmental challenges posed by pollutants such as chemical gases, bacterial and microbial, textile dyes, pharmaceutical antibiotics, agricultural pesticides, and toxic heavy metals in water and air contaminations. It elaborates the applications of 2D nanomaterials in the context of technologies such as sensing and detection to monitor pollutants, as well as photocatalysis and adsorption for the removal of pollutants.

Features:

- Elaborates the applications of 2D nanomaterials in the context of sensing and detection to monitor pollutants, as well as photocatalysis and adsorption for the removal of pollutants.
- Focuses on environmental pollutants detection, removal or remediation, and monitoring device fabrications.
- Discusses materials of specific dimension (2D).
- Covers both water and air remediation.
- Includes photocatalytic degradation and antimicrobial disinfection.

This book is aimed at graduate students and researchers in chemical and civil engineering, materials science, and nanomaterials.

Advanced Two-Dimensional Nanomaterials for Environmental and Sensing Applications

Edited by

Peter R. Makgwane, Naveen Kumar, and David E. Motaung

CRC Press
Taylor & Francis Group
Boca Raton London New York

CRC Press is an imprint of the
Taylor & Francis Group, an **informa** business

Designed cover image: https://www.shutterstock.com/image-photo/scanning-electrong-microscopy-sem-image-graphene-2267423953

First edition published 2025
by CRC Press
2385 NW Executive Center Drive, Suite 320, Boca Raton FL 33431

and by CRC Press
4 Park Square, Milton Park, Abingdon, Oxon, OX14 4RN

CRC Press is an imprint of Taylor & Francis Group, LLC

ISBN: 978-1-032-45880-9 (hbk)
ISBN: 978-1-032-56736-5 (pbk)
ISBN: 978-1-003-43694-2 (ebk)

DOI: 10.1201/9781003436942

Typeset in Times
by codeMantra

Contents

About the Editors

Peter R. Makgwane is currently a Full Professor at the University of South Africa (UNISA) Institute of Catalysis and Energy Solutions (ICES) and an Extraordinary Professor of Chemistry at the University of the Western Cape (South Africa). He previously worked as a Principal Scientist (2011–2023) at the Council for Scientific and Industrial Research (CSIR, South Africa) in catalysis and advanced energy materials. He has a PhD in Chemistry specializing in heterogeneous catalysis from Nelson Mandela University (South Africa, 2010). He was a guest editorial board member for the *Journal of Nanotechnology and Nanoscience* (2014) and *MDPI Catalyst* (2021) in nanocatalysis and photocatalysis thematic issues. He published books for several renowned publishers on advanced materials spheres with diverse applications in environmental remediations, energy, and chemical conversions. Professor Makgwane is a well-published scholar of articles and book chapters in heterogeneous catalysis and photocatalysis with focused applications in renewable chemicals, energy, and environmental remediation.

Naveen Kumar is currently an Associate Professor in the Department of Chemistry, Maharshi Dayanand University, Rohtak (India), and has 18 years of research experience. He is actively engaged in the field of material chemistry with focused research expertise in semiconductor photocatalysis, nanocomposite materials, and electrochemical sensing. He has published more than 110 articles in the journals of high repute. He also worked at Universitat Politècnica de València, Valencia (Spain), in an international research project entitled "Development of a new generation CIGS based solar cells" [NANICIS-269279] in 2013 and 2014. He has presented various research papers at various national and international conferences and also delivered invited lectures at various institutes. Prof. Kumar has published two books as editor with renowned publishers.

David E. Motaung is a Full Professor of Physics at the University of the Free State (South Africa). He previously worked as the Principal Scientist for the Council for Scientific and Industrial Research (CSIR, South Africa) in 2007–2019 and also as a Full Professor of Physics (2019–2021) at the University of Limpopo, South Africa. He obtained his PhD in Physics from the University of the Western Cape, South Africa, in 2011. He has been rated in the Top 2% of Scientists in the World by Stanford University in 2022 and 2023. He has published more than 130 research articles in peer-reviewed journals including three review articles. He has also published more than 15 book chapters and edited one book on metal oxide semiconductors for gas chemical sensing and environmental monitoring.

Contributors

Dana Susan Abraham
Department of Chemistry
Central University of Kerala
Kasaragod, India

Heba Ameen
Materials Testing and Surface Chemical Analysis Laboratory, Chemistry Division
National Institute for Standards
Giza, Egypt

Hosakote Shankara Anusha
Department of Environmental Sciences
JSS Academy of Higher Education and Research
Mysuru, India

Nour F. Attia
Gas Analysis and Fire Safety Laboratory, Chemistry Division
National Institute for Standards
Giza, Egypt

Vellaichamy Balakumar
Department of Chemistry
Sri Ramakrishna College of Arts & Science
Coimbatore, India

Swati Bansal
Department of Chemistry
Maharshi Dayanand University
Rohtak, India

Margandan Bhagiyalakshmi
Department of Chemistry
Central University of Kerala
Kasaragod, India

Mandla B. Chabalala
Department of Physics, School of Science, CSET
University of South Africa
Johannesburg, South Africa

Vongani Chauke
Centre for Nanostructures and Advanced Materials (CeNAM)
Council for Scientific and Industrial Research (CSIR)
Pretoria, South Africa

Steve R. Dima
Next Generation Enterprises and Institutions Cluster
Council for Scientific and Industrial Research
Pretoria, South Africa
and
Department of Physics
University of Venda
Thohoyandou, South Africa

Sally E. A. Elashery
Department of Chemistry, Faculty of Science
Cairo University
Giza, Egypt

Lerato Hlekelele
Centre for Nanostructures and Advanced Materials (CeNAM)
Council for Scientific and Industrial Research (CSIR)
Pretoria, South Africa

Chen Hou
School of Environmental Science and Engineering
Shaanxi University of Science and Technology
Xi'an, PR China

Izaz ul Islam
Department of Chemistry
Abdul Wali Khan University
Mardan, Pakistan

Malefetsane P. Khesuoe
Department of Chemistry
Cape Peninsula University of Technology
Cape Town, South Africa

Suresh Babu Naidu Krishna
Department of Biomedical and Clinical Technology
Durban University of Technology
Durban, South Africa

Naveen Kumar
Department of Chemistry
Maharshi Dayanand University
Rohtak, India

Ponnaiah Sathish Kumar
Department of Physics and Chemistry
Daegu Gyeongbuk Institute of Science and Technology (DGIST)
Daegu, South Korea

Suresh Kumar
Department of Chemistry
Kurukshetra University
Kurukshetra, India

Modjadji R. Letsoalo
Department of Physics
University of Venda
Thohoyandou, South Africa

Potlako John Mafa
Institute for Nanotechnology and Water Sustainability, College of Science, Engineering and Technology
University of South Africa
Johannesburg, South Africa

Eric N. Maluta
Department of Physics
University of Venda
Thohoyandou, South Africa

Peter R. Makgwane
Institute for Catalysis and Energy Solutions, College of Science, Engineering and Technology
University of South Africa
Johannesburg, South Africa

Mangaka C. Matoetoe
Department of Chemistry
Cape Peninsula University of Technology
Cape Town, South Africa

Lindani Mdlalose
Centre for Nanostructures and Advanced Materials (CeNAM)
Council for Scientific and Industrial Research (CSIR)
Pretoria, South Africa

Tshegofatso M. Modungwe
Department of Physics, School of Science, CSET
University of South Africa
Johannesburg, South Africa

Sefako J. Mofokeng
Department of Physics, School of Science, CSET
University of South Africa
Johannesburg, South Africa

Teboho P. Mokoena
Department of Physics
University of the Free State (Qwaqwa Campus)
Phuthaditjhaba, South Africa

Fokotsa V. Molefe
Department of Physics
Tshwane University of Technology
Pretoria, South Africa

Luyanda L. Noto
Department of Physics, School of Science, CSET
University of South Africa
Johannesburg, South Africa

Fredrick O. Okumu
Department of Physical Sciences
Jaramogi Oginga Odinga University of Science and Technology
Bondo, Kenya

Sharangouda J. Patil
Department of Zoology
NMKRV College for Women (Autonomous)
Bengaluru, India

Jijoe Prabagar Samuel
Department of Environmental Sciences
JSS Academy of Higher Education and Research
Mysuru, India

Revanasiddappa
Department of Chemistry
PES University
Bengaluru, India

Bogala Mallikharjuna Reddy
Center for Research, Innovation, Development, and Applications (CRIDA)
Jaiotec Labs (OPC) Private Limited,
Amaravati, Andhra Pradesh, India

Keerthana Sahadevan
Department of Chemistry
Central University of Kerala
Kasaragod, India

Ankita Saini
Department of Applied Sciences, Ganga Institute of Technology and Management, Kablana,
Jhajjar, Haryana- 124104, India

Mohit Saini
Department of Chemistry
Kurukshetra University
Kurukshetra, India

Govindaraj Sangami
Department of Chemistry
Sri Ramakrishna College of Arts & Science
Coimbatore, India

Soham Sarkar
Department of Computer Science and Engineering
PES University, Electronics city Campus
Bengaluru, India

Bhavna Saroha
Department of Chemistry
Kurukshetra University Kurukshetra
Haryana, India

Keiko Sasaki
Department of Earth Resources and Engineering
Kyushu University
Fukuoka, Japan

Katekani Shingange
DSI/Mintek Nanotechnology Innovation Centre
Johannesburg, South Africa
and
Department of Physics
University of the Free State
Bloemfontein, South Africa

Harikaranahalli Puttaiah Shivaraju
Department of Environmental Sciences
JSS Academy of Higher Education and Research
Mysuru, India
and
Center for Water, Food and Energy, GREENS trust
Karnataka, India

Rabbia Shoaib
Department of Chemistry
Government College University Faisalabad
Faisalabad, Pakistan

Makgamathe J. Sithole
Department of Physics, School of Science, CSET
University of South Africa
Johannesburg, South Africa

Pebetsi Thokwane
Department of Physics, School of Science, CSET
University of South Africa
Johannesburg, South Africa

Mari Vinoba
Kuwait Institute for Scientific Research
Safat, Kuwait

Elakkiya Venugopal
PG and Research Department of Biotechnology
Kongunadu Arts and Science College
Coimbatore, India

Divya Vinod
Department of Environmental Sciences
JSS Academy of Higher Education and Research
Mysuru, India

Cong Wang
School of Environmental Science and Engineering
Shaanxi University of Science and Technology,
Xi'an, PR China

Lan Wang
School of Environmental Science and Engineering
Shaanxi University of Science and Technology
Xi'an 710021, PR China

Wei Zhang
School of Environmental Science and Engineering
Shaanxi University of Science and Technology
Xi'an, PR China

Preface

The world has over a century been undergoing industrial evolution, which is accompanied by the conversion of natural raw materials into society's needy products for their improved living standards. This has brought about a huge problem related to the generation of complex pollutants matrix entering the ecosystem. The generated pollutants at the product's production stage and also later at the end-user lifespan of the products are associated with unregulated waste disposal without measures taken to eliminate or minimize their environmental impact consequences. The untreated discharge of many pollutants such as harmful volatile organic compounds, and organic and solid wastes, is a threat to maintaining a healthy sustainable environment. The advancement of various solutions to treat and eradicate these generated wastes has relied heavily on sophisticated material design as the core integration of environmental remediation and monitoring technologies such as adsorption, advanced oxidation processes, and sensing and detection of pollutants. In the past decade, a noticeable breakthrough achievement in new materials design approaches has seen a high rise in layered two-dimensional (2D) materials that offer unique structure properties not accessible to their traditional bulk counterpart with greatly enhanced structure–activity performances. The precise control of achieving multi-stacked or single exfoliated 2D layers has shown to be pivotal in defining their superior functional performances. This book discusses comprehensively the state-of-the-art developments ranging from fundamentals to practical understanding of 2D nanomaterials synthesis, surface structure, and properties integrated with application testing in diverse environmental remediation and protection technologies. This book highlights the interplay relationship between the unique electronic, optical, mechanical, and topological properties of 2D materials manipulated by controlled synthesis approach to maximize their inherent structure effects with high functional performance in achieving efficient and economical technologies for sustainable environmental remediations. Besides, this book covers various emerging 2D materials such as metal oxide semiconductors, carbonaceous (i.e., graphene oxide, graphitic carbon nitride, and metal–organic frameworks), MXenes, layered double hydroxides, and transition metal chalcogenides, including their hybrid composites with diverse applications in sensing, electrochemical detection, photodegradation, and adsorption processes for pollutant monitoring and removal. It consists of 17 chapters written by various experts in advanced materials design and applications, which are classified into three parts. The first part consists of Chapters 1–3 focusing on the fundamental structure properties of 2D materials with extensive emphasis on design, synthesis methodologies, and characterization. The second part discusses the sensing and detection of gaseous, liquids, and solid pollutants with 2D materials covered by Chapters 4–7. The final part of the book focuses on the application of various 2D materials in the removal of volatile gaseous pollutants, radioactive toxic elements, and organic and metal ion pollutants from air, soil, and water contamination. The timely current research progress in 2D materials discussed in this book will appeal to graduate students as well as academic and industrial professionals working in advanced materials with focused applications to achieve sustainable environmental protection.

1 An Introduction to Two-Dimensional (2D) Nanomaterials

Swati Bansal, Naveen Kumar, Peter R. Makgwane, and David E. Motaung

1.1 INTRODUCTION

In the last few decades, nanomaterials had drawn the attention of researchers towards itself not only due to their distinctive features but also due to their applications in various important fields such as medicines, cosmetics, environment preservation and energy. Whenever we talk about nanomaterials, an image of size appears in our mind (i.e., size of 10^{-9}m), but nanotechnology is not just about the size of particles; it is also about how we modify and fabricate these particles to develop more advanced technologies [1]. Nanomaterials can naturally be occurred, can purposefully be prepared or can be occurred as by-products, and their physical and chemical properties are different from their bulk equivalents. The size and shape of nanomaterials affect the properties; for example, their reduced size enhances catalytic properties, while their shape increases their selectivity for catalysis [2]. Different dimensional nanomaterials exist like 0D (nanoparticles and quantum dots), 1D (nanotubes, nanorods, nanofibers and nanopillars), 2D (nanosheets, nanoplates and nanopores) and 3D (nanocomposites and complex graded structures); among which 2D nanomaterials have gained much attention after the isolation of graphene "a 2D ultrathin carbon layer" from graphite, which has exceptionally unique features such as high specific surface area, which results in greater utilization of these materials in biomedical applications such as medication, biosensing and cancer therapeutics. Graphene isolation leads to a new era in the development of 2D nanomaterials; after which various new 2D nanomaterials have been introduced having unique chemical/physical/optical properties, biodegradability and biocompatibility like high mechanical strength; good functionalization capabilities [3]; having applications in optical therapies like PTT (photothermal therapy) and PDT (photodynamic therapy) due to their quick response to external source like light [1,4–6]. Despite all these, their physiological interactions with living tissues are less understood, and hence their biocompatibility cannot be generalized because some are highly compatible in vivo and in vitro and others are not. Different compositions of 2D nanomaterials and their interaction with proteins and cell tissues are also different which will affect them in a different way. 2D nanomaterials can broadly be classified into three main groups – inorganic, organic and hybrid 2D nanomaterials.

Inorganic family includes layered double hydroxides (LDHs), Transition metal oxides (TMOs), Transition metal dichalcogenides (TMDs), Nano clays and MXenes, while covalent organic framework (COF), polymer nanosheets and sequence-defined 2D nanomaterials are associated with family of organic 2D nanomaterials, and hybrid 2D nanomaterials include MOF (metal organic framework), which are derived from both organic and inorganic species [4]. The library of 2D nanomaterials is huge, which exhibited a broad spectrum of properties ranging from conductors-to-semiconductors-to-insulators and from softest to the strongest [7].

DOI: 10.1201/9781003436942-1

1.2 BULK VS 2D MATERIALS: FEATURES AND CONSEQUENCES

The physiochemical properties of 2D nanomaterials are different from their bulk equivalents, which result in greater performance of these materials due to different surface chemistry, quantum-size effect and high-aspect ratio [8]. The following aspects impart unique characteristics to 2D materials:

i. These materials have in-plane strong covalent bonds, thickness of atomic size and high flexibility, high mechanical strength and optical diaphaneity, which enables the utilization of these materials in making future flexible, transparent optoelectronic devices, as well as electrode material with high performance having great energy storage capacity and good conversion applications in batteries and fuel cells in comparison to bulk materials and other dimensional nanomaterials [9–12].
ii. Due to quantum confinement of electron in ultrathin region of 2D, specifically in single-layered material, without much interactions between inter-layers, they have enhanced optical, magnetic and electrical properties [2].
iii. Due to greater exposure of the surface atoms and high specific surface area, the 2D materials are highly desirable in various surface-active applications like photocatalysis, supercapacitors and electrocatalysis. In addition, high surface area, outstanding thermal stability and adsorption capability make 2D nanomaterials promising for constructing functional composites owing to their utilization as templates for the lithographic construction of nanostructures and as fillers to strengthen the resultant nanostructure (like 2D MT&Ds used as promising material for direct growth for noble metal nanostructures) [11,13–17].
iv. Greater fraction of surface atoms enables the easy manipulation of surface through more active-site construction to intensify the intrinsic material properties [18].
v. In comparison to other nanomaterials like nanotubes and nanowires, 2D nanomaterials exhibit tunability to somewhat greater extent due to planar surface, which makes them more compatible with fabricating techniques in device integration and scalable manufacturing [19].

1.3 UNIQUE CHARACTERISTICS OF 2D NANOMATERIALS AND PROPERTIES INFLUENCED

2D nanomaterials have specific features including high anisotropy, mechanical strength, plasmonic, electron confinement and optical properties (Figure 1.1). They have high surface-to-volume ratio, which promotes quick response and helps in the transmission of higher energy. Further, the functional properties of these materials may be tailored with ease by doping, electrodeposition, adsorption and chemical reduction, which promotes more stability, retention ability and permeability than the no-functionalized ones [20].

1.3.1 Thickness of 2D Materials

Thickness and size regulation of 2D materials layers have a significant effect on their chemical properties and little effect on their physical properties. As layer thickness of the material decreases, the flexibility of 2D materials is improved by decrease in flexural stiffness, which is proportional to cube of thickness of materials [21]. The layer thickness also affects the mechanical strength of some materials like in case of graphene; when the layer number is increased from 1 to 8, the strength decreases by 30% or more; but the increase in thickness doesn't affect the mechanical strength of boron nitride (h-BN) nanosheets [22]. With layer thickness reduced to atomic scale, the catalytic properties of these materials are greatly enhanced resulting from changes in electronic state as well as from the generation of in-plane defects by disordering in the structure [12,18]. The atomic-thick

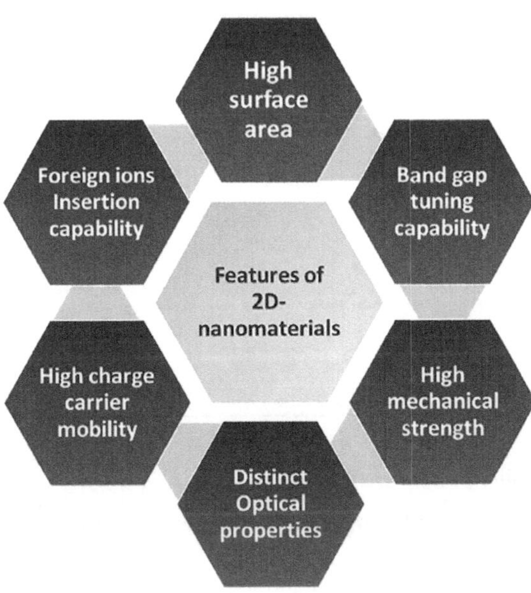

FIGURE 1.1 A representative diagram of characteristic features of 2D-nanomaterials.

Co_3O_4 has abundant active sites enabling the material to absorb CO_2 in large amount, and also due to increased and more dispersed charge density near Fermi level, the electrocatalytic efficiency enhanced to a greater amount; for example, 1.7 nm thick layer of Co_3O_4 shows 1.5 and 20 times greater electrocatalytic activity than 3.51 nm thick layer and bulk materials [23].

The specific area per gram (SAPG) value is highest for graphene among other 2D materials, but as the thickness is reduced from bulk to few layers, the SAPG value increases significantly; for example, an increase of 10–140 m^2g^{-1} is observed in the case of h-BN and 8.4–25 m^2g^{-1} in the case of MoS_2. A high SAPG value and low electrical noise improve the ability to detect analyte even at low concentration [19,24–26]. On fabricating the top-gate transistor with transparent gate electrode, the MoS_2 layer has effect on photodetection ability due to the difference in band gap; for example, the device with triple MoS_2 layer (with a band gap of 1.35 eV) has good photodetection ability of red light, while the single (with a band gap of 1.8 eV) and double layer (with a band gap of 1.65 eV) have good ability in detecting green light [27,28]. Black phosphorous (a most stable allotrope of phosphorous) has a direct band gap of 0.3 eV, but as the thickness is reduced to monolayer, it increases to up to 2 eV, which open the doors for its application in optoelectronics [29].

1.3.2 High Specific Surface Area

2D nanomaterials with greater surface area enable them in loading and delivering therapeutic agents efficiently because of the availability of anchoring sites in larger amount [30]. Due to this feature, 2D materials have great interactions with fluorescent dyes, thus yielding greater quenching efficiency and better performance as fluorescence sensors [18]. The ultrathin nanosheets of TiO_2, ZnO, Co_3O_4 and WO_3 possess specific surface area of 298, 265, 246 and 157 m^2g^{-1}, respectively [31]. Among all the known 2D materials, graphene has the highest specific surface area of 2630 m^2g^{-1} [32], which enables its application in biomedical field like gene transfection, cancer treatment and drug delivery via the delocalized p-orbital electron, which enables this material to bind with aromatic drug molecules via π–π stacking, and the surface area of some 2D nanomaterials is presented in Table 1.1.

TABLE 1.1
Some 2D Nanomaterials with Specific Surface Area

2D Nanomaterials	Specific Surface Area Values ($m^2 g^{-1}$)	References
Graphene	2630	[32]
TiO_2	298	[31]
ZnO	265	[31]
Co_3O_4	246	[31]
WO_3	157	[31]
g-C_3N_4	306	[33]
$Ti_3C_2T_x$	5.154	[34]
In_2S_3	110	[34]

1.3.3 Band Gap Tunability

Material thickness affects the band gap of the materials significantly. By creating defects or by irradiating electron beams on the surface, the thickness of 2D materials can be reduced, i.e., from multilayer to monolayer, which changes the band gap from indirect (present in bulk state) ~1.3 eV to direct (present in monolayer) ~1.8 eV; especially in the case of TMDs (MoS_2) – due to quantum confinement of electron resulting in the better photoluminescence efficiency by a factor more than 10^4 than bulk materials [35–37]. $MoSe_2$ shows changeset in band gap from indirect (~1.41 eV) to direct (~1.58 eV) when the thickness is reduced from multilayer to monolayer resulting in greater photoluminescence efficiency and providing a route to valleytronics [38].

1.3.4 High Mechanical Strength and Flexibility

2D materials possess a high mechanical strength, which is related to stiffness, toughness, strength and stability of materials. Due to high strength, they can bear stress to greater extent. They possess high Young's modulus; for example, monolayer graphene shows extremely high Young's modulus of 1TPa, which can bear more than 25% strain without fracturing [21,39]; monolayer TMDs (MoS_2) shows Young's modulus as high as 270 GPa which is similar to steel; Young's modulus of high crystalline tungsten nitride (WN) is about 390 GPa. Also, 2D nanomaterials possess high flexibility (which makes them to hold strain greater than 10% prior to rupture) due to which they can be used as excellent materials in making flexible sensors and electronics in comparison to traditional semiconductors and metal-based sensors, which are fragile and rigid, limiting their applications as a fruitful material in designing wearable electronics [19,40–42].

1.3.5 Distinctive Optical Properties

The remarkable optical properties of 2D materials include light sensitivity, adsorption, emission and plasmonic effects resulting from their intrinsic properties such as band structure, carrier mobility and density of states [43]. The optical transparency in the visible region requires uniform and nanometer-scale thickness, and due to these characteristics, the thin films of $Ti_3C_2T_z$ exhibit optical transparency of 97%, which is almost equal to graphene (having 97.7% optical transparency) [44]. The sensitivity toward external environment and high surface-to-volume ratio enables 2D materials in making optical sensors, and further X-ray attenuation and high optical properties make these materials a successful agent in phototherapy and radiotherapy [45]. Due to the ability of graphene and its similar 2D materials as fluorescent emitters and quenchers, they can be regarded as great platforms in developing optical biosensors for detecting various biological analytes like ions, proteins, nucleic acid, small molecules and cancer biomarker targets [46].

1.3.6 High Charge Carrier Mobility

The high charge carrier mobility is the property associated with high-speed transistors [7]. 2D materials are known for fast ion and electron transport [47] and hence they have high charge carrier mobility, making them highly useful in various photoconductive applications. Additionally, high electron transport along 2D layer decreases the corrosion rate and hence gives assurance of long-time durability in aqueous electrolyte [23,48]. It was found that graphene has highest charge carrier mobility of $2 \times 10^5 cm^2V^{-1}s^{-1}$, while for the TMDs ($MoX_2$, where M=Mo, W; X= S, Se, and Te), the maximum mobility was found for WS_2 with values of 0.12×10^3 and $0.21 \times 10^3 cm^2V^{-1}s^{-1}$ for electron and hole along the x-direction, and for all others, it was found to be $<100 cm^2V^{-1}s^{-1}$; also, as the chalcogen size increases, the mobility decreases [7]; the α-P (black phosphorous 2D format) exhibits ambipolar charge transport and a high on/off current ratio of $\sim 1 \times 10^4$ due to which they are highly useful in field-effect transistors (FETs)-based devices.

1.3.7 Intercalation of Foreign Species

Intercalation is a phenomenon in which foreign species are reversibly intercalated at the crystal gap, and 2D materials are best known for intercalating the species ranging from atoms-to-ions-to-molecules. This phenomenon is important in redesigning novel 2D-layered materials to improve their physical properties. As in the case of 2D TMDs, intercalation helps in exfoliation and tuning of various physical properties like thermoelectricity, electricity, phase changing for catalysis as well as superconductivity [49]. As in the case of MoS_2, with the thickness reduced to atomic thickness (~2–50 nm) and by intercalation of Li-ion electrochemically, the MoS_2 crystals show a drastic decrease in electrical resistance of sheet and a substantial increase in light transmission (up to 90%, in 4 nm thick lithiated MoS_2) [50]. Besides being their amazing intrinsic properties, intercalation helps in tuning the properties so as to improve the device performance drastically because it increases the doping level to maximum extent possible and is capable of changing the phase of material (ir) reversibly. The intercalation helps in increasing the charge carrier density of 2D nanomaterials (but not all) by shifting the fermi level, which increases the density of states at Fermi level. Also, the intercalated compounds are generally chemically tough as chemically inert and impermeable atomically thick layer 2D materials protect and enclose the unstable intercalants than compounds that are physically or chemically adsorbed onto the surface [51]. Superconductivity is caused by metal intercalation in bulk graphite [52]. Recently, it was found that the Li-ion intercalation in few-layer graphene increases both transmittance and conductivity simultaneously (91.7% transmittance and 3-ohm sq^{-1} sheet resistance) [53].

1.4 SYNTHESIS STRATEGIES FOR 2D NANOMATERIALS

2D nanomaterials synthesis can be broadly classified into two approaches: top-down synthesis and bottom-up synthesis. In **top-down synthesis**, the 2D materials are synthesized via mechanical exfoliation, i.e., by exfoliating the bulk materials into single- or few-layered nanosheets by eliminating weak van der Waals forces between these sheets. The driving force behind these is strong intra-plane covalent bonds and removal of weak inter-plane van der Waals forces. Top-down synthesis involves Mechanical Cleavage [54], Mechanically force assisted liquid exfoliation [55–57], Selective-etching assisted method [58] and Ion-Intercalation assisted method [59].

In **bottom-up synthesis**, the material is synthesized by direct means, i.e., by using precursors through chemical reaction at certain chemical conditions. Chemical Vapor Deposition [60] and Wet-Chemical Methods [61,62] are two typical methods of bottom-up synthesis. In the subsections below, the brief overview of frequently used synthesis methods along with emphasis on their advantages and limitations is explored.

1.4.1 Top-down Synthesis Methods

1.4.1.1 Mechanical Exfoliation

In mechanical exfoliation method, the exfoliation is done with the help of scotch tape (Figure 1.2) in which the bulk crystal is first adhered to scotch tape, and then after peeling off with the help of another adhesive, a thin layer of 2D nanomaterial is attached to tape, which is then transmitted to the targeted substrate [63]. The layer thus obtained is high in quality with little defects, have large lateral dimensions, clean surface, less destructive in nature and remains stable in ambient conditions because the layer is directly exfoliated from bulk not via any chemical reaction. Large-scale production with high quality, hardly controlled size and shape of nanosheets, and constant need for a secondary substrate to support the exfoliated nanosheets are the major issues with this method [11,28,31,64,65]. This method was originally used to exfoliate graphene from graphite, but later on was extended to other materials such as h-BN and TMDs [66].

1.4.1.2 Liquid-Assisted Exfoliation

In liquid-assisted exfoliation technique, the exfoliation of bulk sheets into thin sheets is done in a liquid medium (e.g., dimethyl formamide, N-methyl-pyrrolidone, dimethyl formamide and N-methyl pyrrolidone). The liquid exfoliation has two main means – shear force-assisted liquid exfoliation and sonication-assisted liquid exfoliation. In these, the sonication-assisted exfoliation is more practical and simpler [67]. The solvent used helps in preventing the agglomeration and stacking of resulting nanosheets, and it was found that polymer and surfactant stabilize them during the sonication process. Also, it was found that the organic liquids are highly efficient in exfoliating the sheets. As an example, direct exfoliation of MoS_2 into colloidal nanosheets was carried out in an organic solvent, i.e., dimethyl sulfoxide, and pure water as shown in Figure 1.3. To improve the dispersing ability, a surfactant (sodium cholate) was added to the solution [68]. Moreover, the efficiency can be increased if the surface tension of solvent matches that of the bulk compounds. But compared with the above-mentioned techniques, this is more efficient in preparing ultrathin 2D nanomaterial due to production at large scale and high yield at low cost [56,69]. Along with advantages, there are limitations too – such as the yield obtained of a single-layered 2D material is comparatively low, and the lateral dimension obtained is also low due to strong sonication process [55,70]. In addition, the organic solvents, which are toxic in nature, harm the sheets, and the polymer and surfactants used are also undesirable for further applications. Hence, it is necessary to optimize further the experimental conditions to obtain ultrathin sheets with large lateral dimensions and high yield. Here, in this method, the solvent used acts as an intercalating species, which weakens the force between sheets [11].

1.4.2 Bottom-up Synthesis

The bottom-up synthesis techniques are famous for obtaining high-quality nanosheets. Co-precipitation, sol–gel, chemical vapor deposition, hydrothermal/solvothermal, template-assisted

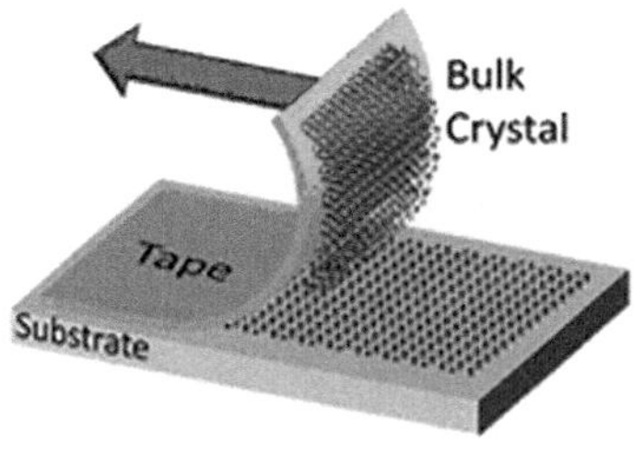

FIGURE 1.2 A representation of mechanical exfoliation via Scotch tape [19].

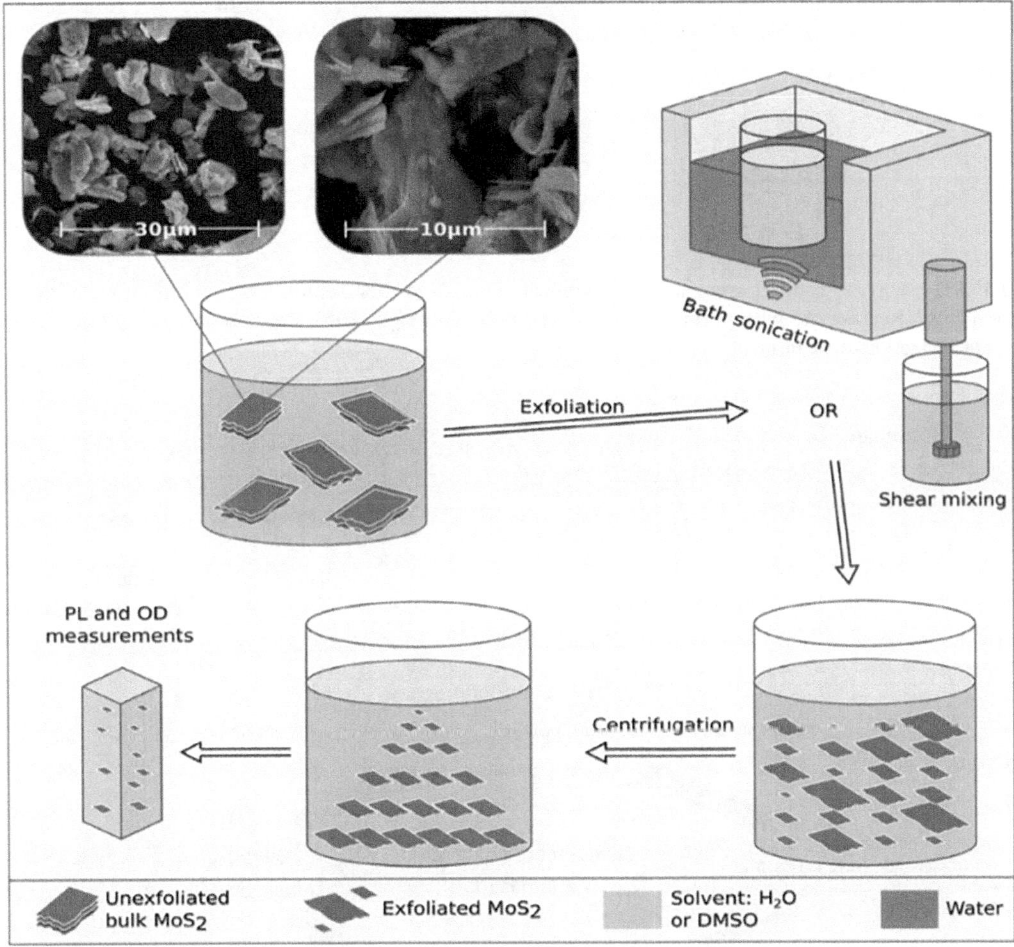

FIGURE 1.3 A representation of synthesis and characterization procedure for MoS_2 nanosheet dispersions. SEM images represent pristine nanomaterial before exfoliation procedure [68].

synthesis, polyol methods, laser ablation, plasma spraying, etc. are various bottom-up synthesis techniques [71]. But in this section, we discuss only few of them.

1.4.2.1 Chemical Vapor Deposition Method

This method is famous for synthesizing ultra-high quality with high purity, low porosity, high controllability and predictability, highly efficient with uniformity, highly stable as well 2D nanosheets onto substrate. Under high temperature and vacuum, the reactive precursors are exposed onto the substrate on which after reacting or decomposing they form a thin layer of 2D nanomaterial (Figure 1.4) [60,65,72,73]. This method is efficient in producing various 2D nanomaterials like TMDs, h-BN, metal oxides and graphene. The disadvantages of this method are high temperature requirement, poisonous gases emission as by-products and requirement of highly expensive instrumentation [65].

1.4.2.2 Wet Chemical Synthesis Technique

Wet chemical synthesis method is effective in synthesizing numerous 2D nanomaterials whether layered or non-layered such as TMDs, graphene, LDHs, MOF, COF, metal oxides, metals, polymers, metal chalcogenides, h-BN and $g\text{-}C_3N_4$ [75]. Co-precipitation [76], Hydrothermal [77], Solvothermal [78] and Sol–Gel are various synthesis strategies of this technique, which gain much

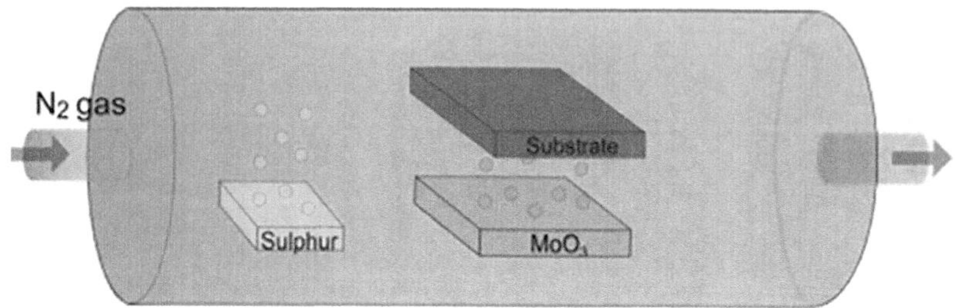

FIGURE 1.4 A typical representation of chemical vapor deposition setup for 2D transition metal dichalcogenides. (Reproduced with permission from Ref. [74], Copyright 2016, Progress in Crystal Growth and Characterization of Materials.)

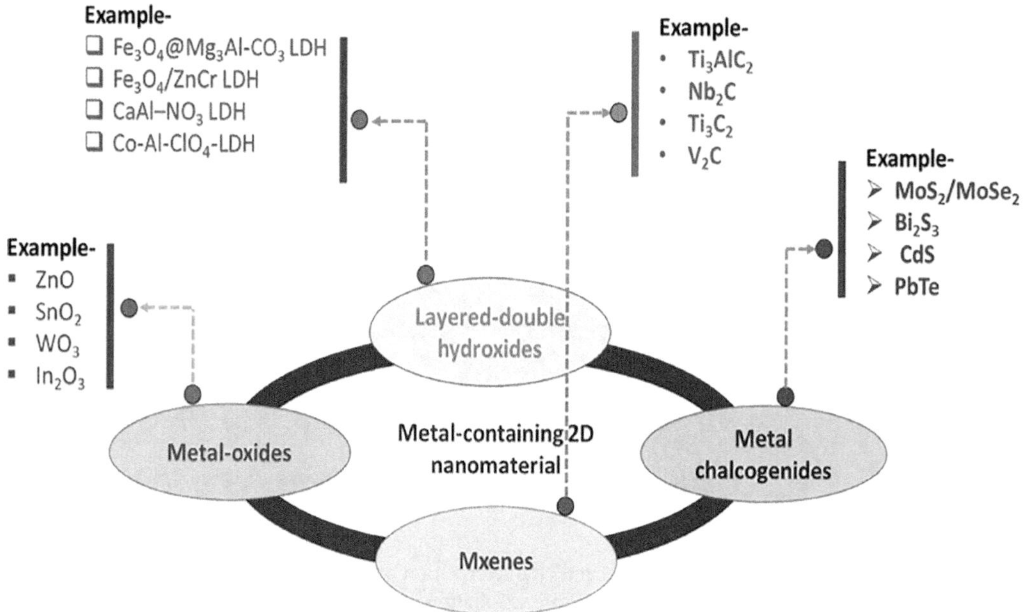

FIGURE 1.5 Various metal containing 2D-nanomaterials along with few examples.

attention due to their easy handling and operational simplicity. In this method, target material is synthesized by certain precursors through chemical reaction in solution assisted with polymers or surfactants. High yield, solution processibility and large-scale production are major advantages of this technique, while difficulty in obtaining single-layer nanosheets and the requirement of surfactants are disadvantages [75].

1.5 METAL CONTAINING 2D MATERIALS

Atomically thin 2D materials are promising materials in novel functional optoelectronic devices including modulators, photodetectors and lasers due to their novel optical and electronic properties. But, light–matter interactions in 2D materials are weak due to atomic-scale thickness, which limits the performances of these devices. Thus, to strengthen the light–matter interactions, metallic nanostructures are to be fabricated [79]. Various metal-containing 2D-nanomaterials as shown in Figure 1.5 are illustrated below.

1.5.1 Metal Oxides

Nowadays, quasi-2D metal oxides got tremendous applications in photonics, nano-sensors, etc., because of their enhanced charge carrier mobility. Also, for FETs, the 2D WO_{3-x} is gaining much importance as a semiconducting material [80] because sub-stoichiometry of this material reduces the band gap and increases the free charge carrier altogether, but after the intercalation of H^+ into this 2D WO_3 material, an increase in the carrier mobility is found by an amount exceeding $319\,cm^2V^{-1}s^{-1}$ than 2D TMDs. So, the conclusion is that this material can be regarded as good material in making various functional FET devices [80]. The metal oxides can be regarded as ideal material in making new electronic devices because of the band gap, which allows them to be electrically conducting, insulating or semiconducting. The amorphous 2D metal oxides have many applications due to their superior performance in catalysis and energy storage [81]. Because of their disordered structure with dangling bonds, ion channels and activity sites, they have different physical and chemical properties than crystalline 2D materials; for example, amorphous 2D SnO_x and NiO have higher photocatalytic performance than their corresponding crystalline counterparts [82]. It was found that the 2D heterostructures that are formed either by van der Waals stacking or by edge covalent bonding generate many properties through synergistic effect of various 2D nanomaterials; for example, on illumination of light of 280 nm, large area heterostructure p-SnO/n-In_2O_3 shows both excellent photoresponsivity and photodetectivity due to narrow band gap at the p-n junction [83].

1.5.2 Metal Chalcogenides

TMDs are MX_2-type materials with M (representing transition metal such as Mo, W, Nb, Re, Ta, Zr, Ga and In) and X (representing chalcogens such as S, Se and Te) having three atomic-layered structure in which transition metal layer is sandwiched between two layers of chalcogens. Depending upon the metal, TMDs known to exist in many structural phases, but the two common structural phases are trigonal prismatic (2H) or octahedral (1T) [84]. The TMDs are semiconducting in nature; due to their nature and defect-free surface, thin exfoliated TMD layers can be regarded as ideal material for FET devices, and TMD-based FET devices possess remarkable electronic [85] and opto-valleytronic properties [86,87]. Recently, it was found that monolayer 2H-$NbSe_2$ has intrinsic superconductivity [88]. The superconductivity is also realized in thin films of 2H-WS_2 and 2H-$MoSe_2$, which are induced by gating [87]. The high strength makes them promising candidates for strain engineering [84].

1.5.3 MXenes

Metal carbides, nitrides and carbonitrides simply known by MXenes are the 2D nanomaterials having nanostructure composed of two or more layers of transition-metal atoms which are packed into honeycomb-like lattice in which carbon or nitrogen are intervened between these layers and occupied octahedral sites [89]. MXenes are generally synthesized by a selective etching technique from the MAX phases (having the formula $M_{n+1}AX_n$, where M represents transition metal, A represents atom from group 12 to 16, and X represents C or N) in which MX bond is stronger than MA bond; so, by the relative difference in their bond energies, the A-layer atoms are selectively etched with the help of strong acid or molten salts [90]. A large number of MAX phases, i.e., >60, are available, but from them, only few MXenes ($M_{n+1}X_nT_x$, where T represents the functional group; e.g., –OH, =O and –F) are found to have practical application [90]. Their unique mechanical, thermal and chemical stability properties make their utilization in various areas like sensors, electronics and catalysis, in electrochemical storage systems like batteries, supercapacitors and electromagnetic interference shielding and in fuel cells thereby creating an interest in researchers to develop more MXenes.

A 2D Ti_3C_2 monolayers were found to be better anode materials for Lithium-Ion Batteries than TiO_2 because of their small open circuit voltage, good electronic conductivity (metallic character) and improved Li storage capacity [91]. MXene-based nanocomposites are also utilized in photocatalytic decontamination of water; for example, $TiO_2/Ti_3C_2T_x$ shows greater adsorption and photocatalytic degradation under UV irradiation (~98% degradation of organic contaminant methyl orange within just 30 minutes) than conventional TiO_2 photocatalyst, which degrades ~77% methyl orange under same conditions [34,92].

1.5.4 Layered Double Hydroxides

LDHs have uniform interlayer galleries, and large number of hydroxyl groups are covalently bonded within the 2D host layers, making them excellent candidates for high-performance membranes [93]. These are also considered as anionic clays and have interesting properties like having intercalated ions with interlayer spacings, ability to intercalate different types of anions (organic, inorganic and biomolecules), swelling properties, high biocompatibility and delivering intercalated anions in a continuous manner [94]. Certain LDHs have more active sites for CO_2 adsorption due to high surface area [95], and the number of surface-active sites has effect on adsorption activity [96]. By increasing the interlayer distance, the more adsorption of CO_2 takes place because it helps in more CO_2 diffusion into the reaction space and thus the reaction with hydroxyl groups to create hydrogen carbonate intermediates [97,98]. This strategy is found to enhance the adsorption activity of LDH catalyst. It was found that Mg-Al LDH nanosheets with C_3N_4 photocatalyst produce more methane from photoreduction of CO_2 than with C_3N_4 alone in an aqueous solution with cocatalyst Pd [99].

1.6 METAL-FREE 2D MATERIALS

The majority of the metal-free 2D materials are composed of elements like hydrogen, nitrogen, carbon and phosphorus, which are the necessary elements in the functioning of human body. Thus, metal-free 2D nanomaterials possess higher biocompatibility and biosafety, which make them more suitable materials for biomedical application and further clinical translation [100]. Also, these materials have applications in electrocatalysis fields, which replace the employment of noble metals making the process cost-effective and favorable for large-scale production of electrocatalysts [101]. Some well-known metal-free 2D nanomaterials as shown in Figure 1.6 are illustrated in the following subsections as presented below.

1.6.1 Carbon Nitride

The graphitic carbon nitride (g-C_3N_4) is a unique polymeric layered 2D nanomaterial that can be considered as N-substituted graphite, but the difference is that the g-C_3N_4 sheets are comprised

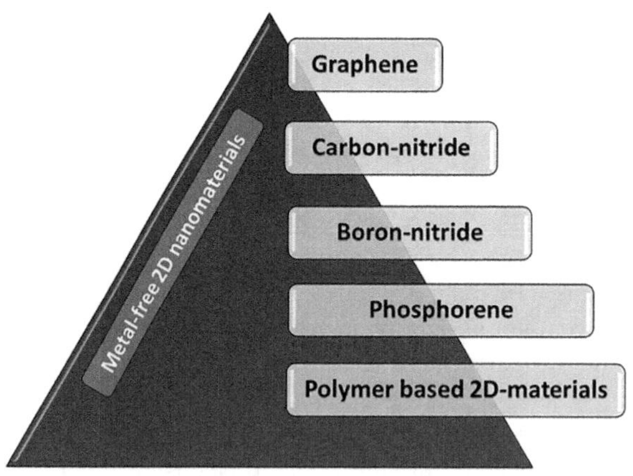

FIGURE 1.6 A graphical representation of metal-free 2D nanomaterials.

of tri-s-triazine units that are bridged by amino groups, with dangling H-atoms in the C-N layer and having periodic vacancies in lattices [33,102,103], and because of surface hydrogen atoms and surface imperfections, the g-C_3N_4 sheets have a wide range of photocatalytic applications [104]. Carbon nitride has five phases, among which g-C_3N_4 has the narrowest band and good light adsorption capability [105]. Recently, among all the known 2D materials, g-C_3N_4 is the carbon material having highest N-content and can exist as stable material in various acidic and alkaline conditions [106]. Despite this, the uniform distribution of triangular nanopores in the layers of tri-s-triazine causes strong in-plane piezoelectricity in g-C_3N_4 [107].

1.6.2 Graphene

It is the single-atom thick layer 2D nanomaterial with a single layer of carbon atoms having sp^2 hybridization arranged in hexagonal pattern in which three valence electrons of carbon are involved in making strong in-plane covalent bonds, while the fourth electron remains in p-orbital for making π-bonds. Graphene shows a versatile range of chemical and physical properties such as large theoretical specific surface area (2630 $m^2 g^{-1}$) [16], high Young's modulus (~1 TPa) [39], low light absorption (~2.3%) [108] and a high carrier mobility of ~10,000 $cm^2 V^{-1} s^{-1}$ at room temperature [66]. It is a zero band-gap semimetal which shows exceptionally high conductivity (~5000 $Wm^{-1} K^{-1}$ [109]) due to the availability of highly mobile electrons, but due to zero-gap, it hinders various applications in bioimaging, optical imaging, logic circuits and FETs [17,110]. Despite the modification in the band gap, it results in degradation of its intrinsic electronic properties [111]. The graphene by itself is non-piezoelectric in nature, but it can be doped by chemical means on the basal planes, which then breaks the inversion symmetry and thus induces piezoelectricity in it [112].

The graphene oxide is an essential derivative of graphene obtained by strong oxidative agents. Various oxygen-containing functional agents like epoxy groups, hydroxy and carboxyl groups were added on the surface of graphene [16]. During the oxidation process, the hybridization changes from sp^2 to sp^3, thus making GO poorer in electrical conductivity. Despite the fact that both graphene and GO have similar structures, there is quite difference in their properties; for example, the graphene is insoluble in water, but GO is quite soluble in water because of the presence of hydrophilic oxygen-containing groups on the surface [113]. Also, graphene oxide is used in harvesting mechanical energy due to high dielectric constant and Young's modulus. The reduction of GO to reduced-GO changes the band gap from insulator to semiconductor, which depends upon the degree of reduction. Due to this, it helps in overcoming the problem of high leakage current, allowing these materials to be promising as interfacial material for triboelectric nanogenerator to increase the output voltage [114,115].

1.6.3 Boron Nitride

h-Boron nitride (h-BN) nanosheets, also called as "white graphene," have similar structure to graphene where it has equal number of boron and nitrogen atoms present in alternative positions in the honeycomb configuration [116] and properties as graphene, but the difference is that it has 1.8% longer lattice constant [117] and is an electric insulator due to its wide band gap than graphene [118]. Also, planar structure cleaves it into flat surface, and ionic bonding makes the surface free of dangling bonds and thus charge traps at the surface [118]. Additionally, h-BN nanosheets are stable at higher temperature than graphene. The thermal and chemical inert abilities enable this as an ideal material for corrosion protection [22]. It has also the ability to promote the photocatalytic activity of the supporting material, thus acting as a cocatalyst. In comparison to graphene on SiO_2 devices, graphene on h-BN devices showed the highest mobility, intrinsic doping, reduced roughness and chemical reactivity [119]. These sheets have wide applications in water treatment and purification because of super-high hydrophobicity power [120]. The band gap of h-BN is about 5.9 eV, and it is thermally stable up to 2000°C [121].

1.6.4 Polymer-Based 2D Materials

The combination of nanomaterials and polymers changes not only the structural and morphological features of the material but also the physiological properties in a significant manner. The addition of even a minute quantity of nanomaterials to polymers matrix enhanced the properties of polymer composites significantly; for example, it has been reported that the addition of graphene or layered silicates to polymer matrix enhances the mechanical properties up to ~200% of the polymer composites [113]. Thus, nanomaterials can be considered as "nanofillers" that bridge the gap between these two, i.e., polymer chemistry and nanotechnology. Thus, a wide variety of polymers–composites are available depending upon the 2D materials like inorganic nanomaterial and LDH, which shows insulating properties that boost the thermal and flame-retardant properties of polymers [94].

i. Graphene-based polymer composite: As graphene has highest surface area and good interaction with polymer matrix, the addition of small amount of this into the polymer matrix greatly influenced its properties like thermal conductivity, electrical conductivity, mechanical properties and dielectric properties; for example, for graphene-Poly(vinylidene fluoride) composite with 0.2 vol% of graphene, the dielectric constant reached 340 at 100 Hz, while with the 1.2 vol% of graphene, the highest dielectric constant was achieved up to 7940 at 100 Hz [122]. A small addition of graphene also increases the thermal stability (e.g., loading of 0.05 wt% of functionalized graphene sheets into polymethyl methacrylate shows an increase of glass transition temperature to 30°C [123].
ii. GO-based polymer composite: Due to rich oxidative functional groups on the surface of graphene oxide, they have good interactions with polymer matrices; thus, only a small loading greatly enhances the properties; for example, by solution mixing technique, the formed GO-Poly(vinyl alcohol) composite exhibits an increase in tensile strength by 76%, i.e., from 49.9 to 87.6 MPa, and high Young's modulus of 3.45 GPa with loading of only 0.7 wt% GO, which was 62% greater than the neat Poly(vinyl alcohol) polymer matrix [124].
iii. h-BN polymer composite: These nanocomposites are used in conditions where insulating and good thermal conducting properties are required, such as thermal interface management and electronic devices. As thermal interface management, in the h-BN-based composites, the increase of loading enhances the properties; for example, polyimide composite films achieved a thermal conductivity of 0.47 $Wm^{-1} K^{-1}$ with the loading of 30 wt% of h-BN [125]. By the loading of 30 and 60 vol% of h-BN, thermal conductivity of 3 and 7 $Wm^{-1} K^{-1}$ was shown by the same polymer composite when prepared by other methods [126].

1.6.5 Phosphorene

Also known as single- or few-layered black phosphorous, it has a puckered layer structure which makes it a promising material in biomedical applications like drug delivery, cell imaging, cancer therapy and biomedical detection. The biomedical applications are attributed to its negligible cytotoxicity and excellent biodegradability in the body. Phosphorenes possess exceptional capabilities like multiple drugs loading and maintaining the bioactivity of loading compounds. The availability of direct and tunable band gap (i.e., when the thickness is reduced from bulk to single layer, the band gap changes from 0.3 to 2 eV) makes it a promising bridging material with other 2D materials; for example, it covers the space in graphene (having zero band gap), TMDs (having large band gap), MXenes (having limited gap) and h-BN (having very large gap) [110].

1.7 CONCLUSIONS AND FUTURE OUTLOOK

The benchmarking of 2D nanomaterials makes them exceptionally different from other nanomaterials and from their bulk parts. These nanomaterials with atomic-level thickness possess distinctive optical properties, high strength, surface area and charge mobility. Along with these features, their physical properties can be enhanced with ease by modifying the band gap and including foreign species. In near future, these nanomaterials are believed to make great contribution toward practical applications in many fields. Currently, among the availability of many synthesis techniques, wet chemical synthesis technique is found to be more efficient due to solution controllability, compatibility and the synthesis of large number of 2D nanomaterials. But no synthesis technique is found to be ideal, which produces 2D nanomaterials with high yield, no defects and large dimensions, limiting their applications in industry. So, focus should be made to develop synthesis techniques which are cost-effective as well. In 2D nanomaterials, the extensive research has been conducted on graphene, TMDs (MoS_2, WS_2) and metal oxides, but similar studies are also possible for other 2D nanomaterials which may impart some better characteristics and applications. Thus, an extensive attention is needed toward 2D class of nanomaterials.

REFERENCES

1. N. Baig, Two-dimensional nanomaterials: A critical review of recent progress, properties, applications, and future directions, *Compos. Part A Appl. Sci. Manuf.* 165 (2023) 107362. https://doi.org/10.1016/j.compositesa.2022.107362.
2. M. Nasilowski, B. Mahler, E. Lhuillier, S. Ithurria, B. Dubertret, Two-dimensional colloidal nanocrystals, *Chem. Rev.* 116 (2016) 10934–10982. https://doi.org/10.1021/acs.chemrev.6b00164.
3. F. Yin, B. Gu, Y. Lin, N. Panwar, S.C. Tjin, J. Qu, S.P. Lau, K.-T. Yong, Functionalized 2D nanomaterials for gene delivery applications, *Coord. Chem. Rev.* 347 (2017) 77–97. https://doi.org/10.1016/j.ccr.2017.06.024.
4. A. Murali, G. Lokhande, K.A. Deo, A. Brokesh, A.K. Gaharwar, Emerging 2D nanomaterials for biomedical applications, *Mater. Today.* 50 (2021) 276–302. https://doi.org/10.1016/j.mattod.2021.04.020.
5. D. Chimene, D.L. Alge, A.K. Gaharwar, Two-dimensional nanomaterials for biomedical applications: Emerging trends and future prospects, *Adv. Mater.* 27 (2015) 7261–7284. https://doi.org/10.1002/adma.201502422.
6. M. Byakodi, N.S. Shrikrishna, R. Sharma, S. Bhansali, Y. Mishra, A. Kaushik, S. Gandhi, Emerging 0D, 1D, 2D, and 3D nanostructures for efficient point-of-care biosensing, *Biosens. Bioelectron. X.* 12 (2022) 100284. https://doi.org/10.1016/j.biosx.2022.100284.
7. S.H. Mir, V.K. Yadav, J.K. Singh, Recent advances in the carrier mobility of two-dimensional materials: A theoretical perspective, *ACS Omega.* 5 (2020) 14203–14211. https://doi.org/10.1021/acsomega.0c01676.
8. Y. Chen, Z. Fan, Z. Zhang, W. Niu, C. Li, N. Yang, B. Chen, H. Zhang, Two-dimensional metal nanomaterials: Synthesis, properties, and applications, *Chem. Rev.* 118 (2018) 6409–6455. https://doi.org/10.1021/acs.chemrev.7b00727.
9. C. Tan, X. Cao, X.-J. Wu, Q. He, J. Yang, X. Zhang, J. Chen, W. Zhao, S. Han, G.-H. Nam, M. Sindoro, H. Zhang, Recent advances in ultrathin two-dimensional nanomaterials, *Chem. Rev.* 117 (2017) 6225–6331. https://doi.org/10.1021/acs.chemrev.6b00558.
10. G. Fiori, F. Bonaccorso, G. Iannaccone, T. Palacios, D. Neumaier, A. Seabaugh, S.K. Banerjee, L. Colombo, Electronics based on two-dimensional materials, *Nature Nanotech.* 9 (2014) 768–779. https://doi.org/10.1038/nnano.2014.207.
11. H. Zhang, Ultrathin two-dimensional nanomaterials, *ACS Nano.* 9 (2015) 9451–9469. https://doi.org/10.1021/acsnano.5b05040.
12. H. Xu, M.K. Akbari, S. Zhuiykov, 2D semiconductor nanomaterials and heterostructures: Controlled synthesis and functional applications, *Nanoscale Res. Lett.* 16 (2021) 94. https://doi.org/10.1186/s11671-021-03551-w.
13. F. Bonaccorso, L. Colombo, G. Yu, M. Stoller, V. Tozzini, A.C. Ferrari, R.S. Ruoff, V. Pellegrini, Graphene, related two-dimensional crystals, and hybrid systems for energy conversion and storage, *Science.* 347 (2015) 1246501. https://doi.org/10.1126/science.1246501.
14. G.R. Bhimanapati, Z. Lin, V. Meunier, Y. Jung, J. Cha, S. Das, D. Xiao, Y. Son, M.S. Strano, V.R. Cooper, L. Liang, S.G. Louie, E. Ringe, W. Zhou, S.S. Kim, R.R. Naik, B.G. Sumpter, H. Terrones, F. Xia, Y. Wang, J. Zhu, D. Akinwande, N. Alem, J.A. Schuller, R.E. Schaak, M. Terrones, J.A. Robinson, Recent advances in two-dimensional materials beyond graphene, *ACS Nano.* 9 (2015) 11509–11539. https://doi.org/10.1021/acsnano.5b05556.
15. S. Stankovich, D.A. Dikin, G.H.B. Dommett, K.M. Kohlhaas, E.J. Zimney, E.A. Stach, R.D. Piner, S.T. Nguyen, R.S. Ruoff, Graphene-based composite materials, *Nature.* 442 (2006) 282–286. https://doi.org/10.1038/nature04969.

16. M.D. Stoller, S. Park, Y. Zhu, J. An, R.S. Ruoff, Graphene-based ultracapacitors, *Nano Lett.* 8 (2008) 3498–3502. https://doi.org/10.1021/nl802558y.
17. X. Chia, M. Pumera, Characteristics and performance of two-dimensional materials for electrocatalysis, *Nat. Catal.* 1 (2018) 909–921. https://doi.org/10.1038/s41929-018-0181-7.
18. F. Yang, P. Song, M. Ruan, W. Xu, Recent progress in two-dimensional nanomaterials: Synthesis, engineering, and applications, *FlatChem.* 18 (2019) 100133. https://doi.org/10.1016/j.flatc.2019.100133.
19. A. Bolotsky, D. Butler, C. Dong, K. Gerace, N.R. Glavin, C. Muratore, J.A. Robinson, A. Ebrahimi, Two-dimensional materials in biosensing and healthcare: From *In Vitro* diagnostics to optogenetics and beyond, *ACS Nano.* 13 (2019) 9781–9810. https://doi.org/10.1021/acsnano.9b03632.
20. A. Jayakumar, S. Mathew, S. Radoor, J.T. Kim, J.-W. Rhim, S. Siengchin, Recent advances in two-dimensional nanomaterials: Properties, antimicrobial, and drug delivery application of nanocomposites, *Mater. Today Chem.* 30 (2023) 101492. https://doi.org/10.1016/j.mtchem.2023.101492.
21. H. Jiang, L. Zheng, Z. Liu, X. Wang, Two-dimensional materials: From mechanical properties to flexible mechanical sensors, *InfoMat.* 2 (2020) 1077–1094. https://doi.org/10.1002/inf2.12072.
22. A. Falin, Q. Cai, E.J.G. Santos, D. Scullion, D. Qian, R. Zhang, Z. Yang, S. Huang, K. Watanabe, T. Taniguchi, M.R. Barnett, Y. Chen, R.S. Ruoff, L.H. Li, Mechanical properties of atomically thin boron nitride and the role of interlayer interactions, *Nat. Commun.* 8 (2017) 15815. https://doi.org/10.1038/ncomms15815.
23. S. Gao, X. Jiao, Z. Sun, W. Zhang, Y. Sun, C. Wang, Q. Hu, X. Zu, F. Yang, S. Yang, L. Liang, J. Wu, Y. Xie, Ultrathin Co_3O_4 layers realizing optimized CO_2 electroreduction to formate, *Angew. Chem. Int. Ed.* 55 (2016) 698–702. https://doi.org/10.1002/anie.201509800.
24. G. Liu, A.W. Robertson, M.M.-J. Li, W.C.H. Kuo, M.T. Darby, M.H. Muhieddine, Y.-C. Lin, K. Suenaga, M. Stamatakis, J.H. Warner, S.C.E. Tsang, MoS_2 monolayer catalyst doped with isolated Co atoms for the hydrodeoxygenation reaction, *Nat. Chem.* 9 (2017) 810–816. https://doi.org/10.1038/nchem.2740.
25. S. Yang, C. Jiang, S. Wei, Gas sensing in 2D materials, Applied Physics Reviews. 4 (2017) 021304. https://doi.org/10.1063/1.4983310.
26. Y.-M. Lin, P. Avouris, Strong suppression of electrical noise in bilayer graphene nanodevices, *Nano Lett.* 8 (2008) 2119–2125. https://doi.org/10.1021/nl080241l.
27. H.S. Lee, S.-W. Min, Y.-G. Chang, M.K. Park, T. Nam, H. Kim, J.H. Kim, S. Ryu, S. Im, MoS_2 Nanosheet phototransistors with thickness-modulated optical energy gap, *Nano Lett.* 12 (2012) 3695–3700. https://doi.org/10.1021/nl301485q.
28. H. Li, J. Wu, Z. Yin, H. Zhang, Preparation and applications of mechanically exfoliated single-layer and multilayer MoS_2 and WSe_2 nanosheets, *Acc. Chem. Res.* 47 (2014) 1067–1075. https://doi.org/10.1021/ar4002312.
29. J. Xiao, M. Long, X. Zhang, J. Ouyang, H. Xu, Y. Gao, Theoretical predictions on the electronic structure and charge carrier mobility in 2D Phosphorus sheets, *Sci. Rep.* 5 (2015) 9961. https://doi.org/10.1038/srep09961.
30. T. Hu, X. Mei, Y. Wang, X. Weng, R. Liang, M. Wei, Two-dimensional nanomaterials: Fascinating materials in biomedical field, *Sci. Bull.* 64 (2019) 1707–1727. https://doi.org/10.1016/j.scib.2019.09.021.
31. Z. Sun, T. Liao, Y. Dou, S.M. Hwang, M.-S. Park, L. Jiang, J.H. Kim, S.X. Dou, Generalized self-assembly of scalable two-dimensional transition metal oxide nanosheets, *Nat. Commun.* 5 (2014) 3813. https://doi.org/10.1038/ncomms4813.
32. J. Zeng, X. Ji, Y. Ma, Z. Zhang, S. Wang, Z. Ren, C. Zhi, J. Yu, 3D graphene fibers grown by thermal chemical vapor deposition, *Adv. Mater.* 30 (2018) 1705380. https://doi.org/10.1002/adma.201705380.
33. P. Niu, L. Zhang, G. Liu, H.-M. Cheng, Graphene-like carbon nitride nanosheets for improved photocatalytic activities, *Adv. Funct. Mater.* 22 (2012) 4763–4770. https://doi.org/10.1002/adfm.201200922.
34. H. Wang, Y. Wu, T. Xiao, X. Yuan, G. Zeng, W. Tu, S. Wu, H.Y. Lee, Y.Z. Tan, J.W. Chew, Formation of quasi-core-shell In_2S_3/anatase TiO_2@metallic $Ti_3C_2T_x$ hybrids with favorable charge transfer channels for excellent visible-light-photocatalytic performance, *Appl. Catal. B Environ.* 233 (2018) 213–225. https://doi.org/10.1016/j.apcatb.2018.04.012.
35. Y.-Y. Chang, H.N. Han, M. Kim, Analyzing the microstructure and related properties of 2D materials by transmission electron microscopy, *Appl. Microsc.* 49 (2019) 10. https://doi.org/10.1186/s42649-019-0013-5.
36. K.F. Mak, C. Lee, J. Hone, J. Shan, T.F. Heinz, Atomically thin MoS_2: A new direct-gap semiconductor, *Phys. Rev. Lett.* 105 (2010) 136805. https://doi.org/10.1103/PhysRevLett.105.136805.
37. A. Splendiani, L. Sun, Y. Zhang, T. Li, J. Kim, C.-Y. Chim, G. Galli, F. Wang, Emerging photoluminescence in monolayer MoS_2, *Nano Lett.* 10 (2010) 1271–1275. https://doi.org/10.1021/nl903868w.

38. Y. Zhang, T.-R. Chang, B. Zhou, Y.-T. Cui, H. Yan, Z. Liu, F. Schmitt, J. Lee, R. Moore, Y. Chen, H. Lin, H.-T. Jeng, S.-K. Mo, Z. Hussain, A. Bansil, Z.-X. Shen, Direct observation of the transition from indirect to direct bandgap in atomically thin epitaxial $MoSe_2$, *Nat. Nanotech.* 9 (2014) 111–115. https://doi.org/10.1038/nnano.2013.277.
39. C. Lee, X. Wei, J.W. Kysar, J. Hone, Measurement of the elastic properties and intrinsic strength of monolayer graphene, *Science.* 321 (2008) 385–388. https://doi.org/10.1126/science.1157996.
40. Q. Zheng, J. Lee, X. Shen, X. Chen, J.-K. Kim, Graphene-based wearable piezoresistive physical sensors, *Mater. Today.* 36 (2020) 158–179. https://doi.org/10.1016/j.mattod.2019.12.004.
41. X. Peng, L. Peng, C. Wu, Y. Xie, Two dimensional nanomaterials for flexible supercapacitors, *Chem. Soc. Rev.* 43 (2014) 3303. https://doi.org/10.1039/c3cs60407a.
42. S. Bertolazzi, J. Brivio, A. Kis, Stretching and breaking of ultrathin MoS_2, *ACS Nano.* 5 (2011) 9703–9709. https://doi.org/10.1021/nn203879f.
43. Y. Zheng, F. Sun, X. Han, J. Xu, X. Bu, Recent progress in 2D metal-organic frameworks for optical applications, *Adv. Opt. Mater.* 8 (2020) 2000110. https://doi.org/10.1002/adom.202000110.
44. A.D. Dillon, M.J. Ghidiu, A.L. Krick, J. Griggs, S.J. May, Y. Gogotsi, M.W. Barsoum, A.T. Fafarman, Highly conductive optical quality solution-processed films of 2D titanium carbide, *Adv. Funct. Mater.* 26 (2016) 4162–4168. https://doi.org/10.1002/adfm.201600357.
45. L. Cheng, X. Wang, F. Gong, T. Liu, Z. Liu, 2D nanomaterials for cancer theranostic applications, *Adv. Mater.* 32 (2020) 1902333. https://doi.org/10.1002/adma.201902333.
46. M. Devi, Application of 2D nanomaterials as fluorescent biosensors, in: L. Singh, D.M. Mahapatra (Eds.), *ACS Symposium Series, American Chemical Society*, Washington, DC, 2020: pp. 117–141. https://doi.org/10.1021/bk-2020-1353.ch006.
47. Q. Guo, N. Chen, L. Qu, Two-dimensional materials of group-IVA boosting the development of energy storage and conversion, *Carbon Energy.* 2 (2020) 54–71. https://doi.org/10.1002/cey2.34.
48. Y. Sun, H. Cheng, S. Gao, Z. Sun, Q. Liu, Q. Liu, F. Lei, T. Yao, J. He, S. Wei, Y. Xie, Freestanding tin disulfide single-layers realizing efficient visible-light water splitting, *Angew. Chem. Int. Ed.* 51 (2012) 8727–8731. https://doi.org/10.1002/anie.201204675.
49. Y. Jung, Y. Zhou, J.J. Cha, Intercalation in two-dimensional transition metal chalcogenides, *Inorg. Chem. Front.* 3 (2016) 452–463. https://doi.org/10.1039/C5QI00242G.
50. F. Xiong, H. Wang, X. Liu, J. Sun, M. Brongersma, E. Pop, Y. Cui, Li Intercalation in MoS_2: In Situ observation of its dynamics and tuning optical and electrical properties, *Nano Lett.* 15 (2015) 6777–6784. https://doi.org/10.1021/acs.nanolett.5b02619.
51. J. Wan, S.D. Lacey, J. Dai, W. Bao, M.S. Fuhrer, L. Hu, Tuning two-dimensional nanomaterials by intercalation: Materials, properties and applications, *Chem. Soc. Rev.* 45 (2016) 6742–6765. https://doi.org/10.1039/C5CS00758E.
52. M.S. Dresselhaus, G. Dresselhaus, Intercalation compounds of graphite, *Adv. Phys.* 51 (2002) 1–186. https://doi.org/10.1080/00018730110113644.
53. W. Bao, J. Wan, X. Han, X. Cai, H. Zhu, D. Kim, D. Ma, Y. Xu, J.N. Munday, H.D. Drew, M.S. Fuhrer, L. Hu, Approaching the limits of transparency and conductivity in graphitic materials through lithium intercalation, *Nat. Commun.* 5 (2014) 4224. https://doi.org/10.1038/ncomms5224.
54. M. Yi, Z. Shen, A review on mechanical exfoliation for the scalable production of graphene, *J. Mater. Chem. A.* 3 (2015) 11700–11715. https://doi.org/10.1039/C5TA00252D.
55. V. Nicolosi, M. Chhowalla, M.G. Kanatzidis, M.S. Strano, J.N. Coleman, Liquid exfoliation of layered materials, *Science.* 340 (2013) 1226419. https://doi.org/10.1126/science.1226419.
56. D. Hanlon, C. Backes, T.M. Higgins, M. Hughes, A. O'Neill, P. King, N. McEvoy, G.S. Duesberg, B. Mendoza Sanchez, H. Pettersson, V. Nicolosi, J.N. Coleman, Production of molybdenum trioxide nanosheets by liquid exfoliation and their application in high-performance supercapacitors, *Chem. Mater.* 26 (2014) 1751–1763. https://doi.org/10.1021/cm500271u.
57. L. Niu, J.N. Coleman, H. Zhang, H. Shin, M. Chhowalla, Z. Zheng, Production of two-dimensional nanomaterials via liquid-based direct exfoliation, *Small.* 12 (2016) 272–293. https://doi.org/10.1002/smll.201502207.
58. M. Naguib, J. Halim, J. Lu, K.M. Cook, L. Hultman, Y. Gogotsi, M.W. Barsoum, New two-dimensional niobium and vanadium carbides as promising materials for Li-ion batteries, *J. Am. Chem. Soc.* 135 (2013) 15966–15969. https://doi.org/10.1021/ja405735d.
59. K. Parvez, Z.-S. Wu, R. Li, X. Liu, R. Graf, X. Feng, K. Müllen, Exfoliation of graphite into graphene in aqueous solutions of inorganic salts, *J. Am. Chem. Soc.* 136 (2014) 6083–6091. https://doi.org/10.1021/ja5017156.

60. Y. Zhang, L. Zhang, C. Zhou, Review of chemical vapor deposition of graphene and related applications, *Acc. Chem. Res.* 46 (2013) 2329–2339. https://doi.org/10.1021/ar300203n.
61. Z. Cai, B. Liu, X. Zou, H.-M. Cheng, Chemical vapor deposition growth and applications of two-dimensional materials and their heterostructures, *Chem. Rev.* 118 (2018) 6091–6133. https://doi.org/10.1021/acs.chemrev.7b00536.
62. Z. Wang, Y. Shen, Y. Ito, Y. Zhang, J. Du, T. Fujita, A. Hirata, Z. Tang, M. Chen, Synthesizing 1T–1H two-phase Mo_{1-x} W_x S_2 monolayers by chemical vapor deposition, *ACS Nano.* 12 (2018) 1571–1579. https://doi.org/10.1021/acsnano.7b08149.
63. K.S. Novoselov, D. Jiang, F. Schedin, T.J. Booth, V.V. Khotkevich, S.V. Morozov, A.K. Geim, Two-dimensional atomic crystals, *Proc. Natl. Acad. Sci. U.S.A.* 102 (2005) 10451–10453. https://doi.org/10.1073/pnas.0502848102.
64. S.Z. Butler, S.M. Hollen, L. Cao, Y. Cui, J.A. Gupta, H.R. Gutiérrez, T.F. Heinz, S.S. Hong, J. Huang, A.F. Ismach, E. Johnston-Halperin, M. Kuno, V.V. Plashnitsa, R.D. Robinson, R.S. Ruoff, S. Salahuddin, J. Shan, L. Shi, M.G. Spencer, M. Terrones, W. Windl, J.E. Goldberger, Progress, challenges, and opportunities in two-dimensional materials beyond graphene, *ACS Nano.* 7 (2013) 2898–2926. https://doi.org/10.1021/nn400280c.
65. X. Li, W. Cai, J. An, S. Kim, J. Nah, D. Yang, R. Piner, A. Velamakanni, I. Jung, E. Tutuc, S.K. Banerjee, L. Colombo, R.S. Ruoff, Large-area synthesis of high-quality and uniform graphene films on copper foils, *Science.* 324 (2009) 1312–1314. https://doi.org/10.1126/science.1171245.
66. K.S. Novoselov, A.K. Geim, S.V. Morozov, D. Jiang, Y. Zhang, S.V. Dubonos, I.V. Grigorieva, A.A. Firsov, Electric field effect in atomically thin carbon films, *Science.* 306 (2004) 666–669. https://doi.org/10.1126/science.1102896.
67. C. Backes, T.M. Higgins, A. Kelly, C. Boland, A. Harvey, D. Hanlon, J.N. Coleman, Guidelines for exfoliation, characterization and processing of layered materials produced by liquid exfoliation, *Chem. Mater.* 29 (2017) 243–255. https://doi.org/10.1021/acs.chemmater.6b03335.
68. M.Y. Lukianov, A.A. Rubekina, J.V. Bondareva, A.V. Sybachin, G.D. Diudbin, K.I. Maslakov, D.G. Kvashnin, O.G. Klimova-Korsmik, E.A. Shirshin, S.A. Evlashin, Photoluminescence of two-dimensional MoS_2 nanosheets produced by liquid exfoliation, *Nanomaterials.* 13 (2023) 1982. https://doi.org/10.3390/nano13131982.
69. Y. Hernandez, V. Nicolosi, M. Lotya, F.M. Blighe, Z. Sun, S. De, I.T. McGovern, B. Holland, M. Byrne, Y.K. Gun'Ko, J.J. Boland, P. Niraj, G. Duesberg, S. Krishnamurthy, R. Goodhue, J. Hutchison, V. Scardaci, A.C. Ferrari, J.N. Coleman, High-yield production of graphene by liquid-phase exfoliation of graphite, *Nat. Nanotech.* 3 (2008) 563–568. https://doi.org/10.1038/nnano.2008.215.
70. J.N. Coleman, M. Lotya, A. O'Neill, S.D. Bergin, P.J. King, U. Khan, K. Young, A. Gaucher, S. De, R.J. Smith, I.V. Shvets, S.K. Arora, G. Stanton, H.-Y. Kim, K. Lee, G.T. Kim, G.S. Duesberg, T. Hallam, J.J. Boland, J.J. Wang, J.F. Donegan, J.C. Grunlan, G. Moriarty, A. Shmeliov, R.J. Nicholls, J.M. Perkins, E.M. Grieveson, K. Theuwissen, D.W. McComb, P.D. Nellist, V. Nicolosi, Two-dimensional nanosheets produced by liquid exfoliation of layered materials, *Science.* 331 (2011) 568–571. https://doi.org/10.1126/science.1194975.
71. P. Karfa, K. Chandra Majhi, R. Madhuri, Synthesis of two-dimensional nanomaterials, in: *Two-Dimensional Nanostructures for Biomedical Technology*, Elsevier, 2020: pp. 35–71. https://doi.org/10.1016/B978-0-12-817650-4.00002-4.
72. Y.-H. Lee, X.-Q. Zhang, W. Zhang, M.-T. Chang, C.-T. Lin, K.-D. Chang, Y.-C. Yu, J.T.-W. Wang, C.-S. Chang, L.-J. Li, T.-W. Lin, Synthesis of large-area MoS_2 atomic layers with chemical vapor deposition, *Adv. Mater.* 24 (2012) 2320–2325. https://doi.org/10.1002/adma.201104798.
73. Q. Ji, Y. Zhang, Y. Zhang, Z. Liu, Chemical vapour deposition of group-VIB metal dichalcogenide monolayers: Engineered substrates from amorphous to single crystalline, *Chem. Soc. Rev.* 44 (2015) 2587–2602. https://doi.org/10.1039/C4CS00258J.
74. S.L. Wong, H. Liu, D. Chi, Recent progress in chemical vapor deposition growth of two-dimensional transition metal dichalcogenides, *Prog. Cryst. Growth Charact. Mater.* 62 (2016) 9–28. https://doi.org/10.1016/j.pcrysgrow.2016.06.002.
75. C. Tan, H. Zhang, Wet-chemical synthesis and applications of non-layer structured two-dimensional nanomaterials, *Nat. Commun.* 6 (2015) 7873. https://doi.org/10.1038/ncomms8873.
76. A. Ghosh, M. Miah, A. Bera, S.K. Saha, B. Ghosh, Synthesis of freestanding 2D CuO nanosheets at room temperature through a simple surfactant free co-precipitation process and its application as electrode material in supercapacitors, *J. Alloys Compd.* 862 (2021) 158549. https://doi.org/10.1016/j.jallcom.2020.158549.

77. A. Rabenau, The role of hydrothermal synthesis in preparative chemistry, *Angew. Chem. Int. Ed. Engl.* 24 (1985) 1026–1040. https://doi.org/10.1002/anie.198510261.
78. Y. Feng, F. Zhou, Q. Deng, C. Peng, Solvothermal synthesis of in situ nitrogen-doped Ti_3C_2 MXene fluorescent quantum dots for selective Cu^{2+} detection, *Ceram. Int.* 46 (2020) 8320–8327. https://doi.org/10.1016/j.ceramint.2019.12.063.
79. S. Yan, X. Zhu, J. Dong, Y. Ding, S. Xiao, 2D materials integrated with metallic nanostructures: fundamentals and optoelectronic applications, *Nanophotonics.* 9 (2020) 1877–1900. https://doi.org/10.1515/nanoph-2020-0074.
80. S. Zhuiykov, E. Kats, B. Carey, S. Balendhran, Proton intercalated two-dimensional WO_3 nano-flakes with enhanced charge-carrier mobility at room temperature, *Nanoscale.* 6 (2014) 15029–15036. https://doi.org/10.1039/C4NR05008H.
81. B. Jia, J. Yang, R. Hao, L. Li, L. Guo, Confined synthesis of ultrathin amorphous metal-oxide nanosheets, *ACS Mater. Lett.* 2 (2020) 610–615. https://doi.org/10.1021/acsmaterialslett.0c00114.
82. T. Yuan, Z. Hu, Y. Zhao, J. Fang, J. Lv, Q. Zhang, Z. Zhuang, L. Gu, S. Hu, Two-dimensional amorphous SnO_x from liquid metal: Mass production, phase transfer, and electrocatalytic CO_2 reduction toward formic acid, *Nano Lett.* 20 (2020) 2916–2922. https://doi.org/10.1021/acs.nanolett.0c00844.
83. M.M.Y.A. Alsaif, S. Kuriakose, S. Walia, N. Syed, A. Jannat, B.Y. Zhang, F. Haque, M. Mohiuddin, T. Alkathiri, N. Pillai, T. Daeneke, J.Z. Ou, A. Zavabeti, 2D SnO/In_2O_3 van der waals heterostructure photodetector based on printed oxide skin of liquid metals, *Adv. Mater. Interfaces.* 6 (2019) 1900007. https://doi.org/10.1002/admi.201900007.
84. S. Manzeli, D. Ovchinnikov, D. Pasquier, O.V. Yazyev, A. Kis, 2D transition metal dichalcogenides, *Nat. Rev. Mater.* 2 (2017) 17033. https://doi.org/10.1038/natrevmats.2017.33.
85. Q.H. Wang, K. Kalantar-Zadeh, A. Kis, J.N. Coleman, M.S. Strano, Electronics and optoelectronics of two-dimensional transition metal dichalcogenides, *Nat. Nanotech.* 7 (2012) 699–712. https://doi.org/10.1038/nnano.2012.193.
86. T. Cao, G. Wang, W. Han, H. Ye, C. Zhu, J. Shi, Q. Niu, P. Tan, E. Wang, B. Liu, J. Feng, Valley-selective circular dichroism of monolayer molybdenum disulphide, *Nat. Commun.* 3 (2012) 887. https://doi.org/10.1038/ncomms1882.
87. W. Shi, J. Ye, Y. Zhang, R. Suzuki, M. Yoshida, J. Miyazaki, N. Inoue, Y. Saito, Y. Iwasa, Superconductivity series in transition metal dichalcogenides by ionic gating, *Sci. Rep.* 5 (2015) 12534. https://doi.org/10.1038/srep12534.
88. M.M. Ugeda, A.J. Bradley, Y. Zhang, S. Onishi, Y. Chen, W. Ruan, C. Ojeda-Aristizabal, H. Ryu, M.T. Edmonds, H.-Z. Tsai, A. Riss, S.-K. Mo, D. Lee, A. Zettl, Z. Hussain, Z.-X. Shen, M.F. Crommie, Characterization of collective ground states in single-layer $NbSe_2$, *Nat. Phys.* 12 (2016) 92–97. https://doi.org/10.1038/nphys3527.
89. G. Deysher, C.E. Shuck, K. Hantanasirisakul, N.C. Frey, A.C. Foucher, K. Maleski, A. Sarycheva, V.B. Shenoy, E.A. Stach, B. Anasori, Y. Gogotsi, Synthesis of Mo_4VAlC_4 MAX phase and two-dimensional Mo_4VC_4 MXene with five atomic layers of transition metals, *ACS Nano.* 14 (2020) 204–217. https://doi.org/10.1021/acsnano.9b07708.
90. M. Naguib, V.N. Mochalin, M.W. Barsoum, Y. Gogotsi, 25th anniversary article: MXenes: A new family of two-dimensional materials, *Adv. Mater.* 26 (2014) 992–1005. https://doi.org/10.1002/adma.201304138.
91. Q. Tang, Z. Zhou, P. Shen, Are MXenes promising anode materials for Li Ion batteries? Computational studies on electronic properties and Li storage capability of Ti_3C_2 and $Ti_3C_2X_2$ (X=F, OH) monolayer, *J. Am. Chem. Soc.* 134 (2012) 16909–16916. https://doi.org/10.1021/ja308463r.
92. K. Rasool, R.P. Pandey, P.A. Rasheed, S. Buczek, Y. Gogotsi, K.A. Mahmoud, Water treatment and environmental remediation applications of two-dimensional metal carbides (MXenes), *Mater. Today.* 30 (2019) 80–102. https://doi.org/10.1016/j.mattod.2019.05.017.
93. J. Hu, X. Tang, Q. Dai, Z. Liu, H. Zhang, A. Zheng, Z. Yuan, X. Li, Layered double hydroxide membrane with high hydroxide conductivity and ion selectivity for energy storage device, *Nat. Commun.* 12 (2021) 3409. https://doi.org/10.1038/s41467-021-23721-9.
94. R. Bayan, N. Karak, Polymer nanocomposites based on two-dimensional nanomaterials, in: *Two-Dimensional Nanostructures for Biomedical Technology*, Elsevier, 2020: pp. 249–279. https://doi.org/10.1016/B978-0-12-817650-4.00008-5.
95. T. Ye, W. Huang, L. Zeng, M. Li, J. Shi, CeO_{2-x} platelet from monometallic cerium layered double hydroxides and its photocatalytic reduction of CO_2, *Appl. Catal. B Environ.* 210 (2017) 141–148. https://doi.org/10.1016/j.apcatb.2017.03.051.

96. Z.W. Seh, J. Kibsgaard, C.F. Dickens, I. Chorkendorff, J.K. Nørskov, T.F. Jaramillo, Combining theory and experiment in electrocatalysis: Insights into materials design, *Science.* 355 (2017) eaad4998. https://doi.org/10.1126/science.aad4998.
97. D. Saliba, A. Ezzeddine, R. Sougrat, N.M. Khashab, M. Hmadeh, M. Al-Ghoul, Cadmium-aluminum layered double hydroxide microspheres for photocatalytic CO_2 reduction, *ChemSusChem.* 9 (2016) 800–805. https://doi.org/10.1002/cssc.201600088.
98. K. Teramura, S. Iguchi, Y. Mizuno, T. Shishido, T. Tanaka, Photocatalytic conversion of CO_2 in water over layered double hydroxides, *Angew. Chem.* 124 (2012) 8132–8135. https://doi.org/10.1002/ange.201201847.
99. J. Hong, W. Zhang, Y. Wang, T. Zhou, R. Xu, Photocatalytic reduction of carbon dioxide over self-assembled carbon nitride and layered double hydroxide: The role of carbon dioxide enrichment, *ChemCatChem.* 6 (2014) 2315–2321. https://doi.org/10.1002/cctc.201402195.
100. C. Lin, H. Hao, L. Mei, M. Wu, Metal-free two-dimensional nanomaterial-mediated photothermal tumor therapy, *Smart Mater. Med.* 1 (2020) 150–167. https://doi.org/10.1016/j.smaim.2020.09.001.
101. Z. Li, Y. Chen, T. Ma, Y. Jiang, J. Chen, H. Pan, W. Sun, 2D metal-free nanomaterials beyond graphene and its analogues toward electrocatalysis applications, *Adv. Energy Mater.* 11 (2021) 2101202. https://doi.org/10.1002/aenm.202101202.
102. X. She, H. Xu, Y. Xu, J. Yan, J. Xia, L. Xu, Y. Song, Y. Jiang, Q. Zhang, H. Li, Exfoliated graphene-like carbon nitride in organic solvents: Enhanced photocatalytic activity and highly selective and sensitive sensor for the detection of trace amounts of Cu^{2+}, *J. Mater. Chem. A.* 2 (2014) 2563. https://doi.org/10.1039/c3ta13768f.
103. B.V. Lotsch, M. Döblinger, J. Sehnert, L. Seyfarth, J. Senker, O. Oeckler, W. Schnick, Unmasking melon by a complementary approach employing electron diffraction, solid-state NMR spectroscopy, and theoretical calculations—structural characterization of a carbon nitride polymer, *Chem. Eur. J.* 13 (2007) 4969–4980. https://doi.org/10.1002/chem.200601759.
104. M.A. Qamar, M. Javed, S. Shahid, M. Shariq, M.M. Fadhali, S.K. Ali, Mohd.S. Khan, Synthesis and applications of graphitic carbon nitride (g-C_3N_4) based membranes for wastewater treatment: A critical review, *Heliyon.* 9 (2023) e12685. https://doi.org/10.1016/j.heliyon.2022.e12685.
105. Y. Wang, L. Liu, T. Ma, Y. Zhang, H. Huang, 2D graphitic carbon nitride for energy conversion and storage, *Adv. Funct. Mater.* 31 (2021) 2102540. https://doi.org/10.1002/adfm.202102540.
106. J. Wu, Y. Chen, J. Wu, K. Hippalgaonkar, Perspectives on thermoelectricity in layered and 2D materials, *Adv. Electron. Mater.* 4 (2018) 1800248. https://doi.org/10.1002/aelm.201800248.
107. M. Zelisko, Y. Hanlumyuang, S. Yang, Y. Liu, C. Lei, J. Li, P.M. Ajayan, P. Sharma, Anomalous piezoelectricity in two-dimensional graphene nitride nanosheets, *Nat. Commun.* 5 (2014) 4284. https://doi.org/10.1038/ncomms5284.
108. R.R. Nair, P. Blake, A.N. Grigorenko, K.S. Novoselov, T.J. Booth, T. Stauber, N.M.R. Peres, A.K. Geim, Fine structure constant defines visual transparency of graphene, *Science.* 320 (2008) 1308–1308. https://doi.org/10.1126/science.1156965.
109. A.A. Balandin, S. Ghosh, W. Bao, I. Calizo, D. Teweldebrhan, F. Miao, C.N. Lau, Superior thermal conductivity of single-layer graphene, *Nano Lett.* 8 (2008) 902–907. https://doi.org/10.1021/nl0731872.
110. M. Qiu, W.X. Ren, T. Jeong, M. Won, G.Y. Park, D.K. Sang, L.-P. Liu, H. Zhang, J.S. Kim, Omnipotent phosphorene: A next-generation, two-dimensional nanoplatform for multidisciplinary biomedical applications, *Chem. Soc. Rev.* 47 (2018) 5588–5601. https://doi.org/10.1039/C8CS00342D.
111. A. Rawat, N. Jena, D. Dimple, A. De Sarkar, A comprehensive study on carrier mobility and artificial photosynthetic properties in group VI B transition metal dichalcogenide monolayers, *J. Mater. Chem. A.* 6 (2018) 8693–8704. https://doi.org/10.1039/C8TA01943F.
112. M.T. Ong, E.J. Reed, Engineered piezoelectricity in graphene, *ACS Nano.* 6 (2012) 1387–1394. https://doi.org/10.1021/nn204198g.
113. W. Liu, B. Ullah, C.-C. Kuo, X. Cai, Two-dimensional nanomaterials-based polymer composites: Fabrication and energy storage applications, *Adv. Polym. Technol.* 2019 (2019) 1–15. https://doi.org/10.1155/2019/4294306.
114. G. Eda, C. Mattevi, H. Yamaguchi, H. Kim, M. Chhowalla, Insulator to semimetal transition in graphene oxide, *J. Phys. Chem. C.* 113 (2009) 15768–15771. https://doi.org/10.1021/jp9051402.
115. R. Que, Q. Shao, Q. Li, M. Shao, S. Cai, S. Wang, S.-T. Lee, Flexible nanogenerators based on graphene oxide films for acoustic energy harvesting, *Angew. Chem.* 124 (2012) 5514–5518. https://doi.org/10.1002/ange.201200773.
116. S. Rana, V. Singh, B. Singh, Recent trends in 2D materials and their polymer composites for effectively harnessing mechanical energy, *iScience.* 25 (2022) 103748. https://doi.org/10.1016/j.isci.2022.103748.

117. L. Liu, Y.P. Feng, Z.X. Shen, Structural and electronic properties of *h* -BN, *Phys. Rev. B.* 68 (2003) 104102. https://doi.org/10.1103/PhysRevB.68.104102.
118. J. Xue, J. Sanchez-Yamagishi, D. Bulmash, P. Jacquod, A. Deshpande, K. Watanabe, T. Taniguchi, P. Jarillo-Herrero, B.J. LeRoy, Scanning tunnelling microscopy and spectroscopy of ultra-flat graphene on hexagonal boron nitride, *Nat. Mater.* 10 (2011) 282–285. https://doi.org/10.1038/nmat2968.
119. C.R. Dean, A.F. Young, I. Meric, C. Lee, L. Wang, S. Sorgenfrei, K. Watanabe, T. Taniguchi, P. Kim, K.L. Shepard, J. Hone, Boron nitride substrates for high-quality graphene electronics, *Nat. Nanotech.* 5 (2010) 722–726. https://doi.org/10.1038/nnano.2010.172.
120. X. Gao, Y. Yao, X. Meng, Recent development on BN-based photocatalysis: A review, *Mater. Sci. Semicond. Process.* 120 (2020) 105256. https://doi.org/10.1016/j.mssp.2020.105256.
121. J. Wang, F. Ma, M. Sun, Graphene, hexagonal boron nitride, and their heterostructures: Properties and applications, *RSC Adv.* 7 (2017) 16801–16822. https://doi.org/10.1039/C7RA00260B.
122. P. Fan, L. Wang, J. Yang, F. Chen, M. Zhong, Graphene/poly(vinylidene fluoride) composites with high dielectric constant and low percolation threshold, *Nanotechnology.* 23 (2012) 365702. https://doi.org/10.1088/0957-4484/23/36/365702.
123. T. Ramanathan, A.A. Abdala, S. Stankovich, D.A. Dikin, M. Herrera-Alonso, R.D. Piner, D.H. Adamson, H.C. Schniepp, X. Chen, R.S. Ruoff, S.T. Nguyen, I.A. Aksay, R.K. Prud'Homme, L.C. Brinson, Functionalized graphene sheets for polymer nanocomposites, *Nat. Nanotech.* 3 (2008) 327–331. https://doi.org/10.1038/nnano.2008.96.
124. J. Liang, Y. Huang, L. Zhang, Y. Wang, Y. Ma, T. Guo, Y. Chen, Molecular-level dispersion of graphene into Poly(vinyl alcohol) and effective reinforcement of their nanocomposites, *Adv. Funct. Mater.* 19 (2009) 2297–2302. https://doi.org/10.1002/adfm.200801776.
125. T.-L. Li, S.L.-C. Hsu, Preparation and properties of thermally conductive photosensitive polyimide/boron nitride nanocomposites, J. Appl. Polym. Sci. 121 (2011) 916–922. https://doi.org/10.1002/app.33631.
126. K. Sato, H. Horibe, T. Shirai, Y. Hotta, H. Nakano, H. Nagai, K. Mitsuishi, K. Watari, Thermally conductive composite films of hexagonal boron nitride and polyimide with affinity-enhanced interfaces, *J. Mater. Chem.* 20 (2010) 2749. https://doi.org/10.1039/b924997d.

2 Defects and Vacancies Effects on the Optoelectronic Properties of Two-Dimensional Nanomaterials

Sefako J. Mofokeng, Teboho P. Mokoena, Mandla B. Chabalala, Pebetsi Thokwane, Luyanda L. Noto, and Makgamathe J. Sithole

2.1 INTRODUCTION

The sustainability and tenacity of various materials at the nanoscale are ongoing research topics. According to International Organization for Standardization (ISO), these materials typically have dimensions of 100 nm or smaller, where their distinctive characteristics allow for innovative applications. The size of a nanometer is one-billionth (10^{-9}) of a meter. The concept of nanoscience was foreseen by renowned physicist Richard Feynman in 1959 at the annual meeting of the American Physical Society (APS). In his talk titled, "There's plenty of room at the bottom – an invitation to start a new branch of physics" at the APS, he persuaded scientists that studying materials at the nanoscale would be more advantageous for various applications in the long run. Although he did not invent the word "nanotechnology," many of his projections in this field have turned out to be accurate because nanotechnology involves the use of materials and structures with dimensions between 1 and 100 nm. What is more intriguing is that it has been discovered that the chemical, physical, mechanical, clinical, and thermodynamical features of materials at the nanoscale are fundamentally and advantageously distinct from those of individual atoms and molecules or bulk matter [1,2]. These nanomaterials have been fabricated into a wide range of geometries, including wires, rods, tubes, thin films, and spheres, to obtain a high surface-to-volume ratio and mechanical strength.

Nanomaterials typically range in size approximately between 1 and 100 nm, as previously mentioned. However, these materials are categorized into different groups based on their dimensions (D), such as zero-dimensional (0D), one-dimensional (1D), two-dimensional (2D), and three-dimensional (3D) nanostructures. Materials with 0D nanostructures have dimensions no larger than 100 nm (e.g., nanospheres, nanoclusters, quantum dots, clusters, solids, and hollow nanoparticles), whereas 1D nanostructured materials are known to have two dimensions measured inside the nanoscale and one dimension outside it (e.g., carbon nanotubes, fibers, nanorods, and nanowires). Furthermore, 2D nanostructured materials are described as containing two dimensions that are not limited to the nanoscale and one dimension that is contained within the nanoscale (for example, nano-textured surfaces, nanofilms, nanolayers, and nanocoatings). Last but not least, 3D nanostructured materials—such as polycrystal nanostructures, nano/composites, pillars, and electromechanical—are thought of as compounds that have no boundaries for the nanoscale in any dimension [3,4], as shown in Figure 2.1.

Currently, 2D nanostructured materials have piqued the interest of researchers due to their degradability, biocompatibility, high specific surface area, rich structural characteristics, anisotropic chemical or physical properties, and versatile functionalities and diverse applications in drug

DOI: 10.1201/9781003436942-2

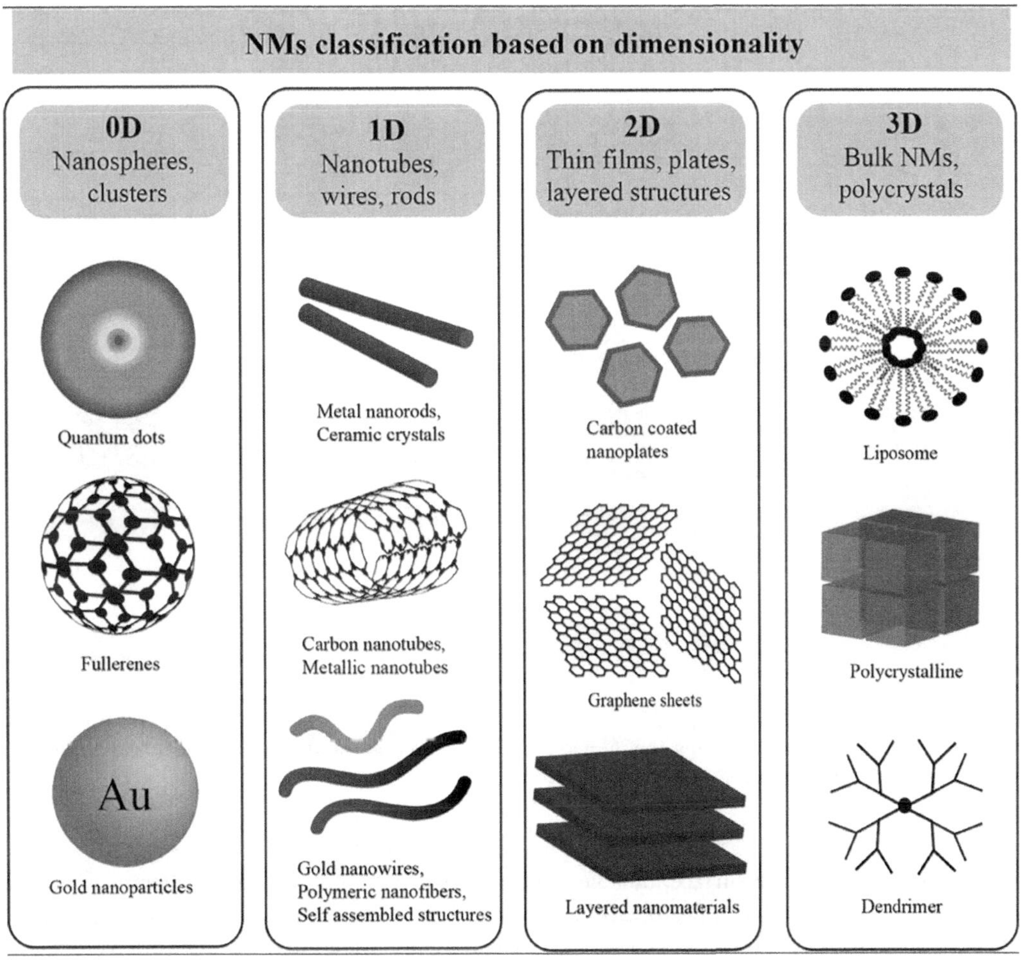

FIGURE 2.1 Classification of nanomaterial according to their dimensions. (Reprinted with permission from Ref. [5], Copyright © 2021 Elsevier.)

delivery, additive manufacturing, bio-engineering, regenerative medicine, bio-sensors, cancer therapeutics, bio-imaging, and others. Another intriguing aspect of 2D nanostructured materials is their ease of synthesis [6–9].

2D nanomaterials are layered materials composed of thin layers with a thickness of at least one atomic layer [11]. Graphene, MXenes, Monoelemental Xenes, Hexagonal boron nitride, Semiconducting transition metal dichalcogenides (STMDs), and 2D COF/MOFs are among the 2D nanomaterials families [9,11]. Figure 2.2 depicts the successful experimental fabrication of several 2D nanomaterials. Many scientists have used these 2D nanomaterials and their future markets in the above-mentioned applications. Remarkably, these 2D nanomaterials with novel physical properties represent a significant opportunity for technological innovation in the next-generation optoelectronic, electronic devices, renewable energy, storage systems, light sources, optical modulators, photodetectors, and processing modules with faster response, higher density, and smaller footprint [10,12,13]. In this context, scientists have recently expressed an interest in learning more about the optoelectronic properties of 2D nanomaterials due to their rich structures and optoelectronic properties. Arsenene, phosphorene, bismuthene, antimonene, and monoelemental materials from group-VA are classified as 2D nanomaterials with optical nonlinearity, incredible structures,

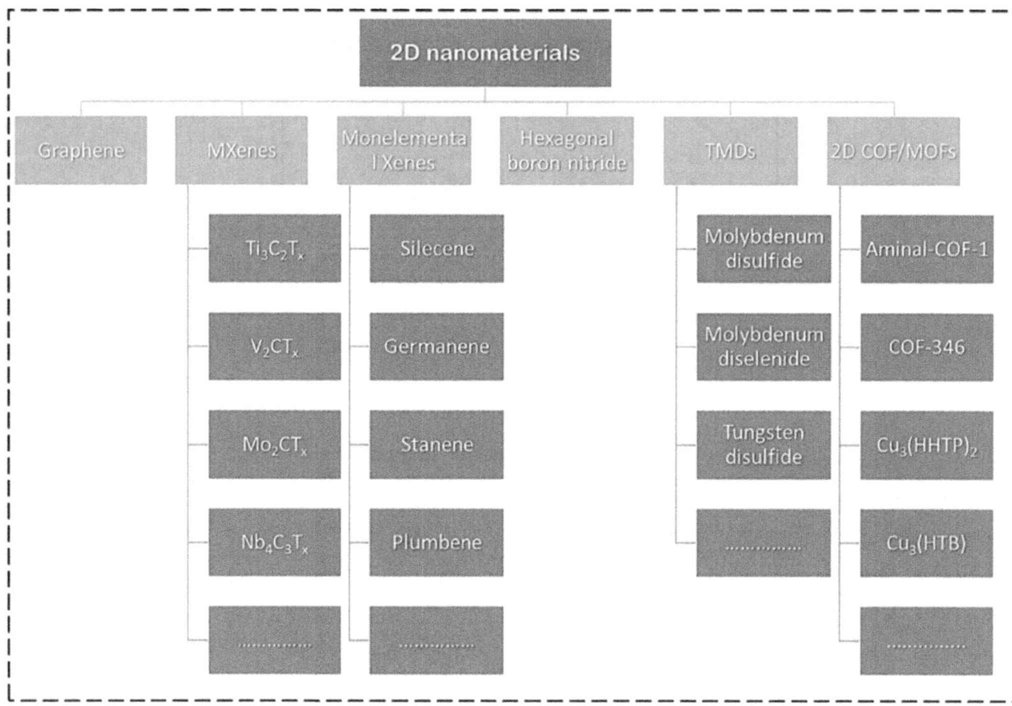

FIGURE 2.2 A diagram depicting the numerous 2D nanomaterial families. (Reprinted with permission from Ref. [10]. Copyright © 2023 Elsevier.)

higher conductivity, large surface, tunable electronic structure, and anisotropic chemical/physical and optoelectronic properties [7–10,12,14]. Because of the aforementioned outstanding properties of these 2D nanomaterials, they are currently selected for a wide range of applications including hydrogen generation for energy storage, optoelectronic devices, cancer treatment, electromagnetic shielding, and water splitting. Extensive research on 2D nanomaterials revealed that some 2D nanomaterials are essential components of bulk layered van der Waals (vdW) heterostructures' crystals. However, because these 2D nanomaterials are naturally thin, they provide more chances for creating functional heterostructures, making them ideal for optoelectronic devices [7,15,16].

2D nanomaterials typically have certain lattice flaws, which vary depending on the synthesis processes used, such as the methods mentioned previously. Depending on the synthesis technique, numerous lattice defects, such as charged impurities, vacancies, grain boundaries, intrinsic defects, extrinsic defects [17], local atomic defects [15], self-interstitials, and vacancies [18], can be introduced into 2D nanomaterials either accidently or deliberately [19]. The benefits of these lattice flaws are that they are more effective in controlling the properties of 2D nanomaterials. Defect engineering is a method of modifying the properties of 2D nanomaterials by eliminating unfavorable defects and adding helpful imperfections in 2D nanomaterials for improved performance of electrical and optoelectronic devices [20]. As a result, we report on the recent effects of different flaws and vacancies on 2D nanomaterials for optoelectronics applications in this chapter.

2.2 SYNTHESIS METHODS FOR 2D NANOMATERIALS

The synthesis of nanomaterials continues to rise in prominence due to the unique optical, magnetic, electrical, mechanical, and chemical properties that nanoparticles have in contrast with bulk materials. Exploration of the distinctive characteristics and phenomena of nanoparticles in order to fully explore their potential applications in science and technology necessitates fabrication and process as main issues in the fields of nanoscience and nanotechnology. Several technical tactics and procedures have been researched to produce nanomaterials. A material's qualities determine how well it operates. The thermodynamics and kinetics of the synthesis influence the qualities in turn, which affect the atomic structure, composition, microstructure, defects, and interfaces [14]. However, the nanostructured material that was ideal for one application might be advantageous in another if it had been produced using entirely distinct processes and techniques than the one employed in the first. Similarly, how a nanomaterial is produced and created determines not just its structure but also its physical form. It is simple to produce and synthesize an amorphous nanostructured thin film using the plasma magnetron sputtering process, for example. However, due to the nature of the approach, it may be challenging to obtain a similar structure utilizing the thermal evaporation method [21]. In this context, it is necessary to address and overcome some of the barriers, such as shape control and particle size distribution (mono-dispersive: all particles of the same size), to build nanostructured materials utilizing any method for modern nanotechnology applications [14,21].

2.3 CLASSIFICATION OF SYNTHESIS METHODS OF 2D NANOMATERIALS

Top-down or bottom-up approaches are most commonly used in the synthesis of 2D nanomaterials. 2D nanostructured materials can be created by mechanically compressing bulk materials. The top-down method is extremely powerful since it is used for the manufacturing of nanomaterials, which involves the reduction of bulk material into the nano-regime and has the capacity to produce reproducible microscale products. However, when the dimensions of the structures to be targeted approach those of the nanoscale, implementation of the concept becomes increasingly problematic. Furthermore, the cost of synthesizing nanoparticles using this method is relatively high [22]. Although top-down approaches are simple to apply, they are ineffective for producing particles with irregular shapes or in extremely small sizes. As a result, the most major disadvantage of utilizing this method is how challenging it is to obtain the appropriate particle size and shape [21]. Researchers, on the other hand, are working diligently and putting out enormous effort to investigate the controlled synthesis of various types of 2D nanostructured materials and their intriguing features. To date, considerable strides have been made in the nano-scale synthesis of 2D materials with regulated compositions, shapes, and sizes with various exfoliation processes. Exfoliation methods for 2D nanomaterials include micromechanical exfoliation, liquid-phase exfoliation (LPE), and ball milling-aided exfoliation, which are all discussed extensively below [10,23,24].

2.3.1 MICROMECHANICAL EXFOLIATION

Micromechanical exfoliation is a popular approach for creating 2D nanomaterials. It is a well-known top-down technique for the simple and low-cost synthesis of 2D nanomaterials. Novoselov and Geim pioneered this strategy in 2004 with the goal of generating graphene. Because of its versatility and inexpensive cost, it has acquired broad application for the manufacturing of 2D materials since that time [25,26]. However, energy must be employed throughout the process to exfoliate the stacked material into nanosheets. Scotch tape is used in the exfoliation technique, which is presently the most common, to exfoliate stacked bulk crystals into nanosheets. The goal of applying mechanical stress/force with the use of Scotch tape is to lower the out-of-plane forces known as vdW without additionally breaking the in-plane rigid covalent bonds inside each layer, resulting in the removal of a just a single or a few layers of 2D nanomaterials. The bulk 2D substance, such as graphite, is first stuck to the adhesive area of the Scotch tape, and then the entire thing is pulled off and split into a single layer or a number of layers by sticking it onto a different sticky surface in the usual process. It is critical to repeat this process several times in order to produce the desired thickness of the 2D nanomaterials, specifically flakes. During the transfer operation, split nanoflakes over Scotch tape are stuck to the appropriate substrate and ripped off from the Scotch tape, dropping a single or several layer nanosheet on the surface of the substrate. However, the major critical challenge associated with micromechanical exfoliation is that production on a large scale is not feasible, the yield is low in comparison to other methods of synthesis that use solutions, there is no means to control the thickness, size, or shape of the nanosheets that are produced, and there is a lack of precision and reproducibility throughout the entire process because it is carried out manually. Furthermore, the purity of the material generated frequently contains remnants of the exfoliating agent. This indicates that micromechanical exfoliation may generate strain on the layer of 2D nanomaterials being formed on a substrate throughout the process. This strain may result in a variety of defects, including atomic flaws, wrinkles or ripples, and tiny corrugation. The electrical performance of optoelectronic devices may be compromised as a result of these flaws [27–31].

2.3.2 Liquid-Phase Exfoliation

The liquid-phase exfoliation (LPE) approach involves the breakdown of large layered 2D structures in liquid media into single- and/or few-layer 2D nanostructures by a range of processes such as ultrasonic treatment, electrochemical, and shear exfoliation. The aforementioned methods have demonstrated significant advancements in the production of nearly any 2D nanomaterial in enormous quantities at a reasonable cost and in a manner that is environmentally conscious. The LPE procedures are extremely promising and highly scalable technologies that are currently easily accessible and can be carried out in mild conditions to manufacture massive quantities of exfoliated 2D materials of exceptional quality in a commercially feasible manner. Direct exfoliation treatments have an added benefit of preserving the 2D materials' inherent physical and electrical properties. This may frequently result in greater quality for the exfoliated nanomaterials, with a relatively small number of flaws and functionalization by chemicals during exfoliation. Direct exfoliation procedures have additionally been shown to be more efficient than indirect exfoliation methods [32].

Figure 2.3 depicts a schematic diagram of the LPE method with (i) bulk layered material as the starting material, (ii) dispersion of layered material in a liquid medium, (iii) ultrasonication, and (iv) final dispersion after the ultracentrifugation process. The LPE process typically consists of two critical steps: dispersion of layered material in a liquid medium and purification. In such a situation, the dispersion of layered material in a liquid medium shows individual particles suspended in a liquid dispersant (Figure 2.3b). Ultracentrifugation is commonly used to separate exfoliated from unexfoliated materials (Figure 2.3c). Submerging in a liquid is one of the most efficient ways to reduce the attraction forces. As a result, the appropriate solvent balances the attractive forces exerted between the sheets of 2D nanomaterials and prevents the 2D nanomaterials from aggregating. However, the most effective solvents for dispersing 2D nanomaterials were those that minimized the amount of interfacial tension between the liquid and the 2D nanomaterials [32–35].

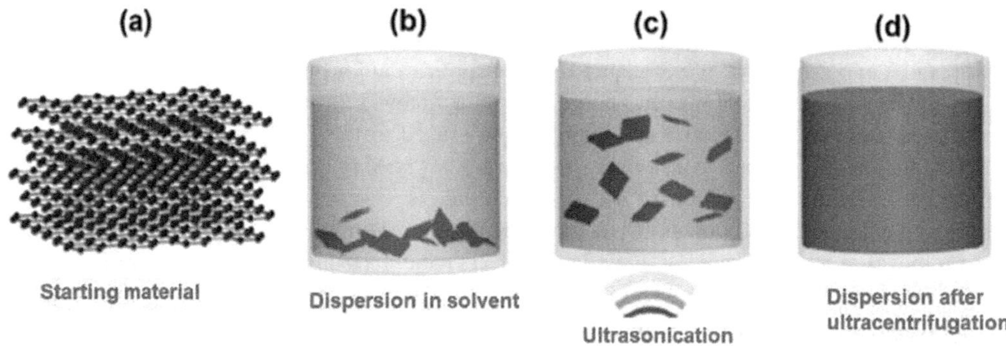

FIGURE 2.3 Schematic representation of LPE process of 2D materials. (Reprinted with permission from Ref. [35], Copyright © 2017 Royal Society of Chemistry.)

Some liquid-based exfoliation methods involve executing chemical reactions on the bulky 2D-powder that make it easier to dissolve in the fluid that is being employed. This is especially important for the processes of generating graphene oxide (GO) from bulk crushed graphite via the process of chemical oxidation, exfoliating transition metal dichalcogenides (TMD)-layered substances/materials through lithium-ion intercalation, resulting in a phase transition from semi-conducting 2H to metal-based 1T phase, and preferential etching of an A-group element in layers of $M_{N+1}AX_n$ (MAX) phases to produce just one-layer MXene nanomaterials. In contrast, bulk 2D materials are exfoliated in aqueous solutions through direct liquid exfoliation, which eliminates the need for chemical interactions. If no chemical interactions occur between the 2D materials and the solvent during the exfoliation process, these methods frequently provide high-quality, faultless 2D nanomaterials with a low number of defects. The produced exfoliated 2D nanomaterials will exhibit appealing material characteristics and will substantially diffuse in the solvent present. Because of this, they can be easily dissolved with other nanomaterial solutions to generate composites and hybrid materials, and they can be cast as a thin film onto any substrate for device and coating applications [32,34,36].

2.3.3 Ball Milling-Assisted Exfoliation

Ball milling is a type of grinding that uses lateral force to grind nanomaterials into extremely fine particles [37–39], in contrast to the sonication or Scotch tape methods of grinding, which predominantly use normal force [40]. The collision between the relatively tiny, hard balls held in a closed container generates a localized high pressure throughout the ball milling process. In many instances, porcelain, flint stones, and stainless steel are used in milling. Certain chemicals can be added to the container during the process to improve the quality of the dispersion and insert functional groups into the surface of the nanomaterial [40,41]. There are some of the elements that can influence the quality of the dispersion such as milling time frame, rotational speed, ball size, and the ratio of 2D nanomaterial to balls [38]. When utilized in the thinning process, low-energy ball milling does not cause substantial damage to the in-plane structure of 2D nanomaterials, and it also produces fewer defects and impurities. Milling at a slow speed for a lengthy period of time guarantees that shear forces dominate, which is appropriate for producing large-scale 2D nanomaterials.

Figure 2.4 depicts the basic mechanism involved in the ball milling-assisted exfoliation process. As shown in Figure 2.4, there are two possible methods for milling-assisted exfoliation: primary and secondary ways. The principal method is known as shear force, which is associated with a good mechanical process for the exfoliation process. The primary method is greatly sought for producing large-sized 2D nanomaterials, such as graphene flakes, whereas the secondary method is related with milling ball vertical impacts or collisions during rolling movements. However, in this secondary approach, huge 2D material flakes can be fractured into small flakes and occasionally even break the crystalline characteristics of structures to generate non-equilibrium or amorphous phases [39]. According to the literature, ball milling at a slow speed for a lengthy period of time reduces the amount of collision or compressive impacts, which limits the fragmentation of large flakes into smaller ones and improves the overall quality of the goods produced by ball milling [33,39]. This process has been used to generate highly curved or closed-shell carbon nanostructures from graphite, to increase the saturation of lithium composition in single-walled carbon nanotubes (SWCNTs), to modify the morphologies of cup-stacked carbon nanotubes, and to generate various carbon nanoparticles from graphitic carbon for use in hydrogen storage applications. Despite the fact that ball milling is simple to use and can be used for powder polymers or monomers, there is a risk of process-induced nanotube damage [42]. These big flakes are prone to being broken by heavy collisions, which may also alter the crystal structure, resulting in a more amorphous bulk. As a result, the number of collisions that occur during the manufacturing of high-quality nanomaterials should be reduced to a minimum [40].

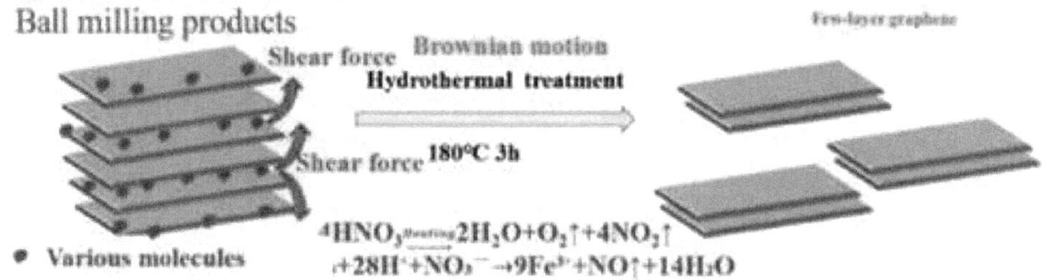

FIGURE 2.4 Schematic illustration of the preparation of 2D nanomaterials via ball milling-assisted exfoliation process. (Reprinted with permission from Ref. [39], Copyright © 2015 Royal Society of Chemistry.)

2.4 BOTTOM-UP TECHNIQUES

Bottom-up manufacturing is an efficient way for creating nanomaterials that are exceedingly thin, of high quality, and have massive lateral dimensions. It is a top-down technique option that has the ability to generate less waste and, as a result, is more cost-effective. The bottom-up technique is the process of developing a nanomaterial from individual components to the larger nanomaterial, either atom-by-atom, molecule-by-molecule, or cluster-by-cluster. Numerous well-known bottom-up methods for producing luminescent nanomaterials have been identified, including the organometallic chemical route, the revere-micelle route, the sol–gel synthesis, colloidal precipitation, hydrothermal synthesis, template-assisted sol–gel, and electrodeposition [21,43,44].

2.4.1 Hydro/Solvothermal Process

The bottom-up methods include hydrothermal and solvothermal methods. Thus, both techniques are considered similar syntheses since they use various solvents such as water (distilled and deionized), ethylene glycol, and polyethylene glycol (PEG). In this instance, solvents are used as the reaction medium to dissolve and recrystallize typically insoluble chemical substances in a high-temperature (~100–1000°C) and high-pressure environment (~1 MPa–1 GPa) [45–48]. Figure 2.5

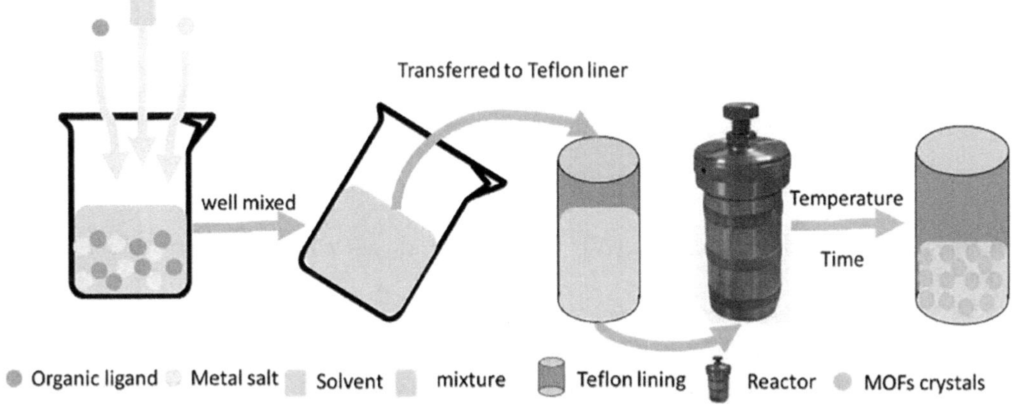

FIGURE 2.5 Schematic diagram of the hydro/solvothermal process. (Reprinted with permission from Ref. [51], Copyright © 2018 MDPI.)

depicts the fundamental mechanism of the hydro/solvothermal synthesis technique. The procedure is often carried out inside an autoclave, which is a form of steel pressure vessel in which temperatures and/or pressures are controlled to obtain the required results. The temperature is raised above the boiling point of the solvent, accomplishing vapor saturation. The hydro/solvothermal method has made a significant contribution to current science and nanotechnology due to its homogeneous precipitation, low cost, easier to synthesize materials in nanoscale, ecologically friendly nature, user-friendliness in terms of scalability, and pure small and uniform end product [21,45]. However, this method can be separated into areas such as hydrothermal synthesis, treatment, and crystal development; the treatment of organic wastes; and the manufacture of functional ceramics powders [21]. Furthermore, the hydrothermal process is separated into categories such as hydrothermal synthesis, treatment, and crystal development; organic waste treatment; and the creation of useful ceramic powder [49]. Hydrothermal methods were widely employed to manufacture several types of MXene-based nanomaterials for the removal of radioactive hazardous contaminants at high temperatures and vapor pressures. This is due to the fact that MXene-based nanomaterials are emerging as viable adsorbents for removing contaminants from the environment [50].

Precursors that match the stoichiometry of the end product are mixed in an organic solvent with reduction and template reagents during the synthesis technique depicted in Figure 2.5. After fully mixing, the ingredients are placed into an airtight autoclave and heated for a lengthy period of time at temperatures ranging from 100 °C to 250 °C. After allowing the products to cool to normal temperature, they are dried, washed, and centrifuged. One of the most prominent benefits of this technology is its capacity to manage the dimensions and form of the goods produced [52]. The hydro/solvothermal technique can also be utilized to create highly crystalline nanomaterials. On the other hand, crystal development information cannot be directly observed, and autoclaves are expensive. One of the method's major limitations is its difficulty to create a big quantity of phosphors in a single process. The comparatively small size of the autoclaves impedes industrial production of phosphors [21,53].

2.4.2 Chemical Vapor Deposition Process

Chemical vapor deposition (CVD) is a versatile and efficient method for creating nanomaterials. This method was an important technique in the field of microelectronics for many decades and remains one of the most appealing approaches utilized today that is capable of overcoming shortcomings provided by advancements in technology. When compared to other methods of synthesis,

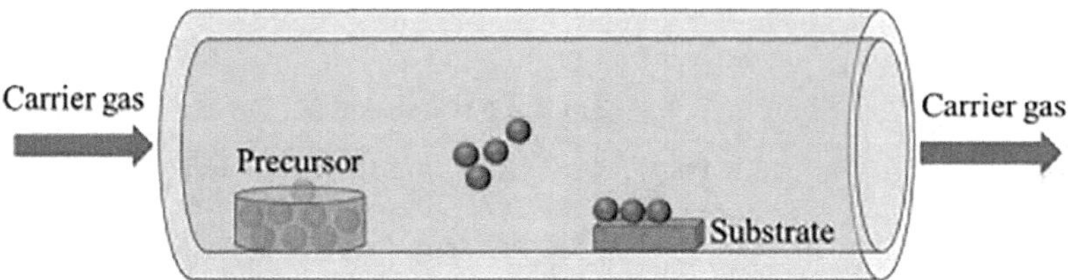

FIGURE 2.6 Schematic diagram of a CVD process setup. (Reprinted with permission from Ref. [57], Copyright © 2021 Elsevier.)

CVD provides the most control in the fabrication of 2D nanomaterials. The CVD process allows for the mass manufacturing of 2D nanomaterials with high crystal clarity, purity, and few imperfections on the substrates. However, in order to conduct additional investigation, this method always necessitates the transfer of nanomaterials from the substrates on which they were deposited. In addition, higher production costs are a disadvantage. Few precursors are hazardous in the sense that they are toxic, combustible, or explosive, and even fewer are inexpensive. The creation of highly poisonous gaseous by-products, as well as the high temperatures at which various CVD techniques deposit their ingredients, limits the substrates that can be used [21,54,55]. However, in order to perform additional examination, the nanomaterials must always be transferred from the substrates on which they were deposited. Furthermore, increasing production costs are a detriment. Few precursors are dangerous in the sense that they are toxic, combustible, or explosive, and even fewer are reasonably priced. Because of the formation of very poisonous gaseous by-products and the high temperatures at which the various CVD procedures deposit their ingredients, the substrates that can be used are limited [55,56].

In this process, the precursors, which can be gases or vapors, can react or decompose on a predetermined substrate, while the chamber is subjected to high temperatures and vacuum, as shown in Figure 2.6. The growth of 2D nanomaterials on the substrate can occur with or without the assistance of catalysts [57]. The controlled delivery of gas-phase reactants, the availability of an enclosed reaction chamber, the discharge of gases, the regulation of reaction pressure, the delivery of an energy source for chemical reactions, the cleaning of exhaust gases to safe and nontoxic levels, and automation process control to increase deposition process stability are all operational requirements for the CVD process. The 2D nanomaterial produced through the CVD method is used in a wide range of practical applications, including electronic, optoelectronic, and solar cell device applications [21].

2.4.3 Physical Vapor Deposition

Physical vapor deposition (PVD) is a vaporization coating technique that involves the transfer of material at the atomic level. This synthesis method typically requires ultra-high vacuum (UHV) settings and all-around regulated, high-purity atomic sources for the development of elemental engineering 2D materials [2]. The process is similar to CVD in several ways, except that in PVD, the precursors, i.e., the material to be deposited, begin in solid form, whereas in CVD, the precursors enter the reaction chamber in gaseous form [58]. The PVC method can be described as follows: (i) the material to be deposited is converted into a vapor by physical means (high-temperature vacuum or gaseous plasma), (ii) the vapor is transported to a low-pressure region from its source to the substrate, and (iii) the vapor condenses on the substrate to form a thin film. As a result, PVD methods are utilized to deposit films ranging in thickness from a few nanometers to thousands of nanometers. They can, however, be utilized to create multilayer coatings, graded composition deposits, extremely thick deposits, and freestanding structures [59]. Sputtering, as opposed to evaporation, is better suited for target materials that are difficult to deposit by evaporation, such as ceramics

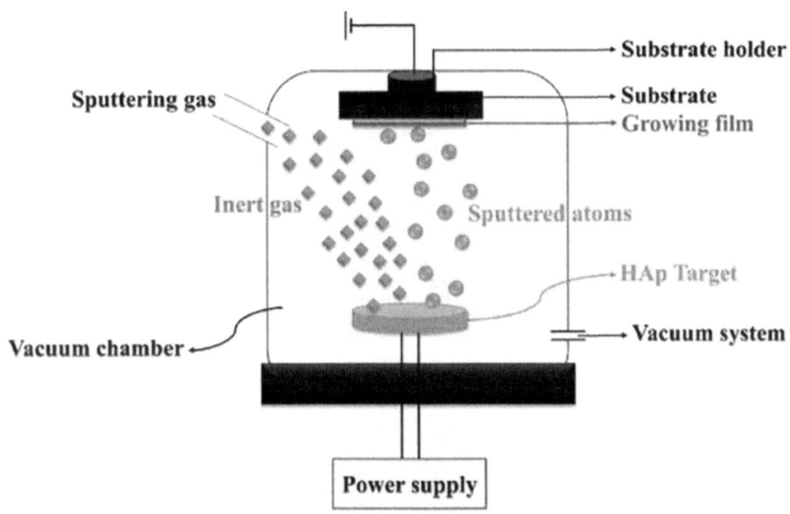

FIGURE 2.7 A schematic representation of a typical PVD process. (Reprinted with permission from Ref. [62], Copyright © 2021 MDPI.)

and refractory metals. Furthermore, sputtered coatings have a higher bonding strength to the substrate than evaporation-deposited coatings [60]. PVD has several advantages, including: (i) coatings formed by PVD may have better properties than the substrate material; (ii) all types of inorganic materials and some types of organic materials can be used; and (iii) the process is more environmentally friendly than many other processes, such as electroplating. However, PVD has some drawbacks, including: (i) issues with coating complex forms; (ii) high process cost and poor production; and (iii) process complexity. Figure 2.7 is a schematic illustration of a typical PVD process. The PVD process often employs sputtering, electron beam evaporation, pulsed laser deposition (PLD), and vacuum arc technologies [58,60–62].

2.5 OPTOELECTRONIC PROPERTIES OF 2D NANOMATERIALS

In addition to the methods discussed above, researchers investigated a variety of methods and strategies for improving the optoelectronic properties of 2D nanomaterials such as graphene, molybdenum disulfide, boron nitride, and black phosphor, which have been established due to their rapid development and growing popularity. Because of the material's ultrathin thickness, these properties are constrained by its intrinsic low light absorption [63,64]. Abdelazeez et al. [65], for example, reported the fabrication of a lateral heterojunction film with a thickness of approximately 1.7 μm for solar cell applications. The thin film was created by depositing 2D molybdenum disulfide (MoS_2) nanosheets (NSs) with selenide (SnSe) in the nanoscale on an indium tin oxide (ITO)/polyethylene terephthalate (PET) substrate using a simple single-step electrophoretic deposition (EPD) procedure. The spectral absorption of the SnSe/MoS_2 heterostructure was found to be higher than that of 2D-MoS_2 and 2D-SnSe nanomaterials alone (see Figure 2.8a). This is owing to the presence of three additional high-intensity peaks in the SnSe/MoS_2 heterostructure, which confirms the enhanced nature of the mixture absorbance behavior. Figure 2.8b displays the voltage versus current density behavior of the constructed SnSe/MoS_2/PET/ITO heterostructure under light (Jph) and dark (Jo) conditions. However, the results demonstrate that the Jo value of the resulting SnSe/MoS_2/PET/ITO heterostructure is quite low. The authors emphasized that the low Jo value is connected to ITO self-conductance and thin films of SnSe/MoS_2. Furthermore, it was discovered that the short-circuit current density (JSC) for the SnSe/MoS_2/PET/ITO photoelectrode is approximately 1.75 mA cm^{-2}

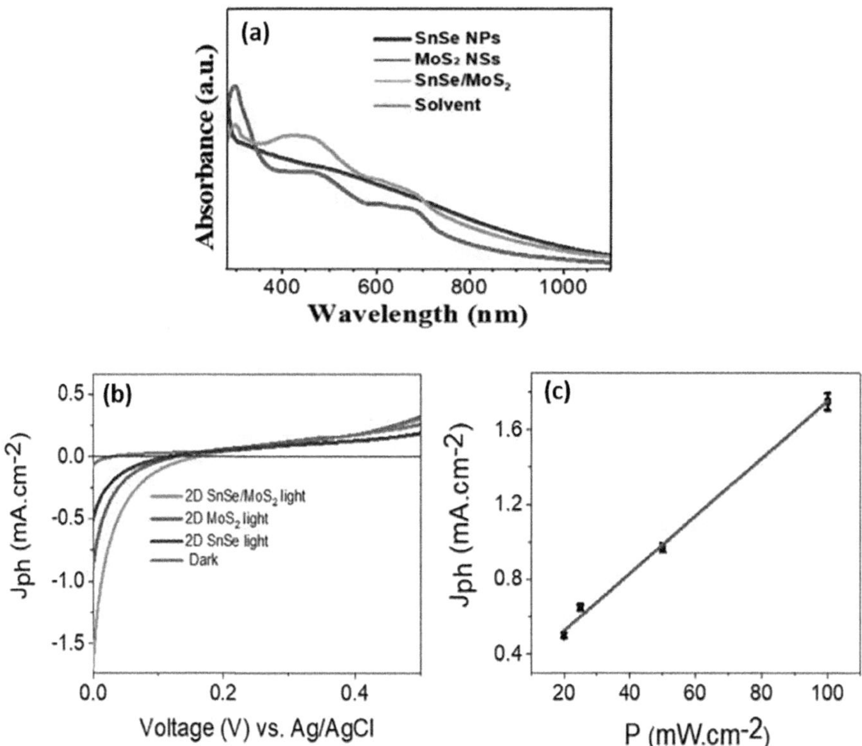

FIGURE 2.8 (a) Optical absorbance, (b) photocurrent density as a function of applied potential under dark or visible light illumination intensity of 100 mW cm^{-2} for MoS_2/PET/ITO, SnSe/PET/ITO, and SnSe/MoS_2/PET/ITO lateral heterostructure flexible electrodes, and (c) response of SnSe/MoS_2/PET/ITO for various intensities of light. (Reprinted with permission from Ref. [65], Copyright © 2022 Elsevier.)

and the open-circuit potential (OCP) is 0.17 V. The created heterostructure thin film, on the other hand, was discovered to respond to light with exceptional efficiency. This was accomplished by varying the intensity of the light from 20 to 100 mW cm^{-2}, resulting in an increase in the value of Jph from 0.52 to 1.75 mA cm^{-2} [65].

The schematic diagram of the constructed solar cell device under light incidence is shown in Figure 2.9a. So far, it has been established that the solar cell generates energy by using photons from the sun. When this solar cell device is exposed to sunlight, photons are received through the 2D-SnSe/MoS_2 lateral heterostructure, which serves as a photocathode to capture photons. The latter produces hydrogen (H_2) gas, whereas the oxygen (O_2) gas is created on the opposite electrode, in this case the Pt-electrode (Figure 2.9b). Although it is known that there is an electron–hole transfer mechanism in solar cell devices, electrons are first transferred from the conduction band (CB) and holes are transmitted from the valence band (VB). The usual transfer electron–hole transfer mechanism in constructed solar cell devices is depicted in Figure 2.9c, in which electrons are moved from the CB of MoS_2 to that of SnSe and subsequently pushed into ITO. As a result, holes migrate from the SnSe valence band to the MoS_2 valence band and ultimately into the electrolyte. Furthermore, the lateral MoS_2/SnSe heterojunction tends to inhibit recombination while accelerating electron–hole pair separation and improving photoelectrochemical cell hydrogen generation performance [65].

Lavanya et al. [66] optimized the molarity of the precursor solution to enhance the photodetection capabilities of the In_2S_3 thin films in terms of responsivity, detectivity, and efficiency. This was accomplished through the low-cost nebulizer-assisted spray pyrolysis (NSP) synthesis method. The results of their research demonstrated that the molarity of the precursor solution does, in fact, have

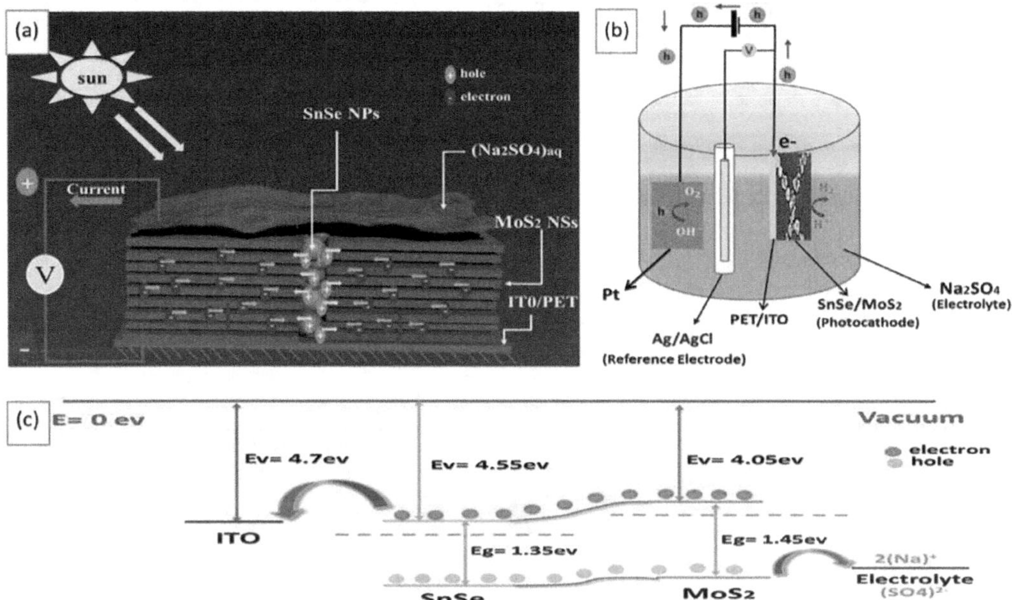

FIGURE 2.9 A schematic diagram depicting the fabrication of photoelectrochemical system for (a) photovoltaic cell, (b) H_2 generation, and (c) the electron transport mechanism within 2D-SnSe/MoS_2 lateral heterostructure thin film. (Reprinted with permission from Ref. [65], Copyright © 2022 Elsevier.)

an impact on the electrical, optical, and photodetection properties. They reported that the films that they prepared showed good absorption in the ultraviolet (UV) region of the electromagnetic spectrum, and the results of their I–V measurements showed a photocurrent of 1.4×10^{-5} A under UV light with a wavelength of 365 nm for an external bias voltage of ±5 V. A sample with a concentration of 0.03 M exhibited a maximum responsivity of 2.74 A/W and a detectivity of 2.65×10^{10} Jones. Additionally, the sample had a high external quantum efficiency (EQE) of 63.9%, indicating that the thin film of In_2S_3 that was prepared can be utilized as an active material in photodetectors [66].

Integration of plasmonic nanostructures with 2D materials is another possible way to improve the optical and electronic properties of 2D nanomaterials. Plasmons focus the light energy into sub-wavelength volumes, which results in a significant improvement in the utilization of light energy by 2D materials. The strong electromagnetic fields formed near the surface of the plasmonic nanostructures enhance both the photoluminescence quantum efficiency of 2D materials and the electron–hole pair excitation. Furthermore, surface plasmons (SPs) generate high-energy electrons that are also referred to as "hot electrons." These electrons are capable of being inserted into the CB of the 2D materials, which encourages photoelectrical response beyond the limitations of band gaps [67]. An et al. [68] investigated the distinctive features of a novel hybrid plasmonic metal-2D MXene photocatalyst. The simple citrate approach was employed to fabricate core–shell Au@Ag-Pd nanoparticles (NPs) with nanocavities as well as Au@Pd NPs with nano-rattle structures. Because of the high number of defects in 2D Ti_3C_2, they discovered that as soon as it was embedded into the core–shell, it boosted the long-term stability of the core–shell NPs, and as a result, the sample has the most prevalent photocurrent density, the greatest efficacy of generation, and the splitting of hotter electrons, among them (Figure 2.10a). This discovery has been confirmed by a research investigation of the luminescence features of Au@Ag/TiO_2/Ti_3C_2 compound because the brightness of the emission was suppressed. Figure 2.10b shows the photoluminescence of Au@Ag/TiO_2/Ti_3C_2 compound. The suppression of Au@Ag/TiO_2/Ti_3C_2 emission clearly demonstrates the successful separation of hot carriers as well as a superior material in the synthesis process. The most significant factor for real-world applications of heterogeneous catalyst-based devices is their ability to be reused and consistency. However, the authors

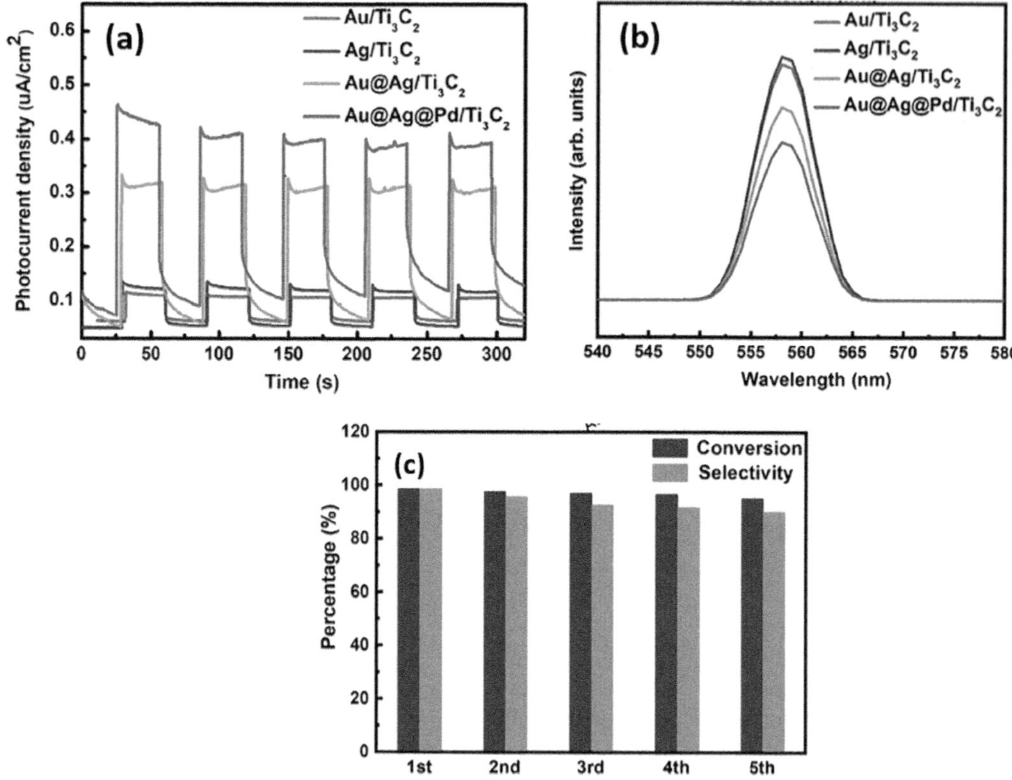

FIGURE 2.10 (a) The transient photocurrent, (b) photoluminescence spectra, and (c) recyclability of Au@Ag@Pd/Ti_3C_2 photocatalyst. (Reprinted with permission from Ref. [68], Copyright © 2023 Elsevier.)

investigated the functioning of the synthesized Au@Ag@Pd/Ti_3C_2 combination roughly five times, as shown in Figure 2.10c. After five consecutive cycles of testing, the Au@Ag@Pd/Ti_3C_2 combination demonstrated stunning photoreduction activities, confirming its outstanding stability. The Au@Ag@Pd/Ti_3C_2 combination is considered as having potential recyclable qualities during the process of reaction due to its excellent stability [68].

2.6 VARIOUS AND POTENTIAL APPLICATIONS OF 2D NANOMATERIALS

2D nanomaterials are ultrathin layered nanomaterials with a sheet-like crystal structure that is bonded by weak vdW forces in an atomic or molecular thickness and infinite dimensions [69]; they are easy to prepare and transfer to any substrate, cost-effective, conducive for constructing heterostructures, and have enhanced integration and suppressed short-channel effects [69,70]. The atomic-level thickness is smaller than other particles' mean free path transport, which includes photons, excitons, and electrons, forcing ballistic transportation than scattering or diffusion [71]; such quantum confinement effect is superlative for fundamental study and electronic applications. The dimension, heterostructures, gating, lighting, intercalation, pressure, and alloying can be tuned to enhance the performance of 2D nanomaterial-based optoelectronic devices [72].

Several 2D nanomaterials find application in optoelectronic devices and that include but not limited to insulators (e.g., boron nitride), semimetals (e.g., $MoTe_2$), semiconductors (e.g., black phosphorus), metals (e.g., 1T-TaS_2), and topological insulators (e.g., Bi_2Se_3) that have optical properties over a large spectral range [73–75]. Figure 2.11 shows specific 2D materials that are commonly used in electronic and optoelectronic devices; these materials include transition metal oxides (TMOs) [76,77], hexagonal boron nitride (h-BN) [78,79], graphene [80,81], antimonene (AM) [82,83], black

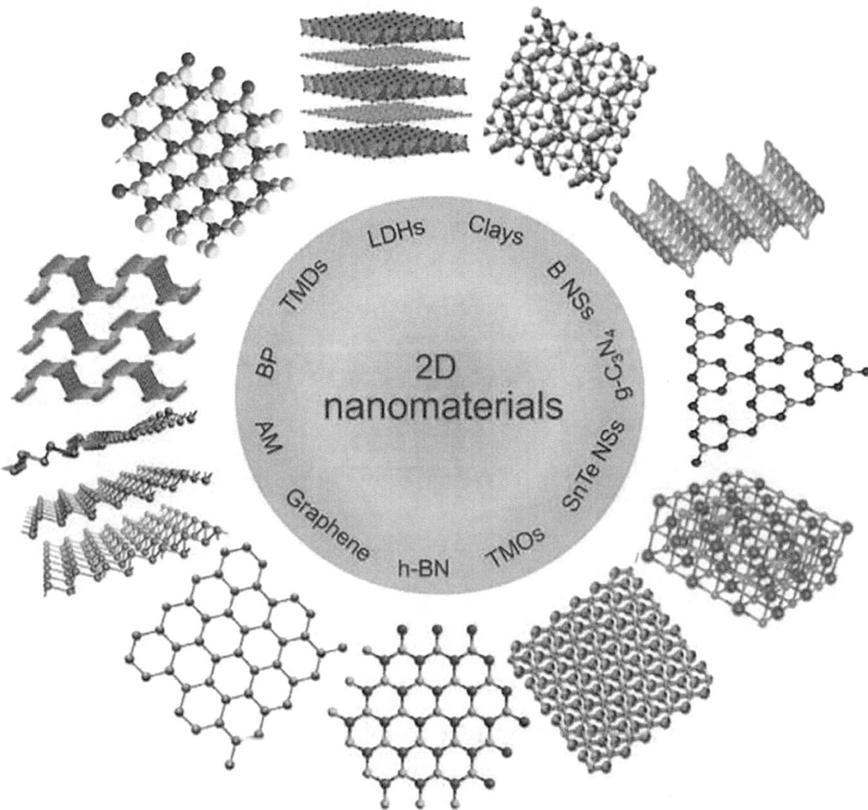

FIGURE 2.11 Schematic of some 2D materials in optoelectronic devices. (Reprinted with permission from Ref. [96], Copyright © 2019 Elsevier.)

phosphorus (BP) [84,85], transition metal dichalcogenides (TMDs) [86,87], layered double hydroxides (LDHs) [88,89], boron nanosheets (BNSs) [90,91], graphitic carbon nitride (g-C_3N_4) [92,93], and tin telluride nanosheets (SnTe NSs) [94,95].

2D nanomaterials and their hybrid systems demonstrate unique exceptional properties such as crystal structure, flexibility, wide band gap coverage, high mobility, durability, speed, efficiency, optical transparency, sensitivity, and integration for potential application in optoelectronic devices [71,97]. Due to these novel properties and their layer dependent nature, 2D materials provide additional advantages in terms of polarization, spectrum range, structural design, and intensity [98]. In addition, TMDs' 2D materials such as ReS_2, MoS_2 and WS_2 provide other advantages which include indirect–direct (~1.0–1.9 eV) band gap transition with bulk thinned to monolayer [74]. The merits of 2D materials are summarized as shown in Figure 2.12.

The novel properties of 2D nanomaterials make them suitable materials with great potential to enhance or even replace other materials in fabrication of high integration and high density low cost arrays for various applications, including optoelectronic devices [99]. These 2D nanomaterials make excellent candidates for future development and application in optoelectronic devices and related areas. This is owing to their rich active sites, large surface area, flexibility, light to heat conversion efficiency, high light absorbing capacity, hydrophilicity, and tunable optical properties [100,101]. In addition, the ability to fabricate 2D nanomaterials with novel properties using environmentally friendly and biodegradable materials reduces their environmental footprint making them the ideal future materials for optoelectronic applications. 2D nanomaterials have thus become critical game changers in all fields of materials and will soon be revolutionary in the energy, environmental, water, and possibly agricultural sectors [10,102].

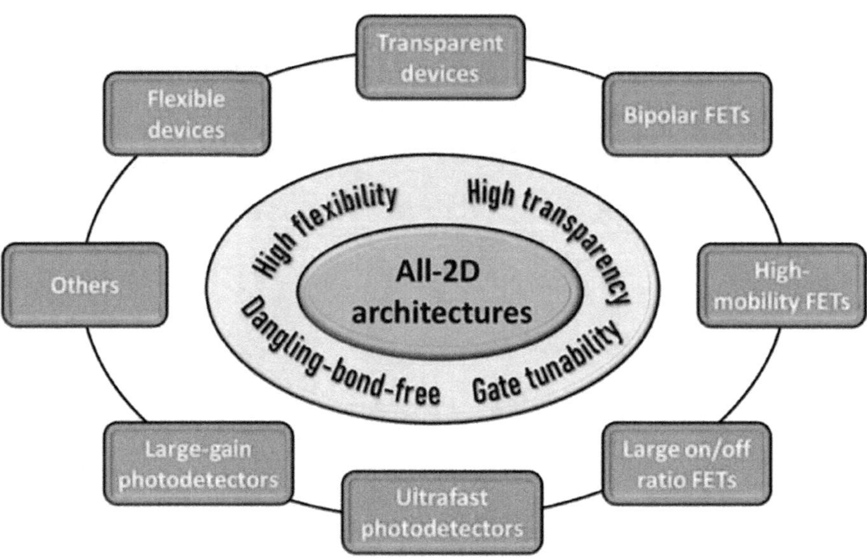

FIGURE 2.12 Merits of 2D materials and applications in electronic and optoelectronic devices. (Reprinted with permission from Ref. [70], Copyright © 2021 Elsevier.)

2.7 SUMMARY AND FUTURE OUTLOOK

The fabrication, properties, and application of 2D nanomaterials have been discussed extensively, particularly for applications in optoelectronic devices. It has been demonstrated that 2D nanomaterials possess excellent optoelectronic properties driven by their defects and vacancies. Additionally, the evolution of fabrication processes, modification, and applicability in various areas makes 2D nanomaterials novel. It is therefore concluded that 2D nanomaterials are the future materials for fabrication of optoelectronic and related devices. The applicability, performance, and properties of materials (including 2D nanomaterials) in various areas are strongly dependent on the fabrication techniques, precursors, and modification procedures. In the near future, aspects such as green synthesis, biocompatibility, biodegradability, recyclability, and reusability will be the driving factors to choose materials that are economic feasible and power saving. In line with climate change, high cost of energy, and increasing demand of high-performance optoelectronic devices, 2D nanomaterials will become the material of choice. Furthermore, we have discussed and summarized several aspects of 2D nanomaterials, which include but not limited to types of 2D nanomaterials, modification methods, properties, advantages and disadvantages, as well as their future prospects. The optical properties and optoelectronic applications of 2D nanomaterials such as transition metal oxides (TMOs), h-BN, graphene, AM, BP, TMDs, LDHs, BNSs, graphitic carbon nitride (g-C_3N_4), and tin telluride (SnTe) nanosheets (NSs) have been discussed. The materials discussed and several others have been found to have potential application in other areas besides optoelectronics. Technological improvement through various means, such as the production and tuning of surface flaws and vacancies in a 2D-lattice and fabrication of heterostructures with 2D materials, is of high importance. The nonlinear optical components' characteristics of these 2D materials must be investigated or explored due to their potential uses in the next-generation devices with outstanding optical switching, optical sensitivity, and so on. The optimization of parameters of the device in relation to applicability and sustainability must also be taken into consideration particularly for greater and novel optical properties. Therefore, this chapter discusses various methodological foundations for designing 2D optical materials, as well as a variety of fabrication procedures' modification strategies. The expansion of the optical application of these materials in the technology and production of optoelectronic devices as well as their overall performance is solely dependent on the first few steps of the process. This significantly improves scientists' ability to examine the relationship between structural traits, physical qualities, optical properties, and economic feasibility of 2D optical nanomaterials.

REFERENCES

1. Zaib, S.; Iqbal, J. Nanotechnology: Applications, techniques, approaches, & the advancement in toxicology and environmental impact of engineered nanomaterials. *Importance Appl. Nanotechnol.* 2019, *8*, 1–10.
2. Noto, L. L.; Mofokeng, S. J.; Molefe, F. V.; Swart, H. C.; Tebele, A. S.; Dhlamini, M. S. Luminescent dynamics of rare earth-doped $CaTiO_3$ phosphors; Elsevier Ltd., 2019. https://doi.org/10.1016/B978-0-08-102935-0.00003-4.
3. El-Kady, M. M.; Ansari, I.; Arora, C.; Rai, N.; Soni, S.; Verma, D. K.; Singh, P.; Mahmoud, A. E. D. Nanomaterials: A comprehensive review of applications, toxicity, impact, and fate to environment. *J. Mol. Liq.*, 2023. https://doi.org/10.1016/j.molliq.2022.121046.
4. Govardhana Reddy, P. V.; Rajendra Prasad Reddy, B.; Venkata Krishna Reddy, M.; Raghava Reddy, K.; Shetti, N. P.; Saleh, T. A.; Aminabhavi, T. M. A review on multicomponent reactions catalysed by zero-dimensional/one-dimensional titanium dioxide (TiO_2) nanomaterials: Promising green methodologies in organic chemistry. *J. Environ. Manage.*, 2021. https://doi.org/10.1016/j.jenvman.2020.111603.
5. Alkaç, İ. M.; Çerçi, B.; Timuralp, C.; Şen, F. Nanomaterials and their classification. In *Nanomaterials for Direct Alcohol Fuel Cells,* Characterization, Design, and Electrocatalysis Micro and Nano Technologies; January 2021; pp. 17–33. https://doi.org/10.1016/B978-0-12-821713-9.00011-1.
6. Murali, A.; Lokhande, G.; Deo, K. A.; Brokesh, A.; Gaharwar, A. K. Emerging 2D nanomaterials for biomedical applications. *Mater. Today*, 2021, 276–302. https://doi.org/10.1016/j.mattod.2021.04.020.
7. Huang, Z.; Liu, H.; Hu, R.; Qiao, H.; Wang, H.; Liu, Y.; Qi, X.; Zhang, H. Structures, properties and application of 2D monoelemental materials (xenes) as graphene analogues under defect engineering. *Nano Today*, 2020. https://doi.org/10.1016/j.nantod.2020.100906.
8. Lyu, Z.; Ding, S.; Du, D.; Qiu, K.; Liu, J.; Hayashi, K.; Zhang, X.; Lin, Y. Recent advances in biomedical applications of 2D nanomaterials with peroxidase-like properties. *Adv. Drug Deliv. Rev.*, 2022, 185, 114269. https://doi.org/10.1016/j.addr.2022.114269.
9. Singhal, J.; Verma, S.; Kumar, S. The physio-chemical properties and applications of 2D nanomaterials in agricultural and environmental sustainability. *Sci. Total Environ.*, 2022. https://doi.org/10.1016/j.scitotenv.2022.155669.
10. Baig, N. Two-dimensional nanomaterials: A critical review of recent progress, properties, applications, and future directions. *Compos. Part A Appl. Sci. Manuf.*, 2023. https://doi.org/10.1016/j.compositesa.2022.107362.
11. Lin, Z.; Carvalho, B. R.; Kahn, E.; Lv, R.; Rao, R.; Terrones, H.; Pimenta, M. A.; Terrones, M. Defect engineering of two-dimensional transition metal dichalcogenides. *2D Mater.*, 2016, 3 (2). https://doi.org/10.1088/2053-1583/3/2/022002.
12. Wang, F. K.; Yang, S. J.; Zhai, T. Y. 2D Bi_2Se_3 materials for optoelectronics. *iScience Rev.*, 2009. https://doi.org/10.1016/j.isci.2021.103291.
13. Sibatov, R. T.; Meftakhutdinov, R. M.; Kochaev, A. I. Asymmetric XMoSiN2 (X=S, Se, Te) monolayers as novel promising 2D materials for nanoelectronics and photovoltaics. *Appl. Surf. Sci.*, 2022, 585. https://doi.org/10.1016/j.apsusc.2022.152465.
14. Patel, J. K.; Patel, A.; Bhatia, D. Introduction to nanomaterials and nanotechnology. *Emerg. Technol. Nanoparticle Manuf.*, 2021, 3–23. https://doi.org/10.1007/978-3-030-50703-9_1.
15. Timpel, M.; Ligorio, G.; Ghiami, A.; Gavioli, L.; Cavaliere, E.; Chiappini, A.; Rossi, F.; Pasquali, L.; Gärisch, F.; List-Kratochvil, E. J. W.; et al. 2D-MoS_2 Goes 3D: Transferring optoelectronic properties of 2D MoS_2 to a large-area thin film. *NPJ 2D Mater. Appl.*, 2021, 5 (1). https://doi.org/10.1038/s41699-021-00244-x.
16. Zhou, X.; Hu, X.; Yu, J.; Liu, S.; Shu, Z.; Zhang, Q.; Li, H.; Ma, Y.; Xu, H.; Zhai, T. 2D layered material-based van der waals heterostructures for optoelectronics. *Adv. Funct. Mater.*, 2018, 28 (14), 1–28. https://doi.org/10.1002/adfm.201706587.
17. Jiang, J.; Xu, T.; Lu, J.; Sun, L.; Ni, Z. Defect engineering in 2D materials: Precise manipulation and improved functionalities. 2019. https://doi.org/10.34133/2019/4641739.

18. Huang, W.; Song, M.; Zhang, Y.; Zhao, Y.; Hou, H.; Huy, L.; Chen, X. Defects-induced oxidation of two-dimensional β -In 2 S 3 and its optoelectronic properties. *Opt. Mater. (Amst).*, 2021, 119, 111372. https://doi.org/10.1016/j.optmat.2021.111372.
19. Qin, H.; Sorkin, V.; Pei, Q.; Liu, Y.; Zhang, Y. Failure in two-dimensional materials: Defect sensitivity and failure criteria. 2019. https://doi.org/10.1115/1.4045005.
20. Lee, H. Y.; Kim, S. Nanowires for 2D material-based photonic and optoelectronic devices. *Nanophotonics*, 2022, 11 (11), 2571–2582.
21. Abid, N.; Khan, A. M.; Shujait, S.; Chaudhary, K.; Ikram, M.; Imran, M.; Haider, J.; Khan, M.; Khan, Q.; Maqbool, M. Synthesis of nanomaterials using various top-down and bottom-up approaches, influencing factors, advantages, and disadvantages: A review. *Adv. Colloid Interface Sci.*, 2022, 300, 102597. https://doi.org/10.1016/j.cis.2021.102597.
22. Pannu, Amandeep Singh (2021) *Synthesis of 2D nanomaterial and their application in opto-electronic devices and sensing.* PhD by Publication, Queensland University of Technology.
23. Korotcenkov, G.; Tolstoy, V. P. Current trends in nanomaterials for metal oxide-based conductometric gas sensors: Advantages and limitations—Part 2: Porous 2D nanomaterials. *Nanomaterials*, 2023, 13 (2). https://doi.org/10.3390/nano13020237.
24. Chen, Y.; Fan, Z.; Zhang, Z.; Niu, W.; Li, C.; Yang, N.; Chen, B.; Zhang, H. Two-dimensional metal nanomaterials: Synthesis, properties, and applications. *Chem. Rev.*, 2018, 118 (13), 6409–6455. https://doi.org/10.1021/acs.chemrev.7b00727.
25. Bhuyan, M. S. A.; Uddin, M. N.; Islam, M. M.; Bipasha, F. A.; Hossain, S. S. Synthesis of graphene. *Int. Nano Lett.*, 2016, 6 (2), 65–83. https://doi.org/10.1007/s40089-015-0176-1.
26. Gumfekar, S. P. Chapter 11- Graphene-based materials for clean energy applications. In *Micro and Nano Technologies*; Bhanvase, B. A., Pawade, V. B., Dhoble, S. J., Sonawane, S. H., Ashokkumar, M., Eds.; Elsevier, 2018; pp. 351–383. https://doi.org/10.1016/B978-0-12-813731-4.00011-4.
27. Nawz, T.; Safdar, A.; Hussain, M.; Lee, D. S.; Siyar, M. Graphene to advanced Mos_2: A review of structure, synthesis, and optoelectronic device application. *Crystals*, 2020, 10 (10), 1–31. https://doi.org/10.3390/cryst10100902.
28. Warner, J. H.; Schäffel, F.; Bachmatiuk, A.; Rümmeli, M. H. Chapter 6- Applications of graphene. In *Graphene*; Warner, J. H., Schäffel, F., Bachmatiuk, A., Rümmeli, M. H., Eds.; Elsevier, 2013; pp. 333–437. https://doi.org/10.1016/B978-0-12-394593-8.00006-0.
29. Warner, J. H.; Schäffel, F.; Bachmatiuk, A.; Rümmeli, M. H. Chapter 4- Methods for obtaining graphene. In *Graphene*; Warner, J. H., Schäffel, F., Bachmatiuk, A., Rümmeli, M. H., Eds.; Elsevier, 2013; pp. 129–228. https://doi.org/10.1016/B978-0-12-394593-8.00004-7.
30. Warner, J. H.; Schäffel, F.; Bachmatiuk, A.; Rümmeli, M. H. Chapter 3- Properties of graphene. In *Graphene*; Warner, J. H., Schäffel, F., Bachmatiuk, A., Rümmeli, M. H., Eds.; Elsevier, 2013; pp. 61–127. https://doi.org/10.1016/B978-0-12-394593-8.00003-5.
31. Warner, J. H.; Schäffel, F.; Bachmatiuk, A.; Rümmeli, M. H. Chapter 5- Characterisation techniques. In *Graphene*; Warner, J. H., Schäffel, F., Bachmatiuk, A., Rümmeli, M. H., Eds.; Elsevier, 2013; pp. 229–332. https://doi.org/https://doi.org/10.1016/B978-0-12-394593-8.00005-9.
32. Alzakia, F. I.; Tan, S. C. Liquid-exfoliated 2D materials for optoelectronic applications. *Adv. Sci.*, 2021, 8 (11). https://doi.org/10.1002/advs.202003864.
33. Witomska, S.; Leydecker, T.; Ciesielski, A.; Samorí, P. Production and patterning of liquid phase–exfoliated 2D sheets for applications in optoelectronics. *Adv. Funct. Mater.*, 2019, 29, 1901126.
34. Amiri, A.; Naraghi, M.; Ahmadi, G.; Soleymaniha, M.; Shanbedi, M. A review on liquid-phase exfoliation for scalable production of pure graphene, wrinkled, crumpled and functionalized graphene and challenges. *FlatChem*, 2018, 8, 40–71. https://doi.org/10.1016/j.flatc.2018.03.004.
35. Hogan, B. T. et al., 2D material liquid crystals for optoelectronics and photonics. *J. Mater. Chem. C*, 2017, 5, 11185–11195. https://doi.org/10.1039/c7tc02549a.
36. Barsoum, M. W. The MN+1AXN Phases: A new class of solids. *Prog. Solid State Chem.*, 2000, 28 (1–4), 201–281. https://doi.org/10.1016/s0079-6786(00)00006-6.
37. Gou, J.; Zhuge, J.; Liang, F. 4-processing of polymer nanocomposites. In *Woodhead Publishing Series in Composites Science and Engineering*; Advani, S. G., Hsiao, K.-T., Eds.; Woodhead Publishing, 2012; pp. 95–119. https://doi.org/10.1533/9780857096258.1.95.
38. Yang, Q.; Zhou, M.; Yang, M.; Zhang, Z.; Yu, J.; Zhang, Y.; Cheng, W.; Li, X. High-yield production of few-layer graphene via new-fashioned strategy combining resonance ball milling and hydrothermal exfoliation. *Nanomaterials*, 2020, 10 (4), 13–16. https://doi.org/10.3390/nano10040667.

39. Yi, M.; Shen, Z. A review on mechanical exfoliation for the scalable production of graphene. *J. Mater. Chem. A*, 2015, 3 (22), 11700–11715. https://doi.org/10.1039/c5ta00252d.
40. Petridis, L. V; Kokkinos, N. C.; Mitropoulos, A. C.; Kyzas, G. Z. Graphene aerogels for oil absorption; Elsevier, 2019; Vol. 30. https://doi.org/10.1016/B978-0-12-814178-6.00008-X.
41. Arao Y.; Tanks J. D.; Aida K., Kubouchi M.. Exfoliation behavior of large anionic graphite flakes in liquid produced by salt-assisted ball milling, *Processes* 2020, 8(1), 28; https://doi.org/10.3390/pr8010028.
42. Sadykov, V. A.; Mezentseva, N. V.; Bobrova, L. N.; Smorygo, O. L.; Eremeev, N. F.; Fedorova, Y. E.; Bespalko, Y. N.; Skriabin, P. I.; Krasnov, A. V.; Lukashevich, A. I.; et al. Advanced materials for solid oxide fuel cells and membrane catalytic reactors; Elsevier Inc., 2018. https://doi.org/10.1016/B978-0-12-814807-5.00012-7.
43. Alam, S.; Asaduzzaman Chowdhury, M.; Shahid, A.; Alam, R.; Rahim, A. Synthesis of emerging two-dimensional (2D) materials – advances, challenges and prospects. *FlatChem*, 2021, 30, 100305. https://doi.org/10.1016/j.flatc.2021.100305.
44. Lei, Y.; Zhang, T.; Lin, Y. C.; Granzier-Nakajima, T.; Bepete, G.; Kowalczyk, D. A.; Lin, Z.; Zhou, D.; Schranghamer, T. F.; Dodda, A.; et al. Graphene and beyond: Recent advances in two-dimensional materials synthesis, properties, and devices. *ACS Nanosci. Au*, 2022, 2 (6), 450–485. https://doi.org/10.1021/acsnanoscienceau.2c00017.
45. Kang, S.; Wang, C.; Chen, J.; Meng, T.; Jiaqiang, E. Progress on solvo/hydrothermal synthesis and optimization of the cathode materials of lithium-ion battery. *J. Energy Storage*, 2023, 67 (May), 107515. https://doi.org/10.1016/j.est.2023.107515.
46. Jeong, Y. R.; Kim, I. H.; Jeong, Y. J. Effects of hydrothermal synthesis conditions on material properties of Cu-Se compounds for thermoelectric and photothermal energy conversion. *Mater. Today Commun.*, 2023, 35, 106324. https://doi.org/10.1016/j.mtcomm.2023.106324.
47. Luo, S.; Ge, Y.; Zhao, M.; Yang, L.; Ren, J. Application of microwave hydrothermal synthesis for the solidification of copper: Effect of heavy metal content and microwave time. *Process Saf. Environ. Prot.*, 2023, 173, 765–774. https://doi.org/10.1016/j.psep.2023.03.066.
48. Kashyap, A.; Singh, N. K.; Soni, M., Soni, A. Chapter 3-Deposition of thin films by chemical solution-assisted techniques. In *Chemical Solution Synthesis for Materials Design and Thin Film Device Applications*; Das, S., Dhara, S., Eds.; Elsevier, 2021; pp. 79–117. https://doi.org/10.1016/B978-0-12-819718-9.00014-5.
49. Zhu, H.; Cao, Y.; Zhang, J.; Zhang, W.; Xu, Y.; Guo, J.; Yang, W.; Liu, J. One-step preparation of graphene nanosheets via ball milling of graphite and the application in lithium-ion batteries. *J. Mater. Sci.*, 2016, 51 (8), 3675–3683. https://doi.org/10.1007/s10853-015-9655-z.
50. Yang, C.; Huang, H.; He, H.; Yang, L.; Jiang, Q.; Li, W. Recent advances in MXene-based nanoarchitectures as electrode materials for future energy generation and conversion applications. *Coord. Chem. Rev.*, 2021, 435, 213806. https://doi.org/10.1016/j.ccr.2021.213806.
51. Bian, Y.; Xiong, N.; Zhu, G. Technology for the remediation of water pollution: A review on the fabrication of metal organic frameworks. *Processes*, 2018, 6 (8). https://doi.org/10.3390/pr6080122.
52. Shi W.; Song S.; Zhang H. Hydrothermal synthetic strategies of inorganic semiconducting nanostructures, *Chem. Soc. Rev.*, 2013, 42, 5714–5743. https://doi.org/10.1039/C3CS60012B
53. Nair, G. B.; Dhoble, S. J. *Fundamentals and Applications of Light-Emitting Diodes: The Revolution in the Lighting Industry*; Woodhead Publishing Series in Electronic and Optical Materials; Woodhead Publishing: Duxford, England; 2020.
54. Creighton J.R.; Ho P., Eds. Chapter 1-Introduction to chemical vapor deposition (CVD), In *Chemical Vapor Deposition*, ASM International, 2001, p. 1.
55. Pottathara, Y. B.; Grohens, Y.; Kokol, V.; Kalarikkal, N.; Thomas, S. Chapter 1- Synthesis and processing of emerging two-dimensional nanomaterials. In *Micro and Nano Technologies*; Beeran Pottathara, Y., Thomas, S., Kalarikkal, N., Grohens, Y., Kokol, V., Eds.; Elsevier, 2019; pp. 1–25. https://doi.org/10.1016/B978-0-12-815751-0.00001-8.
56. Manawi, Y. M.; Atieh, M. A. A review of carbon nanomaterials' synthesis via the chemical vapor deposition (CVD) method. https://doi.org/10.3390/ma11050822.
57. Mittal, M.; Sardar, S.; Jana, A. Chapter 7- Nanofabrication techniques for semiconductor chemical sensors. In *Micro and Nano Technologies*; Hussain, C. M., Kailasa, S. K., Eds.; Elsevier, 2021; pp. 119–137. https://doi.org/10.1016/B978-0-12-820783-3.00023-3.
58. Films, T.; Slof, H. M.; Centre, M. S. Surface fatigue resistance of tool steel coated with thin brittle PVD layers. 2000, 139–144. https://doi.org/10.1016/S0167-8922(00)80119-4.

59. Stoian, M.; Maurer, T.; Lamri, S.; Fechete, I. Techniques of preparation of thin films: Catalytic combustion. *Catalysts*, 2021, 11 (12), 1530.
60. Liu H.; Wang Z. *Perceptual Quality Assessment of Medical Images Encyclopedia of Biomedical Engineering*, Elsevier, Amsterdam, the Netherlands, 2019, pp. 588–596.
61. Mehnath, S.; Kumar, A.; Kumar, S. Biosynthesized/green- synthesized nanomaterials as potential vehicles for delivery of antibiotics/drugs, 1st ed.; Elsevier B.V., 2021; Vol. 94. https://doi.org/10.1016/bs.coac.2020.12.011.
62. Safavi, M. S.; Walsh, F. C.; Surmeneva, M. A.; Surmenev, R. A.; Khalil-Allafi, J. Electrodeposited hydroxyapatite-based biocoatings: Recent progress and future challenges. *Coatings*, 2021. https://doi.org/10.3390/coatings11010110.
63. Gupta, D.; Chauhan, V.; Kumar, R. A comprehensive review on synthesis and applications of molybdenum disulfide (MoS_2) material: Past and recent developments. *Inorg. Chem. Commun.*, 2020, 121, 108200. https://doi.org/10.1016/j.inoche.2020.108200.
64. Hirsch, A.; Hauke, F. Post-graphene 2D chemistry: The emerging field of molybdenum disulfide and black phosphorus functionalization. *Angew. Chemie - Int. Ed.*, 2018, 57 (16), 4338–4354. https://doi.org/10.1002/anie.201708211.
65. Abdelazeez, A. A. A.; Trabelsi, A. B. G.; Alkallas, F. H.; Rabia, M. Successful 2D MoS_2 nanosheets synthesis with SnSe Grid-like nanoparticles: Photoelectrochemical hydrogen generation and solar cell applications. *Sol. Energy*, 2022, 248 (June), 251–259. https://doi.org/10.1016/j.solener.2022.10.058.
66. Lavanya, S.; Kumar, T. R.; Gunavathy, K. V.; Vibha, K.; Shkir, M.; Hakami, J.; Ali, H. E.; Ubaidullah, M. A noticeable improvement in opto-electronic properties of nebulizer sprayed In_2S_3 thin films for stable-photodetector applications. *Micro Nanostruct.*, 2022, 169, 207337. https://doi.org/10.1016/j.micrna.2022.207337.
67. Kalita, D.; Deuri, J. K.; Sahu, P.; Manju, U. Plasmonic nanostructure integrated two-dimensional materials for optoelectronic devices. *J. Phys. D. Appl. Phys.*, 2022, 55 (24). https://doi.org/10.1088/1361-6463/ac5191.
68. An, H.; Zhang, H.; Zhang, K.; Cheng, R.; Ling, Y.; Zhang, X.; Deng, C.; Xu, Z.; Yin, Z. Construction of a novel hybrid plasmonic metal-2D MXene catalyst for plasmon-driven photoreduction of nitroaromatics. *Appl. Surf. Sci.*, 2023, 613, 156055. https://doi.org/10.1016/j.apsusc.2022.156055.
69. Gong, C.; Hu, K.; Wang, X.; Wangyang, P.; Yan, C.; Chu, J.; Liao, M.; Dai, L.; Zhai, T.; Wang, C.; et al. 2D nanomaterial arrays for electronics and optoelectronics. *Adv. Funct. Mater.*, 2018, 28 (16), 1706559–1706682. https://doi.org/10.1002/adfm.201706559.
70. Yao, J. D.; Yang, G. W. All-2D architectures toward advanced electronic and optoelectronic devices. *Nano Today*, 2021, 36, 101026. https://doi.org/10.1016/j.nantod.2020.101026.
71. Yang, F.; Cheng, S.; Zhang, X.; Ren, X.; Li, R.; Dong, H.; Hu, W. 2D organic materials for optoelectronic applications. *Adv. Mater.*, 2018, 30 (2), 1702415–1702442. https://doi.org/10.1002/adma.201702415.
72. Zhao, Q.; Carrascoso, F.; Gant, P.; Wang, T.; Frisenda, R.; Castellanos-Gomez, A. A system to test 2D optoelectronic devices in high vacuum. *J. Phys. Mater.*, 2020, 3 (3). https://doi.org/10.1088/2515-7639/ab8781.
73. Cheng, J.; Wang, C.; Zou, X.; Liao, L. Recent advances in optoelectronic devices based on 2D materials and their heterostructures. *Adv. Opt. Mater.*, 2019, 7 (1), 1800441–1800456. https://doi.org/10.1002/adom.201800441.
74. Wang, X.; Cui, Y.; Li, T.; Lei, M.; Li, J.; Wei, Z. Recent advances in the functional 2D photonic and optoelectronic devices. *Adv. Opt. Mater.*, 2019, 7 (3), 1801274–1801291. https://doi.org/10.1002/adom.201801274.
75. Li, Y.; Li, Z.; Chi, C.; Shan, H.; Zheng, L.; Fang, Z. Plasmonics of 2D nanomaterials: Properties and applications. *Adv. Sci. News*, 2017, 4, 1–25. https://doi.org/10.1002/advs.201600430.
76. Tan, H. T.; Sun, W.; Wang, L.; Yan, Q. 2D transition metal oxides/hydroxides for energy-storage applications. *ChemNanoMat.* 2016, 2, 562–577. https://doi.org/10.1002/cnma.201500177.
77. Yang, T.; Song, T. T.; Callsen, M.; Zhou, J.; Chai, J. W.; Feng, Y. P.; Wang, S. J.; Yang, M. Atomically thin 2D transition metal oxides: Structural reconstruction, interaction with substrates, and potential applications. *Adv. Mater. Interfaces*, 2019, 6 (1), 1–19. https://doi.org/10.1002/admi.201801160.
78. Tan, B.; Wu, Y.; Gao, F.; Yang, H.; Hu, Y.; Shang, H.; Zhang, X.; Zhang, J.; Li, Z.; Fu, Y.; et al. Engineering the optoelectronic properties of 2D hexagonal boron nitride monolayer films by sulfur substitutional doping. *ACS Appl. Mater. Interfaces*, 2022, 14, 16453–16461. https://doi.org/10.1021/acsami.2c01834.
79. Gong, Y.; Xu, Z.; Li, D.; Zhang, J.; Aharonovich, I.; Zhang, Y. Two-dimensional hexagonal boron nitride for building next-generation energy-efficient devices. *ACS Energy Lett.*, 2021, 6, 985–996. https://doi.org/10.1021/acsenergylett.0c02427.

80. Li, X.; Tao, L.; Chen, Z.; Fang, H.; Li, X.; Wang, X.; Xu, J.; Zhu, H. Graphene and related two-dimensional materials: Structure-property relationships for electronics and optoelectronics. *Appl. Phys. Rev.*, 2017, 021306, 1–25. https://doi.org/10.1063/1.4983646.
81. Li, X.; Zhang, X.; Park, H.; Di Bartolomeo, A. Editorial: Electronics and optoelectronics of graphene and related 2D materials. *Front. Mater.*, 2020, 7, 2–4. https://doi.org/10.3389/fmats.2020.00235.
82. Carrasco, J. A.; Congost-escoin, P. Antimonene: A tuneable post-graphene material for advanced applications in optoelectronics, catalysis, energy and biomedicine. *Chem. Soc. Rev.*, 2023, 52, 1288–1330. https://doi.org/10.1039/d2cs00570k.
83. Wang, X.; He, J.; Zhou, B.; Zhang, Y.; Wu, J.; Hu, R.; Liu, L. Zuschriften bandgap-tunable preparation of smooth and large two-dimensional antimonene zuschriften. *Angew. Chem.*, 2018, 130, 8804–8809. https://doi.org/10.1002/ange.201804886.
84. Debnath, P. C.; Park, K.; Song, Y. W. Recent advances in black-phosphorus-based photonics and optoelectronics devices. *Small Methods*, 2018, 2 (4), 1–24. https://doi.org/10.1002/SMTD.201700315.
85. Miao, J.; Zhang, L.; Wang, C. Black phosphorus electronic and optoelectronic devices. *2D Mater.*, 2019, 6 (3), 032003. https://doi.org/10.1088/2053-1583/ab1ebd.
86. Li, S.; Ma, Y.; Ouedraogo, N. A. N.; Liu, F.; You, C.; Deng, W.; Zhang, Y. P-/n-type modulation of 2D transition metal dichalcogenides for electronic and optoelectronic devices. *Nano Res.*, 2022, 15 (1), 123–144. https://doi.org/10.1007/s12274-021-3500-2.
87. Ye, M.; Zhang, D.; Yap, Y. K. Recent advances in electronic and optoelectronic devices based on two-dimensional transition metal dichalcogenides. *Electronics*, 2017, 6 (2), 1–40. https://doi.org/10.3390/electronics6020043.
88. Liu, X.; Zhou, A.; Dou, Y.; Pan, T.; Shao, M.; Han, J.; Wei, M. Ultrafast switching of an electrochromic device based on layered double hydroxide/prussian blue multilayered films. *Nanoscale*, 2015, 7 (40), 17088–17095. https://doi.org/10.1039/c5nr04458h.
89. Liang, R.; Xu, S.; Yan, D.; Shi, W.; Tian, R.; Yan, H.; Wei, M.; Evans, D. G.; Duan, X. CdTe quantum dots/layered double hydroxide ultrathin films with multicolor light emission via layer-by-layer assembly. *Adv. Funct. Mater.*, 2012, 22 (23), 4940–4948. https://doi.org/10.1002/adfm.201201367.
90. Ji, X.; Kong, N.; Wang, J.; Li, W.; Xiao, Y.; Gan, S. T.; Zhang, Y.; Li, Y.; Song, X.; Xiong, Q.; et al. A novel top-down synthesis of ultrathin 2D boron nanosheets for multimodal imaging-guided cancer therapy. *Adv. Mater.*, 2018, 30 (36). https://doi.org/10.1002/adma.201803031.
91. Angizi, S.; Alem, S. A. A.; Hasanzadeh Azar, M.; Shayeganfar, F.; Manning, M. I.; Hatamie, A.; Pakdel, A.; Simchi, A. A comprehensive review on planar boron nitride nanomaterials: From 2D nanosheets towards 0D quantum dots. *Prog. Mater. Sci.*, 2022, 124 (April 2020), 100884. https://doi.org/10.1016/j.pmatsci.2021.100884.
92. Hoh, H. Y.; Zhang, Y.; Zhong, Y. L.; Bao, Q. Harnessing the potential of graphitic carbon nitride for optoelectronic applications. *Adv. Opt. Mater.*, 2021, 9 (16). https://doi.org/10.1002/adom.202100146.
93. Liu, Z.; Wang, C.; Zhu, Z.; Lou, Q.; Shen, C.; Chen, Y.; Sun, J.; Ye, Y.; Zang, J.; Dong, L.; et al. Wafer-scale growth of two-dimensional graphitic carbon nitride films. *Matter*, 2021, 4 (5), 1625–1638. https://doi.org/10.1016/j.matt.2021.02.014.
94. Kumar, V.; Sharma, K.; Sharma, D. K.; Masih, V. G. Study on mechanically alloyed tin telluride screen-printed films for optoelectronic device applications. *Opt. Quantum Electron.*, 2019, 51 (5), 1–9. https://doi.org/10.1007/s11082-019-1843-7.
95. Li, Y.; Ding, T.; Sang, D. K.; Wu, M.; Li, J.; Wang, C.; Liu, F.; Zhang, H.; Xie, H. Evolutional carrier mobility and power factor of two-dimensional tin telluride due to quantum size effects. *J. Mater. Chem. C*, 2020, 8 (12), 4181–4191. https://doi.org/10.1039/c9tc06611j.
96. Hu, T.; Mei, X.; Wang, Y.; Weng, X.; Liang, R.; Wei, M. Two-dimensional nanomaterials: Fascinating materials in biomedical field. *Sci. Bull.*, 2019, 64 (22), 1707–1727. https://doi.org/10.1016/j.scib.2019.09.021.
97. Yadav, S.; Jena, S. R.; M.B, B.; Altaee, A.; Saxena, M.; Samal, A. K. Heterostructures of 2D materials-quantum dots (QDs) for optoelectronic devices: Challenges and opportunities. *Emergent Mater.*, 2021, 4 (4), 901–922. https://doi.org/10.1007/s42247-021-00222-5.
98. Sun, J.; Choi, Y.; Choi, Y. J.; Kim, S.; Park, J. H.; Lee, S.; Cho, J. H. 2D–organic hybrid heterostructures for optoelectronic applications. *Adv. Mater.*, 2019, 31 (34), 1803831–1803861. https://doi.org/10.1002/adma.201803831.
99. Zou, L. R.; Sang, D. D.; Yao, Y.; Wang, X. T.; Zheng, Y. Y.; Wang, N. Z.; Wang, C.; Wang, Q. L. Research progress of optoelectronic devices based on two-dimensional MoS_2 materials. *Rare Met.*, 2023, 42 (1), 17–38. https://doi.org/10.1007/s12598-022-02113-y.

100. Guan, G.; Ye, E.; You, M.; Li, Z. Hybridized 2D nanomaterials toward highly efficient photocatalysis for degrading pollutants: Current status and future perspectives. *Small*, 2020, 16 (19), 1–24. https://doi.org/10.1002/smll.201907087.
101. Lu, B.; Zhu, Z.; Ma, B.; Wang, W.; Zhu, R.; Zhang, J. 2D MXene nanomaterials for versatile biomedical applications: Current trends and future prospects. *Small*, 2021, 17 (46), 1–38. https://doi.org/10.1002/smll.202100946.
102. Chimene, D.; Alge, D. L.; Gaharwar, A. K. Two-dimensional nanomaterials for biomedical applications: Emerging trends and future prospects. *Adv. Mater.*, 2015, 27 (45), 7261–7284. https://doi.org/10.1002/adma.201502422.

3 Methods for the Synthesis of Two-Dimensional Nanomaterials

Structure and Properties

Dana Susan Abraham, Mari Vinoba, and Margandan Bhagiyalakshmi

3.1 INTRODUCTION

Nanotechnology, a young discipline in the 21st century, encompasses exploring and innovating numerous materials as well as developing devices and resources for sustainable well-being. This field relies on engineering in particular and unique nanomaterials endowed with superior capabilities. A glimpse into the prospects of designing devices at the nanoscale was first introduced by Richard Feynman in 1959 during his world-famous speech, "There's Plenty of Room at the Bottom," at the annual meeting of the American Physical Society [1]. Nanomaterials possess one of their dimensions between 1 and 100nm. Based on their morphology, nanomaterials are categorized into four classes: zero-dimensional (0D), e.g., quantum dots; one-dimensional (1D), e.g., nanorods; two-dimensional (2D), e.g., nanosheets; and three-dimensional (3D), e.g., nanowire bundles. Figure 3.1 displays the classification of nanomaterials based on their dimension. Each nanomaterial is endowed with unique qualities by tuning its dimensions, grain size, and shape.

Within this wide range of nanomaterial families, 2D nanomaterials have become one of the intriguing classes of nanomaterials that gained recognition as a captivating group with remarkable features [3]. Their unique characteristics set them apart by rendering scientific research and technological advances. The ultrathin layered structure with a single or few layers of atoms opens the door for engineering their behavior. Ultrathin nature enables an incredibly large surface area compared to volume, which boosts their charge transfer, enhances their reactivity, and allows atomic-level interaction. Current advancements in nanotechnology demonstrate that the properties of nanomaterials will play a significant role in the design of novel technologies in the future. Atomically thin layered structure with massive lateral size imparts with high specific surface area [4]. Chemical and mechanical factors influence their properties with substantial surface-to-volume ratio and quantum effects [5]. Charge carriers and electrons confined within 2D nanomaterials induce quantum confinement, which modulates their optical, thermal, and electronic properties. Notable changes can be witnessed in the optical and electronic properties of 2D materials, which can be attributed to the confinement of electrons and the formation of interfaces as the layers are in close proximity with each other. Despite having an ultrathin structure, these 2D nanomaterials have outstanding mechanical qualities enhanced by strong in-plane covalent bonds [6]. Tailorable band gap, high electrical conductivity, and mobility of carrier ions and electrons attribute to their electric property. Their high flexibility, stretchability, resilience, and strength make them ideal for use in wearable devices and flexible electronics [7]. Optimizing and tuning the properties of 2D nanomaterials based on their application by modulating their composition, size, layer stackings, and functionalizing their surface makes them promising candidates for the future technological advancements.

DOI: 10.1201/9781003436942-3

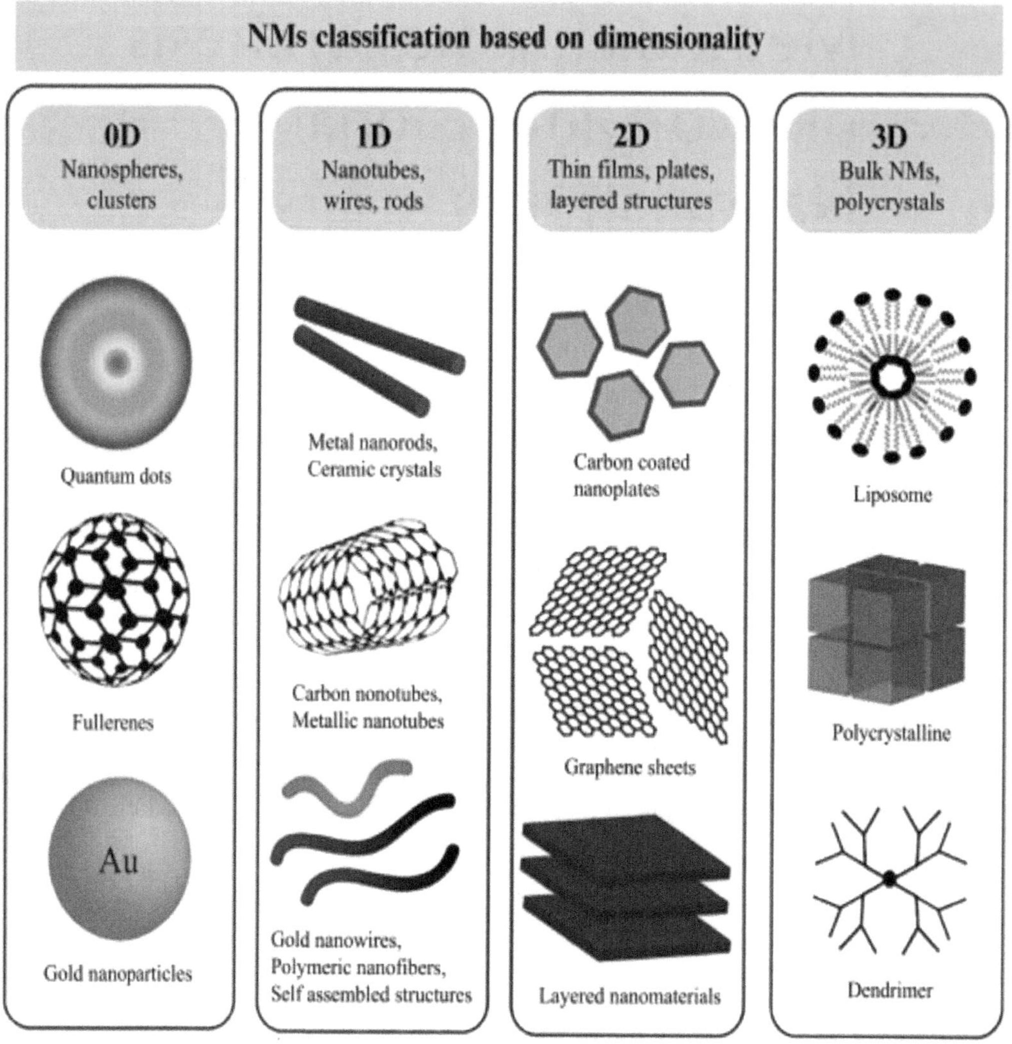

FIGURE 3.1 Classification of nanomaterials based on their morphology. (Reproduced with permission from [2], Copyright © 2017 Elsevier.)

The 2D nanomaterial family involves layered and non-layered 2D nanomaterials [8]. Layered 2D nanomaterials involve stacking their individual atomic layers on top of each other in a repetitive fashion, which is mainly held by the van der Waals force of attraction or strong covalent bonds. These layered 2D nanomaterials are characterized by their appealing properties arising from the interactions between these layers and the confinement of charge carriers within individual layers. In contrast to these layered 2D nanomaterials, the non-layered 2D nanomaterials consist of a single or few layers of atoms arranged randomly.

The firstborn in the family of 2D nanomaterials was graphene in 2004, with an atomically thin layer of carbon atoms arranged in a 2D hexagonal fashion [10]. A groundbreaking experiment was conducted by Konstantin Novoselov, Andre Geim, and their colleagues at the University of Manchester in which they managed to exfoliate graphene from graphite. Graphite is the stacked layer of graphene that was mechanically exfoliated via the Scotch tape method to peel off a very thin atomic layer of graphene. This discovery opened a new era attaining worldwide attention in all fields for its potential properties and enormous applications. Since then, this family has grown

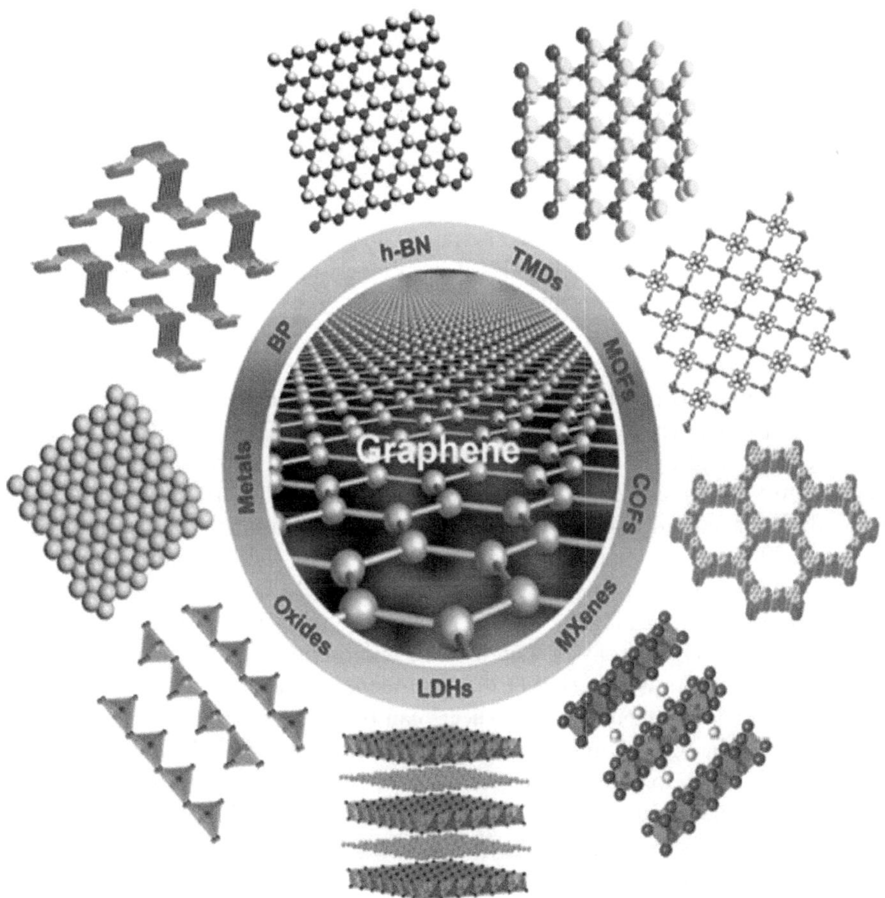

FIGURE 3.2 Illustration of various 2D nanomaterials. (Reproduced with permission from [9], Copyright © 2019 Elsevier.)

with many more members (Figure 3.2) such as graphitic carbon nitride (g-C_3N_4) [11], hexagonal boron nitride (h-BN) [12], transition metal dichalcogenides (TMDCs) [13], black phosphorous (BP) [14], silicene [15], layered double hydroxides (LDHs) [16], 2D metal oxides/sulfides [17], 2D metal–organic framework [18], 2D covalent-organic framework (COF) [19], transition metal carbides or nitrides (MXenes) [20], and 2D perovskites [21]. This chapter briefly introduces the synthesis approaches, structure, and properties of 2D nanomaterial.

3.2 SYNTHESIS APPROACHES FOR 2D NANOMATERIALS

Several research endeavors have been reported, exploring the techniques for synthesizing 2D materials. Briefly, the synthesis approach can be classified into two broad areas: the top-down and the bottom-up approaches.

3.2.1 Top-down Method

This method breaks the bulk material into smaller particles by reducing its size and dimension. The van der Waals force of interaction, which stacks the layers into bulk material, is removed through physical or chemical methods. The physical approach is either by utilizing mechanical force or employing ultrasonic waves. In contrast, the chemical method relies on chemical reactions, ion exchange, or parameter variation. This approach allows engineering material with desired size, morphology, thickness, and properties. Alongside the merits, the potential for introducing impurities and defects during the synthesis and scalability issues are to be considered.

3.2.1.1 Exfoliation Technique

In order to produce nanosheets with distinctive features and a large surface area, the layers of the material must be separated from the bulk crystal. Exfoliation is a frequently used technique, and based on external forces, they are further classified as mechanical exfoliation, liquid-phase exfoliation, and chemical and electrochemical exfoliation.

a. **Mechanical exfoliation**

   Mechanical exfoliation is a widely used method for synthesizing 2D nanomaterial; earlier, it was also known as the "Scotch-tape method" [22]. The bulk material to be exfoliated is cleaned first to be free from contaminants. Then, it is transferred onto a suitable substrate, e.g., Si wafers/ SiO_2, or other substrates with a flat surface and good adhesion properties. A small adhesive tape (Scotch tape/blue Nitto tape) is attached to the bulk sample and pressed firmly to ensure good contact. Then, the tape is peeled off from the surface of the material. This process is repeated several times with fresh tape to ensure the chances of obtaining single- or few-layer flakes. The so-obtained flakes can be characterized by optical microscopy or atomic force microscopy (AFM), high-resolution transmission electron microscopy (HRTEM), and scanning electron microscopy (SEM). As a post-exfoliation treatment, the nanosheets can be additionally treated for annealing or chemical functionalization to modify their properties or improve their quality. This technique entails weakening the van der Waals force of interaction between the layers while keeping the covalent bonds within the layers intact. It acquired the name after the brand of adhesive tape that is used for this method. This method was first employed to synthesize graphene from graphite in 2004. Further, it has been extended to other nanomaterials such as TMDCs [23], h-BN [24], and BP [25]. This method is simple and accessible; however, it is time-consuming, and there can be variations in the application of pressure, resulting in variances in the thickness and size of the obtained nanosheets. Additionally, handling the nanosheets can be challenging and requires careful attention.

b. **Liquid-phase exfoliation**

   Here, the bulk material is dispersed in a liquid solvent, and external forces are applied to exfoliate the layers. The common methods adopted are ball milling, shear mixing, and ultrasonication. This technique is superior to mechanical exfoliation. The bulk material is ground or milled into smaller particles and added to the suitable solvent based on the wetting property and compatibility with the active material [26].

   - Ultrasonication: Ultrasonic waves are applied to agitate the solvent and endorse layer separation. The high-frequency sound waves cause acoustic cavitation, leading to the development and collapse of small bubbles, which generate shear stresses and cause exfoliation.
   - Shear Mixing: Shear stress is applied by high shear homogenization, high-speed mixing, or stirring to exfoliate the layers.
   - Ball Milling: This technique uses a rotating container called a ball mill filled with grinding balls (made of steel, ceramic, or hard agate). These balls act as grinding media, rolling with material, while the walls of the container generate mechanical force that exerts shear forces and break down the particle and mix them into a homogenous form. The bulk material is placed in a high-energy ball mill and ground with medium such as ceramic beads or hardened steel balls. This step involves the continuous collision of the material and friction with the bulk material and the medium generating kinetic energy

that leads to mechanical strain and deformation. This high mechanical force exerted deforms and fragments the bulk material into nanoscale particles. In the case of layered material, this step can lead to the exfoliation of the layers leading to the formation of nanosheets. The weak interlayer forces break and separate the layers. Post-milling treatments, such as annealing, chemical functionalizing, and other processes, can be carried out to obtain the desired nanomaterial. Based on the requirement, researchers need to optimize the milling parameters such as time for milling, rotations per minute, speed, ball-to-powder ratio, and atmosphere. The liquid-phase exfoliation enables large-scale production and control over the thickness of the nanosheets. However, it may introduce defects and impurities into the resulting nanomaterials.

c. **Chemical exfoliation**

Chemical reactions are utilized in separating the layers from the bulk material. Typically, it involves intercalation, followed by exfoliation; in the intercalation step, the bulk material is dispersed in a solvent with compounds or molecules to insert or intercalate between the layers of the material. This weakens the interlayer forces and separates the layers. The additional step is involved exerting an external force, which may be through stirring, sonicating, or thermal expansion. Chemical exfoliation techniques put forward several advantages, including the ability to synthesize nanosheets in bulk, having control over the thickness and size distribution, and the potential for introducing functional groups through different intercalating agents [27].

d. **Electrochemical exfoliation**

The bulk material is dispersed into an electrolyte solution, and an electric field is applied to facilitate the separation of layers [28]. The material is first ground properly into tiny particles to enhance the surface area available for the electrochemical reaction. A suitable electrolyte that can intercalate and exfoliate is chosen and placed as an anodic electrode with a counter-cathodic electrode in the electrolytic solution. The electric field induced the intercalation of electrolyte ions into the layers, resulting in the separation of layers. This method helps to produce nanosheets with large surface area, the potential for scalability, and control over their thickness. However, electrochemical equipment and expertise are essential to perform this technique.

The choice of exfoliation technique is based on the active material and desirable properties. Post-treatment, exfoliation, centrifugation, or filtration is carried out to separate the nanosheets. The selection of solvents, intercalating agents, and process adopted are crucial for achieving high-quality nanosheets. The main advantages of this technique involve the high surface area and tunable thickness for the nanosheets, and further, it is cost-effective compared to other synthesis techniques. However, the introduction of defects and impurities affects their property, time consumption, and scalability challenges of this technique.

3.2.1.2 Nanolithography Technique

Nanolithography is another commonly employed technique that generates precise structure, pattern, and features and tailors desirable properties and functionalities per the application. It helps in precisely positioning and functionalizing at the nanoscale. Nanolithography is used in several ways in the synthesis of 2D nanomaterials:

a. Patterning and fabrication of the device: This technique allows precisely patterning the 2D nanomaterials into desired dimensions, shapes, geometry, and configuration. Photolithography, electron-beam lithography, and dip-pen nanolithography are commonly employed techniques for patterning and fabricating.
b. Selective deposition and growth: Nanolithography can specifically deposit and grow a 2D nanomaterial in a specific region. A patterned layer that is developed as a template using

nanolithography can be used to selectively grow particular material via the help of other methods like molecular beam epitaxy or chemical vapor deposition. In this way, it allows controlled deposition and growth with better spatial resolution.

c. Template formation and assembly: As mentioned above, nanolithography helps develop templates for epitaxial growth and self-orientation of 2D nanosheets to ensure precise alignment and assembly. In this context, nanoimprint lithography and block copolymer lithography are commonly used techniques.
d. Surface patterning and functionalizing: Nanolithography also enables the surface modification and functionalization of 2D nanomaterials. Precise placement of the nanoparticle or functionalizing groups on specific regions of the surface of nanomaterials imparts enhanced reactivity and integrates them into device formation.
e. Etching and Defining edges: Selective etching out a layer and defining the edges of the 2D nanomaterial helps develop quantum confined structures, controlled electronic transport and enhanced catalytic activity.

There are several lithography techniques:

a. **Photolithography**

   It employs light and photo-sensitive substrate materials (photoresist) and is commonly used in the semiconductor industry [29]. A thin layer of photoresist is coated on the material using the spin coating technique; the liquid photoresist is dispensed at the center of the substrate, and as the spinner rotates, the centrifugal force spreads it uniformly across the substrate. After spin coating, it is soft-baked to evaporate the excess solvents, ensuring perfect surface adherence. This is then aligned with a photomask with desired patterns to be transferred onto the photoresist. The aligned substrate with the mask is then exposed to light, which causes chemical reactions in the exposed sites of the photoresist. As a post-exposure treatment, it is again baked to stabilize the chemical reactions and enhance the pattern. Using a developer solution, the so-developed regions on the photoresist are selectively removed. The unexposed regions of the photoresist dissolve in the developer solution, leaving the substrate with the desired pattern. This developed pattern serves as a protective mask for the subsequent steps.

b. **Electron-beam lithography**

   Focused electron beams are used to expose and create patterns on the material selectively. An electron source generates electron beams that are focused and scanned over the resist-coated substrate using electromagnetic lenses. The beam scans precisely over the substrate, which is controlled using a computer-aided design that develops a predefined pattern on the substrate. Similar to the above technique, after the electron beam exposure, it is treated with the developer solution, and the desired pattern is developed on the substrate. Electron-beam lithography offers high resolution down to –10 nm and creates intricate patterns and structures. It enables control over the features and fabrication of devices [30]. However, compared to other lithography techniques, electron-beam lithography is relatively slow and mainly used for low-volume prototyping.

c. **Scanning probe lithography**

   It uses a sharp probe or tip to develop patterns on the substrate. A scanning probe microscope with a sharp probe or tip attached to a cantilever is employed. Depending on the requirement, the probe can be chosen in which Si, silicon nitride, or diamond probes are available. The probe moves in a predefined manner, and it is also controlled by monitoring the deflection of the cantilever. While scanning, different mechanisms can be employed to deposit or modify the surface of the substrate, including a mechanical force/pressure or shear stress can be applied to scratch the surface, electric field application to induce oxidation–reduction processes on the surface of the substrate, thermal energy for localized thermal evaporation or decomposition, and chemical reactions to chemically modify or deposit the materials. The instrument monitors the interaction between the probe and the surface during the scanning process. This technique relies on the probe's controlled movement and its interaction with the surface to selectively modify or deposit the material on the surface. It offers high resolution down to sub-nanometer scale, excellent versatility, and the ability to create high-precision complex patterns. This technique has a wide range of applications, including nano-electrochemistry, surface functionalization, device fabrication, and assembly [31].

d. **Nanoimprint lithography**

It involves the replication of a pattern from a mold or template that contains a desired pattern onto a resist-coated substrate to create nanoscale patterns and structures [32]. The mold can be made of quartz, silicon, or nickel, and the mold pattern can be created using electron beam lithography, focused ion beam, or photolithography. The imprint processes can be thermal or UV nanolithography. In the thermal nanoimprint lithography, the mold is heated above glass transition temperature, and the heated mold is brought in contact with the resist-coated substrate and pressed to facilitate the flow and deformation of the material. While in UV nanoimprint lithography, UV light initiates the photochemical reaction. The photochemical reaction may cause crosslinking or hardening of the exposed sites. Nanoimprint lithography offers many advantages, such as high-resolution patterning, scalability, patterning large areas in a single step, and cost-effectiveness. Limitations include high-quality molds, chances for defect formation, and challenges in releasing the mold without any damage to the resisting material.

e. **Dip-pen lithography**

It's a scanning probe lithography technique that uses a sharp pen-like tip that is coated with molecules of interest, which can be any nanoparticle, organic compounds, or polymers, to form nanoscale patterns on the substrate material [33]. The coating on the tip is achieved by immersing the tip in the solution containing the molecule of interest. When scanned over the substrate surface, this coated tip interacts with the molecules and transfers them from the tip to the substrate, creating a pattern. The movement of the tip is precisely controlled and allows for the designing of desired patterns. The deposition rate is slow, and the stability of the coated molecules limits long-duration patterning; further controlling the amount of ink transferred from the tip to the substrate is also challenging and results in the variable quality of the pattern.

f. **Block copolymer lithography**

This technique leverages the self-assembling phenomena of block copolymers to create nanoscale patterns on the substrate. A block copolymer has two or more chemically distinct polymer chains that are linked by covalent bonds [34]. The choice of block copolymer is crucial as it determines the size and type of resultant patterns. Interaction between the block copolymer and the substrate is achieved by chemical treatment or deposition of a thin layer to form a preferential substrate. The block copolymer is initially dissolved in a solvent and spin-coated onto the substrate. As the solvent evaporates, the block copolymer forms a very thin film. The coated substrate is subjected to annealing, which allows the block copolymer domains to rearrange and align in a well-defined pattern. This self-assembled block copolymer can act as a template. Achieving a perfect long-range order and controlling the defects in the self-assembled pattern challenges its application.

All the nanolithography techniques offer scalability and high-resolution patterning at the nanometer range. Researchers can tailor the morphology, structure, composition, and properties of the 2D nanomaterials for enhanced performance and applications.

3.2.2 Bottom-up Method

This method involves the construction of nanoscale structures or their assembly from the individual atomic and molecular levels to the 2D nanomaterial scale. Controlled growth or self-assembly of these building blocks takes place to form the larger structure with desired properties. This method is characterized by its controlled formation and organization of the building blocks into desired materials with specific structures and properties. Key points of this approach involve its precise control of the size, shape, composition, and structure. Self-assembling the building units into well-defined structures is driven by the electrostatic force of attraction, van der Waals force, and hydrogen bonding. This self-assembling property helps to attain nanoscale structures without any external intervention. Hierarchical structures can be engineered by this technique, where these building blocks organize into multilayered structures. The hierarchical assembly provides enhanced functionalities and properties. It also offers scalability, especially in the case of solution-based methods, that they can be scaled up to bulk quantities by optimizing the parameters such as precursor concentration, solvent, temperature, and reaction time. Atomic-scale control is possible in certain bottom-up methods such as Molecular Beam Epitaxy (MBE) and Atomic Layer Deposition; these techniques allow precise arrangement of atoms and molecules with uniformity and precision. Versatility is also highlighted by the fact that this method permits the synthesis of nanomaterials with a wider range, diverse structure, composition, and functionalities, with different building blocks, altering their surface chemistry, intercalating doping agents, and tailoring the catalyst as per the application. It also enables the integration of multiple functionalities, as well as the incorporation of various functional groups to have synergy. Overall, this method helps in precisely engineering nanomaterials with control of the nanoscale features, opening opportunities for advanced applications. There are several types of bottom-up techniques, including:

3.2.2.1 Chemical Vapor Deposition

Chemical Vapor Deposition (CVD) technique involves the growth of layers or thin films on a heated substrate surface by a controlled reaction of precursor gases containing desired elements [35]. The precursor gases are elevated to high temperatures, or thermal decomposition is carried out to release reactive species, such as molecular fragments or metal atoms. These reactive species are adsorbed onto the heated substrate surface, which undergoes chemical reactions and rearrangements to give a 2D structure. This growth typically takes place as layers deposit in a controlled fashion. The key advantage of this method itself is its controlled growth; optimizing the parameters such as precursor concentration, flow rate, temperature, pressure, and reaction time can result in the deposition of material with desired composition and layer thickness. Scalability, high crystallinity, purity, and versatility are other merits of this method. The CVD method is reported for synthesizing 2D nanomaterials such as graphene and transition metal dichalcogenides. In contrast, this approach challenges the complexity of equipment such as high vacuum pumps, muffle furnace, high-temperature furnaces, and precise gas flow control systems. Setting up and maintaining these types of equipment require resources and exerts, as well as choice of substrate material and thermal stress on them, which limits the temperature-sensitive compounds and substrates.

3.2.2.2 Molecular Beam Epitaxy

Molecular Beam Epitaxy (MBE) is the bottom-up technique known for its atomic-scale control. This technique is sophisticated, especially for synthesizing atomic-level precise, high-quality thin films and nanostructures [36]. It involves the growth of crystalline structure by the deposition of atoms or molecules in an ultra-high vacuum environment onto the substrate. This ultra-high vacuum environment ensures controlled growth without contaminants and impurities. In this chamber, effusion cells evaporate the solid source material or crack the molecular gas into beams directed toward the substrate material. The substrate is heated during deposition to enhance the crystallinity of the growing film. The directed beams impinge onto the substrate leading to the deposition with precise control of the composition and thickness. Reflection of high-energy electron diffraction is used to in situ monitor the crystal structure and the morphology during its growing step. The MBE technique offers layer-by-layer growth, precise atomic-scale control, interface quality, and minimum impurities and defects. Equipment complexity, time consumption, and limited substrate compatibility are the main challenges of this method.

3.2.2.3 Solution-Based Method

The solution-based method is a very commonly used technique in synthesizing 2D nanomaterials. This method involves dispersing the precursors into solvent and allowing controlled chemical reactions. The precursor can be metal complexes, inorganic salts, or organic molecules; these are dissolved in appropriate solvents based on the nature of the precursor and its solubility. Once the precursor is wholly dissolved in the solvent, the controlled chemical reaction is initiated, with the addition of other reagents, variation in pH, temperature, pressure, etc. Capping agents, stabilizing agents, surfactants, etc., may be added to acquire the nanomaterial's desired morphology, dispersibility, size, and shape. The reaction leads to the aggregation of precursor species on the nucleation sites, forming a cluster. As a post-treatment, the nanomaterials obtained are centrifuged, washed, or annealed to remove residual impurities and by-products [37]. Several solution-based methods are employed, such as:

a. **Sol–gel synthesis**

   This technique involves the conversion of sol (precursor colloidal suspension) into a gel matrix by controlled hydrolysis and condensation reactions. The initial step is the preparation of the sol, where the precursors (metal salts, metal alkaloids, and organic compounds) are dissolved in an appropriate solvent, commonly alcohol, with some amount of water [38]. The hydrolysis step involves the reaction between water molecules and precursor molecules, forming hydroxyl groups. M-O bonds are formed during this step. M-O-M bonds are also formed by the condensation process leading to the formation of a gel structure or a three-dimensional network. The aging process follows the hydrolysis and condensation step, where the sol transforms into a stable gel network via the continuous condensation of the M-O bonds. The gel formation step can be initiated at room or elevated temperatures. After the gelation process, it is subjected to the drying step, where the liquid phase is eliminated by evaporation, freeze drying, or supercritical fluid extraction. The resultant material is called aerogel or xerogel, depending on the drying method. The dried gel can be annealed to enhance crystallinity and thermal and structural stability. This method allows homogenization at the molecular level, which helps in attaining highly pure material with uniform distribution of catalysts, dopants, and additives. It also offers to tailor the composition and structure by altering the precursor ratio, solvent composition, and reaction parameters to achieve the desired properties. Meanwhile, during the annealing step, the sol–gel material may crack or shrink due to thermal stress and evaporation of the solvent leading to structural defects. Despite this limitation, the sol–gel method is widely used, offering good control over structure, composition, and properties.

b. **Hydrothermal and solvothermal synthesis**

   This solution-based method involves high temperature and pressure conditions in aqueous (hydrothermal) or organic solvent (solvothermal). This technique is mainly used to synthesize crystalline metal oxides and inorganic nanomaterials. The precursor compound is dissolved in water or a particular solvent and transferred into a stainless steel autoclave that can withstand high pressures. The reaction vessel is placed in the oven with temperatures ranging from 100°C to 220°C. The dissolved reactants undergo chemical reaction under high temperature and pressure, forming nuclei that continue to grow as larger crystalline nanoparticles intercalating precursor species from the solution. In some cases, phase transformation also takes place during this step. After this, the autoclave is allowed to cool naturally to room temperature, and the synthesized products are collected by centrifugation or filtration process. These particles can be washed again and dried. The advantages of the hydrothermal and solvothermal processes involve their homogeneity and purity. It often enables the formation of high crystalline nanoparticles and the versatility in the synthesis procedure. Scalability is limited to the limited capacity of the reaction vessel and complex reaction conditions which are to be optimized. Despite these challenges, it is widely employed by researchers for the synthesis of high-purity and quality nanomaterial [39].

c. **Co-precipitation method**

This method involves the simultaneous precipitation of the precursor species present in the solution. Two or more precursors will be present in the solvent; solubility and stability of the precursor in the solvent determine the success of the method [40]. This mixed precursor solution is subjected to the reaction by adding a precipitating agent, varying the pH, or altering temperature, where it undergoes nucleation. Stabilizing agents or surfactants are usually added to prevent agglomeration and control the particle size. Once precipitation is complete, the resultant nanoparticle is separated via centrifugation, washed, dried, and annealed for further applications. It is relatively the simplest and most cost-effective technique with minimum equipment. Sometimes, nanoparticles undergo agglomeration that can affect their performance.

The solution-based method is the most widely employed technique in synthesizing 2D nanomaterials because of its scalability, cost-efficiency, versatility, control over size and shape, and morphology. Limitations include the potential for contaminants, precursor compatibility, the need for post-synthesis treatment, and growth kinetics.

3.2.2.4 Ion-Exchange Technique

This technique involves the intercalation of ions into the crystal lattice of the bulk material or exchanging an already present ion with another ion [41]. Ions have been exchanged between the substrate (solid or template) and the precursor solution. The solution is prepared with preferred ions to be exchanged, whether with metal ions or with desired organic molecules. Cations with smaller ionic radii, such as Li^+, Na^+, and K^+, are used to create interlayer spacing, enhancing the interspace and weakening the van der Waals force of attraction between the layers. The precursor solution and the substrate are brought together for the interaction and exchange processes. When sonicated, these intercalated ions react with the solvent molecules and generate hydrogen gas, promoting exfoliation efficiency. This leads to the formation or growth of the 2D nanomaterial with a similar lattice structure. This step is followed by centrifugation to separate the nanosheets from the solvent. Based on the charge of ions, there are typically three types: cation, anion, and complex ion-exchange methods. This method's advantages include scalability, versatility, control over the properties, and compatibility with other synthesis techniques. While substrate compatibility, structural constraints, and mismatch between precursor and substrate ions can cause defects and strain in the material. To attain the desired nanomaterial, the choice of precursor ions and their concentration have to be optimized.

3.2.2.5 Atomic Layer Deposition

It involves thin film deposition with atomic-level control on the substrate. It is a sequential process where each precursor molecule is allowed to move one after the other. Common substrates such as silicon wafers, glass, polymers, and metals are used. Precursor gas is first introduced into the reaction chamber, which reacts with the substrate's surface functional groups, leading to the formation of a chemisorbed monolayer of the precursor. An inert gas such as nitrogen is purged into the reaction chamber, followed by a second precursor. The reaction is initiated by temperature, UV light, or plasma, and this cycle is repeated until the desired film thickness is achieved. Film growth is self-limiting due to the sequential nature of this process. This method makes precise thickness control, high uniformity, and quality films possible [42]. Demerits include equipment complexity, cost, limited material selection, and slow deposition rate.

3.3 STRUCTURE OF 2D NANOMATERIALS

Generally, 2D nanomaterials exhibit a planar or layered structure with a nanometer scale thickness. The quantum confinement effect, large surface area, and reduced dimensionality deliver its unique properties. The atomic layer arrangement distinguishes 2D nanomaterials from other materials. 2D nanomaterials generally have a planar or layered structure where atoms are arranged periodically within the planar structure. Each atom must be stacked over the other in a laminar manner through the weak van der Waals interaction force. Common lattice symmetry involves trigonal (BN), hexagonal (graphene), and square (transitional metal dichalcogenides) [43]. Surface chemistry and edge structure can influence the interaction with other materials and environments. Structural irregularities, defects, and vacancies can arise in the synthesis process of 2D nanomaterials, which can have beneficial and detrimental effects depending on the application. The interlayer force of attraction between the layered 2D nanomaterial is generally the weak van der Waals force of attraction; this weak interaction enables interlayer sliding and exfoliation, and in fact, the attraction force leads to restacking and agglomeration. 2D nanomaterials also offer the formation of the heterostructure, stacking various 2D layers together. This stacking enables tailored properties and functionalities. 2D nanomaterials exhibit various morphologies, including nanosheets, ribbons, and flakes, and can be fabricated as per the application.

3.4 PROPERTIES OF 2D NANOMATERIALS

The unique properties of 2D nanomaterials are depicted in Figure 3.3 and detailed below:

- Ultrathin nature: The thickness of 2D nanomaterial is generally a few atomic layers, which delivers them a high surface-to-volume ratio and the ability to interact with the surroundings. The thinnest form of 2D nanomaterial is a monolayered structure, having a single layer of atoms; typically, these have one atom thickness. 2D nanomaterials can also exist as a few-layer structure, multilayer, or bulk material. These few-layer structures retain properties of monolayers while influenced by interlayer interactions. In multilayer or bulk,

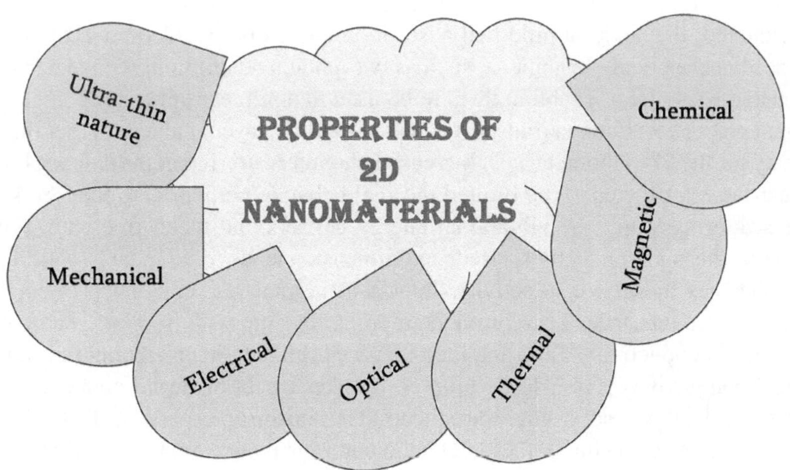

FIGURE 3.3 Major properties of 2D nanomaterials.

these interlayer interactions are more pronounced, resembling a 3D material while preserving the properties of 2D material due to its reduced dimensionality. The thickness of 2D nanomaterial has a significant role in determining the property and application [44].

- Mechanical properties: 2D nanomaterials showcase remarkable mechanical properties, despite being ultrathin, which is attributed to their bonding and atomic-scale structure. 2D nanomaterials are found to withstand large deformations, possessing a high Young's modulus and their ability to resist bending and compression. Even then, some of these 2D nanomaterials possess high elasticity and flexibility. 2D nanomaterials display various fracture mechanisms. For instance, graphene has a brittle fracture behavior owing to its strong covalent bonding, while MoS_2 displays more ductile fracture from its sliding nature. Mechanical anisotropy and variation in the direction of crystallography is another advantageous behavior that 2D nanomaterials exhibit. The weak van der Waals force of interaction that exists between the interlayers contributes to their adhesive and cohesive behavior. Overall, the quantum confinement and the surface effects account for these mechanical properties [45].
- Electrical properties: The ultrathin nature, quantum confinement effects, and unique electronic band structures account for the electrical properties of the 2D nanomaterials. Graphene has exceptional electrical conductivity, low resistivity, and high electronic mobility. While black phosphorous and transition metal dichalcogenides possess finite band gaps to control the electron flow and hence display semiconducting behavior. As the dimension is reduced, the electronic behavior can be quantized. The quantum confinement effect plays a major role in determining the properties of 2D nanomaterials. Thermoelectric properties are displayed by bismuth telluride and antimonene. By modulating the thickness and composition of the 2D nanomaterial, their band gaps can be tuned to engineer materials that possess specific electronic properties. Regarding carrier mobility, 2D nanomaterials exhibit high carrier mobility and fast and efficient charge transport enabling high-speed operation [46].
- Optical properties: Tunability of the band gap structure allows adsorption and emission and exhibits photoluminescence. This property is employed in bio-imaging, light-emitting diodes, and optoelectronics. The ultrathin nature and the planar geometry deliver them optical waveguiding phenomena. The plasmonic effect enables enhanced light adsorption, strong light–matter interactions, and localized electromagnetic field enhancements. Quantum confinement and plasmonic effect enable them to be used in surface-enhanced spectroscopy, sensing, and photodetection. Quantum confinement also influences the optical behavior of 2D nanomaterials; as the electronic band structure and the density of states are quantized, they exhibit modified adsorption and emission. Certain 2D nanomaterials offer nonlinear optical phenomena, such as two-photon adsorption, second harmonic generation and Kerr effect, enabling them to be used in nonlinear optics [47].
- Thermal properties: Phonon transport across the atomic layers induces higher thermal conductivity for the 2D nanomaterials. Increased phonon scattering at the defects, boundaries, and interfaces attributes to the limited thermal behavior compared to their bulk material. These scatterings limit the vibrational energy carriers and mean free path, minimizing the overall thermal conductivity. Reduced dimension leads to boundary scattering, which also influences thermal conductivity. Interface thermal resistance also affects the conductivity; these interfaces also introduce more scattering, with high interfacial resistance and reduced conductivity. Tailoring the interfaces can help in enhancing the conductivity. Thermal conductivity can also be tailored by altering the material composition, doping percentage, defects, and strains introduced. The anisotropic phonon dispersion and the phonon–phonon scattering in these 2D nanomaterials induce anisotropic thermal conductivity. Understanding the thermal stability needed and integrating the devices can enable the efficient utilization of these materials for thermal applications [48].

- Magnetic properties: Magnetic behavior of the 2D nanomaterials is far different from their bulk counterparts arising from the unique atomic arrangements and reduced dimensionality. Magnetic ordering and anisotropy effects are displayed in 2D nanomaterials, magnetic ordering such as ferrimagnetism, ferromagnetism, and antiferromagnetism, and the reduced dimensionality introduces anisotropy, which is utilized in the spintronics and the magnetic storage applications. Spin–orbit coupling effects are also witnessed in 2D nanomaterials in the presence of heavy elements or spin–orbit interactions. This coupling can further influence the magnetic anisotropy and the relaxation time that leverages the spintronics and quantum computing phenomena. Understanding the domain configuration and structure is crucial in developing magnetic memory devices. Variations in the magnetic field and temperature can induce phase transitions and modulate the magnetic behavior, which can be utilized for tunable magnetic devices, actuators, and sensors. Magnetic heterostructures can also be engineered to enhance the magnetic phenomena, spin–valley coupling, spin filtering, and spin-polarized transportation [49].
- Chemical stability and reactivity: These factors depend on the composition, structure, and surface chemistry of 2D nanomaterials. Chemical stability depends on the functional groups, defects, exposure to temperature, environment, etc. Chemical reactivity is accounted by the exposed edges, dangling bonds, and higher surface areas, making them susceptible to chemical transformations. Surface functionalization can also enhance selective binding, adsorption, band alignment, and interfacial reactions. Environmental factors such as humidity, temperature, and exposure to other chemicals and gases also influence the material's chemical stability and reactivity. Precisely optimizing the parameters can give rise to desired 2D nanomaterial, unlocking its potential for a wide range of applications.

3.5 CONCLUSION AND FUTURE OUTLOOK

The research on 2D nanomaterials holds promise for discovering more novel materials with diverse applications in energy harvesting, storage, medical applications, the food industry, etc. More explorations into the chemical and physical properties can enable the fabrication of next-generation functional nanomaterials. Utilizing available resources and dwelling more to the core of the 2D nanomaterial family, altering the composition, tuning its size, shape, morphology, and band structure, and tailoring interlayer distance and interactions can help to achieve high-quality 2D nanomaterial. The challenges in synthesis and processing of 2D nanomaterials are yet to be overcome.

REFERENCES

1. Feynman R.P., Plenty of room at the bottom, *APS Annual Meeting*, 1959. https://doi.org/10.1109/84.128057.
2. Buzea C., Pacheco I., Nanomaterials and their classification, *EMR/ESR/EPR Spectroscopy for Characterization of Nanomaterials* (2017) 3–45. https://doi.org/10.1007/978-81-322-3655-9.
3. Zhang H., Cheng H.-M., Ye P., 2D nanomaterials: Beyond graphene and transition metal dichalcogenides, *Chemical Society Reviews* 47(16) (2018) 6009–6012. https://doi.org/10.1039/C8CS90084A.
4. Bonaccorso F., Colombo L., Yu G., Stoller M., Tozzini V., Ferrari A.C., Ruoff R.S., Pellegrini V., Graphene, related two-dimensional crystals, and hybrid systems for energy conversion and storage, *Science* 347 (6217) (2015) 1246501. https://doi.org/ 10.1126/science.1246501.
5. Gupta A., Sakthivel T., Seal S., Recent development in 2D materials beyond graphene, *Progress in Materials Science* 73 (2015) 44–126. https://doi.org/10.1016/j.pmatsci.2015.02.002.
6. Fiori G., Bonaccorso F., Iannaccone G., Palacios T., Neumaier D., Seabaugh A., Banerjee S.K., Colombo L., Electronics based on two-dimensional materials, *Nature Nanotechnology* 9(10) (2014) 768–779. https://doi.org/10.1038/nnano.2014.207.
7. Boschetto G., Xu T., Yehya M., Thireau J., Lacampagne A., Charlot B., Gil T., Todri-Sanial A., Exploring 1D and 2D nanomaterials for health monitoring wearable devices, *2021 IEEE International Conference on Flexible and Printable Sensors and Systems (FLEPS)*, IEEE, 2021, pp. 1–4. https://doi.org/10.1109/FLEPS51544.2021.9469864.

8. Tao P., Yao S., Liu F., Wang B., Huang F., Wang M., Recent advances in exfoliation techniques of layered and non-layered materials for energy conversion and storage, *Journal of Materials Chemistry A* 7(41) (2019) 23512–23536. https://doi.org/10.1039/C9TA06461C.
9. Hu T., Mei X., Wang Y., Weng X., Liang R., Wei M., Two-dimensional nanomaterials: Fascinating materials in biomedical field, *Science Bulletin* 64(22) (2019) 1707–1727. https://doi.org/10.1016/j.scib.2019.09.021.
10. Novoselov K.S., Geim A.K., Morozov S.V., Jiang D.-E., Zhang Y., Dubonos S.V., Grigorieva I.V., Firsov A.A., Electric field effect in atomically thin carbon films, *Science* 306(5696) (2004) 666–669. https://doi.org/10.1126/science.1102896.
11. Rono N., Kibet J.K., Martincigh B.S., Nyamori V.O., A review of the current status of graphitic carbon nitride, *Critical Reviews in Solid State and Materials Sciences* 46(3) (2021) 189–217. https://doi.org/10.1080/10408436.2019.1709414.
12. Jiang X.-F., Weng Q., Wang X.-B., Li X., Zhang J., Golberg D., Bando Y., Recent progress on fabrications and applications of boron nitride nanomaterials: A review, *Journal of Materials Science & Technology* 31(6) (2015) 589–598. https://doi.org/10.1016/j.jmst.2014.12.008.
13. Manzeli S., Ovchinnikov D., Pasquier D., Yazyev O.V., Kis A., 2D transition metal dichalcogenides, *Nature Reviews Materials* 2(8) (2017) 1–15. https://doi.org/10.1038/natrevmats.2017.33.
14. Ren X., Lian P., Xie D., Yang Y., Mei Y., Huang X., Wang Z., Yin X., Properties, preparation and application of black phosphorus/phosphorene for energy storage: A review, *Journal of Materials Science* 52 (2017) 10364–10386. https://doi.org/10.1007/s10853-017-1194-3.
15. Zhao J., Liu H., Yu Z., Quhe R., Zhou S., Wang Y., Liu C.C., Zhong H., Han N., Lu J., Rise of silicene: A competitive 2D material, *Progress in Materials Science* 83 (2016) 24–151. https://doi.org/10.1016/j.pmatsci.2016.04.001.
16. Nalawade P., Aware B., Kadam V., Hirlekar R., Layered double hydroxides: A review (2009). http://nopr.niscpr.res.in/handle/123456789/3482.
17. Kumbhakar P., Gowda C.C., Mahapatra P.L., Mukherjee M., Malviya K.D., Chaker M., Chandra A., Lahiri B., Ajayan P., Jariwala D., Emerging 2D metal oxides and their applications, *Materials Today* 45 (2021) 142–168. https://doi.org/10.1016/j.mattod.2020.11.023.
18. Chakraborty G., Park I.-H., Medishetty R., Vittal J.J., Two-dimensional metal-organic framework materials: Synthesis, structures, properties and applications, *Chemical Reviews* 121(7) (2021) 3751–3891. https://doi.org/10.1021/acs.chemrev.0c01049.
19. Liang R.-R., Jiang S.-Y., Ru-Han A., Zhao X., Two-dimensional covalent organic frameworks with hierarchical porosity, *Chemical Society Reviews* 49(12) (2020) 3920–3951. https://doi.org/10.1039/D0CS00049C.
20. Naguib M., Barsoum M.W., Gogotsi Y., Ten years of progress in the synthesis and development of MXenes, *Advanced Materials* 33(39) (2021) 2103393. https://doi.org/10.1002/adma.202103393.
21. Williams O.F., Guo Z., Hu J., Yan L., You W., Moran A.M., Energy transfer mechanisms in layered 2D perovskites, *The Journal of Chemical Physics* 148(13) (2018) 134706. https://doi.org/10.1063/1.5009663.
22. Randviir E.P., Brownson D.A., Banks C.E., A decade of graphene research: Production, applications and outlook, *Materials Today* 17(9) (2014) 426–432. https://doi.org/10.1016/j.mattod.2014.06.001.
23. Heyl M., List-Kratochvil E.J., Only gold can pull this off: Mechanical exfoliations of transition metal dichalcogenides beyond scotch tape, *Applied Physics A* 129(1) (2023) 16. https://doi.org/10.1007/s00339-022-06297-z.
24. Zomer P., Dash S., Tombros N., Van Wees B. A transfer technique for high mobility graphene devices on commercially available hexagonal boron nitride, *Applied Physics Letters* 99(23) (2011) 232104. https://doi.org/10.1063/1.3665405.
25. Qing Y.A., Li R., Li S., Li Y., Wang X., Qin Y., Advanced black phosphorus nanomaterials for bone regeneration, *International Journal of Nanomedicine* (2020) 2045–2058. https://doi.org/10.2147/IJN.S246336.
26. Manna K., Huang H.-N., Li W.-T., Ho Y.-H., Chiang W.-H., Toward understanding the efficient exfoliation of layered materials by water-assisted cosolvent liquid-phase exfoliation, *Chemistry of Materials* 28(21) (2016) 7586–7593. https://doi.org/10.1021/acs.chemmater.6b01203.
27. Ozin G.A., Arsenault A., Nanochemistry: A chemical approach to nanomaterials, Royal Society of Chemistry 2015. https://hdl.handle.net/11511/63488.
28. Lei Y., Ossonon B.D., Chen J., Perreault J., Tavares A.C., Electrochemical characterization of graphene-type materials obtained by electrochemical exfoliation of graphite, *Journal of Electroanalytical Chemistry* 887 (2021) 115084. https://doi.org/10.1016/j.jelechem.2021.115084.

29. Gates B.D., Xu Q., Stewart M., Ryan D., Willson C.G., Whitesides G.M., New approaches to nanofabrication: Molding, printing, and other techniques, *Chemical Reviews* 105(4) (2005) 1171–1196. https://doi.org/10.1021/cr030076o.
30. Tseng A.A., Chen K., Chen C.D., Ma K.J., Electron beam lithography in nanoscale fabrication: Recent development, *IEEE Transactions on Electronics Packaging Manufacturing* 26(2) (2003) 141–149. https://doi.org/10.1109/TEPM.2003.817714.
31. Garcia R., Knoll A.W., Riedo E., Advanced scanning probe lithography, *Nature Nanotechnology* 9(8) (2014) 577–587. https://doi.org/10.1038/nnano.2014.157.
32. Guo L.J., Nanoimprint lithography: Methods and material requirements, *Advanced Materials* 19(4) (2007) 495–513. https://doi.org/10.1002/adma.200600882.
33. Liu G., Hirtz M., Fuchs H., Zheng Z., Development of dip-pen nanolithography (DPN) and its derivatives, *Small* 15(21) (2019) 1900564. https://doi.org/10.1002/smll.201900564.
34. Bates C.M., Maher M.J., Janes D.W., Ellison C.J., Willson C.G., Block copolymer lithography, *Macromolecules* 47(1) (2014) 2–12. https://doi.org/10.1021/ma401762n.
35. Ji Q., Zhang Y., Zhang Y., Liu Z., Chemical vapour deposition of group-VIB metal dichalcogenide monolayers: Engineered substrates from amorphous to single crystalline, *Chemical Society Reviews* 44(9) (2015) 2587–2602. https://doi.org/10.1039/C4CS00258J.
36. Huynh S.H., Diep N.Q., Le T.V., Wu S.K., Liu C.W., Nguyen D.L., Wen H.C., Chou W.C., Le V.Q., Vu T.T., Molecular beam epitaxy of two-dimensional GaTe nanostructures on GaAs (001) substrates: Implication for near-infrared photodetection, *ACS Applied Nano Materials* 4(9) (2021) 8913–8921. https://doi.org/10.1021/acsanm.1c01544.
37. Kang J., Sangwan V.K., Wood J.D., Hersam M.C., Solution-based processing of monodisperse two-dimensional nanomaterials, *Accounts of Chemical Research* 50(4) (2017) 943–951. https://doi.org/10.1021/acs.accounts.6b00643.
38. Sutka A., Mezinskis G., Sol-gel auto-combustion synthesis of spinel-type ferrite nanomaterials, *Frontiers of Materials Science* 6 (2012) 128–141. https://doi.org/10.1007/s11706-012-0167-3.
39. Feng S.-H., Li G.-H., Hydrothermal and solvothermal syntheses, in *Modern Inorganic Synthetic Chemistry*, Elsevier 2017, pp. 73–104. https://doi.org/10.1016/B978-0-444-63591-4.00004-5.
40. Kim Y.I., Kim D., Lee C.S., Synthesis and characterization of $CoFe_2O_4$ magnetic nanoparticles prepared by temperature-controlled coprecipitation method, *Physica B: Condensed Matter* 337(1–4) (2003) 42–51. https://doi.org/10.1016/S0921-4526(03)00322-3.
41. Beberwyck B.J., Surendranath Y., Alivisatos A.P., Cation exchange: A versatile tool for nanomaterials synthesis, *The Journal of Physical Chemistry C* 117(39) (2013) 19759–19770. https://doi.org/10.1021/jp405989z.
42. Oviroh P.O., Akbarzadeh R., Pan D., Coetzee R.A.M., Jen T.-C., New development of atomic layer deposition: Processes, methods and applications, *Science and Technology of Advanced Materials* 20(1) (2019) 465–496. https://doi.org/10.1080/14686996.2019.1599694.
43. Parvez K., Two-dimensional nanomaterials: Crystal structure and synthesis, in *Biomedical Applications of Graphene and 2D Nanomaterials*, Elsevier 2019, pp. 1–25. https://doi.org/10.1016/C2017-0-02937-3.
44. Tan C., Cao X., Wu X.-J., He Q., Yang J., Zhang X., Chen J., Zhao W., Han S., Nam G.-H., Recent advances in ultrathin two-dimensional nanomaterials, *Chemical Reviews* 117(9) (2017) 6225–6331. https://doi.org/10.1021/acs.chemrev.6b00558.
45. Li X., Sun M., Shan C., Chen Q., Wei X., Mechanical properties of 2D materials studied by in situ microscopy techniques, *Advanced Materials Interfaces* 5(5) (2018) 1701246. https://doi.org/10.1002/admi.201701246.
46. Lee J.-H., Park S.-J., Choi J.-W., Electrical property of graphene and its application to electrochemical biosensing, *Nanomaterials* 9(2) (2019) 297. https://doi.org/10.3390/nano9020297.
47. Li Y., Li Z., Chi C., Shan H., Zheng L., Fang Z., Plasmonics of 2D nanomaterials: Properties and applications, *Advanced Science* 4(8) (2017) 1600430. https://doi.org/10.1002/advs.201600430.
48. Qiu L., Zhu N., Feng Y., Michaelides E.E., Żyła G., Jing D., Zhang X., Norris P.M., Markides C.N., Mahian O., A review of recent advances in thermophysical properties at the nanoscale: From solid state to colloids, *Physics Reports* 843 (2020) 1–81. https://doi.org/10.1016/j.physrep.2019.12.001.
49. Kumar C.S., *Magnetic Nanomaterials*, John Wiley & Sons 2009. https://doi.org/10.1002/9783527627561.

4 2D Metal Chalcogenides Gas Chemical Sensors

Nour F. Attia, Rabbia Shoaib, Izaz ul Islam, Sally E. A. Elashery, and Heba Ameen

4.1 INTRODUCTION

The tremendous importance of environmental monitoring technologies has increased owing to escalating concerns regarding air pollution caused by industrial chemicals, automobile exhaust, and mega power houses. Hence, the detection of various hazardous gases triggering allergies in humans and impacting both medical treatments and industrial operations has become inevitable. In this regard, the quest to find high-performance gas sensors is a valuable area of technological development. Gas sensors are indispensable tools for ensuring safety in industrial settings, monitoring pollution levels in the environment, in healthcare for diagnosing medical conditions and monitoring patients, and in consumer products and automotive industries [1]. Chemical gas sensors are designed to sense and determine specific gases quantitatively in the environment based on their chemical interactions with sensing materials. These sensors are widely used in various applications, including industrial safety, environmental monitoring, and healthcare. Here is an overview of the technology behind chemical gas sensors. Sensors consist of a recognition part and a transducer. Metal chalcogenides, which are highly responsive to alterations in the surrounding environment, are utilized as semiconductor materials in gas sensors. The general chemical formula for metal chalcogenides is MX_2, where M can be either bi-transition or post-transition metal (Sn, In), and X represents a chalcogen. Multiple crystal phases of both exist with correlated electronic characteristics. Their layered structures are stacked via van der Waals forces [2].

Traditional gas sensors are equipped with power-intensive microheaters for working, while miniaturized two-dimensional (2D) materials diminish the demand for energy-intensive microheaters as they can be operated at room temperature with minimal power consumption, which is a critical requirement for advanced applications. Miniaturization in gas sensor technology is not merely a matter of size reduction but a pivotal factor in enhancing the versatility, accessibility, and energy efficiency of gas sensors. Manufacturing of 2D materials-based gas sensor devices is progressing with the development of 2D materials with chemical vapor deposition (CVD) process [2]. While 1D materials like nanowires have their advantages, the shift toward 2D materials in gas sensing reflects the pursuit of upgraded performance by increasing surface area that enhances the gas interaction with sensing material leading to higher sensitivity, rapid response rate, and versatility in addressing the evolving demands of gas detection applications. Researchers continue to explore and refine the use of 2D materials to advance gas sensing technology [3]. Also, it is also dedicated for other chemical sensors for detection of toxic ions and compounds [4–10].

4.2 SYNTHESIS APPROACHES OF 2D METAL CHALCOGENIDES

2D metal chalcogenides have demonstrated a lot of potential for gas sensing applications. They offer unique physical, chemical, and electrical characteristics that make them suitable for the construction of high-tech precise gas sensors [2]. The synthesis and application of porous 2D nanomaterials

DOI: 10.1201/9781003436942-4

based on metal oxides have been explored to improve the sensitivity and response rate of gas sensors [3]. Additionally, the incorporation of organic functional materials to 2D inorganic materials, such as organic metal chalcogenides (OMCs), graphene, and transition metal, has been investigated to optimize their properties [11]. For instance, the absence of functional groups on the inorganic layer surface in few-layered nanosheets resulted in no response to external gases. The evidential studies of 2D chemiresistive sensor materials with $-NH_2$ groups acting as receptors by completely covering the surface of Ag ($SPh-NH_2$) exhibit superior sensitivity, highest selectivity, and NO_2 response reversibility at room temperature. Further DRIFT analysis confirmed the essential function of $-NH_2$ in the gas sensing ability of Ag ($SPh-NH_2$) [12]. The presence of structured functional organic molecules allows for the control of both the types and the intensity of interactions between these two materials. This regulation facilitates the more straightforward co-assembly of various ordered molecular crystals, with the anticipation of achieving improved performance [13]. The development of 2D metal chalcogenides and their heterostructures with metal oxides holds great potential for the advancement of gas chemical sensors. Gas sensors fabricated on transition metal di-chalcogenides with a general formula MX_2, where M refers to a transition metal (Mo, W, Re, Ta, Hf, Ti, etc.) and X_2 is a chalcogen (S_2, Se_2, and Te_2). Their physical and chemical characteristics have been studied extensively due to their semiconductor behavior, significant surface-to-volume ratio, substantial band gap, elevated absorption capacity, and the presence of active sites for redox reactions. 2D metal chalcogenides (MCs) and 2D GVTMCs can be fabricated using various routes such as CVD, liquid exfoliation, electrochemical exfoliation, and solvothermal process (Figure 4.1). Each of these methods has its advantages and limitations, and researchers have been working to optimize them for specific applications. These synthesis approaches enable the manipulation of dimensions, phases, compositions, defects, and heterostructures of anisotropic MCs [13].

4.2.1 Mechanochemical Synthesis of 2D Gallium Oxyselenide Nanosheets

A recent study demonstrates a novel straightforward approach to synthesize 2D gallium oxyselenide nanosheets through a single-step mechanochemical process via sonication-based exfoliation of bulk GaSe in a specific solvent combination of (75%v/v isopropyl alcohol) IPA with variable concentrations of H_2O_2, followed by centrifugation, ethanol washing, and subsequent material

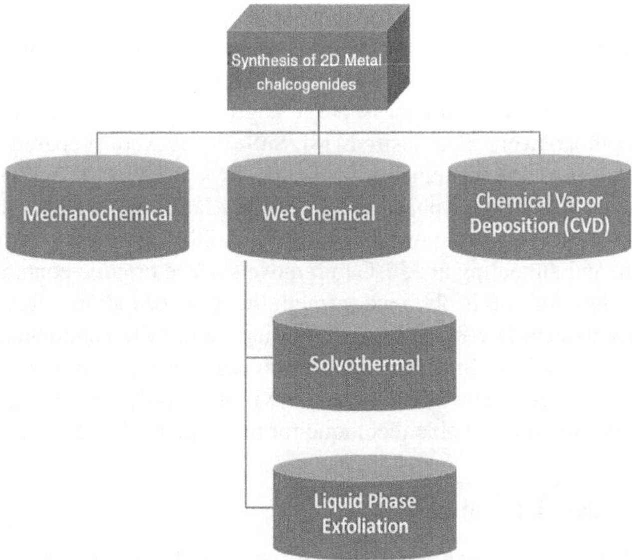

FIGURE 4.1 Schematic diagram representing different synthesis approaches of 2D metal chalcogenides.

characterization and sensor fabrication [14]. While bulk GaSe is being intensely power-driven to exfoliate into 2D atomically thin layers, a key transformation occurred. Partial selenium (Se) atoms within the bulk GaSe were effectively replaced by oxygen (O) atoms in the presence of hydrogen peroxide (H_2O_2) as the oxidant in the exfoliation process. This substitution led to the formation of 2D gallium oxyselenides. Notably, the concentration of H_2O_2 played a crucial role in tailoring the properties of these materials [14].

a. **Wet chemical synthesis**

   These methods involve the synthesis of 2D metal chalcogenides in a liquid or solution medium, including techniques such as hydrothermal/solvothermal pathways, liquid-phase exfoliation, heat-up, hot-injection techniques, and template-mediated synthesis procedures. These methods have been used to synthesize transition metal chalcogenides (TMCs), such as CdS, MoS_2, WS_2, TiS_2, TaS_2, ReS_2, $MoSe_2$, and WSe_2, as indicated in Figure 4.2 [10,14-29].

b. **Solvothermal synthesis**

   A novel approach "in situ topochemical conversion" for the synthesis of Bi_2Se_3/BiOCl layered sensor material involved a solvothermal method. A solution of BiOCl was prepared in a mixture of oleylamine (OLA) and dodecanethiol (DDT) and further hot-injected into another solution containing BiOCl nanosheets. The growth of Bi_2Se_3 on the BiOCl nanosheets was induced by heating. This method allows for the efficient optimization of active edge sites within deposited overlayers and the creation of coherent heterointerfaces. This process involves the transformation of the material's structure while preserving its chemical composition. In this case, layered BiOCl was converted into a Bi_2Se_3/BiOCl heterostructure [15]. By performing the topochemical conversion, hexagonal Bi_2Se_3(001) layers were epitaxially deposited onto the BiOCl (001) basal surfaces. Importantly, this process resulted in the exposure of a significant number of (100) and (110) edge sites in the Bi_2Se_3 layer. This epitaxial growth is crucial for creating well-defined and highly ordered heterointerfaces [15]. The synthesis of WS_2 by hydrothermal method to prepare ultrathin (5 nm) nanosheets was not only successful, but also exhibited p-type attributes in resistive-gas sensors, with a response rate 9.3% to 0.1 ppm NO_2 gas at ambient conditions [19].

c. **Liquid phase exfoliation**

   High-yield suspensions of 2D transition metal dichalcogenides (TMDs) are obtained by exfoliating TMD materials into nanosheets or nanoplatelets with liquid solutions, producing large quantities with high quality. Various techniques for liquid phase exfoliation have been devised, such as shear mixing, mechanical blending and grinding, or a combination of these. These methods could necessitate a lengthy processing period and occasionally produce fewer monolayers than desired [18]. SnS_2 flakes were prepared by wet chemical method and its sensor performance was 10 ppm NO_2 at 120°C [20]. 2D SnS nanoflakes were synthesized via a liquid phase exfoliation method, involving the mechanical mixing of 1 g of SnS micro-sized powder with 200 μL DMF, followed by probe sonication, centrifugation, and annealing at 350°C to remove surface organic contaminants [16]. The p-type conductivity of SnS is obtained through the easy formation of acceptor-like intrinsic Sn vacancy defects [17,18]. Compared to many other 2D nanomaterials, the demonstrated sensitivity values of SnS_2/SnS nanostructured sensing efficiency are higher. Liquid phase exfoliation offers a cost-effective and environmentally friendly approach to obtain TMD flakes, making it a valuable technique for the scalable production of 2D materials.

4.2.2 Chemical Vapor Deposition

CVD is a method that is frequently employed for growing high-quality 2D metal chalcogenides. It involves the reaction of precursor gases on a substrate at elevated temperatures. This method allows for precise control over material thickness and composition, making it suitable for creating uniform

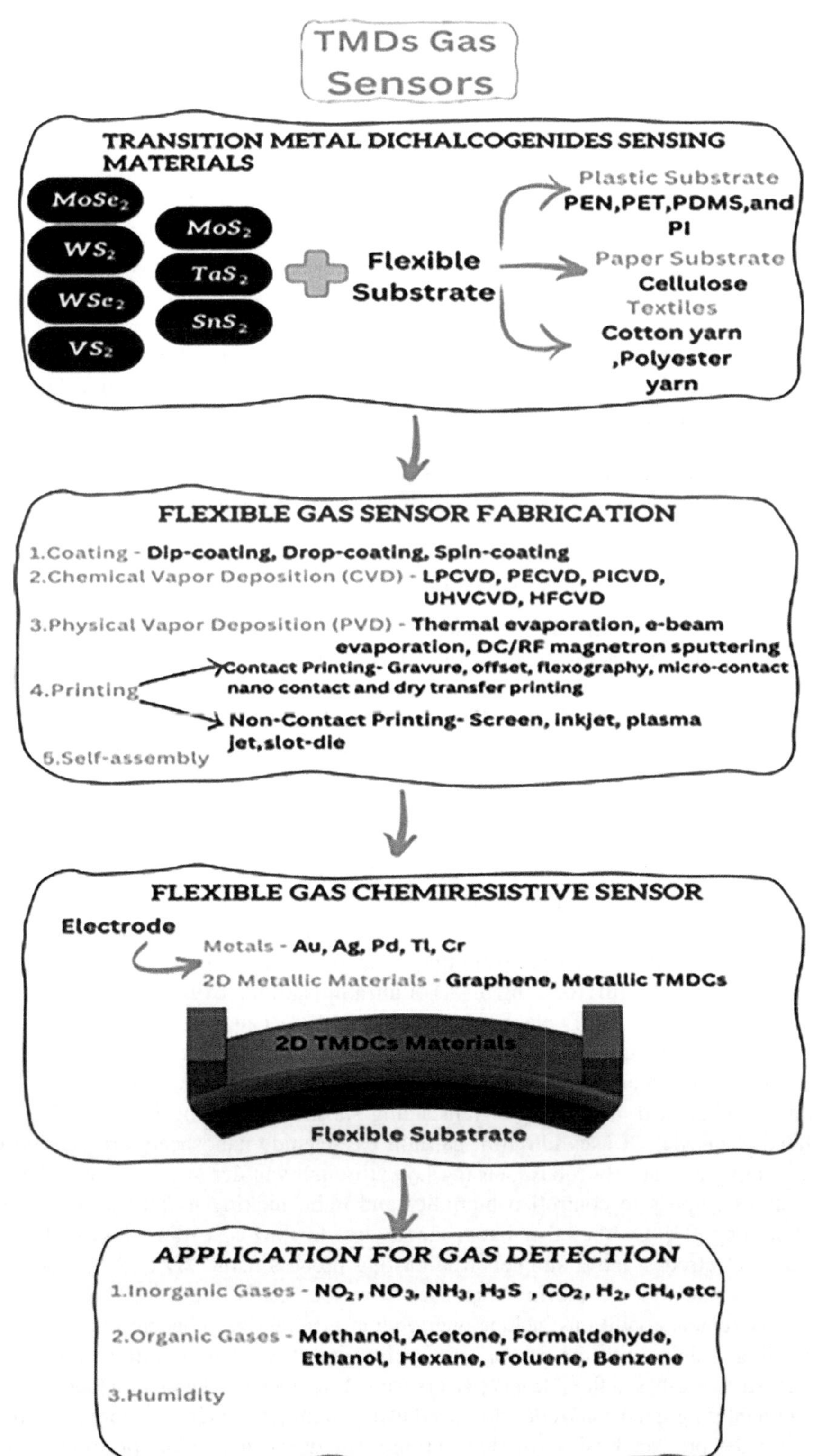

FIGURE 4.2 Transition metal dichalcogenides gas sensors developments, and was adapted from reference [10].

and well-defined sensor materials. The advantage of CVD is its ability to tune the number of layers that are uniformly over a large area. In most of the CVD growth of bulk crystals, TeCl4 is used as a transport reagent. 2D GeSexOy nanosheets, a material for gas sensing, were synthesized using CVD on a Si/SiO_2 substrate with germanium and selenium precursors and controlled oxygen flow [21,22]. Ultrathin layers form via layer-by-layer growth. 2D heterostructure was achieved through a process involving the exfoliation and nucleation of materials [23–28]. The resulting material exhibited excellent room-temperature sensitivity to NO_2 gas, with a recognition limit of 0.986 ppb and a linear response range between 0.05 and 10 ppm [22]. The synthesis of 2D MoS_2 was achieved through a low-temperature CVD process using molybdenum hexacarbonyl (Mo (CO)6) as the precursor and hydrogen sulfide (H_2S) as the reactant gas. The CVD chamber, maintained within a temperature range of approximately 200°C–450°C, facilitated the growth process. Mo (CO)6 was injected directly into the chamber, and H_2S gas, along with additional hydrogen (H_2) and argon (Ar) gases, was flowed in to aid in thermal distribution and chamber pressure stabilization. The synthesis occurred on a silicon dioxide (SiO_2)/silicon (Si) substrate, allowing control of the number of MoS_2 layers by adjusting the process time, typically ranging from 10 to 40 minutes. This method enabled the production of 2D MoS_2 materials with precise control over layer thickness and uniformity (Figure 4.3) [29].

4.3 GAS SENSORS FABRICATION APPROACHES

Fabrication approaches include adding substrate, integrating certain electrode materials, varying nanosheet deposition, enhancing surface roughness, and modifying with functional molecules, metals, metal oxide nanostructures, heterojunction with other 2D materials, etc. Doping metal chalcogenides with atoms including Fe, Ni, and Al can boost their binding attraction for specific gases. Researchers have found that the interaction and charge transmission between the 2D MnPr and adsorbed gas impact material electronic properties and sensing capabilities. Fabrication of material with 2D manganese porphyrin (MnPr) monolayer was done for ultra-sensitive gas sensing of molecules like H_2S, CO, CO_2, SO_2, NO, and NO_2 [23]. For instance, the sensor fabrication process for mechanochemically adjustable in situ 2D oxyselenide of gallium involved creating a transducing substrate with 200 pairs of 10-μm-wide interdigitated electrodes (IDE) through electron-beam lithography on a Si/300 nm SiO_2 substrate, followed by the deposition of a 10 nm Ti layer in conjunction with Au layer of 90 nm thickness for contacts. Gas sensors were then synthesized via drop-casting of 20 μL (0.02 mg mL^{-1}) of ultrathin gallium oxyselenide on the transducing object at 45°C, resulting in a 2D planar gallium oxyselenide nanosheet layer. Sensing measurements were conducted under controlled humidity and light irradiation conditions, with resistance measured using digital multimeter and gas flow maintained at 200 sccm. In this process, sensor fabrication was done with a structured transducing substrate featuring lithographically defined interdigitated electrodes. It uses ultrathin gallium oxyselenide nanosheets drop-casted onto the substrate, forming a relatively robust sensing layer through van der Waals forces. The measurement conditions emphasize controlled humidity and light, making it suitable for precise environmental sensing [23]. On the other hand, Jannat opts for low-cost transducing substrates with Ag–Pd-coated electrodes and a simpler drop-casting process using 2D SnS. It places a strong emphasis on gas selectivity, exploring responses to various gases like NO_2, H_2S, SO_2, CH4, and CO. The measurement conditions include operating at different gas concentrations and temperatures, providing insights into the sensor's versatility under varying conditions [20–28]. Bi_2Se_3/BiOCl heterostructure-based flexible gas sensors were developed via depositing them onto a plastic substrate with interdigitated electrodes produced using an inkjet printer. It is important to note that flexible electrodes produced using simple printing methods, such as inkjet printing, have a lot of space between electrodes and poor resolution. Thus, device performance may suffer as a result of this as opposed to use electrodes fabricated using conventional lithography methods. The surface

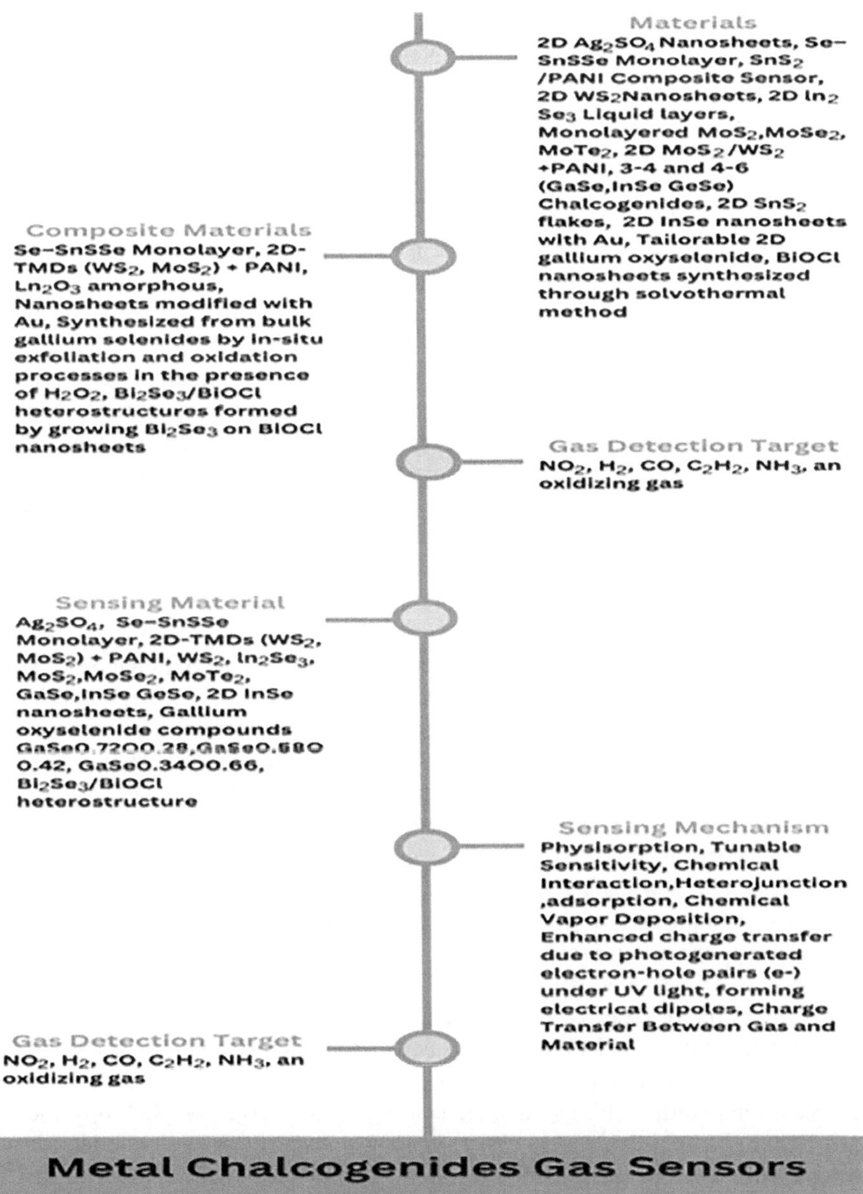

FIGURE 4.3 Recently studied metal chalcogenides for gas sensing applications.

roughness of the BiOCl nanosheet increased due to the formation of the nanoporous Bi_2Se_3 layer, which also improved the interlocking between the stacked hybrid nanosheets [15]. A FET sensor was fabricated with exfoliated WS_2 to improve photoelectrical gas sensing properties [19]. A simpler novel fabrication approach is thinning the FET channels. N-type $MoTe_2$ FETs were fabricated in a similar fashion by tuning the channel thickness [21]. The n-type transfer attitude in the $MoTe_2$ FET channel was attributed to Cl doping from TeCl4, which is used as a transport agent during CVT growth of bulk crystals. Hence, changing from n-type to p-type via declining FET channel depth resulted in a polarity change in ultrathin layers associated with surface states. A remarkable transition occurred in the polarity of the FETs as the channel thickness was reduced from 60 to ~5.6 nm. Initially, FETs exhibited n-type attitude, then ambipolar behavior, and finally, p-type behavior with reducing channel depth. These changes eventually modulated the Schottky barrier height between the semiconductor and metal interface, which in turn affected the charge carrier concentration. $MoTe_2$ FET sensing performance was remarkably enhanced for NH3 by simply adjusting the channel thickness. Enhancing Schottky contacts resulted in enhanced sensitivity in MoS_2 for NH_2, NO_2, and NH3 via $MoTe_2$ at ambient temperature [20,21]. SnS_2 is an n-type semiconductor with a hexagonal crystal structure, whereas SnS is a p-type semiconductor with an indirect band gap of 1.0–1.1 eV. Its poor intrinsic conductivity and weak adsorption qualities prevent it from working at room temperature, even though it is the perfect material for extremely effective NO_2 gas sensors. To address the restrictions of pure SnS_2, researchers have developed NO_2 gas sensors based on SnS_2/SnS p–n heterojunctions. These heterojunction-based sensors achieved a remarkable 660% gas-detecting response to 4 ppm NO_2 at room temperature. Similarly, PbS–SnS_2 was analyzed with the fabrication of Au and pristine. Another interesting fabrication approach was using an extremely sensitive gas sensing ink based on sulfonated reduced graphene oxide (S-rGO) and SnS_2, which has been fabricated for NO_2 and NH3 gas sensing. Interestingly, significant sensitivity to sub-ppm-level NH3 (11% response to 1 ppm) and ppb-level NO_2 (17% response to 125 ppb) was demonstrated by this ink. The exceptional absorption characteristics of SnS_2 enhanced the detection limits of S-rGO significantly, making it fourfold more sensitive to NO_2 gas and 55 times more sensitive to NH3 gas [17] (Table 4.1). Doping of SnS_2 with Mo significantly enhanced the NO_2 detecting response, which was about 23 times greater than pure SnS_2, at a temperature of approximately 150°C. In this fabrication approach, a very tiny n-type SnS_2 nanosheets were decorated with molybdenum (Mo) using the solvothermal technique for NO_2 detection [25].

For hydrogen gas sensing applications of MoS_2, the combinatorial impact of potential energy barrier fluctuations and nanohybrid material led to performance improvements. The fabrication method to achieve success was changing morphology by synthesizing Pd-SnO_2-MoS_2 nanohybrid materials. 2D MoS_2 layers was fabricated by depositing electrode material onto it using thermal evaporation. The electrode materials included Au/Ti, Ag, and Al. Ti layer of 3 nm thickness was used as an adhesion layer when layer of Au was applied as the electrode material. This approach allowed for the creation of gas sensors with different electrode materials to investigate their impact on gas responsivity and selectivity. On the other hand, the responsivity of 3 L of MoS_2 to NO_2 gas increased in case of using Au (12%), Ag (25%), and Al (80%) electrodes (Table 4.1). Additionally, by tuning the device design, fabrication parameters, and operating conditions, a straightforward molybdenum disulfide (MoS_2) field-effect transistor (FET) on a hexagonal boron nitride (hBN) substrate can detect NOx down to concentrations of 6 ppb and potentially much less at ambient temperature (RT) as depicted in Figure 4.4 [26]. For 2L MoS_2, the responsivity toward NO_2 gas increases for Au (63%), Ag (73%), and Al (98%) electrodes. Thus, the use of electrodes with lower work functions resulted in higher responsivity to NO_2 gas. However, when tested with extremely toxic CO gas of 500 ppm and the greenhouse gas CO_2 of 5000 ppm, the 2L MoS_2 gas sensor with Ag electrodes exhibited a response of approximately −15% for each gas. Interestingly, the 2L MoS_2 gas sensor with Au electrodes showed no response to these gases (Table 4.1) [27–29]. Other group reported a scalable hydrothermal synthesis to fabricate MoS_2 nanosheets with SnO_2 nanoparticles resultantly achieving superior gas detecting response (2080.36) to 200 ppm NH3, 27.5 times superior than pure MoS_2 films with rapid response/recovery times (23/1.6 seconds) at 22°C, and excellent selectivity against other gases (CH4, H_2, CO, H_2S, and NO_2) [27]. These remarkable sensing properties were attributed to fabricated thin layers of two-dimensional MoS_2 nanostructures promoting charge-carrier transformation and enhancing the reaction to SnO_2/MoS_2 and NH3 gas [27]. Another highly sensitive tailorable 2D gallium oxyselenide for enhanced optoelectronic NO_2 gas sensing [30] is shown in Figure 4.5 and highlighted in Table 4.1.

TABLE 4.1
Presenting the Reported Sensing Data of Reported 2D Metal Chalcogenides Gas Sensors

Material/System	Properties	Gas of Interest	Notable Findings	Reference
SnS	P-type semiconductor, indirect band gap	NO_2	Low intrinsic conductivity at room temperature	[1]
SnS_2	N-type semiconductor, hexagonal crystal	NO_2, NH_3, CO, and CO_2	Weak adsorption properties at room temperature	[2]
SnS_2/SnS heterojunction	Enhanced NO_2 sensing	NO_2	Achieved 660% gas-detecting response at RT	[24]
S-rGO/SnS_2 gas sensing ink	Enhanced NO_2 and NH_3 sensing	NO_2 and NH_3	Significant sensitivity to ppb-level NO_2	[24]
Doped SnS_2 with Mo	Enhanced NO_2 sensing	NO_2	23 times greater NO_2 detecting response	[25]
MoS_2 for hydrogen sensing	Nanohybrid materials for improved performance	H_2	Synergistic effect of potential energy barriers	[26]
MoS_2 with various electrodes	Impact of electrode materials on gas sensing	NO_2, CO, and CO_2	Electrode materials (Au/Ti, Ag, and Al) impact gas responsivity and selectivity	[29]
Exfoliated WS_2 and $MoTe_2$ FETs	Thinning FET channels, polarity change	NH_3, NO_2, H_2S, CO, and CO_2	Thinning FET channels enhanced gas sensing properties and modulated charge carrier concentration	[21]
Few-layer Ag(SPh-NH_2)	Covered with -NH_2 groups; conductivity of 6×10^{-7} S cm^{-1} at room temperature; band gap of 2.62 eV	NO_3	Excellent repeatability (CV: 6.92%); broad concentration range (0.1–100 ppm); theoretical LOD: 0.9 ppb—Fast response (1.0 minute) and recovery (2.3 minutes). Proposed detection mechanism involving acid–base interactions, hydrogen bonding, and charge transfer	[30]
2D Manganese porphyrin (MnPr)	2D MnPr monolayer	H_2S, CO, CO_2, SO_2, NO, and NO_2	Fabrication of 2D MnPr monolayer enabled ultra-sensitive gas sensing for various gases	[23]

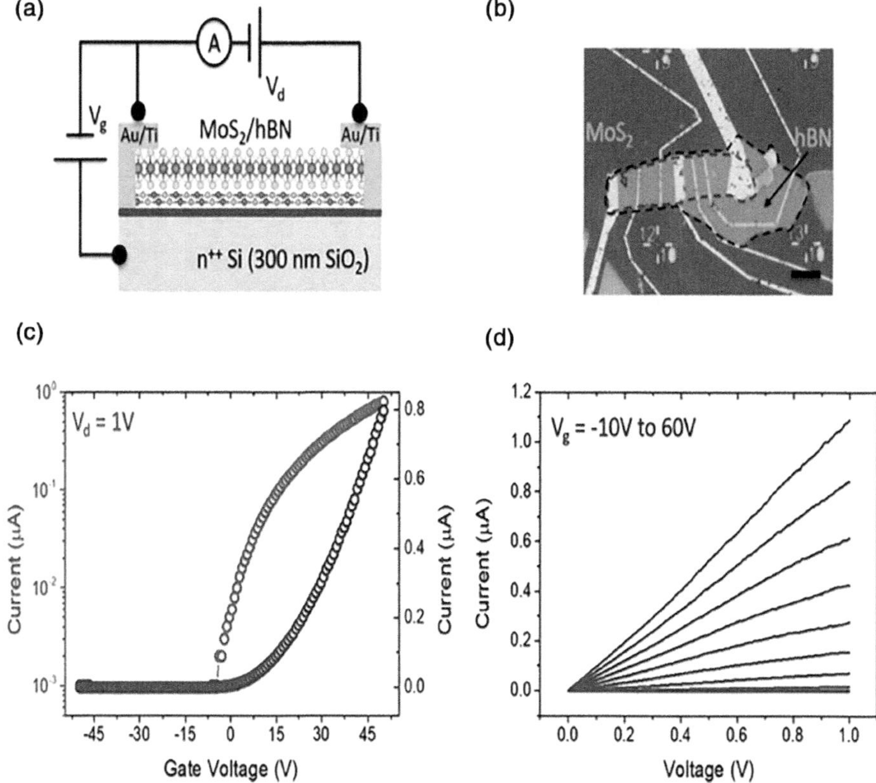

FIGURE 4.4 Schematic and optical image (scale bar 10 μm) of single-layer MoS_2/hBN transistor (a, b). MoS_2/hBN transistor characteristics (c, d). (c) Transfer characteristics at drain voltage (Vd) of 1 V. (d) Series of output characteristic curves with different gate voltages (Vg = −10 V to 60 V). (Reproduced from Ref. [26].)

4.4 CHARACTERIZATION METHODS

Characterization approaches of 2D metal chalcogenides for gas sensors involve various techniques and methods to assess the materials' properties and performance as gas sensors. Here are some common characterization methods.

4.4.1 Structure Characterization

a. **Surface morphology and structural analysis**

Atomic force microscopy (AFM), scanning electron microscopy (SEM), and transmission electron microscopy (TEM) JEOL-2100 are utilized to investigate the morphology and internal structure of the nanosheets. High-resolution TEM (HR-TEM) is utilized to visualize the lattice patterns and confirm epitaxial relationships including layer thickness and defects. For large-sized crystals, using single-crystal X-ray diffraction, one can identify the structure (XRD). XRD is used to determine the crystallographic phases present in the material as indicated in Figure 4.6.

b. **Elemental and chemical and electronic properties analysis**

Dispersive X-ray Spectroscopy (EDX) is a tool used to validate the existence of specific elements (Se, Cl, and Bi) in the hybrid nanosheets. X-ray Photoelectron Spectroscopy (XPS) is used to investigate the binding states of the elements and estimate the composition ratio of elements. The electronic properties of the heterostructure are studied through UV–vis and UV photoelectron spectroscopy (UPS).

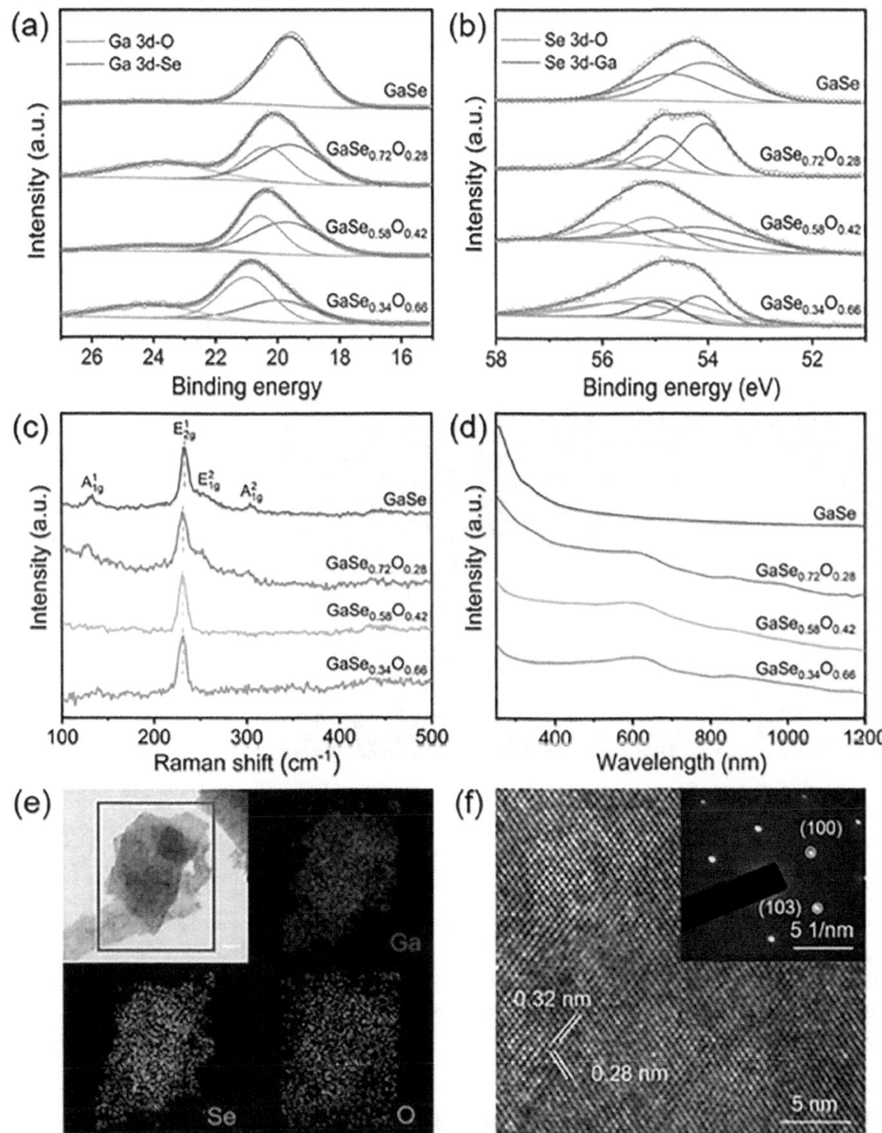

FIGURE 4.5 Chemical composition and structural characterizations of gallium oxyselenide. (a) Ga 2d region and (b) Se 3d region of XPS spectra for gallium oxyselenide. (c) Raman spectrum of gallium oxyselenide from 100 to 500 cm^{-1}. (d) UV–Vis–NIR absorption spectrum of gallium oxyselenide. (e) The TEM image of gallium oxyselenide and corresponding EDS mapping. Scale bar: 20 nm. (f) The HRTEM image with selected area electron diffraction pattern. (Reproduced with from Ref. [30].)

c. **Theoretical and optical properties**

Density functional theory (DFT) calculations are carried out to better understand the electronic properties and charge transfer mechanisms at the heterojunction. The use of DFT calculations can determine stacking fault energies and provide insights into the preference for including such faults in the system. The optical properties of the nanomaterials, such as the band gap and surface defects, are probed using ultraviolet–visible and photoluminescence spectroscopy.

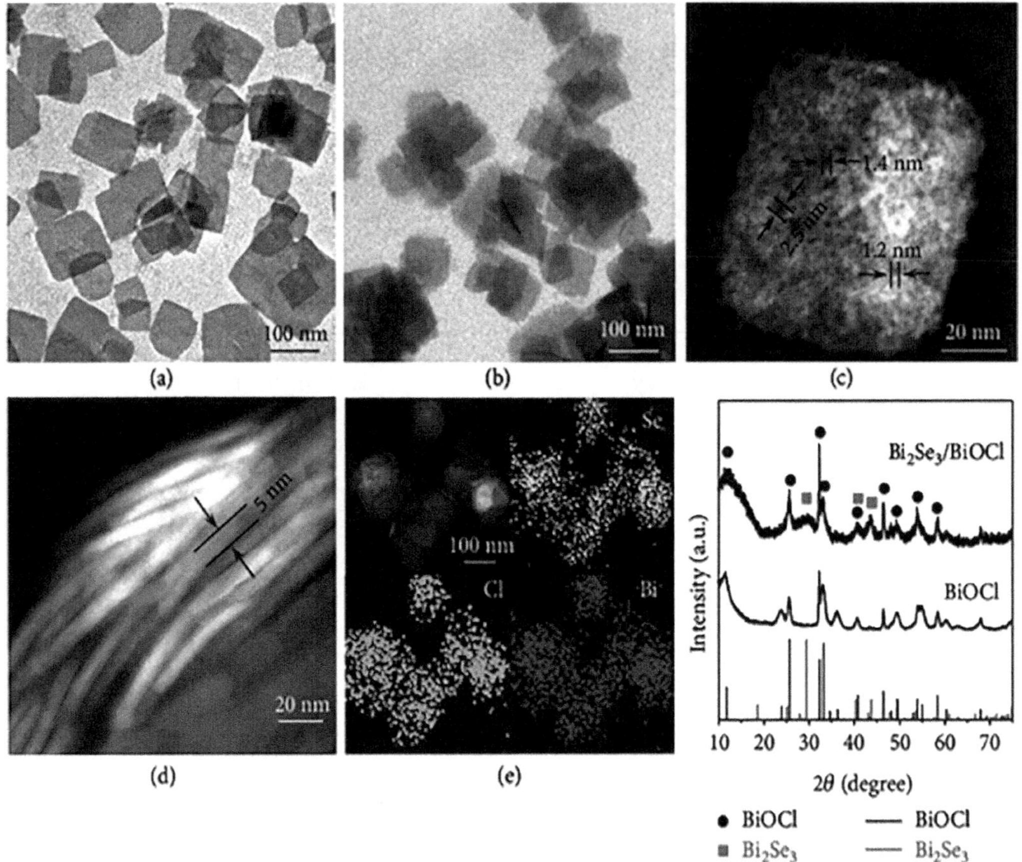

FIGURE 4.6 TEM images of (a) BiOCl nanosheets and (b) Bi_2Se_3/BiOCl heterostructures. (c) STEM image of a Bi_2Se_3/BiOCl heterostructure, revealing a nanoporous Bi_2Se_3 layer with a pore size of 1–3 nm. (d) Side-view STEM image of Bi_2Se_3/BiOCl heterostructures. (e) STEM image and EDX mapping of Bi_2Se_3/BiOCl heterostructures. (f) XRD patterns of BiOCl nanosheets and Bi_2Se_3/BiOCl heterostructures. (Reproduced from Ref. [15].)

4.4.2 Sensing Performance Characterization

Efficiency evaluation for gas sensors typically encompasses a multifaceted set of criteria. The sensor is characterized by its acute sensitivity to target gases with excellent selectivity, rapid response and recovery kinetics while enduring operational stability over extended durations, and minimal energy utilization. These fundamental characteristics shape the development of cutting-edge gas sensing technology [20–29].

a. **Response**

   The response, denoted as ΔX/X0, is characterized by the variation in various parameters regarding the signal observed in the absence of analyte molecules, including current (I), resistance (R), capacitance (C), conductance (G = I/V), light power (P), effective refractive index (RI), and resonant frequency (f), in relation to a particular unit of gas concentration [28]. This can be mathematically expressed as

$$\Delta X/X_0 = \left(X_{gas} - X_0\right) / X_0$$

Here, X represents one of the parameters: I, R, C, G, P, RI, or f, Xgas indicates the signal from the sensor after exposure to the target gas, and X0 denotes the signal baseline observed in the absence of analyte gas. Changes in the concentration of the analyte gas result in diverse responses.

b. **Sensitivity**

Sensitivity represents the sensor's ability to distinguish subtle variations in analyte concentration or mass and can be visualized as the calibration graph's slope, which demonstrates how the sensor response changes for each unit increase in the concentration of the monitored gas. A steep slope indicates high sensitivity, facilitating precise quantification.

In a more concise expression, sensitivity (S) is determined as the response to concentration ratio; hence, S = response/concentration.

Accordingly, a greater sensitivity signifies a superior efficacy of the sensor. Sensitivity is a crucial parameter for a diverse spectrum of applications, including gas leak detection, air quality assessment, medical diagnostics, and industrial safety. In these fields, precise and highly sensitive sensors are indispensable for prompt and accurate measurements. Additionally, in the Internet of Things (IoT) and sensor networks, sensitivity plays a pivotal role in facilitating data-driven decision-making and automation in industries [23–28].

c. **Selectivity**

Selectivity is characterized by the preferential adsorption of specific target gases within a mixed gas environment, while displaying insensitivity to other gases. The selectivity factor or coefficient (K) for the 'target gas' concerning another gas can be quantitatively expressed as

$$K = S_1/S_2$$

In this equation, S1 and S_2 represent the sensitivities of the sensor to the target gas and the other gas, respectively. This coefficient serves as a quantitative measure of the sensor's ability to distinguish and respond selectively to the target gas relative to other gases in the mixture. The Limit of Detection (LOD) stands as a fundamental metric in chemical sensing, serving as a pivotal indicator of the smallest concentration at which an analyte can be reliably detected. LOD is determined using linear regression models, and the following expression calculates the minimum detectable concentration of an analyte in linear sensor systems when there is a threefold increase in signal over noise:

$$LOD = 3 \times RMS_{noise}/S$$

where, in the absence of the analyte gas, RMSnoise is the root-mean-square deviation of noise at the baseline. S stands for sensitivity, which is basically the response curve's linear region's slope.

Light irradiation can enhance selectivity in sensors by enabling wavelength-specific interactions, inducing unique photochemical reactions, and providing spectral fingerprints for analytes, ultimately allowing for the discrimination of target substances amidst complex mixtures. Fluorescence-based sensors are designed to detect a specific biological molecule, and upon exposure to light of a precise wavelength, the target molecule fluoresces with a unique spectral signature, enabling its distinct identification in a complex biological sample while remaining insensitive to another molecules present. An example of a 2D metal chalcogenide sensor is molybdenum disulfide (MoS_2) employed for detecting hydrogen gas (H_2), where the MoS_2 2D structure enables a significant change in electrical conductivity upon exposure to H_2, allowing for precise and selective measurement of hydrogen concentration.

d. **Response and recovery rate**

The response and recovery rates for sensors are often characterized by exponential decay or rise processes. The general formula for these processes can be expressed as:

$$\text{Response Rate}(k_r) = -\ln(0.1)/\tau_r$$

$$\text{Recovery Rate}(k_c) = -\ln(0.9)/\tau_c$$

In this case, τc is the time constant for the recovery, τr is the time constant for the response, and kr is the rate constant for the response.

Response and recovery kinetics refer to the time-dependent behavior of a sensor when exposed to changes in its surrounding environment, particularly in response to the presence or variation in the concentration of a specific analyte, i.e., the durations needed to reach 90% of the sensor's reaction time (response) when the analyte is introduced and to recover 90% of the original value of a sensor when the analyte is removed or reduced. A sensor capacity to consistently and reproducibly deliver results over a specified period determines its stability. The dynamic performance of sensor is analyzed by its swift response and recovery rate as time-sensitive data is crucial for its applicability in various fields. The sensor's responsiveness can be influenced by factors such as the type of gas analyte, sensor material properties, operating conditions (temperature and pressure), humidity, exposure time, and the presence of interfering gases, all of which collectively determine the sensor's speed and effectiveness in detecting and recovering from analyte exposure. For instance, a room-temperature sensor performance, typically designed for normal indoor conditions, may deteriorate or become inaccurate when exposed to high temperatures outside its specified operating range, potentially leading to erroneous readings, damage, or a shorter sensor lifespan [20–29].

e. **Mechanical durability testing**

The flexible sensor mechanical durability is tested by repetitive bending, affirming its ability to maintain its sensing performance throughout the bending process. Mechanical durability verifies their quality, reliability, and suitability for real-world applications to ensure that the sensors can offer reliably accurate data over time, which is crucial for maintaining performance and minimizing costs associated with failures or replacements. This quality assumes paramount importance when sensors are subjected to harsh, corrosive, or high-temperature environments [15].

4.5 SENSING MECHANISM OF 2D GAS SENSORS

4.5.1 Adsorption-Driven Sensing

The adsorption mechanism involves the adsorption of target molecules onto the sensor's surface such as O_2-, O-, and O_2-. At their operating temperatures, the metal oxide surfaces attract and bind oxygen-negative ions. The adsorption process leads to the formation of a negatively charged surface layer on metal oxide, increasing its conductivity, see reactions (4.1 and 4.2). Taking the reduction of CO as an example [28]

$$CO_g + O^- \rightarrow CO_{2(g)} + e^- CO_g + O^- \rightarrow CO_{2(g)} + e^- \quad (4.1)$$

$$\begin{aligned} CO + 2O^- &\rightarrow CO_3{}^{2-} \rightarrow CO_2 + \frac{1}{2}O_2 + 2e^- \\ CO + 2O^- &\rightarrow CO_3{}^{2-} \rightarrow CO_2 + \frac{1}{2}O_2 + 2e^- \end{aligned} \quad (4.2)$$

SnS_2 flakes and ultrathin WS_2 sensors exhibited selective sensitivity for NO_2 due to strong physical affinity between them.

4.5.2 Charge Transfer Mechanism

In contrast to conventional metal oxides, graphene and similar layered inorganic materials primarily rely on charge transfer processes for gas sensing. These materials can act as either charge acceptors or donors when exposed to different gases. The interaction between the sensing materials and the adsorbed gases involves charge transfer reactions, resulting in varying directions and numbers of charge transfer, causing modifications to the material's resistance. The material resistance of these sensing materials returns to its initial state when gas molecules desorb from them when they are exposed to air again. Mono- and a few-layered $MoSe_2$ sensors exhibited higher selectivity for NH3 due to charge transfer between them. Take n-type MoS_2 as an example. Recently, other have been examined the charge transfer relationships between various gas molecules and monolayer MoS_2 [29]. Among the gas molecules adsorbed are O_2, H_2O, NH3, NO, NO_2, and CO. When these gases interact with the monolayer MOS_2, charge density difference images appear [29].

A recent study involved measuring the energy difference between the valence band maximum (VBM) and the lowest unoccupied molecular orbital (LOMO) of NO_2 molecules in various gallium oxyselenide compounds and exhibited the highest NO_2 sensing response by $GaSe_{0.58}O_{0.42}$ that reached 82.2%. Furthermore, $GaSe_{0.58}$ $O_{0.42}$ demonstrated fully reversible responses to 10 ppm NO_2 at room temperature when exposed to various light wavelengths, including UV, green, and red light. This response and recovery time assess gas sensing performance, as it indicates that the material can repeatedly and reliably detect NO_2 gas. Examining the specifics of the process of the gallium oxyselenide-based gas sensor revealed that when exposed to light, photogenerated electrons form a dipole with absorbed NO_2 molecules, improving interfacial charge transfer and increasing the sensitivity of the sensor. The gas sensor barrier height variations and differences in nitrogen (N2) and NO_2 environments are explored through Mott–Schottky plots, which are explained by the interaction between gallium oxyselenide and NO_2. It is expected that the LOD for NO_2 is 3.5 ppb. Overall, increased humidity decreases the sensor response, most likely due to water molecules occupying NO_2 adsorption sites, while the sensor exhibits long-term stability crucial for practical use [30,31]. Exceptional gas sensing performance by Bi_2Se_3/BiOCl heterostructures is also associated with its charge transfer-based sensing mechanism. Sensor faster response time attributed to the emergence of hybrid electronic states at the epitaxial interface. DFT revealed higher NO_2 adsorption affinity toward Bi_2Se_3 compared to BiOCl, which was further enhanced by the formation of a junction at the epitaxial interface between BiOCl and Bi_2Se_3, along with the interdiffusion of electrons and holes, creating a charge depletion region and a built-in potential barrier. The hole concentration in BiOCl is increased upon exposure to NO_2 gas, leading to the expansion of both the depletion region and energy barrier, resulting in a significant improvement in gas sensing, as indicated in Figure 4.7 [15].

The performance of the Se/$GeSe_xO_y$ heterostructure gas sensor was compared to that of pure Se and GeSexOy in response to NO_2 (10 ppm) under red light irradiation. The maximum response magnitude attained was 27.3% when exposed to 10 ppm of NO_2, and an ultra-low LOD of 0.986 ppb was achieved under red light excitation. The Se/$GeSe_xO_y$ heterostructure showed the uppermost response of 27.3% compared to pure Se (18.9%) and $GeSe_xO_y$ (20.4%). This enhanced response in the heterostructure is ascribed to the increased number of photogenerated electrons and holes, enabling charge transfer at the gas–material interface. The $GeSe_xO_y$-based sensor exhibited the shortest response and recovery times, possibly as a result of the junction interface's sluggish carrier diffusion of Se/$GeSe_xO_y$. In another study of Au– and Pt–Pd_2Se_3 systems, the monolayered Pd_2Se_3 acts as the donor, while Au and Pt atoms function as the acceptor. However, in the case of Co–, Cu–, Cr–, Fe–, Ni–, Mn–, and Sc–Pd_2Se_3 systems, the Pd_2Se_3 monolayer takes on the role of the acceptor, with the transition metal (TM) acting as the donor. With the exception of the Au–Pd_2Se_3 system, all the other systems exhibit evident charge transfer, suggesting the formation of covalent bonds between the transition metal and the Pd_2Se_3 monolayer in these systems. In conclusion, the gas sensing mechanism involves the adsorption of molecules of the target gas, the generation and separation

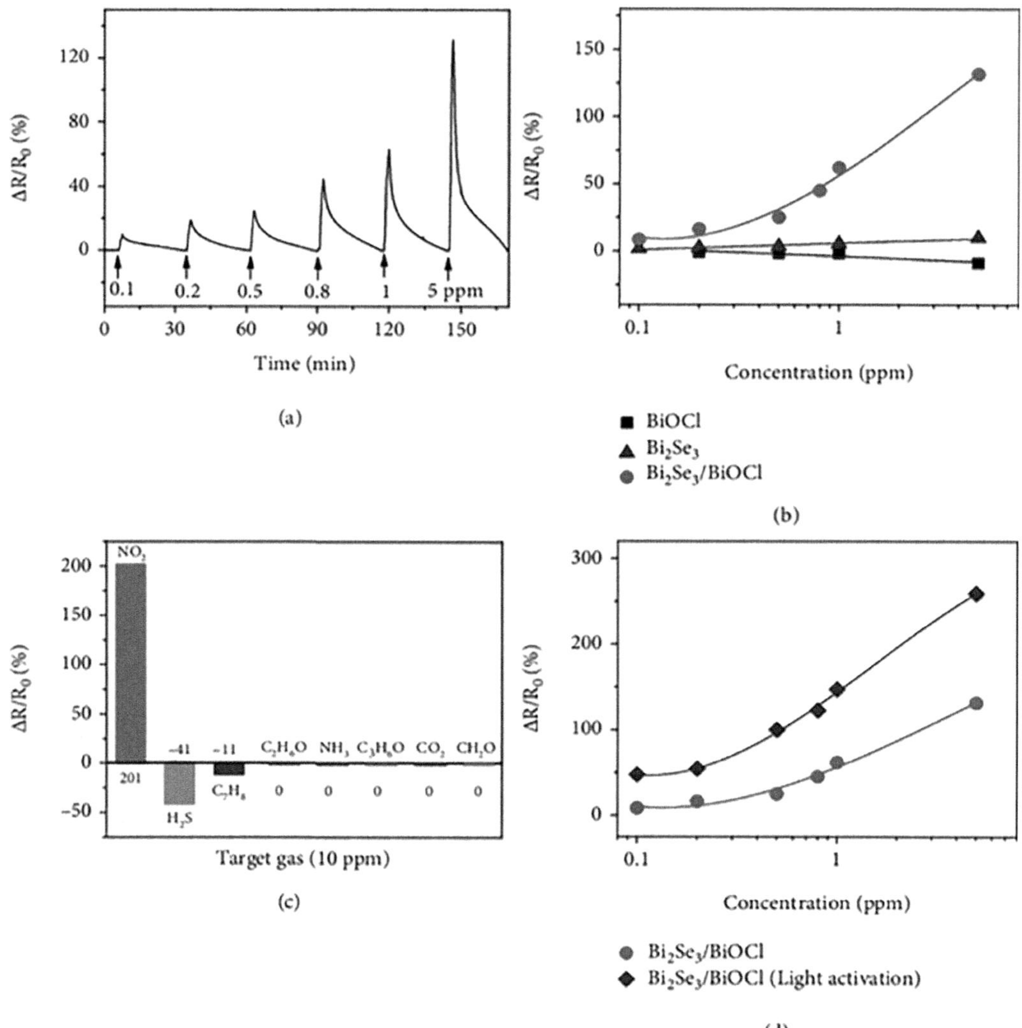

FIGURE 4.7 (a) Dynamic response–recovery curve of the sensor fabricated from Bi_2Se_3/BiOCl heterostructures in response to NO_2 gas with increasing gas concentration at room temperature. (b) Sensing response vs. NO_2 concentration for BiOCl-, Bi_2Se_3-, and Bi_2Se_3/BiOCl-based sensors. (c) Response of Bi_2Se_3/BiOCl heterostructures upon exposure to 10 ppm NO_2, H_2S, C_7H_8, C_2H_5OH, NH_3, $(CH_3)2CO$, CO_2, and HCHO at room temperature. (d) Sensing response vs. NO_2 concentration for Bi_2Se_3/BiOCl-based sensor with and without a 365 nm light irradiation (8.31 μW). (Reproduced from Ref. [15].)

of charge carriers, changes in electrical resistance, and the quantification of the gas response. This mechanism allows for the selective detection of gases like NO, NO_2, and NH_3, and it is dependent on factors such as the nature of the gas, light irradiation, and the heterojunction structure.

4.5.3 Types of Sensors Based on Operational Mechanisms

a. **Chemiresistors**

Chemiresistors represent the extensively studied categories of gas sensing devices. Their operational principle is adjusted on the adsorption of gas molecules onto the gas sensing material surfaces, leading to a modification in its electrical resistance. Then quantifying this change in resistance, these devices are capable of effectively detecting specific gases [32]. Chemiresistors offer several advantages, including ease of manufacturing, straightforward operation, cost-effectiveness, reusability, low power consumption, and the potential for miniaturization, making them a practical choice for gas sensing applications [33]. Typically, a chemiresistor comprises sensing materials fabricated on an inert substrate, along with two metallic electrodes or interdigitated electrodes. This configuration allows chemiresistors to detect a range of gases, including O_2, CO, H_2, NOx, SOx, C12, and certain organic vapors [34].

b. **Field-effect transistors**

The FET presents another noteworthy gas sensing device, garnering significant attention in research circles. This is primarily attributed to its uncomplicated fabrication process, high sensitivity in gas detection, and portability. Furthermore, FETs have been effectively scaled down to the nanoscale, enhancing their versatility and applicability [35]. In its standard configuration, a FET sensor comprises a sensing semiconductor serving as the channel material and two metal electrodes designated as the source and drain electrodes. These electrodes facilitate the modulation of channel conductance by applying distinct bias voltages to the gate electrode via a thin dielectric layer. Gas detection is achieved by measuring changes in drain current before and after exposure to target gases while maintaining a constant voltage. The electronic structure of the sensing material undergoes alterations upon adsorption of gas molecules, resulting in modifications in its conductance. This type of gas sensing device has been successfully employed for the detection of various gases, including but not limited to CO, NO, NH_3, NO_2, SO_2, H_2, hydrogen cyanide (HCN), ethanol, and 2,4-dinitrotoluene [36].

c. **Schottky diodes**

Schottky devices fabricated from layered semiconductor heterojunctions have found utility in the development of electronic devices, such as the Barristor [37], and the creation of sensing devices. This is largely attributed to the atomically tiny layered materials, including graphene and its derivatives [38]. These devices function via adsorbing gas molecules onto the sensing material surface. This adsorption process leads to the modulation of the Fermi level of semiconductor heterojunctions and, crucially, modification of their user interface height of the Schottky barrier (SBH). The SBH is directly influenced by the density of adsorbed gas molecules, and consequently, the reverse current flowing across the device undergoes changes commensurate with shifts in SBH. As a result, Schottky diode gas sensors demonstrate remarkable sensitivity and are relatively easy to fabricate, making them valuable tools for gas detection applications.

d. **Surface acoustic wave**

Surface acoustic wave (SAW) devices have garnered significant interest as gas sensors, primarily because of their little size, durability, cost-effectiveness, exceptional sensitivity, and versatility in detecting a wide range of gases. The underlying principle of SAW gas sensors involves alterations in the properties, whether they are chemical or physical, of the SAW sensor's surface or the region adjacent the surface, as a result of gas molecules' interaction with sensor materials. These alterations, in turn, affect the acoustic wave's velocity or attenuation. Most SAW gas sensors operate based on two primary mechanisms: mass loading of the surface wave or modifications brought on by the adsorption of gas molecules in the sensing materials' conductance. The characteristics of the sensing materials utilized frequently determine the selectivity of SAW sensors. Additionally, their responsiveness can be enhanced by incorporating chemically selective materials or coatings [39].

e. **Conductometric sensors**

Among the most common types of gas sensing equipment are conductometric semiconducting gas sensors. These sensors find widespread use in gas detection applications, particularly in atmospheric conditions. They are favored for their ease of production, user-friendly operation, and the ability to detect a wide array of gases. The capacity of conductometric semiconducting gas sensors to interact reversibly with gas molecules at the surface of the sensitive material is one of its distinguishing features, as seen in Figure 4.8 [40]. These interactions are influenced by various factors, including the fundamental properties of the sensing materials, surface area, surface additives, as well as environmental conditions such as temperature and humidity. Changes in mass, optical characteristics, reaction energy, capacitance, conductivity, or work function can all be used to quantify the changes brought about by these interactions. Among the many important parameters of conductometric semiconducting gas sensors, sensitivity stands out as one that may be adjusted and enhanced by these surface reactions.

f. **Impedance sensors**

An impedance sensor belongs to the category of mixed potential sensors and operates based on monitoring changes in impedance. This type of sensor involves the application of a sinusoidal voltage, with subsequent measurement of the resulting current to calculate impedance, typically in the frequency domain. The impedance sensor leverages solid-state impedance spectroscopy to assess the sensing response across a spectrum of frequencies, ranging from sub-hertz to megahertz. It has shown significant promise in achieving precise gas detection at extremely low concentrations, even down to single parts per million. The design of impedance sensors is notably straightforward. The sensing electrodes typically consist of materials such as noble metals, metal oxides, and more recently, layered materials. An interesting aspect is that they do not necessitate a separate reference air electrode [41]. Impedance sensors have proven to be an efficient tool for detecting low concentrations of several gases, such as humidity, NOx, CO, and hydrocarbons.

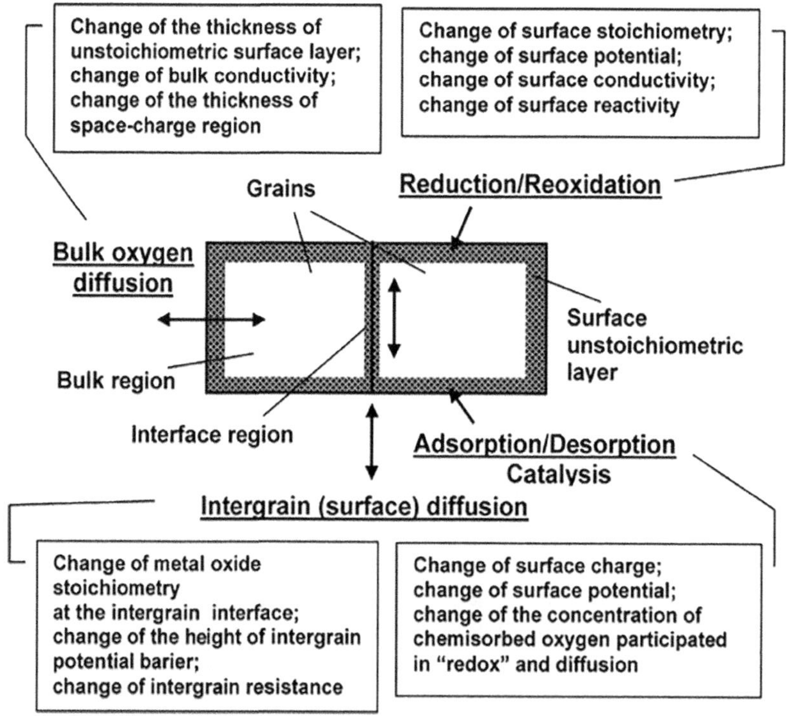

FIGURE 4.8 Diagram illustrating processes taking place in metal oxides during gas detection and their consequences for polycrystalline metal oxides properties. (Reproduced with permission from Ref. [40], Copyright © 2007 Elsevier.)

g. **Surface work function**

A class of gas sensors known as surface work function (SWF) change transistors depends on changes in the sensing material's SWF brought on by gas adsorption [42].

h. **Heterojunction sensors**

Heterojunction-type sensors represent an innovative category of gas sensors aimed at addressing the limitations associated with single-material gas sensors, such as their limited selectivity and insensitivity to low-concentration gases. This is achieved by constructing a heterostructure interface. A heterojunction gas sensor comprises two distinct semiconducting materials, often involving a combination of p-type and n-type semiconductors. The sensor gas sensing performance is significantly improved by exploiting reactions occurring at the interface between these two components. Since its first publication in 1979 [20], the idea of a heterojunction gas sensor has been the subject of much investigation into its potential to detect a wide range of gases, such as CO, H_2, H_2O, NO_2, and C_2H_5OH. The structure of the interfacial barrier varies when gases are adsorbed onto the heterojunction's surface from either side. These changes then affect the heterojunction's charge transfer properties, which affects the current flow characteristics across a rectifying heterojunction [43].

In p–n interfaced junctions, the sensing performance is influenced by the areal coverage of dissimilar materials and the interfacial bonds; higher areal coverage dominates the sensing response while minimizing interfacial states is recommended to enhance charge transfer efficiency. The p–n interfaced junctions serve as fundamental components in gas sensors, enabling the detection of a wide range of gases with distinct electron properties. These junctions consist of one p-type semiconductor, which is rich in "holes," and one n-type semiconductor, which is abundant in electrons. When gas molecules adsorb onto the semiconductor surface, they modify charge carrier concentrations within the depletion region formed at the junction interface, leading to variations in electrical conductivity. The p–n interfaced junctions are remarkably versatile, capable of detecting both electron-deficient and electron-rich gases, making them invaluable in applications ranging from environmental monitoring to industrial safety and medical diagnostics. Their operation relies on the movement of charge carriers and surface chemistry interactions, offering a powerful means to sense gases across diverse fields and industries [43–45].

In n–n and p–p interfaced junctions, heterostructures, especially those based on metal oxide nanomaterials, have been proposed to improve gas sensing performance by enhancing the transfer efficiency of interfacial charge through band bending and electron-accumulation layer formation at the interface, yet further analysis of interface states is needed for a comprehensive understanding of their behavior. The n–n interfaced junctions in gas sensors are formed by connecting two n-type semiconductors, which are rich in electrons. As gas molecules adsorb on the semiconductor surface, they influence the width of the electron-accumulation layer (EAL) created at the junction interface, resulting in changes in electrical conductivity. This shift in conductivity forms the basis for gas detection, with different gases eliciting distinct conductance responses, allowing n–n interfaced junctions to selectively detect electron-deficient analyte gases [43–45]. Conversely, p–p interfaced junctions are composed of two p-type semiconductors, which are rich in "holes" or positive charge carriers. When gas molecules adsorb on their surfaces, the width of the electron-depletion layer (EDL) and hole-accumulation layer (HAL) is modified, leading to changes in electrical conductivity and enabling the specific detection of electron-rich gases. Both n–n and p–p interfaced junctions rely on variations in conductivity due to gas interaction, but their sensitivity and selectivity are tailored to distinct types of analyte gases, enhancing their utility in gas sensing applications.

4.6 APPLICATIONS

2D MCs are promising materials for gas sensing applications due to their tunable band structures in contrast to other materials, such as pristine graphene with deficient selectivity and gapless semimetal. The most observed 2D MCs include layered metal di-selenides ($MoSe_2$, WSe_2,

$ReSe_2$, GaSe, $PdSe_2$, etc.) and metal disulfides (MoS_2, WS_2, SnS_2, ReS_2, GaS, etc.) for various gas sensing applications. Recently, newly discovered MC materials with novel fabrication approaches proved to have superior characteristics. For instance, a heterostructure-based gas sensor, Bi_2Se_3/BiOCl, is an alluring option for diverse practical applications with a LOD as low as 1.6 ppb and the ability to maintain approximately 95% of its original response toward 5 ppm NO_2 even after repeated bending. This property surpassed other 2D flexible sensors. Another NO_2 gas detecting sensor $GaSe_{0.58}\ O_{0.42}$ exhibited a fully reversible response to 10 ppm NO_2 at RT on exposure to various light wavelengths, including UV, green, and red light. This response–recovery rate outsmarts gas sensing performance as compared to others in repeatedly and reliably detecting NO_2 for assessing air pollution. The broad wavelength response makes it suitable for wearable and portable gas sensors. Further analysis proving its long-term stability is evident for deployment in city infrastructure and industrial settings. In optoelectronic NO_2 gas sensing, 2D Ag_2SO_4 nanosheets, another new heterostructure, exhibited sensitive and reversible response at the ppb level upon blue-light illumination. Pristine $PdSe_2$ Monolayer, Rh-$PdSe_2$ Monolayer (Doped), Ru-$PdSe_2$ Monolayer (Doped), and $PdSe_2$ are studied extensively for various gas detections. Rh-$PdSe_2$ (SO_2 adsorbed with a band gap of 0 eV) and Ru-$PdSe_2$ seem to be more ideal for SO_2 and SO3 detection, respectively, and superior adsorption energy implies a stable bonding with SO_2 [46]. Pentagonal $PdSe_2$ exhibited strong adsorption and sensitivity for NO_2 gas molecules. Experimentally, $PdSe_2$/Si heterostructure gas sensor showed high selectivity and response to NO_2 at room temperature.

Pentagonal few-layer $PdSe_2$ flakes were used for field emission measurements, exhibiting high field emission current and potential for field emission applications further used in consumer electronics, industrial safety, medical and aerospace (Figure 4.9) [45–48]. Furthermore, a PbS-SnS_2 nanocomposite that was pure and ornamented with Au was created for room-temperature ethanol detection. PbS-SnS_2 nanocomposite was treated with varying concentrations of Au (0.5, 1, and 1.5 wt%) in order to determine the ideal level of Au-decoration as illustrated in Figure 4.10 [47]. On the other hand, the Se-SnSSe monolayer is employed for gas sensing applications, particularly for detecting H_2, CO, and C2H_2, with tunable sensitivity under negative strains.

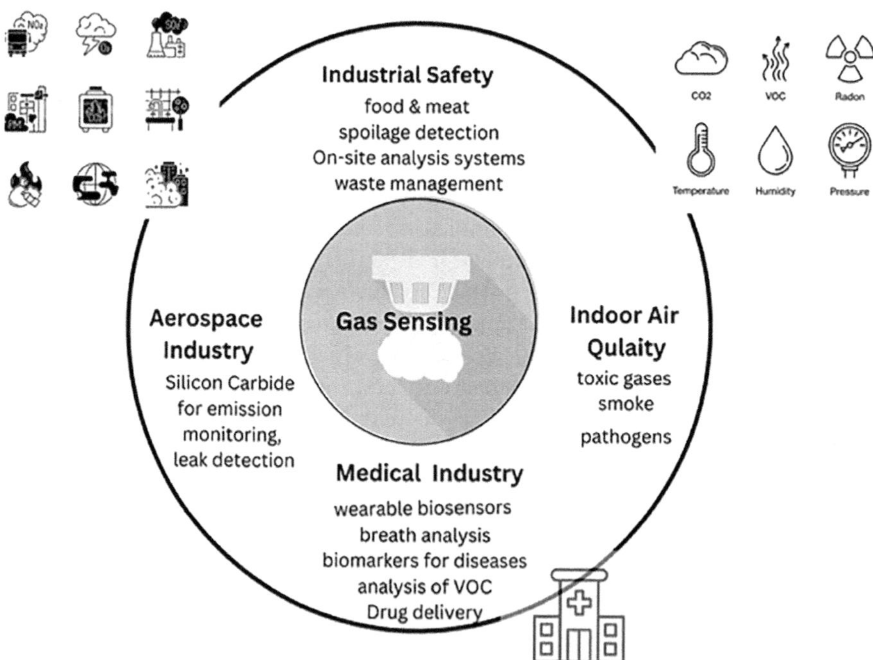

FIGURE 4.9 Schematic diagram representing the applications of MCs gas sensors.

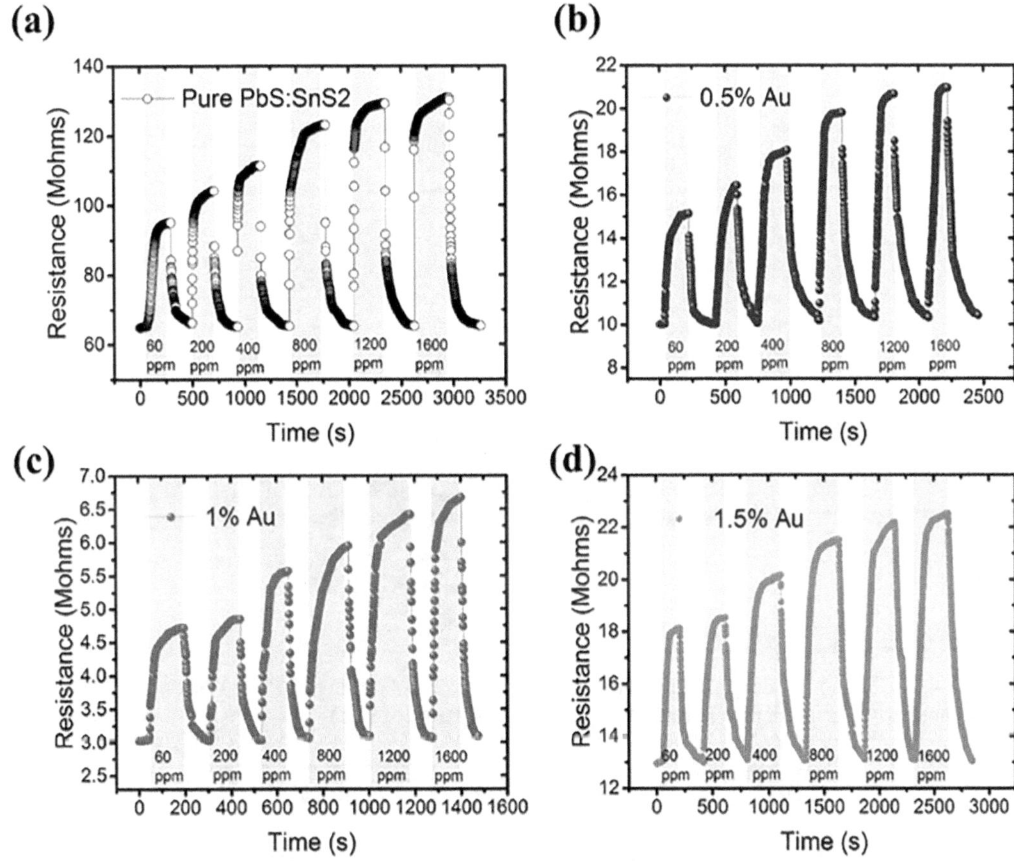

FIGURE 4.10 Dynamic resistance curves of (a) pristine PbS/SnS_2 sensor and Au-decorated PbS/SnS_2 sensors with different amounts of Au: (b) 0.5 wt%, (b) 1 wt%, and (c) 1.5 wt% to different concentrations of ethanol at room temperature (25°C). (Reproduced with permission from Ref. [47], Copyright © 2021 Elsevier.)

It exhibits high sensitivity and selectivity for these gases, with fast response and recovery times. Lastly, the SnS_2/PANI composite sensor finds its application in NH_3 gas detection, showcasing high sensitivity and rapid response/recovery times. The composite material promises accurate NH_3 detection at room temperature, addressing health and safety concerns associated with ammonia exposure.

Adhering to this trend, recently, PtO_2 nanoparticles (NPs) were coated on boron nitride nanotubes (BNNTs) to produce a heterojunction that improved room temperature (RT, 28°C) gas sensing toward dry NH3 gas varied from 0.05 to 10 ppm. By using boron oxide-assisted chemical vapor deposition (BOCVD), the BNNT was fabricated. Pt NPs were deposited via radiofrequency sputtering, and BNNTs were subsequently deposited through slurry casting and annealing to yield PtO_2-BNNT heterostructures [49]. Hence, Figure 4.11 shows the outstanding sensing performance toward ammonia [49].

It is worth noting that investigating 2D TMD materials and their combinations with metal oxides for sensor applications presents a wealth of options. Thus, new materials and their hybrids with creative designs will be a viable choice in the future. 2D TMD sensors have received less attention than graphene-based gas sensors. Furthermore, research on the functionalization of TMDs via metal oxide hybridizations is still in its infancy and needs further design and engineering to afford alternative sensing actions (Figure 4.12) [50]. In this sense, the insights for a heterostructure design could be guided by simulation results before an experimental research.

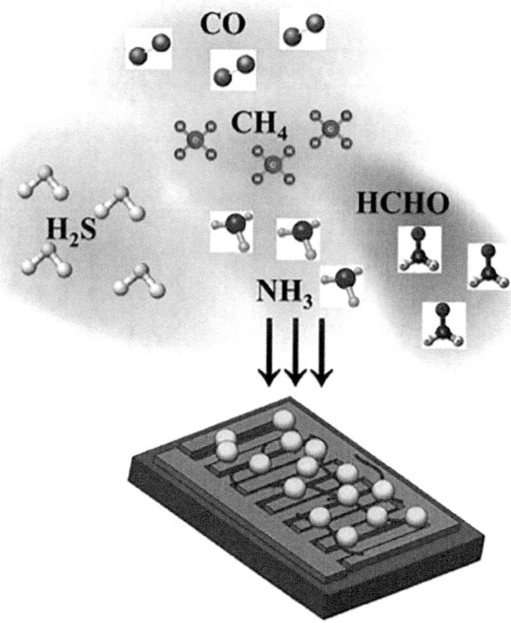

FIGURE 4.11 Schematic diagram illustrating the sensing action of developed PtO2-BNNT heterostructures. (Reproduced with permission from Ref. [49], Copyright © 2024 Elsevier.)

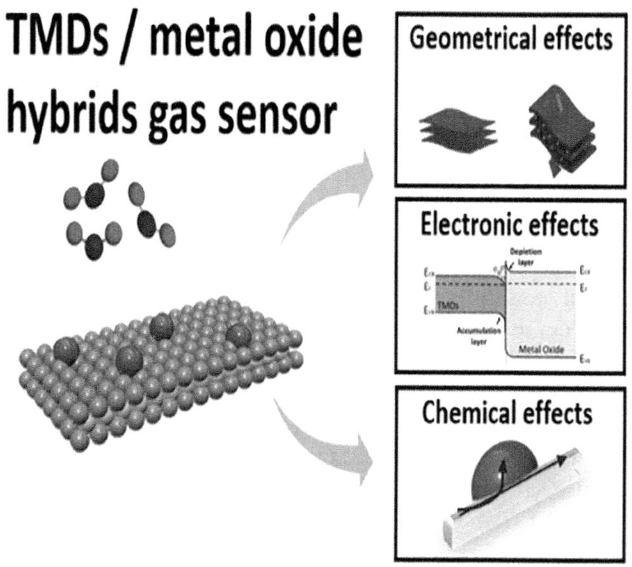

FIGURE 4.12 Schematic diagram displaying the avenues of TMDS/metal oxide hybrid systems for precise gas sensing. (Reproduced with permission from Ref. [50], Copyright © 2018 American Chemical Society.)

4.7 CONCLUSION AND FUTURE OUTLOOK

A typical problem faced by most gas sensing nanomaterials is their narrow selectivity. Comparing these two-dimensional 2D materials to spectroscopy-based sensing systems, there is still a great deal of room for improvement, even if they have attained a convincing degree of selectivity through favored gas adsorption, surface reaction, intercalation, and carrier doping. Transitioning from laboratory-scale synthesis to large-scale production in gas sensing technology while maintaining the desired properties and efficiency poses a complex task. For instance, cultivating high-quality MoS_2 and WS_2 layers, which typically necessitate extremely high temperatures exceeding 800°C, is a complex process. These high-temperature growth methods are incompatible with conventional electronic device preparation techniques. Therefore, the quest for research and development persists in addressing these challenges and discovering more suitable approaches. The influence of fluctuations in ambient temperature and relative humidity on room temperature sensors cannot be ignored, as it has a significant impact on the selectivity of the sensors. To enhance the selectivity of the sensors, the combination of sensor array fabrication and data processing has proven to be a successful approach, and this technique has been widely adopted in various commercial electrical sensors. 2D material-based gas sensors have the potential to reduce power consumption and offer improved sensitivity at ambient temperature. However, they still do not show a high enough level of sensing response at low gas concentrations, which leaves them vulnerable to environmental perturbations. In simpler terms, factors like humidity, temperature, and strain can induce significant alterations in the sensing response, thereby impacting the reliability and durability of these sensors. Hence, there is an urgent requirement to investigate and discover innovative 2D gas sensing materials that possess superior sensitivity, selectivity, and stability. Furthermore, the development of new approaches to further enhance the sensitivity of 2D material gas sensors to operate at room temperature in order to minimize the sintering and power consumption at higher temperatures is desired.

REFERENCES

1. Li T., Yin W., Gao S., Sun Y., Xu P., Wu S., Kong H., Yang G., Wei G. The combination of two-dimensional nanomaterials with metal oxide nanoparticles for gas Sensors: A review. *Nanomaterials* 12 (2022) 982.
2. Goel N., Jarwal D.K., Hu Y., Zhang J., Kumar M. Strategic review on chemical vapor deposition technology-derived 2D material nanostructures for room-temperature gas sensors. *Journal of Materials Chemistry C* 11 (2023) 774–801.
3. Galstyan V., Moumen A., Kumarage G.W.C., Comini E. Progress towards chemical gas sensors: Nanowires and 2D semiconductors. *Sensors and Actuators B-Chemical* 357 (2022) 131466.
4. Mohamed M.E.B., Attia N.F., Elashery S.E.A. Greener and facile synthesis of hybrid nanocomposite for ultrasensitive iron (II) detection using carbon sensor. *Microporous Mesoporous Materials* 313 (2021) 110832.
5. Frag E.Y., El-Zaher N.A., Elashery S.E.A. Carbon thick sheet potentiometric sensor for selective determination of silver ions in X-ray photographic film. *Microchemical Journal* 155 (2020) 104750.
6. Elashery S.E.A., Attia N.F., Oh H. Design and fabrication of novel flexible sensor based on 2D Ni-MOF nanosheets as a preliminary step toward wearable sensor for onsite Ni (II) ions detection in biological and environmental samples. *Analytica Chimica Acta* 1197 (2022) 339518.
7. Elashery S.E.A., Attia N.F., Mohamed G.G., Omar M.M., Tayea H.M.I. Hybrid nanocomposite based graphene sensor for ultrasensitive clomipramine HCl detection. *Electroanalysis* 33 (2021) 2361–2371.
8. Kim S., Cho S.Y., Son K., Attia N.F., Oh H. A metal-doped flexible porous carbon cloth for enhanced CO_2/CH_4 separation. *Separation Purification Technology* 277 (2021) 119511.
9. Attia N.F., Jung M., Park J., Cho S.Y., Oh H. Facile synthesis of hybrid porous composites and its porous carbon for enhanced H2 and CH_4 storage. *International Journal of Hydrogen Energy* 45 (2020) 32797–32807.
10. R. Kumar, N. Goel, M. Hojamberdiev, M. Kumar. Transition metal dichalcogenides-based flexible gas sensors. *Sensors and Actuators* A 303 (2020) 111875.
11. Kang J.-Y., Koo W.-T., Jang J.-S., Kim D.-H., Jeong Y.J., Kim R., Ahn J., Choi S.-J., Kim I.-D., 2D layer assembly of Pt-ZnO nanoparticles on reduced graphene oxide for flexible NO_2 sensors. *Sensors and Actuators B: Chemical* 331 (2021) 129371.
12. Jiang H., Cao L., Li Y., Li W., Ye X., Deng W., Jiang X., Wang G., Xu G. Chem. Organic "receptor" fully covered few-layer organic–metal chalcogenides for high-performance chemiresistive gas sensing at room temperature. *Chemical Communications* 56 (2020) 5366–5369.
13. Li Y., Shu J., Huang Q., Chiranjeevulu K., Kumar P.N., Wang G.E., Deng W.H., Tang D., Xu G. 2D metal chalcogenides with surfaces fully covered with an organic "promoter" for high-performance biomimetic catalysis. *Chemical Communications* 55 (2019) 10444–10447.

14. Tang T., Li Z., Cheng Y.F., Xie H.G., Wang X.X., Chen Y.L., Cheng L., Liang Y., Hu X.Y., Hung C.M., Hoa N.D., Yu H., Zhang B.Y., Xu Kai, Ou J.Z. In-situ mechanochemically tailorable 2D gallium oxyselenide for enhanced optoelectronic NO_2 gas sensing at room temperature. *Journal of Hazardous Materials* 451 (2023) 131184.
15. Wang Z., Dai J., Wang J., Li X., Pei C., Liu Y., Yan J., Wang L., Li S., Li H., Wang X., Huang X., Huang W. Realization of oriented and nanoporous bismuth chalcogenide layers via topochemical heteroepitaxy for flexible gas sensors. *Research (Wash D C)* 2022 (2022) 9767651.
16. Jannat A., Haque F., Xu K., Zhou C., Zhang B.Y., Syed N., Mohiuddin Md., Messalea K.A., Li X., Gras S.L., Wen X., Fei Z., Haque E., Walia S., Daeneke T., Zavabeti A., Ou J.Z. Exciton-driven chemical sensors based on excitation dependent photoluminescent two dimensional SnS. *ACS Applied Materials & Interfaces* 11(2019) 42462–42468.
17. Vidal J., Lany S., d'Avezac M., Zunger A., Zakutayev A., Francis J., Tate J. Band-structure, optical properties, and defect physics of the photovoltaic semiconductor SnS. *Applied Physics Letters* 100 (2012) 032104.
18. Nicolosi V., Chhowalla M., Kanatzidis M.G., Strano M.S., Coleman J.N. Liquid exfoliation of layered materials. *Science* 340 (2013) 1226419.
19. Huo N., Yang S., Wei Z., Li S.-S., Xia J.-B., Li J. Photoresponsive and gas sensing field-effect transistors based on multilayer WS_2 nanoflakes. *Scientific Reports* 4 (2014) 5209.
20. Ou J.Z., Ge W., Carey B., Daeneke T., Rotbart A., Shan W., Wang Y., Fu Z., Chrimes A.F., Wlodarski W., Russo S.P., Li Y.X., Kalantar-zadeh K. Physisorptionbased charge transfer in two-dimensional SnS_2 for selective and reversible NO_2 gas sensing. *ACS Nano* 9 (2015) 10313–10323.
21. Rani A., DiCamillo K., Khan M.A.H., Paranjape M., Zaghloul M.E. Tuning the Polarity of $MoTe_2$ FETs by varying the channel thickness for gas-sensing applications. *Sensors* 19 (2019) 2551.
22. Tang T., Li Z., Cheng Y.F., Xu K., Xie H.G., Wang X.X., Hu X.Y., Yu H., Zhang B.Y., Tao XW., Hung C.M., Hoa N.D., Chen G.Y., Li Y.X., Ou J.Z. Single-step growth of p-type 1D Se/2D GeSexOy heterostructures for optoelectronic NO_2 gas sensing at room temperature. *Journal of Materials Chemistry A* 11 (2023) 6361–6374.
23. Jin Z., Shin S., Kwon D.H., Han S.J., Min Y.S. Novel chemical route for atomic layer deposition of MoS_2 thin film on SiO2/Si substrate. *Nanoscale* 6 (2014) 14453–14458.
24. Huang J.K., Pu J., Hsu C.L., Chiu M.H., Juang Z.Y., Chang Y.H., Chang W.H., Iwasa Y., Takenobu T., Li L.J. Large-area synthesis of highly crystalline WSe_2 monolayers and device applications. *ACS Nano* 8 (2014) 923–930.
25. Wu C.R., Chang X.R., Chu T.W., Chen H.A., Wu C.H., Lin S.Y. Establishment of 2D crystal heterostructures by sulfurization of sequential transition metal depositions: Preparation, characterization, and selective growth. *Nano Letters* 16 (2016) 7093–7097.
26. Ali A., Koybasi O., Xing W., Daniel, Wright N., Varandani D., Taniguchi T., Watanabe K., Mehta B.R., Belle B.D. Single digit parts-per-billion NOx detection using MoS_2/hBN transistors. *Sensors Actuators A Physical* 315 (2020) 112247.
27. Wang W., Zhen Y., Zhang J., Lia Y., Zhong H., Jia Z., Xiong Y., Xue Q., Yan Y., Alharbic N.S., Hayat T. SnO_2 nanoparticles-modified 3D-multilayer MoS_2 nanosheets for ammonia gas sensing at room temperature. *Sensors Actuators B Chemical* 321 (2020) 128471.
28. Jimenez-Cadena G., Riu J., Rius F.X. Gas sensors based on nanostructured materials. *Analyst* 132 (2007) 1083–1099.
29. Yue Q., Shao Z., Chang S., Li J. Adsorption of gas molecules on monolayer MoS_2 and effect of applied electric field. *Nanoscale Research Letters* 8 (2013) 425.
30. Tang T., Li Z., Cheng Y.F., Xie H.G., Wang X X., ChenY.L., Cheng L., Liang Y., Hu X.Y., Hung C.M., Hoa N.D., Yu H., Zhang B.Y., Xu K., Ou J.Z. In-situ mechanochemically tailorable 2D gallium oxyselenide for enhanced optoelectronic NO_2 gas sensing at room temperature. *Journal of Hazardous Materials* 451 (2023) 131184.
31. Zhou J., Xue K., Liu Y., Liang T., Zhang P., Zhang X., Zhang W., Dai Z. Highly sensitive NO_2 response and abnormal P-N sensing transition with ultrathin Mo-doped SnS_2 nanosheets. *Chemical Engineering Journal* 420 (2021) 127572.
32. Hubble L.J., Cooper J.S., Sosa-Pintos A., Kiiveri H., Chow E., Webster M.S., Wieczorek L., Raguse B. High-throughput fabrication and screening improves gold nanoparticle chemiresistor sensor performance. *ACS Combinatorial Science* 17 (2015) 120–129.
33. Davis C.E., Ho C.K., Hughes R.C., Thomas M.L. Enhanced detection of m-xylene using a preconcentrator with a chemiresistor sensor. Sensors and Actuators B: Chemical 2 (2005) 207–216.
34. Ammu S., Dua V., Agnihotra S.R., Surwade S.P., Phulgirkar A., Patel S., Manohar S.K., Am J. Flexible, all-organic chemiresistor for detecting chemically aggressive vapors. *Journal of the American Chemical Society* 134 (2012), 4553–4556.

35. Zhan B., Li C., Yang J., Jenkins G., Huang W., Dong X. Graphene field-effect transistor and its application for electronic sensing. *Small Nano-Micro* 10 (2014) 4042–4065.
36. Wu S., Wang G., Xue Z., Ge F., Zhang G., Lu H., Qiu L. Organic field-effect transistors with macroporous semiconductor films as high-performance humidity Sensors. *ACS Applied Materials & Interfaces.* 9 (2017) 14974–14982.
37. Yang H., Heo J., Park S., Song H.J., Seo D.H., Byun K., Kim P., Yoo I., Chung H., Kim K. Graphene barristor a triode device with a gate-controlled schottky diode. *Science* 336 (2012) 1140–1143.
38. Singh A., Uddin Md.A., Sudarshan T., Koley G. Tunable reverse-biased graphene/silicon heterojunction schottky diode sensor. *Small* 10 (2014) 1555–1565.
39. Wohltjen H. Mechanism of operation and design considerations for surface acoustic wave device vapour sensors. *Sensors and Actuators* 5 (1984) 307–325.
40. Korotcenkov G. Metal oxides for solid-state gas sensors: What determines our choice? *Materials Science and Engineering* 139 (2007) 1–23.
41. Woo L.Y., Martin L.P., Glass R.S., Wang W., Jung S., Gorte R.J., Murray E.P., Novak R.F., Visserc J.H. Effect of electrode composition and microstructure on impedancemetric nitric oxide sensors based on YSZ electrolyte. *Journal of the Electrochemical Society* 155 (2007) J32.
42. Bayram F., Khan D., Li H., Hossain M.M., Koley G. Piezotransistive GaN microcantilevers based surface work function measurements. *Japanese Journal of Applied Physics* 57 (2018) 040301.
43. Traversa E., Bearzotti A., Miyayama M., Yanagida H. Influence of the electrode materials on the electrical response of ZnO-based contact sensors. *Journal of the European Ceramic Society* 18 (1998) 621–631.
44. Jiang H., Cao L., Li Y., Li W., Ye X., Deng W., Jiang X., Wang G., Xu G. Organic 'receptor' fully covered few-layer organic–metal chalcogenides for high-performance chemiresistive gas sensing at room temperature. *Chemical Communications* 56 (2020) 5366–5369.
45. Liang Q., Chen Z., Zhang Q., Wee A.T.S. Pentagonal 2D transition metal dichalcogenides: $PdSe_2$ and beyond. *Advanced Functional Materials* 32 (2022) 2203555.
46. Chettri B., Sharma A., Das S.K., Sharma B. First principle study of Rh/Ru doped pentagonal $PdSe_2$ for detection of SO_2 and SO_3 gas. *Materials Today: Proceedings* 58, Part 2 (2022) 696–701.
47. Roshan H., Kuchi P.S., Sheikhi M.H. Enhancement of room temperature ethanol sensing behavior of PbS–SnS_2 nanocomposite by Au decoration. *Materials Science in Semiconductor Processing* 127 (2021) 105742.
48. Singh J., Thareja R., Malik P., Kakkar R. Size-dependent structural and electronic properties of stoichiometric II–VI quantum dots and gas sensing ability of CdSe quantum dots: A DFT study. *Journal of Nanoparticle Research* 24 (2022) 33.
49. Sharma B., Karuppasamy K., Srivastava A.K., Alfantazi A., Sharma A. Highly sensitive and selective nanoengineered PtO2-BNNT heterostructures for ppb level ammonia gas sensing. *Sensors and Actuators B: Chemical* 400 (2024) 134818.
50. Lee E., Yoon Y.S., Kim D.-J. Two-dimensional transition metal dichalcogenides and metal oxide hybrids for gas sensing. *ACS Sensor* 3 (2018) 2045–2060.

5 Two-Dimensional Nanomaterials for Electrochemical Detection of Organic Wastewater Pollutants

Mangaka C. Matoetoe, Malefetsane P. Khesuoe, and Fredrick O. Okumu

5.1 INTRODUCTION

The scarcity of portable water, industrialization, agricultural practices and other socio-economic human water utilities in domestic settings are increasing exponentially as the population increases. Intensified water usage and population increase adversely affect the wastewater composition and put strain on water purification facilities. The compromised water constitutes a variety of different materials in terms of composition, amounts and impact. These materials are inorganic, microorganisms and organic. Organic contaminants are present in wastewater as a result of poor purification processes, incorrect disposal, improper usage and metabolism [1,2]. Contaminants' presence affects appearance, odor and properties of water. Hydrogen sulfide is responsible for the bad odor and taste; bioaccumulation of some metals has toxic effect; an excessive presence of nutrients causes eutrophication and oxygen depletion; and microorganisms can be pathogenic to marine life. Some non-biodegradable organic materials are carcinogenic and endocrine disruptors, and they have persistent impact since they can stay longer in the environment [3,4]. Waterborne illness has resulted in stricter control measures. It has been reported that non-biodegradable organic materials together with the heavy metals are bio-accumulative and are named priority pollutants [5]. However, the use of many of persistent organic pollutants has been banned, while others are used under strict regulatory measures for environmental protection [5]; hence, they can still be present in trace amounts in the environment [6]. Organic pollutants summary and their effects are depicted in Figure 5.1.

The presence of these organic materials in the water has serious health effects (Figure 5.1b). Therefore, water monitoring is a necessity to ensure water quality, improve health and adhere to environmental policies. Hence, there is a need for accurate and precise analytical techniques for organic pollutants analysis.

5.2 ANALYSIS OF ORGANIC POLLUTANTS IN WASTEWATERS

Analysis of trace levels of organic pollutants is crucial to ensure compliance with the maximum allowed residual levels and assess impacts of accumulated contents. The availability of robust analytical methods is the key to the success of this exercise [7]. Dominating organic pollutants' analyses are liquid chromatography-based methods, such as high-performance liquid chromatography (HPLC) and HPLC commonly hyphenated with spectroscopic techniques. The progress of novel analytical approaches, such as Coupled liquid chromatography to different forms of mass spectrometry (LC-MS), tandem MS (MS^2), or (LC-MS^2), has lowered the detection limits to low

DOI: 10.1201/9781003436942-5

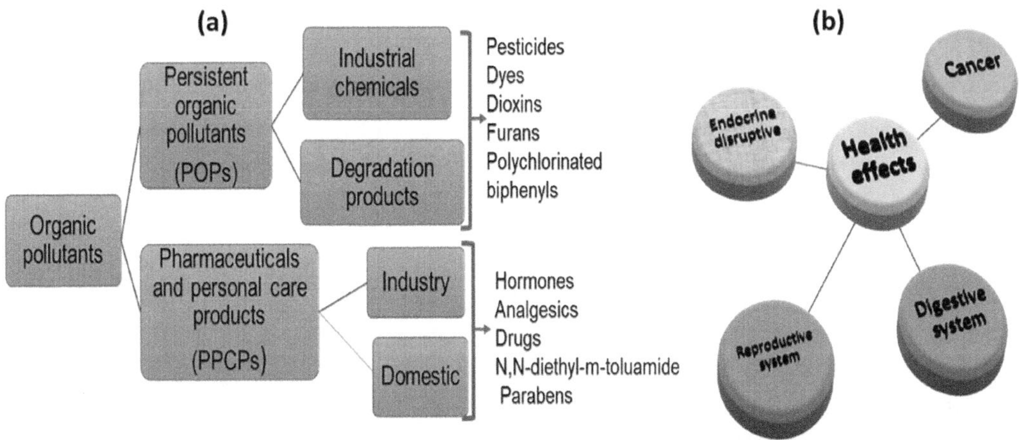

FIGURE 5.1 Summary of (a) organic pollutants and (b) their health effects.

concentrations (μg ng L^{-1}) of these materials in very complex matrixes of liquid (surface and ground waters) and/or solid waste [8,9] due to their reliability and accuracy. However, these methods require expensive instrumentation as well as costly experimental execution due to the requirement for pure organic solvents. They usually need complex pre-treatment sample procedures and skilled operators [10]. Hence, their usage is limited to on-site analysis [11]. Mitigation of these chromatography techniques problems resulted in extensive research on the development of various sensing techniques, such as essays (colorimetric, immunosorbent), biosensors (fluorescence, electrochemical, surface plasmon resonance) and competitive enzyme-linked microfluidic immunoassays [12]. Among these techniques, electrochemical showed promising results.

5.2.1 Electrochemical Analysis of Organic Wastewater Pollutants

Electroanalysis of organic chemicals has been used for decades. However, the application was limited by the availability of electrodes. Electrochemical techniques provide accuracy, high sensitivity, quick response time and cost-effectiveness, making them an ideal alternative for the on-site analysis of organic pollutants [13,14]. A number of common organic functional groups present in several organic pollutants are electroactive. Table 5.1 details these functional groups and the pollutants that contain them. Almost all electroactivity mechanisms of organic compounds involve hydrogen. Thus, the analysis of organic compounds is pH dependent. Hence, electroanalytical methods for organic compounds require intensive optimization of the supporting electrolyte, as well as electroanalytical method parameters such as starting potential, pulse amplitude and scan rate. The development of sensing and NMs technology has widely increased the application of electrochemical analysis. Application of common solid electrodes in monitoring oxidation or reduction processes of organic compounds (drugs, dyes and parasiticides) has been hampered by their low electron transfer kinetics and high overpotential. Hence, modified electrodes are widely used in various analytical chemistry research fields [15,16,17]. These techniques can be classified based on the type of technique used, transducer and the recognition substances used for selective identification of target analyte. The summary of classification is detailed in Figure 5.2.

Revolutions in electrochemical sensors bear the possibility of field and environmental applications of real-time detection (rapid test) since they can be developed in miniaturized formats, making them portable and inexpensive to run because they do not require much technical expertise and use easy-to-synthesize materials [18–20]. Redox activities of the mentioned functional groups can

TABLE 5.1
Electroactive Organic Functional Groups Found in Organic Wastewater Pollutants

Functional Group	Structures	Organic Pollutant
Nitroso, amines, and amides	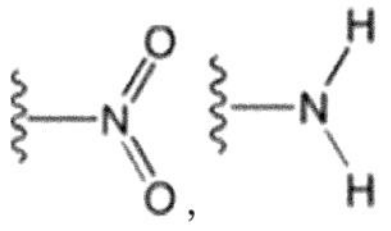 [nitro and amine structures]	Dyes and pharmaceuticals
Alkyl halides	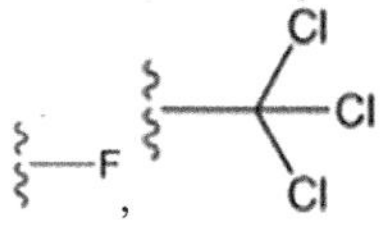 [–F, –CCl3 structures]	PCBs and organochloride pesticides (OCPs)
Carbonyl, carboxylic acid, ketones and aldehydes	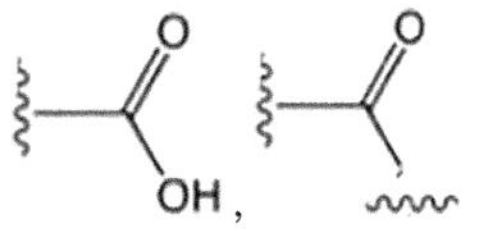 [carboxylic acid and ketone structures]	Analgesics
Carbon phosphorus/ carbon sulfur	-R-P=C-, R-S=0	Dyes
Phenyl, aromatic hydrocarbons	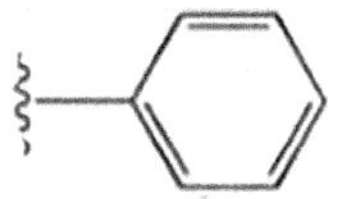 [phenyl structure]	Polychlorobiphenyls, dyes, and Poly(m-aminobenzene sulfonic acid) (PABS)
Olefins and polyolefins	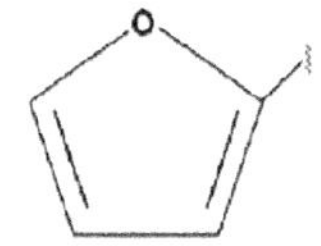 [furan structure]	Furans
Esters and peresters	R-O-R	Parabens and dioxins
Quinones	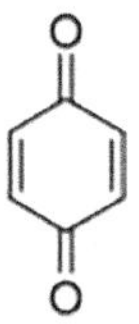 [quinone structure]	Hydroquinones and mercaptans

be used to determine concentration of pollutants. Central to electrochemical sensors' selectivity in the analysis of low concentration pollutants is the characteristic low and high oxidation peaks and overpotential of organic molecules, which are essential in real specimens [21]. Presently, there are some electroactive substances with comparable oxidation/reduction voltage, which can be utilized in determining exact mixtures [22].

Recently, electrodes with NMs have enabled the determination of various organic molecules containing electroactive groups. The modifiers could greatly enhance the response and selectivity for analytes [23,24]. NMs, composites and conjugates have been reported as electrode modifiers. Often, modifiers to fabricate the modified electrodes were organic small molecules (proteins, enzymes and antibodies), polymer, metal nanoparticles, carbon nanotubes and graphene [23,24]. Among these, electrodes grafted with nanoparticles of carbon, polymer and metal are stable, sensitive and have fast response. Classification of sensors according to bioreceptor, technology and electrochemical techniques used is shown in Figure 5.2. The impact of technology in widening the applicability of electrochemistry in organic pollutants analysis through various electroanalytical techniques is summarized in Figure 5.2. These sensors have been used in pharmaceuticals [25], pesticides [26], and dyes and polychlorobiphenyls [27]. Among the commonly used NMs are the 2D NMs such as graphene and its derivatives, metals (mono- and bimetallic), metal oxides and molecularly imprinted polymers (MIPs).

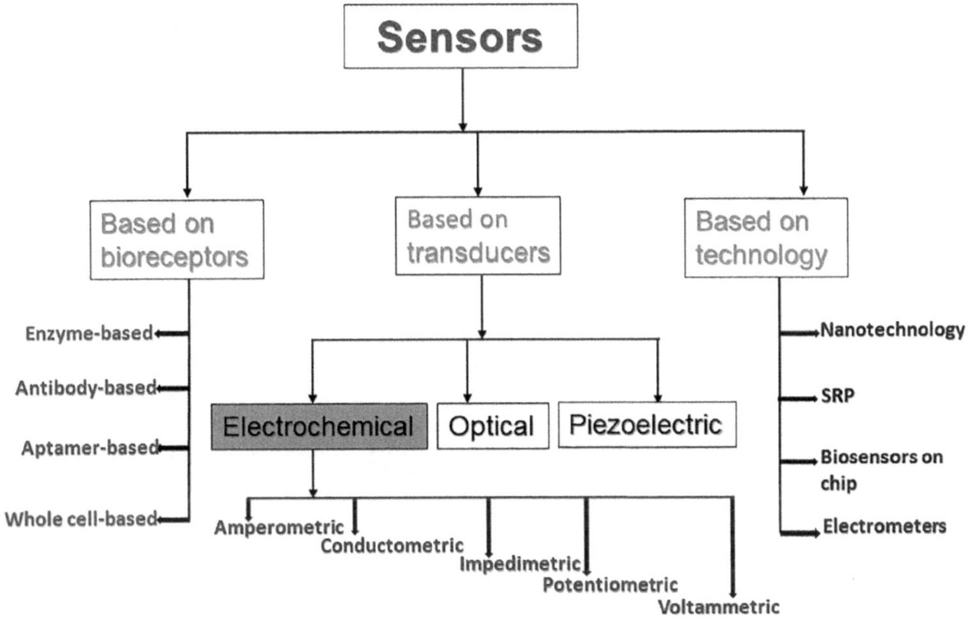

FIGURE 5.2 Classification of common sensors detailing the electrochemical techniques used in analysis.

5.2.2 Application of 2D Nanomaterials in Electrochemical Detection

Compared to other dimensional (zero-dimension=0D, one-dimension=1D and three-dimension=3D) NMs, the 2D NMs have the following advantages: a large surface-to-volume ratio [28,29] and more interaction sites for analytes as a result of their planar structure in addition to their large surface area, [30]. Conductivity of these materials is higher with an easily tunable electronic configuration and band gap [31]. The 2D NMs are more compatible with commercial metal electrodes which have large lateral sizes.

Graphene, one of the commonly used 2D NMs has been widely used for the detection studies of various pharmaceuticals and pesticides. These materials have equally exhibited remarkable results in electroanalysis. Thanks to their remarkable mechanical, thermal, electrical, magnetic and optical properties, they promise to revolutionize electronics. Among the outstanding characteristics of Carbon 2D NMs properties are high mobility of charge carriers, strength resulting from carbon–carbon bonds, the tunable energy structure and band gap using material composition variation, as well as the presence of vacancies and impurities. Occasionally, under favorable conditions, quantum effects manifest. These materials can be topological insulators and high-temperature superconductors. The unique electrochemical properties of 2D NMs strongly indicate a potential in creation of completely new types of devices for technologies in electronics, nanophotonics and quantum [29,30]. Classification based on compounds of 2D NMs that are frequently utilized in electrochemical analysis of organic pollutants is shown in Figure 5.3.

The 2D NMs can act as both the recognition elements and transducer functionalizing material [32]. As a recognition element, in a sensor, the analyte is detected directly through the nanomaterial. As a functionalizing material, it modifies the transducer by enhancing its electrical, electro-activity, and reactivity properties.

5.2.3 Fabrication of Sensors

In a nutshell, biosensors for any analyte have followed the same developmental trends with regard to recognition elements and transducer development. This is attested by a number of reviews [31] that detail advancements in pesticides biosensors. The reviews reveal that in terms of the recognition

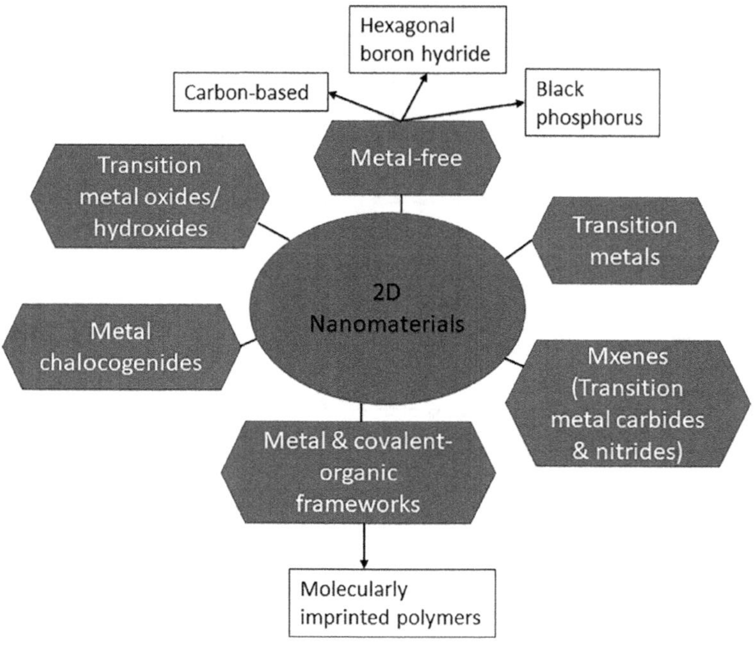

FIGURE 5.3 Classification of 2D NMs used in electrochemical analysis of organic pollutants.

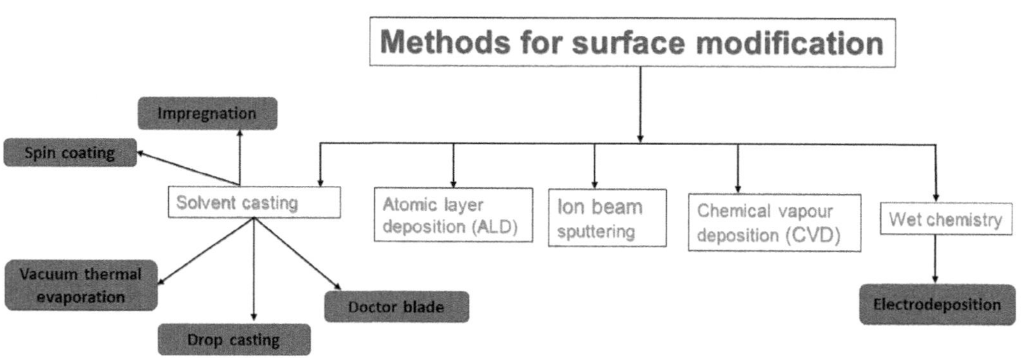

FIGURE 5.4 Methods of surface modifications in sensor development.

elements, the revolutions have been from use of enzymes, antibodies, aptamers, NMs, and lastly MIPs. The last two have not been exhausted for many analytes and are still under extensive research, but they are the promising routes especially the MIPs. The reports also reveal that transducers are more based on the electrochemical approaches in recent years than the previous reliance on the optical and piezoelectric approaches. Nanotechnology has taken the biosensor development to another level with improved mechanical, chemical, electrical, optical, and electrochemical properties [33,34] (Figure 5.4).

5.3 ELECTROANALYTICAL APPLICATION IN SELECTED ORGANIC POLLUTANTS

5.3.1 Electroanalysis of Organic Dyes Pollutants

The major sources of dye pollution are textiles, tanning, and food industry effluents due to common usage of dyes to color materials such as food, medicine, paper, leather, fabric, fiber, plastic, and textile [35,36]. These organic dyes are natural and synthetic materials that contain chromophores or organic functional groups, as depicted in Figure 5.5. Azo dyes ($-N \equiv N-$) are reported to contribute about 60%–70% of all dyes pollutants from food and textile industries [37]. Their widespread

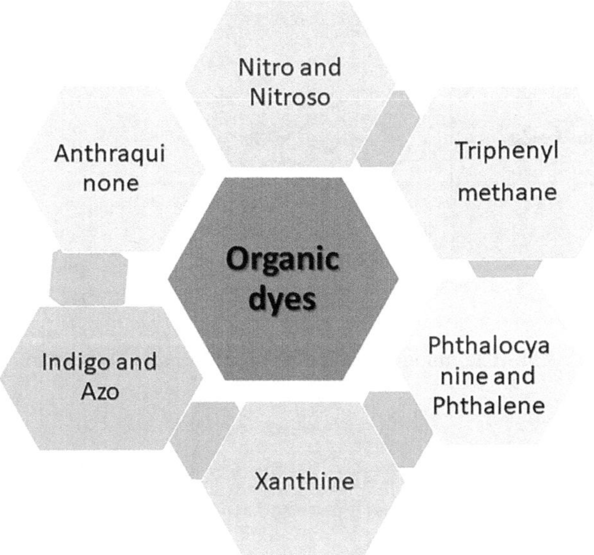

FIGURE 5.5 Electroactive organic chromophores in organic dyes.

applicability is due to their color fastness, chemical stability, wide variety of colors and light resistance. Hence, azo dyes are used in food, lac-dye and textile industry on synthetic and natural fibers, as well as in other industrial fields such as polygraphy, rubber, leather and plastic materials [38]. The application of azo dyes involves two organic chemicals: a binding reagent and an azo [39].

Due to its color, a frequently used technique for quantitative analysis of dyes is UV–visible spectroscopy. However, recently, the identification, characterization and quantification of organic compounds using electrochemistry, especially voltammetry, has increased. The attraction on azo dyes is because they are excellent electron transfer mediators [38]. Dyes electropolymerization has been used in the making of devices such as memory and optoelectronic [40], optical data storage [41], photoswitchable devices [39], nonlinear optical devices [37] and sensors [42]. In sensors, azo dyes have been used as markers for analysis of herbicides [40], reduction of pesticide and environmental pollution [35]. Several reports on electrochemical studies of food dyes such as amaranth, carminic [43], Sudan red [44], Sunset yellow and tartrazine [45] showed dominance of differential pulse voltammetry (DPV) technique, followed by square wave voltammetry (SWV). Among 2D NMs, carbon-based 2D NMs in their pristine or modified form are predominately used in establishing the dye compounds performance. These carbon-based materials are carbon paste, graphite and multiwalled carbon nanotubes (MWCNTs). Metals, metal oxide and polymer modifications are also frequently used. MWCNTs are used in bioactive and electroactive materials sensing because they have an elevated mechanical potency, good electrical conductivity, high surface area, good stability, exceptional electronic behaviors, low cost and produce good voltammograms [44]. Metal and metal oxide functionalization improves electroactivity, while polymer functionalization usually improves biocompatibility. MWCNTs functionalized with amino acid-based polymer, in particular polymerized glutamic acid [poly(GA)], have shown unique properties ideal for its application in sensors. Noticeably, the material is stable, nontoxic, has biomedicinal activity, show biocompatibility and its sensor resulted in excellent reproducibility, sensitivity and strong adherence on substrate with more conducting bridges surface sites [46].

A typical azo dye that has frequently been electrochemically analyzed is sodium salt of 5,5-indigo disulfonic acid, commonly known as indigo carmine (IC), with an IUPAC name disodium;2-(3-hydroxy-5-sulfonato-1H-indol-2-yl)-3-oxoindole-5-sulfonate. Its frequent analysis is due to the widespread application in the detection of ozone and superoxide, medical capsules pH indicator as well as a colorant in food and denim fabric products [47]. Despite the

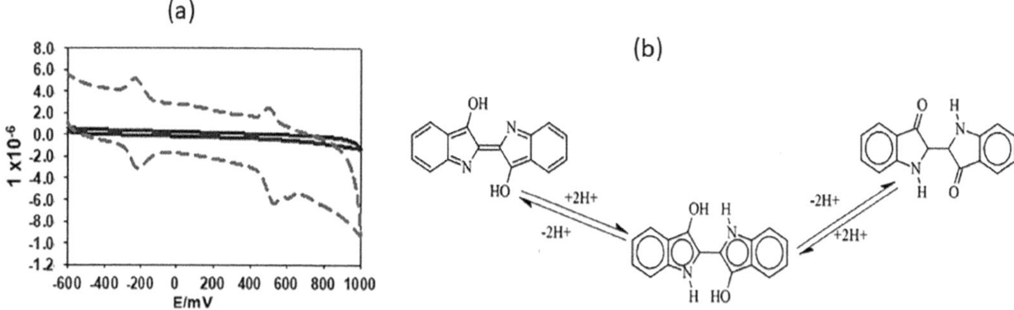

FIGURE 5.6 (a) Typical cyclic voltammogram of indigo in pH 7.0 and (b) the redox reactions involved.

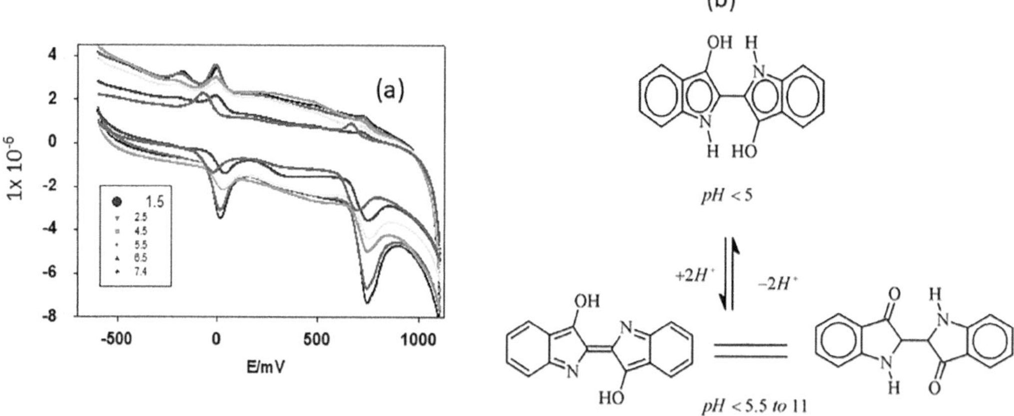

FIGURE 5.7 (a) Effect of pH on cyclic voltammogram and (b) different forms of indigo in various pH conditions.

recommended normal living body dosage range of 100–300 mg kg^{-1}, it has been used in medicine. Its reported medicinal usage has been in cancer (gastric) [48], transurethral resection and vesico-ureteral reflux [49,50], surgery and chemotherapy [51].

The electrochemistry of IC through cyclic voltametric (CV) technique has elucidated electrochemical thermodynamics and kinetics of its redox reactions [52]. Various types of electrodes usage applied in exploring the performance of IC have been reported [53]. Cyclic voltammogram involved a redox reaction as a result of gain or loss of hydrogen ions as shown in Figure 5.6. Electrochemical analysis of indigo performed fairly well when compared to other techniques, with adequate linear ranges and lower detection limits.

As a result of the involvement of hydrogen, pH effect monitoring is a must in electrochemical method development of azo dyes. Figure 5.7b shows the effect of pH on indigo analysis. From Figure 5.7, it is evident that the potential and current are both affected by pH, which can be explained by changes in structure under different pH conditions, as indicated in Figure 5.7b. Hence, for accurate results, it is vital to determine the optimum conditions for each organic dye before analysis.

5.3.2 Electroanalysis of Pesticides, Polychlorinated Biphenyls and other Pollutants

Pesticides are substances used for destroying organisms or undesired plants (also known as weeds) that are harmful and/or sabotage the cultivated plants or to animals in order to increase the yield in agricultural production [53]. Pesticides are extensively used in agriculture, and about one-third of the agricultural production is saved globally by their application [1]. They are categorized either in terms of the targeted pest (fungi, weed or insects) as fungicides, herbicides or insecticides, or according to their chemical makeup such as carbamates, organochlorines and organophosphates [14,54,55]. Their excessive usage however can be a source of ecosystem (air, soil and water) contamination [56,57].

Polychlorinated biphenyls (PCBs) are man-made organic compounds exhibiting 2–10 chlorine atoms attached to 2 benzene rings, the biphenyl, and about 209 PCBs constitute the family as individual congeners (PCB 1–209) [58]. They also exist as a mixture of congeners referred to as Aroclors [59]. They were used as dielectric fluids and insulators for transformers and capacitors, heat transfer fluids, coolants and lubricants, and in preventing fires and -explosions [59]. Other persistent organic pollutants (POPs) include phthalic acid esters (PAEs) or phthalate esters, commonly called phthalates, polybrominated diphenyl ethers (PBDEs) and polycyclic aromatic hydrocarbons (PAHs).

PAEs are organic chemicals with the chemical structure of dialkyl or alkyl aryl esters of 1,2-benzenedicarboxylic acid [60]. They got their extensive application as additives in plastic polymers. They were not forming the polymer chain backbones; hence, they could easily be released into the environment [61]. PBDE together with polybrominated biphenyls (PBBs) is a class of unruly halogenated compounds used as a flame retardant for consumer goods such as electrical equipment, construction materials, coatings, textiles and polyurethane foam (furniture padding). They have similar structure to PCBs [62,63]. PAHs are organic compounds consisting of a least two aromatic rings [64]. They are mainly generated into the environments from incomplete combustion of fossil fuels and other organic compounds. They are also by-products of forests, garbage and oil burning.

These POPs have also been detected using electrochemical sensing. Table 5.2 presents a comprehensive summary of these POPs with details of the electrode composition employed, electrochemical

TABLE 5.2
Electrochemical Analysis of POPs Using 2D NMs, Detailing the Transducers and Analytical Parameters

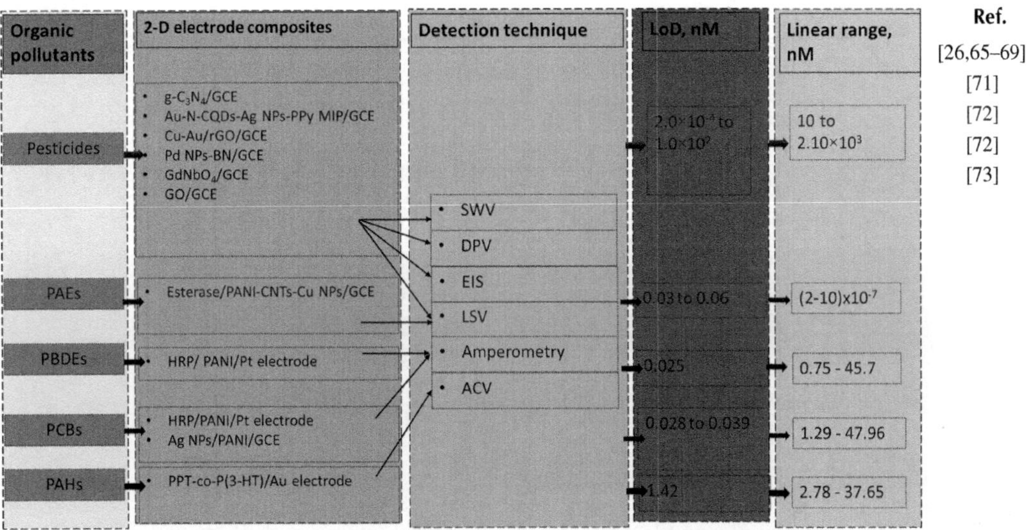

Ref.
[26,65–69]
[71]
[72]
[72]
[73]

ACV, alternating cyclic voltammetry; Ag NPs, silver nanoparticles; Au, gold; BN, boron nitride; CNTs, carbon nanotubes; Cu, copper; Cu NPs, copper nanoparticles; DPV, differential pulse voltammetry; EIS, electrochemical impedance spectroscopy; GCE, glassy carbon electrode; $GdNbO_4$, gadolinium niobate; g-C_3N_4, graphitic carbon nitride; GO, graphene oxide; HRP, horseradish peroxidase; LSV, linear sweep voltammetry; N-CQDs, nitrogen-doped carbon quantum dots; PAEs, phthalic acid esters; Pd NPs, palladium nanoparticles; PAHs, polycyclic aromatic hydrocarbons; PANI, polyaniline; PBDEs, polybrominated diphenyl ethers; PCBs, polychlorinated biphenyls; PPT-co-P(3-HT), poly(propylene thiophenoimine)-co-poly(3-hexylthiophene); PPy MIP, polypyrrole molecularly imprinted polymer; Pt, platinum; rGO, reduced graphene oxide; SWV, square wave voltammetry.

detection technique and the overall performance of the detection methodology (limit of detection, LoD, and linear range). It can be noticed that pesticides, among the POPs, are the most that have been detected using 2D NMs-based electrochemical approaches. Their detections have explored various forms and combinations of the 2D NMs and a variety of electrochemical techniques. Their analyses also date to recent years. These then give deduction that their detection has attracted much attention from the analytical community, and this can be as a result of their immense usage with resultant residual amounts introduced into the wastewater. Their analytical methodologies finally exhibit relatively good analytical characteristics in general with low LoDs and a wide concentration linear range.

5.3.2.1 Detection Criteria for the Pollutants

The specific detection criteria for each of the chemical pollutants are governed by the functional groups present in their structures (Table 5.3). The functional groups render the pollutants electroactive response as alluded earlier. Bearing the electroactive behavior, pollutants undergo either electrochemical oxidization or reduction on the electrode surface with observable resultant anodic or cathodic peak currents during analysis. The nitro groups are responsible for the electrochemical reduction detection of pesticides [65,69,74], while hydroxyl groups facilitate the electrochemical oxidation process [67]. PAEs are said to undergo a reduction electrochemical

TABLE 5.3
Electroactive Organic Functional Groups Found in POPs and Examples of POPs Containing them

Functional groups	POPs
Chloride, Cl^-	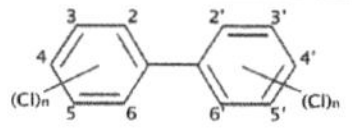 PCBs
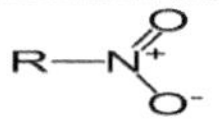 Nitro ($-NO_2$)	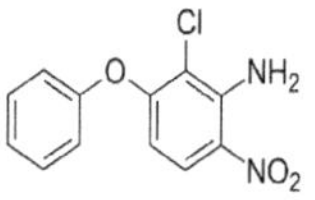 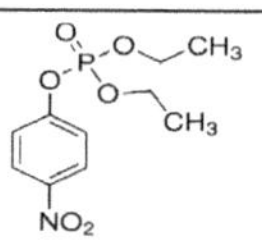 Aclonifen; Paraoxon ethyl (pesticides)
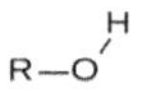 Hydroxyl (-OH)	 4-chlorophenol (pesticide)
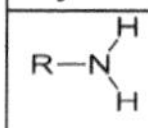 Amino ($-NH_2$)	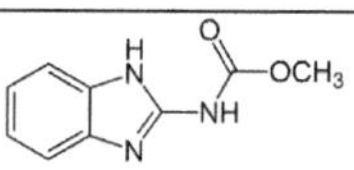 Carbendazim (pesticide)
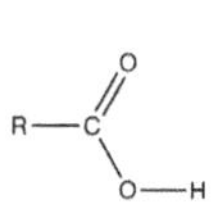 Carboxyl (-COOH)	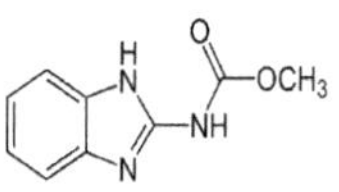 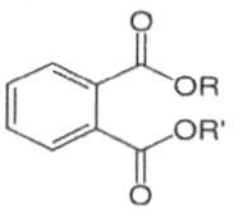 Carbendazim (pesticide); PAEs
 Aromatic structure: delocalized π electrons	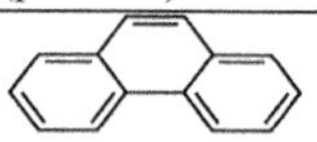 Phenanthrene (PAHs)

process even though they have carboxyl groups that are within the structure of the compound unlike pesticides [70]. PAHs' aromatic structure possesses delocalized π electrons, which enhances their electrochemical oxidation [73].

An example of electrochemical redox process is given in Figure 5.8a where PCB 28 undergoes reduction that is enhanced by the functional chlorine atoms. The process involves the removal of chloride ions (decomposition of the C–Cl bond) and electron transfer (ET) (peaks 1 and 2) on the surface of electrode. The reduction mechanism is depicted in Figure 5.8b, which follows either stepwise (i) or concerted (ii) [75]. This redox experience by the organic compounds is enhanced by the catalytic and fast electron transfer nature of the 2D NMs that are used to modify the electrode surfaces, and these electrochemical processes are mostly said to be diffusion controlled.

Evidently, the 2D NMs have been used in the development of electrochemical sensors as recognition elements and as transducer modifying agents in POP detection, as shown in Table 5.2. The analyses of pesticides have employed only the NMs functionalizing the electrode surface. These materials directly interact with the analyte while at the same time converting the physicochemical interactions into a detectable signal. For the other target pollutants, NMs function mostly to enhance the electrode transduction since the recognition enzymes (esterase and HRP) are present. With the use of these enzymes, the electrochemical detection is based on the inhibition of the enzyme activity by the analyte. The inhibition effect is indirectly proportional to the current signal or the analyte concentration [76].

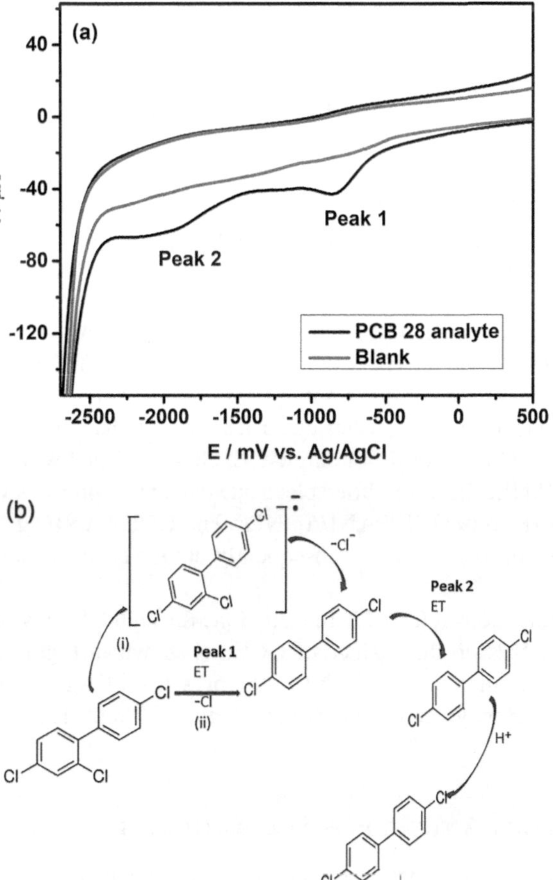

FIGURE 5.8 (a) CV for electrochemical reduction of PCB 28 and (b) reduction mechanism of PCB 28.

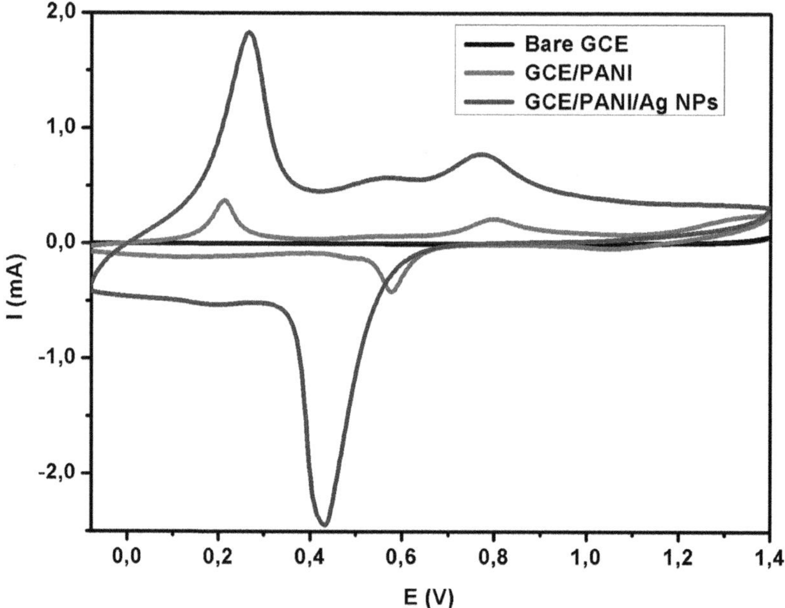

FIGURE 5.9 CV showing enhanced current response of PANI doped with Ag NPs modified GCE relative to bare and PANI-modified GCE.

NMs can be synergized to advance the performance (responsiveness, stability, etc.) of the sensor by improving the electrical, electro-active and surface area of the transducing electrode. This can be explained by referring to the pesticide detections presented in Table 5.2, in which combinations of several materials were deployed on glassy carbon electrode. They are gold, nitrogen-doped carbon quantum dots, silver nanoparticles and molecularly imprinted polypyrrole (Au/N-CQDs/Ag NPs-PPy MIP/GCE), copper-gold-doped reduced graphene oxide (Cu-Au/rGO/GCE) and palladium nanoparticles-hybridized boron nitride (Pd NPs-BN/GCE). Other combinatory examples are molybdenum disulfide (MoS_2) and diamond (D) or gold (Au) nanoparticles (NPs) (MoS_2-D NPs or Au NPs) [77] and nitrogen fluoride-doped MoS_2 hybridized with Ag NPs (NF-MoS_2/Ag NPs) [76].

This modified electrode characteristic synergistic effect is exemplified as depicted in Figure 5.9, where a GCE was successively modified with polyaniline (PANI) and PANI doped with Ag NPs and then assessed the electrode composites electrochemical behavior by cyclic voltammetry (CV). The depiction shows the CV responses of bare GCE, GCE modified with PANI and GCE modified with Ag NPs-doped PANI. The unmodified electrode current response is very low and increases in the order: GCE<GCE/PANI<GCE/PANI/Ag NPs. The GCE/PANI/Ag NPs with enhanced electrical and electroactive properties were subsequently employed in sensor fabrication for PCB 28 analysis [27].

The electrode modifications referred here were done through CV polymerization in the presence of pre-synthesized Ag NPs on the surface of GCE, as shown in Figure 5.10. Figure 5.10a is the polymerization of aniline monomer to PANI in the presence of Ag NPs in acidic medium (HCl), Figure 5.10b is the typical polymerization path and Figure 5.10c is the incorporation of HCl and Ag NPs into PANI films.

5.3.3 Electrochemistry Application in Pharmaceuticals

Pharmaceuticals are one of the classes of emerging contaminants (ECs) in the environment. The ECs' substances are known or suspected chemicals that have adverse ecological or human health effects [78,79]. They are often classified together with personal care products as pharmaceuticals and

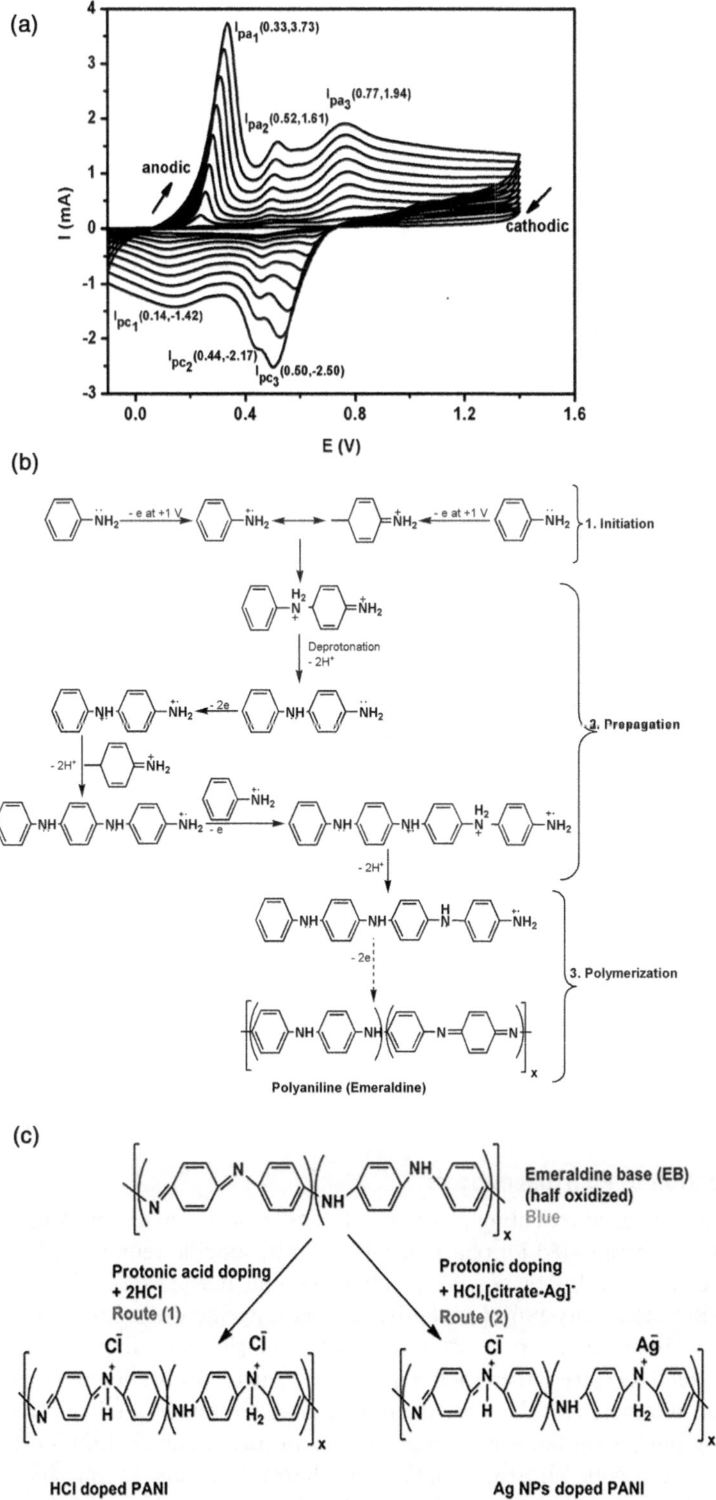

FIGURE 5.10 Oxidative electropolymerization of aniline monomer to PANI in the presence of Ag NPs (a), polymerization mechanism (b), and PANI doping by HCl and Ag NPs (c).

personal care products (PPCPs), and their classification is according to their medicinal usage (oral and topical). The Pharm pollutants common classifications are antibiotics (ATB), analgesics, anti-inflammatory diuretics, anti-microbials, antiseptics, lipid regulators, steroids, stimulant drugs, beta blockers, hormones and illicit drugs [80]. In addition, some organic waste released from poultry and livestock activities are also considered as ECs [81].

The widespread usage of pharmaceuticals as a result of population increases, and health remediation improvements are the major contributors of these drugs presence in sewage systems and the environment. The spread to water bodies, especially aquatic, is a result of sewage leakages, effluents entering the aquatic environment from wastewater treatment plants (WWTPs) as well as the improper dumping of unused or expired medication [82,83]. The consequences of the presence of PPCPs in living organisms are based on exposure time, water matrix (e.g., water stream) and pollutant concentration. PPCPs may induce metabolic changes depending on their accumulation and changes in organisms [84]. The most dramatic reported consequence of PPCPs to organisms is the fish feminization (in Salmo trutta) [85,86] due to their endocrine disruption properties, resulting in reproduction and developmental changes in certain fish species (*Gobiocypris rarus*, *Oryzias latipes* and *Danio rerio*). Feminization has been reported to be caused by very low concentrations of environmental estrogens (estradiol, estrone and ethinyl estradiol) [87]. Other adverse effects of some PPCPs are an increase in the antibiotic resistance of pathogenic microorganisms [88]. Therefore, the presence of PPCPs in the aquatic and terrestrial ecosystems affects the food quality, thus increasing the risks for human health. Notably, ATBs' environmental contamination and residency atmospheres, such as air, water and soil, are a global danger to human health [89]. Excessive utilization of ATBs in the food-constructing to improve food security resulted in their high concentration above the required levels found within food. Hence, there is a steady increase in the development of drug-resilient pathogens or "super microbes" [90]. Food supply ATB contamination brings a significant challenge to food security assignments [91].

The removal of pharmaceuticals from wastewater has mostly depended on the nature of pollutant (e.g., chemical structure, concentration and solubility) and the technology used to treat the wastewater. Bioprocessing with activated sludge and anaerobic digestion are dominant technologies for wastewater treatment. This process is ineffective in PPCPs degradation [92]. Partial degradation of diclofenac, ibuprofen and triclosan has been reported [93]. However, this was time-consuming due to the low degradation rates [86]. However, unexpected degradation of diatrizoate under anaerobic conditions has been reported [94]. These degradation of diatrizoate was achieved at a long hydraulic retention time by microorganisms in the presence of a specific ratio of Cu (II)/S(IV). Although biodegradation of PPCPs using either aerobic or anaerobic conditions has limitations, the biodegradability potential of other PPCPs via these processes can be utilized at increasingly high costs.

5.3.3.1 Detection of Pharmaceuticals

Electrochemical advanced oxidation processes (EAOPs) have recently been shown as a potential technology that can be applied for pharmaceuticals fast, specific removal [95]. Electrochemical sensors for detection of ATBs have several advantages such as ease in fabrication, rapid identification and better selectivity [96]. Currently, sensors utilizing different electrochemical modes (electrochemical impedance spectroscopy, potentiometric and amperometric) are actively researched to effectively strengthen their application. The incorporation of NMs within the electrochemical sensors increases the sites for indication enlargement and selectivity to the target analyte, thereby improving the sensor applicability in trace analysis [97]. Of interest in detection studies are the functional groups present in pharmaceuticals, mainly the -NH- group like in most antivirals like nevirapine (NVP) [98]. Numerous earlier reports examined the role of NMs toward electrochemical sensing of ATBs [99] among others. The dominating NMs recently have been the 2D materials, which are basically carbon based and can be modified using various modifiers, as shown in Table 5.4.

TABLE 5.4
Pharmaceutical Contaminants in Wastewater Detected by Voltammetric Techniques with 2D NMs in Bold

Analyte	Transducer	Electrochemical Technique	LOD (nM)	Ref.
Hydroquinone (HQ)	NRCu-rGO/GCE	DPV and CV	4.9×10^1	[100]
Catechol (CC)			5.2×10^1	
Resorcinol			6.0×10^1	
HQ	Oxidized graphite nanoplatelets	CV	4.0×10^2	[101]
CC			6.0×10^2	
HQ	Ag NPs/MWCNT/GCE	SWV	1.0×10^3	[102]
Sulfadiazine	$MnCO_2O_4$/P-CN	DPV	3	[103]
Diclofenac	GCE/APTES-Amino AT-Silica	SWV	–	[104]
Nevirapine	MWCNTs/Ag-Pt NPs/GCE	DPV	2.1×10^1	[95]
Nitrofurantoin	WS_2 nanosheets	DPV	6.0×10^1	[105]
Ketoprofen	Carbon based paper	CV and DPV	–	[106]

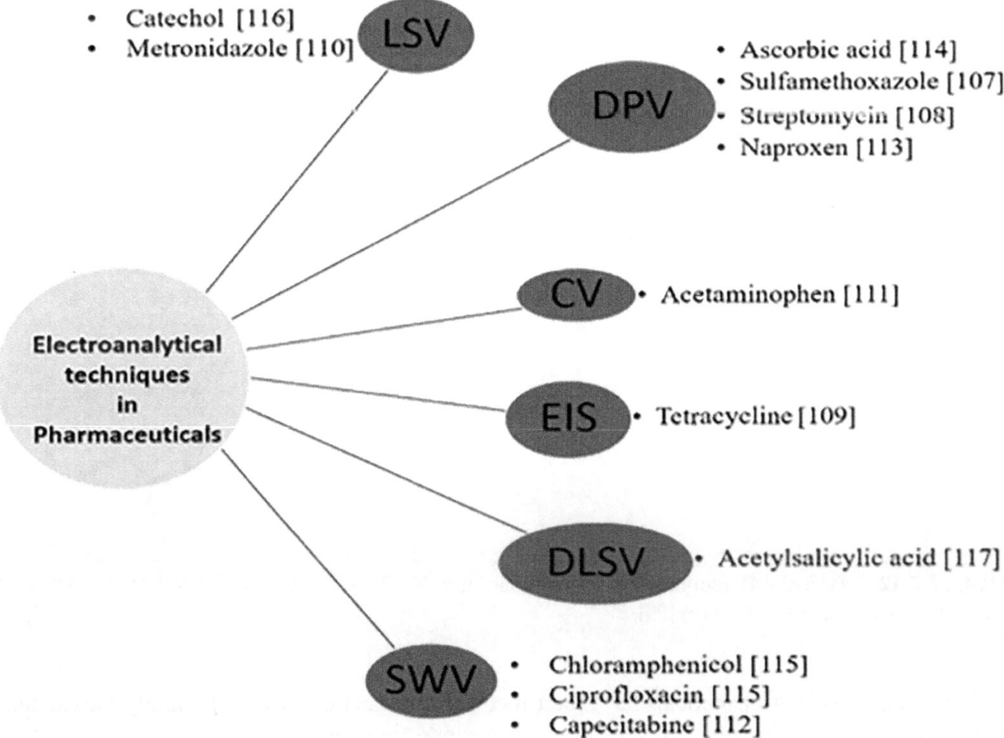

FIGURE 5.11 Electroanalytical techniques commonly used in pharmaceutical detection.

From Table 5.4, we can clearly argue that past detection studies with various NMs have reported high sensitivity with wider linear ranges on lower detection limits. The detection of pharmaceutical drugs with 2D NMs has involved different analytes using various techniques. Some of the notable electroanalytical techniques used in past studies include those mentioned in Figure 5.11.

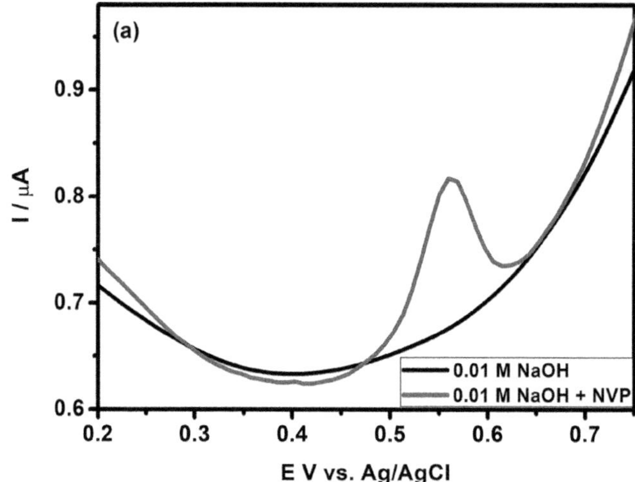

(b)

FIGURE 5.12 Typical differential pulse voltammogram of NVP electrooxidation in NaOH at 20 mV s^{-1} (a) and the redox reaction involved (b) [99].

However, most pharmaceuticals have their mechanism of action involving mainly the exchange of electrons via oxidation processes under suitable supporting electrolytes. For instance, Okumu et al. [99] developed a robust detection platform for an anti-HIV drug, Nevirapine, where MWCNTs/Ag-Pt NPs 3:1/GCE was used to electrooxidize NVP at 0.01 M NaOH as shown in the DPV profile in Figure 5.12a, with a corresponding mechanism of action depicted in Figure 5.12b.

In the presence of hydroxide ions, the NVP was deprotonated to radical ions using 2-electron transfer. This electrochemical process occurred with the same number of protons and electrons. A similar mechanism was reported in the electrochemical reaction process for NVP oxidation at MWCNTs/Ag-Pt NPs 3:1/GCE [99].

5.4 CONCLUSION AND FUTURE OUTLOOK

Presently, 2D NMs' excellent electrical properties contribute to high sensitivity, selectivity and versatility, making them preferred for electrochemical sensing and their extensive utilization in the analysis of organic waste materials. The application of these materials in organic pollutants analysis is dominated by the utilization of modified electrode sensors, which are mainly carbon-based materials. The majority of the analysis is laboratory based, where efficient detection in different media with good detection limits and high sensitivity was obtained. Future research might concentrate on functionalization of nanomaterials to tailor their transduction, biocompatibility and binding bridge properties, thereby improving sensing performance. The wide varieties of organic functional groups redox activities will be explored and used in the development of devices that can be able to be used at the point of interest. Reports on real materials analysis and onsite analysis are limited; therefore, future research will concentrate on the development of devices that can be user friendly, sensitive, selective and versatile for out-of-the-laboratory application. The physicochemical properties of 2D nanomaterials and their ease of functionalization play a pivotal role in their sensing ability.

REFERENCES

1. Hussain N., Bilal M., Iqbal H.M.N. Carbon-based nanomaterials with multipurpose attributes for water treatment: Greening the 21st-century nanostructure materials deployment, *Biomaterials and Polymers Horizon*. 1(2021) 48–58. https://doi.org/10.37819/bph.001.01.0131.
2. Rana D.S., Thakur N., Thakur S., Singh D. Electrochemical determination of hydrazine by using MoS_2 nanostructure modified gold electrode, *Nanofabrication*. 6(2021) 79–86. https://doi.org/10.37819/nanofab.007.190.
3. Henze M., Comeau Y. Wastewater characterization. In: *Biological Wastewater Treatment: Principles Modelling and Design*. (2008) 33–52. https://doi.org/10.2166/9781780408644_001.
4. Akpor O., OtohinoyiD.A., Olaolu T.D., Aferiye J.B.I. Pollutants in wastewater effluents: Impacts and remediation, *International Journal of Environmental Research and Earth Science*. 3(2014) 50–59.
5. Shon H.K., Vigneswaran S., Snyder S.A. Effluent organic matter (EfOM) in wastewater: Constituents, effects, and treatment, *Critical Reviews in Environmental Science and Technology*. 36(2006) 327–374. https://doi.org/10.1080/10643380600580011.
6. Snyder E.M., Pleus R.C., Snyder S.A. Pharmaceuticals and EDCS in the US water industry - An update, *Journal ofAmerican WaterWorksAssociation*. 97(2005) 32–36. https://doi.org/10.1002/j.1551-8833.2005.tb07506.x.
7. Tümay S.O., Senocak A., Sari E., Sanko V. A new perspective for electrochemical determination of parathion and chlorantraniliprole pesticides via carbon nanotube-based thiophene-ferrocene appended hybrid nanosensor, *Sensors and Actuators, B: Chemical*. 345(2021). https://doi.org/10.1016/j.snb.2021.130344.
8. Petrovic M., Gonzalez S., Barcelo D. Analysis and removal of emerging contaminants in wastewater and drinking water, *TrAC: Trends in Analytical Chemistry*. 22(2003) 685–696. https://doi.org/10.1016/S0165-9936(03)01105-1.
9. Abinaya M., Rajakumaran R., Chen S.M., Karthik R., Muthuraj V. In situ synthesis, characterization, and catalytic performance of polypyrrole polymer incorporated Ag_2MoO_4 nanocomposite for detection and degradation of environmental pollutants and pharmaceutical drugs, *ACS Applied Materials & Interfaces*. 11(2019) 38321–38335. https://doi.org/10.1021/acsami.9b13682.
10. Kappelhoff B.S., Rosing H., Huitema A.D., Beijnen J.H. Simple and rapid method for the simultaneous determination of the non-nucleoside reverse transcriptase inhibitors efavirenz and nevirapine in human plasma using liquid chromatography, *Journal of Chromatography B*. 792(2003) 353–362. https://doi.org/10.1016/S1570-0232(03)00325-8.
11. Marzolini C., Beguin A., Telenti A., Schreyer A., Buclin T., Biollaz J., Decosterd L.L.A. Determination of lopinavir and nevirapine by high performance liquid chromatography after solid-phase extraction: Application for the assessment of their transplacental passage at delivery, *Journal of Chromatography B: Analytical Technologies in the Biomedical and Life Sciences*. 774(2002) 127–140. https://doi.org/10.1016/S1570-0232(02)00169-1.
12. Rahi A., Karimian K., Heli H. Nanostructured materials in electroanalysis of pharmaceuticals, *Analytical Biochemistry*. 497(2016) 39–47. https://doi.org/10.1016/j.ab.2015.12.018.

13. Annamalai J., Vasudevan N. Detection of phthalate esters in PET bottled drinks and lake water using esterase/PANI/CNT/Cu NPs based electrochemical biosensor, *Analytica Chimica Acta.* 1135(2020) 175–186. https://doi.org/10.1016/j.aca.2020.09.041.
14. Asif M., Sahid H., Ayub K., Ans M., Mahmood T. A first principles study on electrochemical sensing of highly toxic pesticides by using porous C_4N nanoflake, *Journal of Physics and Chemistry of Solids.* 160(2022) 110345. https://doi.org/10.1016/j.jpcs.2021.110345.
15. Boopathy G., Keerthi M., Chen S.M., Umapathy M.J., Kumar B.N. Highly porous nickel molybdate@ graphene oxide nanocomposite for the ultrasensitive electrochemical detection of environmental toxic pollutant catechol, *Materials Chemistry and Physics.* 239(2020) 121982. https://doi.org/10.1016/j.matchemphys.2019.121982.
16. Gholivand M.B., Akbari A. A novel voltammetric sensor for citalopram based on multiwall carbon nanotube/(poly(p-aminobenzene sulfonic acid)/b-cyclodextrin), *Materials Science and Engineering: C.* 62(2016) 480–488. https://doi.org/10.1016/j.msec.2016.01.066.
17. Gholivand M.B., Mohammadi-Behzad L. An electrochemical sensor for warfarin determination based on covalent immobilization of quantum dots onto carboxylated multiwalled carbon nanotubes and chitosan composite film modified electrode, *Materials Science and Engineering: C. Material for Biological Applications.* 57(2015) 77–87. https://doi.org/10.1016/j.msec.2015.07.020.
18. Ronkainen N.J., Brian H., Heineman W.R. Electrochemical biosensors, *Chemical Society Reviews.* 39(2010) 1747–1763. https://doi.org/10.1039/B714449K.
19. Majdinasab M., Yaqub M.,. Rahim A, Catanante G., Hayat A., Marty J.L. An overview on recent progress in electrochemical biosensors for antimicrobial drug residues in animal-derived food, *Sensors (Switzerland).* 17(2017) 1–21. https://doi.org/10.3390/s17091947.
20. Zheng Q., Chen Y., Fan K., Wu J., Ying Y. Exploring pralidoxime chloride as a universal electrochemical probe for organophosphorus pesticides detection, *Analytica Chimica Acta.* 982(2017) 78–83. https://doi.org/10.1016/j.aca.2017.06.006.
21. Rebelo P., Costa-Rama E., Seguro I., Pacheco J.G., Nouws H.P.A., Cordeiro M.N.D.S., Delerue-Matos C. Molecularly imprinted polymer-based electrochemical sensors for environmental analysis, *Biosensors and Bioelectronics.* 172(2021) 112719. https://doi.org/10.1016/j.bios.2020.112719.
22. Karimi-Maleh H., Karimi F., Fu L., Sanati A.L., Alizadeh M., Karaman C., Orooji Y. Cyanazine herbicide monitoring as a hazardous substance by a DNA nanostructure biosensor, *Journal of Hazardous Materials.* 423(2022) 127058. https://doi.org/10.1016/j.jhazmat.2021.127058.
23. Radi A. Accumulation and trace measurement of chloroquine drug at DNA-modified carbon paste electrode, *Talanta.* 65(2005) 271–275. https://doi.org/10.1016/j.talanta.2004.05.024.
24. Ma J.Y., Zhang Y.S., Zhang X.H., Zhu G.B., Liu B., Chen J.H. Sensitive electrochemical detection of nitrobenzene based on macro-/meso-porous carbon materials modified glassy carbon electrode, *Talanta.* 88(2012) 696–700. https://doi.org/10.1016/j.talanta.2011.11.067.
25. Fabre B., Samorı C., Bianco A. Immobilization of double functionalized carbon nanotubes on glassy carbon electrodes for the electrochemical sensing of the biotin–avidin affinity, *Journal of Electroanalytical Chemistry.* 665(2012) 90–94. https://doi.org/10.1016/j.jelechem.2011.11.029.
26. Ilager D., Malode S.J., Shetti N.P. Development of 2D graphene oxide sheets-based voltammetric sensor for electrochemical sensing of fungicide, carbendazim, *Chemosphere.* 303(2022) 134919. https://doi.org/10.1016/j.chemosphere.2022.134919.
27. Khesuoe M.P., Okumu F.O., Matoetoe M.C. Development of a silver functionalised polyaniline electrochemical immunosensor for polychlorinated biphenyls, *Analytical Methods.* 8(2016) 7087–7095. https://doi.org/10.1039/c6ay01733a.
28. Wang H., Feng H., Li J. Graphene and graphene-like layered transition metal dichalcogenides in energy conversion and storage, *Small.* 10(2014) 2165–2181. https://doi.org/10.1002/smll.201303711.
29. Wang H., Yuan H., Hong S.S, Li Y., Cui Y. Physical and chemical tuning of two-dimensional transition metal dichalcogenides, *Chemical Society Reviews.* 44(2015) 2664–2680. https://doi.org/10.1039/c4cs00287c.
30. Sun Y., Lei F., Gao S., Pan B., Zhou J., Xie Y. Atomically thin tin dioxide sheets for efficient catalytic oxidation of carbon monoxide, *Angewandte Chemie International Edition.* 52(2013) 10569–10572. https://doi.org/10.1002/anie.201305530.
31. Xia F., Wang H., Xiao D., Dubey M., Ramasubramaniam A. Two-dimensional material nanophotonics, *Nature Photonics.* 8(2014) 899–907. https://doi.org/10.1038/nphoton.2014.271.
32. Verma N., Bhardwaj A. Biosensor technology for pesticides - A review, *Applied Biochemistry and Biotechnology.* 175(2015) 3093–3119. https://doi.org/10.1007/s12010-015-1489-2.

33. Ronkainen N.J., Brian H., Heineman W.R. Electrochemical biosensors, *Chemical Society Reviews.* 39(2010) 1747–1763. https://doi.org/10.1039/b714449k.
34. Raja I.S., Vedhanayagam M., Preeth D.R., Kim C., Lee J.H., Han D.W. Development of two-dimensional nanomaterials based electrochemical biosensors on enhancing the analysis of food toxicants, *International Journal of Molecular Sciences.* 22(2021) 1–24. https://doi.org/10.3390/ijms22063277.
35. Liu Q. Pollution and treatment of dye waste-water, *IOP Conference Series: Earth and Environmental Science.* 514(2020) 052001. https://doi.org/10.1088/1755-1315/514/5/052001.
36. Cheng W., Chen H., Liu C., Ji C., Ma G., Yin M. Functional organic dyes for health-related applications, *VIEW.* 1(2020) 20200055. https://doi.org/10.1002/VIW.20200055.
37. Yazdanbakhsh M.R., Ghanadzadeh A., Moradi E. Synthesis of some new azo dyes derived from 4-hydroxy coumarin and spectrometric determination of their acidic dissociation constants, *Journal of Molecular Liquids.* 136(2007) 165–168. https://doi.org/10.1016/j.molliq.2007.03.005.
38. Copaciu F., Coman V., Vlassa M., Opriş M. Determination of some textile dyes in wastewater by solid phase extraction followed by high-performance thin-layer chromatography, *Journal of Planar Chromatography.* 25(2012) 509–515. https://doi.org/10.1556/jpc.25.2012.6.4.
39. Yi C., Tan X., Bie B., Ma H., Yi H. Practical and environment-friendly indirect electrochemical reduction of indigo and dyeing, *Scientific Reports.* 10(2020). https://doi.org/10.1038/s41598-020-61795-5.
40. Hareesha N., Manjunatha J.G., Amrutha B.M. Pushapanjali P.A., Charithra M.M., Subbaiah N.P. Electrochemical analysis of indigo carmine in food and water samples using a poly(glutamic acid) layered multi-walled carbon nanotube paste electrode, *Journal of Electronic Materials.* 50(2021) 1230–1238. https://doi.org/10.1007/s11664-020-08616-7.
41. Edwin D.S., Manjunatha J.G., Raril C., Girish T., Ravishankar D.K., Arpitha H.J. Electrochemical analysis of indigo carmine using polyarginine modified carbon paste electrode, *Journal of Electrochemical Science and Engineering.* 11(2021) 87–96. http://dx.doi.org/10.5599/jese.953.
42. Bessegato G.G., Brugnera M.F., Zanoni M.V.B. Electroanalytical sensing of dyes and colorants, *Current Opinion in Electrochemistry.* 16(2019) 134–142 https://doi.org/10.1016/j.coelec.2019.05.008.
43. Alizadeh M., Demir E., Aydogdu N., Zare N., Karimi F., Kandomal S.M., Rokni H., Ghasemi Y. Recent advantages in electrochemical monitoring for the analysis of amaranth and carminic acid as food colour, *Food and Chemical Toxicology.* 163(2022) 112929. https://doi.org/10.1016/j.fct.2022.112929.
44. Shen Y., Mao S., Chen F., Zhao S., Su W., Fu L., Zare N., Karimi F. Electrochemical detection of Sudan red series azo dyes: Bibliometrics based analysis, *Food and Chemical Toxicology.* 163(2022) 112960. https://doi.org/10.1016/j.fct.2022.112960.
45. Kaya S.I., Cetinkaya A., Ozkan S.A. Latest advances on the nanomaterials-based electrochemical analysis of azo toxic dyes Sunset Yellow and Tartrazine in food samples, *Food and Chemical Toxicology.* 156(2021) 112524. https://doi.org/10.1016/j.fct.2021.112524.
46. Fan Z., Cheng P., Liu M., Li D., Liu G., Zhao Y., Ding Z., Chen F., Wang B., Tan X., Wang Z., Han J. Poly(glutamic acid) hydrogels crosslinked via native chemical ligation, *New Journal of Chemistry.* 41(2017) 8656–8662. https://doi.org/10.1039/c7nj00439g.
47. Ammar S., Abdelhedi R., Flox C., Arias C., Brillas E. Electrochemical degradation of the dye indigo carmine at boron-doped diamond anode for wastewaters remediation, *Environmental Chemistry Letters.* 4(2006) 229–233. https://doi.org/10.1007/s10311-006-0053-2.
48. Ikeda K., Sannohe Y., Araki S., Inutsuka S. A new trial in endoscopic diagnosis for stomach cancer: Intra-arterial dye (IAD) method, *Gastrointestinal Endoscopy.* 26(1980) 1–4. https://doi.org/10.1016/s0016-5107(80)73247-9.
49. Song J.E., Kim S.K. The use of indigo carmine in ureteral operations, *The Journal of Urology.* 98(1967) 669–670. https://doi.org/10.1016/s0022-5347(17)62953-7.
50. Altok M., Sahin A.F., Gokce M.I., Ekin G.R., Divrik R.T. Ureteral orifice involvement by urothelial carcinoma: Long term oncologic and functional outcomes, *International Brazilian Journal of Urology.* 43(2017), 1052–1059. https://doi.org/10.1590/S1677-5538.IBJU.2017.0218.
51. Fujita M., Kuroda C., Hosomi N., Inoue E., Kuriyama K., Ohhigashi H., Kishimoto S., Ishikawa O., Nakaizumi A. Dye-injection method for placement of an infusion catheter in regional hepatic chemotherapy, *Journal of Vascular and Interventional Radiology.* 6(1995) 119–123. https://doi.org/10.1016/S1051-0443(95)71074-0.
52. Fao W. Codex Alimentarius Commission, Safety of Colors, GSFA MPLs of Indigo Carmine. (2015) p. 132.
53. Rhouati A., Majdinasab M., Hayat A. A perspective on non-enzymatic electrochemical nanosensors for direct detection of pesticides, *Current Opinion in Electrochemistry.* 11(2018) 12–18. https://doi.org/10.1016/j.coelec.2018.06.013.

54. Marty J., Garcia D., Rouillon R. Biosensors: Potential in pesticide detection, *Trends in Analytical Chemistry*. 14(1995) 329–333. https://doi.org/10.1016/0165-9936(95)97060-E.
55. Vamvakaki V., Chaniotakis N.A. Pesticide detection with a liposome-based nano-biosensor, *Biosensors and Bioelectronics*. 22(2007) 2848–2853. https://doi.org/10.1016/j.bios.2006.11.024.
56. Songa E.A., Okonkwo J.O. Recent approaches to improving selectivity and sensitivity of enzyme-based biosensors for organophosphorus pesticides: A review, *Talanta*. 155(2016) 289–304. https://doi.org/10.1016/j.talanta.2016.04.046.
57. Mills S.A., Thal D.I., Barney J.A. Summary of the 209 PCB congener nomenclature, *Chemosphere*. 68(2007) 1603–1612. https://doi.org/10.1016/j.chemosphere.2007.03.052.
58. Erickson M.D., Kaley R.G. Applications of polychlorinated biphenyls, *Environmental Science and Pollution Research*. 18(2011) 135–151. https://doi.org/10.1007/s11356-010-0392-1.
59. Srivastava A., Sharma V.P., Tripathi R., Kumar R., Patel D.K., Mathur P.K. Occurrence of phthalic acid esters in Gomti River Sediment, India, *Environmental Monitoring and Assessment*. 169(2010) 397–406. https://doi.org/10.1007/s10661-009-1182-4.
60. Zhang C., Zhou J., Ma T., Guo W., Wei D., Tan Y., Deng Y. Advances in application of sensors for determination of phthalate esters, *Chinese Chemical Letters*. 34(2023). https://doi.org/10.1016/j.cclet.2022.07.013.
61. Siddiqi M.A., Laessig R.H., Reed K.D. Polybrominated diphenyl ethers (PBDEs): New pollutants–old diseases, *Clinical Medicine & Research*. 1(2003) 281–290. https://doi.org/10.3121/cmr.1.4.281.
62. Centers for Disease Control and Prevention, Polybrominated diphenyl ethers (PBDEs) and polybrominated biphenyls (PBBs) Factsheet, National Biomonitoring Program, Division of Laboratory Sciences. (2017). https://www.cdc.gov/biomonitoring/.
63. Yan D., Wu S., Zhou S., Tong G., Li F., Wang Y., Li B. Characteristics, sources and health risk assessment of airborne particulate PAHs in Chinese cities: A review, *Environmental Pollution*. 248(2019) 804–814. https://doi.org/10.1016/j.envpol.2019.02.068.
64. Shetti N.P., Malode S.J., Vernekar P.R., Nayak D.S., Shetty N.S., Reddy K.R., Shkla S.S., Aminabhavi T.M. Electro-sensing base for herbicide aclonifen at graphitic carbon nitride modified carbon electrode - Water and soil sample analysis, *Microchemical Journal*. 149(2019). https://doi.org/10.1016/j.microc.2019.103976.
65. Kıran R., Yola M.L., Atar N. Electrochemical sensor based on Au @ nitrogen-doped carbon quantum dots @ Ag core-shell composite including molecular imprinted polymer for metobromuron recognition, *Journal of Electrochemical Society*. 166(2019) H691–H697. https://doi.org/10.1149/2.0451914jes.
66. Yang Y., Ma N., Bian Z. Cu-Au / rGO nanoparticle based electrochemical sensor for 4- chlorophenol detection, *International Journal of Electrochemical Science*. 14(2019) 4095–4113. https://doi.org/10.20964/2019.05.04.
67. Renganathan V., Balaji R., Chen S., Kokulnathan T. Coherent design of palladium nanostructures adorned on the boron nitride heterojunctions for the unparalleled electrochemical determination of fatal organophosphorus pesticides, *Sensors and Actuators B Chemical*. 307(2020) 127586. https://doi.org/10.1016/j.snb.2019.127586.
68. Gopi P.K., Mutharani B., Chen S., Chen T., Eldesoky G.E., Ali M.A., Wabaidur S.M., Shaik F., Tzu C.Y. Electrochemical sensing base for hazardous herbicide aclonifen using gadolinium niobate ($GdNbO_4$) nanoparticles-actual river water and soil sample analysis, *Ecotoxicology and Environmental Safety*. 207(2021) 111285. https://doi.org/10.1016/j.ecoenv.2020.111285.
69. Annamalai J., Vasudevan N. Detection of phthalate esters in PET bottled drinks and lake water using esterase/PANI/CNT/CuNP based electrochemical biosensor, *Analytica Chimica Acta*. 1135(2020) 175–186. https://doi.org/10.1016/j.aca.2020.09.041.
70. Nomngongo P.N., Ngila J.C., Msagati T.A.M., Gumbi B.P., Iwuoha E.I. Determination of selected persistent organic pollutants in wastewater from landfill leachates, using an amperometric biosensor, *Physics and Chemistry of the Earth*. 50–52(2012) 252–261. https://doi.org/10.1016/j.pce.2012.08.001.
71. Makelane H.R., John S.V., Waryo T.T., Baleg A., Mayedwa N., Rassie C., Wilson L., Baker P., Iwuoha E.I. AC voltammetric transductions and sensor application of a novel dendritic poly(propylene thiophenoimine)-co-poly(3-hexylthiophene) star co-polymer, *Sensors and Actuators, B Chemical*. 227(2016) 320–327. https://doi.org/10.1016/j.snb.2015.12.020.
72. Renganathan V., Balaji R., Chen S.M., Kokulnathan T. Coherent design of palladium nanostructures adorned on the boron nitride heterojunctions for the unparalleled electrochemical determination of fatal organophosphorus pesticides, *Sensors Actuators, B Chemical*. 307(2020) 127586. https://doi.org/10.1016/j.snb.2019.127586.

73. Muthukrishnan A., Boyarskiy V., Sangaranarayanan M.V., Boyarskaya I. Mechanism and regioselectivity of the electrochemical reduction in polychlorobiphenyls (PCBs): Kinetic analysis for the successive reduction of chlorines from dichlorobiphenyls, *Journal of Physical Chemistry C*. 116(2012) 655–664. https://doi.org/10.1021/jp2066474.
74. Song D., Wang Y., Lu X., Gao Y., Li Y., Gao F. Ag nanoparticles-decorated nitrogen-fluorine co-doped monolayer MoS_2 nanosheet for highly sensitive electrochemical sensing of organophosphorus pesticides, *Sensors and Actuators, B: Chemical*. 267(2018) 5–13. https://doi.org/10.1016/j.snb.2018.04.016.
75. Blanco E., Rocha L., Luis V., Petit-domínguez M.D., Casero E., Quintana C. A supramolecular hybrid sensor based on cucurbit [8] uril, 2D - molibdenum disulphide and diamond nanoparticles towards methyl viologen analysis, *Analytica Chimica Acta*. 1182(2021). https://doi.org/10.1016/j.aca.2021.338940.
76. Khan S., Naushad M., Govarthanan M., Iqbal J., Alfadul S.M. Emerging contaminants of high concern for the environment: Current trends and future research, *Environmental Research*. 207(2022) 112609. https://doi.org/10.1016/j.envres.2021.112609.
77. Pai C.W., Wang G.S. Treatment of PPCPs and disinfection by-product formation in drinking water through advanced oxidation processes: Comparison of UV, UV/Chlorine, and UV/H_2O_2, *Chemosphere*. 287(2022) 132171. https://doi.org/10.1016/j.chemosphere.2021.132171.
78. Fu Q., Malchi T., Carter L.J., Li H., Gan J., Chefetz B. Pharmaceutical and personal care products: From wastewater treatment into agro-food systems, *Environmental Science and Technology*. 53(2019) 14083–14090. https://doi.org/10.1021/acs.est.9b06206.
79. Madikizela L.M., Tavengwa N.T., Chimuka L. Status of pharmaceuticals in African water bodies: Occurrence, removal and analytical methods, *Journal of Environmental Management*. 193(2017) 211–220. https://doi.org/10.1016/j.jenvman.2017.02.022.
80. Ayman Z., Isik M. Pharmaceutical active compounds in water, Aksaray, Turkey, *Journal of Environmental Science and Health Part A-Toxic/ Hazardous Substances & Environmental Engineering*. 43(2015) 1381–1388. https://doi.org/10.1002/clen.201300877.
81. Adeleye A.S., Xue J., Zhao Y., Taylor A.A., Zenobio J.E., Sun Y., Han Z., Salawu O.A., Zhu Y. Abundance, fate, and effects of pharmaceuticals and personal care products in aquatic environments, *Journal of Hazardous Materials*. 424(2022) 127284. https://doi.org/10.1016/j.jhazmat.2021.127284.
82. Jarque S., Quirós L., Grimalt J.O., Gallego E., Catalan J., Lackner R., Piña B. Background fish feminization effects in European remote sites, *Scientific Reports*. 5(2015) 1–6. https://doi.org/10.1038/srep11292.
83. Patel M., Kumar R., Kishor K., Mlsna T., C.U. Pittman, Mohan D. Pharmaceuticals of emerging concern in aquatic systems: Chemistry, occurrence, effects, and removal methods, *Chemical Reviews*. 119(2019) 3510–3673. https://doi.org/10.1021/acs.chemrev.8b00299.
84. Kalia V.C. 2-Pharmaceutical and personal care product contamination: A global scenario, *Pharmaceuticals and Personal Care Products: Waste Management and Treatment Technology*. (2019) 27–61. https://doi.org/10.1016/B978-0-12-816189-0.00002-0.
85. Zhang X., Yan S., Chen J., Tyagi R.D., Li J. Physical, chemical, and biological impact (hazard) of hospital wastewater on environment: Presence of pharmaceuticals, pathogens, and antibiotic-resistance genes. In: Tyagi R.D., Sellamuthu B., Tiwari B., Yan S., Drogui P., Zhang X. (Eds.), *Current Developments in Biotechnology and Bioengineering*. Elsevier. (2020) 79–102. https://doi.org/10.1016/B978-0-12-819722-6.00003-1.
86. Nadimpalli M.L., Chan C.W., Doron S. Antibiotic resistance: A call to action to prevent the next epidemic of inequality, *Nature Medicine*. 27(2021) 187–188. https://doi.org/10.1038/s41591-020-01201-9.
87. Sun J., Liao X.P., D'Souza A.W., Boolchandani M., Li S.H., Cheng K., Luis Martínez J., Li L., Feng Y.J., Fang L.X., Huang T., Xia J., Yu Y., Zhou Y.F., Sun Y.X., Deng X.B., Zeng Z.L., Jiang H.X., Fang B.H., Tang Y.Z., Lian X.L., Zhang R.M., Fang Z.W., Yan Q.L., Dantas G., Liu Y.H. Environmental remodeling of human gut microbiota and antibiotic resistome in livestock farms, *Nature Communications*. 11(2020). https://doi.org/10.1038/s41467-020-15222-y.
88. Jin Y., Dou M., Zhuo S., Li Q., Wang F., Li J. Advances in microfluidic analysis of residual antibiotics in food, *Food Control*. 136(2022) 108885. https://doi.org/10.1016/j.foodcont.2022.108885.
89. Hidrovo A., Luek J.L., Antonellis C., Malley J.P., Mouser P.J. The fate and removal of pharmaceuticals and personal care products within wastewater treatment facilities discharging to the Great Bay Estuary, *Water Environmental Research*. 94(2022) e1680. https://doi.org/10.1002/wer.1680.
90. Kim S., Rossmassler K., Broeckling C.D., Galloway S., Prenni J., De Long S.K. Impact of inoculum sources on biotransformation of pharmaceuticals and personal care products, *Water Research*. 125(2017) 227–236. https://doi.org/10.1016/j.watres.2017.08.041.

91. Yang J., Luo Y., Fu X., Dong Z., Wang C., Liu H., Jiang C. Unexpected degradation and deiodination of diatrizoate by the Cu (II)/S (IV) system under anaerobic conditions, *Water Research*. 198(2021) 117137. https://doi.org/10.1016/j.watres.2021.117137.
92. Lozano I., Pérez-Guzmán C.J., Mora A., Mahlknecht J., Aguilar C.L., Cervantes-Avilés P. Pharmaceuticals and personal care products in water streams: Occurrence, detection, and removal by electrochemical advanced oxidation processes, *Science of the Total Environment*, 827(2022) 154348. https://doi.org/10.1016/j.scitotenv.2022.154348.
93. Moumen E., Bazzi L., El Hankari S. Metal-organic frameworks and their composites for the adsorption and sensing of phosphate, *Coordination Chemistry Reviews*. 455(2022) 214376. https://doi.org/10.1016/j.ccr.2021.214376.
94. Qian L., Durairaj S., Prins S., Chen A. Nanomaterial-based electrochemical sensors and biosensors for the detection of pharmaceutical compounds, *Biosensors and Bioelectronics*. 175(2021) 112836. https://doi.org/10.1016/j.bios.2020.112836.
95. Okumu F.O., Silwana B., Matoetoe M.C. Application of MWCNT/Ag-Pt nanocomposite modified GCE for the detection of nevirapine in pharmaceutical formulation and biological samples, *Electroanalysis*. 32(2020) 3000–3008. https://doi.org/10.1002/elan.202060374.
96. Khand N.H., Palabiyik I.M., Buledi J.A., Ameen S., Memon A.F., Ghumro T., Solangi A.R. Functional Co_3O_4 nanostructure-based electrochemical sensor for direct determination of ascorbic acid in pharmaceutical samples, *Journal of Nanostructure in Chemistry*. 11(2021) 455–468. https://doi.org/10.1007/s40097-020-00380-8.
97. Sabbaghi N., Noroozifar M. Nanoraspberry-like copper/reduced graphene oxide as new modifier for simultaneous determination of benzenediols isomers and nitrite, *Analytica Chimica Acta*. 1056(2019) 16–25. https://doi.org/10.1016/j.aca.2018.12.036.
98. Zhu X., Wu C., Miao L., Zhang Y., Cheng S., Liang Y., Nan J. Oxygen-defects functionalized graphite nanoplatelets as electrode materials for electrochemical sensing, *Journal of the Electrochemical Society*. 166(2019b) B1400–B1407. https://doi.org/10.1149/2.0271915jes.
99. Goulart L.A., Gonçalves R., Correa A.A., Pereira E.C., Mascaro L.H. Synergic effect of silver nanoparticles and carbon nanotubes on the simultaneous voltammetric determination of hydroquinone, catechol, bisphenol A and phenol, *Microchimica Acta* 185(2018). https://doi.org/10.1007/s00604-017-2540-5.
100. Sriram B., Baby J.N., Hsu Y.F., Wang S.F., Benadict J.X., George M., Veerakumar P., Lin K.C. $MnCo_2O_4$ microflowers anchored on P-doped g- C_3N_4 nanosheets as an electrocatalyst for voltammetric determination of the antibiotic drug sulfadiazine, *ACS Applied Electronic Materials*. 3(2021a) 3915–3926. https://doi.org/10.1021/acsaelm.1c00506.
101. Jiokeng S.L.Z., Tonle I.K., Walcarius A. Amino-attapulgite/mesoporous silica composite films generated by electro-assisted self-assembly for the voltametric determination of diclofenac, *Sensors and Actuators: B Chemical*. 287(2019) 296–305. https://doi.org/10.1016/j.snb.2019.02.038.
102. Fatima T., Husain S., Narang J., Khanuja M., Shetti N.P., Reddy K.R. Novel Tungsten disulfide (WS_2) nanosheets for photocatalytic degradation and electrochemical detection of pharmaceutical pollutants, *Jouranal of Water Process Engineering*. 47(2022) 102717. https://doi.org/10.1016/j.jwpe.2022.102717.
103. Torrinha A., Martins M., Tavares M., Delerue-Matos C., Morais S. Carbon paper as a promising sensing material: Characterization and electroanalysis of ketoprofen in wastewater and fish, *Talanta*. 226(2021) 122111. https://doi.org/10.1016/j.talanta.2021.122111.
104. Chen C., Chen Y.C., Hong Y.T., Lee T.W., Huang J.F. Facile fabrication of ascorbic acid reduced graphene oxide-modified electrodes toward electroanalytical determination of sulfamethoxazole in aqueous environments, *Chemical Engineering Journal*. 352(2018) 188–197. https://doi.org/10.1016/j.cej.2018.06.110.
105. Yin J., Guo W., Qin X., Zhao J., Pei M., Ding F. A sensitive electrochemical aptasensor for highly specific detection of streptomycin based on the porous carbon nanorods and multifunctional graphene nanocomposites for signal amplification, *Sensors and Actuators. B: Chemical*. 241(2017) 151–159. https://doi.org/10.1016/j.snb.2016.10.062.
106. Li C., Zheng B., Zhang T., Zhao J., Gu Y., Yan X., Li Y., Liu W., Feng G., Zhang Z. Petal-like graphene–Ag composites with highly exposed active edge sites were designed and constructed for electrochemical determination of metronidazole, *RSC Advances*. 6(2016) 45202–45209. https://doi.org/10.1039/C6RA01334A.

6 2D Metal Oxides and Their Heterostructures for Gas Chemical Sensing

Katekani Shingange, Steve R. Dima, Modjadji R. Letsoalo, and Eric N. Maluta

6.1 INTRODUCTION

Rapid gas molecule detection at low concentrations is now crucial in a variety of modern applications, including environmental monitoring, interior air quality monitoring, modern farming, food quality monitoring, and medical diagnosis [1–5]. As a result, it has become crucial to develop low-powered, high-performance gas sensors that can identify and track a wide range of gases and vapors, including dangerous and toxic gases and vapors. Typically, gas sensors must adhere to several specifications depending on the applications for which they are designed, including high sensitivity and selectivity, rapid response and recovery, little drift, a low limit of detection, long term stability, and minimal power consumption [6,7]. A variety of sensing technologies have been developed over the past few decades based on various underlying signal transduction mechanisms, such as electrochemical [8,9], thermoelectric [10,11], calorimetric [12,13], and optical devices [14], to address the challenges of low-concentration and quick detection of gas molecules. Among them, electrically transduced devices have drawn a lot of attention due to their ease of use, low cost, ability to monitor signals in real-time, ability to be miniaturized, and compatibility with other common electronic devices [6,15].

To create gas sensors, a range of materials have been investigated, including metal oxides (MOXs) [16,17], carbon nanomaterials [18–20], and conducting polymers [21,22]. Among them, MOXs provide highly accurate, affordable target gas molecule detection. As much as MOXs are an interesting choice of sensing layer material for gas sensors, these materials suffer from a lack of selectivity and high-temperature operation [23]. To address this, researchers have been on a quest to search for innovative gas sensing materials or structures with improved sensitivity and selectivity for operation at low temperatures, preferably room temperature (RT). It has been established that structure dimensions (0D, 1D, 2D, or 3D) of MOX materials can influence the gas sensing performance of the MOX materials [24–28]. The inimitable physical and chemical properties of 2D nanostructured MOX materials, which have lateral dimensions up to several centimeters and thicknesses ranging from a few to tens of nanometers, make them excellent candidates for electrically transducing chemical gas sensors [29–31].

2D MOXs can be split into three groups according to their crystal structure, i.e., layered, lamellar, and non-layered. Layered MOXs include materials assembled from M-O (M: Mn, Mo, or V) octahedral, whereby the in-plane atoms are joined by strong chemical bonding, and the stacking layers are combined by weak van der Waals (vdW) interface [32,33]. Different oxides from Al, Co, Cu, Fe, Gd, Ge, Mn, Ni, and Ti also possess a layered crystal structure with unique planar hexagonal coordination, which is different from their conventional non-polarized crystal structures seen at room temperature [34]. Most ultrathin 2D MOXs do not have a layered crystal structure because all atoms are bound by strong chemical bonds, leading to abundant unsaturated bonds on surfaces or edges, resulting in high-activity and high-energy surfaces such as

DOI: 10.1201/9781003436942-6

ZnO, In_2O_3, WO_3, and TiO_2 [35]. Furthermore, it should be noted that the lamellar MOX consisting of foreign ions incorporated into the atomic layer gaps has also attracted extensive research, of which Perovskite is a typical example [36,37].

MOX-based sensors function based on the well-known resistance change mechanism, which involves adsorption–desorption processes and catalytic reactions occurring on the surface of the MOX, followed by an electronic exchange between adsorbed molecules and the bulk of MOX. A change in resistance/conduction of the MOX sensing layer follows as a result of these reactions, which is proportional to the concentration of the test gas. Plenty of reports for both n- and p-type MOX sensing mechanisms in reducing or oxidizing gases at different environmental conditions have been reported before [38–43]. Normally, surface oxygen species are recognized as active sites that play an important role in the gas sensing process [43]. Oxygen species absorbed on the surface of metal oxides can react with the target molecules, release electrons, and conduct an interfacial redox reaction, resulting in higher oxygen vacancies [44]. However, these MOX gas sensing reactions are not only dependent on the number of oxygen vacancies, but also highly dependent on their local environment, especially the symmetry of oxygen vacancies [45].

Despite considerable progress in the use of 2D MOXs as sensing materials, they still suffer from some inherent limitations. The ability of 2D heterostructures constructed from vdW stacking of 2D vertical heterostructures or edge covalent bonds of 2D lateral heterostructures has proven to not only solve the lack of performance problems of a single 2D MOX-based materials in chemical gas sensors, but also to generate many unique properties through the synergistic advantages of individual 2D MOX-based materials. The selected MOX materials ultimately determine the principles of gas adsorption, reaction, and electrical structure.

Heterostructures are manufactured by integrating the p-type and n-type MOXs, which can combine various properties of the individual materials into a single system. On a microscopic scale, the physical interface between two different materials is called heterojunction. Heterostructures offer combined effects of increased surface reactivity through catalysis/adsorption and charge transfer between heterojunctions, which leads to better gas sensing performance. It is well known from solid-state physics that by contacting two different semiconductor materials at the interface, the Fermi level reaches the same energy, which often leads to charge transfer and depletion layer formation. Although this depletion zone is one of the most notable effects, many other factors may be responsible for improving the gas detection performance of these heterostructures-based gas sensors.

The work in this chapter is structured to discuss current advances in the development of chemical gas sensors using 2D MOX nanostructures and their heterostructures. Within this chapter, we will present an insight into the achievements and limitations of 2D MOX in the field of chemical gas sensors focusing on selected 2D. Experimental and computational perspectives will be used to understand the sensing performance of the 2D MOX-based chemical gas sensors and their functional characteristics. We will first touch on the fundamentals of 2D MOXs, then discuss different types of MOXs and their gas sensing performance looking at both experimental and computational perspectives.

6.2 BASICS OF 2D METAL OXIDES

Based on structural characteristics, 2D MOXs are divided into three categories: layer, lamellar, and non-layer MOXs. The layered MOXs include the traditional layered MOXs (e.g., MoO_3 and V_2O_5), emerging hexagonal MOXs (e.g., TiO_2 and Ni_2O_3), and layered double hydroxides (LDHs). As an example of the traditional layered MOXs, Figure 6.1 shows the structure of MoO_3 displaying a layered structure in which each layer is largely composed of distorted MoO_6 octahedral in an orthorhombic crystal [46].

Layered planar hexagonal MOX includes mono- and few-layered hexagonal, such as TiO_2 and Ni_2O_3 derived from the metal–gas interface [34]. Each atomic layer is composed of a hexagonal

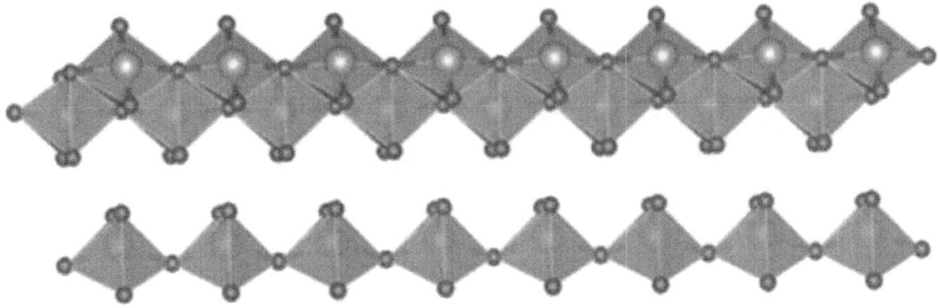

FIGURE 6.1 Crystalline structure of MoO_3. (Produced using software provided by Ref. [47].)

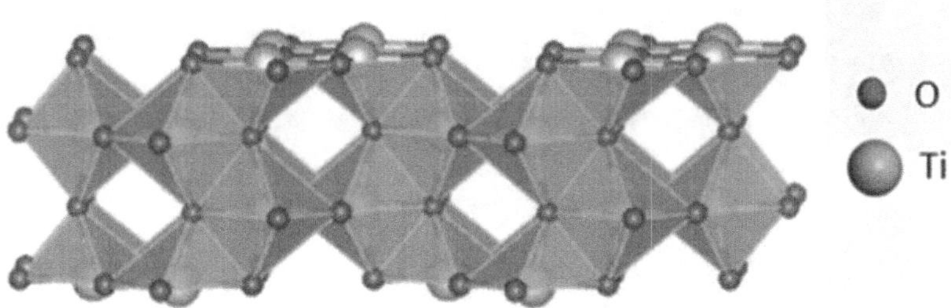

FIGURE 6.2 Crystalline structure of TiO_2. (Produced using software provided by Ref. [47].)

ring that accommodates both the metal and oxygen atoms [48]. Figure 6.2 depicts the crystalline structure of TiO_2 as an example of the layered planar hexagonal MOX.

LDHs are structurally conformed by a consecutive repetition of individual sheets located in parallel spatial planes that are electrostatically bonded by vdW interactions or hydrogen bonds along the perpendicular plane [49]. A schematic interpretation of the LDH structure is displayed in Figure 6.3, which reveals the LDH configuration comprised of stacked lamellas. The chemical composition of LDH materials can be expressed by the following general formula [49]:

$$[M_{1-x}^{2+}M_x^{3+}(OH)_2]_x[A^{n-}]_{x/n}zH_2O$$

where $M^{2+}$and M^{3+} are two metals, M^{2+} can be Mg^{2+}, Ni^{2+}, or Zn^{2+} and M^{3+} can be Al^{3+}, Mn^{3+}, Ga^{3+}, and Fe^{3+}, and A^{n-} is an anion such as Cl^-, CO_3^{2-}, SO_4^{2-}, or RCO_2^-, and x is usually between 0.2 and 0 [49]. LDHs have a lattice structure formed by stacking positively charged brucite-shaped layers, consisting of a divalent metal ion M^{2+} octahedrally surrounded by six $(OH)^-$hydroxyl groups. The substitution of the M^{2+} metal with a trivalent M^{3+} cation gives rise to the periodic repetition of positively charged sheets (lamellas) alternating with charge-counterbalancing A^{n-} ions that allow the electrostatic neutrality of the brucite layers. This charge displacement produces the possibility of dipole–dipole interactions onto the lamella surfaces, enabling interaction with anions or molecules possessing an electrical dipole. Moreover, inside the interlayer space, water molecules are generally accommodated, and a network of hydrogen bonds is present among layers, providing also interacting sites for external molecules. The possibility of intercalating different anions in LDH structure has been largely demonstrated [50], and the possibilities of modifying the chemical, electronic, or optic properties of LDH by changing the anion [51]. In this way, at least in principle, the relative response to various volatile compounds could be tuned by means of anion substitution. LDHs possess a large surface-to-volume ratio and LDH structures can adsorb any kind of molecule [52], making them applicable for chemical gas sensing.

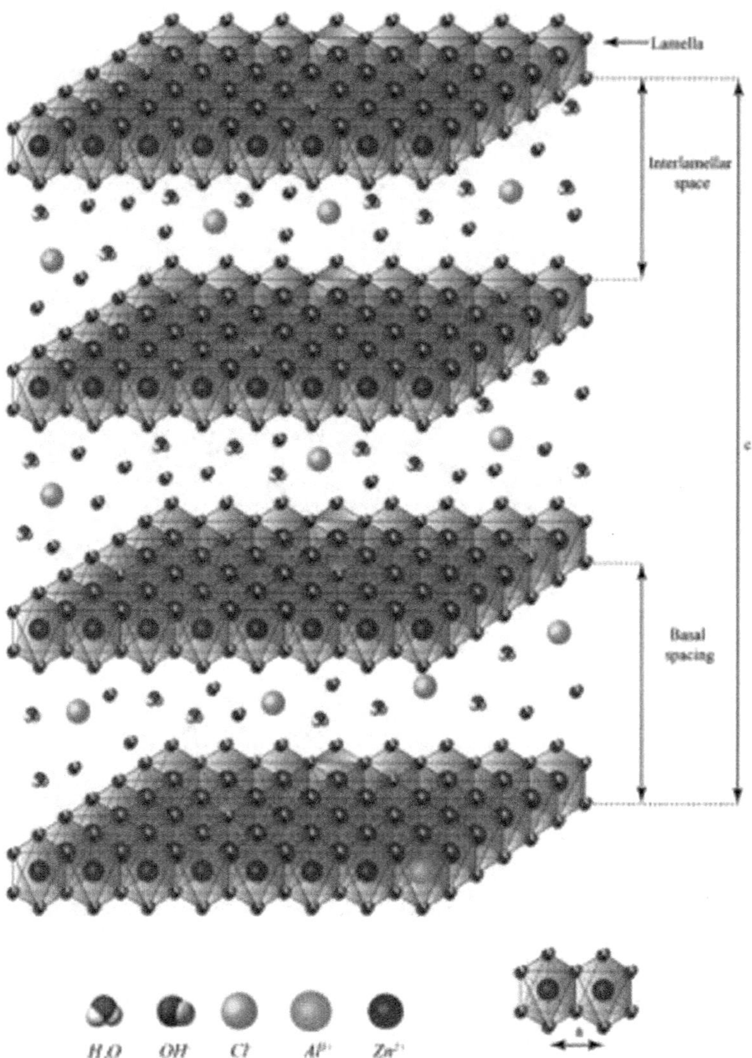

FIGURE 6.3 LDH general structure. (Reproduced with permission from Ref. [53], Copyright © 2017 Elsevier.)

In the lamellar MOXs, atoms or ions are bonded to the oxide layer by weak electrostatic forces. For instance, perovskite $CaTiO_3$ (Figure 6.4) consists of corner-linked TiO_6 octahedra with Ca atoms distributed between the octahedral [54]. At high temperatures, the crystal structure is cubic, but at room temperature, the crystal structure is orthorhombic, with a space group of Pbnm and a Glazer octahedral tilt system of a–a–c+ [55].

The fineness of the atomic scale gives 2D MOX interesting properties such as optical transparency and mechanical flexibility, which can be used in flexible electronics and new optical electronics. Other advantages include high theoretical capacitance, large surface area, and the possibility of oxidation–reduction reactions, making 2D MOX ideal for high-density supercapacitors and batteries [56,57]. Furthermore, their excellent photoelectric properties and high surface chemical reactivity make them suitable for their use as photocatalysts and gas and biosensors [58,59].

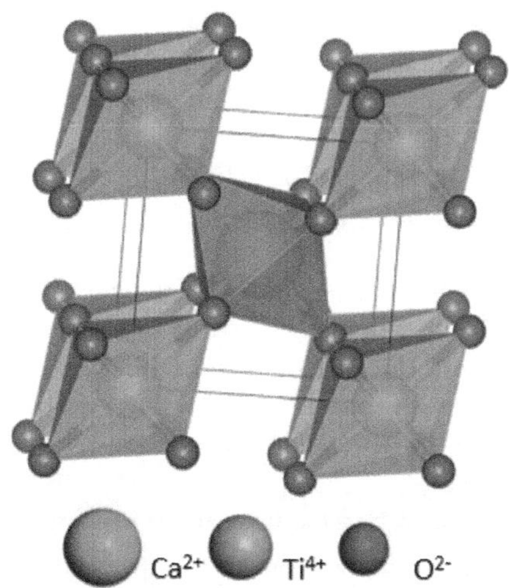

FIGURE 6.4 Crystalline structure of $CaTiO_3$. (Produced using software provided by Ref. [47].)

6.3 THE APPEAL BEHIND 2D METAL OXIDES APPLICATIONS

For MOX-based sensing materials, the morphology influences gas sensing performance as structure dimension (1D, 2D, or 3D) and size affect the material's characteristics, which then influence the sensing performance of the sensing material. 2D MOX nanostructures exhibit a notably greater specific surface area in comparison to their other 0D, 1D, and 3D counterparts. In addition, the gas sensing layers possess distinctive material characteristics due to dimensional confinement, such as the ability to manipulate electrical properties and visible light absorption through quantum size effects that adjust the band gap [60]. Moreover, the mechanical durability and pliability of 2D MOX enable their integration with flexible substrates, exhibiting a high degree of compatibility and facilitating the exploration of novel applications, such as wearable chemical sensors [61,62].

Over the past decade, there has been notable advancement in the fabrication of 2D MOXs ranging from nanosheets, platelets, etc., which have led to a better understanding of the principles behind their engineered synthesis [58,63–65]. This has also resulted in their increased utilization as chemical gas sensing layers. The simplest method is to obtain these 2D MOX structures using various exfoliation techniques from their parent layers, making them different from other low-dimensional nanostructure materials whose synthesis is heavily dependent on crystalline growth.

The thickness of these 2D layers corresponds to the crystal thickness of the individual layer, usually less than 5 nm. The side crystal size of layers of materials along the plane direction remains unchanged, leading to high morphological anisotropy of thickness of sub-nanometer and side dimension up to micrometer. Because the surface-to-volume ratio is close to one, these 2D MOX structures practically expose all components to the surface. This allows the hydrophilic, hydrophobic, surface charge, and surface structure to be adjusted as required for applications [66]. Similarly, there are other techniques such as the top-down approach of soft chemical exfoliation methods that are equally efficient in obtaining 2D MOX structures. In the absence of interlayer interactions and low dimensions, overlaps are observed, improving the surface properties and surface instability of these structures. The description of the structure and properties of any material is essential to understand the resulting behavior and design of unconventional functional materials. This was made possible in recent years by the increase in the capability of the calculation of the First Principle. Similarly, Density Functional Theory (DFT) is widely used to study the intriguing properties of bulk and 2D MOX. However, when working with oxides in 2D form, unique properties begin to develop due to the presence of a large surface area and oxygen atoms on the surface, which leads to significant changes in surface energy that enhance the properties of the material.

The basic atomic structure of MOX determines the nature of conductivity, non-conductivity, and semi-conductivity. The relatively small thickness of 2D material influenced the quantum confinement effect and its effects on electron band structure, charge transport, and optical properties. Due to technological miniaturization, increasing surface is the current need. Consequently, nanostructure MOXs were created by reducing their size to a few nanometers. This change has led to several technological developments toward the development of 2D ultra-thin structures derived from their bulk forms, with significant changes in structure and phase, and improved physiological chemistry properties such as better surface of adsorbents, tuned electronic band structures, and improved photocatalytic behavior.

The predominant avenue of research in the field of gas sensing mechanisms for 2D MOX-based materials pertains to the adsorption of gas molecules. This adsorption may take the form of physisorption or chemisorption, and it is often analyzed through the lens of charge transfer theory. The adsorption of target gases on the surface of 2D MOX-based materials typically results in modifications to the resistance and carrier concentration. In general, 2D MOX materials demonstrate physisorption to gas molecules with relatively low binding energy because of vdW interactions. In the context of chemisorption, the increased binding energy of gas molecules is attributed to the charge transfer mechanism that occurs between the sensing material and the gas molecules. As a result, the recovery process for chemisorption is comparatively prolonged in comparison to physisorption.

It is widely recognized that 2D MOX-based materials exhibit either donor or acceptor behavior toward specific gas molecules. In the context of 2D MOX-based materials, if the Fermi level is situated below the highest occupied molecular orbital (HOMO) state of the adsorbate gas molecules, a phenomenon of charge transfer from the adsorbate to the 2D MOX nanomaterials takes place. This results in an upshift of the Fermi level of the 2D MOX-based materials. In contrast, when the Fermi level of 2D MOX-based materials surpasses the lowest unoccupied molecular orbital (LUMO) state, there is a transfer of charge from the 2D MOX-based materials to the adsorbate, resulting in a reduction of the Fermi level of the 2D MOX-based materials.

The utilization of DFT calculations is feasible for the evaluation of the complete charges transferred and the binding energy that exists between adsorbate and 2D MOX-based materials [67]. In addition, DFT calculations facilitate the determination of the sensing response that occurs when adsorbate and gas molecules interact, based on the calculated total charge and binding energy. It is noteworthy that chemisorption and physisorption can be distinguished by the respective indications of high binding energy and low binding energy [67]. Consequently, it is feasible to anticipate the gas sensing characteristics, such as sensitivity and selectivity toward gas molecules, for the surface of 2D MOX-based materials.

In this section, we will give an overview of current advancements in a variety of chemical gas sensing performances of 2D MOX-based materials in their respective classes, coupled with DFT calculations used to understand the gas molecules' interactions with the surface of the materials. We also compare each of their advantages and weaknesses, placing more emphasis on the overall gas sensor performances to determine the future research directions for better gas sensor performance.

6.4 GAS CHEMICAL SENSING WITH 2D METAL OXIDES

Nanostructured 2D MOX materials including nanoplates, nanosheets, and nanoflakes have advantageous characteristics that cannot be achieved by traditional bulk materials. These characteristics include high surface area and high-surface-to-volume ratio, providing more active sites for gas adsorption, chemisorption, and catalytic activities that control the gas sensing performance. Additionally, for 2D MOX, only one or two crystallographic planes are involved in the sensing performance, thus simplifying the modeling of processes taking place on the MOX sensing material surface, contributing to a better understanding of the influence of sensing performance, and allowing for ways to manipulate better selectivity of the material [68]. Theoretical and experimental results indicate that nanostructured materials with exposed crystal surfaces can exhibit special features in gas sensor applications [27,69,70]. For example, Su et al. [71] fabricated single crystal copper oxide nanoplatelets with a high percentage of {001} facets by a hydrothermal method. X-ray diffraction, Fourier transform infrared spectroscopy, field emission scanning electron microscopy, and high-resolution transmission microscopy were used to characterize the as-prepared materials. From the DFT calculations, it was determined that the {001} facets have the highest surface energy. Due to that, the CuO nanoplatelets revealed high sensing responses to ethanol, acetone, butanol, and isopropanol. In another work by Xue et al. [72], they demonstrated that ZnO nanodishes with exposed (0001) crystal facet sensor exhibited the best response of 49–100 ppm ethanol at 230°C among four as-synthesized samples, while non-customized ZnO was only 28.

Most 2D MOXs have high surface areas and porous structures that provide more gas diffusion channels and more active areas for sensor reactions. Therefore, porous structures improve MOX gas detection properties. Defects in crystal structures, including interstitial and vacancy defects, usually occur in MOXs. In addition to porous structures and high surface areas, surface defects are also considered to be an important factor in MOX gas sensitivity performance. For example, Liu et al. [73] prepared porous ZnO ultrathin nanosheets for the detection of acetylacetone. Their findings revealed that porous ZnO ultrathin nanosheets had excellent selectivity and operational stability, with a response four times higher than the ZnO clusters. They attributed the excellent sensing performance to the high specific surface area and ample surface oxygen vacancy. In another study by Xue et al. [72], the influence of oxygen vacancy (V_O) and zinc interstitial (Zn_i) on the ethanol sensing performance was investigated. The results demonstrated that the ZnO sensor with rich electron donor surface defects Zn_i and V_O displayed greater ethanol gas sensing. These results confirm that surface defects are advantageous for the gas sensing property of MOXs.

Although the advantages of 2D MOX-based materials are interesting, these materials still have some shortcomings when applied in gas sensing. It is difficult to fabricate large and thin 2D MOX nanomaterials using the normal MOX synthesis methods. For a sensor to be able to give a high response, the thickness of the 2D MOX nanostructured material should be comparable with the Debye length (2–5 nm) [39,74,75]; however, this is not the case for 2D MOX materials obtained from traditional synthesis methods; the 2D MOX nanostructured materials are found to be more than 30–50 nm, thus limiting sensitivity. Also, stacking of the 2D MOX nanostructured materials can result in the formation of a denser arrangement, which can hinder gas penetration into the inner voids of the 2D structures, thus decreasing the sensitivity [76]. Therefore, the synthesis of 2D MOX structures must be advanced to be able to yield ultrathin 2D MOXs to maximize their capabilities in gas sensing.

The following sections will discuss different 2D MOX-based materials in detail, as well as their applications in gas sensing, looking at selected examples of 2D-based MOX nanomaterials.

6.5 PREPARATION STRATEGIES OF 2D METAL OXIDES

In general, 2D MOX-based materials are prepared using traditional top-down and bottom-up methods. The top-down synthesis method involves intercalation and exfoliation of bulk-layer MOX in single-layer or few-layer 2D MOX [77]. Bottom-up approaches are a direct growth process that can control the substrate and thickness of desired materials. Wet chemical synthesis, a surfactant-led approach, and self-assembly methods are used in the bottom-up approach [56,78]. In addition, hydrothermal chemical methods can be used to synthesize 2D MOX, with greater yields, easier processing and control, and less energy waste [79,80].

Sun et al. [81] reported a comprehensive and fundamental methodology for the molecular self-assembly synthesis of ultrathin 2D nanosheets of MOXs, such as ZnO, Co_3O_4, and WO_3.

In their work, they proposed a methodology for the surfactant self-assembly of ultrathin 2D MOX nanosheets. The approach involved the formation of inverse lamellar micelles comprising polyethylene oxide–polypropylene oxide–polyethylene oxide (PEO20–PPO70–PEO20, Pluronic P123) surfactant and ethylene glycol (EG) co-surfactant in an ethanol solvent. The molecular assembly process of ultrathin 2D MOX nanosheets from liquid solutions involves the strategic and collaborative self-assembly of MOX precursor oligomers into lamellar structures with the assistance of polymer surfactant molecules. Subsequently, these structures undergo condensation, polymerization, and crystallization to form 2D MOX nanosheets. The synthesized crystallized ultrathin 2D MOX nanosheets were observed to possess limited thickness, a substantial specific surface area, and chemically reactive facets. The UV–vis adsorption spectra of TiO_2, ZnO, Co_3O_4, and WO_3 nanosheets exhibited a blue shift in comparison to their bulk counterparts, indicating a pronounced quantum confinement effect in the thinnest dimension. The phenomenon known as the quantum size effect is observed when the conduction band is raised and the valence band is lowered due to the reduced size of MOX nanoparticles. This is commonly understood to result in a blue shift in the absorption edge.

Different synthesis techniques yield different structures with different functional properties, and therefore, it is important to find or design a synthesis technique that can yield the best characteristics for the application. For gas sensing applications, it would be great to design a synthesis method that can yield materials with high surface area and high carrier mobility. In the following section, the synthesis methods and chemical gas sensing performance of the various MOXs will be discussed.

6.6 GAS SENSING PERFORMANCE

It is important to compare and systematically select the synthesis method according to the type of application. Different methods have different advantages and disadvantages for preparing 2D MOXs. Therefore, it is necessary to choose the method based on its advantages and disadvantages. The following is a discussion on studies for the gas sensing performance applying different 2D MOXs with an emphasis on their methods of fabrication and characteristics correlated with their sensing performance. The experimental and computational perspectives will be considered in the discussion.

6.6.1 Traditional Metal Oxides

Traditional MOXs such as semiconducting metal oxides ZnO [82,83], SnO_2 [84,85], TiO_2 [41,86], CuO [87], and NiO [88] have been extensively studied for chemical gas sensing applications. Particularly in 2D format, traditional MOXs have been synthesized through different techniques. These 2D MOXs have seen applications for the detection of several gases such as NO_2, NH_3, and H_2S [58,89]. Gas sensing mechanisms are known to be based on resistance changes in sensing materials caused by surface oxygen species and target gases. The amount of oxygen adsorbed on the surface is directly affected by the particle size, shape, surface area, surface defects, and interface properties of the sensor material. All these characteristics rely on the method of obtaining the sensing material. For these MOXs, usually the n-type such as ZnO and SnO_2 are favored in contrast to their p-type such as CuO and NiO; this is because even in the same conformations, for instance, rod-shape, the response in p-type MOXs is nearly the square root of the response in n-type [90].

As one of the popular traditional MOXs explored for chemical gas sensing, ZnO is one of the important wide band gap, n-type MOX that is mostly used for chemical gas sensing applications. 2D ZnO structures have been shown to be sensitive to a variety of gases under different conditions. For example, Zhang et al. [91] employed a hydrothermal method to synthesize porous 2D ZnO single-crystal nanosheets with hexagonal wurtzite and mesoporous structures. This was achieved by synthesizing zinc carbonate hydroxide hydrate precursors at a low temperature of 80°C in an environment-friendly hydrothermal method that did not include the use of any surfactant or organic solvent, followed by annealing at 300°C for 0.5h in an air environment. The ZnO nanosheets-based gas sensor unveiled notable sensitivity and rapid response and recovery times when detecting ethanol concentrations within the range of 0.01–1000ppm. The sensor also demonstrated a low detection limit of 10 ppb ($R_a/R_g=3.05$), as well as exceptional selectivity and stability at 400°C. The high ethanol response of the ZnO nanosheets-based sensor was attributed to its large specific surface area, single-crystal structure, plane-contact between nanosheets, and the small thickness (10–40nm thickness). These findings suggest that these nanosheets are highly suitable for the development of ethanol sensors with practical applications.

Oosthuizen et al. [92] used a sonochemical method for the synthesis of p-type CuO nanoplates by the transformation of copper nitrate solution under basic aqueous conditions into CuO nanoplatelets and flower-like nanostructures, without the assistance of any surfactants and additives. Their study demonstrated the effects of reaction time on the diameter of the CuO nanostructured materials, whereby the diameter was found to decrease with increasing reaction time. They also observed that, under the same reaction condition, the modification of the $[Cu(H_2O)_6]^{2+}$ ion complex to $[Cu(NH_3)_4(H_2O)_2]^{2+}$ ion complex, before the addition of NaOH as a base, allowed the surface morphology to change from nanoplatelets to flower-like nanostructures. The as-prepared CuO nanoplatelets were comprehensively characterized in detail using X-ray diffraction, scanning electron microscopy, transmission electron microscopy, Raman spectroscopy, photoluminescence spectroscopy, and Brunauer–Emmett–Teller (BET) surface area analyses. The structural analyses disclosed a size-dependent broadening due to the decrease in platelet size as the reaction temperature was increased. From PL studies, it was observed that the CuO nanostructures displayed several emissions, which were attributed due to various sizes and shapes of CuO nanoplatelets, suggesting that the luminescence property of CuO is dependent on the morphology of the nanomaterials. They further performed a study on the gas sensing performance of the CuO nanoplatelets toward various gases, including CH_4, CO, H_2S, NH_3, and NO_2. Their findings revealed that the surface area, sensing response, and point defects are dependent on the synthesis reaction temperature.

The sensing capabilities of the asymmetric 2D Ga_2O_3 monolayer were examined by Zhao et al. [93], who conducted a study on the adsorption properties of six harmful inorganic gas molecules namely CO, NO, NO_2, NH_3, H_2S, and SO_2, as well as three common ambient molecules namely O_2, CO_2, and H_2O using DFT calculations. The findings indicated that significant variations in adsorption between the upper and lower surfaces were observed due to the inherent internal electric field. The 2D Ga_2O_3 monolayer exhibited distinctive adsorption properties for NO gas molecules. The computed results pertaining to the transport properties of Ga_2O_3 MOSFETs indicated their potential as a viable and recyclable sensor for detecting NO gas. Also, it has been seen that using biaxial strain engineering on a single layer of 2D Ga_2O_3 can make it possible to get gas sensing capabilities that can be changed.

The study conducted by Li et al. [94] utilized DFT computations to examine the adsorption of ten gaseous molecules namely O_2, N_2, NH_3, NO, CO, CO_2, CH_4, C_2H_6, HCHO, and H_2S on three distinct 2D WO_3 nanolayers. They employed a non-equilibrium green function (NEGF) methodology to accurately evaluate the electron transport characteristics, which is a crucial aspect of gas detection. The study's findings also indicated that the 2D WO_3 nanolayers exhibited notable sensitivity and selectivity toward NH_3, NO, HCHO, O_2, and H_2S. The analysis of transport properties, specifically transmission functions and current–voltage (I–V) characteristics, suggested that the electronic device characteristics (ON and OFF) of gas sensors can be effectively achieved through the presence or absence of gas molecules on the 2D WO_3 nanolayers. The binding energies computed indicated that the intermolecular interactions between the gas molecules and 2D WO_3 nanolayers, which were extracted from the (0 0 1) plane of WO_3 bulk crystal, exhibited greater strength compared to those with certain other 2D materials such as graphene and borophene.

6.6.2 Layered Double Hydroxides: Synthesis and Sensing Performance

In recent years, LDHs have been explored for gas sensor applications. This class of materials has attracted attention in the field of gas sensors due to their inherent characteristics such as hierarchical structure, significantly higher stability over a prolonged period, compositional flexibility, anion exchangeability, and a large specific surface that is useful for rapid gas adsorption and desorption [95,96]. LDHs contain anions (e.g., CO_3^{2-} and Cl^-), intercalated between positive layers, which significantly affect their structure. By varying the metallic cations, their ratio, and the interlayer anions, a great number of LDH types can be formed. LDHs structure exhibits a pathway facilitating carrier diffusion; therefore, they received considerable attention for direct potential applications in gas sensors for monitoring a wide class of gases and vapors [42]. Numerous studies related to the synthesis of LDH-based materials have discussed simple and inexpensive methods for synthesis in the laboratory and on an industrial scale [42]. Materials obtained from Pt/Zn/Al LDHs were the first attempts to use LDHs gas sensors whereby changes in the absorbance of all the samples before and after reducing treatments were studied by FT-IR spectra [97]. They reported that due to their memory effect, after calcination, LDHs do not lose their network structure and are able to rebuild it, and this happens after rehydration. Also, the initial structure of LDH does not change after calcination and does not lose its specific surface area. Polese et al. [53] synthesized chlorine-intercalated Zn-Al-LDH and used it for the detection of five volatile compounds, i.e., CO, CO_2, NO, NO_2, and CH_4 at concentrations of 25, 62, 100, 125, 162, and 250 ppm per gas at room temperature (~ 22°C). Each concentration was obtained by diluting in nitrogen CO, CO_2, NO, NO_2, and CH_4 concentrations measured from certified bottles. The sensors were exposed for 60 s to six gas concentrations and then cleaned in wet nitrogen for 500 s. During the entire measurements, the humidity was kept constant at 50% relative humidity (RH) fluxing 100 standard cubic centimeter (sccm) of pure nitrogen through a gas bubbler containing deionized water. The RH level was maintained constant to avoid possible changes in the LDH interlamellar content of water. Finally, every volatile compound was measured in triplicate at room temperature, and the DC sensor resistances were measured by means of an Agilent 34401a multimeter connected to a PC. Under the applied conditions, the LDHs-based sensors showed the ability to differentiate several concentrations with very good short-term stability. The ability of these LDH structures to detect various pollutant volatile compounds at room temperature without the requirements of a heating element is quite interesting in the sensors field as this suggests that the sensors will not require power to operate, thus they are energy-saving. Since this work was one of the very first works on LDHs as sensing materials, the mechanism behind the response could not be fully concluded.

Shinde et al. [30] synthesized a self-assembled Zn-Cr-WO interconnected sheet-like LDH structure to detect gas Cl_2 at room temperature. Exfoliation-restacking-based intercalative hybridization routes have been used to introduce foreign species of polyoxotungstate (POW) in the Zn-Cr-LDH grid, which has led to crucial changes in the chemical composition of host LDH materials. Compared to the original Zn-Cr-LDH, the resulting self-assembled Zn-Cr-WO displayed improved surface characteristics, extended base spacing, and micromesoporous structure. The Zn-Cr-WO sensors were studied at room temperature for various oxidizing and reducing gases, i.e., NO_2, Cl_2, LPG, H_2, H_2S, CO, and NH_3. The selectivity results showed that the Zn-Cr-WO sensor was significantly selective to Cl_2 compared to other interference gases. The lowest POW content of Zn-Cr-WO-1 nanohybrid (Zn-Cr-LDH/POW=0.66) had excellent dynamic response recovery characteristics, with a gas detection limit of 0.1 ppm and a 66.6% response rate to Cl_2. Compared to a variety of Cl_2 sensors such as SnO_2, ZnO, In_2O_3, WO_3, and SWCNT, the LDH-based sensors have shown a lot of promising functions in gas sensor applications, with excellent sensing responses and superior room temperature repeatability [30]. They attributed the great Cl_2 sensing performance to the unique interconnected sheets-like mesoporous microstructure, which facilitated excellent permeability for gas adsorption. The structure also provided enhanced surface area with active surface sites for chemical interaction. Additionally, with expanded interlamellar spacing of 0.73 nm due to the intercalation of foreign POW fractions, Zn-Cr-WO-1 exhibited wider mesopore channels that continuously accelerated the gas molecules adsorption in the inter-gallery space between laminated LDH nanosheets [30]. LDHs have unique physicochemical properties such as high surface area and high porosity, anion and cation exchange capacity, and adjustable structure engineering with high controllability. LDHs have drawbacks including agglomeration and low conductivity, which make their use in gas sensing limited.

6.6.3 Perovskites Oxides: Synthesis and Sensing Performance

Another family of MOX with layered 2D structures is the Perovskites family with a general formula of ABO_3, with A and B cations of different sizes [98]. Perovskites consist of three types of layered phases, which are the Aurivillius (AU) with general formula $[Bi_2O_2]$-$[A_{(n-1)}B_nO_{3n+1}]$, Dion-Jacobson (DJ) with general formula $MA_{(n-1)}B_nO_{(3n+1)}$, and Ruddlesden–Popper (RP) with general formula $A_{n+1}B_nX_{3n+1}$ [99]. In the Ruddlesden–Popper Perovskite (RPP), R is a long chain alkyl or aromatic group that acts as a spacer covering the perovskite layer, and because of isolation, the moister resistibility increases. The number of perovskite layers between spacers is denoted as n [100]. 2D MOX perovskites have several advantages, including air, phase, and thermal stability. Improvements in air stability in 2D perovskites are associated with vdW interactions of capping agent molecules [100].

Despite the superiority of 2D MOX perovskites, the quantum confinement effect increases the band gap of 2D MOX perovskites compared to other dimensions. Furthermore, due to the isolation characteristics of organic space cations, 2D MOX perovskites exhibit high-sensitivity transportation behavior, high inorganic conductivity, and low layers. The physical characteristics of 2D or almost 2D perovskite materials are determined primarily by the bulky organic cations present in the system. These cations play a crucial role in defining the dimensionality and phase behavior of the material, as demonstrated in the studies conducted by Saparov et al. [101]. The electronic characteristics of the resulting materials are solely determined by the properties of the inorganic layers due to the substantial HOMO/LUMO gap exhibited by these organic cations. Maheshwari et al. [102] employed DFT calculations to demonstrate that the incorporation of electron-withdrawing and electron-donating molecules results in the creation of localized states, which can be observed in either the organic or inorganic component. Additionally, it has been demonstrated that the energy levels of the bands located in the organic and inorganic components can be adjusted separately. Modulation of organic cation levels can be achieved through alterations in the electron-withdrawing or -donating properties, while the manipulation of energy levels in the inorganic component can be accomplished by adjusting the number of inorganic perovskite layers.

Recently, various perovskite-based sensors have been reported to detect several gases such as NH_3 [103,104], NO_2 [105], and O_2 [106] and humidity [107]. When a film is exposed to gas/vapor, the gas/vapor can fill the gaps in the film, thereby significantly increasing the film's conductivity. In contrast, when inert gas/vapors reach the perovskite film, the inert gas/vapor molecules remove the target gas/vapor molecules and increase the number of vacancies, thereby reducing the conductivity of the perovskite film [105,107]. Although 3D perovskites have made significant advances in device efficiency, due to their poor stability in water, oxygen, heat, and continuous light exposure, their optical and long-term stability is limited [105,106,108,109]. On the other hand, 2D perovskites are more interesting because of their good stability, unique structure, and excellent optical and electron characteristics [110]. Different findings have previously been reported highlighting the potential for the use of 2D perovskites to produce gas sensors, such as those by Tien et al. [111], whereby they compared 2D-$(PEA)_2PbBr_4$ perovskite with 3D-$MAPbBr_3$ perovskite. They found that the horizontal vapor sensor of 2D-$(PEA)_2PbBr_4$ perovskite was much superior to 3D-$MAPbBr_3$. They attributed high responses to a large ratio of surface to volume and showed a 2D transverse perovskite layer suitable for ethanol detection. Another study by Wang et al. [37] reported on the comparison of 2D $NbMoO_6$, $NbWO_6$, $TaWO_6$, and $TaMoO_6$ nanosheets synthesized through liquid exfoliation and their gas sensing performance toward H_2S and other small molecules of volatile gases. The gas sensor performance toward 5–500 ppm H_2S revealed that, among the four perovskites with similar structures, nanosheets made from n-type $NbWO_6$ showed high sensitivity to H_2S, with response and recovery time of 6 and 30 s at 50 ppm H_2S at a low operating temperature of 150°C. The H_2S sensor performance of the 2D $NbWO_6$ perovskite was attributed to the high surface areas with nearly fully exposed active sites, which significantly facilitated the H_2S gas molecules adsorption–desorption, diffusion, and transmission for high selectivity under low operating temperature. Another reason was the high crystalline framework of the 2D $NbWO_6$ nanosheets that promoted fast transport of electrons between surface and bulk for significantly enhanced response and response dynamics and sensitivity. The electronic transfer associated with the gas sensing process was further explained using the DFT calculations. The DOS of the $NbWO_{6+}H_2S$ showed a new energy level in the conduction band and the band gap was narrowed, which was attributed to the strong bonding adsorption of H_2S on the $NbWO_6$ nanosheets, which decreased the resistance for high selectivity and sensitivity. The adsorption configurations of the (110) plane on the $NbWO_6$ nanosheets were optimized, and no obvious changes in configurations were observed after the adsorption of gas molecules onto the $NbWO_6$ nanosheets. These $NbWO_6$ nanosheets could find application for fast and effective detection of H_2S, especially for the environmental standard set threshold of H_2S which is at 0.1–100 ppm [112].

6.7 HETEROSTRUCTURES OF 2D METAL OXIDES

By growing 2D MOXs monolayers and restacking them into blends with 0D, 1D, 3D, or other 2D nanostructures, different types of heterostructures can be created. Together with the ability to overcome the immanent restrictions of each component material, the many ways that 2D MOX nanostructures can be mixed with other materials create new opportunities to create devices configurations with a flexible electronic property that goes beyond a single material's scope [113,114].

Due to their intriguing features that are absent from bulk semiconductor heterojunction devices, heterostructures of layered 2D MOX nanostructured materials show outstanding potential for applications in many fields. Initial attempts to combine 2D layered nanomaterials with 0D and 1D nanostructures have opened a new area of nanoscale material integration and opportunities to develop new applications for devices that offer exceptional performance [115–118]. Since then, a lot of work has been done on the fabrication of 2D–2D heterostructures made by vertically and laterally stacking several 2D layered nanostructured types. These initiatives have made it possible to control and modify the generation, confinement, and transportation of charge carriers [119,120].

Beyond 2D–2D heterostructures, 2D–3D heterostructures, created by combining unique functionalities of 2D layered materials with standard 3D bulk materials, utilize both the benefits of 3D bulk materials as well as their innovative functions [121]. Although heterostructure and heterojunction are sometimes used interchangeably in the literature, a heterostructure device typically combines several heterojunctions. Normally, when two different materials are connected electrically, a heterojunction is formed at the boundary of the two materials. Charge transport via the interface causes a layer of depletion that results in band alignment because Fermi levels through the heterointerface tend to be balanced at a fixed energy level. The most important step in the heterostructure-based devices used for gas sensing is the adsorption and diffusion of analyte gas toward the heterojunction, which further changes junction properties and improves the sensor response by synergy effects [122,123].

6.8 THE APPEAL BEHIND METAL OXIDES-BASED HETEROSTRUCTURES

Compared to pure MOXs, sensors based on heterojunctions of MOXs show improved gas sensing performance toward the analyte gas. In contact with each other, the transfer of the carrier between the two materials is induced by the inconsistent Fermi levels at the interface. In n–n or n–p heterojunctions, the Fermi values of the two MOXs move to equilibrium, causing the energy bands to bend and create potential barriers between them. The gas sensing performance of the investigated MOXs is mainly explained by redox reactions of the adsorption analyte gas on the surface of the sensor material. The changes in carrier concentration caused by the redox reaction on the composite surface may affect the height of the integrated potential barrier. This process has an additional effect on the resistance or conductivity of the sensor, which is based on n–n or n–p heterojunction as follows [124]:

$$\Delta R \propto \exp\left\{-\frac{\Delta eV_b}{K_B T}\right\} \tag{6.1}$$

where ΔR is the resistance change of the sensor, ΔV_b is the reduction of the height of the potential barrier, k_B is the Boltzmann constant, and T is the temperature [124,125]. As a result, small changes in the height of potential barriers would greatly affect the resistance of the sensor under study, which would improve the gas sensing properties of the heterojunction. In p–n or p–p heterostructures/heterojunctions, the interaction between the target gas and the sensing material also changes the sensor charge carrier, particularly holes, resulting in an additional variation in the thickness of the heterojunction accumulation layer and a more effective modulation of the carrier's conductor channel width. Therefore, sensors based on p–n or p–p heterojunction sensors also have improved the gas detection properties of reduced or oxidizing gases [117,126]. In addition, heterostructures consisting of heterojunctions of MOXs always have a specific surface higher than pure MOXs. The higher surface area allows the gas molecules to diffuse smoothly to the surface, interact more easily with the components, and provide more active sites. Using high specific surface areas increases the size of pores, facilitating the diffusion of gases into sensor materials, and increases the active surface of the internal component of the heterostructure for the adsorption of gases. Gas molecules' adsorption and desorption can also be accelerated in response and recovery processes based on the heterojunction of the MOXs. Thus, the high specific surface area is another positive factor that contributes to improved gas sensing performance in the heterostructure. Compared to the effects of specific surface areas, the study of enhanced gas detection mechanisms by heterojunctions of the MOX sensor materials is more complex. The role of heterojunctions in improving gas sensor performance should be analyzed in detail to fully understand their direct and significant effects on improving gas sensor performance. The gas sensor mechanism of MOX to common reducing and oxidizing gases was extensively discussed in the literature, and the effects of different often studied heterojunctions on improved gas sensor properties of composites were systematically investigated [123,126].

6.9 GAS SENSING PERFORMANCE OF 2D METAL OXIDE HETEROSTRUCTURES

Because it improves reaction and adsorption sites and results in higher catalytic activity than a single material, the development of heterostructures is a preferred method for improving gas sensing performance. When two different materials are combined, such as n–n, p–p, or n–p semiconductors, the interfaces have a synergistic effect that is equal to Fermi energy due to the electron transfer from the highest energy to the lowest energy level that is empty. Due to the band bending caused by different Fermi energies, a potential energy barrier is created in the contact. To overcome this potential energy barrier, electrons must cross the area of depletion (interface). The energy band diagram and heterojunctions manufactured with semiconductors n and p are shown in Figure 6.5. At the interface, an electron depletion layer with higher-energy conductivity band states is formed due to electron loss, as well as an electron accumulation layer with lower-energy conductivity band states [127]. Most of the charge carriers transferred from the higher to the lower state of the valence band at the p–p metal oxide heterojunction interface (Figure 6.5b) are holes containing the higher and lower states of the valence band. Due to its stability in low oxygen conditions and its compatibility with the measurement system, n-type materials are increasingly used [90]. A p–n heterojunction

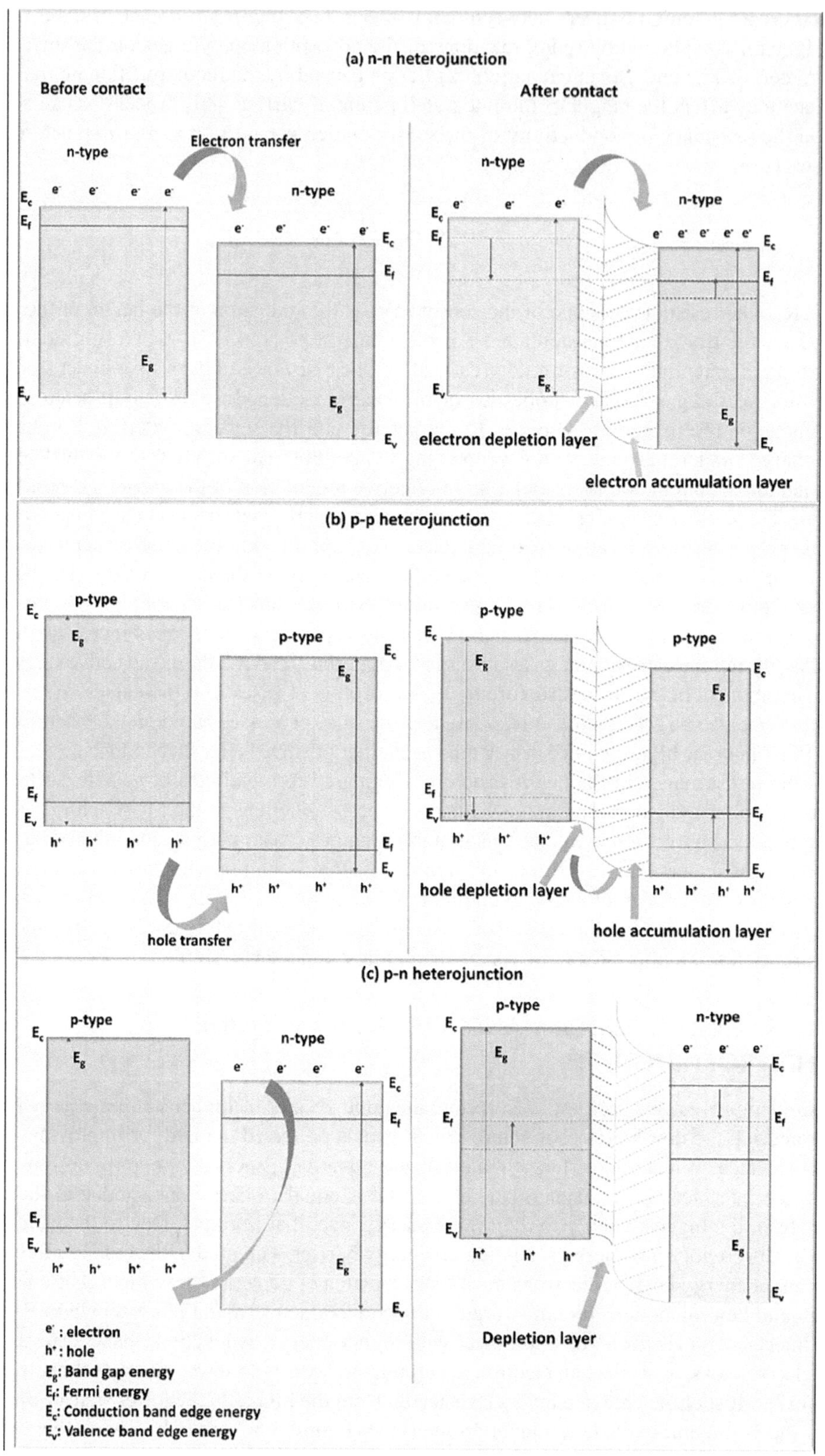

FIGURE 6.5 Types of heterojunctions build from (a) n–n, (b) p–p, and (c) p–n junctions.

can improve the performance of gas sensing. Depending on the backbone material, electron–hole recombination causes electron transfer from n-type to p-type or hole transfer from p-type to n-type at an n–p or p–n heterostructure interface (Figure 6.5c). This allows for more oxygen to be absorbed because of the higher electron density [122]. In comparison to n–n or p–p heterojunctions, these heterostructures have more free electrons [40,122]. There is a space charge area at the p–n interface because n-type materials often have intrinsic Fermi levels that are greater than p-type materials. Heterojunctions can accelerate electron transport and improve oxygen adsorption, resulting in an abundance of oxygen vacancies on the heterostructure surfaces and providing new active sites for higher sensing performances.

6.9.1 Heterojunction p–n Type Sensing Performance

A p–n heterojunction is created by combining a p-type semiconductor material with an n-type semiconductor material. This interface causes a bending of the energy bands and modifies the electronic properties at the junction. For example, Ju et al. [128] reported on the synthesis of ZnO/NiO heterostructures using a pulsed laser deposition technique, whereby they attached p-type NiO nanoparticles onto the surface of n-type ZnO nanosheets. When pure ZnO nanosheets are exposed to air at a high temperature, a depletion layer will form on the surface of the ZnO nanosheet due to the adsorption of oxygen molecules, leading to the high resistance state of sensing materials, as shown in Figure 6.6a and b. When NiO nanoparticles are embedded in the ZnO nanosheet surface, the electrons in ZnO and the holes in NiO diffuse in the opposite direction because of the great gradients of the same carrier concentration. Afterward, the internal electric field is generated in the interface of ZnO/NiO, and the spread of the carrier is finally balanced. After this, the energy band bends into the depletion layer until the system achieves a Fermi (E_F) level equilibrium (Figure 6.6e). Furthermore, the formation of the NiO/ZnO p–n interface with a new depletion layer at the interface increases the resistance of the sensor to further increase the air (Figure 6.6f). However, once the NiO/ZnO sensor is exposed to a reducing Triethylamine (TEA) gas, TEA reacts with oxygen ions absorbed on the surface of the ZnO nanosheet and releases electrons back to ZnO. Thus, the sensor resistance decreases. Furthermore, TEA releases electrons in NiO p-type and electron–hole recombination leads to a decrease in hole concentration. The decrease of holes in NiO results in the increase of electrons and reduces the concentration gradient. The reduction of holes in NiO increases electrons and reduces the concentration gradient of the same carrier on both sides of the p–n connection. As a result, the diffusion of the carriers is weakened, and the interface depletion layer becomes thin. Thus, the resistance of the NiO/ZnO sensor in TEA is even reduced. Figure 6.6g shows a model of the NiO/ZnO-based sensor when exposed to TEA gas. In comparison to the ZnO sensor, the formation of p–n junction in the NiO/ZnO-based sensor greatly increases air resistance and reduces TEA gas resistance. Therefore, based on the sensor response definition ($S = Ra/R_g$), the increase in S to TEA is mainly due to the resistance variation caused by the formation of p–n junction.

6.9.2 Heterojunctions p–p and n–n Types Sensing Performance

The creation of p–p and n–n heterojunctions provides band bending energy levels in an alike method as in p–n heterojunction. For p–n heterojunction, few electrons are involved at the interface due to electron–hole recombination that raises the resistance, whereas at the interface of n–n type heterojunction, there is just the transport of electrons from a material with higher Fermi level to lower Fermi level unoccupied state, making an accretion layer instead of a depletion layer. This accretion layer can be worn out by the successive oxygen adsorption on the surface, which certainly raises the potential energy barrier at the interface and enhances the sensing response.

In a study by Yang et al. [129], n–n type heterostructure based on CeO_2/SnO_2 was prepared and tested for 3H-2B sensing capabilities. Referring to Figure 6.7b, they reported that the Fermi

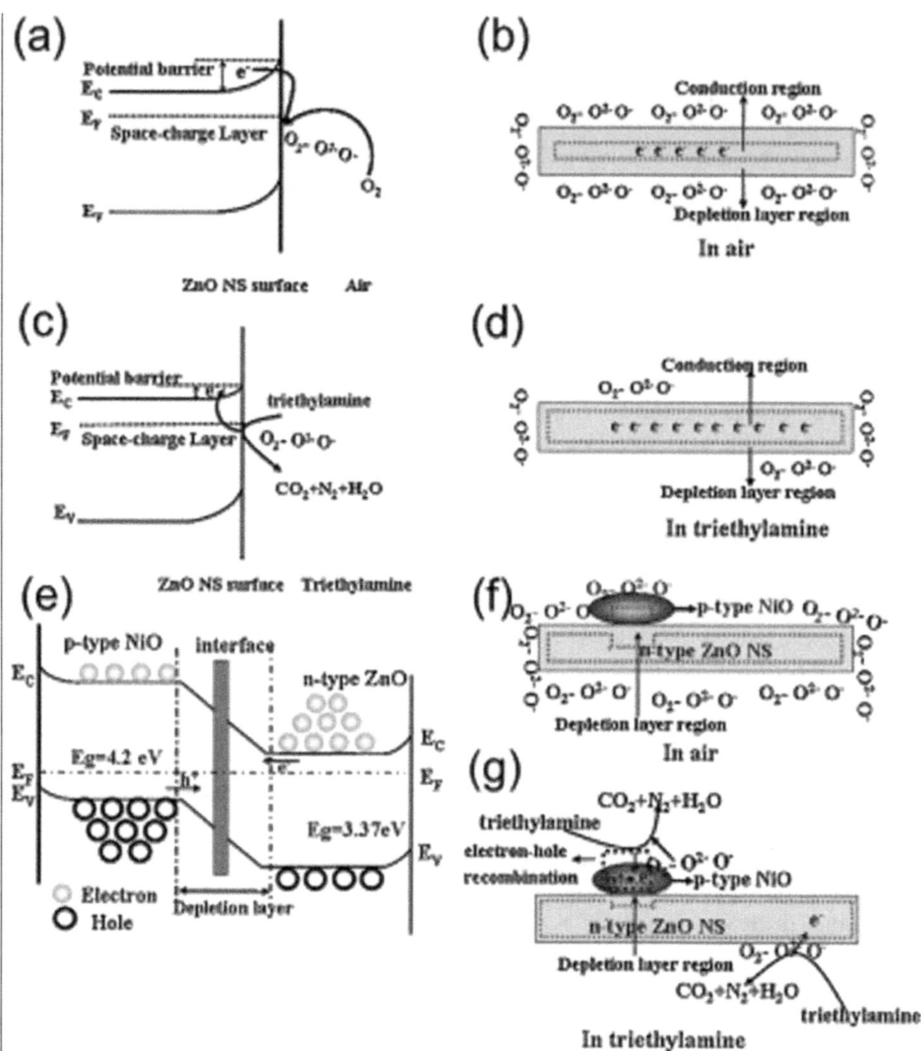

FIGURE 6.6 (a, b) The energy band diagram of ZnO nanosheet and its schematic model in air. (c, d) The energy band diagram of ZnO nanosheet and its schematic model in TEA. (e) The energy band diagram of p-type NiO and n-type ZnO heterostructure. (f, g) Schematic model for the ZNS sensor exposed to air and TEA gas, respectively. (Reproduced with permission from Ref. [128], Copyright © 2014 Elsevier.)

level of CeO_2 was higher than that of SnO_2, resulting in the electron transfer from CeO_2 to SnO_2 until achieving the equilibrium states of their Fermi levels. This would make an n–n type heterojunction at the interface and a depletion layer formed at the interface between CeO_2 and SnO_2 and increase the potential barrier height built into the air due to oxygen adsorption. Normally, the oxygen molecules prefer to be adsorbed on the asymmetric Ce-O-Sn sites to predominantly dissociate to active O^-(ads) species on the surface of the porous CeO_2/SnO_2 nanosheets. When 3H-2B gas is introduced, it interacts with the adsorbed oxygen species on the surfaces, and electrons are released back to the conduction band of CeO_2/SnO_2 nanosheets. The released electrons would decrease the thickness of the depletion layers between CeO_2 and SnO_2, further resulting in a decrease in the height of the potential barrier. This process would increase the conductivity of the sensor and significantly enhance the 3H-2B sensing performance of the CeO_2/SnO_2 nanosheets. Coupled with results from DFT, they showed that the active O^-(ad) species originated from the

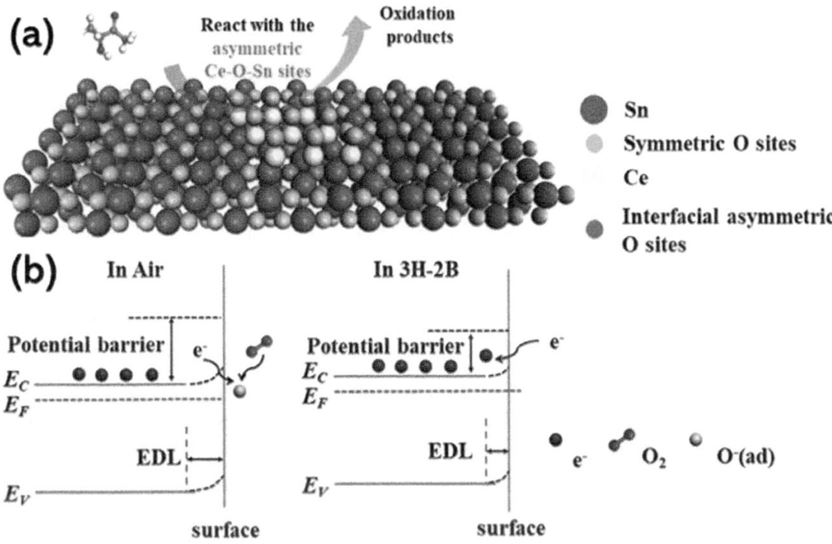

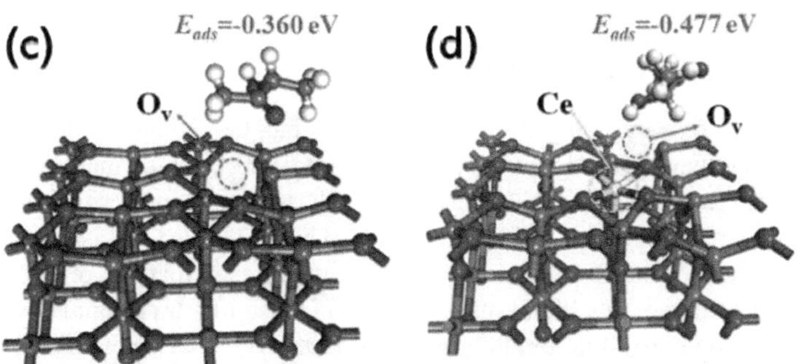

FIGURE 6.7 (a) Demonstration of the surface sensing mechanism of 3H-2B on CeO_2/SnO_2 nanosheets surface, (b) demonstrations of the 3H-2B sensing mechanism band diagrams for CeO_2/SnO_2_400 nanosheets, (c) adsorption configurations of 3H-2B adsorbed on O_v of SnO_2, and (d) O_v near asymmetric Ce-O-Sn sites of CeO_2-SnO_2 interface. (Reproduced with permission from Ref. [129], Copyright © 2022 Elsevier.)

asymmetric Ce-O-Sn sites, as the adsorption energy (E_{ads}) for 3H-2B on (110) plane of SnO_2 was calculated to be −0.477 eV for asymmetric Ce-O-Sn sites and that of symmetric Sn-O-Sn sites was −0.36 eV, as depicted in Figure 6.7c and d, which further enhanced the adsorption and reaction kinetics of 3H-2B on the CeO_2/SnO_2 nanosheets surface, thus resulting in quicker response and recovery times. The porous 2D MOX nanosheet structure in this case endowed the materials with more surface-active sites and a much shorter diffusion pathway, which is highly desirable for the diffusion and adsorption of 3H-2B molecules on the surface of the material, further improving their sensing performance. It was also demonstrated that the local environment of oxygen species influences the sensing performance.

Porous MoO_3/SnO_2 nanoflakes with n–n junctions were also reported by Gao et al. [130] to demonstrate an improved gas sensing property with a higher gas sensor response of 43.5 toward 10 ppm H_2S at 115°C compared with that of the pure SnO_2. The improved H_2S sensing performance was attributed to the formation of n–n junctions, which played an important role in enhancing the sensing characteristics of MoO_3/SnO_2 nanoflakes. Also, the BET surface area of the heterostructure was found to be around 1.8 times larger than that of the pure SnO_2 nanoflakes. Because of this, more gas molecules were able to participate in surface reactions leading to a greater change in the resistance when the heterostructure was exposed to different gas atmospheres. Additionally, the heterostructure nanoflakes possessed larger pore volume than that of the pure SnO_2 nanoflakes and larger pore volume advances gas diffusion, facilitating the enhancement of H_2S sensing performance [130]. Other remarkable achievements in the use of n–n type MOXs for chemical gas sensing have been reported including WO_3@SnO_2 core–shell nanosheets [131], SnO_2:CeO_2 nanosheets [132], and TiO_2/SnO_2 nanosheets [133]. Their sensor performance has been summarized in Table 6.1.

In the p–p type heterojunction, the gas sensing is similar, but the dominant charge carriers are holes and the change in resistance after interaction with analyte gases is reversed in comparison to the n–n type heterojunction. For example, after the heterojunction of NiO-CuO [134] forms at the interface between NiO and CuO (Figure 6.8), a hole depletion layer and a hole accumulation layer form at the interfaces of NiO and CuO. The adsorbed oxygen ions create holes in the hole depletion layer, making them thinner than pure NiO and less resistant. The reaction between H_2S molecules and deposited oxygen ions fills holes and increases the resistance measured when the heterojunction is exposed to H_2S.

Table 6.1 summarizes some of the works found in the literature on 2D MOXs and their heterostructures-based sensing materials. The gas sensing performance including the operating temperature, response, analyte gas and concentration, response, and recovery rate has been summarized. A lot of progress has been made in the chemical gas sensor space, particularly in the use of MOX-based materials, and as evidenced by the referenced works.

6.10 CONCLUSION AND FUTURE OUTLOOK

Over the past two decades, research into 2D MOX materials has been ongoing, and all literature works have paved the way, leaving no doubt about the potential of 2D MOX materials for sensing applications. The research in the sensor technology space is based on a long-term goal of commercializing sensing devices that contain 2D MOX materials. The key industrial aspects to get these devices commercialized are to find 2D MOX development techniques that are reliable and are of low-cost production and scalable. 2D MOX materials are made up of traditional MOXs such as ZnO, TiO_2, and NiO, and the perovskite oxide family and LDHs do have a potential for industrial production based on their characteristics that are beneficial for gas sensing. However, traditional MOXs suffer from high operating temperatures, making them highly energy-consuming. Even though LDHs have unique physicochemical properties such as high surface area and high porosity, anion and cation exchange capacity, and adjustable structure engineering with high controllability, they have drawbacks including agglomeration and low conductivity, which make their use in gas sensing to be limited.

Progress in the manufacturing process of 2D MOX materials has enabled the pairing of 2D structures with different dimensions to form heterostructures. The idea of combining various MOX materials to form heterostructures is relatively recent and stems from the need to improve the selectiveness and other important sensor parameters of chemical gas sensors, such as operating temperature, selectivity, and stability. Heterostructure materials present strong interactions between closely packed interfaces, giving superior performance compared to the construction of single MOX materials. It has also been widely reported that smaller molecules exhibit higher reactions in microstructures of similar sizes, thereby demonstrating higher selectivity. Since wide-band-gap materials generally have fewer reactions, heterostructure materials can be selected to increase surface reactivity through surface–gas interactions. 2D MOX heterostructures-based materials have shown higher possibilities of being used for developing high-performance sensors for detection of different analytical gases. Nonetheless, many discoveries and detailed research remain to be made to fully recognize the opportunities of these materials to realize commercialization. The utilization of DFT

TABLE 6.1
Performance of Gas Sensors Based on 2D Metal Oxides Nanostructures and Their Heterostructures

2D Metal Oxide	Method	Gas	Temp. (°C)	Response	T_{res}/Rec (s)	Ref.
ZnO nanosheets	Hydrothermal	Acetylacetone, 100 ppm	340	191.1	19/94	[73]
ZnO nanosheets	Solvothermal	NO_2, 10 ppm	200	74.68	–	[135]
In_2O_3 nanosheets	Hydrothermal	NO_X, 10 ppm	120	213	4/10	[136]
SnO_2 nanosheets	Hydrothermal	HCHO	240	3	–	[31]
MoO_3 nanosheets	Exfoliation	C_2H_6O	300	33	21/10	[137]
MoO_3, nanosheets	Hydrothermal	DIPA, 10 ppm	217	30.4	4.3/–	[138]
WO_3 nanosheets	Chemical bath deposition	NO_2, 10 ppm	100	460	54/63	[139]
NiO nanosheets	Hydrothermal	NO_2, 20 ppm	250	80%	–	[140]
NiO lotus-root slice-shaped	Hydrothermal	N_2H_4, 100 ppm	92	107.6	50/29.6	[141]
Co_3O_4 nanosheets	Hydrothermal	Acetone, 100 ppm	150	11.4	150-41/300-65	[142]
Co_3O_4	Solvothermal	Xylene, 100 ppm	150	74.5	–	[143]
$NbWO_6$ Perovskite	High-temperature calcination	H_2S, 50 ppm	150	12.5	6/30	[37]
$LaCoO_3$ - ZnO	Sol–gel method	Ethanol,100 ppm	320	55	2.8/9.7	[144]
LDH- Ag/LDH	Glucose reduction	Methanol, 100 ppm	RT	3.35	10/20	[145]
Pd/TiO_2 nanosheets	GO template	H_2, 1000 ppm	230	9.1	1.6/1.4	[146]
WO_3-rGO nanoflakes	Hydrothermal	NO_2, 10 ppm	90	4	10/9	[147]
TiO_2/SnO_2 nanosheets	PLD	TEA, 100 ppm	260	52.3	12/22	[133]
SnO_2:CeO_2 nanosheets	Hydrothermal	Ethanol, 100 ppm	340	44	25/6	[132]
NiO/Co_3O_4 nanosheets	Hydrothermal	Xylene, 100 ppm	140	12.27	22/55	[148]
ZnO-SnO_2 nanosheets	Co-precipitation and decomposition	Ethanol, 50 ppm	240	80	7/42	[149]
CeO_2/ZnO nanosheets	Hydrothermal	Ethanol, 100 ppm	310	90	20/4	[150]
Co_3O_4/MoS_2 nanosheets	In-situ anchored	NH_3, 5 ppm	RT	2.1	105/355	[151]
Pt-$BiVO_4$ nanosheets	Colloidal	Acetone, 100 pm	300	12.5	–	[152]

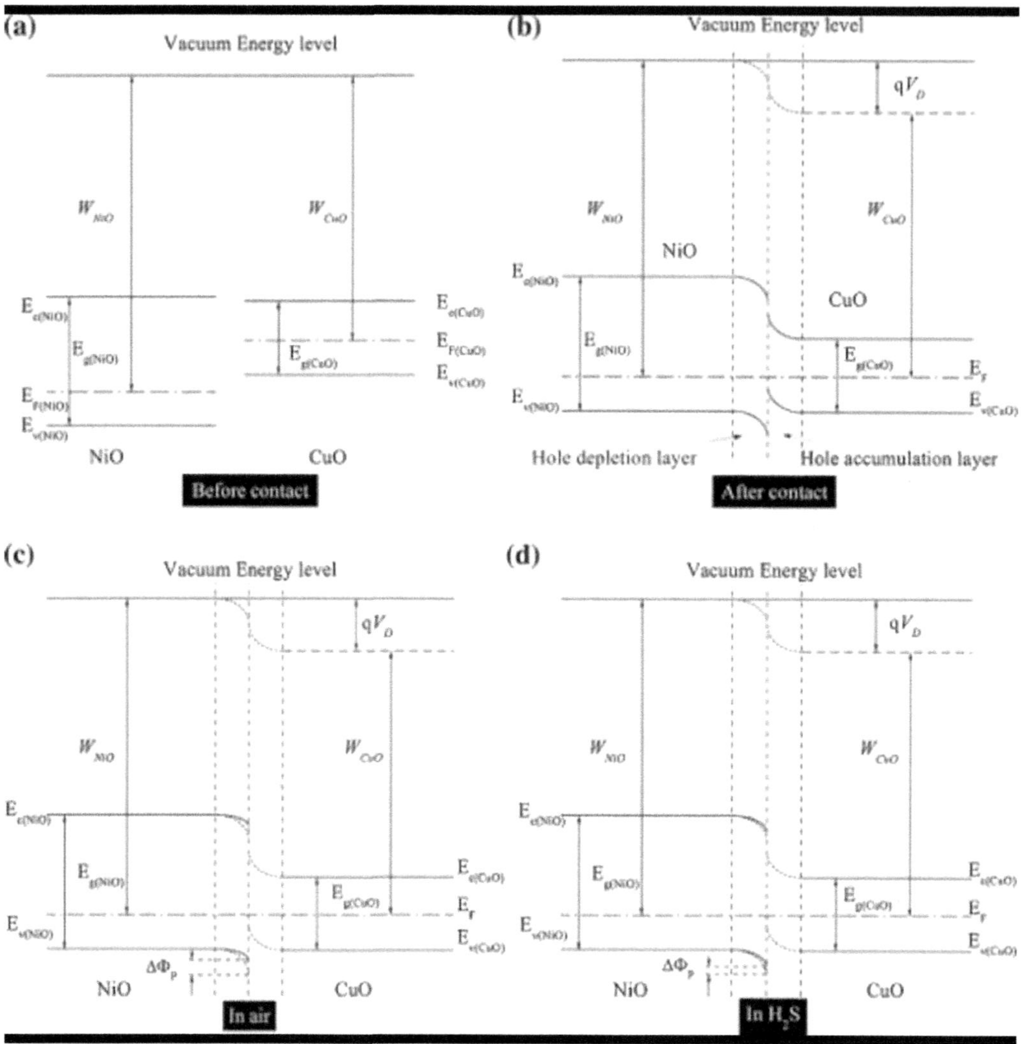

FIGURE 6.8 Schematic diagrams of the energy band structure of NiO and CuO (a) before and (b) after contact. CuO-NiO heterostructures (c) in air and (d) in H_2S. (Reproduced with permission from Ref. [134], Copyright © 2015 Elsevier.)

studies will significantly contribute to the rapid and comprehensive advancement of 2D MOX materials in the field of gas sensing. This is primarily attributed to DFT's capability to offer a profound understanding of atomic geometries and the underlying chemical bonding, thereby serving as a valuable tool for understanding surface–adsorbate interactions through the precise determination of their adoption energy and electronic properties.

ACKNOWLEDGMENTS

The authors would like to acknowledge the Department of Science and Innovation (DSI), MINTEK Advanced Materials Division (AMD), Council for Scientific and Industrial Research (CSIR), and University of Venda, South Africa for financial support.

REFERENCES

1. Kgomo, M.B., Shingange, K., Nemufulwi, M.I., Swart, H.C., Mhlongo, G.H. Belt-like In_2O_3 based sensor for methane detection: Influence of morphological, surface defects and textural behavior. *Materials Research Bulletin*. 158 (2023) 112076. https://doi.org/10.1016/j.materresbull.2022.112076.
2. Nemufulwi, M.I., Swart, H.C., Shingange, K., Mhlongo, G.H. $ZnO/ZnFe_2O_4$ heterostructure for conductometric acetone gas sensors. *Sensors and Actuators B: Chemical*. 377 (2023) 133027. https://doi.org/10.1016/j.snb.2022.133027.
3. Shingange, K., Swart, H.C., Mhlongo, G.H. Enhanced ethanol sensing abilities of fiber-like $La_{1-x}Ce_{x}$-CoO3(0 ≤x≤0.2) perovskites based-sensors at low operating temperatures. *Sensors and Actuators B: Chemical*. 377 (2023) 133012. https://doi.org/10.1016/j.snb.2022.133012.
4. Talwar, V., Singh, O., Singh, R.C. ZnO assisted polyaniline nanofibers and its application as ammonia gas sensor. *Sensors and Actuators B: Chemical*. 191 (2014) 276–282. https://doi.org/10.1016/j.snb.2013.09.106.
5. Thepnurat, M., Chairuangsri, T., Hongsith, N., Ruankham, P., Choopun, S. Realization of interlinked ZnO tetrapod networks for UV sensor and room-temperature gas sensor. *ACS Applied Materials & Interfaces*. 7 (2015) 24177–24184. https://doi.org/10.1021/acsami.5b07491.
6. Meng, Z., Stolz, R.M., Mendecki, L., Mirica, K.A. Electrically-transduced chemical sensors based on two-dimensional nanomaterials. *Chemical Reviews*. 119 (2019) 478–598. https://doi.org/10.1021/acs.chemrev.8b00311.
7. Minh Triet, N., Thai Duy, L., Hwang, B., Hanif, A., Siddiqui, S., Park, K., Cho, C., Lee, N. High-Performance schottky diode gas sensor based on the heterojunction of three-dimensional nanohybrids of reduced graphene oxide–vertical ZnO nanorods on an AlGaN/GaN layer. *ACS Applied Materials & Interfaces*. 9 (2017) 30722–30732. https://doi.org/10.1021/acsami.7b06461.
8. Lazanas, A.C., Prodromidis, M.I. Two-dimensional inorganic nanosheets: production and utility in the development of novel electrochemical (bio)sensors and gas-sensing applications. *Microchimica Acta*. 188 (2021) 1–34. https://doi.org/10.1007/s00604-020-04671-0.
9. Paul, A., Muthukumar, S., Prasad, S. Review—room-temperature ionic liquids for electrochemical application with special focus on gas sensors. *Journal of the Electrochemical Society*. 167 (2019) 037511. https://doi.org/10.1149/2.0112003JES.
10. Shin, W., Goto, T., Nagai, D., Itoh, T., Tsuruta, A., Akamatsu, T., Sato, K. Thermoelectric array sensors with selective combustion catalysts for breath gas monitoring. *Sensors*. 18 (2018) 1579. https://doi.org/10.3390/s18051579.
11. Yu, Y., Hu, Z., Lien, S., Yu, Y., Gao, P. Self-powered thermoelectric hydrogen sensors based on low-cost bismuth sulfide thin films: Quick response at room temperature. *ACS Applied Materials & Interfaces*. 14 (2022) 47696–47705. https://doi.org/10.1021/acsami.2c12749.
12. Kirchner, P., Oberländer, J., Suso, H., Rysstad, G., Keusgen, M., Schöning, M.J. Monitoring the microbicidal effectiveness of gaseous hydrogen peroxide in sterilisation processes by means of a calorimetric gas sensor. *Food Control*. 31 (2013) 530–538. https://doi.org/10.1016/j.foodcont.2012.11.048.
13. Yun, J., Cho, M., Lee, K., Kang, M., Park, I. A review of nanostructure-based gas sensors in a power consumption perspective. *Sensors and Actuators B: Chemical*. 372 (2022) 132612. https://doi.org/10.1016/j.snb.2022.132612.
14. Paliwal, A., Sharma, A., Tomar, M., Gupta, V. Carbon monoxide (CO) optical gas sensor based on ZnO thin films. *Sensors and Actuators B: Chemical*. 250 (2017) 679–685. https://doi.org/10.1016/j.snb.2017.05.064.
15. Kang, X., Yip, S., Meng, Y., Wang, W., Li, D., Liu, C., Ho, J.C. High-performance electrically transduced hazardous gas sensors based on low-dimensional nanomaterials. *Nanoscale Advances*. 3 (2021) 6254–6270. https://doi.org/https://doi.org/10.1039/D1NA00433F.
16. Dey, A. Semiconductor metal oxide gas sensors: A review. *Materials Science and Engineering B*. 229 (2018) 206–217. https://doi.org/10.1016/j.mseb.2017.12.036.
17. Wang, C., Yin, L., Zhang, L., Xiang, D., Gao, R. Metal oxide gas sensors: sensitivity and influencing factors. *Sensors*. 10 (2010) 2106. https://doi.org/10.3390/s100302088.
18. Bannov, A.G., Popov, M.V., Brester, A.E., Kurmashov, P.B. Recent advances in ammonia gas sensors based on carbon nanomaterials. *Micromachines*. 12 (2021) 186. https://doi.org/https://doi.org/10.3390/mi12020186.
19. Kumar, V., Raghuwanshi, S.K., Kumar, S. Recent advances in carbon nanomaterials based SPR sensor for biomolecules and gas detection—A review. *IEEE Sensors Journal*. 22 (2022) 15661–15672. https://doi.org/10.1109/JSEN.2022.3191042.

20. Llobet, E. Gas sensors using carbon nanomaterials: A review. *Sensors and Actuators B: Chemical.* 179 (2013) 32–45. https://doi.org/10.1016/j.snb.2012.11.014.
21. Liu, X., Zheng, W., Kumar, R., Kumar, M., Zhang, J. Conducting polymer-based nanostructures for gas sensors. *Coordination Chemistry Reviews.* 462 (2022) 214517. https://doi.org/10.1016/j.ccr.2022.214517.
22. Pavel, I., Lakard, S., Lakard, B. Flexible sensors based on conductive polymers. *Chemosensors.* 10 (2022) 97. https://doi.org/10.3390/chemosensors10030097.
23. Nikolic, M.V., Milovanovic, V., Vasiljevic, Z.Z., Stamenkovic, Z. Semiconductor gas sensors: Materials, technology, design, and application. *Sensors.* 20 (2020) 6694. https://doi.org/10.3390/s20226694.
24. Lee, S., Galstyan, V., Ponzoni, A., Gonzalo-Juan, I., Riedel, R., Dourges, M., Nicolas, Y., Toupance, T. Finely tuned SnO_2 nanoparticles for efficient detection of reducing and oxidizing gases: The influence of alkali metal cation on gas-sensing properties. *ACS Applied Materials Interfaces.* 10 (2018) 10173–10184. https://doi.org/10.1021/acsami.7b18140.
25. Liang, Y., Yang, Y., Zou, C., Xu, K., Luo, X., Luo, T., Li, J., Yang, Q., Shi, P., Yuan, C. 2D ultra-thin WO_3 nanosheets with dominant {002} crystal facets for high-performance xylene sensing and methyl orange photocatalytic degradation. *Journal of Alloys and Compounds.* 783 (2019) 848–854. https://doi.org/10.1016/j.jallcom.2018.12.384.
26. Wang, T.T., Ma, S.Y., Cheng, L., Xu, X.L., Luo, J., Jiang, X.H., Li, W.Q., Jin, W.X., Sun, X.X. Performance of 3D SnO_2 microstructure with porous nanosheets for acetic acid sensing. *Materials Letters.* 142 (2015) 141–144. https://doi.org/10.1016/j.matlet.2014.12.009.
27. Xiao, Y., Lu, L., Zhang, A., Zhang, Y., Sun, L., Huo, L., Li, F. Highly enhanced acetone sensing performances of porous and single crystalline ZnO nanosheets: High percentage of exposed (100) facets working together with surface modification with Pd nanoparticles. *ACS Applied Materials & Interfaces.* 4 (2012) 3797–3804. https://doi.org/10.1021/am3010303.
28. Zhao, C., Gong, H., Lan, W., Ramachandran, R., Xu, H., Liu, S., Wang, F. Facile synthesis of SnO_2 hierarchical porous nanosheets from graphene oxide sacrificial scaffolds for high-performance gas sensors. *Sensors and Actuators B: Chemical.* 258 (2018) 492–500. https://doi.org/10.1016/j.snb.2017.11.167.
29. Liu, X., Sun, Y., Yu, M., Yin, Y., Du, B., Tang, W., Jiang, T., Yang, B., Cao, W., Ashfold, M.N.R. Enhanced ethanol sensing properties of ultrathin ZnO nanosheets decorated with CuO nanoparticles. *Sensors and Actuators B: Chemical.* 255 (2018) 3384–3390. https://doi.org/10.1016/j.snb.2017.09.165.
30. Shinde, R.B., Padalkar, N.S., Sadavar, S.V., Kale, S.B., Magdum, V.V., Chitare, Y.M., Kulkarni, S.P., Patil, U.M., Parale, V.G., Park, H., Gunjakar, J.L. 2D–2D lattice engineering route for intimately coupled nanohybrids of layered double hydroxide and potassium hexaniobate: Chemiresistive SO_2 sensor. *Journal of Hazardous Materials.* 432 (2022) 128734. https://doi.org/10.1016/j.jhazmat.2022.128734.
31. Yu, H., Yang, T., Zhao, R., Xiao, B., Li, Z., Zhang, M. Fast formaldehyde gas sensing response properties of ultrathin SnO_2 nanosheets. *RSC Advances.* 5 (2015) 104574–104581. https://doi.org/10.1039/C5RA22755K.
32. Apte, A., Mozaffari, K., Samghabadi, F.S., Hachtel, J.A., Chang, L., Susarla, S., Idrobo, J.C., Moore, D.C., Glavin, N.R., Litvinov, D. 2D electrets of ultrathin MoO_2 with apparent piezoelectricity. *Advvanced Materials.* 32 (2020) 2000006. https://doi.org/https://doi.org/10.1002/adma.202000006.
33. Haque, F., Daeneke, T., Kalantar-zadeh, K., Ou, J.Z. Two-dimensional transition metal oxide and chalcogenide-based photocatalysts. *Nano-Micro Letters.* 10 (2017) 23. https://doi.org/10.1007/s40820-017-0176-y.
34. Zhang, B.Y., Xu, K., Yao, Q., Jannat, A., Ren, G., Field, M.R., Wen, X., Zhou, C., Zavabeti, A., Ou, J.Z. Hexagonal metal oxide monolayers derived from the metal–gas interface. *Nature Materials.* 20 (2021) 1073–1078. https://doi.org/10.1038/s41563-020-00899-9.
35. Xie, H., Li, Z., Cheng, L., Haidry, A.A., Tao, J., Xu, Y., Xu, K., Ou, J.Z. Recent advances in the fabrication of 2D metal oxides. *iScience.* 25 (2022) 1–30. https://doi.org/10.1016/j.isci.2021.103598
36. Addabbo, T., Bertocci, F., Fort, A., Gregorkiewitz, M., Mugnaini, M., Spinicci, R., Vignoli, V. Gas sensing properties and modeling of $YCoO_3$ based perovskite materials. *Sensors Actuators B: Chemical.* 221 (2015) 1137–1155. https://doi.org/https://doi.org/10.1016/j.snb.2015.07.079.
37. Wang, J., Ren, Y., Liu, H., Li, Z., Liu, X., Deng, Y., Fang, X. Ultrathin 2D $NbWO_6$ perovskite semiconductor based gas sensors with ultrahigh selectivity under low working temperature. *Advanced Materials.* 34 (2022) 2104958. https://doi.org/10.1002/adma.202104958.
38. Sowmya, B., John, A., Panda, P.K. A review on metal-oxide based pn and nn heterostructured nanomaterials for gas sensing applications. *Sensors International.* 2 (2021) 100085. https://doi.org/https://doi.org/10.1016/j.sintl.2021.100085.
39. Guo, W., Fu, M., Zhai, C., Wang, Z. Hydrothermal synthesis and gas-sensing properties of ultrathin hexagonal ZnO nanosheets. *Ceramics International.* 40 (2014) 2295–2298. https://doi.org/10.1016/j.ceramint.2013.07.150.

40. Miller, D.R., Akbar, S.A., Morris, P.A. Nanoscale metal oxide-based heterojunctions for gas sensing: A review. *Sensors and Actuators B: Chemical.* 204 (2014) 250–272. https://doi.org/10.1016/j.snb.2014.07.074.
41. Tshabalala, Z.P., Shingange, K., Cummings, F.R., Ntwaeaborwa, O.M., Mhlongo, G.H., Motaung, D.E. Ultra-sensitive and selective NH_3 room temperature gas sensing induced by manganese-doped titanium dioxide nanoparticles. *Journal of Colloid and Interface Science.* 504 (2017) 371–386. https://doi.org/10.1016/j.jcis.2017.05.061.
42. Zakaria, S.A., Ahmadi, S.H., Amini, M.H. Chemiresistive gas sensors based on layered double hydroxides (LDHs) structures: A review. *Sensors and Actuators A: Physical.* 346 (2022) 113827. https://doi.org/10.1016/j.sna.2022.113827.
43. Moseley, P.T. Progress in the development of semiconducting metal oxide gas sensors: A review. *Measurement Science and Technology.* 28 (2017) 082001. https://doi.org/10.1088/1361-6501/aa7443.
44. Wetchakun, K., Samerjai, T., Tamaekong, N., Liewhiran, C., Siriwong, C., Kruefu, V., Wisitsoraat, A., Tuantranont, A., Phanichphant, S. Semiconducting metal oxides as sensors for environmentally hazardous gases. *Sensors and Actuators B: Chemical.* 160 (2011) 580–591. https://doi.org/10.1016/j.snb.2011.08.032.
45. Yu, K., Lei, D., Feng, Y., Yu, H., Chang, Y., Wang, Y., Liu, Y., Wang, G., Lou, L., Liu, S., Zhou, W. The role of Bi-doping in promoting electron transfer and catalytic performance of Pt/3DOM-$Ce_{1-x}Bi_xO_{2-\delta}$. *Journal of Catalysis.* 365 (2018) 292–302. https://doi.org/10.1016/j.jcat.2018.06.025.
46. Yang, C.S., Shang, D.S., Liu, N., Shi, G., Shen, X., Yu, R.C., Li, Y.Q., Sun, Y. A synaptic transistor based on quasi-2D molybdenum oxide. *Advanced Materials.* 29 (2017) 1700906. https://doi.org/10.1002/adma.201700906.
47. Momma, K., Izumi, F. VESTA 3 for three-dimensional visualization of crystal, volumetric and morphology data. *Journal of Applied Crystallography.* 44 (2011) 1272–1276. https://doi.org/https://doi.org/10.1107/S0021889811038970.
48. Wang, Y., Ren, B., Zhen Ou, J., Xu, K., Yang, C., Li, Y., Zhang, H. Engineering two-dimensional metal oxides and chalcogenides for enhanced electro- and photocatalysis. *Science Bulletin.* 66 (2021) 1228–1252. https://doi.org/10.1016/j.scib.2021.02.007.
49. Wang, Q., O'Hare, D. Recent advances in the synthesis and application of layered double hydroxide (LDH) nanosheets. *Chemical Reviews.* 112 (2012) 4124–4155. https://doi.org/10.1021/cr200434v.
50. Fahami, A., Al-Hazmi, F.S., Al-Ghamdi, A.A., Mahmoud, W.E., Beall, G.W. Structural characterization of chlorine intercalated Mg-Al layered double hydroxides: A comparative study between mechanochemistry and hydrothermal methods. *Journal of Alloys and Compounds.* 683 (2016) 100–107. https://doi.org/10.1016/j.jallcom.2016.05.032.
51. Khan, A.I., O'Hare, D. Intercalation chemistry of layered double hydroxides: recent developments and applications. *Journal of Materials Chemistry.* 12 (2002) 3191–3198. https://doi.org/10.1039/B204076J.
52. Theiss, F.L., Couperthwaite, S.J., Ayoko, G.A., Frost, R.L. A review of the removal of anions and oxyanions of the halogen elements from aqueous solution by layered double hydroxides. *Journal of Colloid and Interface Science.* 417 (2014) 356–368. https://doi.org/10.1016/j.jcis.2013.11.040.
53. Polese, D., Mattoccia, A., Giorgi, F., Pazzini, L., Di Giamberardino, L., Fortunato, G., Medaglia, P.G. A phenomenological investigation on Chlorine intercalated Layered Double Hydroxides used as room temperature gas sensors. *Journal of Alloys and Compounds.* 692 (2017) 915–922. https://doi.org/10.1016/j.jallcom.2016.09.125.
54. Ru, L., Li, Z., Bai, C., Zhang, N., Wang, H. Effect of $CaTiO_3$ crystal structure and cathodic morphological structure on the electrolysis. *Asian Journal of Chemistry.* 25 (2013) 1814. https://doi.org/10.14233/ajchem.2013.13162.
55. Danaie, M., Kepaptsoglou, D., Ramasse, Q.M., Ophus, C., Whittle, K.R., Lawson, S.M., Pedrazzini, S., Young, N.P., Bagot, P.A.J., Edmondson, P.D. Characterization of ordering in A-site deficient perovskite $Ca_{1-x}La_{2x/3}TiO_3$ using STEM/EELS. *Inorganic Chemistry.* 55 (2016) 9937–9948. https://doi.org/10.1021/acs.inorgchem.6b02087.
56. Kumbhakar, P., Chowde Gowda, C., Mahapatra, P.L., Mukherjee, M., Malviya, K.D., Chaker, M., Chandra, A., Lahiri, B., Ajayan, P.M., Jariwala, D., Singh, A., Tiwary, C.S. Emerging 2D metal oxides and their applications. *Materials Today.* 45 (2021) 142–168. https://doi.org/10.1016/j.mattod.2020.11.023.
57. Raza, A., Zhang, Y., Cassinese, A., Li, G. Engineered 2D metal oxides for photocatalysis as environmental remediation: A theoretical perspective. *Catalysts.* 12 (2022) 1613. https://doi.org/10.3390/catal12121613.
58. Dral, A.P., ten Elshof, J.E. 2D metal oxide nanoflakes for sensing applications: Review and perspective. *Sensors Actuators B: Chemical.* 272 (2018) 369–392. https://doi.org/10.1016/j.snb.2018.05.157.
59. Fine, G.F., Cavanagh, L.M., Afonja, A., Binions, R. Metal oxide semi-conductor gas sensors in environmental monitoring. *Sensors.* 10 (2010) 5502. https://doi.org/10.3390/s100605469.

60. Lin, T., Lv, X., Li, S., Wang, Q. The morphologies of the semiconductor oxides and their gas-sensing properties. *Sensors*. 17 (2017) 2779. https://doi.org/10.3390/s17122779.
61. Singh, E., Meyyappan, M., Nalwa, H.S. Flexible graphene-based wearable gas and chemical sensors. *ACS Applied Materials & Interfaces*. 9 (2017) 34544–34586. https://doi.org/10.1021/acsami.7b07063.
62. Yoon, Y., Truong, P.L., Lee, D., Ko, S.H. Metal-oxide nanomaterials synthesis and applications in flexible and wearable sensors. *ACS Nanoscience Au*. 2 (2022) 64–92. https://doi.org/10.1021/acsnanoscienceau.1c00029.
63. Kanaparthi, S., Govind Singh, S. Highly sensitive and ultra-fast responsive ammonia gas sensor based on 2D ZnO nanoflakes. *Materials Science for Energy Technologies*. 3 (2020) 91–96. https://doi.org/10.1016/j.mset.2019.10.010.
64. Shendage, S.S., Patil, V.L., Vanalakar, S.A., Patil, S.P., Harale, N.S., Bhosale, J.L., Kim, J.H., Patil, P.S. Sensitive and selective NO_2 gas sensor based on WO_3 nanoplates. *Sensors and Actuators B: Chemical*. 240 (2017) 426–433. https://doi.org/10.1016/j.snb.2016.08.177.
65. Timmerman, M.A., Xia, R., Le, P.T., Wang, Y., Ten Elshof, J.E. Metal oxide nanosheets as 2D building blocks for the design of novel materials. *Chemistry–A European Journal*. 26 (2020) 9084–9098. https://doi.org/10.1002/chem.201905735.
66. Neri, G. Thin 2D: The new dimensionality in gas sensing. *Chemosensors*. 5 (2017) 21. https://doi.org/10.3390/chemosensors5030021.
67. Dima, R.S., Tshwane, D.M., Shingange, K., Modiba, R., Maluta, N.E., Maphanga, R.R. Adsorption of NH_3 and NO_2 molecules on Sn-doped and undoped ZnO (101) surfaces using density functional theory. *Processes*. 10 (2022) 2027. https://doi.org/10.3390/pr10102027.
68. Yang, Y., Liang, Y., Wang, G., Liu, L., Yuan, C., Yu, T., Li, Q., Zeng, F., Gu, G. Enhanced gas-sensing properties of the hierarchical TiO_2 hollow microspheres with exposed high-energy {001} crystal facets. *ACS Applied Materials & Interfaces*. 7 (2015) 24902–24908. https://doi.org/10.1021/acsami.5b08372.
69. Chang, J., Ahmad, M.Z., Wlodarski, W., Waclawik, E.R. Self-assembled 3D ZnO porous structures with exposed reactive {0001} facets and their enhanced gas sensitivity. *Sensors*. 13 (2013) 8460. https://doi.org/10.3390/s130708445.
70. Kaneti, Y.V., Yue, J., Jiang, X., Yu, A. Synthesis of ZnO nanoflakes with exposed $(10\bar{1}\bar{0})$ for enhanced gas sensing performance. *Journal of Physical Chemistry C*. 117 (2013) 13153–13162. https://doi.org/10.1021/jp404329q.
71. Su, D., Xie, X., Dou, S., Wang, G. CuO single crystal with exposed {001} facets - A highly efficient material for gas sensing and Li-ion battery applications. *Scientific Reports*. 4 (2014) 5753. https://doi.org/10.1038/srep05753.
72. Xue, Z., Cheng, Z., Xu, J., Xiang, Q., Wang, X., Xu, J. Controllable evolution of dual defect Zni and VO associate-rich ZnO nanodishes with (0001) exposed facet and its multiple sensitization effect for ethanol detection. *ACS Applied Materials & Interfaces*. 9 (2017) 41559–41567. https://doi.org/10.1021/acsami.7b13370.
73. Liu, F., Wang, X., Chen, X., Song, X., Tian, J., Cui, H. Porous ZnO ultrathin nanosheets with high specific surface areas and abundant oxygen vacancies for acetylacetone gas sensing. *ACS Applied Materials & Interfaces*. 11 (2019) 24757–24763. https://doi.org/10.1021/acsami.9b06701.
74. Cao, F., Li, C., Li, M., Li, H., Huang, X., Yang, B. Direct growth of Al-doped ZnO ultrathin nanosheets on electrode for ethanol gas sensor application. *Applied Surface Science*. 447 (2018) 173–181. https://doi.org/10.1016/j.apsusc.2018.03.217.
75. Xiao, J., Liu, P., Liang, Y., Li, H.B., Yang, G.W. Porous tungsten oxide nanoflakes for highly alcohol sensitive performance. *Nanoscale*. 4 (2012) 7078–7083. https://doi.org/10.1039/C2NR32078A.
76. Miao, J., Chen, C., Meng, L., Lin, Y.S. Self-assembled monolayer of metal oxide nanosheet and structure and gas-sensing property relationship. *ACS Sensors*. 4 (2019) 1279–1290. https://doi.org/10.1021/acssensors.9b00162.
77. Tang, L., Tan, J., Nong, H., Liu, B., Cheng, H. Chemical vapor deposition growth of two-dimensional compound materials: controllability, material quality, and growth mechanism. *Accounts of Materials Research*. 2 (2021) 36–47. https://doi.org/10.1021/accountsmr.0c00063.
78. Tang, W., Fan, W., Zhang, W., Yang, Z., Li, L., Wang, Z., Chiang, Y., Liu, Y., Deng, L., He, L. Wet/sonochemical synthesis of enzymatic two-dimensional MnO_2 nanosheets for synergistic catalysis-enhanced phototheranostics. *Advanced Materials*. 31 (2019) 1900401. https://doi.org/10.1002/adma.201900401.
79. Dang, W., Wang, W., Yang, Y., Wang, Y., Huang, J., Fang, X., Wu, L., Rong, Z., Chen, X., Li, X., Huang, L., Tang, X. One-step hydrothermal synthesis of 2D WO_3 nanoplates@ graphene nanocomposite with superior anode performance for Lithium ion battery. *Electrochimica Acta*. 313 (2019) 99–108. https://doi.org/10.1016/j.electacta.2019.04.184.

80. Tripuramallu, B.K., Das, S.K. Hydrothermal synthesis and structural characterization of metal organophosphonate oxide materials: Role of metal-oxo clusters in the self assembly of metal phosphonate architectures. *Crystal Growth & Design*. 13 (2013) 2426–2434. https://doi.org/10.1021/cg4001347.
81. Sun, Z., Liao, T., Dou, Y., Hwang, S.M., Park, M., Jiang, L., Kim, J.H., Dou, S.X. Generalized self-assembly of scalable two-dimensional transition metal oxide nanosheets. *Nature Communications*. 5 (2014) 3813. https://doi.org/10.1038/ncomms4813.
82. Mhlongo, G.H., Shingange, K., Tshabalala, Z.P., Dhonge, B.P., Mahmoud, F.A., Mwakikunga, B.W., Motaung, D.E. Room temperature ferromagnetism and gas sensing in ZnO nanostructures: Influence of intrinsic defects and Mn, Co, Cu doping. *Applied Surface Science*. 390 (2016) 804–815. https://doi.org/10.1016/j.apsusc.2016.08.138.
83. Shingange, K., Mhlongo, G.H., Motaung, D.E., Ntwaeaborwa, O.M. Tailoring the sensing properties of microwave-assisted grown ZnO nanorods: Effect of irradiation time on luminescence and magnetic behaviour. *Journal of Alloys and Compounds*. 657 (2016) 917–926. https://doi.org/10.1016/j.jallcom.2015.10.069.
84. Kumar, R., Mamta, Kumari, R., Singh, V.N. SnO_2-Based NO_2 gas sensor with outstanding sensing performance at room temperature. *Micromachines*. 14 (2023) 728. https://doi.org/10.3390/mi14040728.
85. Motsoeneng, R.G., Kortidis, I., Ray, S.S., Motaung, D.E. Designing SnO_2 nanostructure-based sensors with tailored selectivity toward propanol and ethanol vapors. *ACS Omega*. 4 (2019) 13696–13709. https://doi.org/10.1021/acsomega.9b01079.
86. Tian, X., Cui, X., Lai, T., Ren, J., Yang, Z., Xiao, M., Wang, B., Xiao, X., Wang, Y. Gas sensors based on TiO_2 nanostructured materials for the detection of hazardous gases: A review. *Nano Materials Science*. 3 (2021) 390–403. https://doi.org/10.1016/j.nanoms.2021.05.011.
87. Oosthuizen, D.N., Motaung, D.E., Strydom, A.M., Swart, H.C. Underpinning the interaction between NO_2 and CuO nanoplatelets at room temperature by tailoring synthesis reaction base and time. *ACS Omega*. 4 (2019) 18035–18048. https://doi.org/10.1021/acsomega.9b01882.
88. Mokoena, T.P., Swart, H.C., Hillie, K.T., Motaung, D.E. Colour tuning from violet to blue emission stimulated by various nickel oxide nanostructures: Influence of bias voltage towards volatile organic compounds vapours. *Applied Surface Science*. 542 (2021) 148634. https://doi.org/10.1016/j.apsusc.2020.148634.
89. Yang, S., Jiang, C., Wei, S. Gas sensing in 2D materials. *Applied Physics Reviews*. 4 (2017) 021304. https://doi.org/10.1063/1.4983310.
90. Hübner, M., Simion, C.E., Tomescu-Stănoiu, A., Pokhrel, S., Bârsan, N., Weimar, U. Influence of humidity on CO sensing with p-type CuO thick film gas sensors. *Sensors and Actuator B: Chemical*. 153 (2011) 347–353. https://doi.org/10.1016/j.snb.2010.10.046.
91. Zhang, L., Zhao, J., Lu, H., Li, L., Zheng, J., Li, H., Zhu, Z. Facile synthesis and ultrahigh ethanol response of hierarchically porous ZnO nanosheets. *Sensors and Actuators B: Chemical*. 161 (2012) 209–215. https://doi.org/10.1016/j.snb.2011.10.021.
92. Oosthuizen, D.N., Motaung, D.E., Swart, H.C. In depth study on the notable room-temperature NO_2 gas sensor based on CuO nanoplatelets prepared by sonochemical method: Comparison of various bases. *Sensors and Actuators B: Chemical*. 266 (2018) 761–772. https://doi.org/10.1016/j.snb.2018.03.106.
93. Zhao, J., Huang, X., Yin, Y., Liao, Y., Mo, H., Qian, Q., Guo, Y., Chen, X., Zhang, Z., Hua, M. Two-dimensional gallium oxide monolayer for gas-sensing application. *The Journal of Physical Chemistry Letters*. 12 (2021) 5813–5820. https://doi.org/10.1021/acs.jpclett.1c01393.
94. Li, J., Wu, J., Yu, Y. DFT exploration of sensor performances of two-dimensional WO_3 to ten small gases in terms of work function and band gap changes and I-V responses. *Applied Surface Science*. 546 (2021) 149104. https://doi.org/10.1016/j.apsusc.2021.149104.
95. Qin, Y., Ding, W., Zhao, R. ZIF-8-derived ZnTi-LDHs with unique self-supported architecture and corresponding LDHs/rGO hybrid for gas sensor applications. *Chemical Physics Letters*. 781 (2021) 138965. https://doi.org/10.1016/j.cplett.2021.138965.
96. Sun, H., Chu, Z., Hong, D., Zhang, G., Xie, Y., Li, L., Shi, K. Three-dimensional hierarchical flower-like Mg–Al-layered double hydroxides: Fabrication, characterization and enhanced sensing properties to NO_x at room temperature. *Journal of Alloys and Compounds*. 658 (2016) 561–568. https://doi.org/10.1016/j.jallcom.2015.10.237.
97. Pacuła, A., Socha, R.P., Zimowska, M., Ruggiero-Mikołajczyk, M., Mucha, D., Nowak, P. Application of as-synthesized Co–Al layered double hydroxides for the preparation of the electroactive composites containing N-doped carbon nanotubes. *Applied Clay Science*. 72 (2013) 163–174. https://doi.org/10.1016/j.clay.2012.11.007.

98. Shingange, K., Swart, H.C., Mhlongo, G.H. $LaBO_3$ (B= Fe, Co) nanofibers and their structural, luminescence and gas sensing characteristics. *Physica B: Condensed Matter.* 578 (2020) 411883. https://doi.org/10.1016/j.physb.2019.411883.
99. Li, B., Osada, M., Ozawa, T.C., Sasaki, T. RbBiNb2O7: A new lead-free high-Tc ferroelectric. *Chemistry of Materials.* 24 (2012) 3111–3113.
100. Cao, D.H., Stoumpos, C.C., Farha, O.K., Hupp, J.T., Kanatzidis, M.G. 2D homologous perovskites as light-absorbing materials for solar cell applications. *Journal of the American Chemical Society.* 137 (2015) 7843–7850. https://doi.org/10.1021/jacs.5b03796.
101. Saparov, B., Mitzi, D.B. Organic–inorganic perovskites: Structural versatility for functional materials design. *Chemical Reviews.* 116 (2016) 4558–4596. https://doi.org/10.1021/acs.chemrev.5b00715.
102. Maheshwari, S., Savenije, T.J., Renaud, N., Grozema, F.C. Computational design of two-dimensional perovskites with functional organic cations. *The Journal of Physical Chemistry C.* 122 (2018) 17118–17122. https://doi.org/10.1021/acs.jpcc.8b05715.
103. Jiao, W., He, J., Zhang, L. Synthesis and high ammonia gas sensitivity of $(CH_3NH_3)PbBr_{3-x}I_x$ perovskite thin film at room temperature. *Sensors and Actuators B: Chemical.* 309 (2020) 127786. https://doi.org/10.1016/j.snb.2020.127786.
104. Maity, A., Mitra, S., Das, C., Siraj, S., Raychaudhuri, A.K., Ghosh, B. Universal sensing of ammonia gas by family of lead halide perovskites based on paper sensors: Experiment and molecular dynamics. *Materials Research Bulletin.* 136 (2021) 111142. https://doi.org/10.1016/j.materresbull.2020.111142.
105. Hien, V.X., Hoat, P.D., Hung, P.T., Lee, S., Lee, J., Heo, Y. Room-temperature NO_2 sensor based on a hybrid nanomaterial of methylammonium tin iodide submicron spheres and tin dioxide nanowires. *Scripta Materialia.* 188 (2020) 107–111. https://doi.org/10.1016/j.scriptamat.2020.07.022.
106. Stoeckel, M., Gobbi, M., Bonacchi, S., Liscio, F., Ferlauto, L., Orgiu, E., Samorì, P. Reversible, fast, and wide-range oxygen sensor based on nanostructured organometal halide perovskite. *Advanced Materials.* 29 (2017) 1702469. https://doi.org/10.1002/adma.201702469.
107. Haque, M.A., Syed, A., Akhtar, F.H., Shevate, R., Singh, S., Peinemann, K., Baran, D., Wu, T. Giant humidity effect on hybrid halide perovskite microstripes: Reversibility and sensing mechanism. *ACS Applied Materials & Interfaces.* 11 (2019) 29821–29829. https://doi.org/10.1021/acsami.9b07751.
108. Lang, F., Shargaieva, O., Brus, V.V., Neitzert, H.C., Rappich, J., Nickel, N.H. Influence of radiation on the properties and the stability of hybrid perovskites. *Advanced Materials.* 30 (2018) 1702905. https://doi.org/https://doi.org/10.1002/adma.201702905.
109. Ponchai, J., Srathongsian, L., Amratisha, K., Boonthum, C., Sahasithiwat, S., Ruankham, P., Kanjanaboos, P. Modified colored semi-transparent perovskite solar cells with enhanced stability. *Journal of Alloys and Compounds.* 875 (2021) 159781. https://doi.org/10.1016/j.jallcom.2021.159781.
110. Parikh, N., Tavakoli, M.M., Pandey, M., Kumar, M., Prochowicz, D., Chavan, R.D., Yadav, P. Two-dimensional halide perovskite single crystals: principles and promises. *Emergent Materials.* 4 (2021) 865–880. https://doi.org/10.1007/s42247-021-00177-7.
111. Tien, C., Lee, K., Tao, C., Lin, Z., Lin, Z., Chen, L. Two-dimensional $(PEA)_2PbBr_4$ perovskites sensors for highly sensitive ethanol vapor detection. *Sensors.* 22 (2022) 8155. https://doi.org/10.3390/s22218155.
112. Chen, C., Cai, Q., Luo, F., Dong, N., Guo, L., Qiu, B., Lin, Z. Sensitive fluorescent sensor for hydrogen sulfide in rat brain microdialysis via $CsPbBr_3$ quantum dots. *Analytical Chemistry.* 91 (2019) 15915–15921. https://doi.org/10.1021/acs.analchem.9b04387.
113. Jariwala, D., Marks, T.J., Hersam, M.C. Mixed-dimensional van der Waals heterostructures. *Nature Materials.* 16 (2017) 170–181. https://doi.org/10.1038/nmat4703.
114. Liu, Y., Huang, Y., Duan, X. Van der Waals integration before and beyond two-dimensional materials. *Nature.* 567 (2019) 323–333. https://doi.org/10.1038/s41586-019-1013-x.
115. Boulesbaa, A., Wang, K., Mahjouri-Samani, M., Tian, M., Puretzky, A.A., Ivanov, I., Rouleau, C.M., Xiao, K., Sumpter, B.G., Geohegan, D.B. Ultrafast charge transfer and hybrid exciton formation in 2D/0D heterostructures. *Journal of the American Chemical Society.* 138 (2016) 14713–14719. https://doi.org/https://doi.org/10.1021/jacs.6b08883.
116. Sinha, S., Kim, H., Robertson, A.W. Preparation and application of 0D-2D nanomaterial hybrid heterostructures for energy applications. *Materials Today Advances.* 12 (2021) 100169. https://doi.org/10.1016/j.mtadv.2021.100169.
117. Wang, P., Jia, C., Huang, Y., Duan, X. Van der Waals heterostructures by design: from 1D and 2D to 3D. *Matter.* 4 (2021) 552–581. https://doi.org/10.1016/j.matt.2020.12.015.
118. Xue, S., Wu, G., Li, M., Liu, Z., Deng, Y., Han, W., Lv, X., Wan, S., Xi, X., Yang, D., Dong, A. Generalized assembly of sandwich-like 0D/2D/0D heterostructures with highly exposed surfaces toward superior electrochemical performances. *Nano Research.* 15 (2022) 255–263. https://doi.org/10.1007/s12274-021-3468-y.

119. Withers, F., Del Pozo-Zamudio, O., Mishchenko, A., Rooney, A.P., Gholinia, A., Watanabe, K., Taniguchi, T., Haigh, S.J., Geim, A.K., Tartakovskii, A.I., Novoselov, K.S. Light-emitting diodes by band-structure engineering in van der Waals heterostructures. *Nature Materials.* 14 (2015) 301–306. https://doi.org/10.1038/nmat4205.
120. Zhou, X., Hu, X., Yu, J., Liu, S., Shu, Z., Zhang, Q., Li, H., Ma, Y., Xu, H., Zhai, T. 2D layered material-based van der Waals heterostructures for optoelectronics. *Advanced Functional Materials.* 28 (2018) 1706587. https://doi.org/10.1002/adfm.201706587.
121. Ruzmetov, D., Zhang, K., Stan, G., Kalanyan, B., Bhimanapati, G.R., Eichfeld, S.M., Burke, R.A., Shah, P.B., O'Regan, T.P., Crowne, F.J., Birdwell, A.G., Robinson, J.A., Davydov, A.V., Ivanov, T.G. Vertical 2D/3D semiconductor heterostructures based on epitaxial molybdenum disulfide and gallium nitride. *ACS Nano.* 10 (2016) 3580–3588. https://doi.org/10.1021/acsnano.5b08008.
122. Joshi, N., Braunger, M.L., Shimizu, F.M., Riul Jr, A., Oliveira, O.N. Insights into nano-heterostructured materials for gas sensing: A review. *Multifunctional Materials.* 4 (2021) 032002. https://doi.org/10.1088/2399-7532/ac1732.
123. Singh, A., Sikarwar, S., Verma, A., Chandra Yadav, B. The recent development of metal oxide heterostructures based gas sensor, their future opportunities and challenges: A review. *Sensors and Actuators A: Physical.* 332 (2021) 113127. https://doi.org/10.1016/j.sna.2021.113127.
124. Yang, S., Lei, G., Xu, H., Lan, Z., Wang, Z., Gu, H. Metal oxide based heterojunctions for gas sensors: A review. *Nanomaterials.* 11 (2021) 1026. https://doi.org/10.3390/nano11041026.
125. Yang, S., Lei, G., Lan, Z., Xie, W., Yang, B., Xu, H., Wang, Z., Gu, H. Enhancement of the room-temperature hydrogen sensing performance of MoO_3 nanoribbons annealed in a reducing gas. *International Journal of Hydrogen Energy.* 44 (2019) 7725–7733. https://doi.org/10.1016/j.ijhydene.2019.01.205.
126. Zappa, D., Galstyan, V., Kaur, N., Munasinghe Arachchige, H.M.M., Sisman, O., Comini, E. "Metal oxide -based heterostructures for gas sensors"- A review. *Analytica Chimica Acta.* 1039 (2018) 1–23. https://doi.org/10.1016/j.aca.2018.09.020.
127. Li, Z., Li, H., Wu, Z., Wang, M., Luo, J., Torun, H., Hu, P., Yang, C., Grundmann, M., Liu, X. Advances in designs and mechanisms of semiconducting metal oxide nanostructures for high-precision gas sensors operated at room temperature. *Materials Horizons.* 6 (2019) 470–506. https://doi.org/https://doi.org/10.1039/C8MH01365A.
128. Ju, D., Xu, H., Qiu, Z., Guo, J., Zhang, J., Cao, B. Highly sensitive and selective triethylamine-sensing properties of nanosheets directly grown on ceramic tube by forming NiO/ZnO PN heterojunction. *Sensors and Actuators B: Chemical.* 200 (2014) 288–296. https://doi.org/10.1016/j.snb.2014.04.029.
129. Yang, X., Shi, Y., Gong, F., Chen, J., Jin, G., Guo, Q., Zhang, H., Zhang, Y. Asymmetric interfacial oxygen sites of porous CeO_2-SnO_2 nanosheets enabling highly sensitive and selective detection of 3-hydroxy-2-butanone biomarkers. *Sensors and Actuators B: Chemical.* 371 (2022) 132500. https://doi.org/10.1016/j.snb.2022.132500.
130. Gao, X., Ouyang, Q., Zhu, C., Zhang, X., Chen, Y. Porous MoO_3/SnO_2 nanoflakes with n–n junctions for sensing H_2S. *ACS Applied Nano Materials.* 2 (2019) 2418–2425. https://doi.org/10.1021/acsanm.9b00308.
131. Yuan, K., Zhu, L., Yang, J., Hang, C., Tao, J., Ma, H., Jiang, A., Zhang, D.W., Lu, H. Precise preparation of WO_3@SnO_2 core shell nanosheets for efficient NH_3 gas sensing. *Journal of Colloid and Interface Science.* 568 (2020) 81–88. https://doi.org/10.1016/j.jcis.2020.02.042.
132. Yan, S., Liang, X., Song, H., Ma, S., Lu, Y. Synthesis of porous CeO_2-SnO_2 nanosheets gas sensors with enhanced sensitivity. *Ceramics International.* 44 (2018) 358–363. https://doi.org/10.1016/j.ceramint.2017.09.181.
133. Xu, H., Ju, J., Li, W., Zhang, J., Wang, J., Cao, B. Superior triethylamine-sensing properties based on TiO_2/SnO_2 n–n heterojunction nanosheets directly grown on ceramic tubes. *Sensors and Actuators B: Chemical.* 228 (2016) 634–642. https://doi.org/10.1016/j.snb.2016.01.059.
134. Wang, Y., Qu, F., Liu, J., Wang, Y., Zhou, J., Ruan, S. Enhanced H_2S sensing characteristics of CuO-NiO core-shell microspheres sensors. *Sensors and Actuators B: Chemical.* 209 (2015) 515–523. https://doi.org/10.1016/j.snb.2014.12.010.
135. Sik Choi, M., Young Kim, M., Mirzaei, A., Kim, H., Kim, S., Baek, S., Won Chun, D., Jin, C., Hyoung Lee, K. Selective, sensitive, and stable NO_2 gas sensor based on porous ZnO nanosheets. *Applied Surface Science.* 568 (2021) 150910. https://doi.org/10.1016/j.apsusc.2021.150910.
136. Wang, X., Su, J., Chen, H., Li, G., Shi, Z., Zou, H., Zou, X. Ultrathin In_2O_3 nanosheets with uniform mesopores for highly sensitive nitric oxide detection. *ACS Applied Materials & Interfaces.* 9 (2017) 16335–16342. https://doi.org/10.1021/acsami.7b04395.
137. Ji, F., Ren, X., Zheng, X., Liu, Y., Pang, L., Jiang, J., Liu, S. 2D-MoO_3 nanosheets for superior gas sensors. *Nanoscale.* 8 (2016) 8696–8703. https://doi.org/10.1039/C6NR00880A.

138. Xiao, R., Wang, T., Feng, S., Zhang, X., Cheng, X., Gao, R., Huo, L., Gao, S., Xu, Y. Porous MoO_3 nanosheets for conductometric gas sensors to detect diisopropylamine. *Sensors and Actuators B: Chemical.* 382 (2023) 133472. https://doi.org/10.1016/j.snb.2023.133472.
139. Wang, M., Wang, Y., Li, X., Ge, C., Hussain, S., Liu, G., Qiao, G. WO_3 porous nanosheet arrays with enhanced low temperature NO_2 gas sensing performance. *Sensors and Actuators B: Chemical.* 316 (2020) 128050. https://doi.org/10.1016/j.snb.2020.128050.
140. Hoa, N.D., El-Safty, S.A. Synthesis of mesoporous NiO nanosheets for the detection of toxic NO_2 gas. *Chemistry – A European Journal.* 17 (2011) 12896–12901. https://doi.org/https://doi.org/10.1002/chem.201101122.
141. Feng, S., Yu, H., Zhang, X., Huo, L., Gao, R., Wang, P., Cheng, X., Major, Z., Gao, S., Xu, Y. Ionic liquid-assisted synthesis of 2D porous lotus root slice-shaped NiO nanomaterials for selective and highly sensitive detection of N_2H_4. *Sensors Actuators B: Chemical.* 359 (2022) 131529. https://doi.org/10.1016/j.snb.2022.131529.
142. Zhang, Z., Wen, Z., Ye, Z., Zhu, L. Gas sensors based on ultrathin porous Co_3O_4 nanosheets to detect acetone at low temperature. *RSC Advances.* 5 (2015) 59976–59982. https://doi.org/10.1039/C5RA08536E.
143. Xu, K., Zou, J., Tian, S., Yang, Y., Zeng, F., Yu, T., Zhang, Y., Jie, X., Yuan, C. Single-crystalline porous nanosheets assembled hierarchical Co_3O_4 microspheres for enhanced gas-sensing properties to trace xylene. *Sensors and Actuators B: Chemical.* 246 (2017) 68–77. https://doi.org/10.1016/j.snb.2017.02.071.
144. Qin, W., Yuan, Z., Gao, H., Zhang, R., Meng, F. Perovskite-structured $LaCoO_3$ modified ZnO gas sensor and investigation on its gas sensing mechanism by first principle. *Sensors and Actuators B: Chemical.* 341 (2021) 130015. https://doi.org/10.1016/j.snb.2021.130015.
145. Qin, Y., Gui, H., Ding, W. Enhance methanol-sensing properties based on Ag functionalized monolayer LDH nanosheet: A combined experimental and theoretical calculation investigation. *Physica E: Low-Dimensional Systems and Nanostructures.* 144 (2022) 115409. https://doi.org/10.1016/j.physe.2022.115409.
146. Wang, D., Yang, J., Bao, L., Cheng, Y., Tian, L., Ma, Q., Xu, J., Li, H., Wang, X. Pd nanocrystal sensitization two-dimension porous TiO_2 for instantaneous and high efficient H_2 detection. *Journal of Colloid and Interface Science.* 597 (2021) 29–38. https://doi.org/10.1016/j.jcis.2021.03.107.
147. Hao, Q., Liu, T., Liu, J., Liu, Q., Jing, X., Zhang, H., Huang, G., Wang, J. Controllable synthesis and enhanced gas sensing properties of a single-crystalline WO_3–rGO porous nanocomposite. *RSC Advances.* 7 (2017) 14192–14199. https://doi.org/10.1039/C6RA28379A.
148. Zhang, J., Zhang, K., Liu, S., Liang, X., Zhang, M. Reasonable construction of 2D porous NiO/Co_3O_4 nanosheets for efficient detection of xylene. *Sensors and Actuators B: Chemical.* 377 (2023) 133002. https://doi.org/10.1016/j.snb.2022.133002.
149. Qin, S., Tang, P., Feng, Y., Li, D. Novel ultrathin mesoporous ZnO-SnO_2 n-n heterojunction nanosheets with high sensitivity to ethanol. *Sensors and Actuators B: Chemical.* 309 (2020) 127801. https://doi.org/10.1016/j.snb.2020.127801.
150. Hui, G., Zhu, M., Yang, X., Liu, J., Pan, G., Wang, Z. Highly sensitive ethanol gas sensor based on CeO_2/ZnO binary heterojunction composite. *Materials Letters.* 278 (2020) 128453. https://doi.org/10.1016/j.matlet.2020.128453.
151. Xiong, Y., Liu, W., Wu, K., Liu, T., Chen, Y., Wang, X., Tian, J. Constructing ultrathin defective Co_3O_4/MoS_2 nanosheets based 2D/2D heterojunction toward room temperature NH_3 detection. *Journal of Alloys and Compounds.* 927 (2022) 166962. https://doi.org/10.1016/j.jallcom.2022.166962.
152. Xiao, Y., Hu, S., Liu, Y., Zhang, A., Yao, Z., Tian, Y., Li, H., Ning, Y., Li, F., Qu, F., Yao, D., Zhang, H. Pt-modified $BiVO_4$ nanosheets for enhanced acetone sensing. *Sensors and Actuators B: Chemical.* 389 (2023) 133853. https://doi.org/10.1016/j.snb.2023.133853.

7 Two-Dimensional MXenes-Based Materials in Gas Chemical Sensors

Keerthana Sahadevan, Margandan Bhagiyalakshmi, and Mari Vinoba

7.1 INTRODUCTION

In a wide variety of sectors and applications, gas sensors have grown in significance and have become indispensable in many facets of contemporary civilization. Gas sensors, for instance, are present in many household appliances, such as hygrometers and air purifiers. These sensors are made to detect and show data about the interior humidity and volatile gas concentrations, respectively. They are particularly helpful for identifying dangerous chemicals that can seriously endanger the health and safety of people. Gas sensors are important for finding gas leaks as well, especially around gas pipes and in kitchens, where there is a higher risk of fire and explosion. Gas sensors are even more crucial for detecting harmful gases and ensuring that workers are not exposed to dangerous quantities of these substances in workplaces like laboratories and factories where a variety of chemicals are present.

Gasoline sensors are frequently employed for environmental monitoring on a broader scale, particularly for tracking air pollutants released by companies and cars. To educate the public about air quality, gases including nitrogen oxides, sulfur oxides, and carbon monoxide are frequently monitored. Real-time data on their amounts is frequently posted in public places. Gas sensors are employed in the medical field for a variety of purposes, including the early diagnosis of various disorders. These sensors are made to find particular biomarkers, which are gases found in the body at very low levels. Gas sensors can aid in the early, most effective diagnosis of diseases like cancer and bacterial infections by identifying these indicators.

Typically, gas sensors are a vital technology that is utilized in a variety of applications and industries. These sensors are essential for guaranteeing human health and safety, safeguarding the environment, and improving our understanding of numerous diseases and medical situations since they can detect and monitor gases in real time. Gas sensors must fulfil both performance and practical requirements in order to be used in practical applications. The performance criteria of a gas sensor can be divided into four categories: sensitivity, selectivity, speed, and stability. Sensitivity is essential since gas concentrations might be extremely low in a variety of circumstances. For example, detecting biomarkers like ammonia in exhaled breath requires a minimum detection limit of at least 50 ppb. Selectivity is equally important, as detecting particularly the analyte gas is essential to avoid misleading results. Making materials that can only detect one target molecule, nevertheless, is difficult. In order to study the gas response of marginally adjusted sensing channels, sensing systems use sensing arrays. Additionally, short response and recovery times are important since they allow for prompt and accurate outcomes, particularly in the case of gas leaks. Finally, a gas sensing channel's stability is crucial for maintaining a constant gas response throughout several sensing cycles.

MXenes are a few atomic layers thick inorganic 2D materials composed of transition metal nitrides, carbides, or carbonitrides [1]. Their general formula is $M_{n+1} X_n$, where M represents

DOI: 10.1201/9781003436942-7

transition metals like Sc, V, Zr, Mo, Ti, Nb, Ta, Cr, and Hf and X represents the element carbon or nitrogen [1,2]. This was discovered by Drexel University scientists in 2011 [3]. By intercalation of large organic molecules, sheets were delaminated to form single-layer MXene flakes in 2013 [4]. MXenes are synthesized from MAX phases, which are ternary carbides and nitrides that are layered in structure [5,6]. The MAX phases are with the general formula $M_{n+1}AX_n$. The elements of groups 13 or 14 are denoted by A. They have strong M-X bonds, while the M-A bonds are weaker [7,8]. MXenes are materials possessing characteristics like high conductivity, good fluorescent, plasmonic, and optical properties [9]. They are mainly used in energy storage [9–12], biosensing [13–16], and catalytic application [17]. The good biocompatible nature of the molecules makes them useful for biomedical applications. It can work as a good immobilization matrix that retains the catalytic activity of an immobilized molecule due to the multilayered configuration of MXene.

For the synthesis of MXenes, the MAX phase is used as a template, where the A element is removed using a wet chemical etching process. Etching is done by immersing the MAX phase in hydrofluoric acid (HF) or HF-containing etchants. Etching, delamination, and intercalation are different processes involved in the synthesis process [18]. Selective etching of the MAX phase removes the A layer as the M-A bond is weaker while retaining the stronger mixed metallic-covalent M-X bond.

7.2 MECHANISM OF MXENES SYNTHESIS

After MAX phase formation, etching and exfoliation of the MAX phase result in exfoliated multilayered MXenes [20,21]. Figure 7.1 depicts the exfoliation process of the MAX phase and the formation of MXenes.

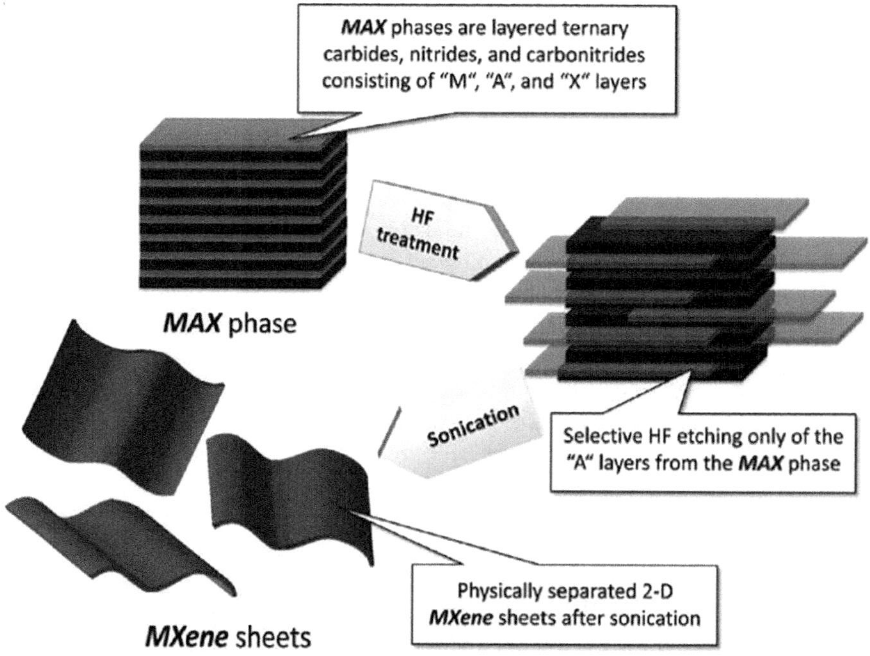

FIGURE 7.1 Exfoliation process of MAX phase and formation of MXenes. (Reproduced with permission from Ref. [19], Copyright © 2012 American Chemical Society.)

7.2.1 Etching and Exfoliation

There are top-down and bottom-up approaches for MXene synthesis [22–24]. In the etching process, $M_{n+1}X_n$ were made intact by etching the A layers from them. The etching process is the oxidation of an element of the MAX phase which poses an oxidation state zero [25]. The M-A bond strength is comparatively weaker than that of the M-X bond, which affects the Gibbs energy parameter and makes the etching process easier.

$$Ti_3AlC_2(s) + 3H^+(aq) \rightarrow Ti_3C_2(s) + Al^{3+}(aq) + 3/2\ H_2(g)$$
$$Al^{3+}(aq) + 3F^-(aq) \rightarrow AlF_3(aq)$$
$$Al^{3+}(aq) + 6F(aq) \rightarrow AlF_6{}^{3-}(aq)$$

When etched with solutions containing HF or HF-containing solutions, the protons H^+ act as the oxidizing agent, while the other by-products of the reaction Al^{3+} are solubilized by the F^- ions. The oxidation of aluminum can also be done electrochemically, where a positive potential is applied, which leads to the etching of Ti_3AlC_2 by the following equation [26,27]:

$$Ti_3AlC_2(s) \rightarrow Ti_3C_2(s) + 3e^- + Al^{3+}(aq)$$

The halide ions or the halogens act as oxidizing agent in the halogen-based etching of aluminum [28]. The halogen- and molten-based etching techniques produce MXenes with functionalities like –Cl, –Br, –I, –NH, –S, –Se, –and Te, and their uniform distribution over the MXene surface [29]. The surface chemistry of MXenes depends on the etching route and the surface coverage of the functionalities. For top-down etching routes of MXene synthesis, according to computer studies, the first adsorption of H and F atoms on Ti weakens the Ti-Al link, causing Al to be removed as soluble AlF_3 and generating an interlayer space to enable further HF and H_2O intercalation for afterward occurring etching [30].

7.2.2 Intercalation and Delamination

The process of separating 2D MXene into different single layers for further application is called delamination. The thicker MXene requires higher adhesion energy compared to the lighter ones. The attractive forces between different layers should be broken down for the delamination process and this becomes challenging because of the polarity and high dipole moment of MXene. Solvents such as dimethyl formamide (DMF), water, N-methyl-2-pyrolidone (NMP), and dimethyl sulfoxide (DMSO) are used for the delamination of MXenes with a surface containing functionalities F, O, and OH.

The intercalation and delamination of etched HF-etched MXene was first done with DMSO. Tetra butyl ammonium hydroxide bases, such as tetra butyl ammonium hydroxide (TBAOH) and tetra methyl ammonium hydroxide (TMAOH), were also used as delaminating agents for other MXenes. Li^+ ions can be used for delamination of MXenes synthesized with HCl and F^- ions. LiCl delamination is not applicable for HF-etched MXenes. The lattice parameter of the multilayer MXenes increases when the ion intercalation happens and swelling of MXenes by water intercalation can also be done for delamination. The delamination by LiCl improves the colloidal stability of the MXene solution due to the stronger interaction of Li^+ ions with the MXene layers [31].

The suitability of MXenes for the application of electrodes and catalysts requires the following properties:

a. Reducing interfacial charge transfer resistance by increasing electrical conductivity.
b. High electrochemical stability, processibility, and hydrophilicity in aqueous electrolytes.
c. Increasing the optimal binding to analytes by tuning and controlling the composition of surface groups.

Both etching methods can be applicable based on the application of the synthesized MXenes. The salt etching process with Lewis acid provides MXene with new surface functionalities, while the composition of T_x functionalities can be controlled by HF, alkali, and halogen-based etching mechanisms. The properties of resulting MXenes also depend on factors such as the structure, stoichiometry, and grain size of the precursor MAX material.

7.3 PROPERTIES OF MXENES

The electrochemical properties of MXene are enhanced by the presence of surface functional groups obtained by suitable etching practices. Surface functionalities such as hydroxyl, fluorine, and oxygen induce hydrophilicity in MXene, which improves its diffusion, electrical conductivity, and ion adsorption properties. Different cationic species can be intercalated into the negatively charged layered structure of MXene by the Coulombic force of attraction, which makes them promising materials for a wide range of applications. Owing to their tunable electronic band structure, MXenes are expected to behave as metals or semiconductors in nature. The presence of certain surface functional groups increases their work function and improves their utility in electronic applications. The presence of surface functional groups on MXene enhances its electrical properties, such as the work function and band structure. The contact angle of MXene with water of 21.5° makes its surface hydrophilic, and it is highly conducting nature (~ 10^4 S cm^{-1} for $Ti_3C_2T_x$) making the material unique to graphene. The stability of MXenes in polar solvents provides the advantage of integration with biomolecules, while their hydrophilic nature makes them biocompatible. It has been reported that organic solvents, such as DMF, PC, ethanol, and NMP, create stable solutions with longer lifetimes. The study also concluded that, when compared with water, organic solvents cause slow degradation with MXene. Figure 7.2 exhibits the dispersion stability of MXene in various solvents.

The size of the intercalated species controlled the spacing between the two MXene nanosheets. The intercalation of MXene with positively charged species makes it hygroscopic, and the MXene layers with intercalated water layers show a spacing of 12.5–15.5 A°. Intercalation of organic molecules, such as stearyl trimethylammonium bromide (STAB) and cetyltrimethylammonium bromide (CTAB), has been reported to create an interlayer spacing of 22.3 A°. Higher interlayer spacing is created by the intercalation of alkylammonium cations and tributylammonium cations, making them suitable for gas sensing applications. Numerous theoretical works have been reported regarding the electronic properties and band structures of different types of MXenes. Most MXenes exhibit a band gap of 0.24–0.18 eV predicting their metallic structure, while a few of the MXenes show semiconducting behavior. MXenes have distinct properties compared with other 2D materials, making them increasingly useful in the fabrication of electrochemical sensors. The obvious benefit of MXene over other 2D materials is its increased electrical conductivity, which accelerates the rate of electron transfer.

MXenes with good solution dispersibility and stability are simple to synthesize. This is critical for the development of electrochemical sensors as the most common method for preparing electrodes is drop casting, which requires the preparation of a good dispersion coating solution in advance. Their biocompatibility with biomolecules, such as enzymes, proteins, and nucleic acids, as well as their non-toxicity, makes them ideal carriers for biosensors and biomedical applications. The unique photothermal conversion ability enables a dual-model detection strategy for the design of electrochemical sensors, which broadens the range of electrochemical sensor signal strategies. MXenes have already proven their potential as electrode materials for batteries, supercapacitors,

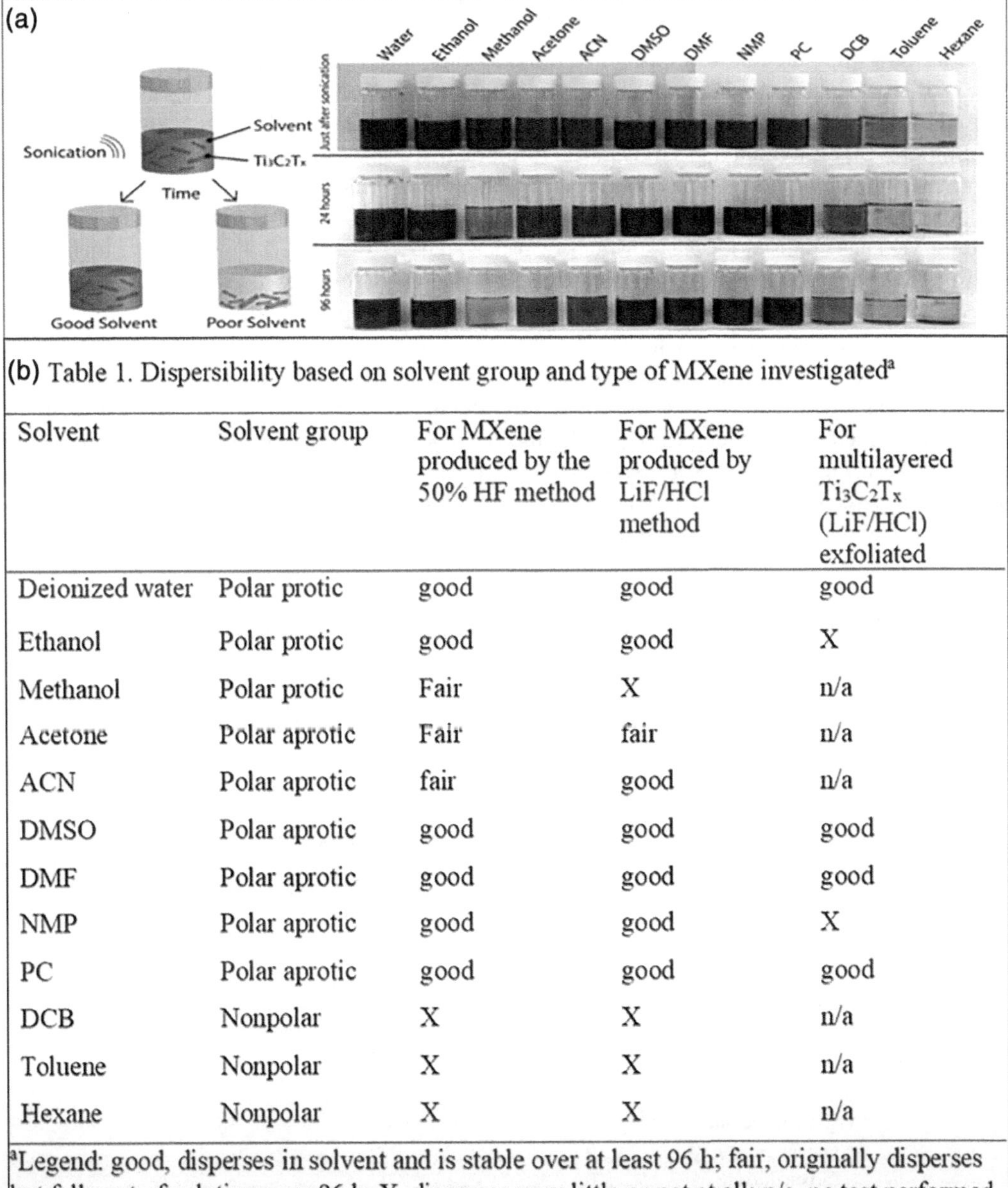

(b) Table 1. Dispersibility based on solvent group and type of MXene investigated[a]

Solvent	Solvent group	For MXene produced by the 50% HF method	For MXene produced by LiF/HCl method	For multilayered $Ti_3C_2T_x$ (LiF/HCl) exfoliated
Deionized water	Polar protic	good	good	good
Ethanol	Polar protic	good	good	X
Methanol	Polar protic	Fair	X	n/a
Acetone	Polar aprotic	Fair	fair	n/a
ACN	Polar aprotic	fair	good	n/a
DMSO	Polar aprotic	good	good	good
DMF	Polar aprotic	good	good	good
NMP	Polar aprotic	good	good	X
PC	Polar aprotic	good	good	good
DCB	Nonpolar	X	X	n/a
Toluene	Nonpolar	X	X	n/a
Hexane	Nonpolar	X	X	n/a

[a]Legend: good, disperses in solvent and is stable over at least 96 h; fair, originally disperses but falls out of solution over 96 h; X, disperses very little or not at all; n/a, no test performed.

FIGURE 7.2 (a) Dispersion stability of MXene in various solvents. (b) Dispersibility of MXene based on solvent type. (Reproduced with permission from Ref. [32], Copyright © 2017 American Chemical Society.)

and sensors because of their exceptional chemical stability, high electrical conductivity, large surface area, layered structure, and environmental friendliness. MXene, which has a high degree of stretchability and biocompatibility, can be an effective substrate for creating platforms with good conductivity that can be used to develop wearable electrochemical sensors, which are used for clinical analysis and healthcare monitoring. The 2D layered structures of MXene and unique surfaces with abundant chemical groups make them excellent candidates for combination with a wide range of functional materials for various purposes.

7.4 MODIFICATIONS OF MXENES FOR GAS SENSING APPLICATION

MXenes have the potential to outperform the conventional 2D materials because of their unique mix of high electrical conductivity and more surface functional groups. The high gas response is caused by the abundance of functional groups, such as –OH, –O, and –F, which can act as efficient binding sites for even diluted gas molecules. Additionally, Ti_3C_2 has shown that MXenes have good electrical conductivity with values as high as 6500 S cm^{-1}. A remarkably low level of electrical noise might be made possible by this extraordinary property. Due to the beforementioned properties, $Ti_3C_2T_x$ MXenes are used for gas sensing of various analytes (Table 7.1). Different parameters that measure the performance of the gas sensor are mainly response time, recovery time, and limit of detection (LOD). The recovery and response time are usually defined by the amount of time it takes for the sensor's resistance to reach certain percentages of its ultimate value. They are critical performance characteristics that indicate how quickly the sensor reacts to changes in gas concentration and how soon it recovers to its original state after being exposed to the analyte gas. The response time reflects the rate at which gas molecules are adsorbed or desorbed on the sensor's surface and represents how rapidly the sensor responds to the gas. It is the time taken by the gas sensor to attain a particular percentage (usually 95%) of its ultimate resistance value after being exposed to a target gas concentration. A faster response time is required for real-time monitoring and detection. The recovery time of a gas sensor can be defined as how rapidly the sensor recovers or desorbs the gas molecules from its surface and returns to its original state. A gas sensor's recovery time (rec) is the time it takes for the sensor to restore to a specified percentage (usually 95%) of its initial resistance value once the gas source is removed.

When real-time measurements are required especially for precise gas sensing applications, both response and recovery times are critical. A gas sensor with a quick reaction and recovery time can

TABLE 7.1
$Ti_3C_2T_X$ MXene for Various Analyte Sensing and their Significance

MXenes	Analytes	Response	Concentration (ppm)	LOD	Ref.
$Ti_3C_2T_X$	Ethanol	0.115	100		[33]
	Methanol	0.143	100		
	Acetone	0.075	100	25 ppm	
	Ammonia gas	0.21	100		
$Ti_3C_2T_X$	Ethanol	1.7	100	100 ppb	[34]
	Acetone	0.97	100	50 ppb	
	Nitrogen dioxide	0.25	100		
	Sulphur dioxide	0.2	100		
	Carbon dioxide	0.1	100		
	Ammonia gas	0.8	100	100 ppb	
	Acetaldehyde	0.88	100		
$Ti_3C_2T_X$	Methanol	2.2	10	50 ppb	[35]
	Acetone	1.4	10	50 ppb	
	Ethanol	1.7	10	50 ppb	
	Ammonia gas	0.7	10		
	TCM	0.1	10,000		
	Water gas	0.5	10,000		
	Nitrogen dioxide	0.9	10		
$Ti_3C_2T_X$	Nitrogen dioxide	4.5 %	5	-	[36]
$Ti_3C_2T_X$	Ethanol	3.5%	0.1%	-	[37]
	Carbon dioxide	0.5%	1%		
$Ti_3C_2T_X$	Ethanol	0.16	100	-	[38]

offer accurate and timely gas concentration readings, making it suited for a wide range of environmental and industrial monitoring applications. The selection of a gas sensor, as well as the optimization of its materials and design, is critical in obtaining optimal response and recovery times. In applications where trace concentration of gases need to be accurately measured, the parameter LOD (limit of detection) is of utmost importance. This parameter is important for assessing the sensitivity and performance of the sensor. It is the lowest concentration of analyte gas that can be reliably detected by the sensor. Mathematically, it can be calculated by the following equation:

$$\text{LOD} = 3.3 * (\sigma / S)$$

where S denotes the sensitivity of the sensor and σ represents the standard deviation of sensor response in the absence of analyte gas. Sensitivity can be obtained from the slope of concentration to sensor response calibration curve.

Functionalization is essential for improving the performance and selectivity of MXene as gas sensors. The surface of MXene can be altered to control how it interacts with target gases and how reactive it is chemically. Strategies for functionalization can include adding functional groups or decorating the surfaces of MXene with nanomaterials, polymers, or metal nanoparticles. Different MXene-based composites and their application in gas sensing are discussed in Table 7.2. The numerous functionalization techniques used for MXenes in gas sensing applications will be covered in detail in this discussion.

Highly sensitive and selective chemiresistive sensors to varied concentrations of ammonia were developed by Tingting and his coworkers. An MXene/SnO_2 heterojunction-based chemiresistive-type sensor was synthesized using a hydrothermal method, where MXene was decorated with SnO_2 nanoparticles. The sensor exhibited improved selectivity to ammonia, with a significantly higher and reproducible response at room temperature compared to the undecorated MXene. This enhanced sensing capability is attributed to the formation of MXene/SnO_2 heterojunctions, which introduce additional electrons at the heterojunction surface, resulting in more sites for NH_3 adsorption. The 2D MXene plays a crucial role in the sensing process due to its high conductivity and selective adsorption properties to ammonia at room temperature [33]. Tai et al. developed a TiO_2/$Ti_3C_2T_x$ bilayer film gas sensor that improved the ammonia sensing capabilities of pure 2D $Ti_3C_2T_x$

TABLE 7.2
MXene Composites for Different Analyte Gas Sensing and its Properties

Composite	Analyte Gas	Response (%)	Recovery Time (seconds)	LOD	Ref.
$Ti_3C_2T_X$/rGO/CuO	Acetone	52.09	7.5	-	[39]
$Ti_3C_2T_X$/ZnO	Nitrogen dioxide	41.90	105	-	[40]
Pd-$Ti_3C_2T_X$	H_2	56	-	-	[41]
Pd-$Ti_3C_2T_X$	H_2	23	161		[42]
$Ti_3C_2T_X$/TiO_2	Hexanal	8.80	461	217	[43]
$Ti_3C_2T_X$/TiO_2	Ethanol	83	-	-	[44]
$Ti_3C_2T_X$/TiO_2	Nitrogen dioxide	16	-	125	[45]
$Ti_3C_2T_X$/TiO_2	Ammonia	40.60	6	5	[46]
FOTS-$Ti_3C_2T_X$	Ethanol	14	139	-	[47]
$Ti_3C_2T_X$/TiO_2	Ammonia	3.10	277	500	[48]
$Ti_3C_2T_X$/WO_3	Ammonia	23.30	228	-	[49]
$Ti_3C_2T_X$/Fe_2O_3	Acetone	16.6	5	560	[50]
$Ti_3C_2T_X$/In_2O_3	Methanol	29.60	3.5	-	[51]
$W_{18}O_{49}$/$Ti_3C_2T_X$	Acetone	11.6	6	170	[52]
CuO/$Ti_3C_2T_X$	Toluene	11.4	10	310	[53]

nanosheets using a rational design method [48]. According to their study, the $TiO_2/Ti_3C_2T_x$ sensor outperformed the pure $Ti_3C_2T_x$ sensor in terms of response values and recovery speed. Figure 7.3 shows the recovery curves of different composite gas sensors to NH_3. The synergistic combination of ZnO nanoparticles and TiO_2 nanosheets, both combined with $Ti_3C_2T_x$, led to the effective attainment of very sensitive NO_2 detection at ambient temperature, which was reported by Li et al. [55].

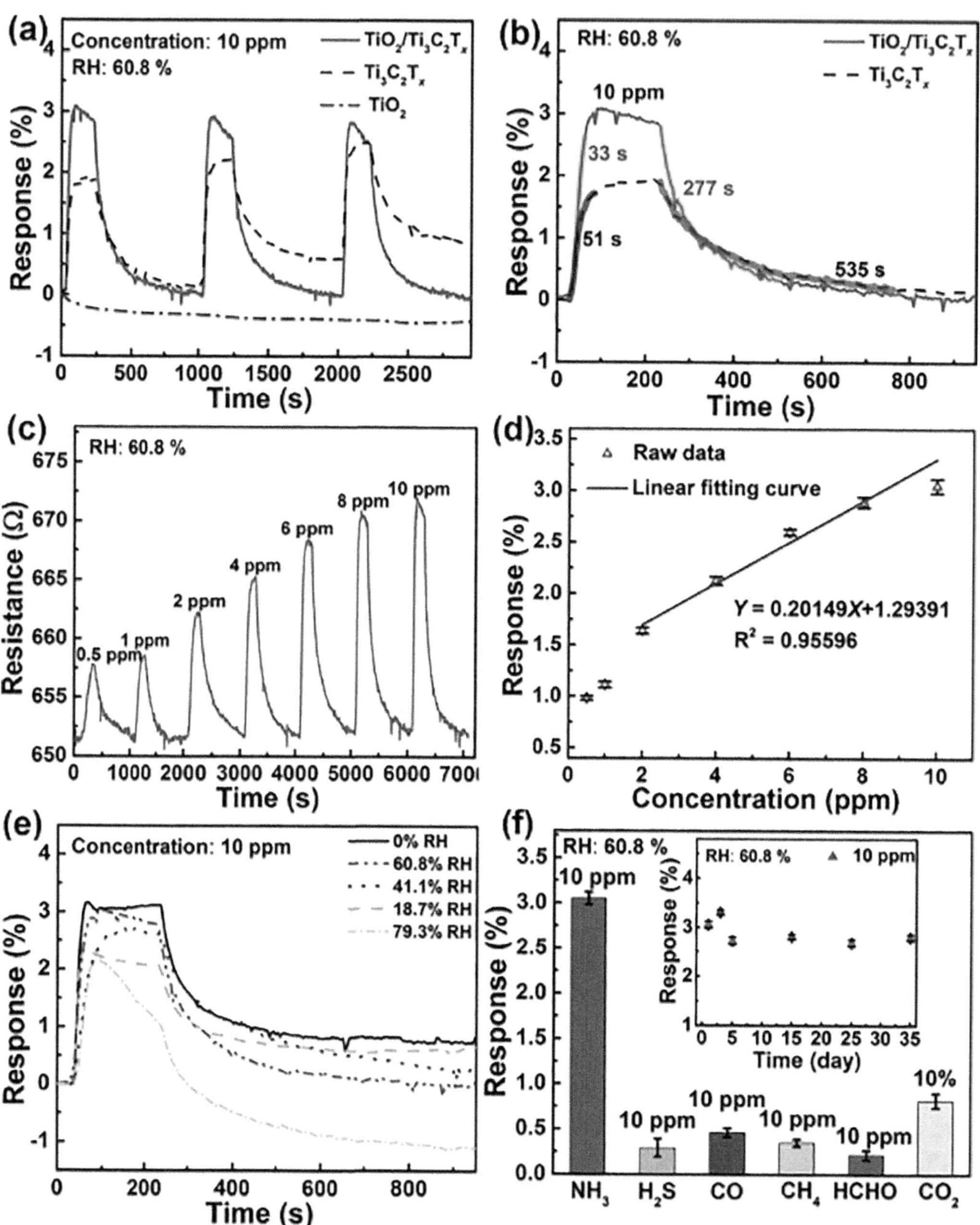

FIGURE 7.3 (a) Recovery curves of different composite gas sensors to NH_3. (b) Recovery times of gas sensors. (c) Dynamic response/recovery curve of the gas sensor to different concentrations of ammonia. (d) Response versus concentration of ammonia. (e) Response–recovery curves at different RH. (f) Responses of the gas sensor to different gases. (Reproduced with permission from Ref. [48], Copyright © 2019 Elsevier.)

The addition of 0D ZnO nanoparticles efficiently prevents adjacent M-TiO_2 nanosheets from stacking on top of one another, creating effective diffusion pathways and open surfaces for target gas molecules. Further, by using 2D M-TiO_2 nanosheets, which have high surface-to-volume ratios, it was possible to increase gas adsorption by guaranteeing the availability of strong active sites. Additionally, the increased detection sensitivity is a result of the various heterojunction interfaces between M-TiO_2 nanosheets and ZnO nanoparticles. The resulting ZnO@M-TiO_2-based gas sensor demonstrates exceptional sensitivity to NO_2 by utilizing the distinct qualities and synergistic effects of various components. Figure 7.4 illustrates the response of the ZnO@M-TiO_2 gas sensor under various conditions.

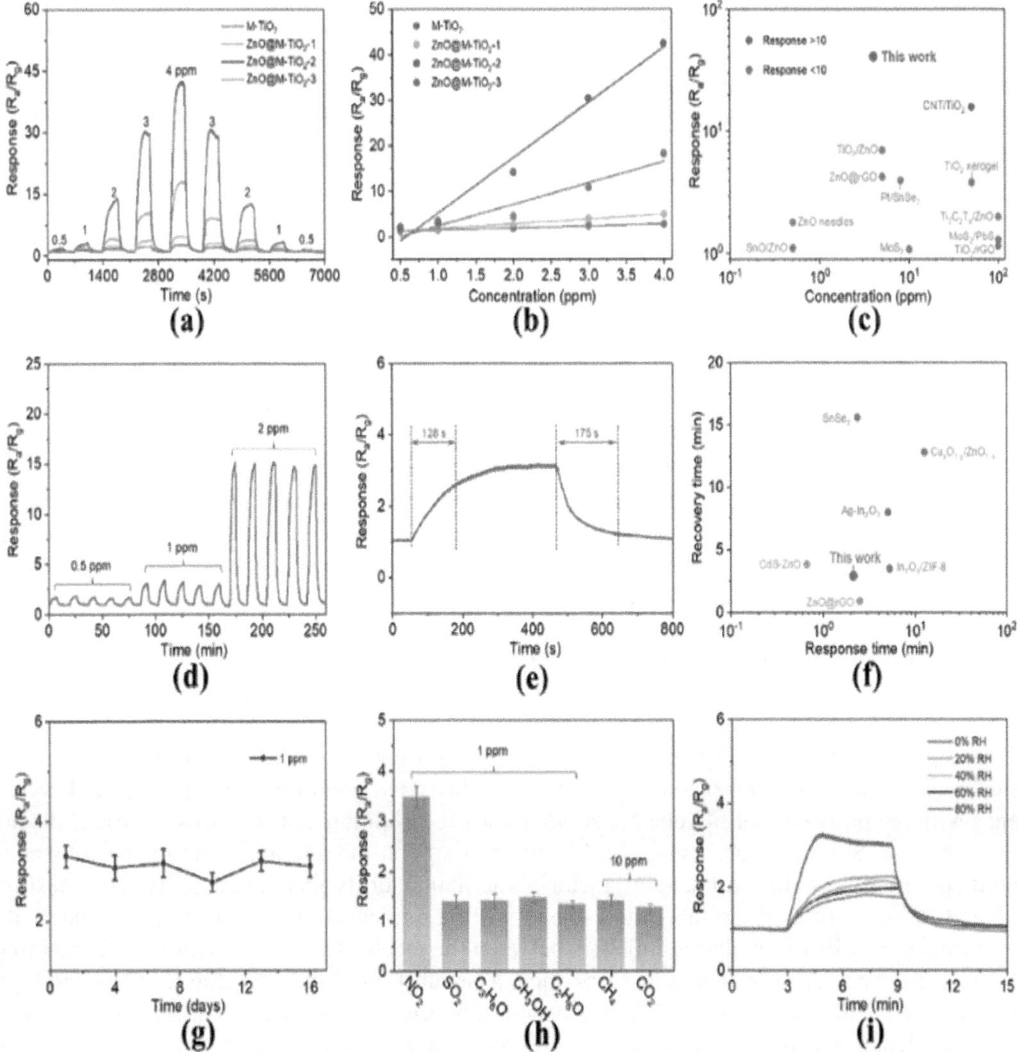

FIGURE 7.4 The response of 2D ZnO@M-TiO_2 gas sensor at various conditions. (a) Response–recovery curves at 40% RH in the presence of different concentrations of NO_2. (b) The relation between the analyte concentration and sensor response. (d) Repeated sensing tests. (e) Response and recovery of the sensor in the presence of NO_2 (1 ppm). (f) Comparison with the reported values in the literature. (g) Stability plot of the sensor. (h) Selectivity and interference plots. (i) Responses of the sensor under different RH conditions in the presence of 1 ppm NO_2. (Reproduced with permission from Ref. [55], Copyright © 2023 Elsevier.)

7.5 FABRICATION AND PREPARATION OF GAS SENSOR

After the synthesis of MXenes and the proper functionalization to form composites of desired properties, the following steps can be taken for the device fabrication. MXene nanosheets must be prepared before being put onto a substrate to create a thin film. Spin coating, drop-casting, spray coating, and vacuum filtration are examples of common methods for MXene film deposition. The choice of deposition technique is influenced by things like substrate compatibility, homogeneity, and film thickness. This step is important for the controlled growth or formation of excellent thin films. Different techniques for film formation are spray coating, drop casting, spin coating, layer-by-layer assembly, vacuum filtration, and chemical vapor deposition.

The spray coating technique involves using a spray nozzle or gun to deposit the solutions of MXene or its composites into the desired substrate. The suspension is made into small droplets from the spray gun and they are continuously deposited onto the substrate, which is then kept for drying. This method is suitable for varied complex composites for a large area deposition. The most used technique is drop-casting, in which an appropriate solvent is used for making the MXene solution which is drop-cast on the suitable substrate. The film will be formed on the substrate after solvent evaporation.

Uniform films can be made by the spin coating technique in which the solution that is dispensed on the suitable substrate is spun at a high speed. The MXene suspension is evenly distributed throughout the substrate by centrifugal force, resulting in the formation of a uniform and thin coating. To regulate film thickness, the spinning speed and duration can be altered. Multilayer films of MXenes with controlled thickness can be made by the layer-by-layer assembly technique. Polyelectrolytes with opposite charge and MXene are alternately deposited on the substrate to form multiple MXene layers.

A porous filter or membrane of MXene can be deposited on the substrate by vacuum filtration. A thin MXene layer will be left on the filter surface after the MXene dispersion is poured onto it and then drawn through the filter by means of suction. After that, the film can be moved from the filter and applied to the selected substrate. By using chemical vapor deposition, MXene-containing gaseous precursors can be delivered into the reaction chamber, where they conduct chemical reactions and deposition of thin films on substrate happens. The preferred film thickness, uniformity requirements coherences with MXene, and scalability characteristics are among the variables that must be taken into consideration when selecting a film deposition procedure. While choosing the best film deposition technique for the fabrication of the sensor, it is crucial to consider the unique properties of MXene composites such as their stability, conductivity, and surface functionalization.

The performance, stability, and selectivity of the sensor can be influenced by the choice of substrate for the sensor fabrication. Excellent conductivity, good thermal stability, and coherence with the properties of MXene are some of the crucial points to take care of while choosing the substrate. Silicon, quartz, glass, polymers, and silicon dioxide are some of the substrates used commonly. Another influencing factor is thermal stability and conductivity. The sensitivity and response time of the fabricated sensor will be influenced by the electrical conductivity of the substrate. Efficient charge transfers between the electrodes and the substrates containing MXene film provide good sensor performance. The substrate should be able to withstand high temperatures while annealing and film deposition, hence the substrate should be thermally stable. Excellent gas sensing performance occurs when there is better contact between the analyte gas and the MXene composite. The substrate surface should be devoid of defects or roughness to facilitate the above factor.

Thin metal foils or polymers with excellent mechanical stability and flexibility as substrates are preferred for easy handling and operation of films. The optical transparency of the substrate is important for optoelectronic gas sensors. Glass and transparent polymers are examples of transparent substrates that allow efficient transmission of light and optical characterization. The substrate

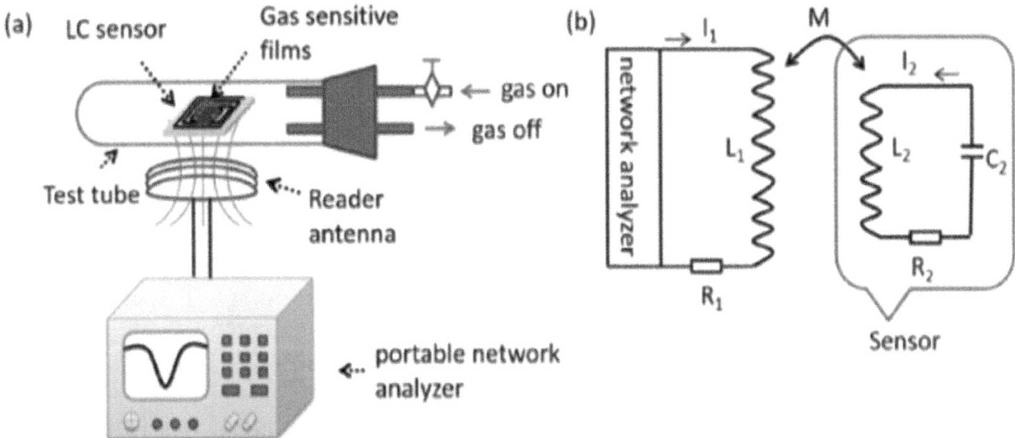

FIGURE 7.5 Gas sensing setup and equivalent circuit of SnO_2 MXene-based gas sensor. (Reproduced with permission from Ref. [54], Copyright © 2020 Elsevier.)

must be accessible in dimensions that match the necessary sensor specifications and methods of fabrication. Additionally, it should be scalable to allow for mass production of MXene-based gas sensors. The choice of the substrate should be based on the requirements of the gas sensor application, and each substrate material has advantages and limits of its own. To guarantee sensor compatibility, dependability, and performance, the substrate parameters must be carefully considered.

Techniques like electron beam lithography or photolithography are used for patterning electrodes on the substrate surface. Designing the gas chamber is a crucial step, and it should ensure compatibility with the substrate, controlled analyte gas flow, and proper sealing in their setup. Figure 7.5 shows the gas sensing setup and equivalent circuit of SnO_2 MXene-based gas sensor. The integration of the fabricated electrode and designed gas chamber into the control systems is the last step in the fabrication of MXene-based gas sensor. The protection of the sensor for its long-term stability from ambient conditions should be also ensured by the encapsulation method. Important parameters like concentration, selectivity, response time, sensitivity, stability, and recovery time will be ensured by testing the sensors response.

7.6 GAS SENSING MECHANISM AND PERFORMANCE OF MXENES

The electron accepting or donating ability of the analyte gas forms the basic mechanism for most of the gas sensors [56]. Compared to common metal oxide gas sensors, which primarily rely on the charge-transfer model caused by electron donation/acceptance from target gases, MXene gas sensing mechanisms are more complex. $Ti_3C_2T_x$ MXene showed a positive response to four electron donor gases in the earliest report on MXene-based gas sensors, which led the authors to classify $Ti_3C_2T_x$ as a p-type semiconductor and explain the positive response by a reduction in the number of majority carriers of $Ti_3C_2T_x$ MXene upon exposure to electron donor gases [57]. Nevertheless, it was discovered in some of the subsequent studies that $Ti_3C_2T_x$ MXene consistently displayed a positive resistance variation in response to oxidizing or reducing types of gases. These positive variations are due to mainly two things:

a. Unlike semiconductors, MXenes exhibit metallic behavior. Gas adsorption onto MXene surfaces can lead to a reduction in the number of carriers, resulting in an increase in channel resistance [48].
b. Upon exposure to gases, the interlayer of MXenes swells, which in turn impedes the transport of electrons out of the plane of the material [42,58].

The content of MXenes in composites decides the response direction of the sensor. When the content of MXenes is high compared to the other material in the composite, the resistance is much smaller as their behavior is more like that of MXene in pure form [49,53]. But on the contrary when the amount of metal oxide is more than the content of MXene, the resistance is higher and behaves like that of pure metal oxide [59]. The sensitivity of sensor can be increased due to the Schottky effect developed between the metal oxide and MXene in the composite [60]. When exposed to four electron donor gases, $Ti_3C_2T_x$ sensors exhibit a sensing behavior that is typical of p-type semiconductors, with the maximum response shown for NH_3. The electrons are transported to $Ti_3C_2T_x$ upon gas molecule adsorption onto functional groups. This decreases the amount of majority carriers and raises sensor resistance. The process of an electron moving from NH_3 to $Ti_3C_2T_x$ specifically goes like this:

$$2NH_3 + 3O^- \rightarrow N_2 + 3H_2O + 3e^-$$

$$NH_3 + OH^- \rightarrow NH_2 + H_2O + e^-$$

There are different gas sensing mechanisms based on the analyte gas and the MXene substrate used. The detection of gases can be through redox reactions, adsorption–desorption, chemiresistive, charge transfer reactions, and surface reactions. The presence of surface terminals like –F, –OH, and –O initiates the charge transfer reactions between the gas molecules and MXene. The charge carrier concentration varies when the surface functional groups donate or accept the electrons from the analyte gas molecules. The mode of interaction can be through Lewis acid–base, hydrogen bonding, or electrostatic interactions. The charge carrier concentration increases when electron transfer happens from the gas molecule to the MXene surface, and when the electron transfer is from MXene film to analyte gas, the carrier concentration decreases. This enhances the electrical conductivity of the sensor. The gas sensor response can be measured quantitatively by the change in resistance or conductivity. The analyte gas concentration is directly proportional to the change in conductivity.

The surface groups on the MXene surface also undergo chemical reactions with the gas molecules. The gas molecules can absorb or penetrate the layers of MXene, which changes the surface chemistry of MXene. Various reactions like acid–base, oxidation, and reduction may occur between MXene and gas molecules. This results in the formation of new chemical species or the release of ions, which affects the conductivity or resistance of the sensor.

Chemiresistive sensing mechanism measures the variation in electrical resistance when the material is exposed to gas molecules. The conductivity of the sensor will be affected as they change the electron carrier concentration. Due to the presence of transition metals, some MXene composites exhibit redox reactions, which allow efficient gas sensing properties.

7.7 STRATEGIES TO ENHANCE GAS SENSING PERFORMANCE OF MXENES

a. The structural arrangement of MXenes has an influence on their gas sensing performance as they tend to restack their layers, which leads to lowering of surface area and active adsorption sites. Ultrasonic spray pyrolysis technique was used by Yang et al. to develop 3D crumpled MXene/ZnO spheres for NO_2 gas sensing. The increased surface area, the formation of p–n heterojunction, and the presence of edges and defects increase the recovery rate of NO_2 sensing from 27.27% to 41.93%. The p–n junction creates a depletion layer, which facilitates the transfer of electrons and holes through drift. Upon adsorption of NO_2 molecules, the molecules acquire electrons, further increasing the width of the depletion layer. This widening of the depletion layer enhances transmission efficiency and induces greater variation in resistance.

b. Modifying the surface of MXene with appropriate functional groups enhances the gas sensing ability of MXene-based sensors. The presence of fluorine functional groups in MXene reduces the gas sensing ability of the sensor. The F^- terminals in MXene were reduced by alkaline sodium hydroxide treatment by Yang et al., thus increasing the amount of oxygen and hydroxyl functional groups.
c. Like their parent MAX phases, MXenes frequently display metallic behavior. Surface terminations, on the other hand, can change the electrical characteristics of MXenes, turning them into semiconductors or even insulators. A Schottky or heterojunction forms at their interface when MXenes and metal oxides are mixed as a result of electron transport from the high Fermi-level side to the low Fermi-level side. The junction barrier height and sensor resistance are exponentially connected, and gas adsorption or desorption can cause a minor change in the barrier height to have a large effect on the sensor resistance. The performance of MXene-based gas sensors can therefore be enhanced by creating hybrid devices with heterojunctions.

7.8 APPLICATIONS OF MXENES-BASED GAS SENSORS

On account of properties like good sensing capability, real-time monitoring, selectivity, and non-invasive monitoring MXenes and their composites are framed for applications like breath analyzers, food quality, and air quality monitoring. Detection of quantity of biomarkers or analyte gas present in the breath of a person is the primary function of a breath analyzer, and these devices are mainly used in diagnosing certain diseases and in alcohol testing. The change in electrical characteristics of MXene composites when they are exposed to the analyte gas in the breath of a person is monitored. Highly sensitive and accurate detection can be possible by correlating the change in conductivity with the concentration of analyte gas. The presence of different gases or biomarkers in the breath is a major challenge and demarcating them from the analyte gas is a major challenge. The issues of false positives and negatives are reduced by the high specific and accurate real-time monitoring property of MXene composites. The proper monitoring of diseases or the diagnosis of diseases by identification of certain biomarkers is the main application of breath analyzers in medical field. Non-invasive MXenes-based breath analysis is more user-friendly and convenient than blood testing or other intrusive techniques. This may result in more frequent monitoring and thus better health management.

Using the 3D MXene architecture, an efficient method for developing high-performance and adaptable VOC sensors was developed. The three-dimensional structure was created by electrospinning an aqueous solution of positively charged polymer. Electrostatic interaction causes negatively charged MXene sheets to self-assemble onto the surfaces of polymer fibers. This 3D structure can build densely connected porous structures that allow gas molecules to easily travel. The resulting sensor demonstrated great sensitivity, a low experimental detection limit, a broad sensing range, significant adaptability, and reversibility for a variety of VOCs in the ppb range, indicating tremendous promise for future sensor development.

The robust nature and gas sensing capabilities are crucial parameters that employed MXenes in the detection of amounts of pollutants and harmful substances in the atmosphere through air quality monitoring. The sensitivity to various gases and volatile organic compounds can be increased by building gas sensing arrays with different layers of MXene. By this way, multiple contaminants can be detected at once and air quality monitored. MXene-based air quality sensors are fabricated for measuring pollution concentrations. Air quality can be improved by making timely interventions in policy decisions and continuous monitoring of pollution trends for improving the quality of air. MXene composites are ideal for portable and wireless air quality monitoring equipment, which can be deployed in a variety of locations, resulting in a more extensive and dispersed data collection network.

Assessing the authenticity, freshness, and safety of the food items is the primary need of food quality control. Volatile compounds generated during food spoilage can be detected by MXene-based gas sensors. It is feasible to identify the freshness and shelf life of perishable food items by monitoring these molecules. MXene sensors can be programmed to detect specific contaminants in food samples, such as pesticides, heavy metals, or adulterants. Keeping tainted or hazardous things out of circulation might help to assure food safety and authenticity. The sensitivity of MXenes to gases has the potential to be utilized to evaluate the quality of food packaging. They can detect leaks in gas sensitive packaging, for example, ensuring that food is packed and protected safely. MXene sensors can be used to monitor food quality and safety during transport and storage in the food supply chain. This can aid in the detection of possible problems and the preservation of food quality from manufacturing to consumption. While the theoretical applications of MXenes seem exciting, practical implementation and commercialization may necessitate further research.

7.9 CHALLENGES AND FUTURE DIRECTIONS FOR MXENE-BASED GAS SENSORS

Finding safer and more effective preparatory techniques for MXenes are necessary since HF etchants are harmful to the humans and environment. Also using the ultrasonication technique to create single layer of MXenes takes time. Finding safer and more effective preparatory techniques is thus important. The majority of MXene-based gas sensors were assembled and fabricated manually. This makes it impossible to accurately manage the thickness and surface areas of films, which makes them unsuitable for mass production. In order to meet the demands of industrial production, there is a need to investigate more advanced technologies. MXenes-based gas sensors possess more complicated gas sensing mechanism than the normal charge transfer mechanism of 2-dimensional materials. Metallic MXenes always exhibit positive resistance fluctuations when exposed to oxidizing or reducing gases due to their high metallic conductivity. The interlayer swelling of MXenes also is crucial in the conductivity change. More theoretical studies should be done to study the mechanism further. The electrical characteristics of MXenes and their gas sensing and their gas sensing capabilities are significantly influenced by the surface groups. Thus, it is crucial for the further development and use of MXene-based gas sensors to synthesize MXenes with regulated surface terminals. The difficulty to withstand high humidity is another crucial challenge faced by MXene-based gas sensors.

7.10 CONCLUSION AND FUTURE OUTLOOK

Due to their special combination of characteristics, such as strong conductivity, numerous surface functional groups, high specific surface area, and the potential for functionalization, MXenes are a promising class of 2D materials for gas sensing applications. However, pristine MXene-based gas sensors have several drawbacks that prevent them from being used in real-world applications. These drawbacks include poor selectivity, a slow response time, and instability in changing environmental conditions. We summarized some of the approaches investigated by researchers to overcome these difficulties and enhance the performance of MXene-based gas sensors. MXene nanosheet stacking can be avoided by building 3D MXene structures through self-assembly, framework support, and the use of reduced graphene oxide (rGO). Gas sensing performance may be enhanced by encouraging the diffusion and adsorption of gas molecules. The extremely reactive surface functional groups of MXenes can operate as covalent connecting sites for the functionalization of self-assembled monolayers as well as active sites for gas sensing. The gas sensing performance of MXene-based sensors can be greatly improved by altering the surface functional groups and adding organic molecules. The addition of noble metals to MXenes can significantly improve the response, sensitivity, and other sensing parameters of gas sensors based on MXenes. This is due to the synergistic effects of chemical sensitization and electronic sensitization. As a result of the faster electron transfer at the interface and faster sensor response/recovery times, the introduction of Schottky junctions or heterojunctions at the interfaces between MXenes and inorganic semiconductors can dramatically increase the resistance variation of gas sensors upon gas exposure. Conducive polymers and MXenes can be used to increase the mechanical flexibility of sensors and to encourage the formation of Schottky junctions/heterojunctions at the interface, which enhances the gas sensing properties and increases wearability for wearable applications. Additionally, with the proper metal-ion intercalation, MXenes can experience gas-induced interlayer swelling, which improves gas sensing performance by changing the sensor resistance when exposed to gas.

REFERENCES

1. Ghidiu, Michael, Maria R Lukatskaya, Meng-Qiang Zhao, and Yury Gogotsi, Conductive two-dimensional titanium carbide 'clay' with high volumetric capacitance, *Nature*, 2014. https://doi.org/10.1038/nature13970.
2. Naguib, Michael, Vadym N Mochalin, Michel W Barsoum, and Yury Gogotsi, 25th anniversary article: MXenes: A new family of two-dimensional materials, Wiley Online Library, 2014. https://doi.org/10.1002/adma.201304138.
3. Naguib, Michael, Murat Kurtoglu, Volker Presser, Jun Lu, and Junjie Niu, Two-dimensional nanocrystals produced by exfoliation of Ti_3AlC_2, Wiley Online Library, 2011. https://doi.org/10.1002/adma.201102306.
4. Mashtalir, Olha, Michael Naguib, Vadym N. Mochalin, Yohan Dall'Agnese, Min Heon, Michel W. Barsoum, and Yury Gogotsi, Intercalation, and delamination of layered carbides and carbonitrides, *Nature Communications*, 2013. https://doi.org/10.1038/ncomms2664.
5. Zhao, Meng-Qiang, Chang E. Ren, Zheng Ling, Maria R. Lukatskaya, Chuanfang Zhang, Katherine L. van Aken, Michel W. Barsoum, et al., Flexible MXene/Carbon nanotube composite paper with high volumetric capacitance, Wiley Online Library, 2015. https://doi.org/10.1002/adma.201404140.
6. Barsoum, Michel W. The $M_{N+1}AXN$ phases: A new class of solids: Thermodynamically stable nanolaminates. *Progress in Solid State Chemistry*, 2000. https://doi.org/10.1016/S0079-6786(00)00006-6.
7. Lukatskaya, Maria R., Joseph Halim, Boris Dyatkin, Michael Naguib, Yulia S. Buranova, Michel W. Barsoum, and Yury Gogotsi, Room-temperature carbide-derived carbon synthesis by electrochemical etching of MAX phases, Wiley Online Library, 2014. https://doi.org/10.1002/ange.201402513.
8. Barsoum, Michel W. *MAX Phases: Properties of Machinable Ternary Carbides and Nitrides*, John Wiley & Sons, Hoboken, NJ, 2013.
9. Li, Xinliang, Zhaodong Huang, Christopher E. Shuck, Guojin Liang, Yury Gogotsi, and Chunyi Zhi, MXene chemistry, electrochemistry and energy storage applications, *Nature Reviews Chemistry*, 2022. https://doi.org/10.1038/s41570-022-00384-8.
10. He, Shiyu, Qizhen Zhu, Razium Ali Soomro, and Bin Xu, MXene derivatives for energy storage applications, *Sustainable Energy & Fuels*, 2020. https://doi.org/10.1039/D0SE00927J.
11. Xiong, Dongbin, Xifei Li, Zhimin Bai, and Shigang Lu, Recent advances in layered $Ti_3C_2T_x$ MXene for electrochemical energy storage, *Small*, 2018. https://doi.org/10.1002/SMLL.201703419.
12. Nan, Jianxiao, Xin Guo, Jun Xiao, Xiao Li, Weihua Chen, Wenjian Wu, Hao Liu, et al., Nanoengineering of 2D MXene-based materials for energy storage applications, Wiley Online Library, 2021. https://doi.org/10.1002/smll.201902085.
13. Chaudhary, Vishal, Virat Khanna, Hafiz Taimoor Ahmed Awan, Kamaljit Singh, Mohammad Khalid, Yogendra Kumar Mishra, Shekhar Bhansali, Chen-Zhong Li, and Ajeet Kaushik. Towards hospital-on-chip supported by 2D MXenes-based 5th generation intelligent biosensors, *Biosensors and Bioelectronics*, 2023. https://doi.org/10.1016/j.bios.2022.114847.
14. Liu, Huiyu, Xiaotong Xing, Yan Tan, and Haifeng Dong, Two-dimensional transition metal carbides and nitrides (MXenes) based biosensing and molecular imaging, *Nanophotonics*, 2022. https://doi.org/10.1515/nanoph-2022-0550.
15. Maleki, Aziz, Matineh Ghomi, Nasser Nikfarjam, Mahsa Akbari, Esmaeel Sharifi, Mohammad Ali Shahbazi, and Mehraneh Kermanian, Biomedical applications of MXene-integrated composites: Regenerative medicine, infection therapy, cancer treatment, and biosensing, *Advanced Functional Materials*, 2022. https://doi.org/10.1002/ADFM.202203430.

16. Lu, Decheng, Huijuan Zhao, Xinying Zhang, Yingying Chen, and Lingyan Feng, New horizons for MXenes in biosensing applications, *Biosensors*, 2022. https://doi.org/10.3390/bios12100820.
17. Kong, Wei, Jinxia Deng, and Lihong Li, Recent advances in noble metal MXene-based catalysts for electrocatalysis, *Journal of Materials Chemistry A*, 2022. https://doi.org/10.1039/D2TA00613H.
18. Lim, Kang Rui Garrick, Mikhail Shekhirev, Brian C. Wyatt, Babak Anasori, Yury Gogotsi, and Zhi Wei She, Fundamentals of MXene synthesis, *Nature Synthesis*, 2022. https://doi.org/10.1038/s44160-022-00104-6.
19. Naguib, Michael, Olha Mashtalir, Joshua Carle, Volker Presser, Jun Lu, Lars Hultman, Yury Gogotsi, and Michel W. Barsoum, Two-dimensional transition metal carbides, *ACS Nano*, 2012. https://doi.org/10.1021/nn204153h.
20. Verger, Louisiane, Chuan Xu, Varun Natu, Hui-Ming Cheng, Wencai Ren, and Michel W. Barsoum, Overview of the synthesis of MXenes and other ultrathin 2D transition metal carbides and nitrides, *Current Opinion in Solid State and Materials Science*, 2019. https://doi.org/10.1016/j.cossms.2019.02.001.
21. Wei, Yi, Peng Zhang, Razium A. Soomro, Qizhen Zhu, and Bin Xu, Advances in the synthesis of 2D MXenes, *Advanced Materials*, 2021. https://doi.org/10.1002/ADMA.202103148.
22. Zeng, Mengqi, Yunxu Chen, Jiaxu Li, Haifeng Xue, Rafael G. Mendes, Jinxin Liu, Tao Zhang, Mark H. Rümmeli, and Lei Fu, 2D WC single crystal embedded in graphene for enhancing hydrogen evolution reaction, *Nano Energy*, 2017. https://doi.org/10.1016/j.nanoen.2017.01.057.
23. Xu, Chuan, Libin Wang, Zhibo Liu, Long Chen, Jingkun Guo, Ning Kang, Xiu-Liang Ma, Hui-Ming Cheng, and Wencai Ren, Large-area high-quality 2D ultrathin Mo_2C superconducting crystals, *Nature Materials*, 2015. https://doi.org/10.1038/nmat4374.
24. Naguib, Michael, Murat Kurtoglu, Volker Presser, Jun Lu, Junjie Niu, Min Heon, Lars Hultman, Yury Gogotsi, and Michel W. Barsoum, Two-dimensional nanocrystals produced by exfoliation of Ti_3AlC_2, Wiley Online Library, 2011. https://doi.org/10.1002/adma.201102306.
25. Magnuson, Martin, and Maurizio Mattesini, Chemical bonding and electronic-structure in MAX phases as viewed by X-ray spectroscopy and density functional theory, *Thin Solid Films*, 2017. https://doi.org/10.1016/j.tsf.2016.11.005.
26. Pang, Sin Yi, Yuen Ting Wong, Shuoguo Yuan, Yan Liu, Ming Kiu Tsang, Zhibin Yang, Haitao Huang, Wing Tak Wong, and Jianhua Hao, Universal strategy for HF-free facile and rapid synthesis of two-dimensional MXenes as multifunctional energy materials, *Journal of the American Chemical Society*, 2019. https://doi.org/10.1021/JACS.9B02578.
27. Yang, Sheng, Panpan Zhang, Faxing Wang, Antonio Gaetano Ricciardulli, Martin R. Lohe, Paul W. M. Blom, and Xinliang Feng, Fluoride-free synthesis of two-dimensional titanium carbide (MXene) using a binary aqueous system, *Angewandte Chemie*, 2018. https://doi.org/10.1002/ANGE.201809662.
28. Jawaid, Ali, Asra Hassan, Gregory Neher, Dhriti Nepal, Ruth Pachter, W Joshua Kennedy, Subramanian Ramakrishnan, and Richard A Vaia, Halogen etch of Ti_3AlC_2 MAX phase for MXene fabrication, *ACS Publications*, 2021. https://doi.org/10.1021/acsnano.0c08630.
29. Kamysbayev, Vladislav, Alexander S Filatov, Huicheng Hu, Xue Rui, and Francisco Lagunas, Covalent surface modifications and superconductivity of two-dimensional metal carbide MXenes, *Science*, 2020. https://doi.org/10.1126/science.aba8311.
30. Srivastava, Pooja, Avanish Mishra, Hiroshi Mizuseki, Kwang Ryeol Lee, and Abhishek K. Singh, Mechanistic insight into the chemical exfoliation and functionalization of Ti_3C_2 MXene, *ACS Applied Materials and Interfaces*, 2016. https://doi.org/10.1021/ACSAMI.6B08413.
31. Naguib, Michael, Vadym N. Mochalin, Michel W. Barsoum, and Yury Gogotsi, 25th anniversary article: MXenes: A new family of two-dimensional materials, *Advanced Materials*, 2014. https://doi.org/10.1002/adma.201304138.
32. Maleski, Kathleen, Vadym N. Mochalin, and Yury Gogotsi, Dispersions of two-dimensional titanium carbide MXene in organic solvents, *Chemistry of Materials*, 2017. https://doi.org/10.1021/acs.chemmater.6b04830.
33. Lee, Eunji, Armin VahidMohammadi, Barton C. Prorok, Young Soo Yoon, Majid Beidaghi, and Dong-Joo Kim, Room temperature gas sensing of two-dimensional titanium carbide (MXene), *ACS Applied Materials & Interfaces*, 2017. https://doi.org/10.1021/acsami.7b11055.
34. Kim, Seon Joon, Hyeong-Jun Koh, Chang E. Ren, Ohmin Kwon, Kathleen Maleski, Soo-Yeon Cho, and Babak Anasori, Metallic $Ti_3C_2T_x$ MXene gas sensors with ultrahigh signal-to-noise ratio, *ACS Nano*, 2018. https://doi.org/10.1021/acsnano.7b07460.
35. Yuan, Wenjing, Kai Yang, Huifen Peng, Fang Li, and Fuxing Yin, A flexible VOCs sensor based on a 3D Mxene framework with a high sensing performance, *Journal of Materials Chemistry A*, 2018. https://doi.org/10.1039/C8TA06928J.

36. Chae, Yoonjeong, Seon Joon Kim, Soo-Yeon Cho, Junghoon Choi, Kathleen Maleski, Byeong-Joo Lee, Hee-Tae Jung, Yury Gogotsi, Yonghee Lee, and Chi Won Ahn, An investigation into the factors governing the oxidation of two-dimensional Ti_3C_2 MXene, *Nanoscale*, 2019. https://doi.org/10.1039/C9NR00084D.
37. Koh, Hyeong-Jun, Seon Joon Kim, Kathleen Maleski, Soo-Yeon Cho, Yong-Jae Kim, Chi Won Ahn, Yury Gogotsi, and Hee-Tae Jung, Enhanced selectivity of MXene gas sensors through metal ion intercalation: In situ X-ray diffraction study, *ACS Sensors*, 2019. https://doi.org/10.1021/acssensors.9b00310.
38. Shuck, Christopher E., Meikang Han, Kathleen Maleski, Kanit Hantanasirisakul, Seon Joon Kim, Junghoon Choi, William E. B. Reil, and Yury Gogotsi. Effect of Ti_3 AlC_2 MAX phase on structure and properties of resultant $Ti_3C_2T_x$ MXene, *ACS Applied Nano Materials*, 2019. https://doi.org/10.1021/acsanm.9b00286.
39. Liu, Miao, Zeyu Wang, Peng Song, Zhongxi Yang, and Qi Wang, Flexible MXene/rGO/CuO hybrid aerogels for high performance acetone sensing at room temperature, *Sensors and Actuators B: Chemical*, 2021. https://doi.org/10.1016/j.snb.2021.129946.
40. Yang, Zijie, Li Jiang, Jing Wang, Fangmeng Liu, Junming He, Ao Liu, Siyuan Lv, et al., Flexible resistive NO_2 gas sensor of three-dimensional crumpled MXene $Ti_3C_2T_x$ /ZnO spheres for room temperature application, *Sensors and Actuators B: Chemical*, 2021. https://doi.org/10.1016/j.snb.2020.128828.
41. Zhu, Zhengyou, Congcong Liu, Fengxing Jiang, Jing Liu, Xiumei Ma, Peng Liu, Jingkun Xu, Lei Wang, and Rui Huang, Flexible and lightweight $Ti_3C_2T_x$ MXene@Pd colloidal nanoclusters paper film as novel H_2 sensor, *Journal of Hazardous Materials*, 2020. https://doi.org/10.1016/j.jhazmat.2020.123054.
42. Dey, Ananya, Semiconductor metal oxide gas sensors: A review, *Materials Science and Engineering: B*, 2018. https://doi.org/10.1016/j.mseb.2017.12.036.
43. Chen, Winston Yenyu, Sz-Nian Lai, Chao-Chun Yen, Xiaofan Jiang, Dimitrios Peroulis, and Lia A. Stanciu, Surface functionalization of $Ti_3C_2T_x$ MXene with highly reliable superhydrophobic protection for volatile organic compounds sensing, *ACS Nano*, 2020. https://doi.org/10.1021/acsnano.0c03896.
44. Pazniak, Hanna, Ilya A. Plugin, Michael J. Loes, Talgat M. Inerbaev, Igor N. Burmistrov, Michail Gorshenkov, Josef Polcak, et al., Partially oxidized $Ti_3C_2T_x$ MXenes for fast and selective detection of organic vapors at part-per-million concentrations, *ACS Applied Nano Materials*, 2020. https://doi.org/10.1021/acsanm.9b02223.
45. Choi, Junghoon, Yong-Jae Kim, Soo-Yeon Cho, Kangho Park, Hohyung Kang, Seon Joon Kim, and Hee-Tae Jung, In situ formation of multiple schottky barriers in a Ti_3C_2 MXene film and its application in highly sensitive gas sensors, *Advanced Functional Materials*, 2020. https://doi.org/10.1002/adfm.202003998.
46. Zhang, Dongzhi, Sujing Yu, Xingwei Wang, Jiankun Huang, Wenjing Pan, Jianhua Zhang, Benjamin Edem Meteku, and Jingbin Zeng, UV illumination-enhanced ultrasensitive ammonia gas sensor based on (001) TiO_2/MXene heterostructure for food spoilage detection, *Journal of Hazardous Materials*, 2022. https://doi.org/10.1016/j.jhazmat.2021.127160.
47. Chen, Winston Yenyu, Sz-Nian Lai, Chao-Chun Yen, Xiaofan Jiang, Dimitrios Peroulis, and Lia A. Stanciu, Surface functionalization of $Ti_3C_2T_x$ MXene with highly reliable superhydrophobic protection for volatile organic compounds sensing, *ACS Nano*, 2020. https://doi.org/10.1021/acsnano.0c03896.
48. Tai, Huiling, Zaihua Duan, Zaizhou He, Xian Li, Jianglong Xu, Bohao Liu, and Yadong Jiang, Enhanced ammonia response of $Ti_3C_2T_x$ nanosheets supported by TiO_2 nanoparticles at room temperature, *Sensors and Actuators B: Chemical*, 2019. https://doi.org/10.1016/j.snb.2019.126874.
49. Guo, Xuezheng, Yanqiao Ding, Delin Kuang, Zhilin Wu, Xia Sun, Bingsheng Du, and Chengyao Liang, Enhanced ammonia sensing performance based on MXene-$Ti_3C_2T_x$ multilayer nanoflakes functionalized by tungsten trioxide nanoparticles, *Journal of Colloid and Interface Science*, 2021. https://doi.org/10.1016/j.jcis.2021.03.115.
50. Liu, Miao, Jun Ji, Peng Song, Min Liu, and Qi Wang, α-Fe_2O_3 Nanocubes/$Ti_3C_2T_x$ MXene composites for improvement of acetone sensing performance at room temperature, *Sensors and Actuators B: Chemical*, 2021. https://doi.org/10.1016/j.snb.2021.130782.
51. Liu, Miao, Zeyu Wang, Peng Song, Zhongxi Yang, and Qi Wang, In_2O_3 Nanocubes/ $Ti_3C_2T_x$ MXene composites for enhanced methanol gas sensing properties at room temperature, *Ceramics International*, 2021. https://doi.org/10.1016/j.ceramint.2021.05.016.
52. Sun, Shibin, Mingwei Wang, Xueting Chang, Yingchang Jiang, Dongzhi Zhang, Dongsheng Wang, Yuliang Zhang, and Yanhua Lei, $W_{18}O_{49}$/$Ti_3C_2T_x$ Mxene nanocomposites for highly sensitive acetone gas sensor with low detection limit, *Sensors and Actuators B: Chemical*, 2020. https://doi.org/10.1016/j.snb.2019.127274.

53. Hermawan, Angga, Biao Zhang, Ardiansyah Taufik, Yusuke Asakura, Takuya Hasegawa, Jianfeng Zhu, Pei Shi, and Shu Yin, CuO nanoparticles/$Ti_3C_2T_x$ MXene hybrid nanocomposites for detection of toluene gas, *ACS Applied Nano Materials*, 2020. https://doi.org/10.1021/acsanm.0c00749.
54. He, Tingting, Wei Liu, Tan Lv, Mingsheng Ma, Zhifu Liu, Alexey Vasiliev, and Xiaogan Li, MXene/SnO_2 heterojunction based chemical gas sensors, *Sensors and Actuators B: Chemical*, 2021. https://doi.org/10.1016/j.snb.2020.129275.
55. Li, Hong-Peng, Jie Wen, Shu-Mei Ding, Jia-Bao Ding, Zi-Hao Song, Chao Zhang, Zhen Ge, Xue Liu, Rui-Zheng Zhao, and Feng-Chao Li, Synergistic coupling of 0D–2D heterostructure from ZnO and $Ti_3C_2T_x$ MXene-derived TiO_2 for boosted NO_2 detection at room temperature, *Nano Materials Science*, 2023. https://doi.org/10.1016/j.nanoms.2023.02.001.
56. Vetter, S., Stefanie Haffer, Thorsten Wagner, and Michael Tiemann, Nanostructured Co_3O_4 as a CO gas sensor: Temperature-dependent behavior, *Sensors and Actuators B: Chemical*, 2015. https://doi.org/10.1016/j.snb.2014.09.025.
57. Anasori, Babak, Chenyang Shi, Eun Ju Moon, Yu Xie, Cooper A. Voigt, Paul R. C. Kent, Steven J. May, Simon J. L. Billinge, Michel W. Barsoum, and Yury Gogotsi, Control of electronic properties of 2D carbides (MXenes) by manipulating their transition metal layers, *Nanoscale Horizons*, 2014. https://doi.org/10.1039/C5NH00125K.
58. Zhang, Yajie, Yadong Jiang, Zaihua Duan, Qi Huang, Yingwei Wu, Bohao Liu, Qiuni Zhao, Si Wang, Zhen Yuan, and Huiling Tai, Highly sensitive and selective NO_2 sensor of alkalized V_2CT MXene driven by interlayer swelling, *Sensors and Actuators B: Chemical*, 2021. https://doi.org/10.1016/j.snb.2021.130150.
59. An, K.H., S.Y. Jeong, H.R. Hwang, and Y.H. Lee, Enhanced sensitivity of a gas sensor incorporating single-walled carbon nanotube–polypyrrole nanocomposites, *Advanced Materials*, 2004. https://doi.org/10.1002/adma.200306176.
60. Van Quang, Vu, Nguyen Van Dung, Ngo Sy Trong, Nguyen Duc Hoa, Nguyen Van Duy, and Nguyen Van Hieu, Outstanding gas-sensing performance of graphene/SnO_2 nanowire schottky junctions, *Applied Physics Letters*, 2014. https://doi.org/10.1063/1.4887486.

8 Two-Dimensional Nanomaterials for the Removal of Radioactive Toxic Environmental Pollutants

Sefako J. Mofokeng, Potlako J. Mafa, Mandla B. Chabalala, Fokotsa V. Molefe, Teboho P. Mokoena, Tshegofatso M. Modungwe, Luyanda L. Noto, and Makgamathe J. Sithole

8.1 INTRODUCTION

Nuclear technologies are appealing globally although their beneficial utilization is discouraged due to human health concerns that include the uncontrolled discharge of radioactive waste to the environment [1]. Some toxic radioactive pollutants/ radionuclides released while using nuclear technology include uranium ($^{235,238}U$) [2], cesium (^{137}Cs) [3], strontium (^{90}Sr) [4], barium ($^{133,140}Ba$) [5], and thorium (232Th) [6] into the environment. The tremendous environmental pollution induced by these radioactive pollutants has raised global concerns that require urgent intervention. However, as the global population continues to increase, we come across a rapid increase in applications that require the use of nuclear technology. Nuclear technology has seen an upsurge in the fields of medicine, energy, research, and industries for drug delivery/medicine, nuclear power generation, and operation of radiation-producing machinery. It is worth noting that wastewater discarded from the aforementioned activities contains persistent emerging pollutants that find their way into the environment. The resulting wastewater or fluids contain organic surfactants, compounds that are phenolic, heavy metal, drug products, toxic solvents, alkalis, antibiotics, bases or salts, and excessively harmful dyes [7], all of which are extremely dangerous. Moreover, water scarcity has raised the demand for wastewater regeneration, which requires the removal of radioactive pollutants prior to utilization [8]. These calls for exploration of means of radioactive waste disposal and isolation from the environment for the benefit of living organisms. Because the use of nuclear technology is well-intentioned and the radioactive materials come as the by-products, scientists are generating nanomaterials to remediate those toxic radioactive pollutants. This is to ensure that every waste is well treated before being released to prevent negative environmental impacts.

Over the past two decades, it has been a huge challenge to find methods of removing environmental pollutants effectively, especially on the internal parts of corroded components and rough surfaces [1]. Multiple routes have been implemented to remediate the environmental pollution induced by radioactive materials such as ion exchange [9], adsorptive filtration [10], capacitive deionization [11], chemical precipitation [12], sedimentation [7], membrane filtration [13], and film coating [14]. However, methods such as precipitation and sedimentation are not cost- and time-effective as they require extensive hard labor work and consume energy. Thus, the aforementioned methods could not be adopted because of their limitations that do not match the global specifications [7]. The methods adopted for the removal of radioactive materials include adsorption-based techniques, which are deemed highly effective in removing long-lived actinides from the environment [15]. In recent

DOI: 10.1201/9781003436942-8

years, there has been an increase in the usage of nanomaterials for the removal of radioactive pollutants from the environment that is widespread in multiple fields such as wastewater purification, as well as agricultural and industrial activities, due to their outstanding properties.

Relative to the bulk or rather microparticles, the nanomaterials have exclusive characteristics, such as stronger adsorption, large surface area, outstanding mobility, and greater reducing ability. In the range of viable nanomaterials, nanoscale zero-valent iron (nZVI) is one of the foremost prevalent engineered nanomaterials employed to eradicate organic contaminants from the environment [16]. It has shown the capability to remove various toxic and radioactive contaminants that could not be degraded into harmless products. Thus, the application of nZVI for the elimination of toxic and radioactive environmental pollutants is hindered by the growing oxidation process and phase transformation that occurs at ultralow concentrations between nZVI and metal ions. In this era, two-dimensional (2D) materials such as graphitic carbon nitride (g-C_3N_4), metal–organic frameworks (MOFs), and carbon-based nanomaterials graphene oxide (GO) with layered nanosheets-like structures are also presented as innovative nanomaterials in the category of the most promising candidates owing to their large specific surface area and great functionalization for enhanced reactivity [17]. Studies showed that pristine carbon-based nanomaterials like GO generate membranes that display intermediate ions selectivity while they possess low water flux rates and structure stability [18]. It has been indicated that creating 2D GO nanocomposite structures by incorporating nanoparticles and polymers improves different properties such as stability, mechanical strength, and water flux [19]. Thus, carbon nanocomposites are incorporated in the novel hybrid membrane technologies due to their potential to resolve limitations of other membrane techniques for the removal of ecological pollutants.

In the present chapter, the utilization of nanomaterials for the elimination of radioactive contaminants from the environment is discussed. Furthermore, it focuses on nanomaterials preparation with preferable layered 2D nanostructures, characterization, applications, and the role of nanomaterials in environmental remediation and their capabilities. We have summarized various technological systems in place for remediating toxic and radioactive pollutants in the environment. Moreover, we provide an outlook of what various nanomaterials can achieve and their drawbacks on various pollutants to the environment.

8.2 SYNTHESIS METHODS

Heavy metals are hazardous and non-biodegradable ecological pollutants that damage the health of the human being. The cleanup of the heavy metal-contaminated water and soil is critical from both an environmental and biological standpoint. Radioactive waste can be created in a variety of phases, including solid, liquid, and gaseous. The radioactivity of these wastes varies depending on the source of production. The use of radionuclides in nuclear reactors, research, agriculture, industry, and medicine has resulted in the generation of radioactive waste, which has a harmful effect on both human health and the environment [20].

With the rapidly growing number of industries, massive amounts of pollutants, both inorganic and organic, are inevitably released into the environment at large, causing pollution and posing a threat to living organisms. The efficient removal of harmful contaminants is critical for environmental and human health protection. To manage the radioactive wastes successfully and protect the environment, new technologies and methods must be developed [21,22]. Currently, nanotechnology offers improved prospects of eliminating pollutants including radionuclides and heavy metal ions. Nanomaterials are given specific preferences because of their unique characteristics such as their broad surface area, productive reactivity, and increased binding sites [23]. There are two well-known strategies for the preparation of these nano-structured materials: top-down and bottom-up approaches [24]. The bottom-up methodology involves arranging smaller molecules into more multifaceted assemblies with the assistance of chemical or physical forces by means of the process of chemical preparation positional assembling, or self-assembling, while the top-down

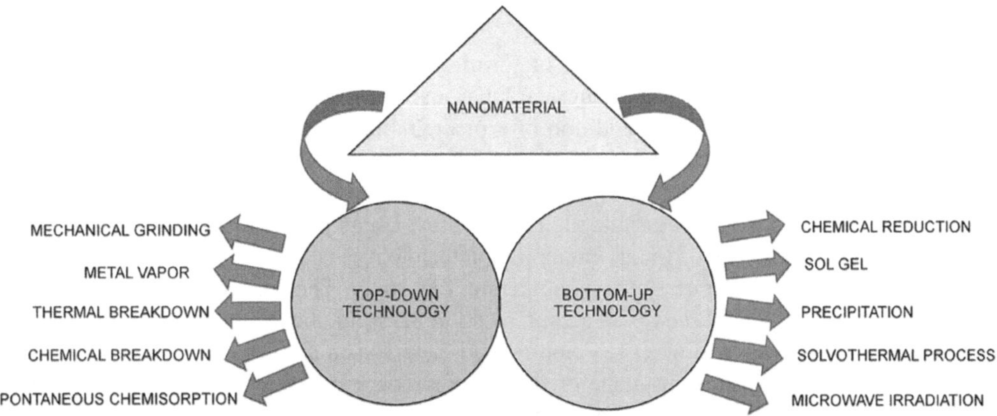

FIGURE 8.1 Various techniques to synthesize nanomaterials. (Reproduced with permission from Ref. [26], Copyright © 2020 Springer.)

technique includes creating a variety of nanostructured materials from large-size (granular or microscale) materials through mechanical- and/or chemical-based steps. A variety of methods for creating nanomaterials with the desired shape, characteristics, and applications have been developed under these methodologies [25]. The methodology includes the catalytic growth, sol–gel process, impregnation, thermal decomposition, co-precipitation, mechanical alloying (sometimes called milling), electrodeposition, laser ablation, and microwave-assisted synthesis [26], as shown in Figure 8.1.

Nanomaterials with distinctive characteristics, such as covalent organic frameworks (COFs), carbon-based material, conjugated microporous polymers (CMPs), and MOFs, have piqued the curiosity of researchers across fields of study. Furthermore, due to their strong chemical stability, regulated pore structure, surface functional group design, low densities, and high surface area, these types of nanostructured materials have been widely used in environmental pollution [27–29]. In this chapter, we will review various commonly used methods for the synthesis of materials for the removal of radioactive harmful compounds utilizing nanomaterials, along with their benefits and drawbacks.

8.2.1 Hydrothermal Method

The term "hydrothermal synthesis" describes the heterogeneous reactions used to prepare inorganic compounds in aqueous systems beyond ambient temperature and pressure. It is mostly a basic method for preparing the nanomaterials using autogenous pressure generated in an autoclave as the reaction medium. Many types of nanomaterials have been successfully synthesized using this approach. Experimentally, hydrothermal treatment at high temperatures results in the formation of regular crystallite patterns due to the distribution of cations throughout the layers. However, the diverse dimensions of morphologies can be successfully created by regulating parameters such as reaction time and aging temperatures [30,31]. Hydrothermal synthesis is considered a more efficient way of dealing with wet solid waste. This is due to the following benefits: minimal energy consumption, gentle reaction conditions, and no moisture requirements for raw materials. As a result, prior to hydrothermal treatment, the waste does not need to be dried [32]. Most importantly, by managing the rate of reaction, temperature, and solvent properties, a hydrothermal technique enables access to more effective powder characteristics such as outstanding purity, stability of phase (stoichiometry), controlled particle size, and narrowed size distribution. However, its biggest downside is its all-time high reaction in terms of pressure and temperature [30,33].

8.2.2 Co-precipitation Method

Chemical co-precipitation is the process of forming solids from a solution by either establishing supersaturated conditions or turning a soluble material into an insoluble form by pH shift, electro-oxidizing potential, or the addition of a precipitation reagent [34,35]. A typical chemical precipitation process consists of four (4) phases, which are (i) precipitation, (ii) flocculation, (iii) sedimentation, and (iv) solid–liquid separation. In contrast to other synthesis methods, the co-precipitation approach does not necessitate high temperatures, high pressures, or expensive precursors. The fundamental advantage of the co-precipitation technology is that it may produce fine-grained nanoparticles in a simple process even in a medium like water. The size and the shape of nanoparticles are established by the kind of extractant, metal molar ratio, the temperature at which the reaction occurs, the pH level, and other reaction-related factors such as stirring rate and basic solution decreasing rate. The sole disadvantage of the precipitation process is that the products obtained appear to be less pure than those acquired by other methods [36,37]. In wastewater, the chemical precipitation method is used to reduce the specific activity of radioactive by co-precipitating the precipitant and the radionuclides in liquid waste. Commonly used precipitates include phosphates, iron salts, and aluminum salts [38].

8.2.3 Sol–Gel Method

The sol–gel procedure is an improved chemical (wet chemical) approach for manufacturing various nanostructures, such as nanoparticles derived from metal oxide. The molecular precursor (frequently a metal alkoxide) undergoes dissolution in water or alcohol before being heated and agitated to form a gel by hydrolysis/alcoholysis. The resulting gel should be dried employing suitable processes based on the required qualities and application of the by-product; for instance, in the event, if the gel is alcoholic, the drying step is achieved by burning away the alcohol. After drying, the gels are pulverized and calcined. Due to the minimal reaction temperature, the sol–gel approach provides inexpensive technology with strong control over chemical composition [37,39].

8.3 2D MATERIALS FOR THE REMOVAL OF RADIOACTIVE TOXIC POLLUTANTS

To date, several therapeutic options for the removal of radioactive hazardous compounds have been developed, including membrane therapy, biological treatment, electrochemical therapies, and improved oxidation processes [40–44]. However, the materials or nanomaterials employed to treat radioactive hazardous compounds must have a high capacity of adsorption and excellent capacity adsorption. Because of its low cost, ease of operation, high removal effectiveness, and accessibility of diverse adsorbent materials, the adsorption process is the most extensively utilized method [40,42]. In addition to MOFs, CMPs and porous aromatic frameworks (PAFs) are also considered promising candidates for the elimination of radioactive toxic materials. However, porous MOF nanomaterials containing plentiful intrinsic pores are extremely effective adsorbents for removing radioactive toxic materials due to their good thermal stability, customizable structure, designable adsorption sites, appropriate pore size, high reproducibility, and high iodine adsorption [45].

Graphene oxide (GO), with its tremendous potential for use in heavy metal adsorption and ease of synthesis from plentiful natural graphite, demonstrated exceptional uranium adsorption characteristics. Furthermore, GO can be produced as surface functionalized GO, reduced GO (RGO), or GO composites. Particularly, the characteristics of GO can be modified through the combination of it with other materials to generate a composite with improved adsorption capabilities. Thus, the production of functionalized magnetic graphene oxide composites (MAG-GO1-NH_2) has an excellent chance for targeted chromium adsorption and removal [46,47]. On the other hand, the produced RGO aerogels doped with ternary Prussian blue analogs (RGOA/TPBAs) with a high specific surface area and great mechanical strength were shown to be active in the removal of radioactive cesium (Cs^+) from effluent [48]. Potassium ferrocyanide is a different compound used for Cs^+ elimination. This compound is used to remove nonradioactive and radioactive liquid waste containing Cs^+ elements from wastewater created by washing contaminated steel ash. In this sense, nickel and iron potassium ferrocyanide are categorized as chemical agents since they are extensively used to effectively eliminate Cs^+ ions from radioactive waste that is water-based. Like other materials or compounds, potassium ferrocyanide exhibited excellent Cs^+ selectivity and capacity to adsorb [49].

Given the functionalization capabilities, chemical stability, and high specific surface area of nanostructures that utilize multiwalled carbon nanotubes (MWCNTs), they are being investigated as adsorption for nitrates, as well as the chemical and physical properties of the adsorbents and adsorption. It was discovered that the amount of nitrate adsorption increased with increasing magnetite content on the surface of MWCNTs [50]. Through the application of the sol–gel process, an Al_2O_3-ZrO_2-CeO_2 nanocomposite with high radioisotope sorption was investigated for the removal of certain radionuclides from radioactive waste effluent. Investigations on these nanocomposites revealed and demonstrated that the removal of ^{134}Cs,^{90}Sr, and^{90}Y was achieved at 94%, 44%, and 8.5%, respectively [51]. Mateo del Rio's group created a silver-functionalized UiO-66 MOF (UiO-66-SO_3H@Ag). This device's use demonstrated the successful detection and removal of radioactive iodine from wastewaters. The key feature of the UiO-66-SO_3H@Ag material is that the adsorbent does not need to be recovered laboriously [52]. A new poly(AA-AM)/ZrW composite was recently developed with an adsorption performance of 86.4% for the removal of radioactive $^{52+154}Eu$ from a model solution. More intriguingly, the composite was found to be capable of immobilizing Eu^{3+} ions or their radioactive isotopes from radioactive waste solutions linked with research, nuclear, industrial, or medical facilities. In practice, the removal efficiency (%) of $^{152+154}Eu$ from tab, groundwater, and the river was 92.2%, 95.2%, and 92.4%, respectively [53].

8.4 CLASSIFICATION OF RADIOACTIVE MATERIALS

The majority of unstable materials attain stability by alterations to the nucleus which can be via spontaneous fission, release of energy/radiation, alpha(α)/beta(β)/gamma(γ) emission particles, or neutron–proton conversion or the opposite process [54]. This is known as radioactive decomposition or transformation, and it is frequently supplemented by the emission of ionizing radiation in the form of β particles, neutrons, or γ rays [55]. Materials that undergo the described stabilization process are called radioactive materials or in short radionuclides.

There are generally three different types of radioactive materials: (i) naturally occurring materials, (ii) man-made materials (only in reactors), and (iii) particle accelerators [56]. Man-made radioactive materials and particle accelerators are naturally not radioactive, but have been made radioactive or the radioactivity of the material is manipulated by human action [57]. Throughout the decay processes of radioactive materials, three types of radiation (α, β, and γ radiation) are emitted [58]. α particles have a limited range and are unable to penetrate through the skin of human beings. β particles can pass through the layers of skin, but not through the human body. The radiation from γ rays can penetrate the entire body [56]. Table 8.1 gives examples of radioactive materials derived from nuclear accidents and their properties. Several substances have radioactive properties that are not included in Table 8.1, such as cobalt, ionizing radiation, radium, radon, thorium, and the famous uranium [57].

Cesium (Cs) is an element that occurs naturally. It is normally in low concentrations alongside other elements in rocks, dirt, and dust. Cs that occurs naturally is not radioactive and is generally referred to as stable Cs. However, the only persistent form of Cs found naturally in the environment is cesium-133 (Cs-133) [59,60]. The fragmentation of uranium in various circumstances or nuclear explosions may result in two types of radioactive Cs: Cs-134 and Cs-137, both of which decompose into non-radioactive elements [61]. As Cs-134 and Cs-137 decay, they generate β and γ particles, respectively. When cesium-134 emits radiation, it takes roughly 2 years for its half to emit radiation,

TABLE 8.1
Radioactive Materials and their Properties

Materials	Type of Radiation	Biological Half-Life	Physical Half-Life	Effective Half-Life	Organs and Tissue Accumulation
Cesium-134	β and γ	70–100 days	2.1 years	64–88 days	Whole body
Cesium-137	β and γ	70–100 days	30 years	70–99 days	Whole body
Iodine-131	β and γ	80 days	8 days	7 days	Thyroid
Strontium-90	B	50 years	29 years	18 years	Bones
Tritium-3	B	10 days	12.3 years	10 days	Whole body
Plutonium-239	α, β, and γ	20 years	24 000 years	20 years	Liver and bones

Reproduced with permission from [59], Copyright © 2016 Elsevier.

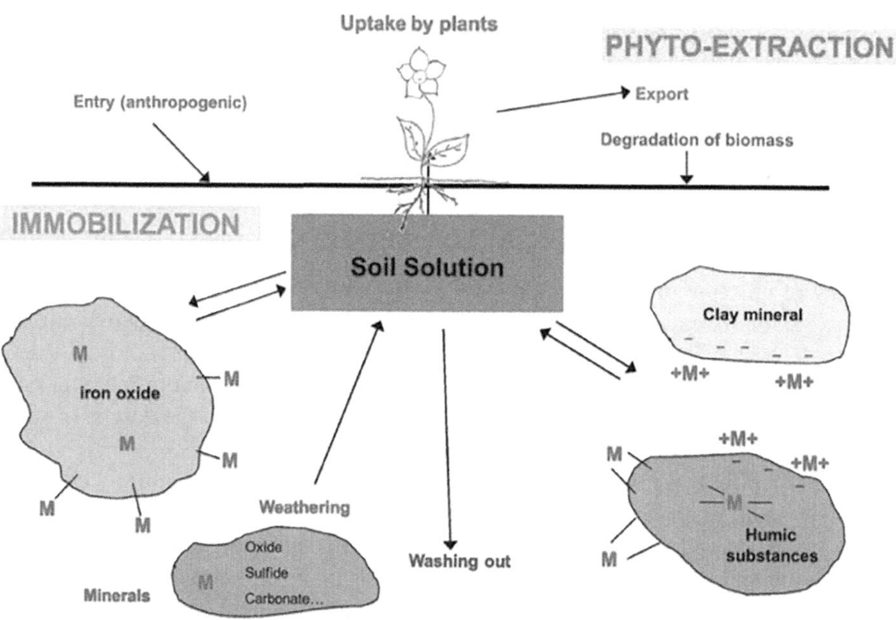

FIGURE 8.2 Availability of cesium to plants from soil. Radioactive Cs^+ enters the soil, is absorbed as a positive metal ion (M^+), and is immobilized by soil constituents such as iron oxides and clay minerals. (Reproduced with permission from Ref. [60], Copyright © 2018 Elsevier.)

whereas cesium-137 takes about 30 years, and this is referred to as the half-life. During the growth stage, radioactive cesium affects organs [62]. Figure 8.2 shows the binding and solubility of Cs in soil. From soil, it can be washed out into ground water and taken up by plants. Hyper-accumulating plants can be used to export cesium from the soil.

Iodine is an inorganic element, which could be extracted from seawater as well as certain rocks and sediments. It is accessible in both radioactive and non-radioactive forms [63]. It is mostly utilized as a sterilizer to clean surfaces, storage areas, bathing soaps, medical bandages, and water purification. In certain situations, iodine is additionally inserted into the table salt to guarantee that everyone gets enough iodine [64–66]. The radioactive form of iodine is mainly created artificially and is utilized in medical diagnostics, therapy, and the treatment of specific disorders. However, radioactive iodine undergoes a rapid (a few days) transformation to a stable element that is not radioactive [67,68]. In contrast to iodine-131, iodine-129 evolves slowly (over millions of years), and the radioactive iodine affects glands and hormones [69].

Strontium (Sr) is an element that can be found in dirt, rocks, dust, oil, and coal. Stable Sr, or simply strontium, is not radioactive [70]. This substance exists in nature in four stable isotopes: Sr-84, Sr-86, Sr-87, and Sr-88. Sr compounds are utilized to make ceramics, glass products, pyrotechnics, paint pigments, fluorescent lights, and medicines [71,72]. The most common radioactive Sr isotope is Sr-90, which is formed in nuclear reactors or during the explosion of nuclear weapons. Sr-90 generates β radiation/particles as it decays. The half-life of Sr-90 is approximately 29 years. This radioactive material affects muscles and bones [73,74]. Humans and animals can get exposed to Sr when they consume food that is contaminated with the substance. Plants get infected with strontium as shown in Figure 8.3.

Under normal conditions, the element Plutonium is a silvery-white material that exists as a solid matter. It is a product or by-product of uranium when it absorbs atomic particles (Figure 8.4). Even though trace amounts of plutonium occur naturally, a large amount has been created in nuclear reactors [75]. Trace plutonium can be found in the environment in numerous forms as a result of past nuclear bomb tests [76]. The most common plutonium isotopes are plutonium-238 and plutonium-239. Plutonium undergoes radioactive decay, whereby energy (radiation) is emitted and a new product is produced [77]. During the decomposition process, plutonium splits into two parts: a small portion known as "alpha" radiation and a big part known as a daughter. The daughter also happens to be a radioactive substance, but it maintains to deteriorate until a nonradioactive daughter is generated. Plutonium radiation can penetrate the body, affecting the immune system, and is carcinogenic [78,79].

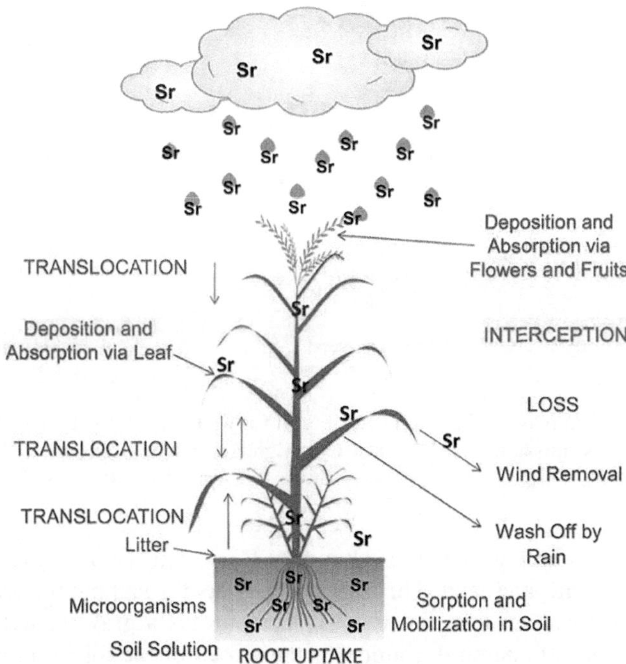

FIGURE 8.3 Schematic illustration of various ways of strontium uptake by plants. (Reproduced with permission from Ref. [71], Copyright © 2019 Elsevier.)

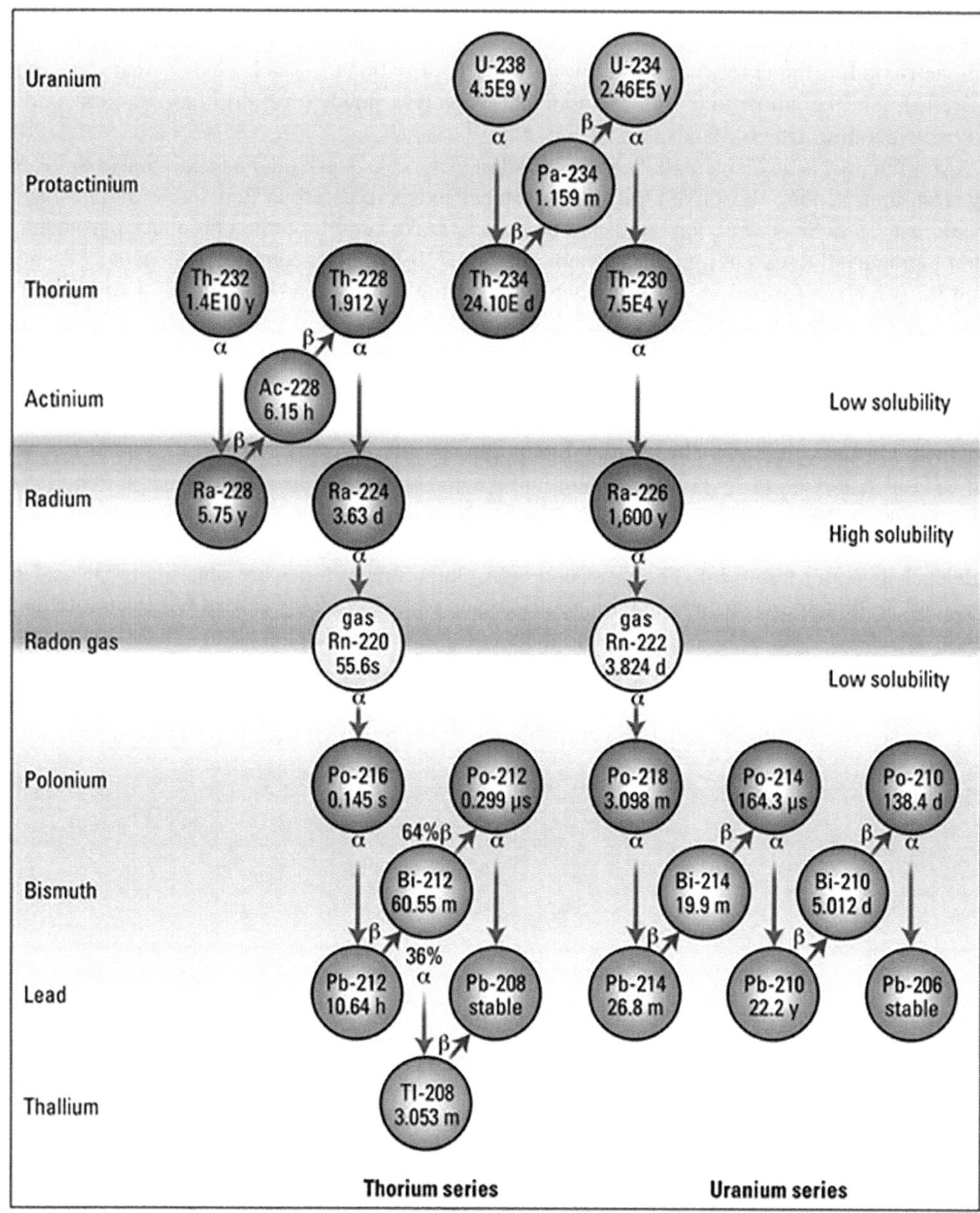

FIGURE 8.4 Decay chains of natural uranium and thorium as well as the formation of other radioactive materials. (Reproduced with permission from Ref. [86], Copyright © 2015 Environmental Health Perspectives.)

Another naturally occurring element is Uranium (U), which is a highly radioactive substance found in water, stones, air, and land. This substance exists throughout the environment as minerals, not as metals. Uranium material is silver-colored has a gray surface and is practically equally sturdy as steel [80]. Natural uranium is composed of the following isotopes: ^{234}U,^{235}U, and^{238}U, each with distinct radioactive qualities [81]. Certain rocks have high sufficient concentrations of mineral uranium to be extracted. After uranium is extracted from the rocks, the residual sand (mill tailings) is frequently found to be rich in compounds and radioactive substances like thorium and radium [82]. The half-lives of uranium isotopes are extremely lengthy, with uranium-234 having a half-life of approximately 200,000 years, uranium-235 having a half-life of 700 million years, and uranium-238 having a half-life of 5 billion years [83,84]. This is why uranium continues to exist throughout the environment and has not decomposed completely. When uranium that is naturally occurring is divided into two distinct components, the fuel component contains more uranium-235 and is commonly referred to as enriched uranium; this is the most radioactive fraction. The other part with low uranium-235 is called depleted uranium, and it is the least radioactive [85]. Figure 8.4 shows the decay chains of natural uranium and thorium as they release radiation in the process and shows other radioactive materials that are produced along the way.

Enrichment of radioactive materials (regardless of their type) raises environmental and human health issues linked with their emission into the environment as either emission from smokestacks, disposal to coal ash ponds, landfills, abandoned mines, or spills [87]. When the organ system (human or animal) is exposed to a hazardous substance such as radiation from radioactive materials, it can suffer from various types of side effects, some of which are fatal [57]. The organ that is affected by the radiation at the lowest dose is called the target organ. Radiation from radioactive materials affects different body parts such as endocrine, renal, cardiovascular, gastrointestinal, lymphoid, hematological, neurological, hepatic, dermal, immunological, musculoskeletal, ocular, reproductive, and respiratory [88,89]. In many cases, particularly long-term exposure can cause cancer, organ failure, and death, the source of contact/contamination could be air, water, soil, or food [90]. The general circulation, human, animal, and environmental exposure to radioactive materials is shown in Figure 8.5 using iodine-131, Cesium-134, and Cesium-137 as examples of radioactive materials.

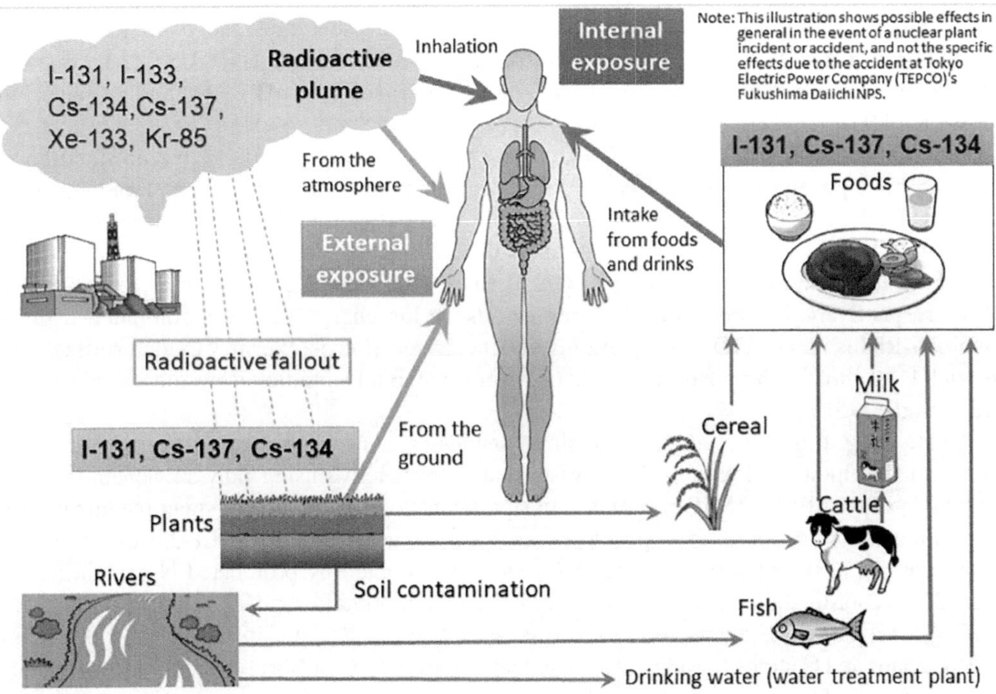

FIGURE 8.5 The effects of radioactive materials on humans, animals, and the environment. (Reproduced with permission from [91], Copyright © 2022 Elsevier.)

8.5 METHODS FOR THE REMOVAL OF RADIOACTIVE TOXIC MATERIALS

New materials and technologies are expected to play a critical role in developing sophisticated nuclear energy systems because of the rapid increase in human demand for nuclear energy. Because of their incredibly lengthy half-life, radionuclides pose a significant problem in the cleanup or decontamination of nuclear waste. Several techniques for removing soluble radionuclides have been proposed by researchers. Membrane dialysis and separation, adsorption, electrocoagulation, photocatalytic reduction, and evaporation are some of the methods used to remove radioactive materials [38,92]. This section will discuss them and pay special attention to those who are using them. The broad range of possible applications for nanomaterials and nanotechnologies, which have lately received considerable attention, could contribute to the creation of safe nuclear energy systems and their disposal. This section investigates the role of nanotechnology in the remediation of nuclear waste.

8.5.1 Adsorption

Adsorption is a theory that describes how an adsorbate from the feed solution is adsorbed onto the outermost layer of an adsorbent. Due to its economical, easy use, and low electrical requirement, reusability, and scalability, it has been commonly used in various treatments. The characteristics and attributes of the adsorbents, on the other hand, are critical in determining the process success. Ion adsorption is enticing as a simple strategy under mild conditions since it can attain highly efficient removal of radioactive ions with no additional energy input [92,93]. The following section discusses various adsorbents used to remove radioactive ions from water.

Wang et al. employed a simple and ecologically friendly method for developing clearly defined magnetic MOFs/graphene oxide (Fe_3O_4@HKUST-1/GO) composite, and they investigated how solution pH, interaction period, and temperature influenced U(VI). According to the results they produced, the Langmuir model and the pseudo-second-order kinetic model can better represent the adsorption process. In this regard, the Fe_3O_4@HKUST-1/GO displayed significant adsorption ability toward U(VI) at an initial solution pH of 4.0 and a temperature of 318 K [94].

Furthermore, Işik et al. conducted a study for the removal of cesium with a novel composite chitosan-bone powder-iron oxide. For Cs^+ ion sorption by these adsorbents, the Langmuir model fitted perfectly. The highest capacity for adsorption calculated by the Langmuir adsorption isotherm for CS-KT was 0.98×10^{-4}mol g^{-1} at 25°C and 1.16×10^{-4}mol g^{-1} at 50°C. For CS-KT-M, the maximum adsorption capacity was 1.79×10^{-4}mol g^{-1} and 2.24×10^{-4}mol g^{-1} at 25°C and 50°C, respectively. Furthermore, the average adsorption energy "E" was computed using the Dubinin–Radushkevich (D–R) equation, and the adsorption isotherm was determined to be around 11 kJ mol^{-1}. The adsorption kinetics were accurately predicted by the pseudo-second-order model [95].

Additionally, this study targeted the simultaneous effective elimination of a radioactive substance metal anions and cations from waste that is radioactive using poly(ethyleneimine) (PEI) customized MIL-101(Cr) (MILP), which is of significant concern for proper spent fuel disposal and environmental protection. The proposed process for ReO_4 adsorption by MILP-3 is outlined in four easy steps: First, the negative-charged ReO_4 ions are attracted by protonated N-containing sites; second, the coordination complex generated by the vacant orbital elements of ReO_4^- and the single pair of electron on the N atoms; third, NHO hydrogen bonds and CNO interactions between N associated groups and O atoms of ReO_4^- play an essential part in ReO_4 adsorption; and fourth, redundant CUSs adsorb ReO_4^- ions through ion-change or Lewis acid/base interactions. Figure 8.6 depicts electrostatic attractions, coordination of ReO4- and amino groups, and host–guest interaction of

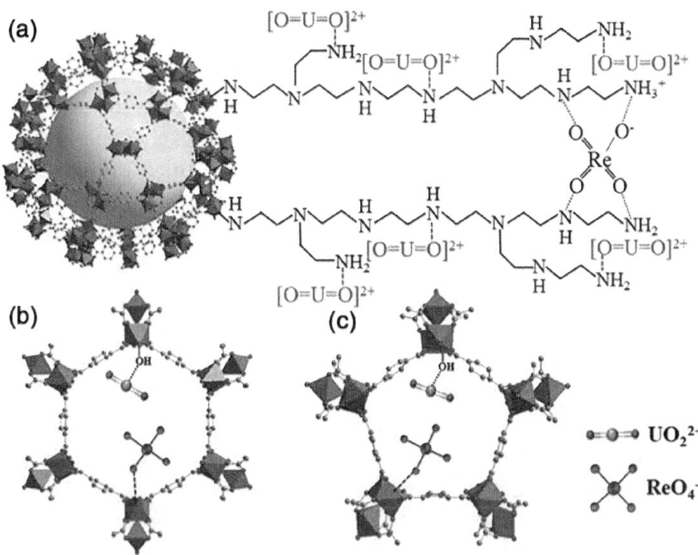

FIGURE 8.6 (a) Representation showing the interaction between UO_2^{2+} and ReO_4– with PEI; Possible host–guest interactions between UO_2^{2+} and ReO_4– with the MIL-101(Cr) cavity. (b) Hexagonal and (c) pentagonal windows. (Reprinted with permission from Ref. [96], Copyright © 2021 Elsevier.)

ReO_4- and cavities. Coordination with CUSs in MIL-101(Cr), creation of NHO hydrogen bonds, and CNO complexation between N-related sites and UO_2^{2+} are all major sources of strong adsorption. The probable mechanism of UO_2^{2+} sorption onto MILP-3 is depicted graphically in Figure 8.6. Finally, the N-containing groups and cavities were among the most critical in the ReO_4^- or UO_2^{2+} adsorption procedure [96].

In general, adsorption methods are exceptionally effective at eliminating radioactive detrimental contaminants; nevertheless, due to the restricted saturated states of the adsorption material and the wide range of surrounding environments, radioactive contaminants could probably be returned to the water-based system, resulting in supplementary pollution on the environment. On top of that, the active area on the surface of the adsorbent, alkalinity, pH level of the solution, and surface energy have a substantial influence on the efficiency of removal of wastes. When choosing an effective adsorbent, consider the cost and accessibility, quicker kinetics, mechanical and chemical stability, and reusability. However, their field uses are limited because of low yield and greater production costs, the risk of nanoparticle leaching, and obstacles in solid–liquid separation [97]. To address the issue of leaching, initiatives were undertaken to immobilize the nano adsorbents employing support matrices such as synthetic polymers, cellulose, chitosan, and inert granular materials. Nonetheless, the immobilization may distort adsorption sites and minimize adsorbent efficient operation. As a result, creating reliable and long-lasting immobilization technologies is essential to improve the utility of nanoscale adsorbents. Co-ions, like bicarbonates, can have a substantial impact on the process of adsorption. As a result, the functionality of higher-alkaline adsorbent materials is in doubt. There is also a shortage of data, and additional research is needed to explore the influence of unidentified background ions on the adsorption approaches. Furthermore, most of the adsorption research is undertaken in laboratory batch reactors, with little information available for assessing the effectiveness of systems at the pilot and industrial scales. An additional decisive component missing from the literature is a comprehensive investigation of the overall sustainability of the system. Encouraging closed-loop recycling and reuse over disposal-based linear approaches is crucial for enhancing treatment process sustainability.

8.5.2 Ion Exchange

Ion exchange is the process of exchanging desired anions or cations from the feed solution for a chemically identical quantity of similar charge ions within the resin's molecular networks [9]. The resin itself undergoes no structural modifications throughout the exchange processes. There are two kinds of exchangers: anion exchangers and cation exchangers. Anion exchangers feature exchange of anions and ions with negative charges functional groups in their feed solutions, whereas cation exchangers have both exchange cations as well as positively charged functional groups in their feed solutions. Furthermore, some amphoteric exchangers are produced from neutrally charged zwitter-ions and are capable of exchanging either cations or anions. As a result, higher valency ions have a preference over low valency ions for exchange, and among ions of identical valency, the ion with a greater atomic number is favored. When selecting the resin (REF), various factors like the initial content of contamination, pH, adsorbate charge, and co-ions must be taken care of. The ion exchange resins are often classified into four categories: strong base anion exchangers (SBA), weak base anion exchangers (WBA), strong acid cation exchangers (SAC), and weak acid cation exchangers (WAC). The SBA is a widely accessible and used for demineralization agent that has a higher reactivity for U(VI) at alkaline pH levels. This greater affinity was owing to the production of uranyl carbonate compounds at alkaline pH, whereby these compounds will be interchanging with SBA ions. Chen et al. focused on the quick and efficient capturing of perrhenate anion from simulated groundwater using a mesoporous silica-supported anion exchanger in their work. The maximal working capacity for Re (VII) was 213 mg g^{-1}. In the presence of 5 mmol L^{-1} of distinct competitive anions, elimination rates of over 90% were reported. The adsorption performance was to be precisely characterized by the Redlich–Peterson model, the Pseudo-second kinetic model, and the Thomas model [98].

8.5.3 Photocatalytic Reduction

Photocatalytic reduction utilizing solar energy as a key driver could be considered as an ecological, green, and economical alternative to the prior methods [99]. Photocatalysis is a sophisticated method of oxidization in which electrons (e^-) jump out of the valence band (EV) and migrate higher into the conduction band (EC), which is set apart by what is called an energy band gap (Eg). When the energy of the light irradiation surpasses the value of Eg, a hole (h^+) is formed in EV. Following that, an e^-/h^+ pair is formed. The created e^- decreases the adsorbed substrate, while the generated h^+ oxidizes it [100,101]. For example, because of its low solubility, U(VI) can be converted to U(IV) by photocatalysis and quickly adsorbed onto the surface of the semiconductor. It is known that radioactive U is present in a variety of chemical valences (U_0, U^{3+}, U^{4+}, and U^{6+}) and that the major species in the natural environment are soluble U^{6+} as well as less soluble U^{4+} [102]. Photocatalysts such as SnO_2, ZnO, Fe_2O_3, TiO_2, ZnS, and CdS are commonly used. By its high stability, minimal cost, non-toxicity, and catalytic performance, TiO_2 is extremely effective [103]. TiO_2 compound is commonly found in three phases: anatase, rutile, and brookite. Along with its high photocatalytic efficiency, the anatase phase is commonly utilized. Ordinary TiO_2 has an Eg value of 3.2 eV, and when excited with UV light at 388 nm, it produces e^-/h^+ pairs [104]. However, efforts have been made to look for alternatives or modify photocatalysts for the enhanced photoactivity, and many new materials are being now used for the photoreduction of radioactive materials.

Deng et al. synthesized non-crystalline TiO_2/g-C_3N_4 compounds using a simple hydrolysis technique, obtaining a non-crystalline TiO_2/30 wt% g-C_3N_4 (TCN-3) compound with a significantly greater surface area (456 m^2g^{-1}). They discovered that the composite had a lower photo-generated e^-/h^+ recombination efficiency cantered on the Z-scheme mechanism among non-crystalline TiO_2 and g-C_3N_4 as expressed by photoelectrochemical and electron paramagnetic resonance investigation. Furthermore, when compared to crystalline TiO_2/g-C_3N_4, the synthesized TCN-3 compound demonstrates outstanding photocatalytic reduction of Re(VII) in STP, with a maximum elimination percentage of 90%, and can be replenished even after eight recycles [105]. Furthermore, Meng et al.

first synthesized metal-free 2D/2D C_3N_4/GO nanosheets before customizing a type-II heterojunction relying on band bending theory to enhance overall uranium extraction efficiency via synergistic adsorption photoreduction engineering. The research shows that even at a low uranium concentration of 10ppm, the removal margin of uranium by 2D/2D C_3N_4/GO heterojunction can reach 96.1% and 93.4% after gamma-ray exposure. This work will lay the groundwork for future work on tailoring the energy band structure of nonmetal-based 2D/2D nanohybrids and enhancing uranium-containing wastewater via adsorption photoreduction designing [106].

Jinna Feng et al. suggested a photoreduction mechanism for Uranium(VI) using Sn-doped In_2S_3 microspheres under visible light. A condition of the photoreduction process was proposed based on experimental results and theoretical knowledge, as shown in Figure 8.7a. They discovered

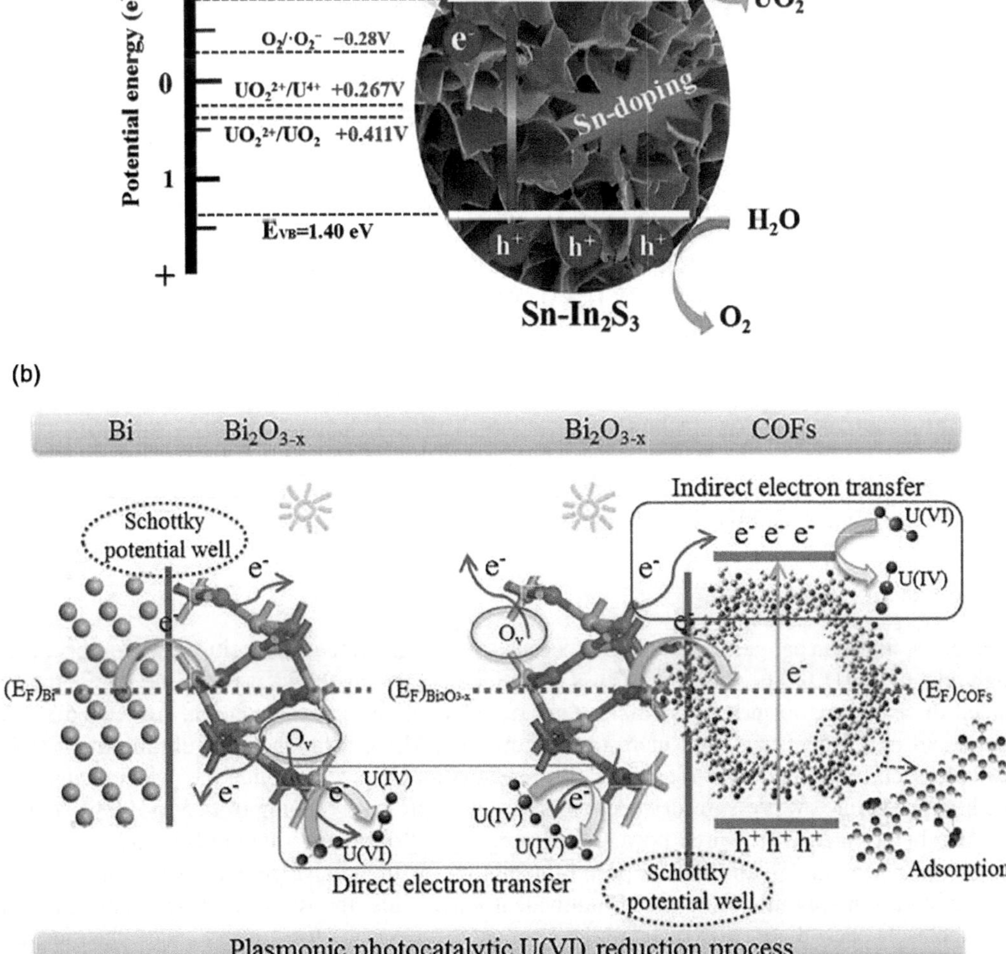

FIGURE 8.7 Possible photoreduction mechanisms of uranium on (a) Sn-In_2S_3 (Reproduced with permission from Ref. [107], Copyright © 2018 Elsevier.) and (b) Bi/Bi_2O_3–x@COFs. Reproduced with permission from Ref. [108], Copyright © 2023 Elsevier.)

that when the photoenergy was significantly higher than the Eg value, Sn-In_2S_3 was excited, and photo-generated e^-/h^+ pairs were formed. In addition, photoinduced electrons (e) have been moved from EV to EC, resulting in the formation of holes (h^+) in VB. Once UO_2^{2-} became adsorbed on the Sn-In_2S_3 surface, photogenerated e^- on the EC of Sn-In_2S_3 may migrate to U(VI), resulting in the corresponding decrease of U(VI) ions to U(IV) and precipitation of U(IV) on the surfaces of Sn-In_2S_3 composite. In the meantime, the pores oxidize moisture to O_2. This configuration enables the photoreduction of U(VI) in a tranquil way. The following are some examples of major processes:

$$\text{Sn-In}_2\text{S}_3 + \text{hv} = \text{Sn-In}_2\text{S}_3\,(\text{e} + \text{h}) \tag{8.1}$$

$$\text{UO}_2^{\;2-} + 2\text{H} + \text{e} = \text{UO}_2\,(\text{s}) \tag{8.2}$$

$$\text{H}_2\text{O} + 2\text{h} = (½)\text{O}_2 + 2\text{H}^+ \tag{8.3}$$

They reported that greater surface area, better absorption of electromagnetic radiation with wavelengths of 500–800 nm, narrowed Eg (~ 2.1 eV) with an upshift of EV, EC potentials, and elongating separation between photogenerated electron–hole pairs (e^-/h^+) were all attributed to Sn-In_2S_3's improved photocatalytic performance [107]. On the other hand, Liu and coworkers developed a plasmonic Bi/Bi_2O_3-x@COFs photocatalyst for photoreduction of U(VI), as indicated in Figure 8.7b. From their work, they established that the presence of oxygen vacancy in Bi_2O_{3-x} decreases the work function and increases Fermi level of Bi_2O_3_x, which is appropriate for Schottky's potential and electron transfer rate improvement [108]. The inclusion of COFs displayed the inclusion of the most important specific binding groups for U(VI). The authors noticed 1411.5 mg g^{-1} removal capacity and 93.9% photoreduction efficiency. The process of photoreduction was confirmed by XPS analysts where U 4f displayed both peaks for U(VI) and U(IV), with U(VI) being more intense, confirming incomplete photoreduction of the targeted U(VI). In the photoreduction process of radioactive materials, various techniques including X-ray diffraction (XRD), X-ray photoelectron spectroscopy (XPS), X-ray absorption near edge structure (XANES), Fourier transform infrared spectroscopy (FTIR), and extended X-ray absorption fine structure (EXAFS) have been used by different researchers with success [109–111].

8.5.4 Membranes

Membrane filtration has gained popularity because of its high removal capacity, small footprint, and high selectivity [112]. The filtering process using membranes involves a pressure-driven method in which the barrier is formed by membrane for pollutants like viruses, bacteria, suspended solids, mono and multivalent ions, and more to pass through. Microfiltration (mF), ultrafiltration (UF), nanofiltration (NF), and reverse osmosis (RO) are the four types of membrane progressions, with the first two being low pressure-driven membranes with pressure ranging from 5 to 100 psi and the last two being intense-pressured powered membranes with pressure ranging from 50 to 1000 psi [113,114]. In general, UF and mF are porous membranes with massive pore diameters that are ineffective at eliminating monovalent and multivalent ions, whilst the RO is a non-porous membrane with a closely packed pore size of 1 nm that eliminates the ions by the process called chemical diffusion. The process is influenced by applied pressure, pH, initial U(VI) concentration, membrane fouling, scaling, pore size, and co-ions [115]. The next section discusses the many membrane processes used to remove radioactive contaminants.

A two-step membrane process consisting of two types of ceramic membranes was proposed. First, a ceramic nanofiltration membrane was used to remove the nuclide from the boric acid.

The purified boric acid solution was then concentrated using a vacuum membrane distillation method using a hydrophobic ceramic membrane. The two-stage NF process removed 99.9% of Co^{2+} ions and 95% of Ag^{+} ions [116].

According to this study, ammonium molybdophosphate ($(NH_4)_3P(Mo_3O_{10})_4{\cdot}3H_2O$) has shown a high selectivity for the cesium ion (Cs+) because of the specific intercalation between Cs^{+} and NH_4^{+}. In this study, the AMP-PAN composite efficiently removed Cs^{+} (95.7%, 94.1%, and 91.3% of 1 mg L^{-1}) from solutions with high ionic strength (400 mg L^{-1} of Na^{+}, Ca^{2+}, or K^{+}). Kinetic and isotherm studies supported the multilayer chemical adsorption process. The maximum adsorption capacities were estimated to be 138.9 ± 21.3 mg g^{-1}. Particularly, the rate-limiting step throughout the filtration was recognized as liquid film diffusion. Eventually, the filtration adsorption process of the AMP-PAN membrane effectively eliminated Cs^{+} from water [117].

Moreover, forward osmosis (FO) was thoroughly studied for concentrating hazardous fluid wastes from hospital radiation treatment facilities. The filtering of both organic and radioactive iodine utilizing FO was initially explored with varied pHs and drawing solutions to find the ideal settings for concentration of FO. FO had an efficient rejection rate of up to 99.3% for all natural and radioactive iodine (125I). This significant rejection rate was reached at an elevated pH, owing to iodine-membrane electric-repellent effects. Enhanced iodine removal by FO was additionally accomplished with a DS with a reverse salt flow sufficient to obstruct iodine transportation. Following that, under these ideal conditions, genuine radioactive medical liquid waste was gathered and concentrated utilizing FO [118].

8.5.5 Iron Electrocoagulation

Electrochemical procedures such as electrocoagulation (EC) can be used to remediate small-size stainless steel parts with simple geometries and static contamination. In general, the components are submerged in solutions with a difference in potential. These decontamination techniques are effective and create little supplementary waste. Electrocoagulation is a method of degrading harmful substances in water that employs a minimal-voltage direct electric current and metal sacrificial electrodes, commonly aluminum/iron (Al/Fe), to generate a greater ion charge that destabilizes the contaminants present. It had previously been reported that iron EC system and discovered that several organic ligands with high binding affinity could precisely sieve UO_2^{2+} and reduce its surface charge. As a result, U(VI) was effortlessly entrapped inside the colloidal ferric oxide/hydroxides produced during EC. The precipitated mixture's U(VI) species could be efficiently concentrated using a Na_2CO_3 and H_2O_2 eluent.

In 2020, Li and colleagues reported that ligand chelation improved an electrocoagulation method to remove and recover U from wastewater. The anode material and chelate have been optimized for the treatment of bulk U-comprising wastewater, and the mechanisms of U precipitation and recycling were extensively probed. By the emission regulation, the uranium was successfully removed in a short time utilizing the optimized Fe anode and Alizarin S as a chelate. Following Alizarin S's selective capture of uranyl ions, the chelation complex will break down and absorb the flocculation precursor, restricting crystal development and reducing interfacial charge to enable precipitation. An oxidizing detergent successfully dissolved uranium in precipitated flocs, and yellowcake heavy uranium amide was obtained via organic digestion and ammonia precipitation. The complete recovery of U proficiency was 89.7% [119].

The remediation of non-degradable radioactive wastes, such as tools and equipment, focuses on the reduction of waste that must be accustomed and stored. The application of EC in the remediation of non-degradable radioactive pollutants from stainless steel comprising U was investigated to determine its technical feasibility. To assess the feasibility of the EC decontamination process, the maximum removal requirements for WO_3 were implemented on plates tainted with $UO_2(NO_3)_2$. In a circular array, U removal rates of 90% were attained in 1 hour at pH 1, 2.4 V, and a distance of 1 cm between anode/cathode. When utilizing the circular A/C design, they were successful in removing 91.4% of U from the stainless steel rod. They proposed that the EC pollutant removal process for stainless steel includes the specific procedure (Figure 8.8) [120].

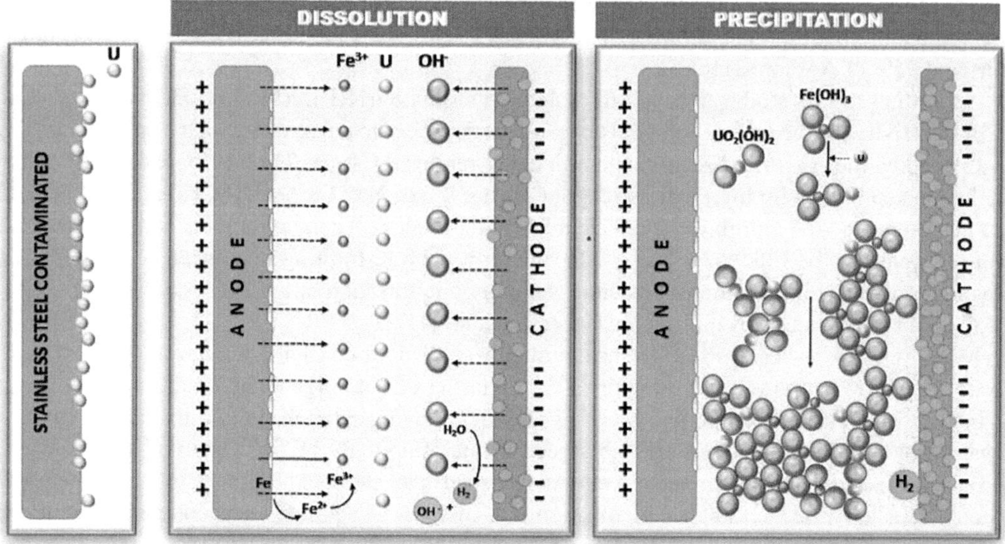

FIGURE 8.8 Mechanisms of uranium removal using electrocoagulation process. (Reproduced with permission from Ref. [120], Copyright © 2019 Elsevier.)

This research team carried out an experimental work on the treatment of radioactive contamination (U) in an electrocoagulation process using multi-objective fuzzy optimization for the Fe- and Al-stainless steel anode/cathode processes. In terms of remnant U-concentration and energy consumption, Cost/benefit solutions revealed that the Fe process (0 mg/L U: 492 USD/g-U) exceeded the Al process (96 mg/L U: 747 USD/g-U). The Fe system was further investigated in a multi-objective assessment because of its practicality in reaching numerous U standard governing constraints. Based on the 30 mg/L MCL, the fuzzy logic decision-making method yielded a 6.1% overall satisfaction at a treatment duration and current density that were 101.6 minutes and 59.9 mA/cm^2, respectively. The following are the results of the fuzzy optimum solution: The U concentration was 5 mg/L, the cumulative uncertainty was 25 mg/L, the energy consumption was 461.7 kWh/g-U, and the operational cost was 60.0 USD/g-U in the United States, 55.4 USD/g-U in South Korea, and 78.5 USD/g-U in Finland [121].

8.6 CONCLUSION AND FUTURE OUTLOOK

Though there are numerous technologies for eliminating radioactive contaminants from the environment, there is no one approach that is adaptable enough to meet all requirements. Each advancement in technology has its pros and cons. As a result, selecting the appropriate technology for a specific ailment is critical to a successful therapy implementation. When establishing a treatment system, the cost of treatment, the effectiveness of removal, contaminants concentration and speciation, sludge formation, feed-flow rate, water chemistry, ease of operation, and other aspects must all be addressed. Finally, because no single current technology is totally efficient in eliminating radioactive pollutants in all scenarios, future attention must be placed on establishing dynamic and cutting-edge technology. The development of viable, inexpensive, and realistically sustainable high-efficiency adsorbents, coagulants, membranes, resins, and catalysts is crucial for the future. Heavy metal testing, including in drinking water, needs a complete water quality monitoring system. Understanding how these contaminants enter groundwater, as well as their fate and transit, can help in the development of effective and comprehensive cleanup strategies. Other than adsorption and photoreduction, there has been little investigation into other methods of eliminating radioactive pollutants.

REFERENCES

1. Roberts, C. J. Management and disposal of waste from sites contaminated by radioactivity. *Radiat. Phys. Chem.*, 1998, 51 (4–6), 579–587. https://doi.org/10.1016/S0969-806X(97)00205-3.
2. Wang, L.; Tao, W.; Yuan, L.; Liu, Z.; Huang, Q.; Chai, Z.; Gibson, J. K.; Shi, W. Rational control of the interlayer space inside two-dimensional titanium carbides for highly efficient uranium removal and imprisonment. *Chem. Commun.*, 2017, 53 (89), 12084–12087. https://doi.org/10.1039/c7cc06740b.
3. Kang, S. M.; Rethinasabapathy, M.; Hwang, S. K.; Lee, G. W.; Jang, S. C.; Kwak, C. H.; Choe, S. R.; Huh, Y. S. Microfluidic generation of prussian blue-laden magnetic micro-adsorbents for cesium removal. *Chem. Eng. J.*, 2018, 341 (January), 218–226. https://doi.org/10.1016/j.cej.2018.02.025.
4. Ortaboy, S.; Acar, E. T.; Atun, G. The removal of radioactive strontium ions from aqueous solutions by isotopic exchange using strontium decavanadates and corresponding mixed oxides. *Chem. Eng. J.*, 2018, 344 (January), 194–205. https://doi.org/10.1016/j.cej.2018.03.069.
5. Fard, A. K.; Mckay, G.; Chamoun, R.; Rhadfi, T.; Preud'Homme, H.; Atieh, M. A. Barium removal from synthetic natural and produced water using MXene as two dimensional (2-D) nanosheet adsorbent. *Chem. Eng. J.*, 2017, 317, 331–342. https://doi.org/10.1016/j.cej.2017.02.090.
6. Li, S.; Wang, L.; Peng, J.; Zhai, M.; Shi, W. Efficient Thorium(IV) removal by two-dimensional Ti_2CT_x MXene from aqueous solution. *Chem. Eng. J.*, 2019, 366 (February), 192–199. https://doi.org/10.1016/j.cej.2019.02.056.
7. Kamran, U.; Rhee, K. Y.; Lee, S. Y.; Park, S. J. Innovative progress in graphene derivative-based composite hybrid membranes for the removal of contaminants in wastewater: A review. *Chemosphere*, 2022, 306 (June), 135590. https://doi.org/10.1016/j.chemosphere.2022.135590.
8. Nain, A.; Sangili, A.; Hu, S. R.; Chen, C. H.; Chen, Y. L.; Chang, H. T. Recent progress in nanomaterial-functionalized membranes for removal of pollutants. *iScience*, 2022, 25 (7), 104616. https://doi.org/10.1016/j.isci.2022.104616.
9. Figueiredo, B. R.; Cardoso, S. P.; Portugal, I.; Rocha, J.; Silva, C. M. Inorganic ion exchangers for cesium removal from radioactive wastewater. *Sep. Purif. Rev.*, 2018, 47 (4), 306–336. https://doi.org/10.1080/15422119.2017.1392974.
10. Kim, Y.; Hyeon, H.; Kon, Y.; Harbottle, D.; Lee, J. W. Chemosphere effective removal of cesium from wastewater via adsorptive filtration with potassium copper hexacyanoferrate-immobilized and polyethyleneimine-grafted graphene oxide. *Chemosphere*, 2020, 250, 126262. https://doi.org/10.1016/j.chemosphere.2020.126262.
11. Xiang, S.; Mao, H.; Geng, W.; Xu, Y.; Zhou, H. Selective removal of Sr (II) from saliferous radioactive wastewater by capacitive deionization. *J. Hazard. Mater.*, 2022, 431 (October 2021), 128591. https://doi.org/10.1016/j.jhazmat.2022.128591.
12. Oh, M.; Lee, K.; Ku, M.; Foster, R. I.; Lee, C. Journal of environmental chemical engineering chemical precipitation – based treatment of acidic wastewater generated by chemical decontamination of radioactive concrete. *J. Environ. Chem. Eng.*, 2023, 11 (5), 110306. https://doi.org/10.1016/j.jece.2023.110306.
13. Wu, E.; Yu, Y.; Hu, J.; Ren, G.; Zhu, M. Piezoelectric-channels in MoS_2-embedded polyvinylidene fluoride membrane to activate peroxymonosulfate in membrane filtration for wastewater reuse. *J. Hazard. Mater.*, 2023, 458 (April), 131885. https://doi.org/10.1016/j.jhazmat.2023.131885.
14. Yin, H. L.; Tan, Z. Y.; Liao, Y. T.; Feng, Y. J. Application of SO_4^{2-}/TiO_2 solid superacid in decontaminating radioactive pollutants. *J. Environ. Radioact.*, 2006, 87 (2), 227–235. https://doi.org/10.1016/j.jenvrad.2005.11.009.
15. Zhang, Y. J.; Zhou, Z. J.; Lan, J. H.; Ge, C. C.; Chai, Z. F.; Zhang, P.; Shi, W. Q. Theoretical insights into the uranyl adsorption behavior on vanadium carbide MXene. *Appl. Surf. Sci.*, 2017, 426, 572–578. https://doi.org/10.1016/j.apsusc.2017.07.227.
16. Wen, T.; Ma, R.; Liu, X.; Song, S.; Wu, B.; Jiang, Z.; Wang, X. Adsorptive and reductive removal of toxic and radioactive metal ions by nanoscale zero-valent iron-based nanomaterials from wastewater; 2021. https://doi.org/10.1016/B978-0-323-85484-9.00001-7.
17. Zaheen, B.; Ahmad, A.; Luque, R.; Hussain, S.; Noreen, R. Inorganic pollutants and their degradation with nanomaterials; INC, 2023. https://doi.org/10.1016/b978-0-12-823551-5.00004-5.
18. Liu, G.; Jin, W. Graphene oxide membrane for molecular separation. *Sci. China Mater.*, 2018, 61, 1021–1026.
19. Jia, F.; Xiao, X.; Nashalian, A.; Shen, S.; Yang, L.; Han, Z.; Qu, H.; Boe, E.; Tfotps, H. B. T. Advances in graphene oxide membranes for water treatment. *Nano Res.* 2022, 15 (7), 6636–6654.
20. Majidnia, Z.; Idris, A. Evaluation of cesium removal from radioactive waste water using maghemite PVA-alginate beads. *Chem. Eng. J.*, 2015, 262, 372–382. https://doi.org/10.1016/j.cej.2014.09.118.

21. Liu, Z.; Ling, Q.; Cai, Y.; Xu, L.; Su, J.; Yu, K.; Wu, X.; Xu, J.; Hu, B.; Wang, X. Synthesis of carbon-based nanomaterials and their application in pollution management. *Nanoscale Adv.*, 2022, 4 (5), 1246–1262. https://doi.org/10.1039/d1na00843a.
22. Wang, Z.; Zhang, L.; Zhang, K.; Lu, Y.; Chen, J.; Wang, S.; Hu, B.; Wang, X. Application of carbon dots and their composite materials for the detection and removal of radioactive ions: A review. *Chemosphere*, 2022, 287 (P3), 132313. https://doi.org/10.1016/j.chemosphere.2021.132313.
23. El-sherif, R. M.; Lasheen, T. A.; Jebril, E. A. Fabrication and characterization of CeO_2-TiO_2-Fe_2O_3 Magnetic nanoparticles for rapid removal of uranium ions from industrial waste solutions. *J. Mol. Liq.*, 2017, 241, 260–269. https://doi.org/10.1016/j.molliq.2017.05.119.
24. Dubey, S.; Banerjee, S.; Upadhyay, S. N.; Sharma, Y. C. Application of common nano-materials for removal of selected metallic species from water and wastewaters: A critical review. *J. Mol. Liq.*, 2017, 240, 656–677. https://doi.org/10.1016/j.molliq.2017.05.107.
25. Swihart, M. T. Vapor-phase synthesis of nanoparticles. *Curr. Opin. Colloid Interface Sci.*, 2003, 8, 127–133. https://doi.org/10.1016/S1359-0294(03)00007-4.
26. Kumar, V.; Katyal, D.; Nayak, S. S. Removal of heavy metals and radionuclides from water using nanomaterials: Current scenario and future prospects. *Environ. Sci. Pollut. Res.*, 2020, 27 (33), 41199–41224. https://doi.org/10.1007/s11356-020-10348-4.
27. Wang, J.; Zhuang, S. Covalent organic frameworks (COFs) for environmental applications. *Coord. Chem. Rev.*, 2019, 400, 213046. https://doi.org/10.1016/j.ccr.2019.213046.
28. Stegbauer, L.; Schwinghammer, K.; Lotsch, B. V. A hydrazone-based covalent organic framework for photocatalytic hydrogen production. *Chem. Sci.*, 2014, 5 (7), 2789–2793. https://doi.org/10.1039/c4sc00016a.
29. Knighten, G. V; Weber, A.; Turner, R. D.; Smith, R. W.; Shen, Y. R.; Fitzgibbon, R.; Lax, B.; Evans, C. L.; Xie, X. S.; Dudovich, N.; et al. Designed synthesis of 3D covalent. *Science*, 2007, 13, 268–273.
30. Gu, P.; Zhang, S.; Li, X.; Wang, X.; Wen, T.; Jehan, R.; Alsaedi, A.; Hayat, T.; Wang, X. Recent advances in layered double hydroxide-based nanomaterials for the removal of radionuclides from aqueous solution. *Environ. Pollut.*, 2018, 240, 493–505. https://doi.org/10.1016/j.envpol.2018.04.136.
31. Gan, Y. X.; Jayatissa, A. H.; Yu, Z.; Chen, X.; Li, M. Hydrothermal synthesis of nanomaterials. *J. Nanomater.*, 2020. https://doi.org/10.1155/2020/8917013.
32. Zang, T.; Wang, H.; Liu, Y.; Dai, L.; Zhou, S.; Ai, S. Fe-doped biochar derived from waste sludge for degradation of rhodamine B via enhancing activation of peroxymonosulfate. *Chemosphere*, 2020, 261, 127616. https://doi.org/10.1016/j.chemosphere.2020.127616.
33. Khan, F. A. *Applications of Nanomaterials in Human Health*, 1st Edition; Springer, 2020. https://doi.org/10.1007/978-981-15-4802-4.
34. Huynh, K. H.; Pham, X. H.; Kim, J.; Lee, S. H.; Chang, H.; Rho, W. Y.; Jun, B. H. Synthesis, properties, and biological applications of metallic alloy nanoparticles. *Int. J. Mol. Sci.*, 2020, 21 (14), 1–29. https://doi.org/10.3390/ijms21145174.
35. Sharma, G.; Kumar, A.; Sharma, S.; Naushad, M.; Prakash Dwivedi, R.; ALOthman, Z. A.; Mola, G. T. Novel development of nanoparticles to bimetallic nanoparticles and their composites: A review. *J. King Saud Univ. - Sci.*, 2019, 31 (2), 257–269. https://doi.org/10.1016/j.jksus.2017.06.012.
36. Chen, D.; Li, Q.; Shao, L.; Zhang, F.; Qian, G. Recovery and application of heavy metals from pickling waste liquor (PWL) and electroplating wastewater (EPW) by the combination process of ferrite nanoparticles. *Desalin. Water Treat.*, 2016, 57 (60), 29264–29273. https://doi.org/10.1080/19443994.2016.1172984.
37. Judith, J. V.; Vasudevan, N. Synthesis of nanomaterial from industrial waste and its application in environmental pollutant remediation. *Environ. Eng. Res.*, 2022, 27 (2), 1–2. https://doi.org/10.4491/eer.2020.672.
38. Ma, H.; Shen, M.; Tong, Y.; Wang, X. Radioactive wastewater treatment technologies: A review. *Molecules*, 2023, 28 (4). https://doi.org/10.3390/molecules28041935.
39. Bokov, D.; Turki Jalil, A.; Chupradit, S.; Suksatan, W.; Javed Ansari, M.; Shewael, I. H.; Valiev, G. H.; Kianfar, E. Nanomaterial by sol-gel method: Synthesis and application. *Adv. Mater. Sci. Eng.*, 2021. https://doi.org/10.1155/2021/5102014.
40. Hwang, S. K.; Kang, S. M.; Rethinasabapathy, M.; Roh, C.; Huh, Y. S. MXene: An emerging two-dimensional layered material for removal of radioactive pollutants. *Chem. Eng. J.*, 2020, 397, 125428. https://doi.org/10.1016/j.cej.2020.125428.
41. Akharawutchayanon, T.; Sopapan, P.; Yotthuan, S.; Gunhakoon, P.; Yubonmhat, K.; Issarapanacheewin, S.; Katekaew, W.; Prasertchiewchan, N. Case studies in chemical and environmental engineering removal efficiency of 137 Cs from radioactively contaminated electric arc furnace dust using different solvents. *Case Stud. Chem. Environ. Eng.*, 2023, 8, 100409. https://doi.org/10.1016/j.cscee.2023.100409.

42. Meng, Y.; Wang, Y.; Ye, Z.; Wang, N.; He, C.; Zhu, Y.; Fujita, T.; Wu, H.; Wang, X. Three-dimension titanium phosphate aerogel for selective removal of radioactive Strontium(II) from contaminated waters. *J. Environ. Manage.*, 2023, 325 (PB), 116424. https://doi.org/10.1016/j.jenvman.2022.116424.
43. Wang, Z.; Chen, K.; Gu, A.; Zhou, X.; Wang, P.; Gong, C.; Mao, P.; Jiao, Y.; Chen, K.; Lu, J.; et al. Journal of solid state chemistry adsorption performance study of bismuth-doped ZIF-8 composites on radioactive iodine in the vapor and liquid phases. *J. Solid State Chem.*, 2023, 325, 124186. https://doi.org/10.1016/j.jssc.2023.124186.
44. Mu, W.; Chen, B.; Yu, Q.; Li, X.; Wei, H.; Yang, Y.; Peng, S. A novel zirconium phosphonate adsorbent for highly efficient radioactive cesium removal. *J. Mol. Liq.*, 2021, 326, 115307. https://doi.org/10.1016/j.molliq.2021.115307.
45. Yu, R. L.; Li, Q. F.; Zhang, T.; Li, Z. Le; Xia, L. Z. Zn, O Co-adsorption based on MOF-5 for efficient capture of radioactive iodine. *Process Saf. Environ. Prot.,* 2023, 174, 770–777. https://doi.org/10.1016/j.psep.2023.04.045.
46. Verma, S.; Kim, K. Graphene-based materials for the adsorptive removal of uranium in aqueous solutions. *Environ. Int.*, 2022, 158, 106944. https://doi.org/10.1016/j.envint.2021.106944.
47. Freire, J. M. A.; Moreira, I. O.; França, A. M. D. M.; Luiz, T. V; Santos, L. P. M.; Lucas, S.; Medeiros, S.; Vasconcelos, I. F. De; Loiola, A. R.; Antunes, R. A.; et al. Functionalized magnetic graphene oxide composites for selective toxic metal adsorption. *Environ. Nanotechnol. Monit. Manag.*, 2023, 20. https://doi.org/10.1016/j.enmm.2023.100843.
48. Li, H.; Zhang, L.; Chen, J.; Lu, M.; Xie, J.; Wang, X.; Han, K.; Li, J.; Lu, J. Journal of water process engineering reduced graphene oxide based aerogels: Doped with ternary prussian blue analogs and selective removal of Cs+from effluent. *J. Water Process Eng.*, 2022, 47, 102741. https://doi.org/10.1016/j.jwpe.2022.102741.
49. Sopapan, P.; Lamdab, U.; Akharawutchayanon, T.; Issarapanacheewin, S.; Yubonmhat, K.; Silpradit, W.; Katekaew, W.; Prasertchiewchan, N. Effective removal of non-radioactive and radioactive cesium from wastewater generated by washing treatment of contaminated steel ash. *Nucl. Eng. Technol.*, 2023, 55 (2), 516–522. https://doi.org/10.1016/j.net.2022.10.007.
50. Lee, H.; Yoo, D.; Jo, S.; Choi, S. Progress in nuclear energy removal of nitrate from radioactive wastewater using magnetic multi-walled carbon nanotubes. *Prog. Nucl. Energy*, 2021, 140, 103893. https://doi.org/10.1016/j.pnucene.2021.103893.
51. Attallah, M. F.; Hassan, H. S.; Youssef, M. A. Synthesis and sorption potential study of Al_2O_3–ZrO_2–CeO_2 composite material for removal of some radionuclides from radioactive waste effluent. *Appl. Radiat. Isot.*, 2019, 147, 40–47. https://doi.org/10.1016/j.apradiso.2019.01.015.
52. Villar, M.; Quesada, S.; Turnes, G.; Ferrer, L.; Palomino, C. Printed device for the removal of radioactive iodine from wastewaters. *Appl. Mater. Today*, 2021, 24, 101130. https://doi.org/10.1016/j.apmt.2021.101130.
53. Helal, A. A.; Breky, M. M. E.; Allan, K. F.; Attallah, M. F. Removal of Eu^{3+} from simulated aqueous solutions by synthesis of a new composite adsorbent material. *Appl. Radiat. Isot.*, 2023, 191 (October 2022), 110543. https://doi.org/10.1016/j.apradiso.2022.110543.
54. Hatra, G. Radioactive pollution: An overview. *Holist. Approach Environ.*, 2018, 8 (2), 48–65.
55. Gilbert, M. R.; Sublet, J. C. PKA distributions: Contributions from transmutation products and from radioactive decay. *Nucl. Mater. Energy*, 2016, 9, 576–580. https://doi.org/10.1016/j.nme.2016.02.006.
56. Thomas, G. A.; Symonds, P. Radiation exposure and health effects - is it time to reassess the real consequences? *Clin. Oncol.*, 2016, 28 (4), 231–236. https://doi.org/10.1016/j.clon.2016.01.007.
57. Adebiyi, F. M.; Ore, O. T.; Adeola, A. O.; Durodola, S. S.; Akeremale, O. F.; Olubodun, K. O.; Akeremale, O. K. Occurrence and remediation of naturally occurring radioactive materials in Nigeria: A review. *Environ. Chem. Lett.*, 2021, 19 (4), 3243–3262. https://doi.org/10.1007/s10311-021-01237-4.
58. Sitaud, B.; Solari, P. L.; Schlutig, S.; Llorens, I.; Hermange, H. Characterization of radioactive materials using the MARS beamline at the synchrotron SOLEIL. *J. Nucl. Mater.*, 2012, 425 (1–3), 238–243. https://doi.org/10.1016/j.jnucmat.2011.08.017.
59. Kakehi, S.; Kaeriyama, H.; Ambe, D.; Ono, T.; Ito, S. ichi; Shimizu, Y.; Watanabe, T. Radioactive cesium dynamics derived from hydrographic observations in the abukuma river estuary, Japan. *J. Environ. Radioact.*, 2016, 153, 1–9. https://doi.org/10.1016/j.jenvrad.2015.11.015.
60. Burger, A.; Lichtscheidl, I. Stable and radioactive cesium: A review about distribution in the environment, uptake and translocation in plants, plant reactions and plants' potential for bioremediation. *Sci. Total Environ.*, 2018, 618, 1459–1485. https://doi.org/10.1016/j.scitotenv.2017.09.298.
61. Kaeriyama, H. Oceanic dispersion of fukushima-derived radioactive cesium: A review. *Fish. Oceanogr.*, 2017, 26 (2), 99–113. https://doi.org/10.1111/fog.12177.

62. Awual, M. R.; Suzuki, S.; Taguchi, T.; Shiwaku, H.; Okamoto, Y.; Yaita, T. Radioactive cesium removal from nuclear wastewater by novel inorganic and conjugate adsorbents. *Chem. Eng. J.*, 2014, 242, 127–135. https://doi.org/10.1016/j.cej.2013.12.072.
63. Ryzhikov, A.; Daou, T. J. Porous sorbents for the capture of radioactive iodine compounds: A review. *RSC Adv.*, 2018, 8, 29248–29273. https://doi.org/10.1039/c8ra04775h.
64. Andersson, M.; Benoist, B. De; Rogers, L. Epidemiology of iodine deficiency: Salt iodisation and iodine status. *Best Pract. Res. Clin. Endocrinol. Metab.*, 2010, 24 (1), 1–11. https://doi.org/10.1016/j.beem.2009.08.005.
65. Martins, C. M. M. R.; Pinheiro, E. S. C.; Gentilini, M.; Benavides, M. L.; Santos, M. V. Efficacy of a high free iodine barrier teat disinfectant for the prevention of naturally occurring new intramammary infections and clinical mastitis in dairy cows. *J. Dairy Sci.*, 2017, 100 (5), 3930–3939. https://doi.org/10.3168/jds.2016-11193.
66. Reijden, O. L. Van Der; Zimmermann, M. B.; Galetti, V. Iodine in dairy milk: Sources, concentrations and importance to human health. *Best Pract. Res. Clin. Endocrinol. Metab.*, 2017, 31 (4), 385–395. https://doi.org/10.1016/j.beem.2017.10.004.
67. Lee, S. L. Complications of radioactive iodine treatment of thyroid carcinoma. *J. Natl. Compr. Cancer Netw.*, 2010, 8 (11), 1277–1287.
68. Rivkees, S. A.; Mazzaferri, E. L.; Verburg, F. A.; Reiners, C.; Luster, M.; Breuer, C. K.; Dinauer, C. A.; Udelsman, R. The treatment of differentiated thyroid cancer in children: emphasis on surgical approach and. *Endocr. Rev.*, 2011, 32, 798–826. https://doi.org/10.1210/er.2011-0011.
69. Kitahara, C. M.; Gonzalez, A. B. De; Bouville, A.; Brill, A. B.; Doody, M. M.; Melo, D. R.; Simon, S. L.; Sosa, J. A.; Tulchinsky, M.; Villoing, D.; et al. Association of radioactive iodine treatment with cancer mortality in patients with hyperthyroidism. *JAMA Intern. Med.*, 2023, 20892 (8), 1034–1042. https://doi.org/10.1001/jamainternmed.2019.0981.
70. Wei, G.; Ma, J.; Liu, Y.; Xie, L.; Lu, W.; Deng, W.; Ren, Z. Seasonal changes in the radiogenic and stable strontium isotopic composition of xijiang river water: Implications for chemical weathering. *Chem. Geol.*, 2013, 343, 67–75. https://doi.org/10.1016/j.chemgeo.2013.02.004.
71. Burger, A.; Lichtscheidl, I. Strontium in the environment: Review about reactions of plants towards stable and radioactive strontium isotopes. *Sci. Total Environ.*, 2019, 653, 1458–1512. https://doi.org/10.1016/j.scitotenv.2018.10.312.
72. Pearce, C. R.; Parkinson, I. J.; Charlier, B. L. A.; Mokadem, F.; Burton, K. W. Reassessing the stable (D88/86Sr) and radiogenic (87Sr/86Sr) strontium isotopic composition of marine inputs. *Geochim. Cosmochim. Acta*, 2015, 157, 125–146. https://doi.org/10.1016/j.gca.2015.02.029.
73. Lei, Y.; Yang, Y.; Li, G.; Liu, Y.; Xu, J.; Xiong, X.; Luo, S.; Peng, T. Demonstration and aging test of a radiation resistant strontium-90 betavoltaic mechanism. *Appl. Phys. Lett.*, 2020, 116, 153901–153905. https://doi.org/10.1063/1.5140780.
74. Chakravarty, R.; Dash, A.; Pilla, M. R. A. Availability of yttrium-90 from strontium-90: A nuclear medicine perspective. *CANCER Biother. Radiopharm.*, 2012, 27 (10), 621–641. https://doi.org/10.1089/cbr.2012.1285.
75. Santschi, P. H.; Xu, C.; Zhang, S.; Schwehr, K. A.; Grandbois, R.; Kaplan, D. I.; Yeager, C. M. Iodine and plutonium association with natural organic matter: A review of recent advances. *Appl. Geochem.*, 2017, 85, 121–127. https://doi.org/10.1016/j.apgeochem.2016.11.009.
76. Kersting, A. B.; Sciences, L.; Livermore, L.; Box, P. O. Plutonium transport in the environment. *Inorg. Chem.*, 2013, 52, 3533–3546.
77. Smith, R. B.; Romero, F.; Vicente, R. Plutonium-238: The fuel crisis. *Braz. J. Radiat. Sci.*, 2021, 09-01A, 1–11.
78. Varga, Z.; Nicholl, A.; Zsigrai, J.; Wallenius, M.; Mayer, K. Methodology for the preparation and validation of plutonium age dating materials. *Anal. Chem.*, 2018, 90, 4019–4024. https://doi.org/10.1021/acs.analchem.7b05204.
79. Witze, A. Desperately seeking plutonium. *Nature*, 2014, 515, 7–9.
80. Goulet, R. R.; Spry, D. J. Uranium. *Homeost. Toxicol. Non-Essential Met.*, 2012, 31 (11), 391–428. https://doi.org/10.1016/S1546-5098(11)31030-8.
81. Kendall, B.; Brennecka, G. A.; Weyer, S.; Anbar, A. D. Uranium isotope fractionation suggests oxidative uranium mobilization at 2.50 Ga. *Chem. Geol.*, 2013, 362, 105–114. https://doi.org/10.1016/j.chemgeo.2013.08.010.
82. Sahu, P.; Mishra, D. P.; Panigrahi, D. C.; Jha, V.; Patnaik, R. L.; Sethy, N. K. Radon emanation from backfilled mill tailings in underground uranium mine. *J. Environ. Radioact.*, 2014, 130, 15–21. https://doi.org/10.1016/j.jenvrad.2013.12.017.

83. Shaki, F.; Zamani, E.; Arjmand, A.; Pourahmad, J. A Review on toxicodynamics of depleted uranium. *Iran. J. Pharm. Res.*, 2019, 18, 90–100. https://doi.org/10.22037/ijpr.2020.113045.14085.
84. Li, J.; Zhang, Y. Remediation technology for the uranium contaminated environment: A review. *Procedia Environ. Sci.*, 2012, 8 (13), 1609–1615. https://doi.org/10.1016/j.proenv.2012.01.153.
85. Quemet, A.; Maloubier, M.; Ruas, A. Contribution of the faraday cup coupled to 10^12 current amplifier to uranium 235/238 and 234/238 isotope ratio measurements by thermal ionization mass spectrometry. *Int. J. Mass Spectrom.*, 2016, 404, 35–39. https://doi.org/10.1016/j.ijms.2016.04.005.
86. Nelson, A. W.; Eitrheim, E. S.; Knight, A. W.; May, D.; Mehrhoff, M. A.; Shannon, R.; Litman, R.; Burnett, W. C.; Forbes, T. Z.; Schultz, M. K. Understanding the radioactive ingrowth and decay of naturally occurring radioactive materials in the environment: An analysis of produced fluids from the marcellus shale. *Environ. Heal. Perspect.*, 2015, 123 (7), 689–696.
87. Lauer, N. E.; Hower, J. C.; Hsu-Kim, H.; Taggart, R. K.; Vengosh, A. Naturally occurring radioactive materials in coals and coal combustion residuals in the United States. *Environ. Sci. Technol.*, 2015, 49 (18), 11227–11233. https://doi.org/10.1021/acs.est.5b01978.
88. Mettler, F. A. Medical effects and risks of exposure to ionising radiation. *J. Radiol. Prot.*, 2012, 32 (1). https://doi.org/10.1088/0952-4746/32/1/N9.
89. Faanu, A.; Ephraim, J. H.; Darko, E. O. Assessment of public exposure to naturally occurring radioactive materials from mining and mineral processing activities of tarkwa goldmine in Ghana. *Environ. Monit. Assess.*, 2011, 180 (1–4), 15–29. https://doi.org/10.1007/s10661-010-1769-9.
90. Morino, Y.; Ohara, T.; Nishizawa, M. Atmospheric behavior, deposition, and budget of radioactive materials from the fukushima daiichi nuclear power plant in march 2011. *Geophys. Res. Lett.*, 2011, 38 (17), 1–7. https://doi.org/10.1029/2011GL048689.
91. Ministry of the environment Government of Japan; National Institutes for Quantum Science and technology. Radiation Exposure, Nuclear Disaster. In BOOKLET to Provide Basic Information Regarding Health Effects of Radiation (2nd edition); 2022; https://www.env.go.jp/en/chemi/rhm/basic-info/1st/.
92. Gandhi, T. P.; Sampath, P. V.; Maliyekkal, S. M. A critical review of uranium contamination in groundwater: Treatment and sludge disposal. *Sci. Total Environ.*, 2022, 825, 153947. https://doi.org/10.1016/j.scitotenv.2022.153947.
93. Cao, Y.; Li, X. Adsorption of graphene for the removal of inorganic pollutants in water purification: A review. *Adsorption*, 2014, 20 (5–6), 713–727. https://doi.org/10.1007/s10450-014-9615-y.
94. Wang, Z.; Zhao, D.; Wu, C.; Chen, S.; Wang, Y.; Chen, C. Magnetic metal organic frameworks/graphene oxide adsorbent for the removal of U(VI) from aqueous solution. *Appl. Radiat. Isot.*, 2020, 162, 109160. https://doi.org/10.1016/j.apradiso.2020.109160.
95. Işık, B.; Kurtoğlu, A. E.; Gürdağ, G.; Keçeli, G. Radioactive cesium ion removal from wastewater using polymer metal oxide composites. *J. Hazard. Mater.*, 2021, 403. https://doi.org/10.1016/j.jhazmat.2020.123652.
96. Li, J.; Zhang, Y.; Zhou, Y.; Fang, F.; Li, X. Tailored metal-organic frameworks facilitate the simultaneously high-efficient sorption of UO^{22+} and ReO^{4-} in water. *Sci. Total Environ.*, 2021, 799, 149468. https://doi.org/10.1016/j.scitotenv.2021.149468.
97. Jeong, C.; Ansari, M. Z.; Hakeem Anwer, A.; Kim, S. H.; Nasar, A.; Shoeb, M.; Mashkoor, F. A review on metal-organic frameworks for the removal of hazardous environmental contaminants. *Sep. Purif. Technol.*, 2023, 305, 122416. https://doi.org/10.1016/j.seppur.2022.122416.
98. Chen, L.; Yin, X.; Yu, Q.; Siming Lu; Meng, F.; Ning, S.; Wang, X.; Wei, Y. Rapid and selective capture of perrhenate anion from simulated groundwater by a mesoporous silica-supported anion exchanger. Microporous Mesoporous Mater., 2019, 274, 155–162. https://doi.org/10.1016/j.micromeso.2018.07.029.
99. Ma, X. J.; Yang, H.; Wang, B.; Wu, L.; Ma, X.; Li, Y. X.; Deng, H. Mechano-synthesized TiO_2/g-C_3N_4composites for rapid photocatalytic removal of perrhenate. *J. Environ. Chem. Eng.*, 2023, 11 (2), 109423. https://doi.org/10.1016/j.jece.2023.109423.
100. Byrne, C.; Subramanian, G.; Pillai, S. C. Recent advances in photocatalysis for environmental applications. *J. Environ. Chem. Eng.*, 2018, 6 (3), 3531–3555. https://doi.org/10.1016/j.jece.2017.07.080.
101. Nguyen, V. H.; Phan Thi, L. A.; Chandana, P. S.; Do, H. T.; Pham, T. H.; Lee, T.; Nguyen, T. D.; Le Phuoc, C.; Huong, P. T. The degradation of paraben preservatives: Recent progress and sustainable approaches toward photocatalysis. *Chemosphere*, 2021, 276, 130163. https://doi.org/10.1016/j.chemosphere.2021.130163.
102. Jiang, X. H.; Xing, Q. J.; Luo, X. B.; Li, F.; Zou, J. P.; Liu, S. S.; Li, X.; Wang, X. K. Simultaneous photoreduction of Uranium(VI) and photooxidation of Arsenic(III) in aqueous solution over g-C_3N_4/TiO_2 heterostructured catalysts under simulated sunlight irradiation. *Appl. Catal. B Environ.*, 2018, 228, 29–38. https://doi.org/10.1016/j.apcatb.2018.01.062.

103. Liu, J.; Zhang, Z.; Dong, Z.; Zhu, X.; Gao, D.; Cheng, Z.; Cao, X.; Wang, Y.; Liu, Y. Metal-Free CQDs Introduced g-C_3N_4 nanosheets with enhanced photocatalytic reduction performance of Uranium (VI). *J. Radioanal. Nucl. Chem.*, 2022, 331 (5), 2093–2104. https://doi.org/10.1007/s10967-022-08264-7.
104. Gong, X.; Tang, L.; Zou, J.; Guo, Z.; Li, Y.; Lei, J.; Liu, H.; Liu, M.; Zhou, L.; Huang, P.; et al. Introduction of cation vacancies and iron doping into TiO_2 enabling efficient uranium photoreduction. *J. Hazard. Mater.*, 2022, 423 (PA), 126935. https://doi.org/10.1016/j.jhazmat.2021.126935.
105. Deng, H.; Wang, X.; Wang, L.; Li, Z.; Liang, P.; Ou, J.; Liu, K.; Yuan, L.; Jiang, Z.; Zheng, L.; et al. Enhanced photocatalytic reduction of aqueous Re(VII) in ambient air by amorphous TiO_2/g-C_3N_4 photocatalysts: Implications for Tc(VII) elimination. *Chem. Eng. J.*, 2020, 401. https://doi.org/10.1016/j.cej.2020.125977.
106. Meng, Q.; Yang, X.; Wu, L.; Chen, T.; Li, Y.; He, R.; Zhu, W.; Zhu, L.; Duan, T. Metal-free 2D/2D C_3N_5/GO nanosheets with customized energy-level structure for radioactive nuclear wastewater treatment. *J. Hazard. Mater.*, 2022, 422, 1–10. https://doi.org/10.1016/j.jhazmat.2021.126912.
107. Feng, J.; Yang, Z.; He, S.; Niu, X.; Zhang, T.; Ding, A.; Liang, H.; Feng, X. Photocatalytic reduction of Uranium(VI) under visible light with Sn-doped In_2S_3 microspheres. *Chemosphere*, 2018, 212, 114–123. https://doi.org/10.1016/j.chemosphere.2018.08.070.
108. Liu, X.; Bi, R. X.; Peng, Z. H.; Lei, L.; Zhang, C. R.; Luo, Q. X.; Liang, R. P.; Qiu, J. D. Synergistic effect of double schottky potential well and oxygen vacancy for enhanced plasmonic photocatalytic U(VI) reduction. *J. Hazard. Mater.*, 2023, 455, 131581. https://doi.org/10.1016/j.jhazmat.2023.131581.
109. Liu, Y.; Gao, Y.; Chen, L.; Li, L.; Ding, D.; Dai, Z. MnOx-decorated oxygen-doped g-C_3N_4 with enhanced photocatalytic activity for efficient removal of Uranium(VI). *Sep. Purif. Technol.*, 2023, 307. https://doi.org/10.1016/j.seppur.2022.122794.
110. Liang, P.; Yuan, L.; Du, K.; Wang, L.; Li, Z.; Deng, H.; Wang, X.; Luo, S. Z.; Shi, W. Photocatalytic reduction of Uranium(VI) under visible light with 2D/1D Ti_3C_2/CdS. *Chem. Eng. J.*, 2021, 420. https://doi.org/10.1016/j.cej.2021.129831.
111. Chen, B.; Zhang, G.; Chen, L.; Kang, J.; Wang, Y.; Chen, S.; Jin, Y.; Yan, H.; Xia, C. Visible light driven photocatalytic removal of Uranium(VI) in strongly acidic solution. *J. Hazard. Mater.*, 2022, 426, 127851. https://doi.org/10.1016/j.jhazmat.2021.127851.
112. Zhang, X.; Gu, P.; Liu, Y. Decontamination of radioactive wastewater: State of the art and challenges forward. *Chemosphere*, 2019, 215, 543–553. https://doi.org/10.1016/j.chemosphere.2018.10.029.
113. Ambashta, R. D.; Sillanpää, M. E. T. Membrane purification in radioactive waste management: A short review. *J. Environ. Radioact.*, 2012, 105, 76–84. https://doi.org/10.1016/j.jenvrad.2011.12.002.
114. Westerling, K. Radioactive waste: Can membranes reduce the fear factor? Wateronline.Com, 2014, 2–4.
115. Huang, Y.; Feng, X. Polymer-enhanced ultrafiltration: Fundamentals, applications and recent developments. *J. Memb. Sci.*, 2019, 586, 53–83. https://doi.org/10.1016/j.memsci.2019.05.037.
116. Chen, X.; Chen, T.; Li, J.; Qiu, M.; Fu, K.; Cui, Z.; Fan, Y.; Drioli, E. Ceramic nanofiltration and membrane distillation hybrid membrane processes for the purification and recycling of boric acid from simulative radioactive waste water. *J. Memb. Sci.*, 2019, 579, 294–301. https://doi.org/10.1016/j.memsci.2019.02.044.
117. Ding, D.; Zhang, Z.; Chen, R.; Cai, T. Selective removal of cesium by ammonium molybdophosphate – polyacrylonitrile bead and membrane. *J. Hazard. Mater.*, 2017, 324, 753–761. https://doi.org/10.1016/j.jhazmat.2016.11.054.
118. Lee, S.; Kim, Y.; Park, J.; Shon, H. K.; Hong, S. Treatment of medical radioactive liquid waste using forward osmosis (FO) membrane process. *J. Memb. Sci.*, 2018, 556, 238–247. https://doi.org/10.1016/j.memsci.2018.04.008.
119. Li, P.; Chen, P.; Wang, G.; Wang, L.; Wang, X.; Li, Y.; Zhang, W.; Jiang, H.; Chen, H. Uranium elimination and recovery from wastewater with ligand chelation-enhanced electrocoagulation. *Chem. Eng. J.*, 2020, 393, 124819. https://doi.org/10.1016/j.cej.2020.124819.
120. Pujol Pozo, A. A.; Bustos, E.; Monroy-Guzmán, F. Decontamination of radioactive metal surfaces by electrocoagulation. *J. Hazard. Mater.*, 2019, 361, 357–366. https://doi.org/10.1016/j.jhazmat.2018.08.061.
121. Choi, A. E. S.; Futalan, C. C. M.; Yee, J. J. Fuzzy optimization for the removal of uranium from mine water using batch electrocoagulation: A case study. *Nucl. Eng. Technol.*, 2020, 52 (7), 1471–1480. https://doi.org/10.1016/j.net.2019.12.016.

9 Two-Dimensional Nanomaterials for the Removal of Gaseous Air Contaminants

Soham Sarkar, Revanasiddappa, and Suresh Babu Naidu Krishna

9.1 INTRODUCTION

Richard Feynman, a famous physicist, gave a speech in 1959 named "There's Plenty of Room at the Bottom" which introduced the ideas that formed nanotechnology. In the modern, fast developing world, the natural environment is confronted with a myriad of issues as a direct result of the progression of industrialization and technological advancement. Because pollution in the environment has negative repercussions, professionals from a variety of fields are emphasizing how urgent it is to make fundamental reforms to save and maintain our natural resources. To solve these environmental challenges and discover novel solutions, significant research efforts are currently being dedicated toward identifying a solution [1,2].

Because a sizeable amount of the air we breathe contains toxins including carbon monoxide, chlorofluorocarbons, and a variety of inorganic wastes, the problem of air pollution has emerged as one of the most pressing environmental issues of our day. The accumulation of these pollutants over the years has led to a higher concentration, which in turn has prompted academics to direct their attention toward eliminating these dangers to the environment. In addition to harming the air we all breathe, pollution can also have a detrimental impact on other components of nature, such as water and soil [3]. Waste materials from industrial processes and the discharge of wastewater that has not been treated into freshwater bodies both contribute substantially to this issue.

The rapid advancement of nanotechnology has led to the emergence of novel approaches to the monitoring and management of pollution in the natural environment. The unique features and abilities of nanomaterials have demonstrated significant progress in addressing pollution-related issues [4,5]. Nanotechnology has been used to create nanomaterials as well as other items such as nanoparticles, nanotubes, and nanorods. Nanomaterials that are two-dimensional have ultrathin layers with a minimum diameter of a single atom. Bulk materials have a much less surface area relative to volume proportion in comparison with nanomaterials, meaning there are more atoms on the surface. The behavior of 2D nanomaterials is altered due to the augmented number of peripheral atoms, which serve a purpose distinct from that of interior atoms. Graphene is one of the most employed and significant 2D materials due to its unique characteristics, which have enabled it to be utilized in many different industries. It is anticipated that other two-dimensional materials will likely possess the same capability as graphene, which has proven successful in numerous applications and industries. However, utilizing other two-dimensional materials will take more time and effort [6–8].

Investigators want to develop techniques to tackle air pollution that are both effective and efficient, and they plan to do it by utilizing the features of nanomaterials. Because of their large surface area relative to their volume, one-of-a-kind electrical and catalytic capabilities, and high surface reactivity, the utilization of 2D nanomaterials offers a great deal of potential. Because of these characteristics, they can absorb and remove gaseous pollutants from the air in an effective manner, hence reducing the negative impacts of air pollution. Overall, the incorporation of nanotechnology

DOI: 10.1201/9781003436942-9

into environmental research paves the way for brand new lines of inquiry that can be pursued to find solutions to the problems caused by air pollution. The utilization of 2D nanomaterials is a promising strategy that shows promise for the removal of gaseous air pollutants in an effective manner. We can work together across academic disciplines and maintain our commitment to research to make the world a better place for future generations to live in.

The purpose of this chapter is to investigate the current status of research on 2D nanomaterials and their possible uses in the regulation of gaseous air pollutants. Exploring the methods of synthesis, characterization techniques, and performance evaluation of these nanoparticles will result in the development of a full understanding of the capabilities that these nanomaterials possess. In addition, the chapter will talk about the difficulties and opportunities that lie ahead in this area, throwing light on the developments that are necessary to obtain long-term and efficient solutions to the problem of air pollution.

9.2 CLASSIFICATION OF 2D NANOMATERIALS

Two-dimensional nanomaterials can be categorized according to their structures, chemical make-ups, and physical characteristics [9]. In the most prevalent classification scheme, graphene-based and non-graphene-based materials are the two categories used to divide materials. Graphene and other graphene-based nanomaterials are composed of a monolayer of carbon atoms that are organized in a hexagonal structure. Nanomaterials that are not based on graphene include transition metal dichalcogenides (TMDs), as well as black phosphorus and other layered materials. Researchers are able to study the possible uses of 2D nanomaterials in a variety of sectors, including electronics, energy, and healthcare, thanks to the classification of 2D nanomaterials (Figure 9.1), which gives a framework for comprehending the unique features of these materials [10].

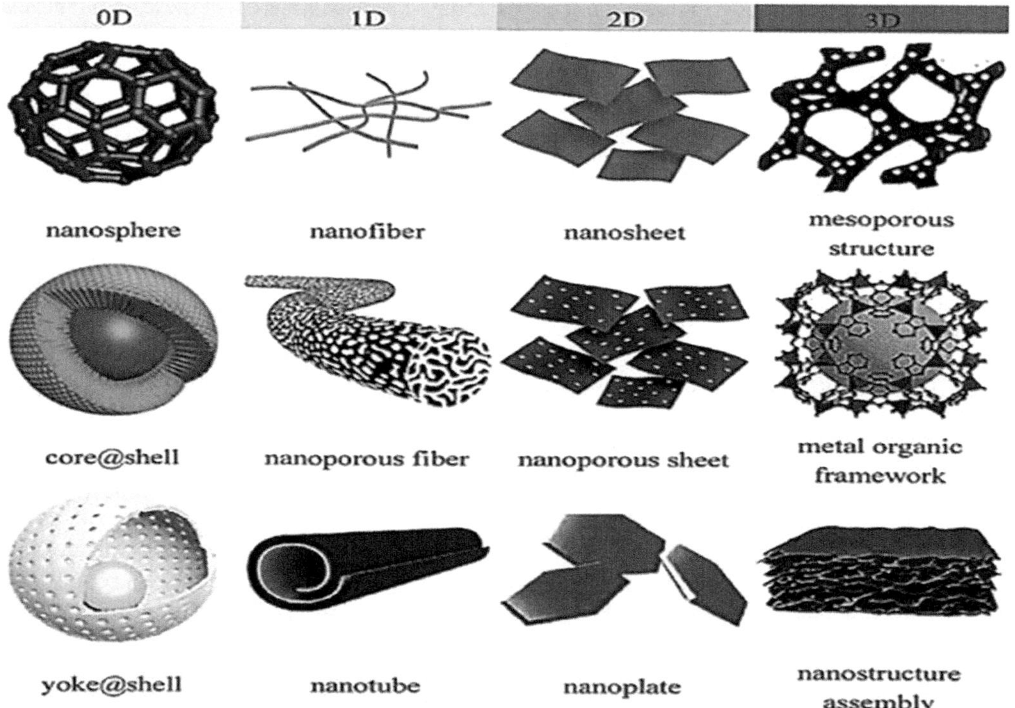

FIGURE 9.1 Categorization of two-dimensional nanomaterials. (Reproduced with permission from [11], Copyright © 2018 Springer.)

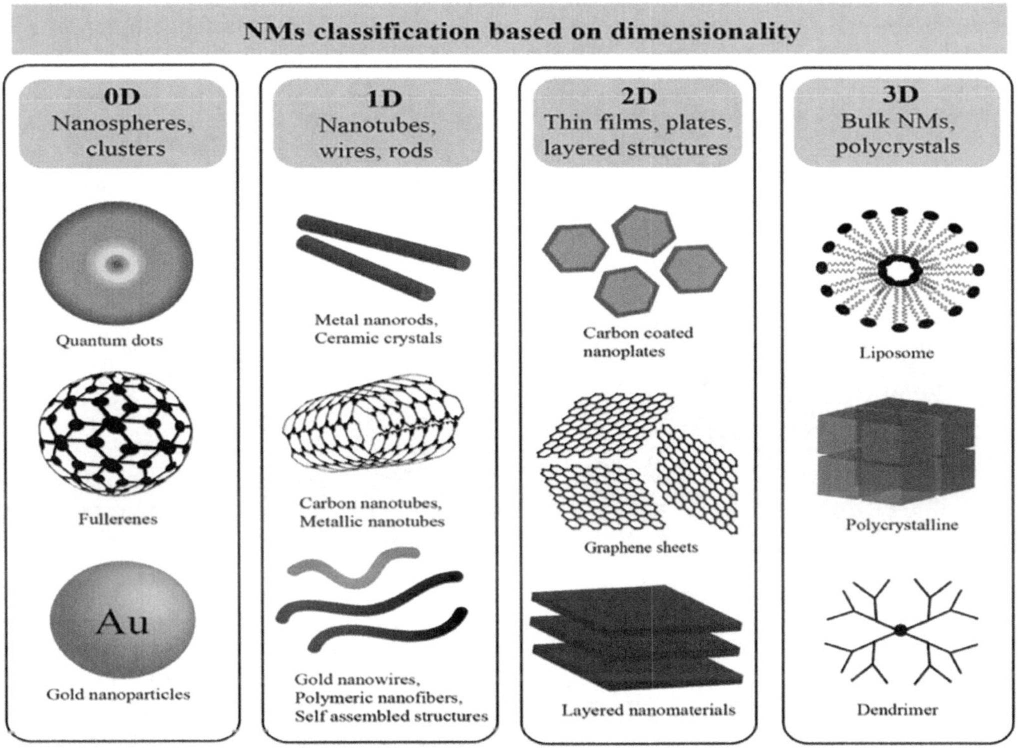

FIGURE 9.2 Nanomaterials categorized based on dimensions. (Reproduced with permission from [12], Copyright © 2018 Springer.)

Figure 9.2 demonstrates how nanomaterials are categorized based on their dimensional number. Nanomaterials are materials with structures and properties that exhibit unique behavior in comparison to their bulk phases due to their high specific surface area. These materials are typically classified into four categories based on their composition.

9.3 SYNTHESIS OF 2D NANOMATERIALS

9.3.1 Synthesis of 2D Non-Layered Nanomaterials via Wet Chemistry

Wet chemistry is used to manufacture non-layered 2D nanomaterials by precisely controlling the precipitation of precursor molecules in a liquid solution or suspension. The nanomaterials have large surface-to-volume ratio, tunable band gaps, and quantum confinement. They are useful in catalysis, sensing, and optoelectronics [13]. Non-layered 2D nanomaterials are used in energy storage, catalysis, sensing, and healthcare. They have a high surface area, unique electronic properties, and are biocompatible. Research on their use is ongoing [14].

9.3.2 2D Templated Synthesis

For the creation of anisotropic nanocrystals (like nanowires), where the growth of the crystal could be constrained within a certain dimension, templated synthesis techniques (Figures 9.3 and 9.4) have been extensively used [15,16]. A new method to create 2D Au nanosheets with stable hcp phase was developed. The nanosheets can be used as seeds to grow other gold nanostructures with different crystal structures. Coating other metals on the nanosheets can also result in further phase transitions and the development of different types of core–shell nanoplates [17–19].

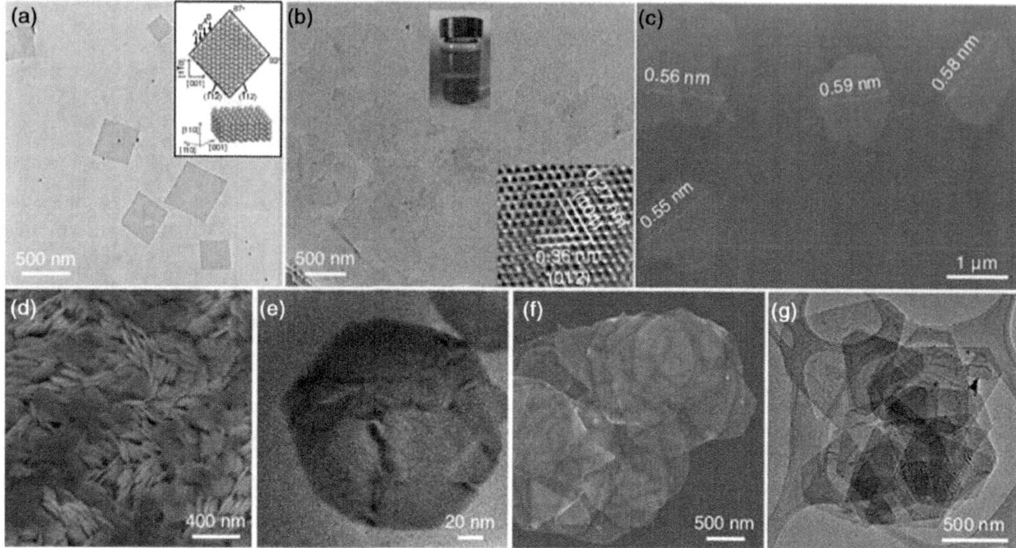

FIGURE 9.3 2D nanosheets prepared by using the 2D templated synthesis techniques. (a) High clotting point (hcp) AuSSs as seen in transmission electron microscopy (TEM). Presented in the inset are crystallographic representations of standard AuSSs, illustrating ABAB stacking in the [001]h direction, with the basal plane aligned with the axis of the [110]h area. Image of α-Fe_2O_3 nanosheets captured by TEM. Inset: α-Fe_2O_3 nanosheets with a Tyndall effect and a HRTEM image. (c) Magnetic resonance imaging (MRI) of α-Fe_2O_3 nanosheets. (d) Scanning electron microscopy and (e) transmission electron microscopy of $CuInS_2$ nanotubes. In (f), the images of CuSe nanosheets are shown using TEM, and in (g), the images of $Cu_{2-x}Se$ nanosheets are shown using SEM. (Reprinted with permission from Ref. [20], Copyright © 2015 Springer.)

Different templates can be used to create unique 2D nanostructures. A new method was developed to create square Au nanosheets with hexagonal-close packed phase. The nanosheets can be used as seeds to grow thick Au sheets or to create core–shell nanoplates with different orientations and structures. CuO nanoplates can be used as a template to create free-standing α-Fe_2O_3 nanosheets [21]. CuO nanoplates can be used to create free-standing α-Fe_2O_3 nanosheets by annealing. Reactive templates, such as CuSe and CuS, can be used to create ternary and quaternary copper-based chalcogenide nanosheets. Hexagonal CuSe nanosheets can be converted into cubic $Cu_{2-x}Se$ nanosheets by heating with CuI cations. Ultrathin hexagonal CuS nanosheets can be transformed into cubic $Cu_{1.9}7S$ nanosheets using the same approach [22].

9.3.3 Wet Chemical Synthesis under High Pressure and Temperature Conditions

Two routinely used methods for producing 2D nanosheets are solvothermal and hydrothermal synthesis by the use of high-temperature aqueous/solvent [23,24]. Solvothermal and hydrothermal methods are used to synthesize 2D nanosheets with desired morphologies and properties. These 2D nanosheets have a wide range of applications, including catalysis, energy storage, and electronic devices. For example, cobalt (II) hydroxide nanosheets can be used as catalysts for the oxygen evolution reaction, while molybdenum disulfide nanosheets can be used as anodes in lithium-ion batteries. Black phosphorus nanosheets can also be used to make field-effect transistors with high performance.

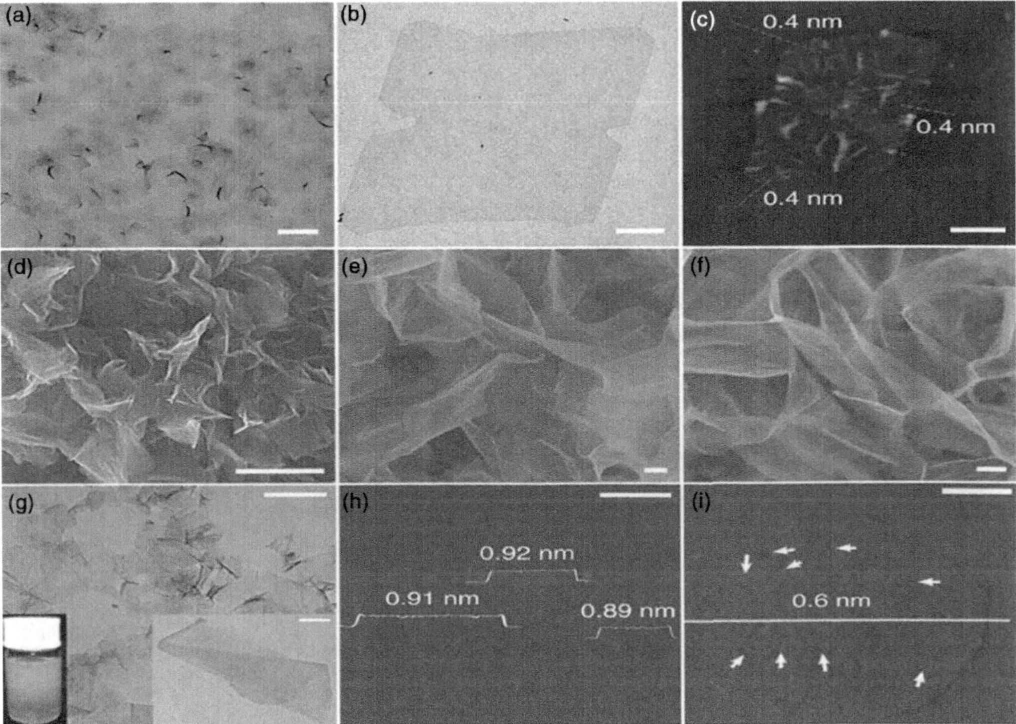

FIGURE 9.4 2D nanosheets prepared by using the aqueous-based chemical synthesis techniques. (a, b) TEM images of Rh nanosheets with PVP caps. (c) AFM image of naked Rh nanosheet. SEM images of two-dimensional nanosheets of TiO_2 (d), ZnO (e), and Co_3O_4 (f; scale bars, 200 nm). (g) TEM image of single layers of ZnSe (scale bar, 500 nm). The inset shows the Tyndall effect on ZnSe single layers and an expanded TEM (scale bar, 100 nm). (h) AFM picture of ZnSe single layers (scale bar, 500 nm). (i) AFM picture of extremely thin CeO_2 sheets with surface pits (scale bars, 100 nm). (Reproduced with permission from Ref. [20], Copyright © 2015 Springer.)

9.4 FABRICATION OF LOW-DIMENSIONAL NANOCRYSTAL ASSEMBLIES VIA SELF-ASSEMBLY TECHNIQUES

Self-assembly of nanocrystals is the spontaneous organization of individual nanocrystals into larger structures in a controlled environment. This can lead to the formation of low-dimensional nanocrystal assemblies with uniform size and shape [25,26].

9.4.1 Methods for Templated Synthesis of Soft Colloids and Other Materials

Soft colloids are synthesized using soft templates, such as micelles or vesicles. Hard templates, such as porous membranes or nanowires, can also be used to synthesize soft colloids, but soft templates are preferred. Soft templates are preferred because they are more flexible and can be easily tailored to the desired structure, as shown in Figure 9.5 [27]. Emulsion templating is a method for synthesizing soft colloids by polymerizing a monomer in an oil-in-water emulsion. The surfactant in the emulsion droplets forms a template for the colloidal particles, which are then released by removing the surfactant. This method can be used to synthesize various types of soft colloids, including polymer particles, vesicles, and capsules [28]. Another method for templated synthesis of soft colloids is

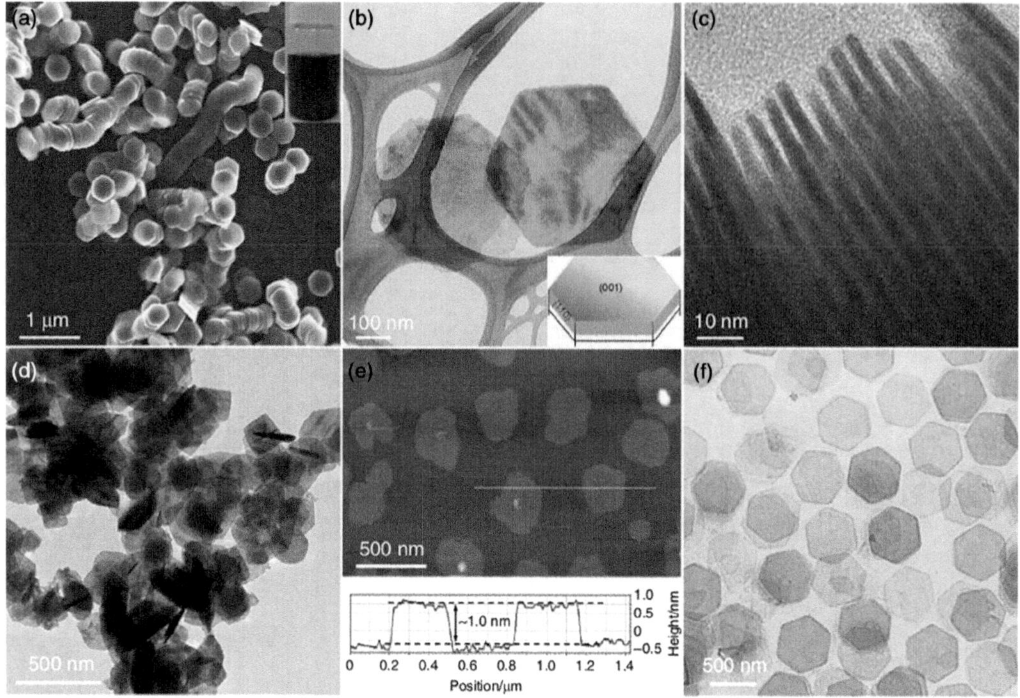

FIGURE 9.5 2D nanosheets prepared by using the soft colloidal templated synthesis and other techniques. (a) Scanning electron micrograph of very thin CuS nanosheets. CuS nanosheets in a colloid solution are shown in the inset. Transmission electron microscopy (TEM) images of (b) flat CuS nanosheets and (c) standing nanosheets on TEM grids. A schematic of a very thin CuS nanosheet is shown in the inset of (b). TEM and AFM pictures of SnSe nanosheets are shown in (d) and (e), respectively. TEM picture of Pd nanosheets (f). (Reproduced with permission from Ref. [20], Copyright © 2015 Springer.)

the self-assembly method [29]. Soft colloids can be synthesized by self-assembling small building blocks under controlled conditions. Electrospinning, layer-by-layer assembly, and lithography are other methods for templated synthesis of soft colloids and other materials.

9.5 ENVIRONMENTAL APPLICATIONS OF 2D NANOMATERIALS WITH NON-LAYERED STRUCTURES

Ultrathin 2D nanomaterials are desirable for surface-active applications due to their large size and thinness, which yields a very high specific surface area. This makes them ideal candidates for efficient catalysts in various catalytic applications (Figure 9.6), and some ultrathin 2D nanosheets have displayed remarkable performance in catalysis applications [30]. Nanotechnology has proven beneficial to the environment in all the three phases which include water treatment, soil remediation, and pollutant gas sensors.

This section will focus on investigating the wide-ranging usage of produced "nanosheets" or "monolayers" (2D nanomaterials), particularly their impressive performance in catalysis and the elimination of gaseous air pollutants.

9.5.1 2D Nanomaterials for the Removal of Gaseous Air Contaminants

2D nanomaterials have been increasingly studied for their potential in air purification applications, specifically for the removal of gaseous air contaminants. The diverse applications are as shown in

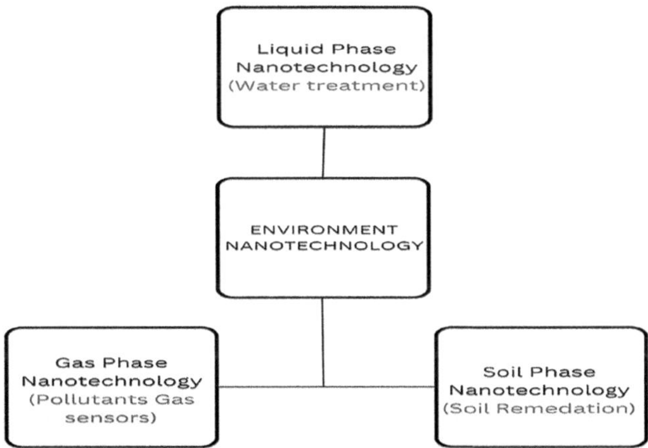

FIGURE 9.6 The use of nanotechnology in various phases of the environment.

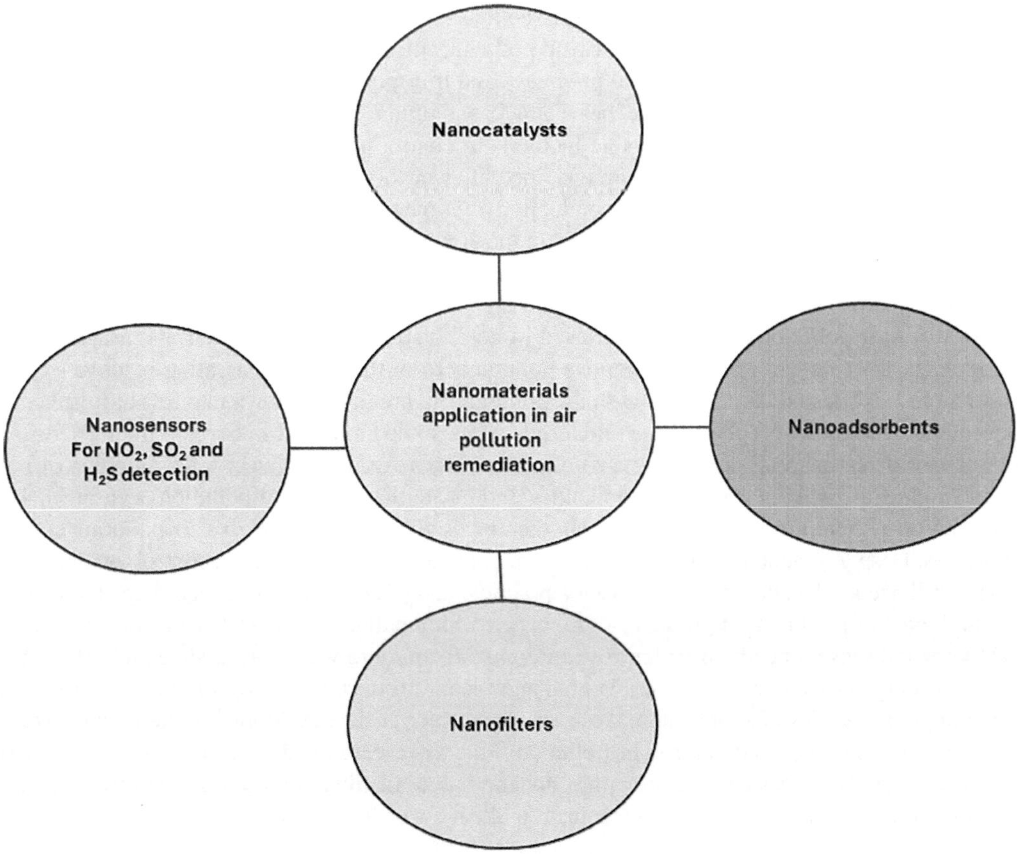

FIGURE 9.7 Application of nanomaterials for air pollution treatment.

Figure 9.7. These materials offer several advantages, including high surface area, tailorable surface chemistry, and exceptional adsorption and catalytic characteristics. Thus, they have exhibited immense capability in eliminating noxious gases, including NO_x, SO_x, CO, and volatile organic compounds (VOCs) [31].

Researchers in nanotechnology firmly believe that the current environmental crisis could be dissolved to a certain extent with the help of nanoscale adsorbents, also known as nano-adsorbents. Nano-adsorbents are named so because of their power to adsorb toxic air contaminants present in air. Various nano-adsorbents are known to specifically detect and adsorb certain air pollutants such as microbes, non-methane organic compounds (NMOCs), H_2S (hydrogen sulfide), SO_2 (sulfur dioxide), and NO_2 (Nitrogen dioxide). Table 9.1. presents different monitoring and treatment techniques used by several researchers in the field of nanotechnology to reduce the threat to environmental elements.

9.5.2 Nano-adsorbents Used for the Process of Air Purification

9.5.2.1 Graphene

Graphene is a layered 2D material with remarkable characteristics that possess exceptional physical and chemical properties [32]. Consisting of a single sheet of carbon atoms that are bonded together in a hexagonal pattern, graphene's unique properties have attracted a great deal of attention from researchers and scientists across a wide range of fields. One of the most promising applications of graphene is in the field of environmental remediation, specifically in controlling air pollution. Air pollution poses a significant environmental issue that adversely impacts the health and well-being of millions of individuals worldwide. Harmful pollutants discharged into the atmosphere from various human activities, such as industrial processes and transportation, can result in life-threatening health conditions, such as lung disease, heart attack, and tumor growth. As a result, there is a critical need to develop efficient and sustainable methods to control and reduce air pollution. Graphene-based materials have demonstrated immense potential in this regard owing to their high surface area, chemical stability, and electrical conductivity. Graphene is known for its exceptional ability to absorb and filter out harmful pollutants from the atmosphere [33]. By designing graphene-based air filters, it is possible to efficiently remove pollutants such as VOCs, nitrogen oxides (NOx), and particulate matter (PM) from the air. Graphene's high electrical conductivity also makes it a suitable material for air pollution control. When used in combination with other materials, graphene can form electrodes that can effectively remove pollutants from the air by generating reactive oxygen species (ROS). These ROS can react with the pollutants to break them down into less harmful components. Furthermore, graphene-based materials can also be employed to develop highly sensitive sensors for detecting and monitoring air pollutants. The sensors can detect trace amounts of pollutants in the air, making it possible to identify and locate the sources of pollution accurately. The information gathered from these sensors can then be used to develop effective air pollution control strategies. Despite graphene's numerous potential applications in air pollution control, several challenges still need to be addressed to make it a practical and effective solution. One such challenge is the high cost of producing high-quality graphene. Additionally, the scalability of graphene-based air filters and sensors needs to be improved to make them more widely accessible and affordable. In conclusion, graphene-based materials hold significant promise in controlling air pollution [34]. The unique properties of graphene make it an ideal material for developing efficient and sustainable methods to reduce and control air pollution. Further research and development are necessary to address the challenges associated with graphene-based air filters and sensors, but the potential benefits for human health and the environment make it a worthy pursuit.

9.5.2.2 Fullerene

Layered fullerenes are a family of carbon-based molecules that have been extensively researched for their unusual properties and potential applications in various fields, including environmental remediation. One of the most promising applications of fullerenes is in the field of air quality management [35]. Air quality management is a significant environmental issue that affects millions of people around the world. The harmful pollutants released into the atmosphere from human activities such as industrial processes and transportation can have serious health problems, such

TABLE 9.1
Diverse Treatment Methods for the Removal of Air Pollutants

Nanoscale Adsorbents	Categories of Nanoparticles	Pollutant Gases of Concern	Removal Strategy
Carbon nanotubes (CNTs)	SWNTs and MWNTs	Oxides of nitrogen (NO and NO_2)	Nitrogen monoxide (NO) and oxygen (O_2) molecules pass through carbon nanotubes (CNTs). NO is oxidized to nitrogen dioxide (NO_2) and then adsorbed on the surface of nitrate species
	CNTs modified with 3-aminopropyltriethoxysilane (APTS), abbreviated as CNTs-APTS	CO_2	The surface of carbon nanotubes (CNTs) has abundant amine groups, which provide numerous chemical sites for carbon dioxide (CO_2) adsorption. This allows CNTs to adsorb more CO_2 gas at low temperatures (20°C–100°C)
	SWNTs/NaClO	Propan-2-ol vapor	Adsorption of molecules onto a surface by van der Waals forces and chemical bonding
	CNTs coated on quartz filters	Organic solvents	The process was mediated by π–π interactions
	Silicon-doped and boron-doped single-walled carbon nanotubes (SWCNTs)	Methyl alcohol and carbon monoxide gases	The electronic properties of SWCNTs significantly improve gas adsorption, regardless of whether it is physisorption or chemisorption
Fullerene	Fullerene B40	CO_2	It has high affinity for CO_2 and can adsorb it through strong chemical bonding
	Boron nitride cage with a fullerene structure	Nitrous oxide	The adsorption and breakdown of nitrous oxide
Graphene	Graphene oxide-reinforced nanocomposites	Carbon dioxide, Ammonia, Sulfur dioxide, Hydrogen sulfide, Nitrogen	The functional groups of graphene oxide (GO) are responsible for the adsorption of gases. The synergistic effect between the metal centers and the GO surface further promotes gas adsorption

as respiratory illnesses, heart problems, and cancer. So, it is essential to develop effective and sustainable methods to control and reduce air pollution. Fullerenes are spherical or ellipsoidal carbon molecules composed of hexagonal and pentagonal carbon rings that have unique structural and electronic properties. These properties make fullerenes highly attractive for various applications, including air pollution control. Fullerene's unique chemical and physical properties have shown great potential in air purification systems. For example, fullerene molecules can absorb gases and pollutants such as nitrogen oxides (NO_x), sulfur dioxide (SO_2), and VOCs due to their high surface area. Fullerene-based air filters can be used to remove air pollutants effectively. When air passes through the fullerene filter, the pollutants are trapped inside the fullerene molecules. The fullerene filter's effectiveness in removing pollutants depends on the fullerene's size, shape, and surface properties. Studies have shown that fullerene cages with diameters between 7 and 15 angstroms are ideal for trapping NO_x, SO_2, and VOCs. Moreover, fullerenes can be modified by adding functional groups to their surface, which can improve their absorption capacity and remove pollutants from air. For example, fullerenes functionalized with hydroxyl (-OH) and carboxylic acid (-COOH) groups have shown excellent results in removing NO_x and VOCs from the air. In addition to air filters, fullerene-based photocatalysts have also shown promising results in air pollution control. Fullerene photocatalysts can remove pollutants from the air by breaking them down into less harmful components when exposed to sunlight. These photocatalysts have been found to be highly effective in removing contaminants like organic solvents, nitrogen oxides, and PM from the air. Fullerenes also have the potential to be used as sensors for detecting and monitoring air pollutants. Fullerene sensors can detect trace amounts of pollutants in the air and provide accurate measurements of their concentrations. The information gathered from these sensors can be used to develop effective air pollution control strategies. Despite the promising potential of fullerenes in air pollution control, several challenges still need to be addressed [36]. One of the most significant challenges is the cost of producing high-quality fullerenes. The production process for fullerenes is complex and expensive, which limits their widespread use. Additionally, the environmental impact of fullerenes needs to be thoroughly investigated to ensure that they do not contribute to other environmental issues. In conclusion, fullerenes hold significant promise in controlling air pollution. The unique properties of fullerenes make them an ideal material for developing efficient and sustainable methods to reduce and control air pollution. Further research and development are necessary to address the challenges associated with fullerene-based air filters, photocatalysts, and sensors, but the potential benefits for human health and the environment make it a worthy pursuit.

9.5.2.3 Metal–Organic Frameworks (MOFs)

Air pollution is a major environmental problem that affects human health and the environment. Various pollutants such as nitrogen oxides (NO_x), sulfur dioxide (SO_2), VOCs, and PM are released into the atmosphere due to human activities such as industrial processes, transportation, and energy production. These pollutants can cause respiratory diseases, heart disease, and cancer, as well as contribute to climate change. Therefore, it is crucial to develop effective and sustainable methods to control and reduce air pollution. Metal–organic frameworks (MOFs) are a promising class of 2D materials that can be used for air pollution control. MOFs are a type of crystalline material composed of metal ions or clusters connected by organic ligands [37]. They have high porosity and surface area, making them excellent candidates for various applications, including air quality management (Figure 9.8). The distinctive properties of MOFs, such as tunable pore size, high selectivity, and reversible adsorption, make them an ideal material for air purification systems. One of the most significant advantages of MOFs is their high surface area. The surface area of MOFs can vary widely from hundreds to thousands of square meters per gram, which allows them to adsorb large amounts of gases and pollutants. MOFs can also be synthesized with tunable pore sizes, allowing for the selective removal of specific pollutants [38]. MOFs have been shown to be effective in removing a variety of air pollutants, including sulfur dioxide, nitrogen oxides, and organic solvents.

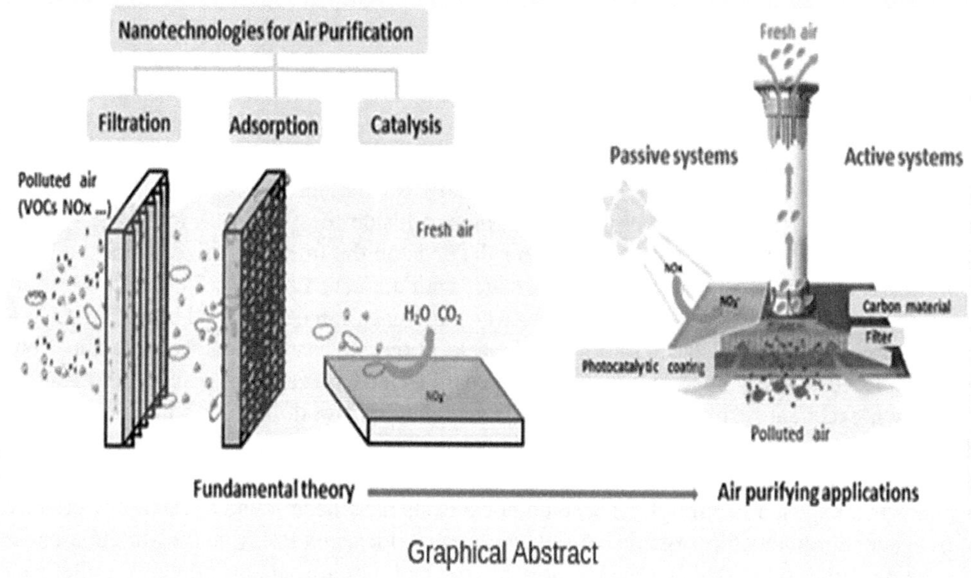

FIGURE 9.8 Air purification by 2D nanomaterials.

For example, researchers have developed MOFs that can selectively remove NO_x from the air. These MOFs have been shown to have high selectivity for NO_x over other gases and can remove up to 99% of NO_x from the air. Similarly, MOFs have been developed that can selectively remove SO_2 from the air. These MOFs have been shown to have high selectivity for SO_2 over other gases and can remove up to 97% of SO_2 from the air. MOFs can also be used for the capture and removal of VOCs. VOCs are a significant contributor to air pollution and can cause respiratory diseases, eye irritation, and other health problems. MOFs have been developed that can selectively capture and remove VOCs from the air. For example, researchers have developed MOFs that can capture and remove benzene, toluene, and xylene from the air. MOFs can be used as adsorbents in air purification systems, such as filters and scrubbers. In these systems, MOFs are used to capture and remove pollutants from the air as they pass through the system. MOFs can also be used as photocatalysts, which are used to break down pollutants under the influence of light. In these systems, MOFs are coated onto a substrate, and when exposed to light, they catalyze the breakdown of pollutants into less harmful components. MOFs also have the potential to be used for the storage and delivery of gases. For example, MOFs can be used for the storage and delivery of hydrogen, which is a clean and renewable fuel. MOFs can also be used for the storage and delivery of carbon dioxide, which can be captured from industrial processes and used for carbon capture and storage (CCS) or converted into useful products. Despite the promising potential of MOFs in air pollution control, several challenges still need to be addressed. One of the most significant challenges is the cost of producing high-quality MOFs. The production process for MOFs is complex and expensive, which limits their widespread use. Additionally, the environmental impact of MOFs needs to be thoroughly investigated to ensure that they do not contribute to other environmental issues.

In conclusion, MOFs are a promising class of materials for controlling air pollution. The unique properties of MOFs, including high porosity, tunable pore size, high selectivity, and reversible adsorption, make them ideal for capturing and removing pollutants from the air. MOFs have been shown to be effective for removal of a wide range of air pollutants, including sulfur dioxide, nitrogen oxides, and organic solvents. They can be used as adsorbents in air purification systems, as photocatalysts for breaking down pollutants, and for the storage and delivery of gases. However, further research is needed to address the challenges of producing high-quality MOFs at a reasonable cost and to ensure that their environmental impact is minimal. Overall, MOFs have the potential to be a significant tool in the fight against air pollution, contributing to improved human health and environmental sustainability [39].

In conclusion, 2D nanomaterials offer great potential for air purification applications. Their high surface area, tunable surface chemistry, and excellent adsorption and catalytic properties make them highly effective in removing harmful gases from the air [40,41]. Future research in this area will likely focus on optimizing the performance of these materials under realistic operating conditions, developing cost-effective and scalable synthesis methods, and exploring their use in integrated air purification systems.

9.6 REVOLUTIONIZING AIR POLLUTION MITIGATION VIA NANOTECHNOLOGY'S IMPACT

The revolution of nanotechnology in bringing a change in the pattern of environmental crisis could be subdivided into three categories, namely:

a. Alteration and purification (refinement): This could be done by making use of nano-adsorbents as discussed in Section 5 or using nanocatalysts in the process of degradation.
b. Waste minimization and Pollution control (sustainable production): This could be done by reducing the number of contaminants released into the air. Nanotechnology has been very proficient in manufacturing environment-friendly products including biodegradable plastics, safe nanocomposites, and many new nanomaterials which have better performance and less toxicity as compared to conventional materials.

c. Analyzing and monitoring the air contaminants can prove to be an important step for minimizing air pollution. Traditional methods used for this purpose are good in terms of their accuracy and precision specifically, but these methods lead to an essential amount of delay in the process and add up to the recurring cost. Also, they are not as specific as nanotechnology today.
d. Furthermore, nanotechnology can be applied to cater to the fuel needs of society by making use of air pollutants which could be developed into fuel. This could in turn leverage people into denying the use of more harmful fossil fuels and reduce their dependency on fossil fuels in turn saving non-renewable resources for our future generations to come while also saving the environment in a single goal. Carbon dioxide (CO_2) could be converted into fuels and chemicals using solar energy, with the potential to produce low-carbon or carbon-neutral product being beneficial. The advancement in the field of nanotechnology has really brought about a change in solar-to-fuel transformation, which is beneficial to the environment and society.

9.7 CONCLUSION AND FUTURE OUTLOOK

This chapter discussed the categorization of nanoparticles as well as their synthesis, with a particular emphasis on two-dimensional nanomaterials and the applications of these nanomaterials in the removal of gaseous air pollutants. We discoursed about the usage of materials with two-dimensional atomic structure, including graphene and transition metal dichalcogenides, focusing on their distinctive qualities and the possible benefits they could offer in the process of pollutant capture. The research that was looked at provides evidence that 2D nanomaterials could adsorb, catalytically convert, and chemically react with gaseous contaminants, which leads to a decrease in the amount of these pollutants in the air. However, there are still a few obstacles that must be dealt with before 2D nanomaterials can be practically implemented for the reduction of air pollution. The synthesis of excellent 2D nanomaterials on a massive scale that could be regulated in terms of their shape, size, and surface properties presents a considerable challenge. To achieve production on a wide scale while maintaining the material's performance and stability, the synthesis methods need to be improved. Another difficulty is in the efficient incorporation of 2D nanomaterials into devices or systems that may be used in everyday life to combat air pollution. It is essential to investigate suitable support materials and device structures in addition to the development of scalable and cost-effective production methods to make the most of 2D nanomaterials in real-world applications. In addition to this, concerns need to be addressed about the recyclability and long-term stability of 2D nanomaterials. To ensure the continuous and effective performance of these materials in pollution collection, it is vital to have a solid understanding of the causes of degradation, the ways in which surfaces might get fouled, and the techniques for regeneration. In the future, developments of two-dimensional nanomaterials for the reduction of air pollution should concentrate on finding solutions to these problems. It is recommended that efforts be made to improve the synthesis procedures and come up with ways of fabrication that are scalable to make it possible to produce high-quality 2D nanomaterials on an industrial scale. In addition, the incorporation of 2D nanomaterials into real-world devices and systems should be investigated, with consideration given to aspects such as longevity, operability, and cost-effectiveness. Furthermore, to build customized materials with greater effectiveness and selectivity, it is vital to have a comprehensive understanding of the fundamental principles driving the interactions between 2D nanomaterials and gaseous contaminants. Novel nanomaterials with improved pollutant capture capabilities can be developed with the help of advanced characterization techniques and computer modeling, both of which can provide valuable insights into the mechanisms involved. In addition, a comprehensive analysis of the consequences for the environment as well as the potential dangers that are linked with the widespread usage of 2D nanomaterials should be carried out. Research should be oriented toward understanding the destiny and behavior of these materials in the environment, as well as their potential consequences on human health and ecosystems. This understanding should be the focus of the research that is conducted.

REFERENCES

1. Sharma A., Makgwane P.R., Lichtfouse E., Kumar N., Bandegharaei A.H., Tahir M. Recent advances in synthesis, structural properties, and regulation of nickel sulfide-based heterostructures for environmental water remediation: An insight review. *Environmental Science and Pollution Research* 30 (2023) 64932–64948. https://doi.org/10.1007/s11356-023-27093-z.
2. Amrane A., Mohan D., Nguyen T.A., Assadi A.A. and Yasin G. eds., 2020. *Nanomaterials for Soil Remediation*. Elsevier, Amsterdam, the Netherlands.
3. Saleem H., Zaidi S.J., Ismail A.F., Goh P.S. Advances of nanomaterials for air pollution remediation and their impacts on the environment. *Chemosphere* 287 (2022) 132083. https://doi.org/10.1016/j.chemosphere.2021.132083.
4. Patel B., Revanasiddappa M., Yallappa S., Rangaswamy D. Iron nanoparticles embedded in polypyrrole and tellurium oxide ternary nanocomposites for electrical conductivity and electromagnetic interference shielding. *Journal of Materials Science: Materials in Electronics* 34 (2023) 1–16. https://doi.org/https://doi.org/10.1007/s10854-023-10298-w.
5. Krishna S.B.N. Emergent roles of garlic-based nanoparticles for bio-medical applications - A review. *Current Trends in Biotechnology and Pharmacy* 15 (2021) 349–360. https://abap.co.in/index.php/home/article/view/104
6. Nguyen-Tri P., Nguyen T.A., Nguyen T.V, Nanomaterial for air remediation: An introduction, in *Nanomaterials for Air Remediation*, Elsevier, 2020, pp. 3–8. https://doi.org/10.1016/B978-0-12-818821-7.00001-4.
7. Occelli F., Lanier C., Cuny D., Deram A., Dumont J., Amouyel P., Montaye M., Dauchet L., Dallongeville J., Genin M. Exposure to multiple air pollutants and the incidence of coronary heart disease: A fine-scale geographic analysis. *Science of the Total Environment* 714 (2020) 136608. https://doi.org/10.1016/j.scitotenv.2020.136608.
8. Ong C.B., Ng L.Y., Mohammad A.W. A review of ZnO nanoparticles as solar photocatalysts: Synthesis, mechanisms and applications. *Renewable and Sustainable Energy Reviews* 81 (2018) 536–551. https://doi.org/10.1016/j.rser.2017.08.020.
9. Murali A., Lokhande G., Deo K.A., Brokesh A., Gaharwar A.K. Emerging 2D nanomaterials for biomedical applications. *Materials Today* 50 (2021) 276–302. https://doi.org/10.1016/j.mattod.2021.04.020.
10. Mu P., Zhou G., Chen C.-L. 2D nanomaterials assembled from sequence-defined molecules. *Nanostructures & Nano-Objects* 15 (2018) 153–166. https://doi.org/10.1016/j.nanoso.2017.09.010.
11. Poh T.Y., Ali N., Mac Aogáin M., Kathawala M.H., Setyawati M.I., Ng K.W., Chotirmall S.H. Inhaled nanomaterials and the respiratory microbiome: Clinical, immunological, and toxicological perspectives. *Part Fibre Toxicol* 15 (2018) 46. https://doi.org/10.1186/s12989-018-0282-0.
12. Poh T.Y., Ali N.A.t.B.M., Mac Aogáin M., Kathawala M.H., Setyawati M.I., Ng K.W., Chotirmall S.H. Inhaled nanomaterials and the respiratory microbiome: Clinical, immunological and toxicological perspectives. *Particle and Fiber Toxicology* 15 (2018) 46. https://doi.org/10.1186/s12989-018-0282-0.
13. Zhang Y., Mei J., Yan C., Liao T., Bell J., Sun Z. Bioinspired 2D nanomaterials for sustainable applications. *Advanced Materials* 32 (2020) 1902806. https://doi.org/10.1002/adma.201902806.
14. Wang S., Liu K., Yao X., Jiang L. Bioinspired surfaces with superwettability: New insight on theory, design, and applications. *Chemical Reviews* 115 (2015) 8230–8293. https://doi.org/10.1021/cr400083y.
15. Fullam S., Cottell D., Rensmo H., Fitzmaurice D. Carbon nanotube templated self-assembly and thermal processing of gold nanowires. *Advanced Materials* 12 (2000) 1430–1432. https://doi.org/10.1002/1521-4095(200010)12:19%3C1430::AID-ADMA1430%3E3.0.CO;2-8.
16. Cao G., Liu D. Template-based synthesis of nanorod, nanowire, and nanotube arrays. *Advances in Colloid and Interface Science* 136 (2008) 45–64. https://doi.org/10.1016/j.cis.2007.07.003.
17. Huang X., Li H., Li S., Wu S., Boey F., Ma J., Zhang H. Synthesis of gold square-like plates from ultrathin gold square sheets: The evolution of structure phase and shape. *Angewandte Chemie International Edition in English* 50 (2011) 12245–12248. https://doi.org/10.1002/anie.201105850.
18. Fan Z., Huang X., Han Y., Bosman M., Wang Q., Zhu Y., Liu Q., Li B., Zeng Z., Wu J., Shi W., Li S., Gan C.L., Zhang H. Surface modification-induced phase transformation of hexagonal close-packed gold square sheets. *Nature Communications* 6 (2015) 6571. https://doi.org/10.1038/ncomms7571.
19. Fan Z., Zhu Y., Huang X., Han Y., Wang Q., Liu Q., Huang Y., Gan C.L., Zhang H. Synthesis of ultrathin face-centered-cubic au@pt and au@pd core-shell nanoplates from hexagonal-close-packed au square sheets. *Angewandte Chemie International Edition in English* 54 (2015) 5672–5676. https://doi.org/10.1002/anie.201500993.
20. Tan C., Zhang H. Wet-chemical synthesis and applications of non-layer structured two-dimensional nanomaterials. *Nature Communications* 6 (2015) 7873. https://doi.org/10.1038/ncomms8873.

21. Cheng W., He J., Yao T., Sun Z., Jiang Y., Liu Q., Jiang S., Hu F., Xie Z., He B., Yan W., Wei S. Half-unit-cell α-Fe_2O_3 semiconductor nanosheets with intrinsic and robust ferromagnetism. *Journal of the American Chemical Society* 136 (2014) 10393–10398. https://doi.org/10.1021/ja504088n.
22. Wu X.J., Huang X., Qi X., Li H., Li B., Zhang H. Copper-based ternary and quaternary semiconductor nanoplates: Templated synthesis, characterization, and photoelectrochemical properties. *Angewandte Chemie International Edition in English* 53 (2014) 8929–8933. https://doi.org/10.1002/anie.201403655.
23. Duan H., Yan N., Yu R., Chang C.R., Zhou G., Hu H.S., Rong H., Niu Z., Mao J., Asakura H., Tanaka T., Dyson P.J., Li J., Li Y. Ultrathin rhodium nanosheets. *Nature Communications* 5 (2014) 3093. https://doi.org/10.1038/ncomms4093.
24. Zhang X., Liu Q., Meng L., Wang H., Bi W., Peng Y., Yao T., Wei S., Xie Y. In-plane coassembly route to atomically thick inorganic-organic hybrid nanosheets. *ACS Nano* 7 (2013) 1682–1688. https://doi.org/10.1021/nn3056719.
25. Schliehe C., Juarez B.H., Pelletier M., Jander S., Greshnykh D., Nagel M., Meyer A., Foerster S., Kornowski A., Klinke C., Weller H. Ultrathin PbS sheets by two-dimensional oriented attachment. *Science* 329 (2010) 550–553. https://doi.org/10.1126/science.1188035.
26. Acharya S., Das B., Thupakula U., Ariga K., Sarma D.D., Israelachvili J., Golan Y. A bottom-up approach toward fabrication of ultrathin PbS sheets. *Nano Letters* 13 (2013) 409–415. https://doi.org/10.1021/nl303568d.
27. Ithurria S., Tessier M.D., Mahler B., Lobo R.P., Dubertret B., Efros A.L. Colloidal nanoplatelets with two-dimensional electronic structure. *Nature Materials* 10 (2011) 936–941. https://doi.org/10.1038/nmat3145.
28. Du Y., Yin Z., Zhu J., Huang X., Wu X.J., Zeng Z., Yan Q., Zhang H. A general method for the large-scale synthesis of uniform ultrathin metal sulphide nanocrystals. *Nature Communications* 3 (2012) 1177. https://doi.org/10.1038/ncomms2181.
29. Ithurria S., Dubertret B. Quasi 2D colloidal CdSe platelets with thicknesses controlled at the atomic level. *Journal of the American Chemical Society* 130 (2008) 16504–16505. https://doi.org/10.1021/ja807724e.
30. Tewari B.B. Engineered nanoparticles in environmental mitigation and green growth, in *Recent Advances in Science and Technology Research*, S.N.U. Yongchun Zhu, (Eds.), Book Publisher International, Guyana, 2020. https://doi.org/10.9734/bpi/rastr/v4.
31. Saleem H., Zaidi S.J., Ismail A.F., Goh P.S. Advances of nanomaterials for air pollution remediation and their impacts on the environment. *Chemosphere* 287 (2022) 132083. https://doi.org/10.1016/j.chemosphere.2021.132083.
32. Huang X., Qi X., Boey F., Zhang H. Graphene-based composites. *Chemical Society Reviews* 41 (2012) 666–686. https://doi.org/10.1039/c1cs15078b.
33. Shende P., Pathan N. Graphene nanoribbons: A state-of-the-art in health care. *International Journal of Pharmaceutics* 595 (2021) 120269. https://doi.org/10.1016/j.ijpharm.2021.120269.
34. Szczęśniak B., Choma J., Jaroniec M. Gas adsorption properties of graphene-based materials. *Advances in Colloid and Interface Science* 243 (2017) 46–59. https://doi.org/10.1016/j.cis.2017.03.007.
35. Yu J.G., Yu L.Y., Yang H., Liu Q., Chen X.H., Jiang X.Y., Chen X.Q., Jiao F.P. Graphene nanosheets as novel adsorbents in adsorption, preconcentration and removal of gases, organic compounds and metal ions. *Science of the Total Environment* 502 (2015) 70–79. https://doi.org/10.1016/j.scitotenv.2014.08.077.
36. Khurram A.R., Rafiq S., Tariq A., Jamil A., Iqbal T., Mahmood H., Mehdi M.S., Abdulrahman A., Ali A., Akhtar M.S., Asif S. Environmental remediation through various composite membranes moieties: Performances and thermomechanical properties. *Chemosphere* 309 (2022) 136613. https://doi.org/10.1016/j.chemosphere.2022.136613.
37. Wang S., McGuirk C.M., d'Aquino A., Mason J.A., Mirkin C.A. Metal-organic framework nanoparticles. *Advanced Materials* 30 (2018) e1800202. https://doi.org/10.1002/adma.201800202.
38. Bathla A., Vikrant K., Kukkar D., Kim K.H. Photocatalytic degradation of gaseous benzene using metal oxide nanocomposites. *Advances in Colloid and Interface Science* 305 (2022) 102696. https://doi.org/10.1016/j.cis.2022.102696.
39. Zhang X., Teng S.Y., Loy A.C.M., How B.S., Leong W.D., Tao X. Transition metal dichalcogenides for the application of pollution reduction: A review. *Nanomaterials (Basel)* 10 (2020). https://doi.org/10.3390/nano10061012.
40. Gulati S., Lingam B.H., Kumar S., Goyal K., Arora A., Varma R.S. Improving the air quality with functionalized carbon nanotubes: Sensing and remediation applications in the real world. *Chemosphere* 299 (2022) 134468. https://doi.org/10.1016/j.chemosphere.2022.134468.
41. Valdés-Madrigal M.A., Montejo-Alvaro F., Cernas-Ruiz A.S., Rojas-Chávez H., Román-Doval R., Cruz-Martinez H., Medina D.I. Role of defect engineering and surface functionalization in the design of carbon nanotube-based nitrogen oxide sensors. *International Journal of Molecular Sciences* 22 (2021). https://doi.org/10.3390/ijms222312968.

10 2D Metal Chalcogenides for Photodegradation and Disinfection of Wastewater Pollutants

Mohit Saini, Ankita Saini, Bhavna Saroha, and Suresh Kumar
Department of Applied Sciences, Ganga Institute of Technology and Management, Kablana, Jhajjar, Haryana- 124104, India

10.1 INTRODUCTION

Because of the industrial revolution, expanding population, protracted droughts, and climate change, water pollution has become a serious problem. Therefore, there is an urgent demand for new solutions for wastewater cleanup. Microorganisms and inorganic/organic pollutants are the two broad categories used to describe water contamination. Carcinogenic substances including arsenic, chromium, lead, and mercury are among the harmful inorganic contaminants. Organic dyes, insecticides, personal care items, detergents, and industrial organic wastes are the key and common organic pollutants. Due to their special physical, chemical, and optical characteristics, nanostructured materials could be suitable for photocatalytic reduction and photodegradation of organic contaminants in wastewater. Due to the increased discharge of dangerous chemical compounds into the environment, the industrial revolution has had a severe influence on the environment [1]. Pesticides, medications, personal care items, petrochemicals, and organic dyes are examples of organic hazardous materials, whereas heavy metals and numerous industrial additives are examples of inorganic hazardous materials. Pollution of the air, land, and water is brought on by hazardous pollutants. A major environmental issue that concerns billions of people globally, especially those in developing nations, is water pollution [2]. There are various covert threats to human health and the ecology posed by hazardous chemicals. These substances are harmful and can harm the organs like lungs, liver, and brain, as well as may cause cancer and neurological disorders. These materials have high operating costs, complex processes, and large sludge generation; in addition, they are difficult to decompose and accumulate as a result [3]. Among the most widely used chemical methods for reducing or oxidizing inorganic pollutants to make them into harmless substances and, consequently, secondary pollutants, are advanced oxidation processes (AOPs), which break down target pollutants into smaller, non-toxic, and biodegradable compounds like carbon dioxide (CO_2) and water (H_2O) [4]. In order to facilitate the aforementioned breakdown processes, active free radicals such as hydroxyl (OH), superoxide anion ($O_2^{-\bullet}$), and sulfate ($SO_4^{-\bullet}$) radicals are generated. Photocatalysis, a nonselective chemical approach, is a promising solution because it may be used for a wide range of hazardous chemicals, such as pathogens and both organic and inorganic pollutants [5].

Photocatalysts are the substances that, when exposed to light, change the pace of a chemical reaction. A semiconductor is used as a catalyst in the process known as photocatalysis [6] and its capability to absorb light energy thereby producing negative electron (e^-) and positive hole (h^+) pairs is known as its photocatalytic capability [7]. When light shines on a semiconductor material, the valence band

DOI: 10.1201/9781003436942-10

(VB) electrons go into the conduction band (CB), where they produce positive holes. The energy band gap of the semiconductor must be equal to or greater than the wavelength of incident light [8]. Redox ability may be found in the photogenerated pairs. The most challenging task in this stage is the recombination of electrons and holes. Before triggering the redox processes, the photogenerated electron–hole pairs might recombine fast [9]. Numerous environmental problems can be solved by the photodegradation of environmental contaminants including organic dyes, medicines, and pesticides via photocatalysts, and also, they are extensively known for their use in the creation of hydrogen and water splitting in energy generation process. In addition, photocatalysts may be utilized to photo reduce heavy metals and destroy a variety of harmful microorganisms, such as fungi, viruses, and bacteria [10]. Photocatalytic environmental remediation is an efficient and affordable technique that uses harmless semiconductors to transform environmental pollutants into inactive form by the use of light of proper wavelengths. This photocatalytic environmental remediation process has four phases. First, contaminants are brought to the surface from the water. Then, they adhered to the semiconductor's surface in the second phase followed by the engagement of pollutants and semiconductors in a photocatalytic process. Lastly, contaminants are broken down and released into the environment [11].

10.2 GENERAL MECHANISM OF PHOTOCATALYTIC REACTION

The photocatalytic rate is primarily determined by the crystal structure of the photocatalyst, which is used in photocatalysis reactions, along with the energy of incident photons. The materials that are employed as catalysts function as a sensitizer for the irradiation of light-stimulated redox processes, based upon their electronic structure. The electrons in the VB will absorb the photon and go to the CB, only if the catalyst's band gap is equal to or less than the energy of the incoming light. The valance band has holes left in it which are crucial and these holes oxidize donor molecules when H_2O reacts with them, producing a powerful hydroxyl radical oxidant. Superoxide ion, a reducing agent, is created when an oxygen molecule absorbs an electron from the CB, as shown in Figure 10.1. Therefore, we may conclude that redox processes are

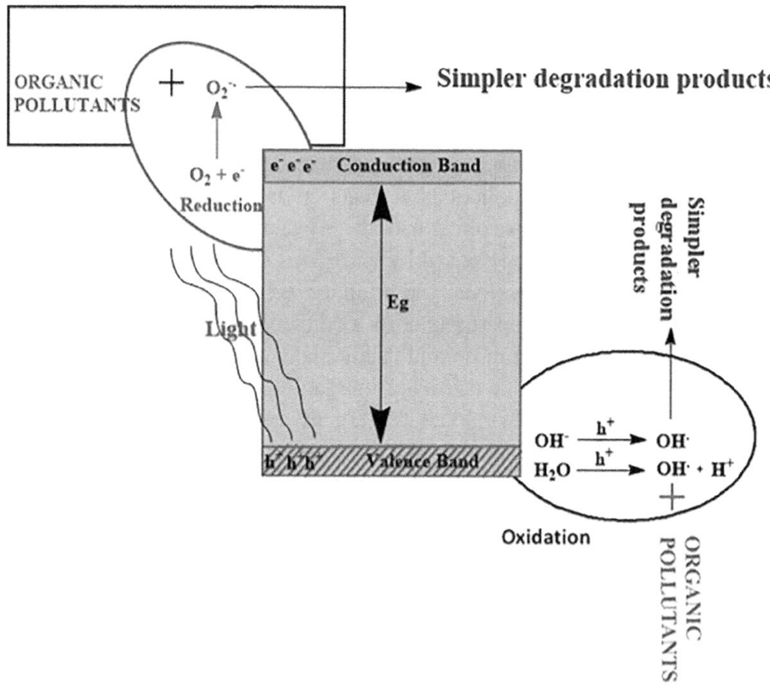

FIGURE 10.1 General mechanism of photocatalytic degradation of organic pollutants.

being triggered by this free electron. Therefore, any substance that comes into touch with the catalyst may undergo an oxidation–reduction process by these pairs of free electrons and holes, changing it into the desired non-toxic compounds [12]. There are two possible mechanisms for this conversion: oxidation and reduction. An electron–hole pair is formed on the catalyst in response to light, and this facilitates the valence electron's promotion to the CB in the oxidation pathway. The catalyst creates a hole that catches the water molecule and oxidizes it to produce the potent hydroxyl radical. Any organic contaminants present will react with these hydroxyls, breaking down into more straightforward, safe compounds. Intermediate radicals from organic molecules and oxygen molecules are involved in a series of chain reactions that continue if oxygen is present during the whole process. When an organic contaminant is present, CO_2 and H_2O are the end products. A pairing reaction lowers the amount of oxygen in the atmosphere during the reduction mechanism. Superoxide ions are created when oxygen interacts with electrons in the CB. The anion can also participate in an oxidation process that produces peroxides before reverting to water by joining with an intermediate. Organic compounds could be degraded more easily than water. As a result, a high organic matter content may enhance photocatalytic activity; however, the existence of holes may eventually cause the rate of carrier recombination to decrease.

10.3 MATERIAL SELECTION FOR NANOMATERIALS AS A PHOTOCATALYST

We choose materials or catalysts that can support both reduction and oxidation processes for photocatalytic reactions. How to select oxidation and reduction reactions taking place at once in photocatalytic process? The answer lies in the selection of materials available for catalysts. Materials are often classified into three fundamental categories namely conductors, insulators, and semiconductors depending upon their electronic characteristics. Both conductors and insulators are not suitable for catalytic activity in reactions as both oxidation and reduction must occur concurrently for a photocatalytic process. For conductors, the VB and CB overlap; as a result, only oxidation reactions are possible in conductors, while in insulators, a high energy band gap is present, which enables the possibility of only reduction reactions. In contrast, semiconductors, which may carry out or promote simultaneous reduction and oxidation events, have a moderate band gap. When light is incident, free electron–hole pairs are produced. To act as a photocatalyst, a semiconductor must have a low recombination rate. Furthermore, semiconductors that exhibit visible light absorption between 350 and 700 nm or band gaps between 1.5 and 3.5 eV are appropriate for photocatalytic action. Because they catalyze processes involving visible light. Semiconductors typically possess a broad spectrum of band gaps; however, for a photocatalyst in the UV–visible range, we only require semiconductors with band gaps between 1.5 and 3.5 eV.

Semiconductors have conductivities that lie in between those of non-conductors/insulators (like ceramics) and conductors (like metals). The types of the semiconductor materials that have received a lot of interest are metal oxides and chalcogenides. Since they are non-toxic, affordable, stable, and photocorrosion resistive, metal oxides such as TiO_2 and ZnO are primarily employed as photocatalysts [13]. The main disadvantage of the use of metal oxides as photocatalysts is about their broad band gap energy, notwithstanding their stability. Only ultraviolet light (about 4% of the solar energy) may be absorbed by them because of the broad band gap energy, which restricts their range of practical uses [14]. There have been several strategies used to create materials having a limited band gap energy [15]. The tiny band gap energy of chalcogenides is widely recognized. At least one electropositive element and one chalcogen anion (S^{2-}, Se^{2-}, or Te^{2-}) combine to form chalcogenides [16]. Chalcogenides are highly sought-after because of their exceptional and desirable properties, including their tiny band gap energy, non-toxicity, biocompatibility, low cost, and simplicity of synthesis [17].

10.4 PROPERTIES AND CLASSIFICATION OF METAL CHALCOGENIDES

Metal-based binary chalcogenides contain one electropositive atom and chalcogen. The works of literature study confirm the effective use of binary chalcogenides for photocatalysis. The photogenerated e^-/h^+ pairs will immediately manifest when photons with sufficient energy $(h\upsilon \geq E_g)$ strike a photocatalyst. As the photogenerated e^- is transferred to the CB, the h^+ will stay in the VB. Superoxide ($O_2^{-}\bullet$) will be produced as a result of the interaction between the adsorbed O_2 molecules on the photocatalyst's surface and the e^- in the CB, as exhibited in Figure 10.1. In addition, the VB's interaction with h^+ will simultaneously result in the production of reactive hydroxyl radicals ($OH^\bullet$) [18]. In binary metal-based chalcogenides, the VB is primarily influenced by the p-orbitals of the chalcogens (S = sulfur, Se = selenides, and Te = tellurides), while the CB is primarily driven by the orbitals of the metal (for example, if the metal selected is a transition metal, then the CB is entirely caused by the d-orbitals of the transition metal). However, research on binary chalcogenides has been less common in recent years, mostly because it is challenging to simultaneously achieve advantageous energy band alignments and charge carrier recombination [19]. Manipulation of surface area, shape, crystal structure modification, inherent flaws, and dopant addition to regulate optical and electrical properties are common methods for improving the performance of binary photocatalysts under visible light irradiation.

The chalcogen group's elements, which include tellurium, selenium, and sulfur, are found in nature. In the earth's crust, sulfur is the most prevalent element with an average of 470 parts per million. Tellurium and selenium are next in order of abundance, with 0.05 and 0.001 parts per million, respectively [20]. These elements are extracted by mining, after which they are purified and/or mixed with other elements to form metal or metal chalcogenides. These chalcogen elements are widely used in numerous sectors, including the tire and chemical industries, which use sulfur extensively [21,22]. While tellurium and selenium are not as abundant as sulfur, their presence is nevertheless significant, particularly when combined with other elements or put through further processing to produce nanomaterials [23,24]. These sulfide-chalcogen-based materials are crucial for a number of products based on optical and electrical functional materials, including batteries, photovoltaics, and photocatalysis [25–28].

Furthermore, unlike metal oxides, metal sulfide semiconductor catalysts have d and sp orbitals in the CB and S 3p orbitals in the VB. These catalysts include a large number of metal cations with electronic configurations of d0, d5, and d10. Because the 3p orbital of S is more negative in energy than the 2p orbital of O, the metal sulfide has a higher VB position than the metal oxide. Additionally, because of the negative CBs, metal sulfides have narrower band gaps than metal oxides [29].

It is appropriate to enhance photogenerated e^-/h^+ pair separation in order to prevent their recombination. Multinary chalcogenides perform better as photocatalysts than binary chalcogenides because they may postpone the mixing together of photogenerated e^-/h^+ couples. Ternary and quaternary chalcogenides are subgroups of multinary metal-based chalcogenides [19]. One electropositive atom along with two chalcogens, either S^{2-} or Se^{2-} or Te^{2-}, or two electropositive atoms and one chalcogen, either S^{2-} or Se^{2-} or Te^{2-}, can be combined to form ternary chalcogenides. But quaternary chalcogenides are capable of combining one electropositive atom with three chalcogens S^{2-}, Se^{2-}, or Te^{2-}; two electropositive atoms with two chalcogens S^{2-}, Se^{2-}, or Te^{2-}; and three electropositive atoms with one chalcogen S^{2-}, Se^{2-}, or Te^{2-}. In addition, compared to binary metal-based chalcogenides, multinary chalcogenides have distinct orbitals that are in charge of the CB and VB [30]. Few examples of different chalcogenides belonging to different categories are presented in Table 10.1. Here, most of the chalcogenides taken into account have the sulfur chalcogen because of the less abundance of Se^{-2} and Te^{-2} in comparison to S^{-2}, thereby making metal sulfide as preferred candidates for the photocatalytic applications.

Several techniques have been attempted to boost chalcogenides' photocatalytic activity. The materials' ability to absorb light is closely linked to their electrical band structure, which may be

TABLE 10.1
Few Examples of Binary and Multinary Chalcogenides

Classification	Sulfides	Selenides	Tellurides
Binary (AX, AX_2, and A_2X_3, where A is the metal and X is the chalcogen ion)	GaS, ZnS, CdS, PbS, MoS_2, SnS_2, Bi_2S_3, and In_2S_3	GaSe, ZnSe, CdSe, $TiSe_2$, $MoSe_2$, and WSe_2	CdTe
Multinary (ABX, ABX_2 and AB_2X_4, where A is the metal, B is the metal, and X is the chalcogen)	$ZnIn_2S_4$, $AgInS_2$, $CuInS_2$, $CuSnS_3$, $CuGaS_2$, $CdIn_2S_4$, and $MgIn_2S_4$	ZnCdSe, $CuInSe_2$, and $CuSnSe_3$	HgCdTe

TABLE 10.2
The Band Gap of Some Metal Sulfide Chalcogenides

Compound	Band Gap (eV)	Ref.
Bi_2S_3	1.52	[38]
$CuInS_2$	1.57	[39]
MoS_2	1.85	[40]
$CdInS_4$	2.26	[41]
$ZnIn_2S_4$	2.29	[42]
SnS_2	2.43	[43]
CdS	2.49	[44]
In_2S_3	2.5	[45]
ZnS	3.6	[46]

changed by increasing the rate at which photogenerated charge carriers are produced and reducing the band gap to the visible area [31]. When the photocatalyst material absorbs light with energy $E \geq E_g$ for a viable photocatalytic reaction, its band gap becomes critical to its catalytic activity. In simpler terms, materials with a broader band gap are limited to absorbing solar energy from the UV or near-UV region. However, materials with a small band gap are also capable of reacting to visible light [32]. Consequently, materials that react to visible light are preferable for efficient solar energy use. With the use of band gap engineering, materials' optical and electrical characteristics may be tailored to efficiently use the solar energy spectrum and achieve effective photocatalytic performance [33].

Table 10.2 represents the band gap of some metal sulfide chalcogenides. For improved visible light absorption, a photocatalyst material's band gap, for instance, should be less than 3 eV. Practical photocatalytic H_2 generation requires a minimal band gap energy of 1.23 eV for water splitting, hence 2 eV is the optimal band-gap energy [34]. In addition to the band gap, the potential of the conduction band minimum (CBM) of any visible-light-driven photocatalyst should have a greater negative potential, the valence band maximum (VBM) shall have a higher positive potential compared to the water oxidation potential (O_2/H_2O), and the hydrogen reduction potential (H^+/H_2) should be less [35]. Through band gap engineering, the edge locations of these two bands may also be changed to increase the catalytic efficiency for the creation of H_2. Consequently, the band gap of the photocatalyst particle directly influences photocatalytic activity. In another investigation, ZnS has a broad band gap of 3.7 eV and is capable of photocatalysis exclusively in the ultraviolet spectrum. Therefore, in order to maximize solar energy harvesting, scientists have tuned Zinc Sulfur (ZnS)'s band gap to make it sensitive to visible light. In order to get over this restriction, Lee et al. doped copper (0%–5%) to compress the band gap of ZnS from 3.65 to 2.89 eV, making it visible-light active for the photosplitting of water [36]. Within the ZnS band gap, there are copper deep trap energy levels that are responsible for this increase in H_2 generation efficiency. At these trap levels, the excited electrons in the ZnS CB slow down because of the slow recombination rate of photogenerated holes, increasing the rate of charge separation. Likewise, by loading metal cations (Au^{3+}, Pd^{2+}) into the framework by ion-exchange treatment, the band gap of the framework is narrowed, increasing the photocatalytic efficacy of GeZnS. This is explained by the fact that the recombination rate of charge carriers is reduced, leading to an improvement in performance of charge separation. Further, Pd@GeZnS and Au@GeZnS structures with excellent electrical conductivity and current density are generated by the addition of metal, which enhances photocatalytic activity for CH_4 production [37]. Similarly, according to the application of interest, the suitable sulfide-based chalcogenide can be chosen, and accordingly, the band gap can be engineered to enhance the photocatalytic efficiency of the chalcogenide photocatalyst.

10.5 STRATEGIES TO ENHANCE PHOTOCATALYTIC ACTIVITIES OF 2D METAL CHALCOGENIDES

One method for addressing the solution of instability under light irradiation and the possibility of photocorrosion or photo-dissolution during photocatalytic processes is doping. Chalcogenides can be modified with both metals and non-metals as dopants to fix their weaknesses. The performance of the photocatalytic system can be enhanced by deficiencies brought about by metal ion dopants that promote charge separation and transfer. Additionally, by preventing spontaneous recombination of the charge carriers, extra trap states may extend the lifespan of electrons and holes. The improvement might be the consequence of the integration of dopants, which resulted in the creation of novel trapping defect sites and a decrease in the rate of e^-/h^+ pair recombination. High doping concentrations, however, can lead to recombination centers, which might lead to a reduction in the quantity of active charge carriers. It may be because the separation between a particle's entrapment sites is getting smaller, which causes the recombination rate to increase [47].

TMCs with two-dimensional (2D) planar architectures have shown potential for use in photocatalytic processes. Because of their greater surface area and strong charge migration across both surfaces, the interfacial charge transfer resistance is reduced, which enhances the kinetics of photocatalytic reactions [48]. Additionally, a 2D planar design permits dominant exposure of a single aspect with a distinctive atomic arrangement [49], which is better suited for the use of photons and the separation of the photogenerated charge pairs. This is because, in the presence of the Fermi-level pinning effect, at the catalyst–electrolyte interface, changes may be seen in the flat band possibility and in the degree of band bending. It is also widely known that some 2D TMCs' thickness and lateral dimensions have a significant impact on their band locations and band gap energies [50]. The VB edge will frequently change toward O_2 oxidation potential, while the CB edge will frequently shift toward H_2 reduction potential. Further, the distinctive 2D layered structure may serve as an effective matrix to produce some particular optical phenomena, such as the plasmonic effect, which may increase the solar spectrum's absorption range [51]. Two categories may be applied to 2D TMCs based on their crystal structure: Initially, there are layered materials with weak van der Waals contact between planes and high lateral chemical bonding inside planes. Second, non-layered materials have 2D sheet-like structures that create atomic bonding in three dimensions and whose development stops after a finite number of layers [52].

10.5.1 LAYERED 2D METAL CHALCOGENIDES

Mo- and W-based chalcogenides (MX_2) have a steady hexagonal crystal structure known as the 2H phase along with semiconducting characteristics. They are typical layered 2D TMCs. A monolayer of MX_2 is typically formed when two planes of hexagonally organized X atoms are covalently connected to M atoms having a hexagonal plane. Around every M atom, six X atoms coordinate prismatically. [53]. It's interesting to note that chemically and electrochemically intercalating alkaline ions can change the stable 2H semiconducting phase into a metastable metallic phase known as the 1T phase [54], where the coordination number of the M atom is changed from 6 to 8, thereby creating enhanced catalytic sites in the substance. One more intriguing feature is that as the thickness of MX_2 is lowered to monolayer [55], the electronic band structure of MX_2 progressively switching from indirect to direct and its band gap energy increases simultaneously, making the hydrogen evolution reaction (HER) more thermodynamically advantageous. Their band gap energies also fall in the perfect range with the purpose of absorbing visible light, between 1.6 and 2.4 eV. Furthermore, in monolayer and few-layer Mo- and W-based TMC, the nonradiative relaxation methods predominate faster exciton recombination due to the indirect–direct band gap transition, which excludes their individual application as photocatalysts [56].

Current theoretical computations and experimental investigations have demonstrated that defect engineering is a useful technique for tuning surface properties, such as the relationship among reactants, intermediates, and vacancy sites, controlling the atom coordination number and electronic structure, increasing carrier mobility and conductivity, and ultimately increasing photocatalyst activity [57–61]. Defect engineering changes the catalyst's geometric and electrical structure to improve light harvesting, speed up carrier separation and migration, and encourage surface reaction [62]. Most semiconductor materials are conductive because of the nonstoichiometric mix of point defects. Conversely, planar and line defects rarely or never change the stoichiometric proportions [63]. Anion vacancies (usually oxygen and halogen vacancies), cation vacancies (metal vacancies), or combined anion and cation vacancies can occur in structurally disordered photocatalysts. Many theoretical calculations and experimental studies have shown that because of the low formation energy, anion vacancies like oxygen, sulfur, and halogen are often present in metal-based catalysts [57,64,65]. To improve photocatalytic performance, anion vacancies engineering plays a critical role by reducing atom coordination numbers, producing more active centers, and efficiently managing the energy band structure and electronic structure of photocatalysts.

Despite the fact that bulk oxygen vacancies lower the band gap energy to support visible light harvesting and partly restrict carrier recombination, surface oxygen vacancies greatly diminish the VB edge to help in the separation of photogenerated carriers. Similar to this, reports of halogen and sulfur anion vacancies in photocatalysts have also been made. By varying the quantity of reducing agent used to produce ZnS with regulated sulfur vacancy concentration, Wang et al. discovered that photocatalytic hydrogen generation activity rises until an appropriate sulfur vacancy content is reached [66]. The narrow band gap and improved photocatalytic performance caused by surface sulfur vacancies in Zn-Cd-S solid solution were discovered by Zhang et al. and were further supported by density functional theory (DFT) calculations [67].

Surprisingly, anion vacancies can modify the energy band structure of a photocatalyst, enhancing its photoresponsiveness. Further, anion vacancies have the ability to form narrow band gaps and mid gap states in semiconductors, which facilitate light harvesting and cause a noticeable color shift in photocatalysts.

Numerous experimental investigations and theoretical computational studies have also indicated that cation vacancies serving as shallow acceptors can speed up hole migration, encourage p-type conductivity, and have a big impact on the photocatalytic process. Catalytic defects give photocatalysts many new features like regulating band structure by shifting the VB maximum upward and the CB minimum downward without creating new intermediate states and promoting the quick migration and separation of photogenerated carriers. Nonetheless, the high formation energy and structural instability brought on by cation defects continue to pose significant difficulties for catalyst production [61].

10.5.2 Non-Layered 2D Metal Chalcogenides

In non-layered 2D TMCs, in contrast to the zinc blende structure, 2D Zn- and Cd-based chalcogenides exhibit a dominant crystal structure of wurtzite and have been extensively used as photocatalysts [68]. Because of the increased surface area and the shifting of the CB edge over the H_2/H_2O potential, they exhibit strong photocatalytic behavior. As an illustration, it is shown that under UV irradiation, 2D ZnS nanosheets with a lateral dimension of 500 nm display a significant rate of deterioration of 0.35 mmol $(gh)^{-1}$ toward methyl orange (MO). Due to its substantial band gap energy, it is inert when exposed to visible light [69]. ZnSe, in contrast, has a significantly narrower band-gap energy of *2.7 eV and is thus more frequently used as a visible light-driven photocatalyst [70]. The same observation can also be observed for CdS chalcogenides. Under visible light irradiation, 2D CdS nanosheets with lateral dimensions of 100–300 nm and a thickness of a few nm exhibit an outstanding H_2 generation rate of *41 mmol $(gh)^{-1}$ [71]. A 3D flower-shaped variant of it that has lateral dimensions between 300 and 800 nm exhibits improved H_2 generation at a rate of *9 mmol $(gh)^{-1}$, which is about three times greater than that of the nanoparticle powder system under similar testing circumstances [72]. This remarkable enhancement can be attributed to both its increased surface area and the CB edge's shift over the H_2/H_2O potential, which may be caused by the quantum confinement effect [30].

10.6 APPLICATIONS OF 2D METAL CHALCOGENIDES PHOTOCATALYSTS IN WASTEWATER TREATMENT

The photocatalytic process involves the use of photocatalyst that are initiated when the right amount of ultraviolet, visible, and infrared light energy is used to decompose the pollutants in water, and it is also a green method of waste treatment. This energy produces an electron–hole pair (e^-/h^+) pair, which in turn produces very reactive reducing and/or oxidizing radicals in the VB and the CB, respectively. By subsequent reactions, these radicals in polluted water break down organic and inorganic contaminants. Due to their appropriate band gap and presence in the visible spectrum, semiconductor nanoparticles are good catalysts for wastewater treatment. Photocatalytic degradation mechanism involving four basic steps is shown in Figure 10.2: in the first step, catalyst reactant diffusion occurs on the surface of catalyst; in the second step, adsorption of reactants occurs on the catalyst's surface; in the third step, basic reaction on surface occurs, in the fourth step, product desorption and regeneration of catalyst occurs.

10.6.1 Factors Affecting Photodegradation Activity of Pollutants in Water Treatment

The following are the elements that are crucial for photocatalytic activity in wastewater treatment. From the literature study, we concluded that the following factors affect the photocatalytic activity of oxide-based photocatalyst, and these generalizations may be utilized for the photocatalytic activity of 2D metal chalcogenides as follows:

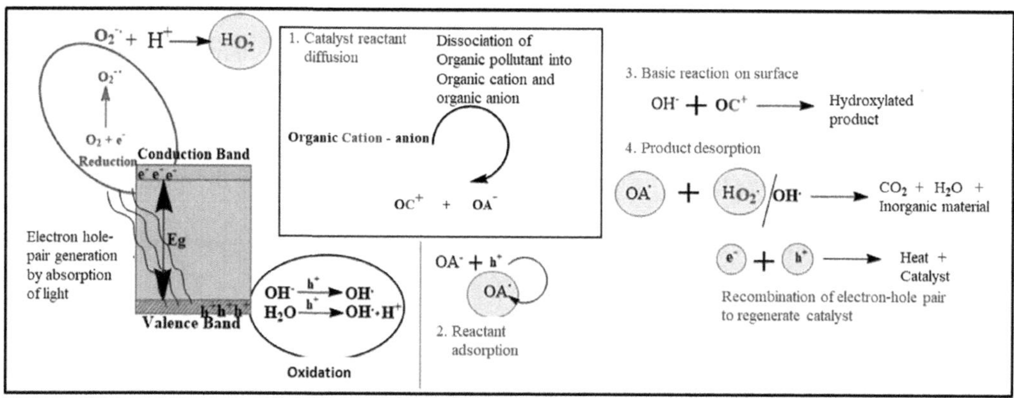

FIGURE 10.2 Photocatalytic mechanism showing four steps of photocatalytic action.

a. **Electron–hole pair recombination rate**
The pace of electron–hole pair division is one of the crucial factors affecting photocatalytic activity. In semiconductor-based catalysts, electron–hole (e^-/h^+) pair has a relatively short lifetime. Utilization of a competent catalyst in a secondary reaction prior to recombination is essential. The ideal conditions for photocatalytic behavior are a wide band gap and little electron recombination [73].

b. **Structure of the catalyst**
The stable structure, increased adsorption power and CB location of a catalyst directly affects its photocatalytic activity. The photocatalytic activity of the photocatalyst is also significantly influenced by its surface morphology. Photocatalysts are utilized in nano form instead of bulk form because of the higher surface-to-volume ratio that allows numerous reactants to react with the catalyst simultaneously [74,75]. This is a result of the tiny size and larger surface area of nanoscale materials that are accessible for the reaction.

c. **pH**
The pH of the solution is crucial for photocatalyst's effectiveness as it changes the surface charges of the photocatalyst and thus influences its photocatalytic activity. The electrostatic interactions between the pollutant and the charge particles may also be observed. The results of the observed photocatalytic activity for semiconductor photocatalyst in dye degradation show that, for pH values below 5 or for acidic media, a lower photocatalyst efficiency is observed, which results from a high proton concentration resulting in a delay in dye degradation owing to fewer available OH^- ions. The photoactivity of the catalyst improves in the pH range of 5–10, and at pH 10 (an alkaline medium), efficiency is at its peak because the OH^- generated interacts with the organic pollutant and breaks it down. The effectiveness decreases for pH ranges 11–13, where there is a significant concentration of hydroxyl ions that are scavenged [75].

d. **Quantity of the catalyst**
The quantity of the catalyst directly affects the photocatalytic activity; therefore, increasing the catalyst will also increase the rate of degradation since more radicals are produced during the degradation process as a result of the increased catalyst quantity. However, once the ideal limit is reached, adding more catalyst does not help the reaction since light cannot easily pass through, leaving the majority of the catalyst surfaces inactive [76].

e. **Intensity of the incident light**
The intensity of incoming light increases the rate of photocatalytic degradation. With an increase in light intensity, the quantum yield rises. Reaction rate divided by absorption rate is known as quantum yield. Wavelength has an impact on response efficacy as well because intense light prevents electron–hole pairs from recombining [77].

f. **Temperature**
Numerous researchers have examined the dependence of photocatalytic efficiency on reaction temperature and discovered that, at 80°C, it drops because the catalyst's absorption rate reduces and the rate of electron and hole recombination increases, leading to a drop in the reaction rate [78].

10.6.2 Photodegradation of Organics with Binary AX_2 and A_2X_3 Structures

Binary AX_2 photocatalysts: The typical formula for 2D transition metal disulfides (TMDs) is AX_2, where X stands for sulfur and A for the IV–VI transition metal group [79]. TMDs have a layered structure that resembles graphene. Each layer of TMDs is made up of three atoms: transition metal A and the sulfur element ligand X, and two chalcogen element layers clip the transition metal element with van der Waals forces [80] (as illustrated in Figure 10.3).

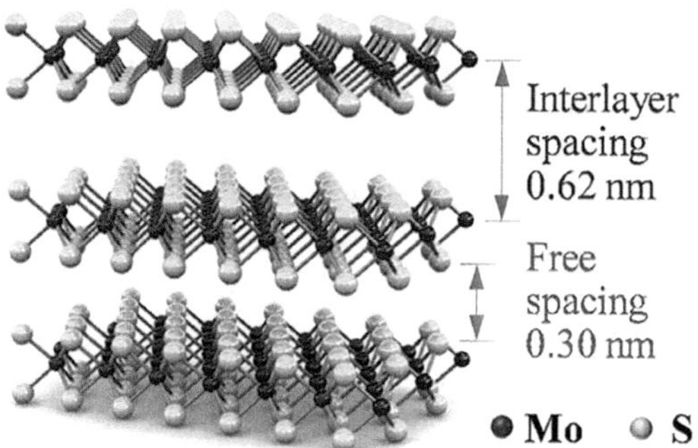

FIGURE 10.3 An illustration of MoS_2 layered structure. (Reprinted with permission from Ref. [81] Copyright 2017 American Chemical Society.)

TMDCs' layered structure may be reduced to the atomic level because of the weak interlayer van der Waals forces. 2D TMD possesses distinct electrical and optical characteristics and a small band gap with a strong carrier migration rate in the visible-near infrared spectrum when the block material is stripped down to the lamellar form. Furthermore, TMDs' crystal edge has a large number of active sites, which makes it a viable material for photocatalysis [82]. 2D MoS_2, which has four crystal structures, has good thermodynamic characteristics and is stable. MoS_2 has a larger solar spectral response range and a substantially lower band gap energy (1.3 eV) in comparison to the band gap of 3.2 eV of conventional anatase TiO_2, which is sensitive to ultraviolet light only. According to reports, the band gap can approach 1.9 eV when MoS_2 is reduced to a single layer. It was also observed that the band gap grows progressively as the number of MoS_2 layers reduces [80,83]. MoS_2 has been extensively researched in the areas of photocatalytic water disinfection, photocatalytic hydrogen production, and pollutant degradation because of its special physical and chemical characteristics. After creating a MoS_2/CdS composite catalyst, Xu et al. discovered that MoS_2 may be utilized as a cocatalyst to help CdS for evolving hydrogen [84]. The composite catalyst's hydrogen evolution process may achieve its ideal state and have 36 times better activity than pure CdS when 0.2 wt% MoS_2 is put on it. Moreover, alternative unstable photocatalysts can be coated with molybdenum disulfide to selectively oxidize organic contaminants. $MoSe_2$ has excellent anti-photocorrosion stability and shares many of the same structural characteristics as MoS_2 [85]. There are several techniques for synthesizing $MoSe_2$ of various morphologies and attributes, including hydrothermal hard template, chemical vapor deposition, and liquid phase stripping, which is a typical approach for creating monolayer $MoSe_2$. By removing the huge block of $MoSe_2$, a few layers of 1T-$MoSe_2$ were created, and this material performed better in the hydrogen generation process than 2H MoS_2 and 2H-$MoSe_2$ [86].

10.6.2.1 Binary A_2X_3 photocatalysts

Having an energy band gap of 1.3–1.7 eV, Bi_2S_3 is a typical layered semiconductor with a strong ability to absorb both visible and ultraviolet light. The ribbon-like structure of Bi_2S_3 is formed by crystal growth that is in line to its layered structure and stretches into limitless space. The Bi_2S_3 nanomaterial's distinct crystal structure gives it nonlinear optical properties in addition to a strong redox capacity under particular light sources [80]. By using a solvent heat technique, Kim et al. created the Bi_2S_3/TiO_2 composite material, which they then used in photocatalytic hydrogen generation, and the proposed model has been represented in Figure 10.4 for catalytic progress.

Since there is a good fit between the two materials' energy bands, the composite 10 wt% Bi_2S_3/TiO_2 can take in all light in the ultraviolet and visible areas, which significantly boosts the efficiency of creating hydrogen [87]. The photogenerated electrons' migration from Bi_2S_3 to ZnS for the

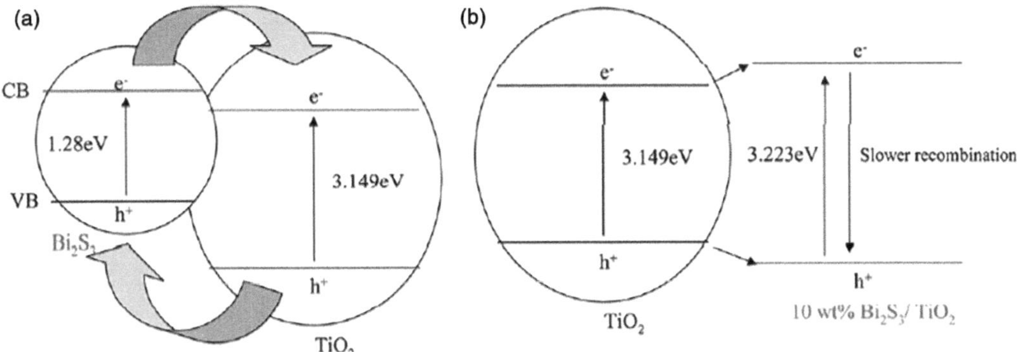

FIGURE 10.4 Proposed models for photocatalytic progress over for TiO_2, Bi_2S_3, and Bi_2S_3/TiO_2 composites: (a) Bi_2S_3 and TiO_2 are separated, and (b) the Bi_2S_3/TiO_2 composite is perfectly formed by the combination between TiO_2 and Bi_2S_3. (Reprinted with permission from Ref. [87] Copyright 2012 Elsevier.)

hydrogen generation reaction during the photocatalytic process is facilitated by the heterogeneous structure of Bi_2S_3/ZnS in the shape of green leaves. In the meanwhile, the carriers are effectively separated by the perforations as they go from ZnS to Bi_2S_3 [88].

10.6.3 Photodegradation of Organics with Ternary ABX_2 and AB_2X_4 Structures

Sulfur-based ternary chalcogenides with the AB_2X_4 structure constitute a novel class of materials with outstanding steadiness in photocatalytic activity. $ZnIn_2S_4$ is among the best-performing photocatalysts for energy and environmental applications in the AB_2X_4 family of compounds. Ternary sulfide chalcogenides are highly dependent on crystal structure, morphology, and optical properties for their photocatalytic action due to their visible region activity and significant chemical stability [89]. Heavy metals, medical wastes, and dyes with a high distinctive color are frequently found in the wastewater released from enterprises [89–91]. With a host of benefits over rival and established procedures, including (i) complete mineralization of the pollutant, (ii) no secondary pollution leading to only tame products like H_2O and CO_2, (iii) low cost, and (iv) ambient temperature and pressure as working conditions, the photocatalytic degradation process is a boon to attenuate the environmental pollution caused by organic pollutants. The following qualities are desired in an ideal photocatalyst material: (i) photoactivity, (ii) an acceptable band gap to use the light energy, (iii) photostability without experiencing photocorrosion, (iv) chemical inertness, (v) affordability, and (vi) non-toxicity. Since binary sulfides are susceptible to photocorrosion, the stability of sulfide catalysts during the photocatalytic degradation process has long been a cause for worry. For example, ZnS and CdS are the most researched binary sulfides for photocatalytic use [92], but they have disadvantages such as a greater band gap and photocorrosion.

$ZnIn_2S_4$-based materials with an $A^{II}B_2^{III}X_4^{VI}$ (where A^{II}=Zn or Cd; B^{III}=Ga or In, and X^{VI} =S) structure are proving to be excellent prospects for a variety of energy and environmental photocatalytic applications [93]. In 2009, Chen et al. [94] presented a precise mechanism for how $ZnIn_2S_4$ breaks down organic dyes. According to the process, when exposed to visible light, $ZnIn_2S_4$ microspheres produce electron–hole pairs. The electrons in the VB being excited to go into the CB as a result of which photon absorption leaves holes (h^+) in the VB. The oxygen molecules that have been adsorbed scavenge these excited electrons at the photocatalyst's surface to produce the superoxide radical anion, or O_2^-. This $O_2^{-\bullet}\rightarrow$ radical anion then reacts with H^+ to form $^{\bullet}HO_2$ radical, which when paired with the trapped electrons can result in significant quantities of $OH^{\bullet}\rightarrow$ radical.

The photocatalytic process takes place based on the following equations (Chen et al. 2009) [89,94]:

$$ZnIn_2S_4 + hv \rightarrow ZnIn_2S_4\left(e_{cb}^- + h_{vb}^+\right) \quad (10.1)$$

$$ZnIn_2S_4\left(e_{cb}^-\right) + O_2 \rightarrow ZnIn_2S_4 + O_2^{-\bullet} + H^+ \rightarrow {}^\bullet HO_2 + O_2^{-\bullet} \quad (10.2)$$

$$ZnIn_2S_4\left(e_{cb}^-\right) + {}^\bullet HO_2 + H^+ \rightarrow H_2O_2 \quad (10.3)$$

$$H_2O_2 + ZnIn_2S_4\left(e_{cb}^-\right) \rightarrow OH^\bullet + OH^- \quad (10.4)$$

$$OH^\bullet, {}^\bullet HO_2, O_2^{-\bullet} \text{ or } h_{vb}^+ + \text{dyes} \rightarrow \text{intermediates} \rightarrow \text{products} \quad (10.5)$$

According to the arrangement of octahedral or tetrahedral cationic sites, $A^{II}B_2^{III}X_4^{VI}$ compounds display three distinct structures, including tetragonal, spinel, and layered [96], while $ZnIn_2S_4$ is the only $A^{II}B_2^{III}X_4^{VI}$ compound that exists in layered structure and has three possible polymorphs, including cubic, hexagonal, and rhombohedral, as shown in Figure 10.5. While the cubic polymorph of $ZnIn_2S_4$ exhibits thermoelectric characteristics, hexagonal polymorphs have been reported to display photoconductivity and photoluminescence. Further, $ZnIn_2S_4$ has a band gap range of 2.3–2.8 eV where S-3p orbitals occupy $ZnIn_2S_4$'s valence band maxima, and the highest occupied molecular orbital (HOMO) is mostly composed of hybridized S-3p and Zn-3d orbitals [97]. And In-5s5p orbitals and S-3p orbitals generate the lowest occupied molecular orbital (LUMO), while Zn-4s4p orbitals located at higher positive energy levels do not participate in the CBM; also, In-5s orbitals, which exist at greater negative energy than S-3p orbitals, do not significantly contribute to the VBM.

Higher photocatalytic activity of chalcogenides is further encouraged by the surface imperfections produced by microwave treatment as they capture photoinduced electrons and extend the lifespan of holes [98]. According to another study [99], $ZnIn_2S_4$ hollow nanotubes with larger specific surface area (compared to 1D nanowires) promoted the improved photodegradation activity.

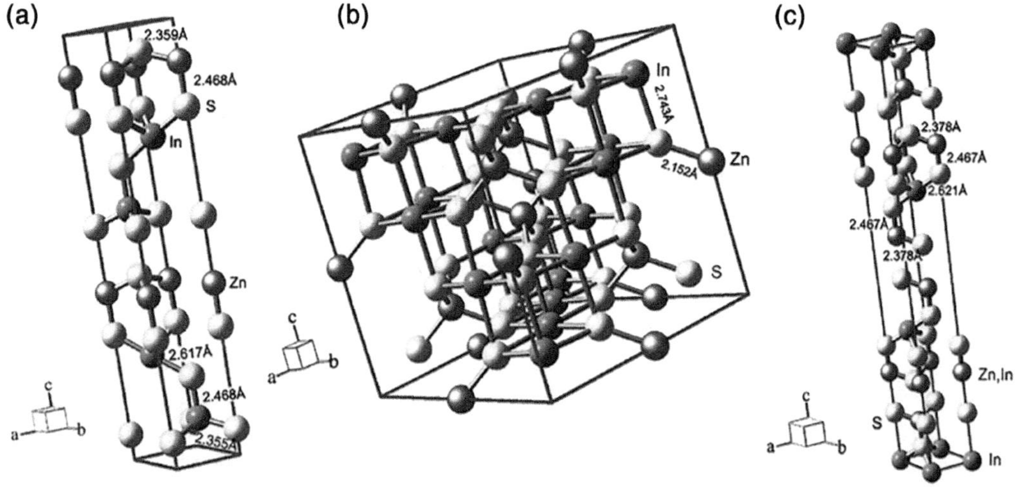

FIGURE 10.5 Crystal structure of $ZnIn_2S_4$ polymorphs: (a) hexagonal, (b) cubic, and (c) rhombohedral. (Reprinted with permission from Ref. [95] Copyright 2011 Elsevier.)

This was demonstrated by the photocatalytic degradation of Methyl Orange dye solution by these materials. $ZnIn_2S_4$ can be combined in a number of ways to photochemically breakdown a wide range of antibiotic medications in aquatic environments. As a result, $ZnIn_2S_4$ has been merged with the high chemical stability and carrier mobility of conductive polymer polypyrrole (PPy) for effective performance. This performance is also observed in photocatalytic degradation of the antibiotic Chloramphenicol (CHL) [100]. The degradation of CHL by the $PPyZnIn_2S_4$ combination was found to be two times greater than that of pure $ZnIn_2S_4$, with mineralization of CHL reaching up to 48.5%. The similar photodegradation studies have also been implemented on five different antibiotics, including erythromycin, chloramphenicol, rifampicin, lincomycin hydrochloride, and tetracycline hydrochloride. It is evident that the composite's greater surface area, visible light absorption, and Z-scheme degrading route contribute to its enhanced performance [101]. In similar modification [102], Sb_2S_3-$ZnIn_2S_4$ core–shell structure degraded tetracycline hydrochloride and created an intimate contact between Sb_2S_3 nanorods and $ZnIn_2S_4$ sheets. In another study, it was found that Ag_3PO_4 nanoparticle-adorned $ZnIn_2S_4$ photocatalyst was also able to degrade tetracycline with an enhanced efficiency of 92.3%. According to the suggested degradation route [91], the electron transport in Ag_3PO_4 provided the stability against photocorrosion.

Figure 10.6 shows another improvement in the photocatalytic degradation of antibiotic tetracycline in water bodies for the composite of g-C_3N_4/$ZnIn_2S_4$ [103]. It was concluded that the electrons from the CB of g-C_3N_4 migrate to the CB of $ZnIn_2S_4$ due to the band edge potentials of these two materials, while the photogenerated holes migrate from the VB of $ZnIn_2S_4$ and get transferred to the VB of g-C_3N_4 in order to promote charge carrier separation. $ZnIn_2S_4$ has a CB potential that is more negative than the O_2/O_2^- reduction potential. So, by a partial transformation of $O_2^{-\bullet}$, $O_2^{-\bullet}$ was formed at the CB of $ZnIn_2S_4$ together with the creation of $OH^\bullet$. Thus, the pollutant was immediately oxidized by the holes at the VB of g-C_3N_4.

Due to fundamental chemical and biological durability, herbicides and pesticides in waterbodies are among those by-products that are challenging to breakdown [104]. A heterojunction formed between $ZnIn_2S_4$ and g-C_3N_4 was exploited for the photocatalytic breakdown of 2,4-dichlorophenoxyacetic acid (2,4-D), which is one of the main constituents used in herbicides [104]. The $ZnIn_2S_4$/g-C_3N_4 composite showed excellent photocatalytic activity with improved charge carrier separation. The composite shows photostability up to five recycles, with O_2^- and h^+ being the main contributing active species in photocatalysis.

Additionally, on continuous exposure to visible light for 4 hours, the composite $ZnIn_2S_4$/g-C_3N_4 made with a 40% hydrothermal addition of g-C_3N_4 had proven remarkable photocatalytic

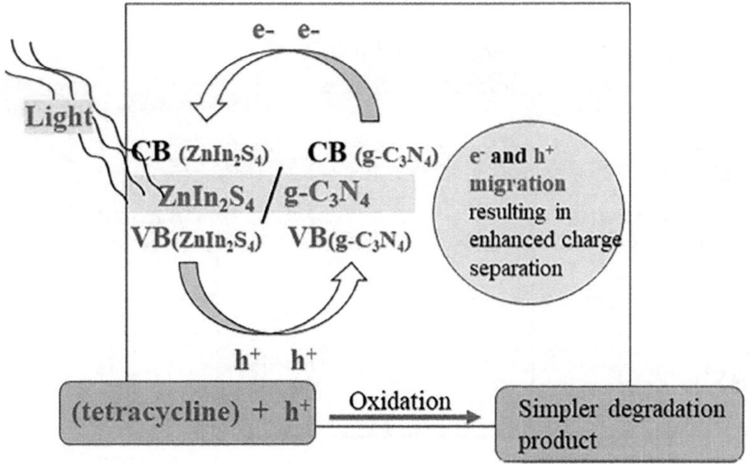

FIGURE 10.6 Photocatalytic degradation of antibiotic tetracycline by g-C_3N_4/$ZnIn_2S_4$.

activity in the degradation of phenol [105]. $ZnIn_2S_4$ and g-C_3N_4's with well-aligned band structure and close contact surfaces are found to be responsible for the increased photoactivity. Due to the extensive use of brominated phenols as pesticides, wood preservatives, and significant pharmaceutical intermediates, brominated phenols are another pervasive contaminant observed in the aquatic environment [106]. Debromination and full mineralization of 2,4,6-tribromophenol have been accomplished with the photocatalyst made with deliberate addition of zerovalent iron nanoparticles with $ZnIn_2S_4$. The results obtained from LCMS method were used to pinpoint the debromination route and confirmed the contribution of both oxidative and reductive debromination reactions.

Similarly, Pd-doped $ZnIn_2S_4$ photocatalyst was used to remove atrazine under the direct sunlight [107]. The enhanced photocatalytic elimination of atrazine up to 90% in 60 minutes is caused by the electrons from the CB of $ZnIn_2S_4$ transferring to vacant bands of Pd^{2+} ion during photocatalysis (Figure 10.7) and r the recombination of electron–hole pairs.

10.6.4 Antimicrobial Activity of 2D Chalcogenides for Microbes Disinfection in Water

The presence of pathogenic bacteria in hospitals and other healthcare facilities is a matter of concern as they obstruct the proper operation of medical instruments, surgical devices, and other equipment's and are the reason for several microbial infections. As a result of antibiotic resistance or multidrug resistance, combating these microbial infections has emerged as a significant concern in science and medicine as well as a significant public health issue. Antibiotic resistance is defined as the presence of naturally occurring resistance that evolved in microbes to antimicrobial agents. Designing materials with appropriate antimicrobial techniques is so crucial. Among other currently known antimicrobial agents, chalcogenide-based materials have emerged as potential antimicrobial agents due to their intrinsic antimicrobial activity and powerful capacity to both kill and inhibit the formation of bacteria. Due to their greater efficacy, lower toxicity, adaptable structure, and tunable band gap energy, chalcogenides (Ag_2S, MoS_2, and CuS) provide fascinating potential for antimicrobial applications [108]. Photocatalytic disinfection utilizes the abundant and free sunlight in order to efficiently inactivate numerous pathogens such as bacteria, viruses, and fungi and perhaps avoid nosocomial infection. Therefore, selecting highly stable photocatalysts with high photocatalytic activity is crucial [108].

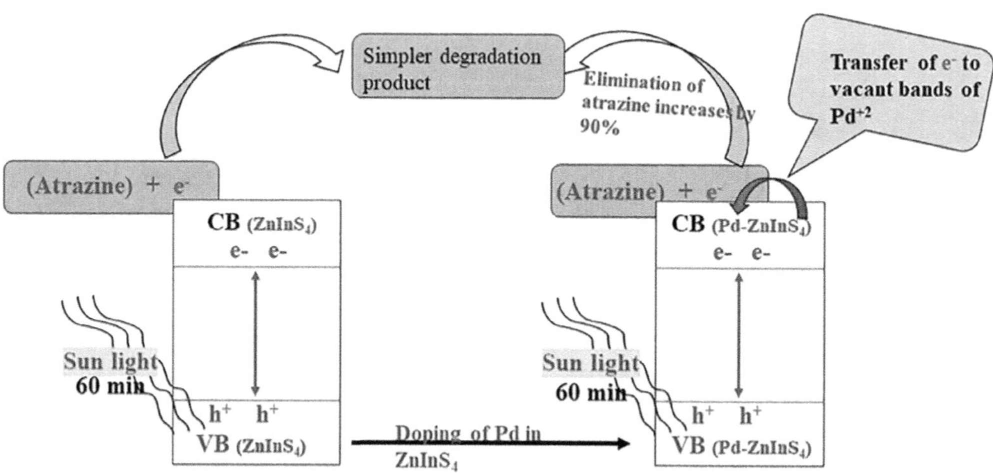

FIGURE 10.7 Enhanced photocatalytic elimination of atrazine by doping of Pd^{+2} in $ZnIn_2S_4$.

Chalcogenides are preferred because of their exceptional and in-demand properties, such as their adjustable band gap energy and electron transport capabilities, non-toxicity, biocompatibility, low cost, and ease of synthesis. In addition, their properties can be altered through a variety of modifications, such as the creation of composites, the addition of metal, and doping [109]. Using the sol–gel/ultrasound technique, Wang and colleagues produced Ag_2S-MgO/GO nanocomposite successfully and was applied to *Bacillus vallismortis*, *Escherichia. coli*, *Aspergillus flavus*, and *Trichoderma viride*. Ag_2S-MgO/GO showed better antibacterial properties than pure MgO and Ag_2S-MgO composite [110]. The betterment was observed due to the presence of large surface area of fine Ag_2S-MgO/GO particles providing greater number of active sites for antibacterial action. Further, morphologically changed Nisin@PEGylated-MoS_2 showed superior antibacterial activity in studies against *Staphylococcus aureus* [111]. The progress noticed is due to the breach of MoS_2 in *E. coli* cells which happens more readily. Moreover, it is well known that because of their outstanding broad-spectrum antimicrobial activity and notable cytotoxicity toward a variety of pathogens, Ag and Ag-based compounds/solids have antibacterial potential. Pure Ag_2S, CdS, and CdS/Ag_2S composites chemically produced using the co-precipitation approach, studied against *Pseudomonas aeruginosa*, *E. coli*, and *S. aureus* for their antibacterial potential [112], and it was observed that CdS/Ag_2S considerably surpassed pure 1D Ag_2S and CdS in terms of antibacterial activity.

Chalcogenides-based materials have the potential to be employed as antimicrobial agents due to excitation and production of e^-/h^+ couples on exposure to light against bacteria. The charge carriers produced by absorption of light can either migrate to the surface of the photocatalyst or undergo recombination process. On photoexcitation, e^- combines with adsorbed O_2 molecules, resulting in the production of ROS [108]. The impact of MoS_2, synthesized by several techniques, including ultrasonic, hydrothermal, and lithium-ion intercalation, giving different morphological structures, was studied for their photoinduced antibacterial activity against *E. coli* [111]. *E. coli* cells on exposure to visible light for 180 minutes in the presence of MoS_2, synthesized by Lithium-ion intercalation approach, showed incredible antibacterial activity. This might be because more ROS were produced to encourage the photoinduced inactivation of *E. coli*. In another example, the inactivation of *S. aureus* under visible light irradiation utilizing MoS_2/α-$NiMoO_4$ composite manufactured using microwave-assisted hydrothermal was also observed [113]. The electron spin resonance measurements and radical trapping studies revealed that the O_2^- radical was the primary active species controlling the bacterial inactivation under visible light irradiations. It was found that, under the influence of visible light, α-$NiMoO_4$ and MoS_2 were both excited, resulting in the production of e^- and h^+ in the CB and VB, respectively, in both materials. Because the VB and CB edges of α-$NiMoO_4$ are higher than those of MoS_2, the photogenerated e^- was transported from α-$NiMoO_4$ to MoS_2, whereas the photogenerated h^+ of MoS_2 migrated to α-$NiMoO_4$, as shown in Figure 10.8.

This makes it possible for the photogenerated e^- and h^+ in this composite material to be separated effectively. Additionally, in the CB of MoS_2 (including those transferred from α-$NiMoO_4$), O_2 suffers reduction by e^- to make O_2^-radical anions, whereas h^+ from α-$NiMoO_4$'s VB reacts with H_2O/OH to produce $OH^•$ radicals.

Regarding the mechanism behind antimicrobial action of chalcogenides, Kannan and colleagues proposed a mechanism in their study of antibacterial properties of Cu_2WS_4 against *E. coli*, *Bacillus subtilis*, *Micrococcus luteus*, *Klebsiella pneumoniae*, and *P. aeruginosa* at two distinct contact times (24 and 48 hours) [114], which is shown in Figure 10.9. They said that the interaction of Cu_2WS_4 with the bacterial membrane leads to a change in the membrane's permeability and causes the leaking of reducing sugars and proteins, which may be the origin of the antibacterial properties ensuing in deactivation of respiratory chain dehydrogenases. Another factor influencing the antibacterial activity is the electrostatic interaction present between the Cu_2WS_4 and the integral membrane proteins on the bacterial surface. Ultimately, similar to metal oxides, nanoparticles of above photocatalyst may cause a large increase in ROS, i.e., Hydroxyl radicals (OH), superoxide anion radicals ($O_2^{-•}$), and hydrogen peroxide (H_2O_2), which then reacts with proteins, lipids, and DNA of microbes [114]. Although the cell membrane is impermeable to $OH^•$ and $O_2^{-•}$, the H_2O_2 may readily get into it.

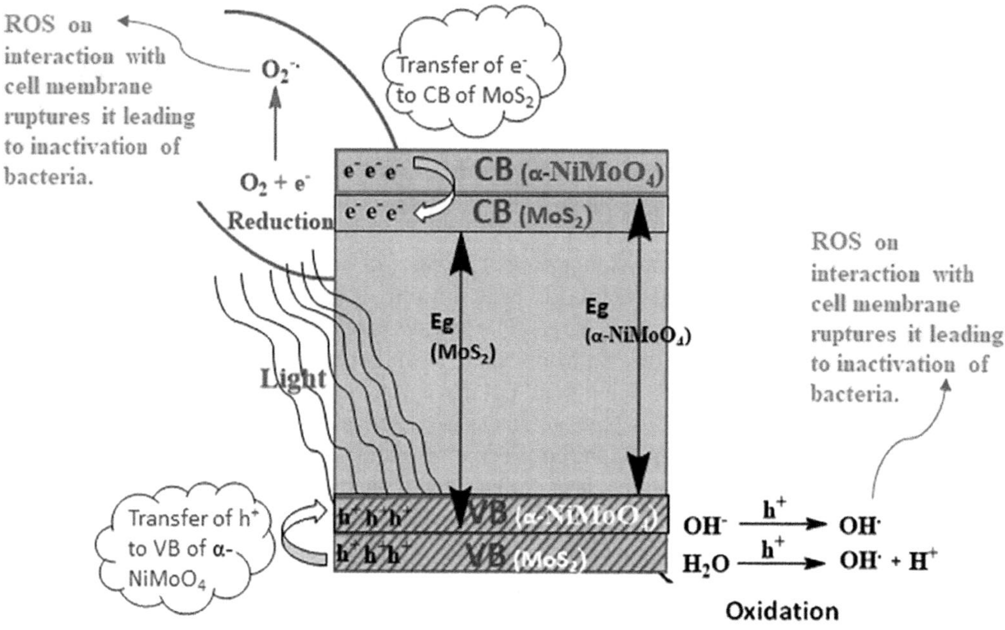

FIGURE 10.8 Transfer of photogenerated e^- to CB of MoS_2 and h^+ to VB of $NiMoO_4$.

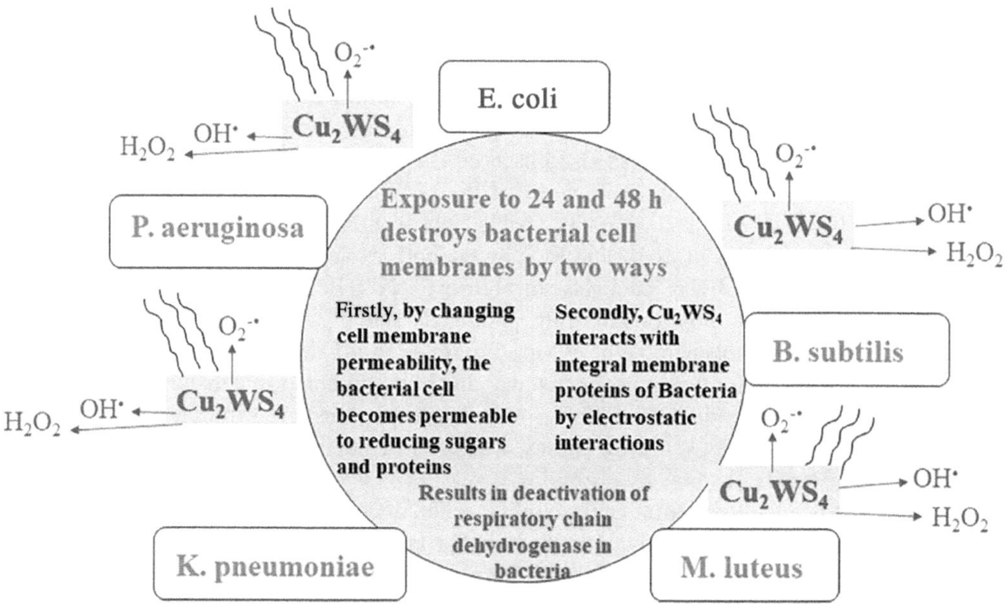

FIGURE 10.9 Kannan and colleagues antibacterial mechanism of Cu_2WS_4.

It is likely that more ROS will be produced when Cu_2WS_4 nanoparticles contact with *E. coli*; this will be a crucial element in boosting the antibacterial activity. Another mechanism proposed by Kokilavani and colleagues for photocatalytic antimicrobial action of MnS/Ag_2WO_4 confirmed the existence of two pathways [115]. First, it may be caused by coming into direct contact with bacterial cells, or it might be caused by the interaction of inferior products with them. The formation of denser metal ions, ROS, and the structure of the NPs are the three main components responsible for the antibacterial activity mechanism of the MnS/Ag_2WO_4. The denser metal ions can penetrate and attack the sulfhydryl moiety of cell membrane or these metal ions can cause lipid peroxidation on reaction with carbonic chains. This finally results in deactivation of protein and dysfunction of DNA molecules.

10.7 SUMMARY

Water pollution, both organic and inorganic, is a global hazard to aquatic life, human health, agriculture, and the economy. Since water contaminants are harmful and non-biodegradable substances are challenging to handle, it is necessary to use efficient, unconventional techniques to remove them. One strategy with a lot of potential is photocatalytic elimination. Because of their electronic structure and optical characteristics, semiconductor materials have a high potential to clean up the environment. They may act as photocatalysts because of their capacity to absorb light and generate free radicals. The core concepts of semiconductor 2D transition metal chalcogenides photocatalysts, their varieties, mechanism, and the variables influencing their effectiveness have all been covered in this study. Electrons and holes are photogenerated as they travel from the VB to the conducting band, leaving free holes. Photocatalysts are stimulated by light, producing photogenerated charge carriers. The photocatalysts' band gap energies (E_g) are helpful in determining the number of excited electrons or e^-/h^+ pairs. Reactive oxygen species (ROS), including superoxide radicals ($O_2^{-\bullet}$) and hydroxyl radicals ($OH^\bullet$), are created when photogenerated electrons and holes interact with oxygen or water. Organic pollutants may be photodegraded by ROS, and charge carriers have the power to decrease or oxidize the target chemical species, rendering them harmless. The bulk of studies are focused on bench-scale operations in a laboratory context based on the existing information regarding the limits of photocatalytic technology in danger reduction. As a result, more research including extensive simulations is needed. Nevertheless, there are still a lot of obstacles to be solved despite the fact that substantial study has been done on the elimination of risks utilizing semiconductor photocatalysts metal chalcogenides. We will be able to better comprehend the potential of 2D nanomaterials for practical catalysis in the future by expanding our knowledge of 2D nanocatalyst manufacture, material design, hybridization, catalytic processes, and applications.

REFERENCES

1. Yu, C., et al., Ag_3PO_4-based photocatalysts and their application in organic-polluted wastewater treatment. *Environmental Science and Pollution Research*, 29 (2022). 18423–18439. https://doi.org/10.1007/s11356-022-18591-7.
2. Shi, Y., et al., Recent progress of silver-containing photocatalysts for water disinfection under visible light irradiation: A review. *Science of the Total Environment*, 804 (2022). 150024. https://doi.org/10.1016/j.scitotenv.2021.150024.
3. Guo, R., et al., Recent advances and perspectives of g–C_3N_4–based materials for photocatalytic dyes degradation. *Chemosphere*, 295 (2022). 133834. https://doi.org/10.1016/j.chemosphere.2022.133834.
4. Ren, G., et al., Recent advances of photocatalytic application in water treatment: A review. *Nanomaterials*, 11 (2021). 1804. https://doi.org/10.3390/nano11071804.
5. Saeed, M., et al., Photocatalysis: An effective tool for photodegradation of dyes—A review. *Environmental Science and Pollution Research* (2022). 1–19. https://doi.org/10.1007/s11356-021-16389-7.
6. Elgohary, E.A., et al., A review of the use of semiconductors as catalysts in the photocatalytic inactivation of microorganisms. *Catalysts*, 11 (2021). 1498. https://doi.org/10.3390/catal11121498.
7. Ameta, R., et al., Photocatalysis, in *Advanced Oxidation Processes for Waste Water Treatment*. Elsevier. (2018). 135–175. https://doi.org/10.1016/B978-0-12-810499-6.00006-1.
8. Jabbar, Z.H. and B.H. Graimed, Recent developments in industrial organic degradation via semiconductor heterojunctions and the parameters affecting the photocatalytic process: A review study. *Journal of Water Process Engineering*, 47 (2022). 102671. https://doi.org/10.1016/j.jwpe.2022.102671.

9. Zhao, H., et al., Photodeposition of earth-abundant cocatalysts in photocatalytic water splitting: Methods, functions, and mechanisms. *Chinese Journal of Catalysis*, 43 (2022). 1774–1804. https://doi.org/10.1016/S1872-2067(22)64105-6.
10. Hasija, V., et al., Progress on the photocatalytic reduction of hexavalent Cr (VI) using engineered graphitic carbon nitride. *Process Safety and Environmental Protection*, 152 (2021). 663–678. https://doi.org/10.1016/j.psep.2021.06.042.
11. Deng, Y., et al., Nanohybrid photocatalysts for heavy metal pollutant control, in *Nanohybrid and Nanoporous Materials for Aquatic Pollution Control*. Elsevier. (2019). 125–153. https://doi.org/10.1016/B978-0-12-814154-0.00005-0.
12. Tahir, M.B., et al., Role of nanotechnology in photocatalysis. *Encyclopedia of Smart Materials*, 2 (2022). 578. https://doi.org/10.1016/B978-0-12-815732-9.00006-1.
13. Rahman, A., et al., Zinc oxide and zinc oxide-based nanostructures: Biogenic and phytogenic synthesis, properties and applications. *Bioprocess and Biosystems Engineering*, 44 (2021). 1333–1372. https://doi.org/10.1007/s00449-021-02530-w.
14. Hojamberdiev, M., et al., Binary flux-promoted formation of trigonal $ZnIn_2S_4$ layered crystals using ZnS-containing industrial waste and their photocatalytic performance for H_2 production. *Green Chemistry*, 20 (2018). 3845–3856. https://doi.org/10.1039/C8GC01746H.
15. Khan, M.M., S.F. Adil, and A. Al-Mayouf, *Metal Oxides as Photocatalysts*. (2015), Elsevier. 462–464. https://doi.org/10.1016/j.jscs.2015.04.003.
16. Bouroushian, M. and M. Bouroushian, Chalcogens and metal chalcogenides. *Electrochemistry of Metal Chalcogenides*, (2010). 1–56. https://doi.org/10.1007/978-3-642-03967-6_1.
17. Choi, Y.I., et al., Fabrication of ZnO, ZnS, Ag-ZnS, and Au-ZnS microspheres for photocatalytic activities, CO oxidation and 2-hydroxyterephthalic acid synthesis. *Journal of Alloys and Compounds*, 675 (2016). 46–56. https://doi.org/10.1016/j.jallcom.2016.03.070.
18. Khan, M.M., D. Pradhan, and Y. Sohn, *Nanocomposites for Visible Light-Induced Photocatalysis*. Vol. 101. (2017): Springer. https://doi.org/10.1007/978-3-319-62446-4.
19. Rahman, A. and M.M. Khan, Chalcogenides as photocatalysts. *New Journal of Chemistry*, 45 (2021). 19622–19635. https://doi.org/10.1039/D1NJ04346C.
20. Yaroshevsky, A., Abundances of chemical elements in the Earth's crust. *Geochemistry International*, 44 (2006). 48–55. https://doi.org/10.1134/S001670290601006X.
21. Gruendken, M., et al., Structure-property relationship of low molecular weight 'liquid' polymers in blends of sulfur cured SSBR-rich compounds. *Polymer Testing*, 87 (2020). 106558. https://doi.org/10.1016/j.polymertesting.2020.106558.
22. Duarte, M.E., et al., The unrevealed potential of elemental sulfur for the synthesis of high sulfur content bio-based aliphatic polyesters. *Polymer Chemistry*, 11 (2020). 241–248. https://doi.org/10.1039/C9PY01152H.
23. Chivers, T. and R.S. Laitinen, Tellurium: A maverick among the chalcogens. *Chemical Society Reviews*, 44 (2015). 1725–1739. https://doi.org/10.1039/C4CS00434E.
24. Zou, Z., et al., Photochemical vapor generation of selenium: Mechanisms and applications. *Trends in Environmental Analytical Chemistry*, 27 (2020). e00094. https://doi.org/10.1016/j.teac.2020.e00094.
25. Scheer, R. and H.-W. Schock, *Chalcogenide Photovoltaics: Physics, Technologies, and Thin Film Devices*. (2011): John Wiley & Sons, Hoboken, NJ.
26. Nie, L. and Q. Zhang, Recent progress in crystalline metal chalcogenides as efficient photocatalysts for organic pollutant degradation. *Inorganic Chemistry Frontiers*, 4 (2017). 1953–1962. https://doi.org/10.1039/C7QI00651A.
27. Zheng, N., et al., Open-framework chalcogenides as visible-light photocatalysts for hydrogen generation from water. *Angewandte Chemie International Edition*, 44 (2005). 5299–5303.
28. Rosenman, A., et al., Review on Li-sulfur battery systems: An integral perspective. *Advanced Energy Materials*, 5 (2015). 1500212. https://doi.org/10.1002/anie.200500346.
29. Muslih, E.Y., B. Munir, and M.M. Khan, Advances in chalcogenides and chalcogenides-based nanomaterials such as sulfides, Selenides, and tellurides, in *Chalcogenide-Based Nanomaterials as Photocatalysts*. (2021), Elsevier. 7–31. https://doi.org/10.1002/aenm.201500212.
30. Chen, S., et al., Crystal and electronic band structure of Cu2ZnSnX4 (X= S and Se) photovoltaic absorbers: First-principles insights. *Applied Physics Letters*, 94 (2009). https://doi.org/10.1016/B978-0-12-820498-6.00002-0.
31. Siddiqui, M.S. and M. Aslam, Chalcogenides as well as chalcogenides-based nanomaterials and its importance in photocatalysis, in *Chalcogenide-Based Nanomaterials as Photocatalysts*. (2021), Elsevier. 33–76. https://doi.org/10.1063/1.3074499.

32. Xu, Y., Y. Huang, and B. Zhang, Rational design of semiconductor-based photocatalysts for advanced photocatalytic hydrogen production: The case of cadmium chalcogenides. *Inorganic Chemistry Frontiers*, 3 (2016). 591–615. https://doi.org/10.1039/C5QI00217F.
33. Guo, Z., et al., Band gap engineering in huge-gap semiconductor $SrZrO_3$ for visible-light photocatalysis. *International Journal of Hydrogen Energy*, 39 (2014). 2042–2048. https://doi.org/10.1016/j.ijhydene.2013.11.055.
34. Dincer, I. and C. Zamfirescu, Hydrogen production by photonic energy, in *Sustainable Hydrogen Production*, (2016). 309–391. https://doi.org/10.1016/B978-0-12-801563-6.00005-4.
35. Glanz, K., Using behavioral theories to guide decisions of what to measure, and why. *International Journal of Hydrogen Energy*, 27 (2010). 991–1022.
36. Lee, G.-J., et al., Sonochemical synthesis of hollow copper doped zinc sulfide nanostructures: Optical and catalytic properties for visible light assisted photosplitting of water. *Industrial & Engineering Chemistry Research*, 53 (2014). 8766–8772. https://doi.org/10.1021/ie500663n.
37. Sasan, K., et al., Open framework metal chalcogenides as efficient photocatalysts for reduction of CO_2 into renewable hydrocarbon fuel. *Nanoscale*, 8 (2016). 10913–10916. https://doi.org/10.1039/C6NR02525K.
38. Li, P., et al., Highly efficient visible-light driven solar-fuel production over tetra (4-carboxyphenyl) porphyrin iron (III) chloride using CdS/Bi_2S_3 heterostructure as photosensitizer. *Applied Catalysis B: Environmental*, 238 (2018). 656–663. https://doi.org/10.1016/j.apcatb.2018.07.066.
39. Deng, F., et al., Novel visible-light-driven direct Z-scheme $CdS/CuInS_2$ nanoplates for excellent photocatalytic degradation performance and highly-efficient Cr (VI) reduction. *Chemical Engineering Journal*, 361 (2019). 1451–1461. https://doi.org/10.1016/j.cej.2018.10.176.
40. Sun, S., et al., Highly correlation of CO_2 reduction selectivity and surface electron Accumulation: A case study of $Au\text{-}MoS_2$ and $Ag\text{-}MoS_2$ catalyst. *Applied Catalysis B: Environmental*, 271 (2020). 118931. https://doi.org/10.1016/j.apcatb.2020.118931.
41. Huang, L., et al., Fabrication of hierarchical Co_3O_4@ $CdIn_2S_4$ p–n heterojunction photocatalysts for improved CO_2 reduction with visible light. *Journal of Materials Chemistry A*, 8 (2020). 7177–7183. https://doi.org/10.1039/D0TA01817A.
42. Wang, S., B.Y. Guan, and X.W.D. Lou, Construction of $ZnIn_2S_4$–In_2O_3 hierarchical tubular heterostructures for efficient CO_2 photoreduction. *Journal of the American Chemical Society*, 140 (2018). 5037–5040. https://doi.org/10.1021/jacs.8b02200.
43. Shown, I., et al., Carbon-doped SnS_2 nanostructure as a high-efficiency solar fuel catalyst under visible light. *Nature Communications*, 9 (2018). 169. https://doi.org/10.1038/s41467-017-02547-4.
44. Zhou, M., et al., Boron carbon nitride semiconductors decorated with CdS nanoparticles for photocatalytic reduction of CO_2. *ACS Catalysis*, 8 (2018). 4928–4936. https://doi.org/10.1021/acscatal.8b00104.
45. Li, H., et al., Construction and nanoscale detection of interfacial charge transfer of elegant Z-scheme $WO_3/Au/In_2S_3$ nanowire arrays. *Nano Letters*, 16 (2016). 5547–5552. https://doi.org/10.1021/acs.nanolett.6b02094.
46. Kočí, K., et al., ZnS/MMT nanocomposites: The effect of ZnS loading in MMT on the photocatalytic reduction of carbon dioxide. *Applied Catalysis B: Environmental*, 158 (2014). 410–417. https://doi.org/10.1016/j.apcatb.2014.04.048.
47. Yang, X., et al., Enhanced photocatalytic activity of Zn-doped dendritic-like CdS structures synthesized by hydrothermal synthesis. *Journal of Photochemistry and Photobiology A: Chemistry*, 329 (2016). 175–181. https://doi.org/10.1016/j.jphotochem.2016.07.005.
48. Luo, B., G. Liu, and L. Wang, Recent advances in 2D materials for photocatalysis. *Nanoscale*, 8 (2016). 6904–6920. https://doi.org/10.1039/C6NR00546B.
49. Nasilowski, M., et al., Two-dimensional colloidal nanocrystals. *Chemical Reviews*, 116 (2016). 10934–10982. https://doi.org/10.1021/acs.chemrev.6b00164.
50 Peng, Y., et al., Flower-like CdSe ultrathin nanosheet assemblies for enhanced visible-light-driven photocatalytic H_2 production. *Chemical Communications*, 51 (2015). 4677–4680. https://doi.org/10.1039/C5CC00136F.
51. Wang, Y., et al., Plasmon resonances of highly doped two-dimensional MoS_2. *Nano Letters*, 15 (2015). 883–890. https://doi.org/10.1021/nl503563g.
52. Tan, C. and H. Zhang, Wet-chemical synthesis and applications of non-layer structured two-dimensional nanomaterials. *Nature Communications*, 6 (2015). 7873. https://doi.org/10.1038/ncomms8873.
53. Acerce, M., D. Voiry, and M. Chhowalla, Metallic 1T phase MoS_2 nanosheets as supercapacitor electrode materials. *Nature Nanotechnology*, 10 (2015). 313–318. https://doi.org/10.1038/nnano.2015.40.

54. Lukowski, M.A., et al., Enhanced hydrogen evolution catalysis from chemically exfoliated metallic MoS_2 nanosheets. *Journal of the American Chemical Society*, 135 (2013). 10274–10277. https://doi.org/10.1021/ja404523s.
55. Xie, J., et al., Controllable disorder engineering in oxygen-incorporated MoS_2 ultrathin nanosheets for efficient hydrogen evolution. *Journal of the American Chemical Society*, 135 (2013). 17881–17888. https://doi.org/10.1021/ja408329q.
56. Song, Y., et al., Synthesis of few-layer MoS_2 nanosheet-loaded Ag_3PO_4 for enhanced photocatalytic activity. *Dalton Transactions*, 44 (2015). 3057–3066. https://doi.org/10.1039/C4DT03242J.
57. Zhang, Y.-C., et al., Role of oxygen vacancies in photocatalytic water oxidation on ceria oxide: Experiment and DFT studies. *Applied Catalysis B: Environmental*, 224 (2018). 101–108. https://doi.org/10.1016/j.apcatb.2017.10.049.
58. Wang, S., et al., Titanium-defected undoped anatase TiO_2 with p-type conductivity, room-temperature ferromagnetism, and remarkable photocatalytic performance. *Journal of the American Chemical Society*, 137 (2015). 2975–2983. https://doi.org/10.1021/ja512047k.
59. Li, H., et al., Oxygen vacancy-mediated photocatalysis of BiOCl: Reactivity, selectivity, and perspectives. *Angewandte Chemie International Edition*, 57 (2018). 122–138. https://doi.org/10.1002/anie.201705628.
60. Pan, L., et al., Constructing TiO_2 pn homojunction for photoelectrochemical and photocatalytic hydrogen generation. *Nano Energy*, 28 (2016). 296–303. https://doi.org/10.1016/j.nanoen.2016.08.054.
61. Zhang, Y.C., et al., Structure-activity relationship of defective metal-based photocatalysts for water splitting: Experimental and theoretical perspectives. *Advanced Science*, 6 (2019). 1900053. https://doi.org/10.1002/advs.201900053.
62. Chen, X., et al., Increasing solar absorption for photocatalysis with black hydrogenated titanium dioxide nanocrystals. *Science*, 331 (2011). 746–750. https://www.science.org/doi/abs/10.1126/science.1200448.
63. Bai, S., et al., Defect engineering in photocatalytic materials. *Nano Energy*, 53 (2018). 296–336. https://doi.org/10.1016/j.nanoen.2018.08.058.
64. Zhang, Y.-C., et al., Unraveling the facet-dependent and oxygen vacancy role for ethylene hydrogenation on Co_3O_4 (110) surface: A DFT+ U study. *Applied Surface Science*, 401 (2017). 241–247. https://doi.org/10.1016/j.apsusc.2017.01.031.
65. Song, J., et al., Oxygen-deficient tungsten oxide as versatile and efficient hydrogenation catalyst. *ACS Catalysis*, 5 (2015). 6594–6599. https://doi.org/10.1021/acscatal.5b01522.
66. Wang, G., et al., Synthesis and characterization of ZnS with controlled amount of S vacancies for photocatalytic H_2 production under visible light. *Scientific Reports*, 5 (2015). 8544. https://doi.org/10.1038/srep08544.
67. Zhang, X., et al., Surface defects enhanced visible light photocatalytic H_2 production for Zn-Cd-S solid solution. *Small*, 12 (2016). 793–801. https://doi.org/10.1002/smll.201503067.
68. Bao, N., et al., Self-templated synthesis of nanoporous CdS nanostructures for highly efficient photocatalytic hydrogen production under visible light. *Chemistry of Materials*, 20 (2008). 110–117. https://doi.org/10.1021/cm7029344.
69. Wu, Q. and F. Sun. Photochemical synthesis of ZnS nanosheet and its use in photodegradation of organic pollutants, in *2010 4th International Conference on Bioinformatics and Biomedical Engineering*. IEEE. (2010). https://doi.org/10.1109/ICBBE.2010.5517517.
70. Feng, B., et al., Growth mechanism, optical and photocatalytic properties of the ZnSe nanosheets constructed by the nanoparticles. *Journal of Alloys and Compounds*, 555 (2013). 241–245. https://doi.org/10.1016/j.jallcom.2012.12.074.
71. Xu, Y., et al., Synthesis of ultrathin CdS nanosheets as efficient visible-light-driven water splitting photocatalysts for hydrogen evolution. *Chemical Communications*, 49 (2013). 9803–9805. https://doi.org/10.1039/C3CC46342G.
72. Xiang, Q., B. Cheng, and J. Yu, Hierarchical porous CdS nanosheet-assembled flowers with enhanced visible-light photocatalytic H_2-production performance. *Applied Catalysis B: Environmental*, 138 (2013). 299–303. https://doi.org/10.1016/j.apcatb.2013.03.005.
73. Smith, T., K. Stevenson, and C. Zoski, *Handbook of Electrochemistry*. (2007): Elsevier, Amsterdam. 73–110. https://doi.org/10.1016/B978-044451958-0.50005-7.
74. Magdalane, C.M., et al., Photocatalytic decomposition effect of erbium doped cerium oxide nanostructures driven by visible light irradiation: Investigation of cytotoxicity, antibacterial growth inhibition using catalyst. *Journal of Photochemistry and Photobiology B: Biology*, 185 (2018). 275–282. https://doi.org/10.1016/j.jphotobiol.2018.06.011.
75. Turki, A., et al., Phenol photocatalytic degradation over anisotropic TiO_2 nanomaterials: Kinetic study, adsorption isotherms and formal mechanisms. *Applied Catalysis B: Environmental*, 163 (2015). 404–414. https://doi.org/10.1016/j.apcatb.2014.08.010.

76. Saikia, L., et al., Photocatalytic performance of ZnO nanomaterials for self sensitized degradation of malachite green dye under solar light. *Applied Catalysis A: General*, 490 (2015). 42–49. https://doi.org/10.1016/j.apcata.2014.10.053.
77. Prasad, G., et al., Photocatalytic inactivation of Bacillus anthracis by titania nanomaterials. *Journal of Hazardous Materials*, 165 (2009). 506–510. https://doi.org/10.1016/j.jhazmat.2008.10.009.
78. Dong, F., et al., Enhancement of the visible light photocatalytic activity of C-doped TiO_2 nanomaterials prepared by a green synthetic approach. *The Journal of Physical Chemistry C*, 115 (2011). 13285–13292. https://doi.org/10.1021/jp111916q.
79. Chhowalla, M., et al., The chemistry of two-dimensional layered transition metal dichalcogenide nanosheets. *Nature Chemistry*, 5 (2013). 263–275. https://doi.org/10.1038/nchem.1589.
80. Xia, D., et al., Band gap engineered chalcogenide nanomaterials for visible light-induced photocatalysis, in *Chalcogenide-Based Nanomaterials as Photocatalysts*. Elsevier. (2021). 135–172. https://doi.org/10.1016/B978-0-12-820498-6.00006-8.
81. Wang, Z. and B. Mi, Environmental applications of 2D molybdenum disulfide (MoS_2) nanosheets. *Environmental Science & Technology*, 51 (2017). 8229–8244. https://doi.org/10.1021/acs.est.7b01466.
82. Guo, W., et al., Colloidal synthesis of $MoSe_2$ nanonetworks and nanoflowers with efficient electrocatalytic hydrogen-evolution activity. *Electrochimica Acta*, 231 (2017). 69–76. https://doi.org/10.1016/j.electacta.2017.02.048.
83. Mak, K.F., et al., Atomically thin MoS_2: A new direct-gap semiconductor. *Physical Review Letters*, 105 (2010). 136805. https://doi.org/10.1103/PhysRevLett.105.136805.
84. Min, Y., et al., Dual-functional MoS_2 sheet-modified CdS branch-like heterostructures with enhanced photostability and photocatalytic activity. *Journal of Materials Chemistry A*, 2 (2014). 2578–2584. https://doi.org/10.1039/C3TA14240J.
85. Shi, Y., et al., Highly ordered mesoporous crystalline $MoSe_2$ material with efficient visible-light-driven photocatalytic activity and enhanced lithium storage performance. *Advanced Functional Materials*, 23 (2013). 1832–1838. https://doi.org/10.1002/adfm.201202144.
86. Gupta, U., et al., Characterization of few-layer 1T-$MoSe_2$ and its superior performance in the visible-light induced hydrogen evolution reaction. *APL Materials*, 2 (2014). https://doi.org/10.1063/1.4892976.
87. Kim, J. and M. Kang, High photocatalytic hydrogen production over the band gap-tuned urchin-like Bi_2S_3-loaded TiO_2 composites system. *International Journal of Hydrogen Energy*, 37 (2012). 8249–8256. https://doi.org/10.1016/j.ijhydene.2012.02.057.
88. Nawaz, M., Morphology-controlled preparation of Bi_2S_3-ZnS chloroplast-like structures, formation mechanism and photocatalytic activity for hydrogen production. *Journal of Photochemistry and Photobiology A: Chemistry*, 332 (2017). 326–330. https://doi.org/10.1016/j.jphotochem.2016.09.005.
89. Ravichandran, J. and S. Singh, A review on potential sulfide-based ternary chalcogenides for emerging photo-assisted water purification applications. *Environmental Science and Pollution Research*, 30 (2023). 1–23. https://doi.org/10.1007/s11356-023-27113-y.
90. Mahlambi, M.M., C.J. Ngila, and B.B. Mamba, Recent developments in environmental photocatalytic degradation of organic pollutants: The case of titanium dioxide nanoparticles—a review. *Journal of Nanomaterials*, 2015 (2015). 5. https://doi.org/10.1155/2015/790173.
91. Zhang, G., et al., A mini-review on $ZnIn_2S_4$-Based photocatalysts for energy and environmental application. *Green Energy & Environment*, 7 (2022). 176–204. https://doi.org/10.1016/j.gee.2020.12.015.
92. Zhang, G., et al., Preparation of $ZnIn_2S_4$ nanosheet-coated CdS nanorod heterostructures for efficient photocatalytic reduction of Cr (VI). *Applied Catalysis B: Environmental*, 232 (2018). 164–174. https://doi.org/10.1016/j.apcatb.2018.03.017.
93. Arif, M., et al., Enhance photocatalysis performance and mechanism of CdS and Ag synergistic co-catalyst supported on mesoporous g-C_3N_4 nanosheets under visible-light irradiation. *Journal of Environmental Chemical Engineering*, 5 (2017). 5358–5368. https://doi.org/10.1016/j.jece.2017.10.024.
94. Chen, Z., et al., Photocatalytic degradation of dyes by $ZnIn_2S_4$ microspheres under visible light irradiation. *The Journal of Physical Chemistry C*, 113 (2009). 4433–4440. https://doi.org/10.1021/jp8092513.
95. Shen, S., et al., Insights into photoluminescence property and photocatalytic activity of cubic and rhombohedral $ZnIn_2S_4$. *Journal of Solid State Chemistry*, 184 (2011). 2250–2256. https://doi.org/10.1016/j.jssc.2011.06.033.
96. Manjon, F.J., I. Tiginyanu, and V. Ursaki, *Pressure-Induced Phase Transitions in AB2X4 Chalcogenide Compounds*. (2014): Springer, Berlin. https://doi.org/10.1007/978-3-642-40367-5.
97. Shen, S., et al., Enhanced photocatalytic hydrogen evolution over Cu-doped $ZnIn_2S_4$ under visible light irradiation. *The Journal of Physical Chemistry C*, 112 (2008). 16148–16155. https://doi.org/10.1021/jp804525q.

98. Chen, Z., et al., Microwave-assisted hydrothermal synthesis of marigold-like $ZnIn_2S_4$ microspheres and their visible light photocatalytic activity. *Journal of Solid State Chemistry*, 186 (2012). 247–254. https://doi.org/10.1016/j.jssc.2011.12.006.
99. Shi, L., P. Yin, and Y. Dai, Synthesis and photocatalytic performance of $ZnIn_2S_4$ nanotubes and nanowires. *Langmuir*, 29 (2013). 12818–12822. https://doi.org/10.1021/la402473k.
100. Gao, B., et al., Identification of intermediates and transformation pathways derived from photocatalytic degradation of five antibiotics on $ZnIn_2S_4$. *Chemical Engineering Journal*, 304 (2016). 826–840. https://doi.org/10.1016/j.cej.2016.07.029.
101. Jo, W.-K. and T.S. Natarajan, Fabrication and efficient visible light photocatalytic properties of novel zinc indium sulfide ($ZnIn_2S_4$)–graphitic carbon nitride (g-C_3N_4)/bismuth vanadate ($BiVO_4$) nanorod-based ternary nanocomposites with enhanced charge separation via Z-scheme transfer. *Journal of Colloid and Interface Science*, 482 (2016). 58–72. https://doi.org/10.1016/j.jcis.2016.07.062.
102. Xiao, Y., et al., Hierarchical $Sb_2S_3/ZnIn_2S_4$ core–shell heterostructure for highly efficient photocatalytic hydrogen production and pollutant degradation. *Journal of Colloid and Interface Science*, 623 (2022). 109–123. https://doi.org/10.1016/j.jcis.2022.04.137.
103. Guo, F., et al., Graphite carbon nitride/$ZnIn_2S_4$ heterojunction photocatalyst with enhanced photocatalytic performance for degradation of tetracycline under visible light irradiation. *Journal of Physics and Chemistry of Solids*, 110 (2017). 370–378. https://doi.org/10.1016/j.jpcs.2017.07.001.
104. Qiu, P., et al., Enhanced visible-light photocatalytic decomposition of 2, 4-dichlorophenoxyacetic acid over $ZnIn_2S_4$/g-C_3N_4 photocatalyst. *Journal of Hazardous Materials*, 317 (2016). 158–168. https://doi.org/10.1016/j.jhazmat.2016.05.069.
105. Liu, H., et al., Fabrication of $ZnIn_2S_4$-gC_3N_4 sheet-on-sheet nanocomposites for efficient visible-light photocatalytic H_2-evolution and degradation of organic pollutants. *RSC Advances*, 5 (2015). 97951–97961. https://doi.org/10.1039/C5RA17028A.
106. Gao, B., et al., Photocatalytic degradation of 2, 4, 6-tribromophenol over Fe-doped $ZnIn_2S_4$: Stable activity and enhanced debromination. *Applied Catalysis B: Environmental*, 129 (2013). 89–97. https://doi.org/10.1016/j.apcatb.2012.09.007.
107. Bo, L., et al., Preparation, activity, and mechanism of $ZnIn_2S_4$-based catalysts for photocatalytic degradation of atrazine in aqueous solution. *Journal of Water Process Engineering*, 36 (2020). 101334. https://doi.org/10.1016/j.jwpe.2020.101334.
108. Khan, M.M., S.N. Matussin, and A. Rahman, Recent development of metal oxides and chalcogenides as antimicrobial agents. *Bioprocess and Biosystems Engineering*, 46 (2023). 1–19. https://doi.org/10.1007/s00449-023-02878-1.
109. Rahman, A., et al., Molybdenum disulfide-based nanomaterials for visible-light-induced photocatalysis. *ACS Omega*, 7 (2022). 22089–22110. https://doi.org/10.1021/acsomega.2c01314.
110. Wang, H., G. Li, and A. Fakhri, Fabrication and structural of the Ag_2S-MgO/graphene oxide nanocomposites with high photocatalysis and antimicrobial activities. *Journal of Photochemistry and Photobiology B: Biology*, 207 (2020). 111882. https://doi.org/10.1016/j.jphotobiol.2020.111882.
111. Wang, P., et al., Antibacterial activity and cytotoxicity of novel silkworm-like nisin@ PEGylated MoS_2. *Colloids and Surfaces B: Biointerfaces*, 183 (2019). 110491. https://doi.org/10.1016/j.colsurfb.2019.110491.
112. Iqbal, T., et al., Facile synthesis and antimicrobial activity of CdS-Ag_2S nanocomposites. *Bioorganic Chemistry*, 90 (2019). 103064. https://doi.org/10.1016/j.bioorg.2019.103064.
113. Ray, S.K., et al., Visible light driven MoS2/α-NiMoO4 ultra-thin nanoneedle composite for efficient Staphylococcus aureus inactivation. *Journal of Hazardous Materials*, 385 (2020). 121553. https://doi.org/10.1016/j.jhazmat.2019.121553.
114. Kannan, S., et al., Antibacterial studies of novel Cu_2WS_4 ternary chalcogenide synthesized by hydrothermal process. *Journal of Solid State Chemistry*, 258 (2018). 376–382. https://doi.org/10.1016/j.jssc.2017.11.005.
115. Kokilavani, S., et al., Decoration of Ag_2WO_4 on plate-like MnS for mitigating the charge recombination and tuned bandgap for enhanced white light photocatalysis and antibacterial applications. *Journal of Alloys and Compounds*, 889 (2021). 161662. https://doi.org/10.1016/j.jallcom.2021.161662.

11 2D Metal–Organic Framework Nanocomposites for Removal of Organic Pollutants

Divya Vinod, Jijoe Prabagar Samuel, Hosakote Shankara Anusha, and Harikaranahalli Puttaiah Shivaraju

11.1 INTRODUCTION

The term "organic pollution" is used when there are substantial quantities of organic chemicals present. Its main sources include domestic sewage, industrial effluents, urban runoff, and agricultural wastewater. Sewage treatment facilities are strategically located near businesses engaged in food processing, paper and pulp production, crop cultivation, and fish farming [1]. When organic contaminants break down, the dissolved oxygen in the receiving water may be depleted more quickly than it can be replaced. This oxygen shortage has a detrimental effect on the aquatic life in the stream. The significant amounts of suspended particles found in wastewater containing organic pollutants limit the penetration of light available to photosynthetic species. The properties of the riverbed are also altered when these particles settle, making it an unfavorable habitat for numerous creatures [2].

Efforts to find effective approaches for eliminating highly hazardous chemical compounds from water have garnered significant attention. Diverse techniques, including coagulation, adsorption, coagulation-assisted filtration, ion exchange, precipitation, reverse osmosis, ozonation, and advanced oxidation processes, including photodegradation, have been investigated to remove organic pollutants from polluted water and wastewater. However, the widespread implementation of these techniques is hindered by their substantial capital and operational expenses. In contrast, ion exchange and reverse osmosis processes offer more favorable outcomes as they not only remove pollutants from wastewater but also allow for the recovery of valuable substances. Nonetheless, the significant initial investment and ongoing operational costs associated with reverse osmosis, ion exchange, and advanced oxidation techniques present challenges in terms of their economic viability [3].

Among the available methods for water treatment, the adsorption and photodegradation process using two-dimensional (2D) metal–organic framework (MOF) stands out as one of the most effective techniques for treating and eliminating organic contaminants in wastewater [4]. With the introduction of graphene, 2D nanomaterial that possesses remarkable chemical and physical characteristics, such as an incredibly huge specific surface area, exceptional optical transparency, and high electric and thermal conductivity, researchers have been captivated by its potential. Additional 2D nanomaterials, including noble metal nanosheets, graphite-carbon nitride (g–C_3N_4), transition metallic dichalcogenides (TMDs), hexagonal boron nitride (h-BN), and layered double hydroxides (LDHs), are also currently attracting further research attention [5–8]. Due to their diverse crystal structures and compositions, these 2D nanomaterials possess a range of characteristics, despite their shared sheet-like form. Unlike three-dimensional (3D) hierarchical nanostructures, zero-dimensional (0D) nanoparticles, one-dimensional (1D) nanorods/wires/tubes, and bulk substances, these nanomaterials exhibit several unique features [9,10]. For instance, 2D nanosheets, which are atomically thin and have excellent strength and transparency, are good candidates for

DOI: 10.1201/9781003436942-11

flexible and clear electrical devices. Due to their high lateral dimensions and ultrathin thickness, which reduce transport barriers and increase flux, 2D nanomaterials are desirable components for ultrathin membranes used in gas separation. The broad surfaces and high surface-to-volume atom ratios make it simpler for reactant molecules to come into contact with the active sites, thereby increasing catalytic activity [11,12].

MOFs are crystalline porous materials formed by combining metal-containing components such as ionic metals or organic ligands clusters like carboxylate ligands as well as added anionic ligands. Due to their customizable shapes, functions, enormous surface area, and ultra-high permeability, materials based on MOFs have recently witnessed rapid growth, making MOF-related development one of the rapidly advancing disciplines in chemistry and materials. For a variety of applications, including gas storage and separation, energy storage and conversion, biomedicine, catalysis, and sensors, MOFs are highly desirable due to their favorable properties [13]. To create MOFs of various sizes, shapes, and compositions, ranging from larger crystals to nanocrystals, a variety of well-established procedures have been devised. The focus of most research efforts has been on crystals, 0D as well as 1D MOF. The growth of MOF crystals should be restricted to the nanoscale in only one direction while maintaining the other two lateral directions, which makes the creation of 2D MOF nanocomposites challenging. Nevertheless, the integration of 2D MOF nanocomposites is anticipated to display exceptional accomplishment in applications such as separation of gas, sensing, storage of energy as well as conversion, and catalysis, thereby introducing new possibilities within the realm of 2D nanomaterials. Consequently, 2D MOF composites, as a novel addition to the 2D nanomaterial family, offer promising avenues for both fundamental research and practical applications [14–19].

This chapter is organized into several sections to comprehensively cover relevant topics. The initial section presents the classification and structural properties of 2D MOF nanocomposites. The subsequent part provides an overview of the synthesis methods employed for creating 2D MOF nanocomposites, including both top-down and bottom-up approaches. Sections 3 and 4 focus on various applications of 2D MOF nanocomposites in adsorption and photodegradation, encompassing non-modified, metal oxide modified, carbon modified, polymer modified, and metal sulfides modified MOF hybrids. Section 5 delves into the utilization of MOF-based membranes for removing organic pollutants from wastewater. Finally, Sections 6 and 7 discuss the recyclability performance and progress in implementing MOFs in commercial aspects of wastewater treatment (WWT).

11.2 CLASSIFICATIONS AND STRUCTURE PROPERTIES OF 2D MOFS

2D-based MOFs can be categorized according to their structural architecture features, connectivity, and composition. The following are some commonly known classifications of 2D MOFs:

- Layered 2D MOFs: These are composed of stacked 2D sheets that are weakly interconnected. These sheets consist of metal clusters or nodes connected by organic linkers. The arrangement of the layers can be random or exhibit varying degrees of disorder. Layered MOFs often involve metal-based frameworks with organic linkers such as carboxylates or imidazolates. Examples of layered MOFs include those based on metal ions like copper, zinc, or nickel [20].
- Metal Chalcogenide MOFs: In Metal Chalcogenide MOFs, 2D nanosheets are formed by coordinating chalcogen (sulfur, selenium, or tellurium) atoms with metal nodes or clusters. The chalcogen atoms serve as connectors between metal centers or as sites for ligand coordination. The presence of chalcogen elements in these MOFs often results in intriguing electrical and optical properties [21].
- Coordination Polymer Networks (CPNs): CPNs are another category of 2D MOFs characterized by the connection of metal nodes through organic ligands. These networks form

highly organized 2D nanosheets with well-defined pore shapes. The compositions and structures of CPNs can vary depending on the metal ions and ligands employed [22].

- Metal–Organic Polyhedral (MOPs): MOPs are 2D MOFs composed of metal clusters interconnected by organic linkers. These metal clusters can form extended 2D nanosheets with porous characteristics and adopt polyhedral shapes. MOPs often exhibit high thermal stability and can be functionalized to impart specific properties [23].

The proportions of a material have significant effects on its characteristics. Due to their drastically different properties from 3D MOFs, 2D MOFs are beneficial in situations where bulk MOFs are inadequate. For example, the electrical conductivity of 3D MOFs is hindered in electrochemical sensor applications due to their porous nature, crystal morphology, and chemical composition [24]. Nevertheless, 2D MOFs can be created by delicately combining ligands and metal ions. It is fascinating to note that the 2D complements of these MOFs exhibit improved conduction due to shorter charge transfer channel lengths [25]. For instance, Ohata et al. [26] found that the conductivity of nanocomposite nickel-hexaminotriphenylene (Ni-HITP) MOF was 0.6 S cm^{-1}. This represents the highest level recorded so far for HITP-based MOFs with a thickness of fewer than 100 nm. Sheberla et al. [27] reported that the conductivity values of the bulk as well as thin film analogs of the MOF remained 2 and 40 Scm^{-1}, respectively. Furthermore, 2D MOFs possess a vast surface area (owing to their high aspect ratio) and abundant dynamic sites, rendering them highly attractive for a range of applications, predominantly in sensing. The increased surface area enhances sensitivity by promoting the adsorption of additional analyte moieties and providing a robust matrix for effective enzyme binding. Moreover, the increased adsorption also improves reaction time. Its 2D design also has the advantage of making the active spots on its surface easily accessible. By efficiently utilizing these sites and reducing diffusion resistance and mass transfer resistance, this improves their catalytic properties. The transfer of energy and charges between fluorophores and quenchers is also facilitated for optical sensors, specifically when there are more available sites [28]. Another enticing quality of these materials is their excellent mechanical adaptability and durability, which find uses in sectors like sensor fabrication and gas separation. These characteristics are made feasible by the structure's shape and the large number of coordination bonds [29]. Figure 11.1 represents the schematic illustration of molecular structure of 2D MOFs.

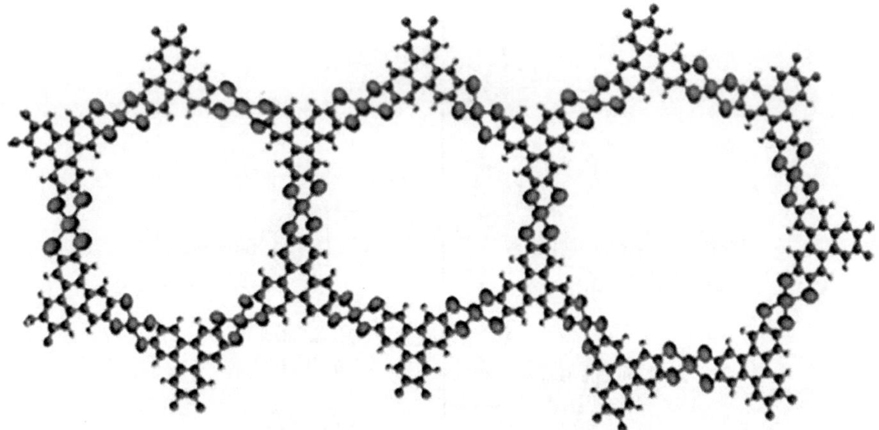

FIGURE 11.1 Schematic illustration of molecular structure of 2D MOFs.

11.3 OVERVIEW OF SYNTHESIS STRATEGIES OF 2D MOFS

For synthesizing 2D MOF nanosheets, a variety of techniques have been developed, including mechanical, sonication, Li-intercalation, chemical, interfacial, three-layer, surfactant-aided, modulated, and sonication exfoliation [30]. The top-down approaches and the bottom-up approaches are two groups into which these techniques can be divided. The bottom-up methods pertain toward the immediate production of 2D MOF nanocomposites using metallic nods and organic ligands, whereas the top-down procedures, as their names suggest, entail the flaking of layered MOFs. The layer-structured MOFs contain modest van der Waals forces or hydrogen bonds among the layers, but strong coordination interactions inside the layers. By using top-down techniques, it is simple to conquer the weak interlayer interactions, which causes layer-structured MOFs to exfoliate and transform into 2D MOF nanocomposites. The subsequent part will give an outline of numerous top-down as well as bottom-up techniques for producing 2D MOF nanocomposites [31]. Figure 11.2. represents the synthesis strategies of 2D MOFs.

11.3.1 Top-Down Method

The term "top-down method" describes the use of physical or chemical processes to separate the entire MOF with multiple layers from it. MOFs are characterized by strong intra-layer coordination bonds and frail interlayer π–π interactions, hydrogen bonds, or van der Waals forces. To create extremely thin 2D MOF nanosheets, one can thus simply use an external force to break van der Waals forces, weak interactions, or hydrogen bonds [32].

a. **Physical exfoliation**
   To create 2D MOF nanocomposites, abrasive exfoliation involves exterior substantial force, such as wet ball grating, sonication, and freeze–defrost methods. The selection of solvent plays a crucial role in the exfoliation efficiency of these physical methods as different solvents can influence the stripping effectiveness in various ways. Common solvents utilized include methanol, water, and acetone. Sonication exfoliation is frequently engaged in a variety of solvents to generate 2D MOF nanocomposites [33]. To create ultrathin

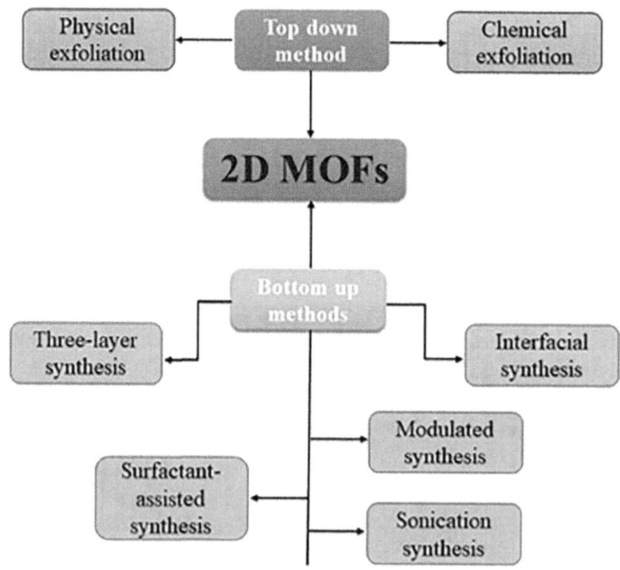

FIGURE 11.2 Synthesis strategies of 2D MOFs.

molecular sieved $Zn_2(bim)_4$ nanomaterial ($Zn_2(bim)_4$ MSNs), Y. Peng et al. [18] utilized sonication in conjunction with wet ball milling. The $Zn_2(bim)_4$ MOF exhibited moderate van der Waals force in its bulk form. To produce the ultrathin $Zn_2(bim)_4$ MSNs, wet ball milling was employed by using methanol and propanol solvents at approximately 60 rpm for a duration of 1 hour. To generate MOF nanomaterial, X. Wang et al. [34] developed a freeze–thaw exfoliation technique. The MOF crystal was dispersed in hexane, followed by a thorough freeze-drying method using liquid nitrogen, which was then thawed in hot water. The bulk MOF crystal experienced peeling caused by the shear stress arising from the difference in volume between the solid and liquid phases.

b. **Chemical exfoliation**

Chemical processes can yield higher results compared to physical processes. These chemical processes involve techniques such as chemical exfoliation and intercalation, which result in the breakdown or physical alteration of the core ligand in the MOF crystal [35]. To enhance the interlayer spacing, Ding et al. [36] initially introduced 4,4′-dipyridyl disulfide to the bulk layer of Zn_2(PdTCPP) MOF. The disulfide linkage was subsequently broken, and the bulk MOF crystals were isolated using trimethyl phosphorus. To achieve a high yield of 57%, the texture of the nanomaterials was regulated by altering the concentration and duration of the trimethyl phosphorus reaction. In summary, the top-down method is viewed as a simple way to produce 2D MOF nanomaterials from multilayered pure MOFs. Yet, there are several drawbacks to this approach, such as low consistency, instability, susceptibility to restacking, and insufficient yield of the products, which hinders their large-scale production [37].

11.3.2 Bottom-Up Method

The bottom-up methods depend on crystal growth as well as produce wafer-like crystals from organic linkages and metals in controlled conditions, indicating that their vertical expansion is restricted while horizontal growth is encouraged. More significantly, the bottom-up methods can produce non-layered MOF sheets with consistent thickness and enhanced yields [24]. We will discuss five different synthesis techniques, including interfacial synthesis, three-layer synthesis, surfactant-assisted synthesis techniques, modulated synthesis, and sonication synthesis.

a. **Interfacial synthesis method**

The interface represents the location where organic ligands and metal ions interact. The liquid/liquid interface and gas/liquid interface are two subtypes of the interfacial synthesis technique. Thus, the growth of 2D MOF nanomaterial is limited to the constrained interface regions. In gas/liquid interfacial synthesis, a small amount of organic diluent containing dispersed ionic metals as well as organic ligands is functionalized to the liquid's exterior. This creates an interface between gas/liquid, and when the organic solvent evaporates, ultrathin MOF nanomaterials are formed. This technique has been widely used for creating MOF nanocomposites. Liquid/liquid interfacial synthesis involves utilizing a pair of immiscible liquids, such as water/dichloromethane or water/ethyl acetate, to dissolve the organic ligands and metal ions. This method allows the production of ultrathin MOF nanomaterials at the interface of the two liquids [19].

By varying the ligand concentration as well as metal ions, the liquid/liquid interfacial synthesis approach enables precise control over the thickness of the synthesized 2D MOF nanocomposites. However, the thickness of these nanocomposites often reaches 100 nm. Consequently, the interface between gas/liquid methods is preferred over liquid/liquid interfacial methods for producing thinner or single-layer MOF nanocomposites [38]. The creation of 2D MOF nanocomposites can be effortlessly achieved through interfacial synthesis. Interfacial synthesis techniques are specifically applicable to MOF crystals produced at ambient temperature. However, the expansion of interfacial area can impose limitations on the yield of MOF. Consequently, this approach is not well-suited for the extensive production of 2D MOF nanocomposites [19].

b. **Three-layer synthesis method**

The diffusion rate and crystal growth rate are both delayed in the three-layer synthesis process, which is commonly employed for creating MOF crystals. The three-layer solvent method, as the name suggests, involves using two mixable diluents with significantly diverse densities, where the lower-density solvent sits atop the higher density. A buffer region is then formed between these two mixed solutions [39].

c. **Surfactant-assisted synthesis method**

Surfactants are employed in the surfactant-assisted synthesis technique of 2D MOF nanocomposites to expedite the synthesis process. Surfactants are amphiphilic compounds that possess both hydrophilic (water-loving) and hydrophobic (water-repelling) regions. By introducing surfactant molecules during the synthesis of MOF nanocomposites, the growth and morphology of the resulting materials can be controlled [15]. The interaction of organic ligands and metal ions used in the creation of MOFs with surfactant molecules can modify their assembly and orientation at the nanoscale. By selecting specific surfactants and adjusting their concentration, the form, dimensions, and structure of the resulting 2D MOF nanocomposites can be altered. The surfactant-assisted synthesis approach offers several advantages, including greater control over the structure of nanocomposite materials, improved dispersion in solvents, and stabilization of the nanoscale morphology. This enables the creation of homogeneous and well-defined 2D MOF nanocomposites with specialized properties for applications such as gas storage, sensing, and catalysis.

d. **Modulated synthesis**

Modulated synthesis of 2D MOF nanocomposites is a method that involves carefully modifying the reactions throughout the synthesis process to produce materials with desired structures and characteristics. In this approach, parameters such as temperature, solvent composition, reactant concentration, and reaction duration are manipulated or altered at specific points or stages during the synthesis. The growth kinetics, crystal size, shape, and surface characteristics of the 2D MOF nanocomposites can be influenced by carefully adjusting these parameters. This variation allows for the creation of unique structures, such as multilayered structures, hierarchical architectures, or nanocomposites with specialized functionalities [40]. Modulated synthesis offers a flexible framework for adjusting the characteristics of 2D MOF nanocomposites based on specific requirements in various applications. It enables the regulation of the nanocomposites' mechanical properties, surface area, surface functionality, and pore size [41]. Additionally, the incorporation of different external molecules or nanomaterials into the MOF framework enhances their potential use in fields such as gas storage, separation, catalysis, and sensing. In general, modulated synthesis provides a versatile and adaptable method for creating 2D MOF nanocomposites with improved characteristics and customized capabilities [42].

e. **Sonication synthesis**

The technique called sonication synthesis utilizes ultrasonic waves to accelerate the creation of 2D MOF nanocomposites. In this method, high-frequency sound waves generated by an ultrasonic bath or probe are applied to the precursors, which may consist of organic ligands and metal ions. The procedure involves combining precursors with a suitable solvent and then subjecting the resulting mixture to sonication [19]. The ultrasonic waves generate significant pressure fluctuations and cavitation bubbles in the solution. When these bubbles collapse abruptly near the surfaces of the precursors, they create localized extreme temperatures and pressures. This phenomenon promotes the dissolution, diffusion, and reaction of the precursors, resulting in the formation of 2D MOF nanocomposites [43].

The production of 2D MOF nanocomposites can benefit from several advantages offered by sonication synthesis. The precursors are effectively mixed and dispersed, enhancing the reaction kinetics and resulting in a more homogeneous final product. Additionally, the sonication procedure can promote the growth of smaller nanoscale crystals with high crystallinity. Moreover, by incorporating functional molecules, nanoparticles, or guest species into the MOF framework, sonication synthesis creates hybrid nanocomposites with customized structural and functional characteristics. Overall, sonication synthesis is a powerful method for the rapid and precise synthesis of 2D MOF nanocomposites. It offers advantages such as improved homogeneity, reaction kinetics, and the ability to incorporate additional substances into the final product [44].

11.4 ADSORPTION OF ORGANIC WATER POLLUTANTS WITH 2D MOFS-BASED COMPOSITES

2D MOFs possess unique surface properties, making them highly effective at adsorbing organic pollutants from water solutions. As an example, 2D MOFs exhibit diverse surface areas, sizes, and pore structures, spanning from 0.3 to 9.8 nm. These characteristics enable them to effectively adsorb a wide range of substances, such as dyes, insecticides, and antibiotics [45]. The ability of 2D MOFs to accommodate various species greatly impacts their adsorption capabilities. Furthermore, MOFs adhere to the isoreticular principle, allowing them to be synthesized with the widest pore openings (98 Å) and the lowest densities ($0.13\,g\,cm^{-3}$). Moreover, their exceptional thermal and chemical stability makes them suitable for post-synthetic covalent organic and metal-complex functionalization. This characteristic enhances their ability to adsorb organic pollutants from aqueous solutions [46]. Figure 11.3 represents the adsorption mechanism of organic water pollutants with 2D MOFs.

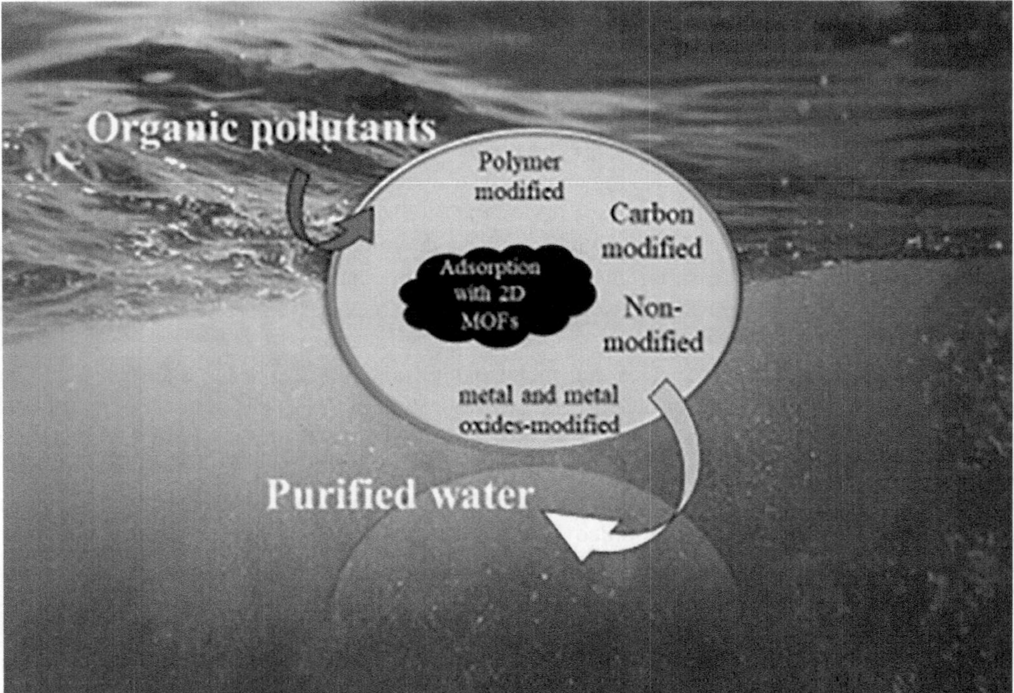

FIGURE 11.3 Adsorption mechanism of organic water pollutants with 2D MOFs.

11.4.1 Adsorption with Non-Modified MOFs

Adsorption is a technique in which particles or molecules bind to the surface of a solid material. MOFs are a group of porous substances composed of metal ions or clusters coordinated with organic ligands. These materials are well-suited for adsorption applications due to their high surface area and well-defined porous framework. The interaction between the MOF surface and the pollutant molecules is essential for the adsorption of organic pollutants using non-modified MOFs. The porous nature of MOFs provides numerous sites where the pollutant molecules can physically adsorb or chemically react with the MOF material [47]. During physical adsorption, organic pollutant molecules are attracted to the MOF surface through weak intermolecular forces such as van der Waals interactions. These forces facilitate the temporary adhesion and retention of the pollutant molecules on the MOF surface. In contrast, chemical interactions involve stronger bonds between the pollutants and the MOF surfaces. Depending on the characteristics of the MOF and the pollutant molecule, these interactions can include hydrogen bonds, π–π stacking, or coordination bonds. Organic contaminants can be efficiently adsorbed onto MOFs due to their large surface area and specific porosity. The extensive number of accessible sites provided by the porous structure enhances the overall adsorption capacity for pollutants to interact with. Additionally, the composition of organic ligands can be modified to adjust the selectivity of MOFs, allowing them to target specific contaminants [48].

In summary, the interaction between organic pollutant molecules and the surface of non-modified MOFs, occurring through physical adsorption or chemical interactions, is the underlying mechanism driving the adsorption of organic pollutants. This technique enables the separation or removal of organic pollutants from various environments, contributing to environmental restoration efforts [49].

11.4.2 Adsorption with Metal and Metal Oxides-Modified MOFs

Incorporating metal species with metal oxides into the MOF structure can enhance the ability of MOFs to adsorb organic contaminants, which is the purpose of using metal- and metal oxide-modified MOFs. The addition of metal ions or clusters during the synthesis process or through post-synthetic modification creates more adsorption sites [50]. Through coordination bonds or complexation interactions, these metal species interact with organic pollutants, often through mechanisms such as Lewis acid–base interactions. These modified MOFs exhibit stronger and more targeted bonding compared to non-modified MOFs, leading to improved adsorption of specific organic contaminants. The presence of metal and metal oxide species on the MOF surface contributes to both adsorption and catalytic activity. This catalytic activity facilitates the removal of organic pollutants from the environment or their transformation into less harmful compounds. The metal species act as catalysts by providing active sites where adsorbed organic molecules can undergo oxidation, reduction, and other relevant transformations [51].

The combination of metal species and metal oxides, along with the large surface area and porosity of MOFs, has significant benefits for the adsorption process. The increased surface area provides more adsorption sites, while the metal species or metal oxides enhance the adsorption capacity, selectivity, and catalytic activity of the MOF material. Overall, the utilization of metal- and metal oxide-modified MOFs offers an effective approach for the degradation and removal of organic contaminants. These modified MOFs are promising candidates for various environmental applications due to their enhanced adsorption characteristics, including increased adsorption capacity, selectivity, and surface catalytic activity [52]. Qin et al. [53] utilized an ion-assisted solvothermal method to synthesize Co-M-MOFs (M=Cu, Mn, Ni, and Zn). The incorporation of Cu^{2+}, Mn^{2+}, Ni^{2+}, and Zn^{2+} into the Co_3O_4 lattice was successfully achieved, resulting in the presence of abundant adsorbed oxygen and oxygen vacancies on the material's surface. This unique composition and surface characteristics contributed to the remarkable performance and stability of the composite material. Krap et al. [54] reported the successful synthesis of $[Ga_2(OH)_2(L)]$ (H_4L=biphenyl-3,3′,5,5′-tetracarboxylic acid), which is referred to as MFM-300(Ga_2) (MFM representing Manchester Framework Material, replacing the NOTT designation). This synthesis involved a solvothermal reaction of Ga $(NO_3)_3$ and H_4L in a mixture of N,N-dimethylformamide (DMF), Tetrahydrofuran (THF), and water with the addition of hydrochloric acid (HCl), which was allowed to proceed for a duration of 3 days.

11.4.3 Adsorption with Carbon-Modified MOFs

Adsorption of organic pollutants by carbon-modified MOFs involves the incorporation of carbon-based components, such as activated carbon, carbon nanotubes, or graphene oxide, into the MOF structure. These materials possess large surface areas and porous structures, providing ample space for the adsorption of organic contaminants. The carbon-based substances play a crucial role in facilitating adsorption on the MOF surface, utilizing weak intermolecular forces like van der Waals interactions and π–π interactions to attract and bind organic molecules. The porous structure and surface chemistry of the carbon materials contribute to their adsorption capabilities, allowing for the effective removal of organic pollutants. By incorporating functional groups onto the surface of the carbon material or applying chemical treatments to enhance its affinity for specific types of organic contaminants, surface modification further enhances the adsorption capabilities of the material. This modification enables more selective interactions and more efficient adsorption. The combination of MOFs as well as carbon-based materials causes synergistic effects, where the MOF framework provides structural stability and the carbon materials introduce additional adsorption sites, thereby increasing the overall adsorption capacity of the material. In addition, carbon-modified MOFs offer the advantage of regenerability and reusability. The deposited organic contaminants on the carbon surface can be easily removed using techniques such as solvent washing or thermal treatment, allowing for repeated cycles of adsorption and desorption. This regenerability makes carbon-modified MOFs a cost-effective and environmentally sustainable option for the efficient removal and remediation of organic contaminants in various environmental applications [55–58]. M. Li et al. [59] conducted a study where they synthesized interfaces of activated carbon (AC) and copper (II)- benzene-1,3,5-tricarboxylate (Cu-BTC) MOF to produce AC/Cu-BTC complexes using the hydrothermal technique to investigate VOCs adsorption capabilities (specifically n-hexane). The AC served as a nucleation substratum as well as Cu-BTC development, subsequently influencing the specific surface area characteristics. As a result, the specific surface area of AC/Cu-BTC was reduced, leading to a decrease in its adsorption capacity.

Y. Sun et al. [60] conducted a study to examine the adsorption capabilities of Cu-BTC MOF composites based on graphene oxide (GO) and reduced graphene oxide (rGO) of GO/Cu-BTC as well as rGO/Cu-BTC for organic adsorption such as methane and ethane. The ability of these composites to adsorb gas particles was found to be affected by the surface characteristics of the graphene. The existence of surface functional clusters on the constituents led to electrostatic repulsion as well as hydrophilicity, enabling their even dispersal and preventing clumping during the preparation procedure. This led to an increase in the specific surface area of the composites. Furthermore, by regulating the reduction degree of rGO, the composite material could achieve selective adsorption of gases. Y. Li et al. [61] successfully produced Cu-BTC@GO and investigated toluene adsorption. They observed that Cu-BTC@GO-5 exhibited significantly higher toluene adsorption capacity compared to traditional AC and zeolite. This enhanced performance can be attributed to two factors. First, Cu-BTC@GO-5 showed an increased specific surface area, which contributed to its improved adsorption capacity. Second, the introduction of GO enhanced the surface dispersibility of Cu-BTC, further enhancing its adsorption properties.

11.4.4 Adsorption with Polymers-Modified MOFs

The process of adsorption with polymer-coated MOFs involves modifying MOFs to enhance their adsorption capabilities. Initially, polymer coatings are applied to the surfaces of MOFs, either during their production or through subsequent treatments. These polymer coatings introduce additional functional groups and adsorption sites, thereby enhancing the efficiency and selectivity of adsorption for the MOFs. In comparison to non-modified MOFs, the functional groups within polymer coatings exhibit stronger and more selective adsorption of organic pollutants. This is achieved through interactions involving hydrogen bonding, electrostatic interactions, and specific chemical bonds. Furthermore, the porous structure of MOFs allows for the circulation and penetration of organic contaminants, facilitating their adsorption. Despite the polymer modification, the enhanced surface area and the porosity of MOFs remain intact. The polymer coatings do not significantly diminish the porosity; instead, they introduce additional adsorption sites within the pores, thereby enhancing the overall adsorption capacity. Furthermore, the functional groups within the polymers can be tailored to target specific types of organic contaminants, offering adjustable selectivity. By selecting polymers with specific chemical characteristics, modified MOFs can preferentially adsorb organic pollutants based on their size, charge, or functional groups. The combination of MOFs and polymer coatings results in synergistic effects, where the polymer coatings enhance adsorption capacity and selectivity, while the MOFs contribute to structural stability. This synergy leads to improved overall adsorption performance of the modified MOFs. Additionally, polymer-modified MOFs offer the advantage of regeneration and reusability. After adsorption, the adsorbed organic contaminants can be effectively removed from the polymer coatings through methods such as solvent washing or thermal treatment, allowing for the recovery of adsorption capacity and enabling the repeated use of modified MOFs [62–64].

In their pioneering work, M.L. Liu et al. [65] introduced an innovative method by incorporating electro-conductive polyaniline (PANI) into the filtration process, thereby enhancing electrocatalysis. The resulting membrane showcased remarkable advantages, including improved dye-solution permeability (204.1 L $m^{-2}h^{-1}$ bar^{-1}) achieved through a synergistic interaction of heightened hydrophilicity, expanded size, and an abundance of adsorption active sites (with an impressive 99.8% removal rate for Congo red), all while maintaining exceptional selectivity. Moreover, the membrane exhibited a substantial nearly two-orders-of-magnitude increase in electrical conductivity, leading to significantly enhanced permeability recovery (87%) and more efficient electro-oxidative removal of pollutants. Additionally, the optimal compatibility between PANI and the other polymer contributed to a robust and crack-free membrane with superior mechanical strength, enabling large-scale production without any compromise in performance. This groundbreaking approach holds great promise for advanced filtration applications in water treatment and beyond infused polyvinylidene fluoride (PVDF) membranes with zirconium 1,4-dicarboxybenzene (UiO-66), chromium (III) terephthalate (MIL-101) MOFs, and faujasite (FAU) zeolite nanocrystals [66]. As a result, the hydrophilic character of the adapted membranes was significantly affected. Moreover, the porosity of the composite membrane was notably increased. While the neat membrane had a porosity of 65%, the created composite membranes exhibited porosity ranging from 65% to 80% and enhanced the porosity, hydrophilicity, and hydrogen bonding capacity of polyethersulfone (PES) membranes during water filtration by incorporating nitrogen-doped GO [67]. This modification led to a significant improvement in rejection capacity, achieving a range of 91.1%–95.6%, evaluated to the pristine PES membrane that only exhibited 88.6% efficiency in rejecting Reactive Red 195 dye.

11.5 PHOTODEGRADATION OF ORGANIC WATER POLLUTANTS WITH 2D-BASED MOF COMPOSITES

Photodegradation happens when there is a parity between the energy levels of the incident light and the band gap energy. In such cases, electrons in the valence band absorb energy and become excited, moving to the conduction band. This occurs when the energy of the illuminating light surpasses the band gap energy. As a result, electrons create a positive hole in the valence band, leading to reduction and oxidation half-reactions involving electrons and holes, respectively. When exposed to light, 2D MOFs exhibit semiconductor-like behavior, making them potentially effective as photocatalysts. When MOFs are irradiated with light, electrons in the material become excited and subsequently undergo electron transfer. Numerous experiments conducted using MOFs for photocatalysis have yielded satisfactory results, indicating their suitability as photocatalysts for purification processes [68–70]. Here, Table 11.1 represents the photodegradation of organic pollutants found in the previous literature.

MOFs can absorb photons from light, leading to the generation of excited states. The resulting electron–hole pairs play a crucial role in subsequent photocatalytic reactions. The photogenerated electron–hole pairs participate in redox processes on the MOF surface, where electrons act as reducing agents and holes act as oxidizing agents. Through multiple reduction and oxidation reactions, this facilitates the degradation of organic contaminants. Furthermore, in the presence of oxygen molecules, photogenerated holes can react with them to produce reactive oxygen species (ROS), including hydroxyl radicals. These high ROS levels enable the immediate oxidation and degradation of organic contaminants. The combined action of ROS and photogenerated electrons leads to the breaking of chemical bonds in the organic contaminants adsorbed on the MOF surface. During this degradation process, organic molecules are transformed into smaller, less harmful compounds or mineralized into carbon dioxide and water. The photocatalytic process follows a catalytic cycle, where photogenerated electrons and holes return to their respective energy levels and participate in subsequent photocatalytic cycles. Overall, the inherent light-absorbing and photocatalytic properties of unmodified MOFs enable the efficient oxidation of organic contaminants through a series of redox reactions and the generation of reactive species.

11.5.1 Photodegradation with Non-Modified MOFs

Employing the inherent light-absorbing and photocatalytic capabilities of non-modified MOFs is the basis of photodegradation using these materials. For instance, Saidi et al. [84] investigated Zr-fumaric-based MOF in which the outcomes of the photocatalytic experiments demonstrate that

TABLE 11.1
Photodegradation of Organic Pollutants Found in the Literature

MOF	Band Gap (eV)	Degradation Efficiency (%)	Time (minutes)	Pollutant	Ref.
MIL-53(Fe)-NH_2@g-C_3N_4/PDI	1.88	78	150	Carbamazepine	[71]
MIL-88A(Fe)@g-C_3N_4@BiOI	1.77	88	180	Acid Blue 92	[72]
Cd-TCAA	1.89	81	171	MB	[73]
$[Cu_9(OH)_6Cl_2(itp)_6(bdc)_3](NO_3)_2(OH)_2$	0.62	92.5	90	RhB	[74]
UiO-66-NH_2@g-C_3N_4@BiOI	2.52	95	180	RhB	[75]
MOF-199	5.43	96	120	Basic Blue 41	[76]
MIL-125-NH_2@Zr (TiZr15)	2.74	100	90	Acetaminophen	[77]
MIL-68(In)-NH_2@GO	2.62	93	120	Amoxicillin	[78]
siloxane-based Cu-MOF	3.86	100	20	Doxorubicin	[79]
MIL-53(Fe)-NH_2@g-C_3N_4/PDI	2.72	90	60	Tetracycline	[71]
MIL-53(Fe)@$AgIO_3$	3.39	70	60	Chlorpyrifos	[80]
UiO-66@HPW, HPW: phosphotungstic acid	3.6	87.2	60	Tetracycline	[81]
UiO-66-NH_2@g-C_3N_4@BiOI	1.78	80	180	Tetracycline	[75]
(MB+BY24+BR14)@In-MOF	1.70	95.5	720	Reactive blue 21	[82]
MIL-53(Fe)@Ag/Ag_3PO_4	2.45	94.8	50	Ponceau BS	[83]

the nonmodified MOF displays efficient photodegradation of MV2B under UV light irradiation. These findings suggest that the non-modified MOF has great potential as a heterogeneous photocatalyst for degrading organic dyes. Arroussi and colleagues conducted an additional experiment utilizing a nickel-based MOF to degrade 4-nitrophenol. The MOF demonstrated an impressive 78% degradation efficiency over a period of 240 minutes [85]. Similarly, R. Li et al. [86] investigated the use of ZIF-8 MOF for tetracycline degradation, achieving a remarkable 90% efficiency within just 90 minutes. Table 11.2 represents a summary of the photodegradation of non-modified MOFs found in the previous literature.

11.5.2 Photodegradation with Doped and Metal Oxide Hybrids of MOFs

The photodegradation of organic pollutants with doped and metal oxide hybrids of MOFs combines the unique properties of doped MOFs and metal oxide nanoparticles. These hybrid materials exhibit strong light absorption across various wavelengths, thanks to the incorporation of metal oxide nanoparticles such as TiO_2 or ZnO. When light is absorbed, electron–hole pairs are created, resulting in the generation of holes in the valence band while the electrons move to the conduction band. This initiates photocatalytic reactions. The excellent charge separation and utilization made possible by the passage of photogenerated electrons and holes from metal oxide nanoparticles to doped MOF components increases photocatalytic activity. Redox reactions take place on the surface of doped MOFs and metal oxide nanoparticles, where electrons behave as reducing agents and holes as oxidizing agents, leading to the degradation and transformation of organic pollutants. ROS, including hydroxyl radicals ($\cdot OH^-$), are generated when photogenerated holes react with oxygen molecules, contributing to pollutant degradation. The high surface area and porous structure of doped MOFs enable the adsorption of organic pollutants, aiding their subsequent degradation. Organic pollutants are degraded through oxidation and reduction reactions as a result of the combined effects of electron–hole pairs, ROS, and adsorption. The synergistic effects of doping MOFs and incorporating metal oxide nanoparticles enhance photocatalytic performance and effectiveness in pollutant degradation. Overall, the photodegradation of organic pollutants with doped and metal oxide hybrids of MOFs involves a range of mechanisms, including light absorption, electron–hole pair generation, charge transfer, redox reactions, ROS generation,

TABLE 11.2
Photodegradation of Non-Modified MOFs Found in the Literature

MOF	Band Gap (eV)	Degradation Efficiency (%)	Time (minutes)	Pollutant	Ref.
Zr MOF	–	90	180	Methyl violet 2b	[84]
Ni- MOF	–	78	240	4-nitrophenol	[85]
MIL-53(Fe)	–	99	60	Methylene blue	[87]
Cu-MOF	2.2	70	60	Acid blue 92	[88]
Co- MOF	–	96.3	50	Basic fuchsin	[89]
MOF-5	2.06	95	90	Moxifloxacin	[90]
Fe-MOF	1.7	95	120	Rhodamine B	[91]
ZIF-8	2.9	90	120	Tetracycline	[86]
Ni-MOF	2.3	100	180	Rhodamine B	[92]
UiO-66-NH_2	2.8	70	180	Paraben	[93]
HKUST-1	3.02	92.4	60	Tetracycline	[94]
ZIF-67	–	97.6	40	Rhodamine B	[95]
UiO-67	–	>85	200	Methylene blue	[96]
HKUST-1	2.57	94	60	Rhodamine B	[94]
ZIF-67	2.95	>90	60	Basic Yellow 28	[97]

adsorption, and degradation, leading to efficient remediation of organic pollutants [98–102]. Metal oxide nanoparticles (ZnO and TiO_2) and Fe-based MOF nanoparticles were used [103]. By observing the photodegradation of Rhodamine B (RhB) dye for ZnO and TiO_2, as well as the removal of Ciprofloxacin antibiotic for Fe-MOF, the study attempted to evaluate the activity of these hybrid materials. The hybrid photocatalysts' capacity to be reused was also proven. Using organic coupling agents to modify the metal oxide nanoparticles' surfaces and then plasma-graft them onto polymer substrates was found to increase their photodegradation effectiveness by up to 20% in separate trials. Amini and colleagues conducted a study involving the synthesis of ZnO and TiO_2 nanoparticles (NPs). These prepared NPs were then employed for the photocatalytic degradation of various dyes, including methylene blue (MB), RhB, and Acridine orange (AO), under solar light irradiation at room temperature. The researchers observed that the photocatalytic activity of the system experienced a significant improvement when ZnO was altered by TiO_2. This enhancement was attributed to the instability and subsequent inactivation of ZnO particles due to their tendency to undergo incongruous dissolution, succeeding in the formation of $Zn(OH)_2$ on the surfaces of ZnO particles [104]. Table 11.3 shows the photodegradation of metal oxide hybrids of MOFs found in the previous literature.

11.5.3 Photodegradation with Carbon-Modified MOFs

It is crucial to modify MOFs with carbon-based materials to enhance their photocatalytic capabilities for the photodegradation of organic contaminants using carbon-modified MOFs. This process involves incorporating carbon components, which include carbon nanotubes, activated carbon, or graphene oxide, into the MOF structure. The wide surface areas and excellent light absorption properties of these carbon-based materials significantly improve the photocatalytic activity of the MOFs. When carbon-modified MOFs are exposed to light, they efficiently absorb photons from various light sources, including UV and visible light, facilitating the photocatalytic reaction. By promoting the electron movement from the valence band to the conduction band upon light absorption, the carbon-containing materials facilitate the generation of electron–hole pairs. The effective separation of these charges plays a vital role in subsequent photocatalytic reactions [117]. On the surface of carbon-modified MOFs, redox processes occur, involving the participation of photogenerated electrons and holes. The electrons act as reducing agents, while the holes act as oxidizing agents, leading to the degradation and transformation of organic pollutants. In the presence

TABLE 11.3
Photodegradation of Metal Oxide Hybrids of MOFs Found in the Literature

MOF	Band Gap (eV)	Degradation Efficiency (%)	Time (minutes)	Pollutant	Ref.
TiO_2	3.2	95	180	Rh B	[105]
CuO-ZnO	2.8	98.1	30	Acid Orange 7	[106]
ZnO/Ni0.9Zn0.1O-91	3.1	97.4	60	MB	[107]
ZrO_2	2.7	100	80	Dexamethasone	[108]
Fe_3O_4-COOH/ZIF-8/Ag/Ag_3 PO_4	3	99.7	90	Diazinon	[109]
Ag-ZnO	0.17	100	30	4-Nitrophenol	[110]
$Fe_3O_4/Co_3O_4–TiO_2$	3.10	>90	300	Tetracycline	[111]
ZnO	3.37	95	180	MB	[112]
WO3/Ag	2.7	96	300	Sulfonamides	[113]
NH_2-UiO-66 /ZnO	2.86	98	60	Malachite green	[114]
TiO_2/BN/Pd	3.09	>90	240	Acetaminophen	[115]
TiO_2@MIL-100 (Fe)	3.05	99	120	MB	[116]

of oxygen molecules (O_2) in the environment, the interaction between photogenerated holes and oxygen leads to the formation of ROS, such as hydroxyl radicals. These ROS possess a high reactivity and play a critical role in the breakdown of organic contaminants [55]. Organic contaminants undergo degradation through the coordinated actions of ROS, holes, and photogenerated electrons. Various reduction and oxidation reactions occur, leading to the breakdown of chemical bonds in the organic pollutants. This process results in either complete mineralization or the transformation of the contaminants into smaller, less harmful compounds [102].

The combination of carbon-based materials and MOFs creates synergistic effects. The photocatalytic capacity of MOFs and the adsorption power of carbon materials work together to enhance the overall efficiency of organic contaminant degradation. Carbon-modified MOFs exhibit higher efficiency and superior photocatalytic performance compared to unmodified MOFs or individual carbon-based compounds [118]. Figure 11.4 represents the photodegradation process of carbon-modified MOFs. Carbon-modified MOFs can be recycled and reused. After the photodegradation process, the degraded organic compounds can be removed from the outer layer of carbon, allowing the MOFs to undergo regeneration. This regeneration step enables carbon-modified MOFs to be used for multiple cycles of photodegradation, making them a practical and cost-effective solution for the remediation of organic pollutants [56]. Thi et al. [119] synthesized MOF with reduced graphene oxide via hydrothermal synthesis and evaluated its photocatalytic degradation efficiency for various organic dyes under 20 minutes of sunlight exposure. The results showed that the MOFs/reduced graphene oxide composite exhibited a degradation rate of over 90%, surpassing the degradation rate achieved when reduced GO was used alone. This significant enhancement in photocatalytic performance can be attributed to the synergistic effect created by the combination of MOFs and reduced graphene oxide. This synergistic effect effectively delays the recombination rate of photogenerated electron–hole pairs and maximizes charge transfer within the system, thereby achieving highly efficient photocatalytic performance. Lv et al. [107] conducted a study focusing on the enhancement of visible light absorption ability of g-C_3N_4 through the utilization of $CoFe_2O_4$/Fe_2O_3. The results demonstrated that the obtained g-C_3N_4@$CoFe_2O_4$/Fe_2O_3-2 composites exhibited remarkable performance in the degradation of various organic pollutants. Specifically, the composite materials exhibited high removal efficiencies for Tetracycline (TC) at 99.7%, Bisphenol A (BPA) at 98.1%, Sulfamethoxazole (SMX) at 94.8%, Diclofenac (DFC) at 97.0%, Ibuprofen (IBP) at 96.1%, and Ofloxacin (OFX) at 96.5% at 80 minutes reaction time. Furthermore, these composites

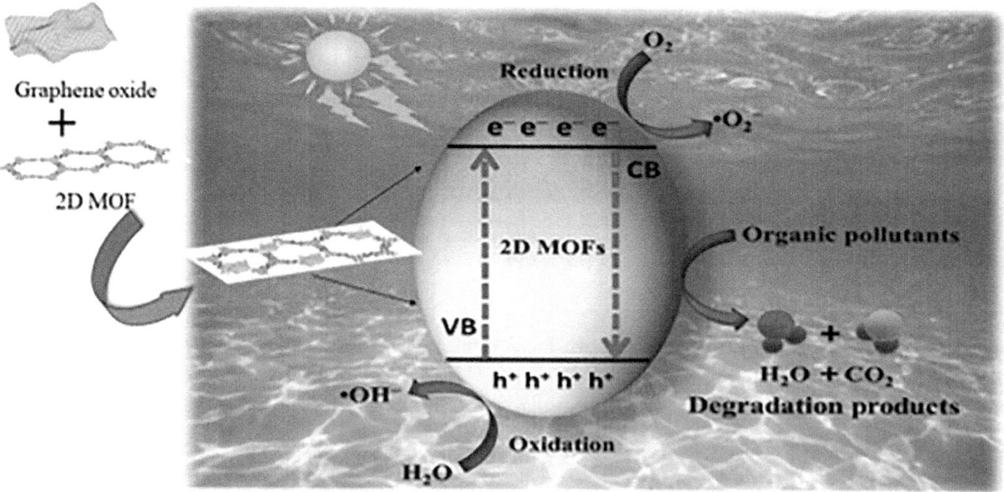

FIGURE 11.4 Schematic illustration of photodegradation with carbon-modified MOFs.

demonstrated good stability and showed great potential for practical application in degradation tests. Overall, this study provided novel insights and ideas for the advancement of photocatalytic oxidation technology in organic pollutant degradation.

11.5.4 Photodegradation with Metal Sulfides and MOFs Hybrids

The photodegradation of organic pollutants using metal sulfides and MOFs hybrids is a promising approach that combines the unique properties of metal sulfides (such as ZnS, CdS, or CuS) and MOFs. These hybrid materials exhibit enhanced photocatalytic capabilities, and their mechanism can be described in detail. Metal sulfides are incorporated into MOF, capitalizing on their semiconducting properties to facilitate efficient light absorption. These metal sulfides have the capability to effectively absorb photons across a broad spectrum, encompassing ultraviolet (UV) and visible light, when subjected to illumination [120].

Light absorption by the metal sulfides leads to the generation of electron–hole pairs. As photons are absorbed, electrons are elevated from the valence band to the conduction band, leaving positively charged holes in the valence band. This process initiates the onset of photocatalytic reactions. The transference of photogenerated electrons and holes from the metal sulfides to the MOF components plays a crucial role in the process [121]. This charge transfer facilitates efficient separation and utilization of electron–hole pairs, thereby enhancing the overall photocatalytic activity of the hybrid material. On the surface of the metal sulfides and MOFs, the photogenerated electrons and holes participate in redox reactions. The holes act as oxidizing agents, while the electrons function as reducing agents. These redox processes greatly contribute to the degradation and transformation of organic pollutants within the system. Additionally, the photogenerated holes react with oxygen molecules to generate ROS, including hydroxyl radicals (OH), in the presence of oxygen. These ROS exhibit high reactivity and play a significant role in the degradation of organic pollutants [90].

The MOF component of the hybrid material, characterized by its large surface area and porous structure, aids in the adsorption of organic contaminants. The surface of the MOF facilitates the absorption of organic molecules, thereby promoting their subsequent oxidation and transformation. Through a combination of oxidation and reduction reactions, the photogenerated electrons, holes, and ROS work together to degrade organic pollutants. This leads to the breakdown of chemical bonds in organic molecules, resulting in their conversion into less harmful compounds or complete mineralization into carbon dioxide and water [122]. One of the key advantages of this hybrid system is the synergistic relationship between metal sulfides and MOFs. While MOFs contribute to adsorption capacity, metal sulfides excel in light absorption and charge transfer. This combination enhances the overall photocatalytic activity and effectiveness in degrading organic contaminants compared to using the individual components separately. Another advantage is the regenerability and reusability of metal sulfide and MOF hybrids. Following photodegradation process, degraded organic substances can be removed from the hybrid material's surface. This allows for the regeneration of its photocatalytic activity, enabling multiple cycles of photodegradation [123]. The regenerability of these hybrids makes them environmentally friendly and economically advantageous for the elimination of organic contaminants. As a result, the photodegradation of organic contaminants using metal sulfides and MOFs hybrids involves a combination of light absorption, charge transfer, redox reactions, ROS formation, adsorption, and degradation mechanisms. The synergy between the hybrid components enhances the photocatalytic efficacy, while the ability to recycle and reuse the hybrids makes them an effective and environmentally friendly method for remediating organic pollutants [124]. H. Liang et al. [125] conducted a study where they synthesized a novel porous loofah-sponge-like ternary heterojunction material, namely g-C_3N_4/Bi_2WO_6/MoS_2 (CN-BM), using a simple method. The researchers evaluated the photocatalytic performance of the optimized sample, CN-BM2, for the degradation of sulfamethoxazole (SMX) under visible light irradiation. The results demonstrated that CN-BM2 exhibited excellent degradation efficiency, achieving over 99% degradation of SMX within 60 minutes. Moreover, the pseudo-first-order kinetic rate constant for

SMX degradation was determined to be 0.089 minutes^{-1}, which was almost three times higher compared to pure g-C_3N_4. These findings highlight the enhanced photocatalytic activity of the CN-BM2 hybrid material compared to the individual components, suggesting its potential application in the efficient degradation of organic pollutants such as SMX under visible light conditions. Gautam et al. [101] conducted a comprehensive review that focused on the synergistic utilization of metal oxides and MOFs for photocatalytic applications, specifically in the degradation of dyes and the removal of antibiotics and pharmaceutical products. The review provided a consolidated and in-depth analysis of the combined use of these materials, highlighting their potential and effectiveness in addressing environmental challenges related to dye pollution and the presence of pharmaceutical compounds.

11.6 MOFS-BASED MEMBRANE FOR THE REMOVAL OF ORGANIC POLLUTANTS FROM WASTEWATER

In this section, we will examine and compare the latest progress in eliminating Emerging Organic Contaminants (EOCs) from water, utilizing both MOFs and materials derived from MOFs. Despite being a relatively new field of study, there is a growing body of fascinating research on this topic. While the removal of organic dyes constitutes more than 75% of the documented cases, there are other concerning EOCs as well. Depending on the category of contamination, EOCs are categorized into four groups: organic dyes, Pharmaceuticals and Personal Care Products (PPCPs), herbicides and pesticides, and industrial compounds/by-products.

a. **Organic dyes**

Dye discharges constitute a significant portion of industrial waste. Annually, 800,000 tonnes of dyes, or 10%–15% of global production, enter the environment as effluent or from drying losses. Dyes are classified into various categories, such as acidic, basic, dispersal, mordant, responsive, sulfur, azoic, and vat dyes, based on their solubility, chemical characteristics, and application in the textile industry. The primary chromophores of these dyes are mainly azo (phenylazobenzene) and anthraquinone [126,127].

Azo dyes and anthraquinone are the dominant chromophores in reactive, dispersion, and vat dyes, comprising approximately 70% and 15%, respectively, of all textiles produced. While the aim of industrial progress in numerous countries is to minimize environmental impact, both production and consumption of textiles continue to increase steadily, with a staggering global annual consumption estimated at 30 million tonnes [128]. The breakdown products of many released dyes, in particular, are hazardous, carcinogenic, or mutagenic to various life forms. Furthermore, a significant number of these dye by-products persist in the environment for extended periods, with some examples like hydrolyzed reactive blue 19 having a half-life of nearly 46 years [129]. Organic dye pollution in water is a major issue that is traditionally treated by isolation or partial/full breakdown.

Researchers have explored various porous materials, such as clays, activated carbons, hydroxyapatite composites, nickel oxide nanoplates, and graphene oxides, to address the issue of dye removal from aqueous environments. Additionally, the potential of MOFs has been investigated for the elimination of different dye colors [130]. H. Li et al. [131] revealed the application of PCN-222(Zr) employed for effective dye adsorption. PCN-222(Zr) is a promising adsorbent due to its large surface area (SBET = 2330 mg gl), hexagonal 1D exposed channels having diameters up to 36, excellent stability, and significant surface area. When MB and MO dyes are together in a solution, their binding rate to PCN-222(Zr) increases by 36.8% (1239 mg g^{-1}) for MB and 73.5% (1022 mg g^{-1}) for MO. The authors tested this technique's real-world application and reusability for dye removal using a small chromatographic column. Certain MOFs when employed as catalysts display remarkable efficiency in the catalytic degradation of organic dyes. When these MOFs are exposed to radiation in the presence of H_2O_2, they can effectively decolorize a dye solution. For example, CuTz-1 exhibits a pore size of 6.4 and has the chemical formula $[Cu_8(3,5\text{-}Ph_2\text{-}tz)_6 (HSO_4)_2 (H_2O)_{5.5}]$, where 3,5-$Ph_2$-tzH represents 3,5-diphenyltriazole. When introduced to water, both MB and RhB can be completely degraded in just 8 and 22 minutes, respectively, with constant rates of photodegradation (k) measured at 0.13 and 0.19 $minute^{-1}$. Recently, researchers have investigated four environmentally friendly Fe-MOFs for the catalytic ozonation of RhB [132]. The degradation kinetics of RhB were calculated as 5.76 $minute^{-1}$ for iron-based MIL-53, 1.82 $minute^{-1}$ for iron-based MIL-88B, and 0.87 $minute^{-1}$ for iron-based MIL-100. Surprisingly, the blend of MIL-53+O_3 showed a RhB elimination rate four times higher compared to ozonation, demonstrating excellent reusability as well as stability for up to five cycles. In an alternative study, Y. Li et al. [133] utilized a composite of the Fe-MOF MIL-88B(Fe) and TiO_2 to enhance Fenton-like degradation capacity. In the presence of H_2O_2 and LED light, this composite achieved a complete destruction of 100% of the MB present in the solution after 2.5 hours.

b. **Pharmaceuticals and Personal Care Products (PPCPs)**

PPCPs constitute a unique category of emerging environmental pollutants because they can elicit physical effects in humans even at minimal concentrations. Currently, significant progress in advanced analytical methods has led to the widespread recognition of many of these micropollutants, along with their continuous introduction into the environment [134]. Due to their extensive usage, limited human metabolic breakdown, and improper disposal, residues of PPCPs are frequently detected in drinking water sources, sewage treatment plants (STPs), and wastewater treatment plants (WWTPs). Moreover, metabolites excreted from PPCPs may further evolve into secondary contaminants and undergo additional modifications once they enter receiving water bodies. PPCPs can enter the environment through various sources, including STPs, industrial facilities, hospitals, surface water runoff via fields, and soil runoff via animal farms or manure [135].

We will focus on the most promising materials, given the abundance of studies on removing PPCPs from water using MOFs, similar to the approach taken with dyes. A study modified the synthesis of Al-based MOF, MIL53(Al), for creation of a mesoporous material for water remediation. The mesoporous form had superior adsorption capacity and was almost four times faster at adsorbing triclosan (TCS) at small concentrations than the microporous MIL-53(Al). This is due to its enhanced surface area and total pore volume (SBET ~1270 vs. 1030 $m^2 g^{-1}$; Vp~ 0.73 vs. 0.51 $cm^3 g^{-1}$). The mesoporous form of MIL-53(Al) exhibits rapid adsorption, capturing 488 mg g^{-1} (99%) of TCS within approximately 10 minutes. In contrast, the microporous MIL-53(Al) requires 60 minutes to achieve a similar degree of TCS adsorption. These discrepancies in adsorption capacity and equilibrium time are likely due to the highly mesoporous structure, which facilitates accelerated mass diffusion and transfer. Following a similar approach, researchers also developed a hierarchically porous UiO66 with varying mesoporous diameters. In comparison to the microporous UiO-66, the 17.3 nm derivative of UiO-66 demonstrates a significant enhancement in PPCP removal. The adsorption capacity of TC in UiO-66-17.3 nm reaches 667 mg g^{-1}, which is 2080% higher than UiO-66@Mn and 430% higher than the value achieved using the original microporous UiO-66 [136,137].

While the majority of studies focus on adsorption for removing PPCPs from water using MOFs, there are some intriguing instances of MOFs being utilized as catalysts in PPCP degradation. One such example involves a highly water-stable Co-based MOF known as bioMOF-11(Co) ($[Co_2(ad)_2(CO_2CH_3)_2]$; ad=adeninate; S_{BET} ~ 1040 $m^2 g^{-1}$; Vp ~ 0.45 $cm^3 g^{-1}$). This MOF was effectively employed as a catalyst in the degradation of sulfachloropyridazine (SCP) and hydroxybenzoic acid (HBA) using peroxymonosulfate (PMS) [138]. Researchers used H_2O_2 as an oxidizing agent and a hydrophobic siloxane-based MOF as a catalyst to completely break down doxorubicin (DOX) after 20 minutes of sunlight exposure. The examination of the remaining water established the safety of the process, as it revealed the formation of the harmless chemical glycerol, and complete elimination of DOX. The MOF exhibited remarkable stability and efficiency in removing DOX from water, as evidenced by the absence of any metal leaching, as confirmed by X-ray fluorescence (XRF) analysis. In a recent study, MIL-101(Cr) and sodium persulfate (PS) were employed for the removal of sulfonamides from contaminated water. Initially, the combined adsorption of MIL-101(Cr) for three sulfonamides, i.e., sulfadimethoxine (SDM), sulfamonomethoxine (SMM), and SCP, was investigated, resulting in a total adsorption capacity of 340.9 mg g^{-1}. Subsequently, by introducing PS into the reaction mixture and raising the temperature to 60°C (to hasten PS disintegration), highly reactive SO_4^- radicals were generated, leading to a maximum degradation efficiency of 97.84% [139].

11.7 RECYCLABILITY PERFORMANCE AND REGENERATION OF MOFS MATERIALS

To achieve successful commercialization of innovative adsorbent materials, recyclability is a crucial requirement. During regeneration, analytes are removed from the MOF to prepare it for future adsorption or reaction sequences. The desorption process must be highly efficient, through minimal energy costs and minimal physical damage, to enable the recycling of adsorbents effectively. A good sorbent should have thermal, mechanical, and hydrolytic stability during adsorption, recycling, and multiple adsorption–desorption cycles.

Numerous studies focus on the reconstruction or regeneration of inundated or disintegrated MOFs back into their original form, which is essential for the feasibility of the procedure and the reduction of waste and operating expenses. The reconstruction process aims to recover the action of the new material after numerous recycling runs, especially when the functioning of a MOF considerably declines due to structural deterioration [140]. Desorption efficiency is influenced by several factors, including the adsorbate molecule polarity, the surface characteristics of the adsorbents, and the temperature and chemical constancy of the adsorbent. A crucial aspect of desorption efficiency in MOFs is their lower isosteric heat of adsorption, as it enhances recycling efficiency [141].

In the regeneration process, desorption parameters like temperature, pressure, and humidity play a crucial role, especially when dealing with adsorbates that have a strong affinity for adsorbents. For example, substances with high molecular weight, such as the nonpolar volatile organic compound (VOC) n-dodecane, can easily penetrate the micropores of adsorbents due to their small kinematic diameter and high boiling point (216°C). Under such circumstances, regeneration becomes particularly challenging as bulky molecules struggle to desorb easily from small micropores [142]. In the regeneration process, the adsorbate sorbed on the adsorbent must be removed using severe treatments, such as enhanced temperatures and pressures. However, these intense desorption conditions often result in a reduced amount of adsorption, limited by the adsorbate percentage that can be effectively desorbed over specific temperature and pressure conditions [143]. The effectiveness of sorbent regeneration is influenced by two main parameters: the enthalpy of adsorption and the kinetics of adsorption/desorption. To maximize energy efficiency, a critical approach is to minimize the input energy required for rejuvenation by carefully adjusting the thermodynamics of the interface between the adsorbent and adsorbate [144,145].

Regeneration methods can be broadly classified into two categories: thermal and non-thermal. Thermal methods encompass conventional techniques such as heating with hot inert gases and steam, as well as unconventional methods like heating with microwave, UV radiation, or electric currents [146]. Each regeneration approach comes with its own set of advantages and disadvantages, making them more suitable depending on the specific limitations of the process at hand [147]. For example, some methods may be time- or energy-intensive, while others may lead to irreversible alterations in the sorbents' structure or their chemical and physical properties [148]. Table 11.4 presented a compilation of MOFs that were regenerated using various methods.

TABLE 11.4
Compilation of MOFs Regenerated Using Various Methods

Regeneration Technique	MOF	Conditions	Ref.
Thermal regeneration	MOF-74	The ability for CO_2 adsorption remained unchanged over ten cycles.	[151]
Temperature swing adsorption (TSA) regeneration	UTSA-280	Under He flow at 80°C	[152]
Vacuum thermal swing adsorption (VTSA) regeneration	TMU	–	[153]
	UiO-66	Under a 42 mT field, 100% CO_2 desorption is achieved	[154]
Magnetic induction swing adsorption (MISA) regeneration	Mg-MOF-74	–	[155]
	ZIF-67	When exposed to UV light with a wavelength between 300 and 650 nm and a temperature of 127°C	[156]
Light-induction swing adsorption (LISA) regeneration	PCN-250	UV light exposure	[157]
	Cu-FINA-1 and Cu-FINA-2	Helium gas purging at room temperature	[158]
Non-thermal regeneration	MUF-16	Dry air purging at a temperature of 20°C	[159]
Purging with inert gas	MOF-801	NaOH (0.1M) for 6 hours	[160]
Solvent or solution treatment regeneration	MIL-101(Cr)-SO_3H	0.1 M HCl washing at 40°C	[161]
VPSA	HKUST-1, and MIL-53(Al)	The adsorption pressure is inputted and the desorption pressure is lower than the pressure of the surrounding environment	[162]
PSA	Cu-BTC	–	[163]
VSA	MUF-15	Helium purging at 70°C	[163]

a. **Thermal regeneration**

This method comprises heating saturated MOFs to deliver the necessary energy for desorbing the reserved adsorbates. Several techniques use this approach with trivial variations, mainly due to the variable behavior of adsorbates on the MOF, which governs the system's response. These adsorbates have been categorized in various publications based on the behavior once held on the MOF and then subjected to heated inert gas. In conventional thermal regeneration, the adsorbent is heated for a predetermined duration, trailed by dampening with an inert gas (such as He, Ar, or N_2) for purging purposes. For example, Cu-BTC honeycomb monoliths have been heated to temperatures ranging from 120°C to 200°C for at least 8 hours [149].

b. **Non-thermal regeneration**

Non-thermal regeneration methods remove adsorbed material from MOFs without heating, unlike thermal regeneration. This is advantageous as it reduces the consumption of energy. Non-thermal regeneration methods include chemical regeneration, purging of inert gas, vacuum swing adsorption, pressure swing adsorption, and vacuum pressure swing adsorption [150].

11.8 PROGRESS IMPLEMENTATION OF MOFS COMMERCIAL ASPECTS IN WASTEWATER TREATMENT

As a result of the extensive efforts invested in the development of MOFs, there are now over 20,000 distinct varieties of MOFs available for various applications. Recently, numerous research teams have been emphasizing the significance of utilizing MOFs in practical applications to address their limitations in WWT. These limitations include reduced structural stability, dispersibility, and solubility. Furthermore, when used in columns, especially in WWT, their exceptionally small particle sizes and insufficient mechanical durability have posed challenges. Therefore, to effectively meet the requirements of industrial waste management, substantial research is still necessary either to modify their structure or to develop alternative designed solutions.

Considerable endeavors have been undertaken in the quest to produce diverse types of porous materials. Consequently, extra water-soluble as well as stable MOFs are engineered, expanding their potential applications. Despite these advancements, a broad spectrum of biological and non-biological sectors has not yet experienced significant benefits from the implementation of this unique class of MOFs. To improve the constancy of various MOFs with different water sorption capabilities, several functionalization techniques have been employed successfully, altering their chemical adsorption characteristics. Within this category, there are several approaches that can be employed, including incorporating Keggin polyoxometalates (POMs) into the MOFs pores (e.g., Cu-MOF), conducting numerous post-synthetic alterations and replacing either the organic linker or the metal in the secondary building units (SBU). The MOF community should give due consideration to the following aspects to enhance the applicability of MOFs in WWT: (i) Focus on achieving enhanced selectivity and the sorption process for specific toxic ions that occur at a fast rate. (ii) Construct MOFs with particle dimensions that permit the unobstructed flow of waste through MOF-based membranes, columns, or thin films. (iii) Ensure decent mechanical potency to endure extreme water pressures. To effectively increase the practical use of MOFs, it is crucial to concentrate on these qualitative elements in laboratory research, with the ultimate goal of scaling up these materials for large-scale WWT applications. In future studies, it is essential to evaluate the execution of water-stable MOFs as well as their complexes, analyze the pH dependency of absorption, and assess selectivity in the presence of competitive components. This evaluation can be achieved through a novel integrated approach that combines kinetic and thermodynamic absorption investigations [164,165].

11.9 CONCLUSION AND FUTURE OUTLOOK

One of the prominent global environmental concerns is the presence of organic pollutants, especially persistent organic pollutants (POPs), in ecosystems. Extensive research indicates a significant rise in the production and usage of these pollutants in recent decades, posing a serious threat to the environment. As a result, there is a strong focus on developing effective methods to eliminate highly hazardous organic chemicals from water and wastewater. Among the available techniques, adsorption and photodegradation have emerged as recognized and efficient methods for removing organic contaminants from water. These methods offer not only high efficiency but also affordability, resulting in the production of high-quality treated effluent. This chapter provides a concise overview of the two primary synthesis methods used to create nanocomposites for this purpose. The top-down method involves physical and chemical exfoliation techniques, while the bottom-up method encompasses three-layer synthesis, interfacial synthesis, modulated synthesis, surfactant-assisted synthesis, and sonication synthesis. Each method offers distinct advantages and enables the fabrication of well-defined 2D MOF nanocomposites. Additionally, the chapter delves into the underlying principles of adsorption and photodegradation exhibited by MOFs, highlighting various 2D MOF-based hybrids that effectively remove organic pollutants from water. It concludes by emphasizing the potential applications of 2D MOF nanocomposites in eliminating organic pollutants from water, considering their mechanisms of adsorption and photodegradation. In summary, this chapter provides a comprehensive overview of the synthesis methods used to produce 2D MOF nanocomposites. It discusses the mechanisms of adsorption and photodegradation, showcasing the potential of these nanocomposites in effectively removing organic pollutants from water with various MOF-based hybrids. Furthermore, the chapter explores the usage of MOF membranes for the removal of organic pollutants, considering the recyclability and regeneration performance of MOF materials. Finally, the chapter concludes with the progress and implementation of MOFs in commercial aspects of WWT.

REFERENCES

1. M.F. Gasim, Z.Y. Choong, P.L. Koo, S.C. Low, M.H. Abdurahman, Y.C. Ho, M. Mohamad, I.W.K. Suryawan, J.W. Lim, W. Da Oh, Application of biochar as functional material for remediation of organic pollutants in water: An overview, *Catalysts*, 12 (2022) p. 210.
2. Y. Wen, G. Schoups, N. Van De Giesen, Organic pollution of rivers: Combined threats of urbanization, livestock farming and global climate change, *Scientific Reports*, 7 (2017) pp. 1–9.
3. P. Shumbula, C. Maswanganyi, N. Shumbula, P. Shumbula, C. Maswanganyi, N. Shumbula, Type, sources, methods and treatment of organic pollutants in wastewater, BoD—Books on Demand (2021) pp. 1–90. doi: 10.5772/intechopen.101347.
4. M. Arif, U. Fatima, A. Rauf, Z.H. Farooqi, M. Javed, M. Faizan, S. Zaman, A new 2D metal-organic framework for photocatalytic degradation of organic dyes in water, *Catalysts*, 13 (2023) p. 231.
5. M. Hossain, X. Duan, B. Zhao, D. Shen, Z. Zhang, P. Lu, J. Li, B. Li, B. Zhao, D.Y. Shen, Z.C. Zhang, P. Lu, M. Hossain, J. Li, B. Li, X.D. Duan, 2D Metallic Transition-metal dichalcogenides: Structures, synthesis, properties, and applications, Wiley Online Library, 31 (2021). doi: 10.1002/adfm.202105132.
6. S. Roy, X. Zhang, A.B. Puthirath, A. Meiyazhagan, S. Bhattacharyya, M.M. Rahman, G. Babu, S. Susarla, S.K. Saju, M.K. Tran, L.M. Sassi, M.A.S.R. Saadi, J. Lai, O. Sahin, S.M. Sajadi, B. Dharmarajan, D. Salpekar, N. Chakingal, A. Baburaj, X. Shuai, A. Adumbumkulath, K.A. Miller, J.M. Gayle, A. Ajnsztajn, T. Prasankumar, V.V.J. Harikrishnan, V. Ojha, H. Kannan, A.Z. Khater, Z. Zhu, S.A. Iyengar, P.A. da S. Autreto, E.F. Oliveira, G. Gao, A.G. Birdwell, M.R. Neupane, T.G. Ivanov, J. Taha-Tijerina, R.M. Yadav, S. Arepalli, R. Vajtai, P.M. Ajayan, Structure, properties and applications of two-dimensional hexagonal boron nitride, *Advanced Materials*, 33 (2021) p. e2101589.
7. X. Gao, P. Wang, Z. Pan, J.P. Claverie, J. Wang, Recent progress in two-dimensional layered double hydroxides and their derivatives for supercapacitors, *ChemSusChem*, 13 (2020) pp. 1226–1254.
8. X. Zhang, X. Yuan, L. Jiang, J. Zhang, H. Yu, H. Wang, G. Zeng, Powerful combination of 2D g-C_3N_4 and 2D nanomaterials for photocatalysis: Recent advances, *Chemical Engineering Journal*, 390 (2020) p. 124475.
9. V. Nicolosi, M. Chhowalla, M.G. Kanatzidis, M.S. Strano, J.N. Coleman, Liquid exfoliation of layered materials, *Sciences*, 340 (2013) p. 1226419.
10. H. Zhang, Ultrathin two-dimensional nanomaterials, *ACS Nano*, 9 (2015) pp. 9451–9469.
11. F. Song, X. Hu, Exfoliation of layered double hydroxides for enhanced oxygen evolution catalysis, *Nature Communications*, 5 (2014) pp. 1–9.
12. H. Li, Z. Song, X. Zhang, Y. Huang, S. Li, Y. Mao, H.J. Ploehn, Y. Bao, M. Yu, Ultrathin, molecular-sieving graphene oxide membranes for selective hydrogen separation, *Science*, 342 (2013) pp. 95–98.
13. Z. Chen, Y. Li, Y. Cai, S. Wang, B. Hu, B. Li, X. Ding, L. Zhuang, X. Wang, Application of covalent organic frameworks and metal–organic frameworks nanomaterials in organic/inorganic pollutants removal from solutions through sorption-catalysis strategies, *Carbon Research*, 2 (2023). doi: 10.1007/s44246-023-00041-9.
14. J. Shan, J. Liao, C. Ye, J. Dong, Y. Zheng, S.Z. Qiao, The dynamic formation from metal-organic frameworks of high-density platinum single-atom catalysts with metal-metal interactions, *Angewandte Chemie - International Edition*, 61 (2022) p. e202213412.
15. Z. Wang, G. Wang, H. Qi, M. Wang, M. Wang, S.W. Park, H. Wang, M. Yu, U. Kaiser, A. Fery, S. Zhou, R. Dong, X. Feng, Ultrathin two-dimensional conjugated metal–organic framework single-crystalline nanosheets enabled by surfactant-assisted synthesis, *Chemical Science*, 11 (2020) pp. 7665–7671.
16. Y. Peng, W. Yang, Metal-organic framework nanosheets: A class of glamorous low-dimensional materials with distinct structural and chemical natures, *Science China Chemistry*, 62 (2019) pp. 1561–1575.
17. T. Rodenas, I. Luz, G. Prieto, B. Seoane, H. Miro, A. Corma, F. Kapteijn, F.X. Llabrés I Xamena, J. Gascon, Metal-organic framework nanosheets in polymer composite materials for gas separation, *Nature Materials*, 14 (2015) pp. 48–55.
18. Y. Peng, Y. Li, Y. Ban, H. Jin, W. Jiao, X. Liu, W. Yang, Metal- organic framework nanosheets as building blocks for molecular sieving membranes, *Science*, 346 (2014) pp. 1356–1359.
19. M. Zhao, Q. Lu, Q. Ma, H. Zhang, Two-dimensional metal–organic framework nanosheets, *Small Methods*, 1 (2017) p. 1600030.
20. W. Zhu, L. Wang, H. Cao, R. Guo, C.W, Introducing defect-engineering 2D layered MOF nanosheets into Pebax matrix for CO_2/CH_4 separation, *Journal of Membrane Science*, 669 (2023) p. 121305.
21. C.L. Storedot, I. Hussain, H. Rabiee, O.R. Ogunsakin, C. Lamiel, K. Zhang, Metal-organic framework-derived transition metal chalcogenides (S, Se, and Te): Challenges, recent progress, and future directions in electrochemical energy storage, *Coordination Chemistry Reviews*, 480 (2023) p. 215030.

22. J. Sun, L. Dai, F. Yao, H. Zhao, J. Bi, W. Xue, J. Deng, C. Fang, Y. Fu, J. Zhu, Poly (triazine imide) ligand based 2D metal coordination polymers: Design, synthesis and application in electrocatalytic water oxidation, *Electrochim Acta*, 401 (2022) p. 139463.
23. B. Garai, A. Mallick, A. Das, R. Mukherjee, R. Banerjee, Self-exfoliated metal-organic nanosheets through hydrolytic unfolding of metal-organic polyhedra, *Chemistry - A European Journal*, 23 (2017) pp. 7361–7366.
24. Y. Zheng, F.Z. Sun, X. Han, J. Xu, X.H. Bu, Recent progress in 2D metal-organic frameworks for optical applications, *Advanced Optical Materials*, 8 (2020) p. 2000110.
25. U. Khan, A. Nairan, J. Gao, Q. Zhang, Current progress in 2D metal–organic frameworks for electrocatalysis, *Small Structures*, 4 (2023) p. 2200109.
26. T. Ohata, A. Nomoto, T. Watanabe, I. Hirosawa, T. Makita, J. Takeya, R. Makiura, Uniaxially oriented electrically conductive metal-organic framework nanosheets assembled at air/liquid interfaces, *ACS Applied Materials & Interfaces*, 13 (2021) pp. 54570–54578.
27. D. Sheberla, L. Sun, M. Blood-Forsythe, S. Leyman Er, C.R. Wade, C.K. Brozek, A. Alán Aspuru-Guzik, M. Dincă, High electrical conductivity in Ni_3 (2, 3, 6, 7, 10, 11-2 hexaiminotriphenylene)$_2$, a Semiconducting metal– organic graphene analogue, *Journal of the American Chemical Society*, 136 (2014) pp. 8859–8862.
28. K. Zhao, W. Zhu, S. Liu, X. Wei, G. Ye, Y. Su, Z. He, Two-dimensional metal–organic frameworks and their derivatives for electrochemical energy storage and electrocatalysis, *Nanoscale Advances*, 2 (2020) pp. 536–562.
29. M.K. Lee, M. Shokouhimehr, S.Y. Kim, H.W. Jang, Two-dimensional metal–organic frameworks and covalent–organic frameworks for electrocatalysis: Distinct merits by the reduced dimension, *Advanced Energy Materials*, 12 (2022) p. 2003990.
30. G. Chakraborty, I.H. Park, R. Medishetty, J.J. Vittal, Two-dimensional metal-organic framework materials: Synthesis, structures, properties and applications, *Chemical Reviews*, 121 (2021) pp. 3751–3891.
31. W. Wang, Y. Yu, Y. Jin, X. Liu, M. Shang, X. Zheng, T. Liu, Z. Xie, Two-dimensional metal-organic frameworks: From synthesis to bioapplications, *Journal of Nanobiotechnology*, 20 (2022) pp. 1–17.
32. Y. Zhou, P. Yan, S. Zhang, Y. Zhang, H. Chang, X. Zheng, J. Jiang, Q. Xu, CO_2 coordination-driven top-down synthesis of a 2D non-layered metal–organic framework, *Fundamental Research*, 2 (2022) pp. 674–681.
33. Y. Peng, Y. Li, Y. Ban, W. Yang, Two-dimensional metal–organic framework nanosheets for membrane-based gas separation, *Angewandte Chemie*, 129 (2017) pp. 9889–9893.
34. X. Wang, C. Chi, K. Zhang, Y. Qian, K.M. Gupta, Z. Kang, J. Jiang, D. Zhao, Reversed thermo-switchable molecular sieving membranes composed of two-dimensional metal-organic nanosheets for gas separation, *Nature Communications*, 8 (2017) p. 14460.
35. Y. Pan, R. Abazari, J. Yao, J. Gao, Recent progress in 2D metal-organic framework photocatalysts: Synthesis, photocatalytic mechanism and applications, *Journal of Physics: Energy*, 3 (2021) p. 032010.
36. Y. Ding, Y.P. Chen, X. Zhang, L. Chen, Z. Dong, H.L. Jiang, H. Xu, H.C. Zhou, Controlled intercalation and chemical exfoliation of layered metal-organic frameworks using a chemically labile intercalating agent, *Journal of the American Chemical Society*, 139 (2017) pp. 9136–9139.
37. W. Liu, R. Yin, X. Xu, L. Zhang, W. Shi, X. Cao, W. Liu, R. Yin, X. Xu, L. Zhang, X. Cao, W. Shi, Structural engineering of low-dimensional metal–organic frameworks: Synthesis, properties, and applications, Wiley Online Library, 6 (2019). doi: 10.1002/advs.201802373.
38. Z. Wang, L.S. Walter, M. Wang, P. Petkov, B. Liang, H. Qi, Nguyen Ngan Nguyen, M. Hambsch, H. Zhong, M. Wang, S.-W. Park, L. Renn, K. Watanabe, T. Taniguchi, Stefan, T. Heine, U. Kaiser, S. Zhou, R. Thomas Weitz, X. Feng, Interfacial synthesis of layer-oriented 2D conjugated metal–organic framework films toward directional charge transport, *ACS Publications*, 34 (2021) pp. 13624–13632.
39. J.E. Chen, Z.J. Yang, H.U. Koh, J. Shen, Y. Cai, Y. Yamauchi, L.H. Yeh, V. Tung, K.C.W. Wu, Current progress and scalable approach toward the synthesis of 2D metal–organic frameworks, *Advanced Materials Interfaces*, 9 (2022) p. 2102560.
40. J. Yang, S. Xiao, X. Cui, W. Dai, X. Lian, Z. Hao, Y. Zhao, J.-S. Pan, Y. Zhou, L. Wang, W. Chen, Inorganic-anion-modulated synthesis of 2D nonlayered aluminum-based metal-organic frameworks as carbon precursor for capacitive sodium ion storage, *Energy Storage Materials*, 26 (2020) pp. 391–399.
41. M. Xu, S. Yuan, X.Y. Chen, Y.J. Chang, G. Day, Z.Y. Gu, H.C. Zhou, Two-dimensional metal-organic framework nanosheets as an enzyme inhibitor: Modulation of the α-chymotrypsin activity, *Journal of the American Chemical Society*, 139 (2017) pp. 8312–8319.
42. Z. Hu, Y. Wang, D. Zhao, Modulated hydrothermal chemistry of metal-organic frameworks, *Accounts of Materials Research*, 3 (2022) pp. 1106–1114.

43. S. Wu, L. Qin, K. Zhang, Z. Xin, S. Zhao, Ultrathin 2D metal–organic framework nanosheets prepared via sonication exfoliation of membranes from interfacial growth and exhibition of enhanced, *RSC Advances*, 9 (2019) pp. 9386–9391.
44. S. Khan, M. Shahid, Throwing light on the current developments of two-dimensional metal–organic framework nanosheets (2D MONs), *Materials Advances*, 2 (2021) pp. 4914–4944.
45. R. Zhao, X. Shi, T. Ma, H. Rong, Z. Wang, F. Cui, G. Zhu, C. Wang, Constructing mesoporous adsorption channels and MOF-polymer interfaces in electrospun composite fibers for effective removal of emerging organic contaminants, *ACS Applied Materials & Interfaces*, 13 (2021) pp. 755–764.
46. M. Beydaghdari, F.H. Saboor, A. Babapoor, M. Asgari, Recent progress in adsorptive removal of water pollutants by metal-organic frameworks, *ChemNanoMat*, 8 (2022). doi: 10.1002/cnma.202100400.
47. F. Jeremias, V. Lozan, S. Henninger, C. Janiak, Programming MOFs for water sorption: amino-functionalized MIL-125 and UiO-66 for heat transformation and heat storage applications, *Dalton Transactions*, 42 (2013) p. 15967.
48. T. Devic, F. Salles, S. Bourrelly, B. Moulin, G. Maurin, P. Horcajada, C. Serre, A. Vimont, J.-C. Lavalley, H. Leclerc, G. Clet, M. Daturi, P.L. Llewellyn, Y. Filinchuk, G. Férey, Effect of the organic functionalization of flexible MOFs on the adsorption of CO_2, *Journal of Materials Chemistry*, 22 (2012) p. 10266.
49. J. Canivet, A. Fateeva, Y. Guo, B. Coasne, D. Farrusseng, Water adsorption in MOFs: Fundamentals and applications (2013) pp. 1–3. doi: 10.1039/c4cs00078a.
50. X. Huang, Y. Ma, H. Lai, Q. Jia, L. Zhu, J. Liu, B. Quan, Conductive substrates-based component tailoring via thermal conversion of metal organic framework for enhanced microwave absorption performances, *Journal of Colloid and Interface Science*, 608 (2022) pp. 1323–1333.
51. J. Huang, W. Luo, X. Yuan, J. Wang, Bimetallic MOF-derived oxides modified $BiVO_4$ for enhanced photoelectrochemical water oxidation performance, *Journal of Alloys and Compounds*, 930 (2023) pp. 167397.
52. M. Li, K. Huang, J.A. Schott, Z. Wu, S. Dai, Effect of metal oxides modification on CO_2 adsorption performance over mesoporous carbon, *Microporous and Mesoporous Materials*, 249 (2017) pp. 34–41.
53. C. Qin, B. Wang, N. Wu, C. Han, Y. Wang, General strategy to fabricate porous co-based bimetallic metal oxide nanosheets for high-performance CO sensing, *ACS Appl Mater Interfaces*, 13 (2021) pp. 26318–26329.
54. C.P. Krap, R. Newby, A. Dhakshinamoorthy, H. García, I. Cebula, T.L. Easun, M. Savage, J.E. Eyley, S. Gao, A.J. Blake, W. Lewis, P.H. Beton, M.R. Warren, D.R. Allan, M.D. Frogley, C.C. Tang, G. Cinque, S. Yang, M. Schröder, enhancement of CO_2 adsorption and catalytic properties by Fe-doping of $[Ga_2(OH)_2(L)]$ (H_4L=Biphenyl-3,3′,5,5′-tetracarboxylic Acid), MFM-300(Ga_2), *Inorganic Chemistry*, 55 (2016) pp. 1076–1088.
55. Y. Gong, Y. Wang, H. Zhang, X. Zhao, K. Xiao, T. Guo, J. Zhang, H. Shao, Y. Wang, G. Yu, MOF-derived nitrogen doped carbon modified g-C_3N_4 heterostructure composite with enhanced photocatalytic activity for bisphenol A degradation with peroxymonosulfate under visible light irradiation, *Applied Catalysis B-Environmental*, 233 (2018) pp. 35–45.
56. G. Xu, Y. Zhang, D. Peng, D. Sheng, Y. Zhang, Y. Tian, D. Ma, MOF derived carbon modified porous TiO_2 mixed-phase junction with efficient visible-light photocatalysis for cyclohexane oxidation, *Materials Research Bulletin*, 146 (2022) p. 111602.
57. H. Ye, W. Na, W. Gao, H. Wang, Carbon-modified CuO/ZnO catalyst with high oxygen vacancy for CO_2 hydrogenation to methanol, *Energy Technology*, 8 (2020). doi: 10.1002/ente.202000194.
58. G. Zeng, Z. Yu, M. Du, N. Ai, W. Chen, Z. Gu, B. Chen, Enhanced CO_2 adsorption on activated carbon-modified HKUST-1 composites, *ChemistrySelect*, 3 (2018) pp. 11601–11605.
59. M. Li, W. Huang, B. Tang, F. Song, A. Lv, X. Ling, Preparation of a composite material AC/Cu-BTC with improved water stability and n-hexane vapor adsorption, *Journal of Nanomaterials* (2019). doi: 10.1155/2019/5429264.
60. Y. Sun, M. Ma, B. Tang, S. Li, L. Jiang, Graphene modified Cu-BTC with high stability in water and controllable selective adsorption of various gases, *Journal of Alloys and Compounds*, 808 (2019) p. 151721.
61. Y. Li, J.-P. Miao, X. Sun, J. Xiao, Y. Li, H. Wang, J. Xiao, Z. Li, Mechanochemical synthesis of Cu-BTC@GO with enhanced water stability and toluene adsorption capacity, *Chemical Engineering Journal*, 298 (2016) pp. 191–197.
62. A.R. Abbasi, M. Yousefshahi, K. Daasbjerg, Non-enzymatic electroanalytical sensing of glucose based on nano nickel-coordination polymers-modified glassy carbon electrode, *Journal of Inorganic and Organometallic Polymers and Materials*, 30 (2020) pp. 2027–2038.

63. D.W. Kang, M. Kang, D.W. Kim, H. Kim, Y.H. Lee, H. Yun, J.H. Choe, C.S. Hong, Engineered removal of trace NH_3 by porous organic polymers modified via sequential post-sulfonation and post-alkylation, *Advanced Sustainable Systems*, 5 (2021). doi: 10.1002/adsu.202000161.
64. Y. Belmabkhout, A. Sayari, Isothermal versus non-isothermal adsorption-desorption cycling of triamine-grafted pore-expanded MCM-41 mesoporous silica for CO_2 capture from flue gas, *Energy and Fuels*, 24 (2010) pp. 5273–5280.
65. M.L. Liu, L. Li, Y.X. Sun, Z.J. Fu, X.L. Cao, S.P. Sun, Scalable conductive polymer membranes for ultrafast organic pollutants removal, *Journal of Membrane Science*, 617 (2021), p. 118644.
66. M. Dehghankar, T. Mohammadi, M.T. Moghadam, M.A. Tofighy, Metal-organic framework/zeolite nanocrystal/polyvinylidene fluoride composite ultrafiltration membranes with flux/antifouling advantages, *Materials Chemistry and Physics*, 260 (2021) p. 124128.
67. V. Vatanpour, S.S. Mousavi Khadem, A. Dehqan, M.A. Al-Naqshabandi, M.R. Ganjali, S. Sadegh Hassani, M.R. Rashid, M.R. Saeb, N. Dizge, Efficient removal of dyes and proteins by nitrogen-doped porous graphene blended polyethersulfone nanocomposite membranes, *Chemosphere*, 263 (2021) p. 127892.
68. M. Liu, P. Ye, M. Wang, L. Wang, C. Wu, J. Xu, Y. Chen, 2D/2D Bi-MOF-derived $BiOCl/MoS_2$ nanosheets S-scheme heterojunction for effective photocatalytic degradation, *Journal of Environmental Chemical Engineering*, 10 (2022) p. 108436.
69. M. Govarthanan, R. Mythili, W. Kim, S.A. Al-Farraj, Sulaiman Ali Alharbi, Facile fabrication of (2D/2D) MoS_2@MIL-88(Fe) interface-driven catalyst for efficient degradation of organic pollutants under visible light irradiation, *Journal of Hazardous Materials*, 414 (2021) p. 125522.
70. Y. Qian, F. Zhang, H. Pang, A review of MOFs and their composites-based photocatalysts: Synthesis and applications, *Advanced Functional Materials*, 31 (2021) p. 2104231.
71. Y. Li, Y. Fang, Z. Cao, N. Li, D. Chen, Q. Xu, J. Lu, Construction of g-C_3N_4/PDI@MOF heterojunctions for the highly efficient visible light-driven degradation of pharmaceutical and phenolic micropollutants, *Applied Catalysis B: Environmental*, 250 (2019) pp. 150–162.
72. S. Gholizadeh Khasevani, M.R. Gholami, Engineering a highly dispersed core@shell structure for efficient photocatalysis: A case study of ternary novel BiOI@MIL-88A(Fe)@g-C_3N_4 nanocomposite, *Materials Research Bulletin*, 106 (2018) pp. 93–102.
73. L. Xia, J. Ni, P. Wu, J. Ma, L. Bao, Y. Shi, J. Wang, Photoactive metal–organic framework as a bifunctional material for 4-hydroxy-4′-nitrobiphenyl detection and photodegradation of methylene blue, *Dalton Transactions*, 47 (2018) pp. 16551–16557.
74. T.-R. Zheng, L.-L. Qian, M. Li, Z.-X. Wang, K. Li, Y.-Q. Zhang, B.-L. Li, B. Wu, A bifunctional cationic metal–organic framework based on unprecedented nonanuclear copper(II) cluster for high dichromate and chromate trapping and highly efficient photocatalytic degradation of organic dyes under visible light irradiation, *Dalton Transactions*, 47 (2018) pp. 9103–9113.
75. Q. Liang, S. Cui, J. Jin, C. Liu, S. Xu, C. Yao, Z. Li, Fabrication of BiOI@UIO-66(NH_2)@g-C_3N_4 ternary Z-scheme heterojunction with enhanced visible-light photocatalytic activity, *Applied Surface Science*, 456 (2018) pp. 899–907.
76. N.M. Mahmoodi, J. Abdi, Nanoporous metal-organic framework (MOF-199): Synthesis, characterization and photocatalytic degradation of Basic Blue 41, *Microchemical Journal*, 144 (2019) pp. 436–442.
77. A. Gómez-Avilés, M. Peñas-Garzón, J. Bedia, D.D. Dionysiou, J.J. Rodríguez, C. Belver, Mixed Ti-Zr metal-organic-frameworks for the photodegradation of acetaminophen under solar irradiation, *Applied Catalysis B: Environmental*, 253 (2019) pp. 253–262.
78. C. Yang, X. You, J. Cheng, H. Zheng, Y. Chen, A novel visible-light-driven In-based MOF/graphene oxide composite photocatalyst with enhanced photocatalytic activity toward the degradation of amoxicillin, *Applied Catalysis B: Environmental*, 200 (2017) pp. 673–680.
79. C. Racles, M.F. Zaltariov, M. Silion, A.M. Macsim, V. Cozan, Photo-oxidative degradation of doxorubicin with siloxane MOFs by exposure to daylight, *Environmental Science and Pollution Research*, 26 (2019) pp. 19684–19696.
80. A.A. Oladipo, R. Vaziri, M.A. Abureesh, Highly robust $AgIO_3$/MIL-53 (Fe) nanohybrid composites for degradation of organophosphorus pesticides in single and binary systems: Application of artificial neural networks modelling, *Journal of the Taiwan Institute of Chemical Engineers*, 83 (2018) pp. 133–142.
81. S. Subudhi, S. Mansingh, G. Swain, A. Behera, D. Rath, K. Parida, HPW-anchored UiO-66 metal-organic framework: A promising photocatalyst effective toward tetracycline hydrochloride degradation and H_2 evolution via Z-scheme charge dynamics, *Inorganic Chemistry*, 58 (2019) pp. 4921–4934.
82. Q. Li, Z.-L. Fan, D.-X. Xue, Y.-F. Zhang, Z.-H. Zhang, Q. Wang, H.-M. Sun, Z. Gao, J. Bai, A multi-dye@MOF composite boosts highly efficient photodegradation of an ultra-stubborn dye reactive blue 21 under visible-light irradiation, *Journal of Materials Chemistry A*, 6 (2018) pp. 2148–2156.

83. F.A. Sofi, K. Majid, Enhancement of the photocatalytic performance and thermal stability of an iron based metal–organic-framework functionalised by Ag/Ag_3PO_4, *Materials Chemistry Frontiers*, 2 (2018) pp. 942–951.
84. M. Saidi, A. Benomara, M. Mokhtari, L. Boukli-Hacene, Sonochemical synthesis of Zr-fumaric based metal-organic framework (MOF) and its performance evaluation in methyl violet 2B decolorization by photocatalysis, *Reaction Kinetics, Mechanisms and Catalysis*, 131 (2020) pp. 1009–1021.
85. A. Arroussi, H. Gaffour, M. Mokhtari, L. Boukli-Hacene, Investigating metal-organic framework based on nickel (II) and benzene 1,3,5-tri carboxylic acid (H3BTC) as a new photocatalyst for degradation of 4-nitrophenol, *International Journal of Environmental Studies*, 77 (2020) pp. 137–151.
86. R. Li, W. Li, C. Jin, Q. He, Y. Wang, Fabrication of ZIF-8@TiO_2 micron composite via hydrothermal method with enhanced absorption and photocatalytic activities in tetracycline degradation, *Journal of Alloys and Compounds*, 825 (2020) p. 154008.
87. J.J. Du, Y.P. Yuan, J.X. Sun, F.M. Peng, X. Jiang, L.G. Qiu, A.J. Xie, Y.H. Shen, J.F. Zhu, New photocatalysts based on MIL-53 metal-organic frameworks for the decolorization of methylene blue dye, *Journal of Alloys and Compounds*, 190 (2011) pp. 945–951.
88. E. Akbarzadeh, H.Z. Soheili, M. Hosseinifard, M.R. Gholami, Preparation and characterization of novel Ag_3VO_4/Cu-MOF/rGO heterojunction for photocatalytic degradation of organic pollutants, *Materials Research Bulletin*, 121 (2020) p. 110621.
89. X.Y. Lu, M.L. Zhang, Y.X. Ren, J.J. Wang, X.G. Yang, Design of Co-MOF nanosheets for efficient adsorption and photocatalytic degradation of organic dyes, *Journal of Molecular Structure*, 1288 (2023) p. 135796.
90. A. Anum, M.A. Nazir, S.M. Ibrahim, S.S.A. Shah, A.A. Tahir, M. Malik, M.A. Wattoo, A. ur Rehman, Synthesis of Bi-metallic-sulphides/MOF-5@graphene oxide nanocomposites for the removal of hazardous moxifloxacin, *Catalysts*, 13 (2023) p. 984.
91. K. Vinothkumar, S. Jyothi, C. Lavanya, M. Sakar, S. Valiyaveettil, R. Geetha Balakrishna, Strongly co-ordinated MOF-PSF matrix for selective adsorption, separation and photodegradation of dyes, *Chemical Engineering Journal*, 428 (2022) pp. 1385–8947.
92. Z. Sun, X. Wu, K. Qu, Z. Huang, S. Liu, M. Dong, Z. Guo, Bimetallic metal-organic frameworks anchored corncob-derived porous carbon photocatalysts for synergistic degradation of organic pollutants, *Chemosphere*, 259 (2020) p. 127389.
93. M. Peñas-Garzón, M.J. Sampaio, Y.L. Wang, J. Bedia, J.J. Rodriguez, C. Belver, C.G. Silva, J.L. Faria, Solar photocatalytic degradation of parabens using UiO-66-NH_2, *Separation and Purification Technology*, 286 (2022) p. 120467.
94. T. Zhou, G. Liao, Y. Qiao, C. Sun, J. Jian, X. Xue, J. Shi, G. Che, Efficient removal of organic pollution via photocatalytic degradation over a TiO_2@HKUST-1 yolk-shell nanoreactor, *Journal of Molecular Liquids*, 385 (2023) p. 122383.
95. Z. Yang, H.S. Xie, W.Y. Lin, Y.W. Chen, D. Teng, X.S. Cong, Enhanced adsorption-photocatalytic degradation of organic pollutants via a ZIF-67-derived Co-N codoped carbon matrix catalyst, *ACS Omega*, 7 (2022) pp. 40882–40891.
96. R.N. Amador, M. Carboni, D. Meyer, Sorption and photodegradation under visible light irradiation of an organic pollutant by a heterogeneous UiO-67–Ru–Ti MOF obtained by post-synthetic exchange, *RSC Advances*, 7 (2016) pp. 195–200.
97. M. Zabihi, A. Motavalizadehkakhky, PbS/ZIF-67 nanocomposite: novel material for photocatalytic degradation of basic yellow 28 and direct blue 199 dyes, *Journal of the Taiwan Institute of Chemical Engineers*, 140 (2022) p. 104572.
98. Mian Zahid Hussain, A. Schneemann, R.A. Fischer, Y. Zhu, Y. Xia, MOF derived porous ZnO/C nanocomposites for efficient dye photodegradation, *ACS Applied Energy Materials*, 1 (2018) pp. 4695–4707.
99. Y. Zhang, J. Zhou, X. Chen, Q. Feng, W. Cai, MOF-derived C-doped ZnO composites for enhanced photocatalytic performance under visible light, *Journal of Alloys and Compounds*, 777 (2019) pp. 109–118.
100. N. Liu, W. Huang, X. Zhang, L. Tang, L. Wang, Y. Wang, M. Wu, Ultrathin graphene oxide encapsulated in uniform MIL-88A(Fe) for enhanced visible light-driven photodegradation of RhB, *Applied Catalysis B: Environmental*, 221 (2018) pp. 119–128.
101. S. Gautam, H. Agrawal, M. Thakur, A. Akbari, H. Sharda, R. Kaur, M. Amini, Metal oxides and metal organic frameworks for the photocatalytic degradation: A review, *Journal of Environmental Chemical Engineering*, 8 (2020) p. 103726.
102. Q.-Q. Chang, Y.-W. Cui, H.-H. Zhang, F. Chang, B.-H. Zhu, S.-Y. Yu, C-doped ZnO decorated with Au nanoparticles constructed from the metal–organic framework ZIF-8 for photodegradation of organic dyes, *RSC Advances*, 9 (2019) pp. 12689–12695.

103. K. Li, Y. De Rancourt De Mimérand, X. Jin, J. Yi, J. Guo, Metal oxide (ZnO and TiO_2) and Fe-based metal-organic-framework nanoparticles on 3D-printed fractal polymer surfaces for photocatalytic degradation of organic pollutants, *ACS Applied Nano Materials*, 3 (2020) pp. 2830–2845.
104. M. Amini, M. Ashrafi, Photocatalytic degradation of some organic dyes under solar light irradiation using TiO_2 and ZnO nanoparticles, *Nanochemistry Research*, 1 (2016) pp. 79–86.
105. J. Abdi, M. Yahyanezhad, S. Sakhaie, M. Vossoughi, I. Alemzadeh, Synthesis of porous TiO_2/ZrO_2 photocatalyst derived from zirconium metal organic framework for degradation of organic pollutants under visible light irradiation, *Journal of Environmental Chemical Engineering*, 7 (2019) p. 103096.
106. B. Abdollahi, A. Najafidoust, E. Abbasi Asl, M. Sillanpaa, Fabrication of ZiF-8 metal organic framework (MOFs)-based CuO-ZnO photocatalyst with enhanced solar-light-driven property for degradation of organic dyes, *Arabian Journal of Chemistry*, 14 (2021) p. 103444.
107. S.W. Lv, J.M. Liu, N. Zhao, C.Y. Li, F.E. Yang, Z.H. Wang, S. Wang, MOF-derived $CoFe_2O_4/Fe_2O_3$ embedded in g-C_3N_4 as high-efficient Z-scheme photocatalysts for enhanced degradation of emerging organic pollutants in the presence of persulfate, *Separation and Purification Technology*, 253 (2020) p. 117413.
108. A. Ghenaatgar, R. M.A. Tehrani, A. Khadir, Photocatalytic degradation and mineralization of dexamethasone using WO_3 and ZrO_2 nanoparticles: Optimization of operational parameters and kinetic studies, *Journal of Water Process Engineering*, 32 (2019) p. 100969.
109. M. Naimi Joubani, M.A. Zanjanchi, S. Sohrabnezhad, The carboxylate magnetic – Zinc based metal-organic framework heterojunction: Fe_3O_4-COOH@ZIF-8/Ag/Ag_3PO_4 for plasmon enhanced visible light Z-scheme photocatalysis, *Advanced Powder Technology*, 31 (2020) pp. 29–39.
110. H. Mou, C. Song, Y. Zhou, B. Zhang, D. Wang, Design and synthesis of porous Ag/ZnO nanosheets assemblies as super photocatalysts for enhanced visible-light degradation of 4-nitrophenol and hydrogen evolution, *Applied Catalysis B*, 221 (2018) pp. 565–573.
111. M.M. Abutalib, H.M. Alghamdi, A. Rajeh, O. Nur, A.M. Hezma, M.A. Mannaa, Fe_3O_4/Co_3O_4–TiO_2 S-scheme photocatalyst for degradation of organic pollutants and H_2 production under natural sunlight, *Journal of Materials Research and Technology*, 20 (2022) pp. 1043–1056.
112. B. Chen, G. Ma, D. Kong, Y. Zhu, Y. Xia, Atomically homogeneous dispersed ZnO/N-doped nanoporous carbon composites with enhanced CO_2 uptake capacities and high efficient organic pollutants removal from water, *Carbon*, 95 (2015) pp. 113–124.
113. W. Zhu, J. Liu, S. Yu, Y. Zhou, X. Yan, Ag loaded WO_3 nanoplates for efficient photocatalytic degradation of sulfanilamide and their bactericidal effect under visible light irradiation, *Journal of Hazardous Materials*, 318 (2016) pp. 407–416.
114. S. Sayegh, F. Tanos, A. Nada, G. Lesage, F. Zaviska, E. Petit, V. Rouessac, I. Iatsunskyi, E. Coy, R. Viter, D. Damberga, M. Weber, A. Razzouk, J. Stephan, M. Bechelany, Tunable TiO_2 –BN–Pd nanofibers by combining electrospinning and atomic layer deposition to enhance photodegradation of acetaminophen, *Dalton Transactions*, 51 (2022) pp. 2674–2695.
115. M. Danish, M. Saud Athar, I. Ahmad, M.Z.A. Warshagha, Z. Rasool, M. Muneer, Highly efficient and stable Fe_2O_3/g-C_3N_4/GO nanocomposite with Z-scheme electron transfer pathway: Role of photocatalytic activity and adsorption isotherm of organic pollutants in wastewater, *Applied Surface Science*, 604 (2022).
116. R. Hejazi, A.R. Mahjoub, A.H.C. Khavar, Z. Khazaee, Fabrication of novel type visible-light-driven TiO_2@MIL-100 (Fe) microspheres with high photocatalytic performance for removal of organic pollutants, *Journal of Photochemistry and Photobiology A: Chemistry*, 400 (2020) p. 112644.
117. C. Jing, Y. Zhang, J. Zheng, S. Ge, J. Lin, D. Pan, N. Naik, Z. Guo, In-situ constructing visible light CdS/Cd-MOF photocatalyst with enhanced photodegradation of methylene blue, *Particuology*, 69 (2022) pp. 111–122.
118. W.-J. Zhou, L.-X. Ma, L.-Y. Li, M. Zha, B.-L. Li, B. Wu, C.-J. Hu, Synthesis of a 3D Cu(II) MOF and its heterostructual g-C_3N_4 composite showing improved visible-light-driven photodegradation of organic dyes, *Journal of Solid State Chemistry*, 315 (2022) p. 123520.
119. Q.V. Thi, M.S. Tamboli, Q. Thanh Hoai Ta, G.B. Kolekar, D. Sohn, A nanostructured MOF/reduced graphene oxide hybrid for enhanced photocatalytic efficiency under solar light, *Materials Science and Engineering: B*, 261 (2020) p. 114678.
120. D. Gao, Y. Zhang, H. Yan, B. Li, Y. He, P. Song, R. Wang, Construction of UiO-66@MoS_2 flower-like hybrids through electrostatically induced self-assembly with enhanced photodegradation activity towards lomefloxacin, *Separation and Purification Technology*, 265 (2021) p. 118486.
121. F. Chen, Y. Li, M. Zhou, X. Gong, Y. Gao, G. Cheng, S. Ren, D. Han, Smart multifunctional direct Z-scheme In_2S_3@PCN-224 heterojunction for simultaneous detection and photodegradation towards antibiotic pollutants, *Applied Catalysis B-Environmental*, 328 (2023) pp. 122517.

122. C. Zhang, L. Ai, J. Jiang, Solvothermal synthesis of MIL–53(Fe) hybrid magnetic composites for photoelectrochemical water oxidation and organic pollutant photodegradation under visible light, *Journal of Materials Chemistry A*, 3 (2015) pp. 3074–3081.
123. Ch.V. Reddy, K.R. Reddy, V.V.N. Harish, J. Shim, M.V. Shankar, N.P. Shetti, T.M. Aminabhavi, Metal-organic frameworks (MOFs)-based efficient heterogeneous photocatalysts: Synthesis, properties and its applications in photocatalytic hydrogen generation, CO_2 reduction and photodegradation of organic dyes, *International Journal of Hydrogen Energy*, 45 (2020) pp. 7656–7679.
124. M.Z. Hussain, Z. Yang, Z. Huang, Q. Jia, Y. Zhu, Y. Xia, Recent advances in metal–organic frameworks derived nanocomposites for photocatalytic applications in energy and environment, *Advanced Science*, 8 (2021) p. e2100625.
125. H. Liang, J. Guo, M. Yu, Y. Zhou, R. Zhan, C. Liu, J. Niu, Porous loofah-sponge-like ternary heterojunction g-C_3N_4/Bi_2WO_6/MoS_2 for highly efficient photocatalytic degradation of sulfamethoxazole under visible-light irradiation, *Chemosphere*, 279 (2021) p. 130552.
126. O.J. Hao, H. Kim, P.C. Chiang, Decolorization of wastewater, *Critical Reviews in Environmental Science and Technology*, 30 (2000) pp. 449–505.
127. M. Hassaan, A. El Nemr, A.H.-A.J., Health and environmental impacts of dyes: Mini review, *Journal of Environmental Science and Engineering*, 1 (2017) pp. 64–67.
128. T. Maneerung, J. Liew, Y. Dai, S. Kawi, C. Chong, C.-H. Wang, Activated carbon derived from carbon residue from biomass gasification and its application for dye adsorption: Kinetics, isotherms and thermodynamic studies, *Bioresource Technology*, 200 (2016) pp. 350–359.
129. J.E. Aguiar, J.A. Cecilia, P.A.S. Tavares, D.C.S. Azevedo, E.R. Castellón, S.M.P. Lucena, I.J. Silva, Adsorption study of reactive dyes onto porous clay heterostructures, *Applied Clay Science*, 135 (2017) pp. 35–44.
130. K. Niinimäki, L. Hassi, Emerging design strategies in sustainable production and consumption of textiles and clothing, *Journal of Cleaner Production*, 19 (2011) pp. 1876–1883.
131. H. Li, X. Cao, C. Zhang, Q. Yu, Z. Zhao, X. Niu, X. Sun, Y. Liu, L. Ma, Z. Li, Enhanced adsorptive removal of anionic and cationic dyes from single or mixed dye solutions using MOF PCN-222, *RSC Advances*, 7 (2017) 16273–16281.
132. D. Yu, M. Wu, Q. Hu, L. Wang, Chencheng Lv, L. Zhang, Iron-based metal-organic frameworks as novel platforms for catalytic ozonation of organic pollutant: Efficiency and mechanism, *Journal of Hazardous Materials*, 367 (2019) pp. 456–464.
133. Y. Li, J. Jiang, Y. Fang, Z. Cao, D. Chen, N. Li, Q. Xu, J. Lu, TiO_2 nanoparticles anchored onto the metal-organic framework NH_2-MIL-88B(Fe) as an adsorptive photocatalyst with enhanced fenton-like degradation of organic pollutants under visible light irradiation, *ACS Sustainable Chemistry & Engineering*, 6 (2018) pp. 16186–16197.
134. A.J. Ebele, M. Abou-Elwafa Abdallah, S. Harrad, Pharmaceuticals and personal care products (PPCPs) in the freshwater aquatic environment, *Emerging Contaminants*, 3 (2017) pp. 1–16.
135. A.B.A. Boxall, M.A. Rudd, B.W. Brooks, D.J. Caldwell, K. Choi, S. Hickmann, E. Innes, K. Ostapyk, J.P. Staveley, T. Verslycke, G.T. Ankley, K.F. Beazley, S.E. Belanger, J.P. Berninger, P. Carriquiriborde, A. Coors, P.C. DeLeo, S.D. Dyer, J.F. Ericson, F. Gagné, J.P. Giesy, T. Gouin, L. Hallstrom, M. V. Karlsson, D.G. Joakim Larsson, J.M. Lazorchak, F. Mastrocco, A. McLaughlin, M.E. McMaster, R.D. Meyerhoff, R. Moore, J.L. Parrott, J.R. Snape, R. Murray-Smith, M.R. Servos, P.K. Sibley, J.O. Straub, N.D. Szabo, E. Topp, G.R. Tetreault, V.L. Trudeau, G. Van Der Kraak, Pharmaceuticals and personal care products in the environment: What are the big questions? *Environmental Health Perspectives*, 120 (2012) pp. 1221–1229.
136. O.R. Price, G.O. Hughes, N.L. Roche, P.J. Mason, Improving emissions estimates of home and personal care products ingredients for use in EU risk assessments, *Integrated Environmental Assessment and Management*, 6 (2010) pp. 677–684.
137. Y. Yang, Y.S. Ok, K.-H. Kim, E.E. Kwon, Y.F. Tsang, Occurrences and removal of pharmaceuticals and personal care products (PPCPs) in drinking water and water/sewage treatment plants: A review, *Science of the Total Environment*, 596–597 (2017) pp. 303–320.
138. M. Azhar, P. Vijay, M.O. Tadé, M.O. Tadé, S. Wang, Submicron sized water-stable metal organic framework (bio-MOF-11) for catalytic degradation of pharmaceuticals and personal care products, *Chemosphere*, 196 (2018) pp. 105–114.
139. J. Peng, E. Wu, N. Wang, X. Quan, M. Sun, Q. Hu, Removal of sulfonamide antibiotics from water by adsorption and persulfate oxidation process, *Journal of Molecular Liquids*, 274 (2019) pp. 632–638.
140. S. Gadipelli, W. Travis, W. Zhou, Z. Guo, A thermally derived and optimized structure from ZIF-8 with giant enhancement in CO_2 uptake (2012) (undefined 2014). doi: 10.1039/C4EE01009D.

141. T. Islamoglu, S. Goswami, Z. Li, A.J. Howarth, O.K. Farha, J.T. Hupp, Postsynthetic tuning of metal-organic frameworks for targeted applications, *Accounts of Chemical Research*, 50 (2017) pp. 805–813.
142. I. Ahmed, N.A. Khan, Z. Hasan, S.H. Jhung, Adsorptive denitrogenation of model fuels with porous metal-organic framework (MOF) MIL-101 impregnated with phosphotungstic acid: Effect of acid site inclusion, *Journal of Hazardous Materials*, 250–251 (2013) pp. 37–44.
143. J. Faccini, S. Ebrahimi, D.J. Roberts, Regeneration of a perchlorate-exhausted highly selective ion exchange resin: Kinetics study of adsorption and desorption processes, *Separation and Purification Technology*, 158 (2016) pp. 266–274.
144. K. Sumida, D.L. Rogow, J.A. Mason, T.M. McDonald, E.D. Bloch, Z.R. Herm, T.H. Bae, J.R. Long, Carbon dioxide capture in metal-organic frameworks, *Chemical Reviews*, 112 (2012) pp. 724–781.
145. G. Majano, O. Martin, M. Hammes, S. Smeets, C. Baerlocher, J. Pérez-Ramírez, Solvent-mediated reconstruction of the metal-organic framework HKUST-1 ($Cu_3(BTC)_2$), *Advanced Functional Materials*, 24 (2014) pp. 3855–3865.
146. K. Vellingiri, Y.X. Deng, K.H. Kim, J.J. Jiang, T. Kim, J. Shang, W.S. Ahn, D. Kukkar, D.W. Boukhvalov, Amine-functionalized metal-organic frameworks and covalent organic polymers as potential sorbents for removal of formaldehyde in aqueous phase: Experimental versus theoretical study, *ACS Applied Materials & Interfaces*, 11 (2019) pp. 1426–1439.
147. T. Dutta, K.-H. Kim, Richard, Yong Jin Kim, D.W. Boukhvalov, Metal-organic framework and Tenax-TA as optimal sorbent mixture for concurrent GC-MS analysis of C1 to C5 carbonyl compounds, *Scientific Reports*, 8 (2018) p. 5033.
148. K. Vellingiri, J.E. Szulejko, P. Kumar, E.E. Kwon, K.-H. Kim, A. Deep, D.W. Boukhvalov, R.J.C. Brown, Metal organic frameworks as sorption media for volatile and semi-volatile organic compounds at ambient conditions, *Scientific Reports*, 6 (2016) p. 27813.
149. P. Küsgens, A. Zgaverdea, H.G. Fritz, S. Siegle, S. Kaskel, Metal-organic frameworks in monolithic structures, *Journal of the American Ceramic Society*, 93 (2010) pp. 2476–2479.
150. L. Hashemi, M.Y. Masoomi, H. Garcia, Regeneration and reconstruction of metal-organic frameworks: Opportunities for industrial usage, *Coordination Chemistry Reviews*, 472 (2022) p. 214776.
151. T.M. McDonald, J.A. Mason, X. Kong, E.D. Bloch, D. Gygi, A. Dani, V. Crocellà, F. Giordanino, S.O. Odoh, W.S. Drisdell, B. Vlaisavljevich, A.L. Dzubak, R. Poloni, S.K. Schnell, N. Planas, K. Lee, T. Pascal, L.F. Wan, D. Prendergast, J.B. Neaton, Cooperative insertion of CO_2 in diamine-appended metal-organic frameworks, *Nature*, 519 (2015) pp. 303–308.
152. R.-B. Lin, L. Li, H.-L. Zhou, H. Wu, C. He, S. Li, R. Krishna, J. Li, W. Zhou, B. Chen, Molecular sieving of ethylene from ethane using a rigid metal–organic framework, *Nature Materials*, 17 (2018) pp. 1128–1133.
153. M.Y. Masoomi, M. Bagheri, A. Morsali, P.C. Junk, High photodegradation efficiency of phenol by mixed-metal–organic frameworks, *Inorganic Chemistry Frontiers*, 3 (2016) pp. 944–951.
154. M.M. Sadiq, H. Li, A.J. Hill, P. Falcaro, M.R. Hill, K. Suzuki, Magnetic induction swing adsorption: An energy efficient route to porous adsorbent regeneration, *Chemistry of Materials*, 28 (2016) pp. 6219–6226.
155. H. Li, M.M. Sadiq, K. Suzuki, R. Ricco, C. Doblin, A.J. Hill, S. Lim, P. Falcaro, M.R. Hill, Magnetic metal-organic frameworks for efficient carbon dioxide capture and remote trigger release, *Advanced Materials*, 28 (2016) pp. 1839–1844.
156. J. Espín, L. Garzón-Tovar, A. Carné-Sánchez, I. Imaz, D. Maspoch, Photothermal activation of metal-organic frameworks using a UV-Vis light source, *ACS Applied Materials & Interfaces*, 10 (2018) pp. 9555–9562.
157. H. Li, M.R. Martinez, Z. Perry, H.C. Zhou, P. Falcaro, C. Doblin, S. Lim, A.J. Hill, B. Halstead, M.R. Hill, A Robust metal–organic framework for dynamic light-induced swing adsorption of carbon dioxide, *Chemistry - A European Journal*, 22 (2016) pp. 11176–11179.
158. X.-Q. Wu, J.-H. Liu, T. He, P.-D. Zhang, J. Yu, J.-R. Li, Understanding how pore surface fluorination influences light hydrocarbon separation in metal–organic frameworks, *Chemical Engineering Journal*, 407 (2021) p. 127183.
159. O.T. Qazvini, S.G. Telfer, MUF-16: A robust metal-organic framework for pre- and post-combustion carbon dioxide capture, *ACS Applied Materials & Interfaces*, 13 (2021) pp. 12141–12148.
160. T.L. Tan, P.A. Krusnamurthy, H. Nakajima, S.A. Rashid, Adsorptive, kinetics and regeneration studies of fluoride removal from water using zirconium-based metal organic frameworks, *RSC Advances*, 10 (2020) pp. 18740–18752.

161. Z. Li, M. Ma, S. Zhang, Z. Zhang, L. Zhou, J. Yun, R. Liu, Efficiently removal of ciprofloxacin from aqueous solution by MIL-101(Cr)-HSO_3: The enhanced electrostatic interaction, *Journal of Porous Materials*, 27 (2020) pp. 189–204.
162. I. Majchrzak-Kucęba, D. Bukalak-Gaik, Regeneration performance of metal–organic frameworks: TG-Vacuum tests, *Journal of Thermal Analysis and Calorimetry*, 125 (2016) pp. 1461–1466.
163. J. Xiao, L. Fang, P. Bénard, R. Chahine, Parametric study of pressure swing adsorption cycle for hydrogen purification using Cu-BTC, *International Journal of Hydrogen Energy*, 43 (2018) pp. 13962–13974.
164. N.C. Burtch, H. Jasuja, K.S. Walton, Water stability and adsorption in metal-organic frameworks, *Chemical Reviews*, 114 (2014) pp. 10575–10612.
165. S. Rojas, P. Horcajada, Metal-organic frameworks for the removal of emerging organic contaminants in water, *Chemical Reviews*, 120 (2020) pp. 8378–8415.

12 Two-Dimensional Nanomaterials for Antibacterial and Antimicrobial Activity for Water Contaminants Disinfection

Bogala Mallikharjuna Reddy

12.1 INTRODUCTION

The Earth is suffering greatly as a consequence of human activity. With ever expanding industrialization and rapid human population growth, the energy issues and deterioration of the environment have emerged as major concerns for mankind in the twenty-first millennium [1]. Pathogenic microorganisms, viruses, and bacteria, as well as inorganic and organic wastes, have emerged as major pollutants of the environment that pose a great threat to the well-being of humanity [2]. For instance, every year, enterotoxigenic *E. coli* (*Escherichia coli*), a bacteria that is responsible for causing diarrhea, mostly spreads through the contaminated water, and kills approximately 1.3 million children globally below 5 years of age [3]. Water that is clean and safe is considered to be vital to every species on Earth. For survival, all animals on Earth need clean and healthy water [4]. However, sewage discharge may not be eliminated, leading to the continuous contamination of accessible sources of freshwater. According to the previous evaluation, inadequate sanitation and the absence of safe drinking water contributed to many deaths [5]. In terms of overall real statistics, the number of people killed by pathogen infections (through contaminated water) has overtaken the number of people killed by terrorism, war, and weapons of mass destruction [6]. In accordance with the WHO (World Health Organization) reports released in 2015, in spite of tremendous initiatives, 10% of the world population still remains without a sufficient supply of safe and clean water to drink [7]. According to the World Health Organization, roughly one-sixth of the world's population (1.1 billion people) do not have clean water access to consume (WHO, 2004). The consequences are horrible: diarrhea kills more than 2.2 million individuals per year, the vast majority of whom are young children under 5 years of age [8]. The control of microorganisms and disinfection of wastewater are critical. Accessibility to reliable, safe, and clean water remains a major concern for all humans, making the development of effective procedures for polluted water purification vital [9]. As a consequence, the advancement of successful bacteria and virus disinfection technologies is crucial.

The most serious type of water contamination is antibacterial and microbial contamination, which causes many fatal diseases such as diarrhea, cholera, hepatitis, and intestinal parasite infections [4]. The ongoing enforcement of water quality standards has put further emphasis on the establishment of more effective water disinfection systems. Chlorination, ozonation, and irradiation using UV (ultraviolet) light have long been regarded as established chemical processes for large-scale purification of water [4]. These solutions, nevertheless, have obvious drawbacks. For instance, chlorine, the most widely used chlorination reagent, may react with fulvic acid to form a range of halogenated organic compounds, and these disinfection by-products (DBPs) have been

DOI: 10.1201/9781003436942-12

linked to the weight loss and genogenic aberrations in the cardiovascular and nervous systems [10,11]. While current disinfection procedures for drinking water purification may efficiently eradicate microbial pathogens, recent studies indicate a conflict between efficient disinfection and the formation of dangerous DBPs. Ozone, free chlorine, and chloramines, which are commonly used in the wastewater treatment industry, can react with natural water elements to produce DBPs, some of which are cancer-causing agents [8]. Throughout the disinfection process, the chemicals used for ozonation and chlorination may generate by-products that possess the capability to result in cancer [1,12]. Moreover, ozone's considerable environmental toxicity and hazardous qualities prohibit its broader application. Despite the fact that UV disinfection is a "green" technique, since no toxic by-products are produced, certain microorganisms are inherently resistant to the UV light irradiation [1]. Besides these challenges, the existing disinfection systems frequently consume large amounts of energy, which is generated by important nonrenewable fossil fuels such as petroleum and coal. Moreover, the costly exploitation of these natural resources may result in substantial degradation of the environment [13]. Considering the potential challenges of safety, resource consumption, and environmental pollution, the discovery of novel materials that are low cost, energy efficient, effective, and eco-friendly for wastewater treatment remains extremely important.

Novel procedures for the treatment of water and wastewater systems based on nanosized materials are being explored throughout the past decade or so [14]. The rapid growth of nanotechnology has prompted tremendous curiosity in the environmental uses of nanomaterials [8]. Nanotechnology contributes to the growth of future water supply systems. Nanoengineered materials possess the ability to adsorb and/or breakdown many water contaminants (dyes, pesticides, metal ions, plastics, pharmaceuticals, organic pollutants, etc.) and they exhibit attractive antibacterial properties against aquatic microbes (protozoa, viruses, bacteria, etc.) [14]. Because of useful characteristics such as large specific surface area, quantum confinement effect, facile modification, and high reactivity, nanomaterials have captured the interest of researchers in the areas of adsorbents, catalysis, biomedicine, sensors, disinfection, and environmental remediation [1,8,15–21]. Numerous nanoscale materials, including nanoscale metal oxides (such as titanium dioxide), carbon nanotubes, fibers, noble metals, zeolites, and enzymes, have found usage in overcoming the environmental issues [22]. As shown in Figure 12.1, all these kinds of nanomaterials can be classified into 0D, 1D, 2D, or 3D NMs. Metal oxides, graphene-based hybrids, carbon nanotubes/graphitic carbon nitride (CNT/g-C_3N_4) composites, metal–organic frameworks (MOFs), and other readily available NMs are some examples of NMs.

Two-dimensional nanoparticles (2D NMs) are intriguing and prospective antibacterial materials for water treatment operations among the nanomaterials mentioned above. In contrast, high conductivity, good charge mobility, and the enormous specific area (surface) of some 2D layered NMs (like graphene) permit the development of a huge number of surface-active sites, easy adhesion to bacteria, and efficient generation of photoinduced electron (e^-)/hole (h^+) separated pairs [23]. Furthermore, due to their distinctive morphologies, 2D nanomaterials are excellent platforms for the development of catalyst-well-dispersed systems with functional properties [24]. Given these advantages, extensive efforts have been undertaken to investigate the antibacterial and antimicrobial activities of 2D nanomaterials. The Li group investigated the most recent advances in 2D antibacterial nanomaterials in this area [8]. They thoroughly investigated several antimicrobial NMs for disinfection of water, covering mechanism elucidation, and evaluated their strengths and weaknesses for additional studies. Metal cation release, ROS (reactive oxygen species) generation, and physical contact were proposed as important antibacterial mechanisms. Both assessments were entirely focused on antimicrobial nanomaterials such as nanosilver, TiO_2, ZnO, MgO, fullerenes, and carbon nanotubes [1]. Despite the rapid advancements in novel antibacterial and antimicrobial NMs, an up-to-date overview is essential on this research topic.

The current book chapter focuses on recent advances in 2D antibacterial and antimicrobial nanomaterials. In addition to new findings, the critical antibacterial mechanisms are explained, and key parameters influencing antibacterial activity are reviewed. There is also discussion of the

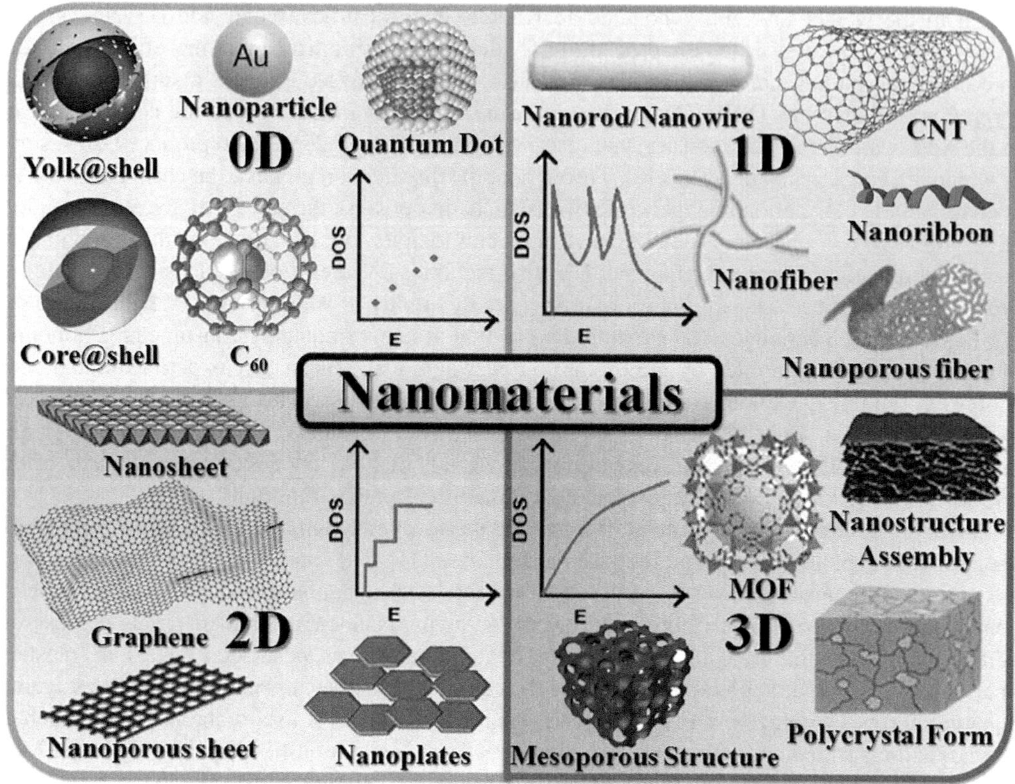

FIGURE 12.1 Classification of nanomaterials (0D, 1D, 2D, and 3D).

classification of various 2D nanomaterials, in addition to their structural features and antibacterial properties. The primary areas to investigate for future antibacterial 2D nanomaterial design include material recyclability, toxicities management, and efficiency enhancement. A complete summary of current advances in 2D antimicrobial nanomaterials is provided, as well as considerations of the factors impacting disinfection performance. This study will help in the production of 2D NMs with antibacterial and antimicrobial activities for the water pollutants treatment in the future.

12.2 WATER CONTAMINATION

12.2.1 Emerging Water Contaminants

Wastewater is thought to contain a broad range of pollutants (inhibitors, surfactants, pharmaceuticals, and so on) with widely varying origins and properties (Figure 12.2) [14]. These new poisons are constantly discharged into the environment as a result of their use in industry, agriculture, medical treatment, consumer products, and domestic activity. Because the majority of the microcontaminants and their by-product metabolites are weakly degradable in nature and just little of them may be removed by wastewater treatment methods, they can be found in wastewater and aquatic habitats even after treatment in all over the world [25]. Agricultural and livestock runoff, agricultural reuse of sewage and wastewater sludge as fertilizer, irrigation with treated wastewater, leakage from septic tanks, landfills, and industrial wastewater systems are all notable sources of the micropollutants [26]. These microcontaminants are regularly found in tiny/trace amounts (varying from ng L^{-1} to g L^{-1}) in water. As a result, discovering and measuring them is a major undertaking [25, 26]. Ul-Islam et al. investigated advanced magnetic nanoparticles for organic pollutant adsorption and degradation [27]. The use of cellular isolates (from plants, algae, fungus, and bacteria) as green

FIGURE 12.2 A schematic showing the contaminants of wastewater.

reagents for nanomaterial production is now recognized as a low-cost, sustainable, energy-efficient, and environmentally friendly alternative to water and wastewater treatment. Gautam et al. recently investigated the manufacture and use of biogenic NMs in various applications [28]. Likewise, Ali et al. investigated microbe-derived biogenic NMs for treatment of water and wastewater, whereas Hennebel et al. investigated palladium NM biosynthesis [29,30].

Pharmaceuticals are a significant class of growing environmental micropollutants [31]. Each year, a large number of pharmaceuticals are manufactured, consumed, and released into the environment, and their accumulation in water bodies (groundwater, seawater, surface water, and treated sewage effluents) is widely reported worldwide, posing a major concern because they negatively affect the growth, reproduction, and behavior of non-target organisms, mainly aquatic organisms. Pharmaceutical micropollutants have been associated to a number of environmentally toxicological effects [31]. Around the world, various pharmaceutical substances for veterinary and human health purpose have been recommended (10,000 in the United States and 5000 compounds in Europe) [32]. Over 200 pharmaceutical compounds have been discovered in river waters all over the world [31]. Fekadu et al. recently looked at the presence of medicines in aquatic and freshwater environments in

Europe and Africa. Surprisingly, numerous drugs were detected at quantities that surpassed stated objectives for ecotoxicity [32]. Any product featuring healthcare or medical functions for humans and/or animals is referred to as PPCPs (pharmaceutical and personal care products) [33,34]. Because of their widespread consumption, low human metabolic aptitude, inappropriate disposal, and the inability of wastewater treatment plants to adequately remove them, PPCPs have been detected in water bodies at concentrations ranging from ng L^{-1} to g L^{-1}. At low quantities, PPCPs can impact human health and the environment [33]. To avoid negative effects on aquatic creatures, Hopkins and Blaney addressed the necessity for enhanced treatment of wastewater using polycyclic musks, antimicrobials, and UV filters [34].

Regardless of the well-known toxicity and adverse consequences of organic micropollutants (such as polycyclic aromatic hydrocarbons (PAHs), solvents, pesticides, bisphenol A, organochlorine pesticides, alkyl phenols, polychlorinated biphenyls, and polybromodiphenyl ethers), their manufacturing, use, and distribution are expected to rise in the years to come. Surprisingly from 1907 to 2008, about 33 million chemical and inorganic molecules were synthesized, with roughly 4000 new compounds being put on the list per every day of the year. Pesticides (fungicides, insecticides, molluscicides, herbicides, rodenticides, and nematocides) are sprayed in about 4.6 million tons per year, with a large portion making its way to water resources [31]. The current farming system's misuse of artificial pesticides and fertilizers has damaged the environment. Pesticides from household waste, industrial effluents, agricultural activities, and other irrelevant/ relevant resources enter the environment through air, runoff, and other agents [35]. Pollutants enter the water ecosystem via an array of channels, including spills in transportation, leaks from landfills, warehouses, and industrial supplies [36]. Pesticides can enter crop soil by the wind, irrigation water, and rain. They subsequently percolate through surface water courses, enter groundwater via runoff and soil fissures, and eventually reach wastewater treatment plants (Figure 12.3) [37]. Aside from the ones mentioned above, many pesticides are widespread compounds; due to their poor bioabundance, they

FIGURE 12.3 An illustration of pesticide circulation into the water ecosystem.

persist in sediments and soils [38]. As a result, pesticides are exceedingly hazardous and can harm human beings and non-targeted organisms, as well as devastate living habitats and ecosystems [39].

Sewer networks and urban runoff systems transport organisms from the urban microbiome to water recipients on a regular basis, creating an impression of the urban fingerprint in polluted rivers [40]. The authors stressed the importance of emphasizing research on microbiomes present in municipal water because of their beneficial effects on both human and environmental health. Vittecoq et al. stressed the necessity of effective water treatment and surveillance in the Mediterranean basin for sickness control [41]. More than 50 novel agents of infection have been discovered in the recent past decades, with bacteria constituting around 10% of them [42]. Every year, over 200 million people are killed by non-fatal and waterborne diseases [31]. *Pseudomonas aeruginosa*, *Legionella pneumophila*, *Mycobacterium avium*, and other non-tuberculous mycobacteria are some of the opportunistic diseases causing an expanding waterborne illness problem with a significant overall economic impact. Falkinham et al. investigated the characteristics of opportunistic residential water pathogens in an effort to alert the drinking water population [43]. According to La Rosa et al., viruses have the potential to become rising pathogens due to their biology (ability to infect new hosts and adapt to new environments). They focused on aquatic viruses as likely rising agents [44].

While it was once expected that Fleming's 1928 discovery of antibiotics would result in the abolition of bacterial illnesses, antibiotic misuse in recent decades has reduced bacteria's resistance to antibiotics, leading to the drug potency fall or reduction [45–47]. Antibiotics are more progressively utilized in livestock and human therapy; nevertheless, they are poorly metabolized and end up in wastewater and soil via human and animal urine. Each year, 63,151 tons of 250 different varieties of antibiotics are used in healthcare for humans and animals, with approximately 70% of them never being processed or assimilated in the human or animal body, resulting in their regular release into the ecosystems [31]. Certain antibiotics are only partially transformed in sewage and wastewater treatment units and thus are constantly released into bodies of water, creating a significant class of growing water pollutants [14]. As a result, treating bacterial infections has grown increasingly challenging [48]. The development of antibacterial drug strategies and the prevention of resistance to drugs are important areas of scientific study. Classical techniques (ozonation, chlorination, sand filtration, and UV irradiation) announced the end of waterborne epidemics in the industrialized world more than a century ago. Despite this, waterborne outbreaks continue to occur at disturbingly high rates because many bacteria are resistant to UV irradiation therapy, and successful disinfection requires long irradiation duration and a high dose [49]. DBPs from chlorination decontamination increase the risk of cancer, and UV irradiation can even harm the central nervous system and produce stronger stimulation, leading to more significant oxidative stress, higher gene expression, and faster ARG binding transfer [50]. The downside of conventional water purification processes' high energy usage is especially obvious in light of the present energy crisis [51]. Traditional decontamination methods in developing countries rely primarily on energy generated by fossil fuels such as petroleum and coal, both of which are dwindling and may cause significant environmental issues [52]. The interplay of increasing human energy consumption and severe environmental issues provides an obstacle to standard water disinfection systems [53]. Moreover, existing water disinfection techniques demand massive instruments, restricting the scale of water treatment plants. As a consequence, the development of new materials for water filtration that are practical, eco-friendly, highly effective, easily accessible, and energy-efficient is important [54].

12.2.2 Nanotechnology for Water Disinfection

Because of the rapid advancement of nanoscience and technology, major attempts have lately been undertaken to develop adaptable nanomaterials as novel antibacterial agents against a broad range of bacterial pathogens based on their unique physicochemical features [55, 56]. Due to their large membrane permeability, benign biocompatibility, and potential for many antibacterial actions, NMs are unlikely to generate bacterial resistance than traditional antibiotics [57]. It is a possible method to leverage modern nanomaterial technologies to update wastewater treatment plants, transform micropollutants into less harmful molecules, or even mineralize them. Photocatalysis, ozonation, sonolysis, electrochemical oxidation, Fenton catalysis, and similar reactions are examples of AOPs (advanced oxidation processes) that are based on the high production of ROS and may be utilized as a pre- or post-treatment for a biological process [14]. As an effective antibacterial therapy, nanomaterials are gaining favor. They are recognized as very promising avenues for addressing existing ailments [58]. Ul-Islam et al. investigated magnetic nanoparticle advancements in organic contaminants adsorption and breakdown [27]. Cellular extracts (from fungi, bacteria, plants, and algae) have been acknowledged as a sustainable, low-cost, energy-efficient, and ecologically friendly alternative for treatment of water and wastewater. Gautam et al. recently investigated the manufacture and use of biogenic NMs in various applications [28].

In addition, Ali et al. investigated microbe-derived biogenic NMs for water and wastewater treatment, whereas Hennebel et al. presented on palladium nanoparticle biogenesis [29,30]. Due to the creation of ROS, the release of toxic metal ions, and the loss of cell membrane integrity upon direct contact, several unique nanomaterials (for example, nano-Ag, nano-TiO_2, nano-ZnO, and carbon nanotubes) have significant antibacterial capabilities. These nanoparticles provide an effective alternative to traditional disinfectants while avoiding the formation of potentially toxic by-products of disinfection [59]. Choudhury and Hashmi have reported article on the environmental viability in the treatment of wastewater using novel nanostructured photocatalysts [60]. Zhao et al. examined metal-free catalysts for the production of sulfate radicals (graphene, carbon nanotubes, mesoporous carbon, activated carbon fiber, activated carbon, and nanodiamond) [61]. Saqib et al. studied the enhancement of TiO_2 photocatalysts with the use of rare earth metals [62]. Beltrán et al. investigated solar photocatalytic ozonation to emphasize the importance of the hybrid procedure as an environmentally friendly water treatment technology for reducing emerging contaminants [63]. Tsydenova et al. investigated the viability of simultaneous pathogen and chemical contaminant abatement utilizing solar-enhanced AOPs [64]. Duan et al. described metal-free carbocatalysis in AOPs as a green remediation choice for metal-based processes, which are described by poor stability and metal leaching [65]. Wols and Hofman-Caris investigated the photochemical reaction parameters of photochemical AOPs for organic micropollutant degradation in water [66].

12.3 2D NANOMATERIALS

After the graphene discovery in 2004, researchers have been interested in 2D NMs [67]. Among various forms of nanomaterials, 2D NMs are interesting due to their promising antibacterial capabilities. 2D NMs are nanoscale (1–100 nm) substances in which electrons move freely in only two dimensions (planar motion) [68]. The large specific surface area, high conductivity, and charge mobility of certain 2D layered nanomaterials (e.g., graphene) enable for many surface active sites, facile adhesion to microorganisms, and effective generation of photoinduced electron (e^-)/hole (h^+) segregated pairs [23, 69]. Furthermore, due to their distinctive shapes, 2D nanomaterials are excellent substrates for the fabrication of catalyst-distributed systems with functionalities [24]. Given these advantages, numerous studies have been conducted to investigate the antibacterial activity of 2D nanomaterials. In this context, the Hossain group thoroughly examined the most recent advances in 2D antibacterial nanomaterials [70]. They thoroughly investigated several antimicrobial nanoparticles for water disinfection, including mechanism elucidation, and evaluated their strengths and weaknesses for additional studies. Also, 2D NMs with a high surface area and better surface functionalization allow for close interaction with bacterial membranes, which improves antibacterial activity. Moreover, antibacterial treatments based on 2D NMs may be used at lower doses than normal antibiotics, overcoming resistance and lowering other negative consequences to some extent [71]. Thus, comparatively the 2D nanomaterials may be preferred over the traditional antibiotics against bacteria and microorganisms in wastewater treatment.

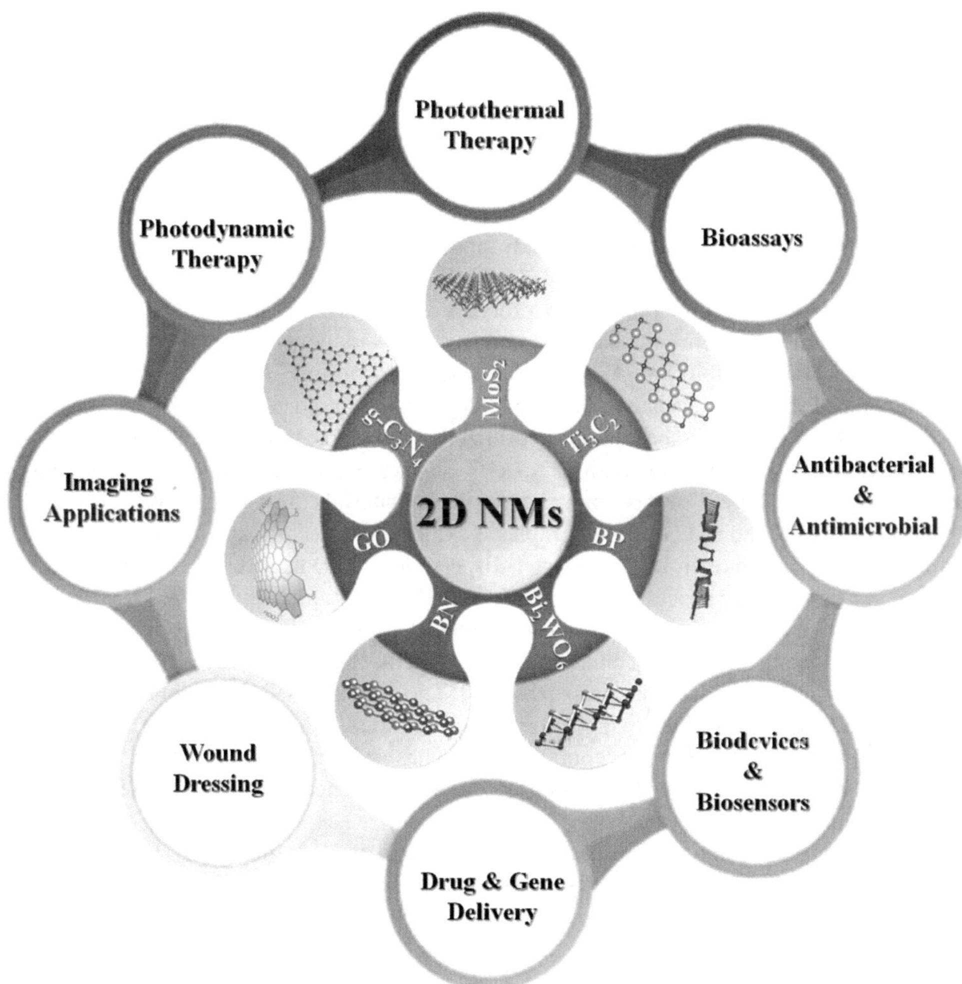

FIGURE 12.4 Commonly used 2D NMs and their bioapplications.

12.3.1 Classification of Antibacterial 2D NMs

As illustrated in Figure 12.4, typical 2D NMs include graphene oxide (GO), graphene, reduced graphene oxide (rGO) and its derivatives, graphitic carbon nitride (g-C_3N_4), transition metal dichalcogenides (TMD—MoS_2, WS_2) and their derivatives, titanium carbide (MXenes), bismuth-based materials, BN, and black phosphorus (BP) [45]. Due to their distinctive structures and photoelectrical properties, these 2D NMs offer significant potential for biological applications. A broad overview of these 2D NMs and their uses in medicine is depicted in Figure 12.4. In recent years, there has been a significant quantity of research focused on 2D NMs and their antimicrobial uses. Researchers have examined the prospective antibacterial utilization of 2D NMs such as metal-based nanocomposites, transition-metal dichalcogenides/oxides (TMD/Os), C_3N_4, MXenes, BP, and other 2D NMs, after inspiration from graphene's antibacterial properties.

12.3.1.1 Graphene-Based 2D NMs

Because of the sp^2 hybrid atomic orbitals, graphene is a 2D NM composed of carbon atoms organized in a hexagonal honeycomb lattice [72]. In this configuration, the p-s orbital of each carbon atom in graphene is vertical to the direction of the layer, resulting in a broad polyatomic-bond network across the layer and configurable electrical, thermal, optical, and mechanical properties, as well as an enormous specific surface area [73]. Graphene is created using chemical vapor deposition (CVD) and thermal stripping. The Hummers oxidation technique is used for the production of graphite oxide [74]. A reduction process was used to prepare rGO flakes. Because graphene has a zero band gap and readily undergoes oxidation, it is frequently mixed with other semiconductors and metal nanostructures to boost its efficiency, producing composite materials that can be utilized in an array of utilization [75]. Graphene, in specific, has been exploited in photocatalysis due to its high electrical conductivity (106 Sm^{-1}) and advantageous role in absorbing or moving electrons in nanocomposites, which aids in better photocatalytic activity [76,77]. To improve the antimicrobial activity of graphene-based materials, many procedures are currently being used: producing a heterojunction structure, changing the size and number of layers, constructing a composite material, and surface functionalization [78]. Antibacterial carbon-containing NMs comprise fullerenes, 1D carbon nanotubes, 2D graphene nanosheets, and 3D graphene-based networks [79]. Graphene-based materials, a 2D single-atomic-sheet of sp^2-hybridized carbon atoms, have been widely employed in disinfection due to their low toxicity, high specific surface area, higher electric conductivity, and good water dispersity [80]. The most prevalent of these materials are Gt (graphite), Gr (graphene), GO (graphene oxide), and rGO (reduced graphene oxide). GO is a carboxyl, hydroxyl, and epoxide-containing oxidized form of Gr [81]. The water-dispersible GO 2D nanosheets may be produced utilizing a modified Hummer process based on acid-based exfoliation synthesis. GO has the strongest antibacterial activity among the graphene-based material family, according to research [82]. The physicochemical features of GO, such as lateral size, surface charges, and functional groups, have a large influence on its antibacterial properties [83]. The antibacterial effect of GO 2D nanosheets is proportional to their size. Surprisingly, very pure GO was found to have no effect on bacterial development [84]. There are further reports that rGO has greater cytotoxicity than GO while having lower antibiotic properties, which could be due to differences in charge density and surface functional groups [85].

Until now, efforts are being done to improve the disinfection efficacy of 2D graphene-based nanosheets. Qi and colleagues developed supramolecular carbohydrate-functionalized graphene nanosheets. The technique accomplished bacterial adhesion and, as a result, high-efficiency antibacterial killing by utilizing the multivalent action of sugar ligands [86]. Zou et al. generated wrinkled 2D GO nanosheets with hierarchical structures and multiscale lengths via vacuum-based filtration of GO suspensions across a porous filter membrane. They discovered that the antibacterial capabilities of GO films are related to their surface roughness. Adequate roughness (as determined by the R_q value) ensured maximum binding between 2D GO nanosheets and bacteria, which is required for optimal antibacterial activity. Additionally, the numerous sharp edges and better carrier transit efficiency contribute to successful bacterial death [87]. Decorating 2D graphene nanosheets with metal particles is one approach for increasing its antibacterial characteristics [88]. When 2D GO is coupled with Ag NPs and TiO_2 nanorods, the optical absorption range expands while the rate of e^-/h^+ pair recombination decreases. As a result, the photocatalytic disinfection efficacy of GO-TiO_2-Ag nanocomposites has been considerably enhanced [89]. Deng et al. created graphene-CdS (G-CdS) nanocomposites using a solvothermal method. G-CdS can absorb visible light and has a high inactivation ability due to the well-dispersion of CdS on the GO surface and efficient charge transfer [90]. Gao et al. also employed a two-phase assembling approach to make GO-CdS composites. Uniformly disseminated CdS NPs on wrinkled single-layered 2D GO nanosheets generated more surface reactive sites. Due to excellent e^-/h^+ separation and strong bonds between GO and CdS, outstanding photostability and antibacterial performance are achieved [91].

In addition, the large size of 2D GO sheets allows for simple recycling after use. By depositing $CoFe_2O_4$ on 2D GO-Ag sheets, the same function was proven [92]. To achieve a better catalytic heterojunction, Bai et al. invented a PEG-functionalized 2D GO-CuS nanocomposite with improved characteristics (hydrophilicity and a decreased e^-/h^+ recombination rate). The synergistic effect might explain why near-infrared (NIR) chemo-photothermal disinfection is more effective [93]. Zeng et al. discovered a possible antibacterial system by co-decorating TiO_2 and carbon dots (C-dots) on graphene. TiO_2 aggregation on 2D rGO nanosheets was prevented by the presence of glucose. The inclusion of C-dots on the rGO surface improved charge transfer and resulted in a good ROS generation yield [94]. As a result, 2D graphene-based (mainly GO) antibacterial nanomaterials have attracted considerable attention of the researchers. The modified Hummers approach is the most commonly utilized GO synthetic method. Because of their intrinsic antibacterial capability and large size effect, 2D GO nanosheets are an excellent supportive substrate for photocatalysts, reducing catalyst aggregation and increasing antibacterial activity. 2D GO sheets are frequently large in size, allowing for post-use recycling and avoiding subsequent contamination.

12.3.1.2 Graphitic Carbon Nitride

Like graphene, graphitic carbon nitride (g-C_3N_4) has a planar 2D sheet structure [95, 96]. It is composed of two basic units: triazine rings and 3-s triazine rings. The layers are joined via van der Waals interactions to form a 2D nanosheet [97]. Because of its ease of synthesis and functionalization, stable electronic field emission, and regulated forbidden band, g-C_3N_4 has received popularity in the photocatalysis literature. Furthermore, g-C_3N_4 is both thermally and chemically stable. It stabilizes the structure and functions well at high temperatures, as well as in acidic and alkaline situations. Furthermore, the material is non-polluting and does not contribute to secondary contamination. In the generic g-C_3N_4 manufacturing strategy, several nitrogen (N)-rich precursors (e.g., urea, dicyandiamide, thiourea, and melamine) are used in a variety of processes, including solvothermal, solid-phase reaction, thermal polymerization, and electrochemical deposition [98]. As a result, g-C_3N_4 has received a considerable deal of attention in a range of scientific fields. However, due to its large particle size, poor water solubility, larger band gap (i.e., light absorption wavelengths shorter than 460 nm), low conductivity, and other limitations, its practical application is limited. As a result, new studies showed that doping, surface modification, composite structure, and defect creation can boost the efficiency of photocatalytic reactions [99].

Carbon nitride (C_3N_4) was invented by Berzelius in the 1830s, and Liebig named it "melon" after him [100]. The 2D g-C_3N_4 nanosheets, a burgeoning nitride-based polymeric semiconductor material with laminar structure and narrow band gap, have recently attracted enormous attention in the antibacterial field due to their large surface area, highly efficient photocatalytic performance, adjustable electron band structure, and good chemical stability [52]. Wu et al. used an in situ reduction approach to generate an antibacterial Ag/polydopamine/2Dg-C_3N_4 bio-photocatalyst [101]. On the other hand, employing Ag NPs and polydopamine via photo-irradiation increased photocatalysis of Ag/polydopamine/2Dg-C_3N_4, which was followed by the generation of abundant ROS (particularly OH). Alternatively, light has the potential to accelerate the discharge of Ag ions. Finally, because of the synergistic effect of Ag ions and photocatalytic therapy (PCT) of g-C_3N_4 nanosheets, the Ag/polydopamine/2Dg-C_3N_4 nanosheets have shown potent antibacterial properties. This bio-photocatalyst has outstanding biocompatibility as well as exceptional long-term antibacterial action. Wang et al. developed a novel metal-free heterojunction by coating cyclooctasulfur (α -S_8) with 2D rGO and 2D g-C_3N_4 nanosheets for photocatalytic disinfection under visible light in order to increase the use of g-C_3N_4 as photocatalysts [102]. They observed that photocatalytic bacteria inactivation of this nano-heterojunction is mostly driven by photogenerated ROS in an aerobic environment or electron transfer produced oxidative stress under anaerobic conditions. Surprisingly, the S_8 (core)/rGO (inner shell)/g-C_3N_4 (outer shell) (CNRGOS_8) arrangement had a stronger antimicrobial effect (considering aerobic conditions), whereas the S_8(core)/g-C_3N_4 (outer shell) (RGOCNS_8) had an excellent antibacterial ability in anaerobic environments.

Moreover, Li and colleagues constructed 2D g-C_3N_4 nanosheets-functionalized hydrophilic composite membranes and discovered that changing the 2D g-C_3N_4 nanosheets considerably boosted the antibacterial capacity [103]. Wang et al. discovered that vanadate quantum dots-inserted 2D g-C_3N_4 (vanadate QDs/g-C_3N_4) nanosheets performed faster photocatalytic disinfection with profuse ROS than naked g-C_3N_4 nanosheets [104]. In addition, Chen's groups created a novel all-organic self-assembled semiconductor photocatalytic nanomaterial C_3N_4/PDINH heterostructure for photocatalytic antibacterial applications by recrystallizing PDINH (perylene-3,4,9,10-tetracarboxylic diimide) on the plane of 2D C_3N_4 nanosheets in situ to reduce heavy metal toxicity [105]. Interestingly, they discovered that the heterostructure's absorption spectrum was significantly enlarged from UV to NIR, enhancing the photocatalytic capability to generate more ROS for increased bactericidal action while exhibiting low toxicity to healthy tissue cells. Simultaneously, the C_3N_4/PDINH heterostructure may cure the diseased region and stimulate wound regeneration in vivo. Although C_3N_4-based photocatalysts have been described in a range of biological disciplines, their photocatalytic efficacy is still insufficient to meet the demands of practical applications. As a result, future improvements in the catalytic efficacy of 2D C_3N_4 nanosheets are highly anticipated.

2D g-C_3N_4, a new metal-free polymeric semiconductor material, has received greater attention due to its appealing chemical and thermal stability, ease of fabrication, low cost, and tiny band gap [106]. As a consequence, the graphite-substituted 2D layered g-C_3N_4 was suggested. Thermal polymerization of low-cost precursors such as melamine, urea, dicyandiamide, and thiourea may be utilized to create bulk g-C_3N_4 [107]. Conversely, the photogenerated charges' tiny specific area of surface and long migration distance limited the photocatalytic performance of bulk g-C_3N_4. Alternatively, the razor-thin thickness of 2D g-C_3N_4 nanosheets can dramatically reduce the diffusion distance of photogenerated electrons while also adding active sites for catalysis [108]. Nevertheless, the photogenerated e^-/h^+ couples' photocatalytic efficiency was limited by their rapid recombination and poor molar extinction coefficient in visible light [109]. As a consequence, methods for increasing photocatalytic efficiency remain in great demand. Zhao et al. created wrinkled atomic single-layer (SL) 2D g-C_3N_4 (SL 2D g-C_3N_4) by ultrasonic exfoliation. The synthesized SL 3D g-C_3N_4 displayed outstanding photocatalytic disinfection activities under visible light irradiation because of the improved photogenerated charges separation efficiency resulting from the short migration distance paired with the low electron transfer resistance [110]. A different method is to make nitride-based nanocomposites. In this regard, Wang et al. developed sandwich-like metal-free nanocomposite photocatalysts including 2D RGO, heterojunction 2D g-C_3N_4 nanosheets, and α-S_8. Based on their composition sequence, the photocatalysts were named $CNRGOS_8$ and $RGOCNS_8$. RGO has a smooth surface with spherical sulfur crystals when it is positioned in the outer layer. On the 2D g-C_3N_4 nanosheets surface, sulfur crystals have cross-linked cylindrical morphologies. Both materials have a red-shifted light absorption spectrum and a strong photocatalytic antibacterial activity [102].

In addition, the metal heterostructured Au/g-C_3N_4 nanocomposite has high disinfection efficiency [111]. Ma et al. also used thermal polymerization and light reduction to generate an Ag/g-C_3N_4 heterostructure. The concentration of Ag NPs has an effect on disinfection efficacy, and they observed that 3% is the optimal ratio. The hybrid effect of Ag and g-C_3N_4 significantly raised the oxidation potential while inhibiting the rate of recombination of the material's e^-/h^+ pairs [112]. Li et al. encased lamellar g-C_3N_4 in micron-sized spherical flocculent anatase TiO_2. Due to its enhanced light absorption efficiency, this nanocomposite displayed improved photocatalytic bacterial inactivation capabilities [113]. Li et al. used an in situ solvothermal approach to construct a Bi_2MoO_6/g-C_3N_4 heterojunction in another investigation. The increased specific area of the nanocomposite surface facilitates additional active site regions and accelerates the flow of photogenerated carriers, which is advantageous for photocatalytic decontamination [114]. Because of its numerous advantages, 2D g-C_3N_4 nanosheets have received a great deal of interest. The most frequent method for preparing pristine g-C_3N_4 is bulk exfoliation. Adding heterojunctions is a common method used by researchers to address the flaws of pristine g-C_3N_4 [115]. The g-C_3N_4 virus possesses the capability to inactivate not just bacteria but also viruses such as MS2 [116]. More research should be conducted to develop more efficient synthetic processes.

12.3.1.3 Metal-Based Antibacterial Nanomaterials

Metal nanoparticles (NPs)-based bactericidal materials, such as Ag, Cu, and Ni NPs, have sparked considerable interest and investigation [1]. These metal NPs have a poorer bacteria-killing efficacy than 2D NMs because of their reduced specific area of the surface and fewer surface-active sites [117]. Building a 2D nanocomposite by depositing metal NPs onto the surface of 2D nanomaterials is one way for manufacturing high-performance antibacterial chemicals [118]. Mo and colleagues generated Pd@Ag nanosheets by lowering the Ag^+ concentration on the Pd nanosheets (Pd NSs) surface. The nanosheets created had a hexagonal plate-like shape and an average diameter of 85 nm. When subjected to NIR light at 808 nm, Pd@Ag nanosheets had an excellent photothermal conversion effect and could release bacteria cell membranes damaging Ag^+ [119]. Photothermal conversion combined with Ag^+ release results in possible antibacterial activity against Gram-negative bacteria with thinner cell walls (*E. coli*). Wang et al. developed GO@Au@Ag 2D nanosheets by depositing Ag on Au-doped BSA-GO nanosheets. They revealed that the 2D nanosheets formed are more antimicrobial than pure Au@Ag NPs. The high concentration of Ag NMs near bacteria, together with the polyvalent effect, is thought to have resulted in strong interactions between nanocomposites and bacteria, resulting in successful bacteria removal [120].

2D metallic NMs such as ZnO, MgO, and TiO_2 have showed remarkable antibacterial capabilities and attracted tremendous consideration because of their favored photocatalytic activities [1]. The reasons for this are a high density of surface active sites, a large specific surface area, and a short migration distance for photogenerated charges from the bulk to the material surface [121]. Ultrathin TiO_2 nanosheets are the most accessible antibacterial materials among 2D metallic nanocomposites due to their low cost, promising stability, excellent disinfection efficacy, nontoxicity, and robust oxidizing activity [122]. TiO_2 has been classified into three phases (brookite, anatase, and rutile) [123, 124]. TiO_2 anatase was widely employed as a photocatalyst and was acknowledged as a critical semiconductor [125]. Rutile (TiO_2) has a visible light absorption range; however, the enhanced recombination rate of photoinduced e^-/h^+ pairs restricts its usefulness when compared to anatase (TiO_2) [126]. Conversely, brookite (TiO_2) nanocrystals displayed the highest photocatalytic activity among them, which may be due to the orthorhombic structure of brookite [127]. However, the difficulties in getting pure TiO_2 brookite nanoparticles slowed their development [128]. As an outcome, anatase (TiO_2) is still the most used photocatalyst. Wang et al. exfoliated layered titanate precursor to form TiO_2 nanosheets with lateral dimensions of hundreds of nanometers. TiO_2 nanosheets (TNS) can also break bacterial cell membranes, resulting in bacterial inactivation, according to the researchers [129]. Because of its relatively broad band gap and high recombination rate of photoinduced e^-/h^+ pairs, pure TiO_2 nanosheets are not commonly employed [130]. One technique in this area could be to dope TiO_2 with cocatalysts. Mn- and Co-doped TiO_2 nanosheets were used to inactivate MS2 coliphages, with a population drop rate of up to 99.9% [131]. To improve disinfection performance, Ma et al. used a solvothermal technique to generate ultrathin Fe_3O_4-TiO_2 nanosheets (Fe_3O_4-TNS). Photogenerated electrons can be captured by the excellent conductor Fe_3O_4, preventing e^-/h^+ pair recombination [132]. The generated nanoparticles demonstrated great antibacterial activity as well as strong visible light area absorption. Magnetic separation enables the recovery of Fe_3O_4-TNS nanosheets from water, allowing photocatalyst reuse and minimizing secondary environmental damage. Wang et al. also generated B, F-codoped TiO_2 nanosheets in a one-pot hydrothermal process employing $Ti(OBu)_4$ and HBF_4. Their findings demonstrated that codoped nanosheets outperformed single-doped nanosheets in photocatalytic activity [133]. Zhang et al. discovered a heterostructure composed of $Na_2W_4O_{13}$ flake and a WO_3/PdO mixture. The heterojunction formation has been shown to efficiently limit e^-/h^+ pair recombination and widen the absorption to the visible light range, resulting in better photocatalytic disinfection efficiency [134].

Antibacterial activity has been demonstrated for metallic NMs like Ag, Ti, Au, Cu, Mg, Zn, and Ni [55]. Nonetheless, the bactericidal action of these metallic nanoparticles is diminished when compared to 2D NMs because of their lower specific area of surface and fewer surface-active sites.

As a result, to overcome the drawbacks of metal alone and get high-performance antibacterial action, either coat 2D nanosheets with metallic nanoparticles or construct 2D metallic nanosheets [55]. Mo et al. used this method to create Pd@Ag 2D nanosheets by reducing Ag^+ ions on the Pd nanosheets surface with formaldehyde. Photothermal therapy (PTT) and Ag^+ ions created under NIR light irradiation may have a synergistic antibacterial effect [119]. The amount of Ag^+ released from Pd@ Ag 2D nanosheets was measured using ICP-MS (inductively coupled plasma mass spectrometry), and the findings indicated that NIR irradiation may considerably boost silver release compared to non-irradiated controls. Interestingly, bacteria viability demonstrated that only NIR lasers or Pd@ Ag 2D nanosheets effectively killed the bacteria; however, Pd@Ag nanosheets after NIR irradiation may provide efficient bacteria-killing action, as approximately 100% of bacteria were killed after a 10-minute synergistic treatment. When exposed to NIR light, Pd@Ag 2D nanosheets with excellent photothermal conversion effect may not only generate heat to kill bacteria but also disintegrate the bacterial membrane with freed Ag ions. 2D nanosized metallic oxides such as TiO_2, CuO, and ZnO have lately demonstrated good antibacterial performance and received widespread interest because of their beneficial photocatalytic activities, benign biosafety, strong oxidizing power, and rich surface-active sites [55]. Ultrathin TiO_2 nanosheets are most efficient antibacterial agents, owing to its numerous benefits such as low cost, minimal toxicity, exceptional stability, and robust oxidizing activity. In general, TiO_2 nanosheets have three-phase structures: brookite, rutile, and anatase. For example, anatase (TiO_2) nanosheets have been widely employed as a photocatalyst [55]. Arnab et al. developed MoS_2-TiO_2 nanohybrids with important antibacterial capability to examine TiO_2's in-depth capability [135]. In addition, using solvothermal approach and reverse lamellar micelles, Ma et al. developed ultrathin Fe_3O_4-TiO_2 2D nanosheets (2D Fe_3O_4-TNS) for efficient antibacterial use [132]. Fe_3O_4-TNS, in contrast, could successfully capture photoinduced electrons and accelerate electron–hole pair separation. However, when exposed to sunshine, Fe_3O_4-TNS displayed superior antibacterial activity against *E. coli.* Wang et al. also developed a bifunctional layered Cu_2S nanoflowers/biopolymer electrospun membrane for tumor therapy and healing of wounds [136]. Though these trials indicated that metal-based nanomaterials have great antibacterial potential, more research into their cytotoxicity and immune response effect on human cells is required. As a result, more in vivo research on the biological function of metal-based antibacterial nanoparticles is required before they may be used in clinical sterilization.

12.3.1.4 MXenes

MXene is a layered 2D metal carbide or nitride structure [137]. The majority of their composition is made up of transition metals (such as Ti, V, Cr, and Nb), IIIA/IVA group elements (such as Si and Al), and carbon or nitrogen. Its chemical formula is $M_{n+1}AX_n$ (n = 1, 2, or 3). MXenes, like graphene, have a huge specific area of surface and excellent conductivity (9.88 x 10^5 Sm^{-1}), but they also benefit from flexible component adjustment and controlled minimum nanolayer thickness [138]. As they are formed through etching, MXenes have a single or few layers (width < 1 nm). Their lateral dimensions range from nanometers to micrometers. MXene surfaces modified with functional groups such as hydroxyl, oxygen, or fluorine are hydrophilic and have numerous applications in medicine [139]. They are also eliminated by the human body because of their nontoxicity, biocompatibility, and degradability. MXenes are also extremely sensitive to NIR light. As a result, surface modifications are used to improve their photocatalytic-based antibacterial applications [140].

MXenes, a novel class of 2D materials, were recently shown to have outstanding antibacterial characteristics. Rasool et al. used $Ti_3C_2T_x$ powders to produce thin, transparent, and hydrophilic $Ti_3C_2T_x$ nanosheets (T refers to the surface termination, such as O, OH, or F) [141]. $Ti_3C_2T_x$ nanosheets have a greater surface area than Ti_3AlC_2 and multilayer 2D $Ti_3C_2T_x$. Hydrogen bonding between the oxygenate groups of 2D $Ti_3C_2T_x$ MXene and cell membranes inhibits the bacterial growth. Furthermore, the high hydrophilicity and cation surface improve bacterial contact. As a consequence, in terms of antibacterial activity, 2D $Ti_3C_2T_x$ outperforms virgin 2D GO nanosheets. Surprisingly, due to the synergistic action of $Ti_3C_2T_x$ and TiO_2/C produced on the surface, the aged $Ti_3C_2T_x$ MXene membrane displayed considerably better antibacterial properties than the freshly generated membrane [141]. Studies on Ti_2C and Ti_3C_2 MXenes revealed that the atomic structure of MXenes has considerable role on their ability to kill bacteria [142]. The fabrication of silver/MXene ($Ti_3C_2T_x$) 2D nanosheets increased $Ti_3C_2T_x$ MXene's antibacterial activity [143]. However, additional investigation into the antimicrobial performance of distinct MXenes with different transition metals or surface terminations is required. MXenes are a novel type of 2D NMs containing transition metal carbides/nitrides with the general formula $M_{n+1}A_nT_x$ (M for transition metal, A for carbon or nitrogen, and T for surface-terminating functional groups such as -OH, -F, or -O) [144]. Over 70 different varieties of MXenes are reported, with $Ti_3C_2T_x$ constituting the most distinctive MXenes material [145]. Due to their ultrathin lamellar structure and unusual physiochemical properties, $Ti_3C_2T_x$ MXenes nanosheets have demonstrated outstanding antibacterial capabilities in recent studies [55, 145].

Rasool et al., for instance, used ultrasonication in argon (Ar) gas to form $Ti_3C_2T_x$ MXenes colloidal suspension and first shown their outstanding antibacterial action against *Bacillus subtilis* and *E. coli* [141]. Unlike GO, the $Ti_3C_2T_x$ MXenes 2D nanosheets showed distinct dose-dependent bactericidal efficacy, with up to 98% of bacteria killed after 4 hours of incubation with 200 g mL^{-1} $Ti_3C_2T_x$. Based on the SEM and TEM images, they believed that the bactericidal mechanism of $Ti_3C_2T_x$ MXenes 2D nanosheets was due to the synergism of sharper edge-produced membrane damage and electron mobility rendered oxidative stress. They next utilized vacuum-assisted filtration to make $Ti_3C_2T_x$ MXenes 2D nanosheets wrapped in polyvinylidene fluoride (PVDF) membranes, which they tested for antibacterial action against *E. coli* and *B. subtilis* [141]. When compared to fresh $Ti_3C_2T_x$/PVDF membranes, aged $Ti_3C_2T_x$/PVDF membranes were more active in boosting overall antibacterial properties due to the synergistic effect of $Ti_3C_2T_x$ 2D nanosheets and the growth of anatase (TiO_2) nanocrystals with sharp edges. Shamsabadi et al. also discovered that colloidal $Ti_3C_2T_x$ MXenes nanosheets have antibacterial properties that vary with size and exposure time. They discovered that direct contact with the pointed ends of $Ti_3C_2T_x$ MXenes 2D nanosheets and the bacterial membrane damages the membrane significantly [146]. Considering that the chemical behavior of MXenes is determined by the transition metals (such as Mo, Nb, V) and surface-terminating functional groups, unique physicochemical properties are directly associated to antibacterial activity. As a consequence, fine-regulated ultrathin MXenes nanosheets will find widespread application in antibacterial surfaces and applications in medicine.

12.3.1.5 Black Phosphorus

Phosphorus has received a considerable interest in recent decade because of its widespread availability and environmentally friendly features [45]. So far, phosphorus allotropes such as BP, amorphous red phosphorus (RP), Hittorf's RP, and fibrillar RP have been identified [45]. Among these, BP stands out from other allotropes due to its 2D structure, which has attracted a great interest in the literature. The most sophisticated techniques for preparing BP include mechanical extraction, liquid extraction, and in situ CVD. Every phosphorus atom has a pair of unshared electrons due to the valence electrons and configuration of phosphorus [147]. As a consequence, it has the potential to be a photocatalytic antibacterial, and BP's antibacterial physical puncture approach is also beneficial. BP, an emerging player in the 2D nanomaterials family, has received a lot of interest since 2014 [55]. 2D BP nanosheets have been receiving considerable interest because of their remarkable thermal/optical/electrical performance and the poor van der Waals forces interactions for easy exfoliation into ultrathin 2D nanosheets [55]. Because of their metal-free semiconductor properties and thickness-dependent band gap, BP nanosheets offer extraordinarily broad light absorption over full visible areas [55]. Therefore, under visible light, 2D stacked BP NSs may be utilized to create $^1O_2$, which can subsequently be used in photodynamic therapy (PDT) [55]. In this regard, Mao et al. created a BP-based hybrid hydrogel (CS-BP) therapeutic system for reproducible PDT decontamination and healing of wounds [148]. They captured $^1O_2$ using ESR (electron spin resonance) spectra and discovered that only the CS-BP hydrogel could rapidly produce $^1O_2$ when subjected to visible light. Furthermore, when placed under sunlight (simulated conditions), the CS hydrogel displayed moderate bactericidal efficacy, whereas the CS-BP killed much of the bacteria (*S. aureus* and *E. coli*) in less than 10 minutes. Moreover, even after four challenges with high concentrations of *S. aureus*, the CS-BP hydrogel has favorable reusable antibacterial effectiveness. Huang et al. also synthesized thin-layer BP@ silk fibroin NSs with slight solution-refinement employing the silk fibroin (exfoliating agent), confirming that BP@ silk fibroin NS could not only be fabricated into multiple forms but also showed excellent PTT decontamination and wound repair capability [149]. Given the preceding excellent results, BP-based NMs show great promise for future material science, biomedicine, and biotechnology.

12.3.1.6 Transition Metal Dichalcogenides

TMDs are a new category of 2D NMs [45]. The formula of this substance is MX_2, and it has three atomic layers. Three chalcogenide atomic layers (S, Se, and Te) are connected by a transition metal atomic layer [45]. In TMDs, van der Waals interactions connect two adjacent chalcogenide layers, enabling a 2D structure to be preserved. Their ultrathin atomic single- or multilayer structure produces semiconductivity and semimetallic magnetism. Because of their distinctive optical, mechanical, and electrical properties, TMDs are broadly utilized in optoelectronics, catalysis, biomedical engineering, and solar energy batteries [150]. MoS_2 has significant catalytic activity throughout its edge, which, as opposed to the base surface of TMDs, functions as the catalytic center. However, the thermodynamic reactions that occur on the surface of substrate limit the activation of active surface sites [45]. Accordingly, increasing the photocatalytic antibacterial efficiency of TMDs necessitates atomically controlling the surface structure, which includes increasing the defective proportion and specific surface area to expose more edge regions. TMD nanocomposites have stimulated extensive research in the last few years, and they have become crucial building blocks for solar cells, electronic devices, sensors, and photothermal cancer treatment systems [1]. Nanostructured WX_2 (X=S, Se) and MoS_2 were evaluated to be exceptional antimicrobial materials among all TMDs, and disinfection efficiency was greatly improved. Until now, synthetic approaches for TMD have been widely established. High-quality TMDs can be produced via chemical exfoliation or CVD [1].

Exfoliation techniques comprise liquid phase exfoliation using suitable solvents or surfactants, as well as chemical exfoliation by basal-plane modification [1]. Bang et al. developed a simple and high-yield exfoliation technique for WX_2 (X=Se, S) using single-stranded DNA (ssDNA). The incorporation of ssDNA into WX_2 nanosheets can prevent aggregation and increase exfoliation efficiency. They underlined that the as-synthesized 2D nanomaterials have a significant amount of specific surface area, which facilitates the formation of surface active sites [151]. Furthermore, the antibacterial effects of MoS_2 nanomaterial have been studied. Yang et al. revealed that ceMoS_2 (chemically exfoliated MoS_2) with stacked layer structures have strong antibacterial activity because of their 2D planar structure, high specific surface area, and superior electrical properties. The 2D structure of ceMoS_2 allows for efficient interaction with bacteria, leading to bacterium-death membrane and oxidative stress [152]. Using the lithium intercalation exfoliation procedure, Pandit et al. created neutral, positively, and negatively charged MoS_2 nanosheets. These MoS_2 nanosheets demonstrated significant wide spectrum of antibacterial properties [153]. The researchers developed antibacterial Fe_3O_4-modified water-dispersible MoS_2 nanocomposites. Magnetic behavior enhancement using Fe_3O_4 nanoparticles can boost MoS_2 photothermal conversion efficiency and thus antibacterial efficacy.

Additionally, polyethylene glycol-modified MoS_2 (PEG-MoS_2) was produced using a one-pot hydrothermal method. The composite is shaped like a flower and has great aqueous dispersibility. The as-prepared PEG-MoS_2 nanocomposite exhibits considerable NIR absorbance as well as peroxidase-like activity, capable of catalyzing H_2O_2 to generate OH. •OH produced can then damage the cell membrane of bacteria. By combining the •OH generating and photothermal effects, the as-synthesized nanocomposite displayed outstanding disinfection performance. The studies also demonstrated that the antibacterial system can be used in vivo to clean wounds [154]. According to Alimohammadi et al., vertically aligned nanosheets show better antibacterial activity than randomly oriented nanosheets [155]. Cui and coworkers reduced the domain size to 3–10 layers to develop few-layered vertically aligned MoS_2 (FLV MoS_2) films. The band gap of this FLV-MoS_2 raised (1.3–1.55 eV). Moreover, the FLV-MoS_2 nanosheets have a high e^-/h^+ transport rate, contributing to a high ROS generation yield and photocatalytic killing of bacteria with high efficiency [156]. Metal heterojunctions produced on FLV-MoS_2 with Cu or Au NPs exhibited rapid electron movement, suppression of photogenerated e^-/h^+ pair recombination, and a considerable ROS emission. As a consequence, a satisfactory ROS generation yield was obtained, indicating that efficient bacterial killing is possible. Because of its unusual graphene-like features, which are made up of a "sandwich" structure of "B-A-B" or "B-A-O" via van der Waals interactions, ABX (A: transition metal, B: chalcogen), with the typical chemical formula for 2D TMDC/Os, has been intensively investigated in the nanotechnology area in the recent years [55, 157]. The ones that have received the most focus include archetypal TMDC/Os, $MoSe_2$, MoS_2, MoO_2, WS_2, and WO_{3-x} [55]. Due to their exclusive characteristics linked with their 2D ultrathin atomic layer structure and enormous surface area, MoS_2 nanosheets, for example, have shown significant promise in the healthcare field [158].

Because of their layer-dependent band gap, MoS_2 nanosheets with very broad photodetection capacity may be utilized as a candidate for photothermal/photocatalytic antibacterial applications [158]. The authors also used this property to build biocompatible 808 nm laser-mediated NO-releasing MoS_2-BNN_6 nanovehicles for low-cost, rapid, and effective antibacterial therapy [55]. MoS_2 nanosheets exhibiting excellent photothermal conversion efficiency and hyperthermia might regulate NO delivery and release after 808 nm laser irradiation. PTT and NO release synergistically resulted in outstanding germicidal ability that killed *Enterococcus faecalis* and *E. coli*. In addition, the test results showed that MoS_2-BNN_6 nanovehicles with PTT/NO synergetic antimicrobial properties could cause considerable damage to DNA. The antibacterial strategy's practical significance was validated by healing of wounds (in vivo) in mice. Likewise, Liu et al. developed a few-layered vertically aligned MoS_2 (FLV-MoS_2) film that can capture the whole visible light spectrum for efficient photocatalytic water disinfection [159]. Because of the increased band gap of MoS_2 from 1.3 (bulk material) to 1.55 eV (FLV-MoS_2), FLV-MoS_2 may generate considerable ROS for bacteria inactivation in water when exposed to visible light. They also discovered that incorporating Cu or Au into FLV-MoS_2 films may aid in electron–hole pair separation and catalyze ROS production reactions, allowing FLV-MoS_2 to inactivate >99.99% of bacteria in less than 20 minutes. Furthermore, Cheng et al. used CVD to construct a pyramid-type MoS_2 (pyramid MoS_2) on transparent glass for very effective water disinfection when exposed to visible light [160]. They also found that the pyramid MoS_2 has a lower band gap, allowing it to capture a broader spectrum of sunlight and produce more ROS for bacteria killing. MoS_2/$MoSe_2$ nanosheets have peroxidase-like activity in addition to photoinduced germicidal effect.

12.3.1.7 Bismuth-Based Materials

Bismuth (Bi)-based compounds have attracted wide attention due to their large stability, nontoxicity, wide energy band structure, and outstanding photodegradation performance against pollutants. Bismuth halide oxide, bismuth oxide, bismuth molybdate, and bismuth tungstate (Bi_2WO_6) are all being investigated as photocatalysts [161]. The bonding of bismuth and oxygen is possible due to the hybridization of the Bi(6s) and O(2p) orbitals. Their distinctive atomic layer configuration assists in electronic conduction. Also, their adjustable physical and electronic characteristics permit Bi-based NMs to improve photocatalytic activity throughout a wide temperature range [162]. Consequently, the following research directions are now being explored in order to increase the photocatalytic antibacterial properties of Bi-based materials: doping, or introducing defects, altering the size, semiconductor bonding, conjugated structural surface modification, and metal deposition. When compared to 1D and 3D nanomaterials, 2D BiOCl-based nanomaterials displayed higher catalytic activity because of huge specific area of their surface [163]. However, their drawbacks of UV spectrum absorption and low overall quantum efficiency cannot be solved. Metal NPs deposition is one remedy for the aforementioned BiOCl nanosheet flaws. BiOCl-Ag nanocomposites were produced by depositing Ag NPs on the BiOCl NS surface [164]. The generated nanosheets have a red-shifted light absorption range due to the surface plasmonic resonance effect of Ag NPs. The inclusion of Ag NPs can also boost the charge separation efficiency and quantum yield of nanosheets throughout photocatalytic processes. Each of these variables contributes to increased disinfection efficiency.

12.3.1.8 Boron Nitride

The two-dimensional crystal of BN is composed of N and B atoms. High-temperature stability, strength, thermal conductivity (22 $Wm^{-1} K^{-1}$), a low expansion coefficient ($8.1 \times 10^{-6} K^{-1}$), high resistivity (14.4 cm), and corrosion resistance are all advantages of hexagonal BN [165]. They are made using hydrothermal synthesis, chemical gas phase synthesis, benzene thermal synthesis, carbothermal synthesis technology, and self-propagating technology [166]. The properties of BN NS and their applications complement those of graphene, and they have advantages in terms of thermal and chemical stability. The biocompatibility of BN is influenced -by its size, shape, structure, and surface groups. Unsaturated boron atoms on the edge or surface of nanosheets promote bacterial mortality. In the realm of photocatalytic antibacterial NMs, the large thermal conductivity and specific area of BN surface are widely used to create heterojunctions by combining with metals, semiconductors, and metal oxides to advance antibacterial activity [167]. BN-based materials may also be utilized as antibacterial reagents. Amorphous BN (a-BN), hexagonal BN (h-BN), and cubic BN (c-BN) are the three known crystalline formations of BN [168]. The most attention has been paid to h-BN. It is an atomic-thick 2D material with an iso-structure comparable to graphene that has a broad band gap as well as remarkable thermal and chemical stability. Other methods for manufacturing BN nanosheets have been devised, including CVD and chemical and thermal exfoliation [168]. Due to carbon's sp^2 hybridization, the edges of h-BN show high chemical activity while having a low binding energy. As a consequence, h-BN is an excellent building scaffold for a broad variety of functional materials [168,169]. The investigation of h-BN-based nanoparticles' antibacterial activity is in its early phases. Gao et al. used a microwave to create Ag NP (SNP) embedded h-BN in a one-pot technique, resulting in uniform growth of SNP. The proposed h-BN nanomaterial displayed outstanding antimicrobial properties [170]. Furthermore, Roy et al. created Ag NP-adorned BN nanosheets (AgNP-BNNS), which displayed substantial antibacterial activity. They believed that the antibacterial entity is Ag NPs, whereas the substrate is BN nanosheets [171]. Therefore, it is necessary to develop new synthetic techniques for BN-related materials and conduct additional study on their antibacterial characteristics.

12.3.1.9 Other 2D Nanomaterials

Other novel 2D nanomaterials that have been identified as prospective antibacterial agents include laponite (Lap), layered double hydroxides (LDHs), group III–VI compounds (In_2Se_3, Bi_2Se_3, and Sb_2Se_3), hexagonal BN, and RuO_2 [55]. Wang et al., for example, modified the cation ratio to create several morphologies of lysozyme-modified LDHs and discovered that the bloom flower form of LDHs not only could load more lysozyme but also cling to more bacteria, illustrating superior antibacterial activity and wound repair abilities [172]. Ghadiri et al. also developed a Laponite/mafenide/alginate (Lap/Maf/Alg) film with good antibacterial and wound healing capabilities [173]. Additionally, Zhu et al. used a solvent exfoliation method to produce large quantities of In_2Se_3 nanosheets. Their findings showed that In_2Se_3 nanosheets had good photothermal performance, which results in significant antibacterial activity [174]. Similarly, Miao et al. used liquid exfoliation technique to develop unique atomically thin antimony Sb_2Se_3 nanosheets capable of killing bacteria via physical contact destruction and short-term hyperthermia sterilization when exposed to laser light [175]. Thus, other researchers may be interested in various antibacterial properties of these new 2D nanomaterials because of their good bacteria inhibition efficiencies.

Many advances have been made in the improvement of nanotechnology-based antibacterial nanoagents (graphene, noble metals, organic polymers, and so on) for combating multidrug resistance in bacteria through physical contact damage, oxidative stress, photoinduced antibacterial (such as PTT, PCT, and PDT, controlled drug/metallic ions release, and multi-mode synergistic antibacterial) [55]. The 2D graphene and graphene-based 2D NMs have shown promising roles in biomedicine (antibacterial activity, tumor diagnosis, therapy, and neurological disorder treatment) in the last decade due to their intriguing biochemical properties such as easy preparation and functionalization, high specific surface area, and good biocompatibility [55]. In addition, graphene-based nanomaterials research has aided in the identification of new two-dimensional nanomaterials beyond graphene (2D NBG) [78]. It's grateful that in recent decade, 2D NBG showed appealing applications such as optical/electronic devices, biomedicine, and catalysis, which may be due to graphene-like physical and chemical properties such as large specific surface area, ultrathin 2D nature, remarkable light-to-heat capability, exceptional photocatalytic features, facile surface modification, and comparatively safe biocompatibility [55]. According to recent studies, 2D NBG has a substantial antibacterial effect. Moreover, these antibacterial nanoparticles based on graphene-like nanomaterials displayed a broad variety of antimicrobial mechanisms [176]. Several novel 2D nanomaterials have been introduced, such as transition metal dioxides (TMDOs), transition metal dichalcogenides (TMDs), graphitic carbon nitride (g-C_3N_4), BP, LDHs, and transition metal carbides and nitrides (MXenes) [177]. At the present time, there is a lot of research being done on 2D NM antibacterial applications, which includes the direct interaction mechanism between 2D NM and bacteria [55].

12.3.2 Antibacterial and Antimicrobial Mechanisms of 2D Nanomaterials

In studying the antibacterial action of 2D NMs (metals and metal oxides, as well as newly produced NMs), exogenous and endogenous antimicrobial approaches are applied. These 2D NMs interact with microbial cells in a variety of ways because of their different 2D structures and plethora of surface functional groups. They also have quantum mechanical properties such as changeable thickness and energy band gaps. 2D NMs hinder bacterial action by their intrinsic antibacterial properties, but they also have an indirect influence on microbes by their reaction in specific settings. The main antibacterial mechanisms (endogenous and exogenous modes) documented in the literature are depicted in Figure 12.5. The nanoparticles may directly interact with microbial cells in ways that harm them, such as disrupting transmembrane electron transfer, disrupting/penetrating

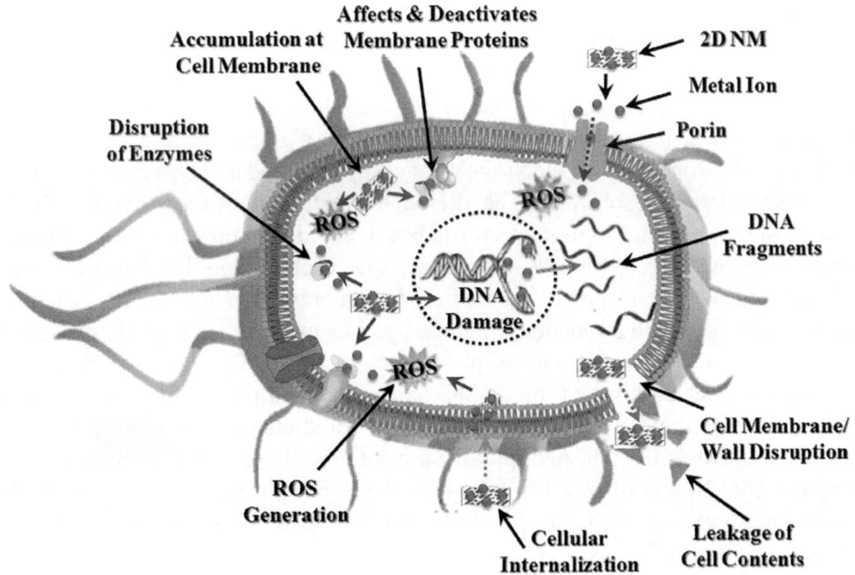

FIGURE 12.5 A depiction of various antibacterial mechanism modes of 2D NMs.

the cell envelope, or oxidizing cell components, or they may produce secondary products that cause damage (ROS or dissolved heavy metal ions).

12.3.2.1 Cell Membrane/Wall Disruption

The cell wall of bacteria can be found in the cell membrane's outermost layer [45]. Although they share the functional peptidoglycan component, the surface constituents of Gram-positive bacteria's cell membrane and Gram-negative bacteria's cell membrane differ considerably [45]. The membrane of bacterial cells serves the following purposes: (i) maintaining cell shape and increasing mechanical strength; (ii) inhibiting mechanical and osmotic damage; (iii) assisting in cell mobility and proliferation; and (iv) conferring the specific antigenic properties of antibiotics against bacteriophages [45]. When nanosheets interact with microbes, their pointed edges contribute to bacterial membrane destruction. DNA, phospholipids, and proteins leak into the cell as the bacterial membrane degrades. By physical means of action, antibacterial 2D NMs have the subsequent benefits: (i) they possess complete eradicating role on a broad variety of bacteria; (ii) their large stability thwart the development of bacteria for an extended period of time; and (iii) they would not result in secondary pollution. A rising variety of 2D NMs, including MoS_2, WS_2, GO, and BP, have shown to cause physical rupture capabilities and are used in physical contact sterilization in recent years [82]. Lu et al. studied the interaction of several GO inclinations against microorganisms [178]. A magnetic field was employed to generate three different GO nanosheet orientations (random, vertical, and planar). Bacteria on random and planar GO kept their shape, whereas bacteria on vertical GO shriveled considerably, indicating a loss of viability and cell membrane damage. The rough edges of graphene nanosheets were shown to disrupt the bacterial membrane in this process. Dye fluorescence analysis demonstrated that membrane rupture in the GO/lipid vesicle NMs was resulted from physical destruction of the bilayer structure of the GO NSs, instead of oxidation by ROS species. Particular 2D NMs have been shown to have physical penetration bactericidal properties, while the precise mechanism is unknown.

There are three types of NM insertion into bacteria: First, gravity causes bacteria to come into touch with the edge of the nanosheet. Wei et al., for example, built and tested the antibacterial activities of horizontally and vertically oriented graphene on silicon (Si) and silicon dioxide (SIO) substrates against diverse bacteria [179]. According to the results of the trials, the specific antibacterial mechanisms of different bacteria species vary. Gram-positive bacteria used physical penetration as a defense strategy. Because of their gravity, the bacteria were easily punctured by the vertically oriented graphene layer, shattering the bacterial membrane and deactivating the germs. Electron transfer was determined to be the antibacterial mechanism in Gram-negative bacteria. The bacteria membrane surface had higher negative charges because the respiratory proteins in the microbial membrane generated a negative charge. Furthermore, because graphene has a high conductivity and can be used as an electron acceptor, electrons were easily captured and transferred from the microbial layer to the underneath substrate. The silicon substrate system outperformed the insulating silica substrate in terms of antibacterial performance. It could be more clearly established that graphene caused a continuous electron loss in gram-negative bacterial biofilm, thus leaving it inactive.

Second, nanosheets can adsorb to the bacterial membrane and disrupt its structure via van der Waals interactions and hydrophobic interactions. For example, Cui et al. studied the sterilizing properties of g-C_3N_4 treated with N plasma (N-g-C_3N_4) [180]. The researchers discovered that direct physical contact between the g-C_3N_4 nanosheets and the cell membrane produced cell rupture. When the nanosheets were close to the phospholipid molecules on the cell membrane, the Coulomb contact between the phospholipids with positively charged amino groups and the negatively charged N-g-C_3N_4 nanosheets drove the nanosheets into the phospholipid bilayer. Following that, the nanosheets were easily integrated into the phospholipid bilayer, resulting in N-g-C_3N_4 with strong physical antibacterial properties.

Finally, it has been demonstrated that disruptive extraction causes membrane deformation and integrity loss. Tu et al., for instance, studied the chemical process of graphene-induced *E. coli* cell membrane disintegration and discovered that, in addition to severe insertion and rupture, lipid molecule extraction played a part in the process [181]. Computer models and TEM images were employed to verify the presence of direct phospholipid extraction from lipid membranes. The peculiar 2D structure of graphene, which contains sp^2 carbon atoms, contributed to the formation of an extraordinarily strong dispersion contact between graphene and lipid molecules, leading to a strong adhesion. As a result of the rearrangement of hydrophobic tails, the isolated lipids performed synergistic movements on the graphene surface to maximize the hydrophobic contact.

12.3.2.2 Oxidative Stress

It can be explained as an imbalance in microorganisms' oxidative and antioxidative activity in response to environmental stressors. The formation of oxidative stress oxidizes and destroys the molecules on membrane surface, disrupts critical cell functions, impedes bacterial metabolism, and eventually leads to bacterial deactivation [45]. Therefore, it is a popular antibacterial strategy. 2D NMs, as shown in Figure 12.5, mediate two oxidative stress types. The first is ROS-induced oxidative stress. When bacteria are stimulated, the internal redox reaction becomes imbalanced, resulting in an excess buildup of active chemicals that consist of hydrogen peroxide (H_2O_2), hydroxyl radical (OH), superoxide anion ($\bullet O_2^-$), and singlet molecular oxygen ($^1O_2$) [45]. The second kind of oxidative stress is ROS-independent oxidative stress that happens when nanomaterials come into close contact with bacterial membranes through electron transfer or redox processes, causing damage or oxidizing membrane structure and its surface contents instead of producing ROS [45].

Karunakaran et al. tested the underlying cause of ROS-dependent oxidative stress by exfoliating and modifying 2H-MoS_2 nanosheets with different surfactant molecules having thiol functionalities [182]. In the main mechanism, positively charged and functionalized 2H-MoS_2 interacted with the bacterial surface to generate ROS. The conduction band (CB) of 2H-MoS_2 was lower than the redox potential of ROS creation, contributing to ROS generation. A glutathione (GSH) depletion experiment validated the cells' ROS production. Gurunathan et al. discovered that GO and rGO exhibit antibacterial effects against *P. aeruginosa* [183]. GO and rGO both increased oxidative stress in cells by disrupting the balance of oxidant and antioxidant activity. ROS accumulated and harmed biological components like DNA and proteins. Another 2D NM (BP) is employed in killing bacteria by boosting the creation of oxidative stress. Xiong et al. looked into antibacterial performance of BP NSs following the exfoliation [184]. They assessed intracellular ROS levels by measuring fluorescence intensity, revealing that the main bactericidal mechanism of BP was to generate ROS-dependent oxidative stress. The BP nanosheet group, in particular, had more obvious green fluorescence than the untreated group, suggesting that BP increased ROS formation in bacteria, leading to bacterial mortality. Furthermore, the BP nanosheets displayed time- and concentration-dependent bactericidal efficacy against two bacterial strains, Gram-positive *B. subtilis* and Gram-negative *E. coli*. Because *E. coli*'s cell wall was thinner, it was more sensitive to BP nanosheet treatment at an earlier stage (6 h), hence the growth inhibition and bactericidal effects on *E. coli* were more pronounced than those on *B. subtilis*. Given *E. coli*'s proclivity to self-heal, BP's antibacterial actions against it were partially reversible, with the effect wearing off after 12 hours.

In contradiction to ROS-dependent mechanisms, ROS-independent mechanisms are direct oxidation reactions. Pandit et al. produced ce-MoS_2 by chemically exfoliating and modifying MoS_2 with different thiol ligands. The surface of the material was then charged differentially and had distinct hydrophobicities, resulting in highly effective antibacterial activity against microbes and related biofilms [153]. Their research demonstrated that the bacterial surface was effectively connected to MoS_2 nanosheets because of the positive charge, worsening the bacterium's response to oxidative stress. Long-chain alkanes bind hydrophobically with bacterial cell membranes, depolarizing them to varying degrees, modifying their surface potential, restricting bacterial metabolic processes, and causing bacterial death. In addition, they may be able to limit the formation of bacterial biofilms while still inhibiting the formation of pre-existing biofilms. Kim et al. investigated GOs and MoS_2 composite systems with improved antibacterial activity because of improved oxidation ability and electrical conductivity [88]. In terms of ROS-independent oxidation, the GO-MoS_2 nanocomposite performed better. That is, it directly oxidized the bacterial cell structure, producing membrane damage and intracellular component leaking, and thereby eradicating the organism. Aside from direct contact with bacteria, 2D NMs can wrap them and render them inactive. Rasool et al. studied $Ti_3C_2T_x$'s antibacterial activities in colloidal suspension [185]. $Ti_3C_2T_x$ has good antibacterial properties due to its high hydrophilicity, which boosted interaction with germs, according to their findings. When the $Ti_3C_2T_x$ concentration grew, the bacteria were surrounded by nanoscale $Ti_3C_2T_x$ flakes, which formed aggregates. The high reducing activity and active surface of $Ti_3C_2T_x$ could be used to directly inactivate adherent bacteria. Finally, hydrogen bonds established by $Ti_3C_2T_x$ oxygen-containing groups and lipopolysaccharides on membrane inhibited the growth of the bacteria by preventing the intake of food.

12.3.2.3 Photocatalysis

Photodynamic treatment is a new technique for curing bacterial infections that makes advantage of the photocatalytic properties of photosensitive materials [45]. When photosensitive material is activated by light, electrons gain energy and transfer from the ground state to the excited state, as shown in Figure 12.6. The excited electrons react with water or oxygen to form extremely reactive $^1O_2$ and •OH, which immediately oxidize bacteria, causing cell damage and even cell death [186]. This technology has received a lot of attention, and the use of 2D NMs in the antibacterial industry is growing [45]. Chong et al. demonstrated the mechanism of GO's antibacterial activity under the influence of simulated sunlight [187]. According to ROS tests, GO produced only a few $^1O_2$ under dark conditions, and its oxidative activity was minimal. When exposed to light, photoinduced electron–hole pairs were produced on the active sites on the GO surface. These photoinduced electrons expedited the GO reduction, which ultimately introduced carbon-centered radicals, damaging the bacterial antioxidant defense system and destroying microorganisms. When exposed to visible light, Huang et al. observed that graphitic carbon nitride polymer was an effective and recyclable catalyst that allowed electrons and holes to interact with the bacterial membrane, leading to bacterial inactivation [188].

$$e^- + O_2 \rightarrow O_2^- \quad (12.1)$$

$$2H^+ + 2O_2^- \rightarrow H_2O_2 + O_2 \quad (12.2)$$

$$2e^- + 2H^+ + O_2^- \rightarrow H_2O_2 \quad (12.3)$$

$$h^+ + H_2O \rightarrow H^+ + \bullet OH \quad (12.4)$$

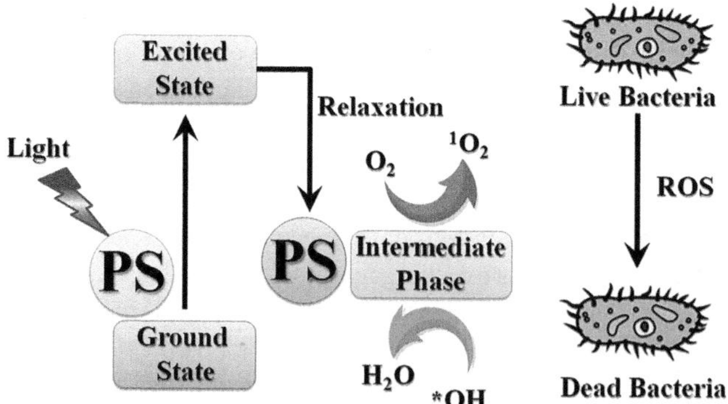

FIGURE 12.6 A display of steps involved in antibacterial activity of photosensitive 2D NMs.

Using *E. coli* as a model, Zhao et al. discovered that single-layer g-C_3N_4 exhibits exceptional disinfection efficacy due to the effective generation of ROSs [110]. Using the Ag/g-C_3N_4 nanocomposite as an example, the disinfection process typically consists of the three stages indicated below. To begin with, H_2O, O_2, and bacterial cells adhere to the Ag/g-C_3N_4 surface. Second, electrons (e^-) are transferred from the valence band (VB) to the CB, resulting in the formation of holes (h^+) in the VB. Third, when e^- interacts with adsorbed O_2, superoxide radicals (O_2), H_2O_2 are formed (Equations 12.1–12.3), which can cause cell wall and membrane damage [4,106]. These investigators additionally found that coating the g-C_3N_4 surface with Ag stimulates the formation of more $\bullet O_2$ than pure g-C_3N_4 [106]. An investigation of the n-WO_3 photocatalyst indicated the production of O_2 (Equation 12.1) and hydroxide radicals (OH) (Equation 12.4). Finally, disinfection activity can be increased by raising the specific surface area, lowering the layer thickness, depositing uniformly dispersed cocatalysts onto photocatalysts, and inhibiting e^-/h^+ pair recombination. This innovative disinfection methodology based on photocatalysis is promising since it not only may be used in disinfecting water more efficiently, but it may also assist to minimize contamination of the environment.

12.3.2.4 Photothermal Disinfection

Photothermal disinfection is the conversion of solar energy to heat produced by nanoparticles upon light adsorption. Following photon energy absorption, this mechanism, which has been observed in GNRs, involves the generation of heat via electron–phonon interactions and surface plasmonic resonance [189]. Despite this, photothermal disinfection can only use a small fraction of the solar spectrum. Only NIR and infrared light can pass through since organisms reflect the bulk of UV–vis light. Unfortunately, due to the low energy intensity of mid- and far-infrared light, not all infrared light spectrums can be employed for photothermal disinfection. NIR light is the most effective for photothermal disinfection. GNRs, Cu_7S_4 nanorods, and MoS_2 exhibit high NIR absorbance and photothermal conversion efficiency. When GNRs are subjected to NIR, the bacterial cell walls are irreversibly damaged and lysed, allowing biological components to spill. Both Gram-negative and Gram-positive bacteria have been discovered to be susceptible to photothermal nanosized antiseptics. The absence of potentially harmful by-products is the most enticing aspect of photothermal disinfection. However, the efficacy of photothermal disinfection is lower than that of other disinfection methods, such as metal ion disinfection and photocatalytic decontamination [4]. To raise disinfection efficacy even further, exceptionally efficient photothermal disinfection approaches, specifically the creation of new semiconductors with a suitable band gap structure and improved absorbance of NIR light, are still required. Tuning the structure and shape of nanoparticles, as well as heterojunction formation and elemental doping, may be effective strategies for developing 2D NMs for effective photothermal decontamination.

12.3.2.5 Photo-Fenton Disinfection

So far, water disinfection has depended on homogeneous and photo-Fenton techniques. The major Fenton reaction is explained by Equations (12.5) and (12.6). The extremely reactive radical OH generated can directly cause DNA damage, leading to bacterial cell death [190]. However, the main Fenton reaction produces just a little amount of $\bullet OH$ (Equations 12.5 and 12.6). To overcome this constraint, ferrous or ferric iron was combined with hydrogen peroxide (H_2O_2) and subjected to light (Equation 12.7), and the ESR spin trap technique was used to demonstrate an accelerated generation of ROS [4,191]. A photochemical technique can renew Fe^{2+} species after irradiation. In addition, the regenerated Fe^{2+} would react with H_2O_2 to form the $\bullet OH$ radical (Equation 12.5) [192]. Because of the synergistic action of H_2O_2 and light, the photo-Fenton process results in higher intracellular H_2O_2 concentrations and hence plays an important role as photocatalysis [77]. The presence of Fe^{3+} facilitates the formation of Fe^{3+}-organic complexes, allowing the photo-Fenton process to occur at near neutral pH, and the synthesis of organic-Fe^{3+}-bacteria congeries results in the localization of ROS to the membrane [190].

$$H_2O_2 + Fe^{2+} \rightarrow OH^- + \bullet OH + Fe^{3+} \quad (12.5)$$

$$H_2O_2 + Fe^{3+} \rightarrow H^+ + \bullet O_2H + Fe^{2+} \quad (12.6)$$

$$h\nu + Fe(OH)^{2+} \rightarrow OH + Fe^{2+} \quad (12.7)$$

$$O_2^- + Cu^{2+} \rightarrow O_2 + Cu^+ \quad (12.8)$$

$$H_2O_2 + Cu^+ \rightarrow OH^- + \bullet OH + Cu^{2+} \quad (12.9)$$

Ruales-Lonfat et al. conducted extensive research on photo-Fenton disinfection. According to these researchers, Fe^{3+} breaks the cellular membrane, while Fe^{2+} enters the cell to create ROS [193]. Cu^{2+} in particular is excellent for the Fenton reaction (Equations 12.8 and 12.9) [4]. Javid et al. developed Cu:C nanocomposite thin films and showed that the antibacterial properties were dependent on the concentration of released cupric ions. A comparison of three different types of Cu:C nanocomposites found that smaller NMs improve disinfection due to their increased volume ratio and surface activity. Furthermore, increasing the concentration of Cu^{2+} ions boosts antibacterial activity and entirely suppresses bacterial growth [194]. This technology could enable for the effective utilization of nanosized materials in water disinfection; however, due to the restricted variety of transition metals, their usage would be limited [4].

12.3.2.6 Other Disinfection

In addition, Hong and colleagues' study revealed the effectiveness of electrochemical decontamination of water [195]. Using Ag NW-CC nanocomposites, these researchers developed a filter system for an electrical gravity filtering device. The researchers discovered that the sharp nanoedges of Ag NWs create a strong electric field, causing the transmembrane potential to reach the threshold value in a short amount of time. Following that, a large number of pores form on the cell membranes, causing the cell death [4]. In addition, Singh et al. recently found that employing laser-induced graphene layers as an electrode resulted in the electrochemical formation of H_2O_2, and they hypothesized that strong electric currents induce physical damage to bacteria [196].

12.3.3 Water Contaminants Disinfections by NMs

Waterborne infections are caused by consuming polluted water and functioning as a passive carrier of the infecting agent [14]. Human and/or animal fecal contamination is the most common source of disinfection. Bacteria, viruses, and protozoa feed upon them, grow, and cause many emerging aquatic infectious diseases. Traditional bacterial markers of fecal contamination (*E. coli* and *Enterococci*) are well established to be ineffective for detecting the presence and determining the amount of viruses and protozoa [197].

12.3.3.1 Bacteria

Bacteria can thrive in an extensive variety of environments; therefore, they can be found all around the biosphere. *E. coli*, *Pseudomonas*, *Shigella*, *Salmonella*, *M. avium*, *Vibrio cholerae*, *Campylobacter*, *Helicobacter*, *Legionella*, and other bacterial agents have been linked to waterborne diseases [14]. Akhil et al. showed approximately 100% suppression of *S. aureus* and *P. aeruginosa* at 100 mg L^{-1} of ZnO NPs, whereas Raghupathi et al. observed a similar outcome for *S. aureus* inhibition at 5 mM of ZnO NPs [198,199]. ZnO NPs have also been demonstrated to diminish biofilms of *S. aureus*, *P. aeruginosa*, and *E. coli*, an action that could be due to the production of ROS. Silver nanoparticles (AgNPs) were discovered to suppress methicillin-resistant *S. aureus* biofilm development [14]. AgNPs, with a diameter of 10 nm, have improved antibacterial activity against *Methylobacterium* spp., most likely owing to the discharge of Ag^+ ions. Surprisingly, at the same concentration and surface chemical composition, the antibacterial performance of 10 nm particles was greater than that of 100 nm particles [14]. *Gordonia* sp., a bacterium species that has been shown to be beneficial in natural water bodies and wastewater treatment, were inactivated by AgNPs and slowly released silver ions. The highest AgNPs dose led to the greatest average log inactivation for *Gordonia* sp., which was consistent with AgNPs dosage (1–8 mg L^{-1}) against *E. coli* [200, 201].

When the two agents were present simultaneously, the breakdown of organic substances and bacterial disinfection (*E. coli*, *Salmonella typhimurium*, and *Shigella sonnei*) in water by heterogeneous photocatalysis with TiO_2 and near-neutral photo-Fenton were less efficient [14]. Suri et al. studied the disinfection capabilities of TiO_2, Pt-TiO_2, and Ag-TiO_2 photocatalysts against *E. coli* under artificial light and sunlight, and discovered that the Ag-TiO_2 photocatalyst was more successful in both water and wastewater environments [202]. Under sunlight, an Ag/TiO_2 nanofiber membrane deactivated 99.9% of *E. coli* in 30 minutes [14]. Venieri et al. used solar light to exhibit considerable antibacterial effects against *E. coli* and *Klebsiella pneumonia* using Co-Mn- and binary Mn/Co-doped TiO_2 catalysts [203]. Fakhri et al. developed composites of ZnO QDs furnished CuO nanosheets and TiO_2 quantum dots filled WO_3 nanosheets, illustrating improved photocatalytic breakdown of neurotoxins in addition to deactivation of Gram-negative and Gram-positive bacteria, *E. faecalis* and *Micrococcus luteus*, respectively [204]. Hemdan et al. discovered one-of-a-kind nanoporous copper aluminosilicate (CAS) material. The antimicrobial properties and minimum inhibitory concentration of this promising material against broad variety of target microorganisms (*E. coli*, *Salmonella enterica*, *Listeria monocytogenes*, *P. aeruginosa*, *E. faecalis*, *Candida albicans*, *Staphylococcus aureus*, and *Aspergillus niger*) were determined. The authors reported that all bacteria investigated were completely inactivated in 20–40 minutes, emphasizing the potential applicability of nanoporous CAS for water disinfection [205]. Zhan et al. devised a facile method for swiftly eliminating dangerous microorganisms (bacteria and viruses) from water using modified core–shell Fe_3O_4-SiO_2-NH_2 nanoparticles. The clearance efficiencies of Gram-positive bacteria (*S. aureus* and *B. subtilis*) and Gram-negative bacteria (*P. aeruginosa*, *Salmonella*, and *E. coli*) were 93.4%, 97.4%, 95.1%, 90.1%, and 90.1%, respectively [206].

Jin et al. synthesized bactericidal paramagnetic nanoparticles (Fe_3O_4@CTAB) by modifying Fe_3O_4 nanoparticles with the antibacterial agent cetyltrimethylammonium bromide (CTAB). The researchers claimed that these nanoparticles had a significant potential for water disinfection, proving that they can kill more than 99% of *B. subtilis* and *E. coli* bacteria in 60 minutes [207]. SnO_2/PSi/NH_2 nanocomposite has been shown to have antibacterial action by killing *S. aureus* and *E. coli*. The core of the nano-adsorbent was SnO_2 nanoparticles, which were coated with mesoporous silica and modified with 3-aminopropyl triethoxysilane [14]. Khaydarov et al. present a new process for producing nano-photocatalysts using colloidal nanocarbon-metal composition (NCMC) with Ti as metal. These nano-photocatalysts effectively killed *E. coli* bacteria in water within 10–30 minutes [208]. Under visible light, Pandiyan et al. discovered that SnO_2-doped nanocomposites (sulphonated CNT and GO with SnO_2 dopant) displayed significant dose-dependent bactericidal effect against *Puccinia graminis* and *E. coli* isolated from a household wastewater treatment facility [209]. Also, SnO_2-doped nanomaterials have already been demonstrated to have potent antibacterial action against a wide variety of bacteria (*E. coli*, *S. aureus*, *P. aeruginosa*, *L. monocytogenes*, *Salmonella typhi*, *Trichoderma viride*, and *B. subtilis*) [14].

Hassouna et al. looked at the antibacterial characteristics of kaolin clay and its modified forms, which included carbon nanotubes and silver nanoparticles. They studied bacterial strains from different water matrices (surface, subsurface, tap water, and wastewater). They discovered that AgNPs-loaded clay (at 0.1 mg L^{-1}) showed the most pronounced antibacterial activity against *Salmonella* spp. (90%), *E. coli*, *K. pneumonia*, *Shigella flexneri* (80%), and *Klebsiella aerogenes* (70%) after 2 hours of exposure [210]. At the same concentration, the antibacterial effect of CNTs-loaded clay was 70% for *Salmonella* spp. and *K. pneumonia*, and 60% for *E. coli* strains. Antimicrobial characteristics of g-C_3N_4-based photocatalysts for the disinfection of water have been widely researched.

The antibacterial effect was apparent after approximately 10 minutes of exposure to visible light [14]. Another study discovered that GO/g-C_3N_4 exhibited antimicrobial effects [211]. Significant *E. coli* inactivation (97.9% of 10^7 CFU mL^{-1}) was achieved in 2 hours under visible light. Metal-free antimicrobials have been proposed as promising alternatives for water treatment and microbiological regulation [14].

12.3.3.2 Viruses

There are more than 140 different types of enteric viruses observed in human stool and urine [14]. The vast majority of viruses transferred by the fecal-oral path is non-enveloped and has great stability in the environment, with significant rising and re-emerging etiological agents. Surprisingly, current treatment procedures of wastewater are inefficient at removing these viruses, which can be found in both treated and untreated wastewater. Members of the *Caliciviridae*, *Adenoviridae*, *Hepeviridae*, *Picornaviridae*, and *Reoviridae* families are among the most important human harmful viruses [197]. When pathogenic enteric viruses are prevalent in ambient waters, they pose a significant risk to human health (e.g., 23–25 nm for MS2, 28–30 nm for *Hepatitis A* virus, 27–30 nm for poliovirus, 65–85 nm for adenovirus, and 35–39 nm for norovirus). *Adenoviruses* and *coliphages* MS2 and FX174 have been investigated as markers of viral contamination in fecally polluted water. Experimental results based on coliphages, in contrast, should be interpreted with caution because it cannot predict the behavior of diverse types of viruses. The MS2 phage was utilized as a viral indicator in the vast majority of the investigations, with only a few studies using *Poliovirus-1*.

Cheng et al. found that commercial nano $P_{25}TiO_2$ was better than a range of other nanomaterials for phage elimination (using MS2 as a model virus) than carbon nanotubes, graphene, nano ZnO, nano Ni, nano Fe_3O_4, and nano TiO_2-anatase [212]. The clearance efficiency rose with increasing P_{25} concentration in the 0–1000 mg L^{-1} range, while higher concentrations had little effect. Under certain conditions (P_{25} concentration, irradiation dose, and transmembrane pressure), MS2 clearance efficiencies of up to 100% may be achieved. These results support prior findings that P_{25} has higher adsorption and photocatalysis performance when compared to other nanomaterials and that the nanomaterial-membrane coupling system is suitable for virus elimination. Zheng et al. used visible light photocatalysis to demonstrate the enhanced disinfection capacity of Cu-TiO_2 nanofibers for bacteriophage (f_2) and the host (*E. coli*). As expected, the virus was more resistant to photocatalytic oxidation than bacteria; viral inactivation was hindered in a virus/host bacteria hybrid system; and ROS were required for virus deactivation [213]. Zhan et al. observed capture efficiency of 76.7% and 81.5% over bacteriophage f_2 and *Poliovirus-1*, respectively, using amine-functionalized magnetic Fe_3O_4-SiO_2-NH_2 nanoparticles [206]. Liga et al. showed that silver doping TiO_2 nanoparticles were an effective approach for increasing TiO_2 photocatalytic activity for virus inactivation; this was tested in aqueous settings for bacteriophage MS2 deactivation [214]. In terms of virucidal action against MS2, g-C_3N_4 outperformed other visible light active photocatalysts such as Ag@AgCl, Bi_2WO_6, and $NeTiO_2$ [14]. It was found that under visible light, 8-log of MS2 were inactivated without regeneration after 6 h; with procedure optimization, this time could be reduced to 4 hours, implying the potential utilization of g-C_3N_4 for water virus removal. The g-C_3N_4/EP composite (carrier of expanded perlite-EP) totally inactivated 8-log *E. coli* and MS2 after 3 and 4 hours of visible light irradiation, respectively. Surprisingly, viruses from real source water samples can be inactivated after 7 hours without regrowth. These findings emphasized the antibacterial effects of water-surface floating photocatalytic composites for source water decontamination [14].

12.3.3.3 Protozoa

Some of the most common waterborne diseases include *Cryptosporidium*, *Giardia*, *Cyclospora*, *Acanthamoeba*, *Isospora*, and other newly and re-emerging members of the protozoan parasite family. The majority of the papers reviewed were about *Cryptosporidium parvum*. Abebe et al. used physical filtration in a porous ceramic filter medium to explore the free effects of Ag salt and AgNPs on *C. parvum*, in addition to the removal of this protozoan illness with global consequences for the well-being of humans. They found that the efficiency of removal ranged from 96.4% to 99.2%. They showed that employing silver-impregnated ceramic water filters, both physical filtering and silver nanoparticle disinfection assisted in the treatment of *C. parvum*, with physical filtration likely contributing more than silver disinfection [215]. In another study, Darwish et al. produced a composite by embedding nanospheres of Ag onto aragonitic cuttlefish bone -immobilized samarium-doped zinc oxide (Sm-doped ZnO) nanorods. They found that Ag@Sm-doped ZnO/cuttlefish bone had a high biocidal activity toward pathogenic parasites and bacteria in both light and dark environments. The nanocomposite, in specific, displayed increased disinfection efficiency against *P. aeruginosa* (60%), *S. aureus* (80%), and *Schistosoma mansoni cercariae* (100%), which was coupled with gradual cercarial body deconstruction. Additionally, the nanocomposite has an exterminating impact on adult *S. mansoni* microbes, causing approximately 100% cell death and significant cell surface fragmentation [216]. Hussein et al. have demonstrated that MgO NPs had a significant effect on *Cyclospora oocysts.* MgO NPs have anti-*Cyclospora cayetanensis* activity on both sporulated and unsporulated oocysts [217]. Sunnotel et al. concentrated on using TiO_2 photocatalysis to treat surface water contaminated with *Cryptosporidium oocysts.* They discovered that photocatalytic deactivation of *C. parvum* oocysts occurred in both surface water (73.7%, 180 minutes) and buffer solution (78.4%, 180 minutes). There was no discernible disinfection in the absence of TiO_2 in the dark or under UV light irradiation [218]. Table 12.1 outlines several 2D NMs that have been described in the literature, the types of microorganisms involved, various antibacterial methods, and their disinfection effectiveness performance characteristics. The antibacterial activity mechanisms and performance of two-dimensional nanomaterials differ and are particular to the type of 2D NMs and microbes studied for water disinfection investigations.

TABLE 12.1
A List of 2D NMs, Their Antibacterial Mechanism and Disinfection Performance

Nanomaterial	Microorganism	Mechanism	Performance	Ref.
Ag-$CoFe_2O_4$-GO	*Staphylococcus aureus*, *Escherichia coli*	ROS production driven by light	*S. aureus* and *E. coli* were inactivated in 89.8% and 95.3% of cases, respectively	[92]
Ag/Ag_3PO_4/ $BiPO_4$	*E. coli*	Photoinduced ROS disinfection and Ag+ release	*E. coli* disinfection of >99% within 20 minutes of visible light irradiation	[232]
Ag/g-C_3N_4	*E. coli*	ROS production driven by light	After 90 minutes of visible light irradiation, *E. coli* cells were completely distorted	[106]
Ag/g-C_3N_4	*E. coli*	ROS production driven by light	After 90 minutes of visible light irradiation, the *E. coli* cell is completely distorted	[151]
Ag–AC	*E. coli*	Releasing of Ag+ ions	Bacterial cell inactivation is complete in 25 minutes	[220]
Ag–ACF	*S. aureus*, *E. coli*	Releasing of Ag+ ions	Bacterial death is complete in 3.5 minutes	[221]
Ag–MgO	*E. coli*	Production of ROS	Within 25 minutes, *E. coli* is completely inactivated	[226]
B, F-codoped TiO_2 nanosheet	*E. coli*	Disinfection due to photocatalysis	Bacteria are inactivated in 15 minutes when exposed to visible light	[133]
Bi_2MoO_6–RGO	*E. coli*	ROS production driven by light	Within 3 hours of being exposed to visible light, there was a 5-log reduction in *E. coli* count	[231]

(Continued)

TABLE 12.1 (*Continued*)
A List of 2D NMs, Their Antibacterial Mechanism and Disinfection Performance

Nanomaterial	Microorganism	Mechanism	Performance	Ref.
$Bi_2MoO_6/g\text{-}C_3N_4$	*E. coli*	ROS production driven by light	6-log disinfection of *E. coli* after 2.5 hours of visible light irradiation	[114]
$Bi_2MoO_6/g\text{-}C_3N_4$	*E. coli*	ROS production driven by light	Under 2.5 hours of visible light irradiation, the *E.coli* infection rate reached 6 log	[114]
BiOCl–Ag	*B. subtilis, E. coli*	ROS production driven by light	After 2 hours of visible light irradiation, *E. coli* and *B. subtilis* were completely inactivated	[164]
BiOCl–Ag nanocomposite	*B. subtilis, E. coli*	ROS production driven by light	After 2 hours of visible light irradiation, 100% of *B. subtilis* and *E. coli* were killed	[164]
ce-MoS_2 nanosheet	*E. coli*	Membrane oxidative stress resulted from ROS	Bacteria are killed 90% after 2 hours of contact	[152]
Cu/FLV-MoS_2 nanosheet	*E. coli*	ROS production driven by light	Bacterial inactivation is complete after 20 minutes of sun exposure	[159]
Cu/FLV-MoS_2 nanosheet	*E. coli*	ROS production driven by light	Bacterial death is complete after 20 minutes of sunlight exposure	[159]
Cu/Ni	*B. subtilis, E. coli*	Direct contact	Within 3 hours, the bacterial count was reduced by 99%	[222]
Cu_7S_4	*S. aureus, E. coli*	Photothermal	Bacterial inactivation after 10 minutes of exposure to sunlight	[229]
Fe_3O_4-TiO_2 nanosheet	*S. aureus, E. coli*	ROS production driven by light	Solar irradiation for 2 hours killed up to 93.7% and 87.2% of *S. aureus* and *E. coli,* respectively	[132]
Fe_3O_4–TiO_2 nanosheet	*S. aureus, E. coli*	ROS production driven by light generation	After 2 hours of solar irradiation, *E. coli* (87.2%) and *S. aureus* (93.7%) were killed	[132]
G/Ag NP–MS	*S. aureus, E. coli*	Releasing of Ag^+ ions and disinfection by ROS	Inactivation of 99.8% *E. coli* and 99.3% *S. aureus*	[219]
g-C_3N_4	MS2 virus particles	ROS production driven by light	MS2 inactivation is complete after 6 hours of visible light exposure	[116]
g-C_3N_4	MS2 virus particles	ROS production driven by light	MS2 was totally deactivated (> 6 hours) of visible light irradiation	[116]
g-C_3N_4/TiO_2	*E. coli*	ROS production driven by light	Within 3 hours of being exposed to visible light, *E. coli* was completely inactivated	[113]
g-C_3N_4/TiO_2	*E. coli*	ROS production driven by light	*E. coli* was completely inactivated after 180 minutes of visible light irradiation	[113]
GNR–MNP	*Enterococcus faecalis, E. coli*	Photothermal effect	Within 12 minutes, *E. coli* (99%) and *E. faecalis* (95%) are killed	[189]
GO	*Mycobacterium smegmatis, S. aureus, E. coli*	Oxidative stress	*M. smegmatis* has the best antibacterial activity	[87]
GO@Au@Ag nanosheet	*E. coli*	Releasing of Ag+ ions	Damage to *E. coli* intracellular components	[120]
GO@Au@Ag nanosheet	*E. coli*	Releasing of Ag+ ions	*E. coli* intracellular component damage	[120]

(*Continued*)

TABLE 12.1 (*Continued*)
A List of 2D NMs, Their Antibacterial Mechanism and Disinfection Performance

Nanomaterial	Microorganism	Mechanism	Performance	Ref.
GO–Ag_3PO_4	*S. aureus, E. coli*	Releasing of Ag+ ions and photoinduced ROS	After 0.5 hours of visible light irradiation, *S. aureus* and *E. coli* are completely killed	[233]
GO–CdS	*B. subtilis, E. coli*	ROS production driven by light	Within 25 minutes of being exposed to visible light, *E. coli* and *B. subtilis* were completely inactivated	[91]
GO–CdS	*E. coli*	ROS production driven by light	Under visible light irradiation, about 100% of *E. coli* are killed after 25 minutes	[91]
GO–TiO_2–Ag	*E. coli*	ROS production driven by light	Within 90 minutes of solar irradiation, the *E. coli* count dropped by 8.24 log	[89]
m-Bi_2O_4	*E. coli*	ROS production driven by light	Bacterial inactivation is complete after 120 minutes of light exposure	[227]
m-$BiVO_4$	*Pseudomonas aeruginosa*	ROS production driven by light	After 2 hours of visible light irradiation, the sterilization rate was 99.9%	[230]
MNP–CPQN	*S. aureus, E. coli*	Direct contact	Within 5 minutes, *E. coli* and *S. aureus* were 99.9% completely eliminated	[234]
Pd@Ag nanosheet	*S. aureus, E. coli*	Photothermal effect and releasing of Ag+ ions	After 10 minutes of NIR irradiation, nearly all bacteria were destroyed	[119]
SL g-C_3N_4	*E. coli*	ROS production driven by light	Bacterial inactivation is complete after 4 hours of exposure to visible light	[110]
SL g-C_3N_4	*E. coli*	ROS production driven by light	Bacteria are completely killed after 4 hours of visible light exposure	[110]
$Ti_3C_2T_x$ MXene	*B. subtilis, E. coli*	Oxidative stress	Bacterial cell viability is reduced by more than 98% within 4 hours	[185]
$Ti_3C_2T_x$ MXene	*B. subtilis, E. coli*	Oxidative stress	Bacterial cell viability is reduced by 98% within 4 hours of exposure	[185]
TiO_2 and C-dots co-decorated GO	*E. coli*	ROS production driven by light	In 1 hour, solar light inactivated 1.03 log *E. coli*	[236]
TiO_2 film coated on the TLM alloy	*S. aureus, E. coli*	ROS production driven by light	Up to 90% after 80 minutes of visible light irradiation	[235]
TiO_2 nanosheet	*E. coli, E. faecalis*	ROS production driven by light	Bactericidal activity in the dark after post-illumination	[129]
TiO_2 nanosheet/PdO	*E. coli*	ROS production driven by light	90% *E. coli* deactivation in 8 hours	[224]
TiO_2 nanotube	*Salmonella typhimurium, P. aeruginosa, E. coli*	Oxidative stress	Bacterial inactivation of 99.9%–99.99%	[223]
TiO_2/CdS	*E. coli*	ROS production driven by light	Within 10 minutes of being exposed to visible light, 99.9% of *E. coli* are killed	[228]
TiO_2–CuS	*E. coli*	ROS production driven by light	Within 100 minutes, *E. coli* is completely inactivated	[225]
WX_2 (X=Se, S) nanosheets	*E. coli*	Oxidative stress that is not caused by ROS	82.3%±1.7% Bacterial viability decline	[106]

12.3.4 Factors Affecting Microbial Enzymatic Degradation

Despite the advantages listed above, most 2D NMs have poor inherent photocatalytic performance due to low photoelectric conversion efficiency or rapid coupling of light-generated electrons and holes, resulting in poor antibacterial efficiency when exposed to light. Several methods have been identified for increasing photocatalytic antibacterial effects. Recent studies in the literature have revealed methods for boosting the photocatalytic efficiency of 2D NMs, hence improving their antibacterial characteristics [237].

12.3.4.1 Effect of Size or Thickness of 2D Nanomaterials

The shape and size of 2D NMs are closely related to their performance [45]. Traditional multilayer bulk nanocrystalline Bi_2WO_6, for example, has a band gap of 2.9 eV. When Bi_2WO_6 was produced as a monolayer with CTAB (cetyl trimethyl ammonium bromide), the energy band gap was reduced to 2.7 eV. The created Aurivillius oxide monolayer, Bi_2WO_6, displayed a new sandwich structure $[BiO]^+$-$[WO_4]_2$-$[BiO]^+$, which approached a normal heterojunction shape [238]. Since it was a one-atom-layer nanosheet, several unsaturated Bi atoms (on the surface) were utilized as active sites. As a result, when exposed to light, the active sites produced holes while the interlayer produced electrons, allowing for fast charge separation. As an outcome, the brominated surface modification decreased the band gap of the material by 0.2 eV while enhancing the intensity of light uptake. Likewise, the unique atomic structure was similar to a heterojunction, which facilitated in the separation of photogenerated carriers, significantly increasing photocatalytic activity. By varying the amount of layers in a given monolayer 2D NM, the structure can be adjusted to form a porous structured 2D NM with a higher surface area, lowering the band gap and enhancing light uptake. Bulk g-C_3N_4 photocatalytic performance is suboptimal due to its larger band gap (2.7 eV), high exciton dissociation energy, and restricted electron mobility. The first issue to address was the development of a layering and exfoliation method, which can be advantageous to the entire process.

In another study, Kang et al. produced high-quality g-C_3N_4 nanosheets at room temperature using a simple bacterial etching technique based on exfoliation [239]. The bacteria-treated g-C_3N_4 surface has a 2D structure (BT-CN-2D) as well as a porous structure. As a result, its specific area of surface rose significantly, whereas the energy band gap was decreased to 2.11 eV. Several unpaired electrons were concentrated in BT-CN-2D, and the electrons were quickly transferred. Consequently, the photocatalytic performance of BT-CN-2D was four times that of bulk g-C_3N_4, and its antibacterial stability was improved. Apart from shape, improving the thickness of 2D NMs may enhance their antibacterial performance dramatically [45]. For example, a single thin layer of g-C_3N_4 with 0.5 nm thickness, generated by heat etching and ultrasonic stripping, displayed good photocatalytic activity [110]. As a consequence, the 2D conduction path was reduced, lowering charge transfer resistance. When coupled, the short transfer distance and small charge transport resistance promoted effective separation of photogenerated charges and boosted photocatalytic efficiency. Thus, the photocatalytic activity is affected by the number of layers of 2D NMs [45]. Liu et al. developed FLV-MoS_2 film by vertically stacking multiple layers of MoS_2 [159]. MoS_2's energy band gap was improved (1.3–1.55 eV) upon reducing its thickness to a few layers or one layer. The improvement in the energy band gap enhanced visible light absorption considerably. When the photocatalytic disinfection rates of FLV-MoS_2 and horizontal MoS_2 were compared, the vertical arrangement had a higher photocatalytic disinfection rate since the vertical thin layer structure reduced the electron transfer distance, elevated hole separation efficiency, had greater in-plane conductivity, and brought out more active edge sites on the FLV-MoS_2 film.

12.3.4.2 Effect of Surface Modification

2D NMs exhibit weak photocatalytic performance due to their low electrochemical activity and easy agglomeration, limiting potential photocatalytic antibacterial applications. The alteration of organic substances, i.e., small organic molecules, macromolecules, or antimicrobial peptides, gives 2D NMs a variety of features that have a major effect on their bactericidal applications [45]. For example, perylene-3,4,9,10-tetracarboxylic diimide (PDINH) has been recrystallized on the g-C_3N_4 nanosheets surface in one new investigation [105]. Recrystallization of PDINH improved the g-C_3N_4 crystal structure. The lattice spacing was lowered to 0.475 nm due to strong intermolecular interactions. PEI altered the electronic structure of g-C_3N_4 by increasing the gap between light-excited electrons and holes, leading to ROS (O_{2-} and H_2O_2) production. The modified g-C_3N_4 photocatalytic activity is double that of the untreated group. PEI interacted with the negatively charged membrane surface as a result of these characteristics, which enhanced bacterial adherence to g-C_3N_4 and increased ROS suppression. Likewise, antimicrobial peptides are known to have a considerable impact on the hydrophilicity of materials. Samak et al. generated cyclic dodecapeptide (AMP) with different arginine (Arg) to tryptophan (Trp) ratios, which was immobilized on a nanocomposite (RGO/MnO_2) hybridized by flaky RGO and MnO_2 crystals [240]. The cationic nature of AMP boosted electrostatic interactions with the negative surfaces of *P. aeruginosa* and improved adhesion to phospholipids, resulting in an amphiphilic structure with flexibility at the membrane–water interfaces. Thus, its antibacterial activities were improved.

Organic photosensitizers in 2D NMs boost absorption in the visible NIR region, improving their light excitation ability significantly. Organic photosensitizers absorb light and promote electron transitions from the ground state to the short-lived singlet excited state and then to the long-lived triplet state. In the triplet state, electrons undertake direct reduction processes with the substrate or electronic disintegration. When an organic photosensitizer contacts bacteria, unsaturated phospholipids in the cell membrane are damaged [45]. Organic photosensitizers additionally interact with O_2 via energy transfer to create $^1O_2$. The extraordinarily active $^1O_2$ swiftly combines with biomolecules on bacterial membranes such as amino acids, peptides, lipids, receptors, and enzymes, thus causing damage [45]. For example, Shi et al. were the first researchers to combine vanadyl phthalocyanine (VOPc) with g-C_3N_4 using a phase approach [241]. VOPc is a photosensitizer that increases g-C_3N_4 absorption from visible to NIR light. Subsequently, g-C_3N_4's optoelectronic activity and photocatalytic performance were considerably improved. Li et al. fabricated a hybrid system comprised of MoS_2, IR780, and arginine-glycine-aspartate-cysteine [242]. IR780 is a photosensitizer that binds to the MoS_2 layer (negatively charged) electrostatically. It altered the separation of electrons and holes formed by light by modifying the electrical structure of the 2D NMs via atomic-level interaction. Furthermore, when exposed to NIR light, it produced $^1O_2$, which has antimicrobial properties. Zhang et al. used a straightforward polymerization technique to modify GO using sodium anthraquinone-2-sulfonate (AQS) [243]. AQS displayed a broad variety of bioactivities in composite materials as a photosensitizer, whereas GO provided a huge surface area. Both were linked together through π–π bond network. According to the experimental findings, AQS-GO provides good photocurrent response, indicating greater charge transport performance and e^--h^+ pair isolation efficacy. The holes and electrons produced on the AQS's CB and VB reacted with oxygen and H_2O to form $\bullet O_2^-$ and $\bullet OH$, respectively. Moreover, the holes in the VB of GO immediately responded to *E. coli*. As a result, in terms of light-induced antibacterial properties, the composite system outperformed single components.

12.3.4.3 Effect of Surface Defects

During the preparation and storage of 2D NMs, defective structures may be produced [45]. The presence of flaws represents the capability to enhance the number of unsaturated active sites. A catalytic reaction happens preferably on coordinated unsaturated active sites on the material's surface, including steps, edges, and protrusions, throughout heterogeneous catalysis [45]. The most common types of defects in 2D NMs that promote photocatalytic activity are depicted in Figure 12.7 [45]. There are four types of defects in 2D NMs. Nonmetallic element vacancy is the first type. The most prevalent defect elements are carbon (C), nitrogen (N), oxygen (O), and sulfur (S). Guirguis et al., for example, synthesized graphene with carbon edge defects by growing vertically aligned graphene nanosheets on carbon fibers using electron cyclotron resonance microwave plasma CVD [244].

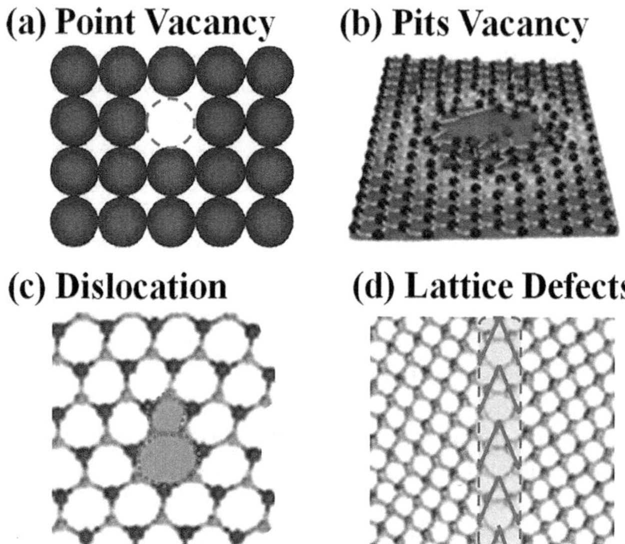

FIGURE 12.7 Common defects in 2D NMs for improving photocatalytic activity.

The second category is cation vacancies, such as metal cation holes. Using the solvothermal technique, Zhang et al. constructed Bi_2WO_6 nanosheets with a large fraction of crystal faces, and "Bi-O" dimer vacancy pairs on the (100) high-energy plane [245]. The third type includes pits or cavities. Atoms on neighboring surface break out from crystal lattice in ultrathin 2D NMs due to the small atomic escape energy. Guan et al. developed ultrathin nanosheets with large solar photocatalytic performance by fully exposing (001) facets [246]. In addition, as the nanosheet thickness was decreased to the atomic size, extra atoms escaped, resulting in the shift of defects from the separated one-atom defect "VBi" to the polyatomic (three-vacancy) defect "VBiVOVBi". The fourth type includes lattice faults such as dislocation, distortion, and disorder. During the synthesis of Bi_2MoO_6, Dai et al. used metal cerium (Ce) to cause ion substitution and a charge compensation reaction mechanism to produce "O" vacancies, modifying the crystal structure [247]. These faults affected the local electronic structure of 2D NMs, altering their physical and chemical properties and ultimately boosting their photocatalytic efficiency.

12.3.4.4 Effect of Heteroatom Doping

Heteroatom doping is an efficient procedure for improving the physical and chemical properties of 2D NMs [45]. Whether it replaces atoms (unit cell) or penetrates the crystal structure via heteroatom doping, changes in the 2D NMs properties are caused. Doping atoms efficiently modify the material's band gap width and electrical structure, altering photocatalytic activity [45]. Carbon (C), nitrogen (N), oxygen (O), and sulfur (S) are common heteroatom elements used for doping in 2D NMs [45]. Xiong et al. developed C-doped $(BiO)_2CO_3$ (CBCO) nanosheets and self-assembled microspheres using a simple hydrothermal method with glucose as a carbon source [248]. As the concentration of doped carbon element in crystal lattice of BCO raised a localized state appeared on the VB's edge, and both the potential of CB and the energy band gap of CBOC dropped, which resulted in increased visible light absorption. In addition, carbon doping produced the formation of more defect sites in the crystal lattice, which might be exploited as active site centers to improve electron–hole isolation. Likewise, the layered surface structure of the nano-microspheres permitted repeated scattering and reflection of incident light, increasing light absorption efficiency. This structure also assisted the diffusion of reaction intermediates and products, resulting in a quicker reaction rate. The photocatalytic performance of CBCO was efficiently and steadily raised, much above that of undoped BCO. Zhang et al. introduced a highly efficient porous O-doped g-C_3N_4 photocatalytic catalyst [249]. The catalyst guaranteed that melamine was thoroughly linked to its hydrolysate by forming uniform supramolecular assemblies throughout the hydrothermal process. On the surface of a g-C_3N_4 network, nanopores and oxygen (O) doping were evenly distributed. The porosity framework, together with the addition of heteroatoms (O), increased the amount of active regions and altered the electronic structure, narrowing the energy band and enhancing light-harvesting capabilities. Feng et al. created S-doped h-BN using a simple heat-treatment procedure and employed doping to change the band structure and interlayer gap [250]. The efficiency of the optimized S-doped h-BN was significantly improved over the original h-BN due to the sulfur (S) doping. The energy band gap was altered by the amount of S-doped in the h-BN. The hybridization was aided by p- and d-orbitals of S atoms, resulting in the improvement of novel VB and CB edges in h-BN. Consequently, the CB potential was lowered, the energy band gap was narrowed, the h-BN absorption spectrum was red-shifted, and light absorption performance increased. Because of the increased interaction between layers generated by many polar B-S bonds, the interlayer distance of h-BN was reduced. The presence of many S-doped sites and a shorter layer stacking distance may have been the key reasons in increasing charge mobility and surface reactivity, resulting in higher photocatalytic activity.

12.3.4.5 Effect of Heterojunction Formation

The usage of heterojunction structures can significantly increase the photocatalytic action of materials. Metal or semiconductor 2D NMs are the material classifications for heterojunction structures [45]. The combination of precious metals (Au or Ag) with 2D NM substantially increases photoreactivity due to two fundamental characteristics: the Schottky junction and LSPR (localized surface plasmon resonance) [45]. At the Schottky junction, precious metals capture light-emitting electrons from 2D NMs and then take part in the reduction reaction [251]. When the 2D NM gets into close proximity with the metal, their E_{Fermi} changes simultaneously and settles to a uniform constant value. As a result, the 2D NM energy band bends and deforms, showing a possible energy difference. The Schottky barrier prevents electrons from the 2D NM trapped by the metal from returning. Thus, the Schottky connection speeds up photogenerated electron transfer, boosting separation efficiency from holes, photogenerated carrier dynamics, and the photocatalytic efficiency of 2D NM [45]. A heterojunction of semiconductor photocatalysts and 2D NM, in addition to the previously mentioned approach, was constructed to produce high antibacterial activity [252]. Heterojunctions that make close contact between dissimilar semiconductors are attractive to researchers because they enhance electron transmission, indirectly improving electron–hole isolation [45]. Based on their material types and band gaps, heterojunctions are categorized into four types: conventional, p–n, Z-scheme, and S-scheme heterojunctions. Each of them has a unique effect on the charge transport of 2D NMs [45].

12.4 CHALLENGES OF 2D NANOMATERIALS FOR WATER TREATMENT

There are numerous difficulties to address for the appropriate utilization of antimicrobial nanoparticles in drinking water treatment, particularly nanomaterial dispersion and retention, in addition to antimicrobial activity and environmental sustainability [8,14].

12.4.1 Toxicity

Although nanoparticles are useful instruments for water treatment methods, their production procedures often employ hazardous and volatile compounds, resulting in severe additional pollution [28]. One of the biggest disadvantages of using metal- and metal oxide-based NMs is their potential toxicity, in addition to their by-products and the costs involved with their cleanup [253]. It must be noted that biological systems have not evolved alongside the artificial nanomaterials that are currently being produced and disseminated throughout the environment. As a consequence, several manmade nanomaterials have been discovered to have a broad variety of ecotoxicological effects on various organisms (plants, bacteria, fish, invertebrates, and so on) [22]. Toxicity study should preferably concentrate on many organisms in food chains/pyramids. However, the majority of study is limited to a small number of persons. Moreover, rather than functionalized NMs, they use unreasonably large concentrations of fresh NMs. The investigation of nanomaterials in the form in which they get released into the environment, e.g., embedded in a matrix, and/or are finally exposed to living organisms (e.g., after chemical and/or biological alterations caused by an array of environmental factors), is essential for an accurate evaluation of their impact on the environment [22]. Interestingly, the molecular structure and size of 2D NMs, which make them desirable, are also linked to their potential toxicity. Immobilization of nanoparticles on supporting medium or reactor surfaces may lessen toxicity issues; this process may also have good outcomes such as greater activity and less agglomeration [254].

12.4.2 Operating Conditions

In natural water matrices, 2D NMs interact physically and chemically in complicated ways, leading to their oxidation vs. reduction, dissolution vs. precipitation, dispersion vs. aggregation, and complexation with environment chemicals [14]. The absence of residues is the primary disadvantage of most nanomaterial-based decontamination techniques [255]. Under normal conditions, some nanomaterials are intrinsically unstable and agglomerate. The main constraint is agglomeration, which can have a major influence on the material's reactivity. Preventing agglomeration and improving monodispersity and stability are essential difficulties for an effective nanomaterial [253]. Mass transfer constraints and high-pressure drop have been identified as the primary challenges of powder catalyst applications. One technique that has already been demonstrated to solve these issues is the use of structured catalysts with carbon nanomaterials as an active stage in the catalytic ozonation process [14]. Though numerous nanoparticles have been extensively modified and have significant disinfection abilities, they have yet found real-world uses, owing primarily to the complex manufacturing procedures [14]. MOFs' inherent flaws, including a lack of catalytically active sites and poor thermal and chemical stability, hinder their widespread adoption. MOF synthesis techniques are currently complex, time-consuming, and expensive, with scaling up considered as a tough task. As a result, better metal- and metal oxide-NM composites based on MOFs have been produced. As heterogeneous catalysts, MOF composites are still in the early phases of development. One of the future breakthroughs is in the production of MOFs [14]. New nanotechnologies for sewage and water treatment must be accepted by society, commercially viable, and environmentally safe [255]. To overcome the limitations of poor isolation after use, sorption site blockage, and colloidal nanomaterial leaching, 2D NMs were immobilized on a supporting matrix [256].

12.4.3 Reusability

Regeneration and recyclability are two of the most essential problems concerning the cost-effectiveness of nanomaterials. Most studies consider regeneration to be high if nanomaterials retain most of their original characteristics after 3–10 cycles; however, these are laboratory-scale research with no economic consideration [257]. The recyclability of numerous NMs were described in earlier reports, but their efficacy deteriorates over time and they finally cease to be active [253]. The long-term reusability of nanomaterials boosts their cost-effectiveness. Numerous studies on regenerated nano-adsorbents, magnetically detachable multifunctional nanomaterials, and catalysts that may be reused numerous times have been reported [255]. A number of variables for particular nanomaterials should be modified to optimize their regeneration properties, and similarly, immobilization of photocatalysts on support matrices or as in nanocomposites would improve their ability to recover, reusability, and financial viability [256].

12.4.4 Others

Nanomaterial preservation is important not just for the expense of nanomaterial loss, but additionally and more importantly, for the possible consequences of nanoparticles on human health and ecosystems [8]. Nanomaterial preservation is important not just for the expense of nanomaterial loss, but additionally and more importantly for the possible consequences of nanoparticles on human health and ecosystems. This is primarily attributable to the greater diversity of the class of 2D NMs, and the maximum allowable amount varies for each type of 2D NM, as well as for different countries throughout the world. Nanomaterials will not be used in huge-scale water treatment systems until their human toxicity is thoroughly explored and methods to keep the nanomaterials in the treatment system are devised. Since the majority of nanomaterial antimicrobial activity research has been done in comparatively easy and hygienic solutions, the retention of their antimicrobial activities in natural or wastewater, whose elements may interfere with nanomaterial–microbe interactions, is unclear. Likewise, subsequent research on the generation of 2D NMs must address the following essential issues: (i) long-term efficacy and fouling caused by direct contact, (ii) economic feasibility and sustainability, (iii) development of hybrid compounds with enhanced stability and low toxicity, (iv) optimized 2D nanomaterial design with enhanced antibacterial activity, (v) large-scale synthesis/reproducibility, and (vi) assessment of the negative human and environmental consequences [210].

12.5 CONCLUSION AND FUTURE OUTLOOK

This book chapter describes various types of two-dimensional antibacterial nanomaterials used in water disinfection, in addition to their probable antibacterial processes. 2D nanomaterials outperform 1D nanomaterials in antibacterial activity due to their higher specific surface area, more surface-active sites, improved chemical stability, and conductivity. Because of their excellent qualities, they can be found in a wide range of antibacterial applications, including water disinfection, wound disinfection, and dental care and therapy. Various approaches have been utilized to manufacture 2D antibacterial nanomaterials of varied sizes, surface areas, morphologies, and topologies. These variables had an effect on their antibacterial performance. Nanomaterials with higher specific surface areas have higher surface reactivities, making them suited for contact with bacteria. Smaller dimensions allow them to penetrate faster through cell membranes. The toxicity of the released metal ions to attack bacterial cells, the induction of stress on cells via direct contact, Fenton process disinfection, and photocatalytic ROS generation are the primary antibacterial mechanisms of nanosized materials, though the use of the generated ROSs for disinfection has raised significant issues. To boost antibacterial efficiency, the size, specific surface area, and water solubility of a nanomaterial should all be addressed. Solar light absorption ranges, electron transformation process performance, and electron–hole pair recombination prevention are all critical features of photocatalytic decontamination. Most antibacterial activity studies use bacterial species as models, but viruses, fungi, and even parasites have only been studied to a limited extent. Significant research is needed to investigate the disinfecting capacities of antibacterial chemicals aimed at viruses, fungi, and even parasites. Because ROS play a significant role in bacterial mortality, increasing the generation of ROS could be critical in enhancing nanoscale antibacterial materials. Furthermore, metal ion toxicity must be considered, and the use of metal nanoparticles in water disinfection is controversial. As a result, metal ion removal is crucial. In addition, various practical methods for easy recycling of nanoparticles after their use in water disinfection have been presented. However, coordinated efforts are still needed to develop recycling processes that are simple to execute. Finally, the rapid development of water disinfection methods should contribute to the goal of a clean and safe water supply in future.

CONFLICT OF INTEREST

The author declares no conflict of interest.

ACKNOWLEDGMENTS

The author is thankful to CRIDA, Jaiotec Labs (OPC) Private Limited, Amaravati, AP, India for the access to necessary facilities for the completion of this work.

REFERENCES

1. Miao, H., Teng, Z., Wang, C., Chong, H. and Wang, G., 2019. Recent progress in two-dimensional antimicrobial nanomaterials. *Chemistry–A European Journal*, 25(4), pp. 929–944.
2. Manna, J., Goswami, S., Shilpa, N., Sahu, N. and Rana, R.K., 2015. Biomimetic method to assemble nanostructured Ag@ZnO on cotton fabrics: application as self-cleaning flexible materials with visible-light photocatalysis and antibacterial activities. *ACS Applied Materials & Interfaces*, 7(15), pp. 8076–8082.
3. Santosham, M., Chandran, A., Fitzwater, S., Fischer-Walker, C., Baqui, A.H. and Black, R., 2010. Progress and barriers for the control of diarrhoeal disease. *The Lancet*, 376(9734), pp. 63–67.
4. Miao, H., Teng, Z., Wang, S., Xu, L., Wang, C. and Chong, H., 2019. Recent advances in the disinfection of water using nanoscale antimicrobial materials. *Advanced Materials Technologies*, 4(5), p. 1800213.
5. Prüss, A., Kay, D., Fewtrell, L. and Bartram, J., 2002. Estimating the burden of disease from water, sanitation, and hygiene at a global level. *Environmental Health Perspectives*, 110(5), pp. 537–542.
6. Bartram, J., Lewis, K., Lenton, R. and Wright, A., 2005. Focusing on improved water and sanitation for health. *The Lancet*, 365(9461), pp. 810–812.
7. UNICEF and WHO, Progress on sanitation and drinking water, http://apps.who.int/iris/bitstream/10665/177752/1/9789241509145_eng. pdf?ua=1 (accessed: April 2023).
8. Li, Q., Mahendra, S., Lyon, D.Y., Brunet, L., Liga, M.V., Li, D. and Alvarez, P.J., 2008. Antimicrobial nanomaterials for water disinfection and microbial control: Potential applications and implications. *Water Research*, 42(18), pp. 4591–4602.
9. Schwarzenbach, R.P., Escher, B.I., Fenner, K., Hofstetter, T.B., Johnson, C.A., Von Gunten, U. and Wehrli, B., 2006. The challenge of micropollutants in aquatic systems. *Science*, 313(5790), pp. 1072–1077.
10. Parker, K.M., Zeng, T., Harkness, J., Vengosh, A. and Mitch, W.A., 2014. Enhanced formation of disinfection byproducts in shale gas wastewater-impacted drinking water supplies. *Environmental Science & Technology*, 48(19), pp. 11161–11169.
11. Surendhiran, D., Sirajunnisa, A. and Tamilselvam, K., 2017. Silver–magnetic nanocomposites for water purification. *Environmental Chemistry Letters*, 15, pp. 367–386.
12. Shannon, M.A., Bohn, P.W., Elimelech, M., Georgiadis, J.G., Marinas, B.J. and Mayes, A.M., 2008. Science and technology for water purification in the coming decades. *Nature*, 452(7185), pp. 301–310.
13. Turki, A., Guillard, C., Dappozze, F., Ksibi, Z., Berhault, G. and Kochkar, H., 2015. Phenol photocatalytic degradation over anisotropic TiO_2 nanomaterials: Kinetic study, adsorption isotherms and formal mechanisms. *Applied Catalysis B: Environmental*, 163, pp. 404–414.
14. Kokkinos, P., Mantzavinos, D. and Venieri, D., 2020. Current trends in the application of nanomaterials for the removal of emerging micropollutants and pathogens from water. *Molecules*, 25(9), p. 2016.
15. Reddy, B.M., 2022. Nanomedicine and nanovaccinology tools in targeted drug delivery. In *Nanovaccinology as Targeted Therapeutics*, pp. 21–52. https://doi.org/10.1002/9781119858041.ch2.
16. Bogala, M.R., 2022. Three-dimensional (3D) printing of hydroxyapatite-based scaffolds: A review. *Bioprinting*, 28, p. e00244.
17. Jeshurun, A., Mohammad, I., Behara, S. and Bogala, M.R., 2021. Structural and optical properties of (Y^{3+}, Tb^{3+})-codoped sodium bismuth titanate nanoparticles. *Materials Today Chemistry*, 20, p. 100476.
18. Irfan, M., Jeshurun, A., Baraneedharan, P. and Reddy, B.M., 2022. A photoluminescence study of nitrogen-doped carbon quantum dots/hydroxyapatite (NCQDs/HAp) nanocomposites. *Materials Technology*, 37(8), pp. 768–779.
19. Irfan, M., Suprajaa, P.S., Praveen, R. and Reddy, B.M., 2021. Microwave-assisted one-step synthesis of nanohydroxyapetite from fish bones and mussel shells. *Materials Letters*, 282, p. 128685.
20. Irfan, M., Suprajaa, P.S., Baraneedharan, P., Reddy, B.M., 2020. A comparative study of nanohydroxyapetite obtained from natural shells and wet chemical process. *Journal of Materials Science and Surface Engineering*, 7, pp. 938–43.
21. Bogala, M.R., 2021. Phyto-fabricated metal oxide nanoparticles as promising antibacterial agents. In: Pal, K. (ed.) *Bio-manufactured Nanomaterials: Perspectives and Promotion* (pp. 279–297). Cham: Springer International Publishing.

22. Karn, B., Kuiken, T. and Otto, M., 2011. Nanotechnology and in situ remediation: A review of the benefits and potential risks. *Ciência&SaúdeColetiva*, 16(1), pp. 165–178.
23. Xiang, Q., Yu, J. and Jaroniec, M., 2012. Graphene-based semiconductor photocatalysts. *Chemical Society Reviews*, 41(2), pp. 782–796.
24. Low, J., Cao, S., Yu, J. and Wageh, S., 2014. Two-dimensional layered composite photocatalysts. *Chemical Communications*, 50(74), pp. 10768–10777.
25. Wang, Y., Fan, L., Khan, S.J. and Roddick, F.A., 2020. Fugacity modelling of the fate of micropollutants in aqueous systems—Uncertainty and sensitivity issues. *Science of the Total Environment*, 699, p. 134249.
26. Barbosa, M.O., Moreira, N.F., Ribeiro, A.R., Pereira, M.F. and Silva, A.M., 2016. Occurrence and removal of organic micropollutants: An overview of the watch list of EU Decision 2015/495. *Water Research*, 94, pp. 257–279.
27. Ul-Islam, M., Ullah, M.W., Khan, S., Manan, S., Khattak, W.A., Ahmad, W., Shah, N. and Park, J.K., 2017. Current advancements of magnetic nanoparticles in adsorption and degradation of organic pollutants. *Environmental Science and Pollution Research*, 24, pp. 12713–12722.
28. Gautam, P.K., Singh, A., Misra, K., Sahoo, A.K. and Samanta, S.K., 2019. Synthesis and applications of biogenic nanomaterials in drinking and wastewater treatment. *Journal of Environmental Management*, 231, pp. 734–748.
29. Ali, I., Peng, C., Khan, Z.M., Naz, I., Sultan, M., Ali, M., Abbasi, I.A., Islam, T. and Ye, T., 2019. Overview of microbes based fabricated biogenic nanoparticles for water and wastewater treatment. *Journal of Environmental Management*, 230, pp. 128–150.
30. Hennebel, T., De Corte, S., Verstraete, W. and Boon, N., 2012. Microbial production and environmental applications of Pd nanoparticles for treatment of halogenated compounds. *Current Opinion in Biotechnology*, 23(4), pp. 555–561.
31. Ahmad, J., Naeem, S., Ahmad, M., Usman, A.R. and Al-Wabel, M.I., 2019. A critical review on organic micropollutants contamination in wastewater and removal through carbon nanotubes. *Journal of Environmental Management*, 246, pp. 214–228.
32. Fekadu, S., Alemayehu, E., Dewil, R. and Van der Bruggen, B., 2019. Pharmaceuticals in freshwater aquatic environments: A comparison of the African and European challenge. *Science of the Total Environment*, 654, pp. 324–337.
33. Yang, Y., Ok, Y.S., Kim, K.H., Kwon, E.E. and Tsang, Y.F., 2017. Occurrences and removal of pharmaceuticals and personal care products (PPCPs) in drinking water and water/sewage treatment plants: A review. *Science of the Total Environment*, 596, pp. 303–320.
34. Hopkins, Z.R. and Blaney, L., 2016. An aggregate analysis of personal care products in the environment: Identifying the distribution of environmentally-relevant concentrations. *Environment International*, 92, pp. 301–316.
35. Ahmed, G., Anawar, H.M., Takuwa, D.T., Chibua, I.T., Singh, G.S. and Sichilongo, K., 2015. Environmental assessment of fate, transport and persistent behavior of dichlorodiphenyltrichloroethanes and hexachlorocyclohexanes in land and water ecosystems. *International Journal of Environmental Science and Technology*, 12, pp. 2741–2756.
36. Marican, A. and Durán-Lara, E.F., 2018. A review on pesticide removal through different processes. *Environmental Science and Pollution Research*, 25, pp. 2051–2064.
37. Raffa, C.M. and Chiampo, F., 2021. Bioremediation of agricultural soils polluted with pesticides: A review. *Bioengineering*, 8(7), p. 92.
38. Odukkathil, G. and Vasudevan, N., 2013. Toxicity and bioremediation of pesticides in agricultural soil. *Reviews in Environmental Science and Bio/Technology*, 12, pp. 421–444.
39. Jariyal, M., Gupta, V.K., Jindal, V. and Mandal, K., 2015. Isolation and evaluation of potent Pseudomonas species for bioremediation of phorate in amended soil. *Ecotoxicology and Environmental Safety*, 122, pp. 24–30.
40. McLellan, S.L., Fisher, J.C. and Newton, R.J., 2015. The microbiome of urban waters. *International Microbiology: The Official Journal of the Spanish Society for Microbiology*, 18(3), p. 141.
41. Vittecoq, M., Thomas, F., Jourdain, E., Moutou, F., Renaud, F. and Gauthier-Clerc, M., 2014. Risks of emerging infectious diseases: Evolving threats in a changing area, the mediterranean basin. *Transboundary and Emerging Diseases*, 61(1), pp. 17–27.
42. Vouga, M. and Greub, G., 2016. Emerging bacterial pathogens: the past and beyond. *Clinical Microbiology and Infection*, 22(1), pp. 12–21.
43. Falkinham III, J.O., Pruden, A. and Edwards, M., 2015. Opportunistic premise plumbing pathogens: increasingly important pathogens in drinking water. *Pathogens*, 4(2), pp. 373–386.

44. La Rosa, G., Fratini, M., della Libera, S., Iaconelli, M. and Muscillo, M., 2012. Emerging and potentially emerging viruses in water environments. *Annalidell'Istitutosuperiore di sanità*, 48, pp. 397–406.
45. Li, B., Luo, Y., Zheng, Y., Liu, X., Tan, L. and Wu, S., 2022. Two-dimensional antibacterial materials. *Progress in Materials Science*, 130, p.100976.
46. Fleming, A., 1929. On the antibacterial action of cultures of a penicillium, with special reference to their use in the isolation of B. influenzae. *British Journal of Experimental Pathology*, 10, pp. 226–36.
47. Smith, P.A., Koehler, M.F., Girgis, H.S., Yan, D., Chen, Y., Chen, Y., Crawford, J.J., Durk, M.R., Higuchi, R.I., Kang, J. and Murray, J., 2018. Optimized arylomycins are a new class of Gram-negative antibiotics. *Nature*, 561(7722), pp. 189–194.
48. Baker, S., Thomson, N., Weill, F.X. and Holt, K.E., 2018. Genomic insights into the emergence and spread of antimicrobial-resistant bacterial pathogens. *Science*, 360(6390), pp. 733–738.
49. Koivunen, J. and Heinonen-Tanski, H., 2005. Inactivation of enteric microorganisms with chemical disinfectants, UV irradiation and combined chemical/UV treatments. *Water Research*, 39(8), pp. 1519–1526.
50. Hu, Z.T., Chen, Y., Fei, Y.F., Loo, S.L., Chen, G., Hu, M., Song, Y., Zhao, J., Zhang, Y. and Wang, J., 2022. An overview of nanomaterial-based novel disinfection technologies for harmful microorganisms: Mechanism, synthesis, devices and application. *Science of the Total Environment*, 837, p. 155720.
51. Logan, B.E. and Elimelech, M., 2012. Membrane-based processes for sustainable power generation using water. *Nature*, 488(7411), pp. 313–319.
52. Ong, W.J., Tan, L.L., Ng, Y.H., Yong, S.T. and Chai, S.P., 2016. Graphitic carbon nitride (g-C_3N_4)-based photocatalysts for artificial photosynthesis and environmental remediation: Are we a step closer to achieving sustainability? *Chemical Reviews*, 116(12), pp. 7159–7329.
53. Cao, S., Low, J., Yu, J. and Jaroniec, M., 2015. Polymeric photocatalysts based on graphitic carbon nitride. *Advanced Materials*, 27(13), pp. 2150–2176.
54. Malato, S., Fernández-Ibáñez, P., Maldonado, M.I., Blanco, J. and Gernjak, W., 2009. Decontamination and disinfection of water by solar photocatalysis: recent overview and trends. *Catalysis Today*, 147(1), pp. 1–59.
55. Mei, L., Zhu, S., Yin, W., Chen, C., Nie, G., Gu, Z. and Zhao, Y., 2020. Two-dimensional nanomaterials beyond graphene for antibacterial applications: current progress and future perspectives. *Theranostics*, 10(2), p. 757.
56. Miller, K.P., Wang, L., Benicewicz, B.C. and Decho, A.W., 2015. Inorganic nanoparticles engineered to attack bacteria. *Chemical Society Reviews*, 44(21), pp. 7787–7807.
57. Li, Y., Liu, X., Tan, L., Cui, Z., Jing, D., Yang, X., Liang, Y., Li, Z., Zhu, S., Zheng, Y. and Yeung, K.W.K., 2019. Eradicating multidrug-resistant bacteria rapidly using a multi functional g-C_3N_4@ Bi_2S_3 nanorod heterojunction with or without antibiotics. *Advanced Functional Materials*, 29(20), p. 1900946.
58. Lin, L.C.W., Chattopadhyay, S., Lin, J.C. and Hu, C.M.J., 2018. Advances and opportunities in nanoparticle-and nanomaterial-based vaccines against bacterial infections. *Advanced Healthcare Materials*, 7(13), p. 1701395.
59. Qu, X., Brame, J., Li, Q. and Alvarez, P.J., 2013. Nanotechnology for a safe and sustainable water supply: Enabling integrated water treatment and reuse. *Accounts of Chemical Research*, 46(3), pp. 834–843.
60. Choudhury, I. and Hashmi, S., 2020. *Encyclopedia of Renewable and Sustainable Materials*. Elsevier, Amsterdam.
61. Zhao, Q., Mao, Q., Zhou, Y., Wei, J., Liu, X., Yang, J., Luo, L., Zhang, J., Chen, H., Chen, H. and Tang, L., 2017. Metal-free carbon materials-catalyzed sulfate radical-based advanced oxidation processes: A review on heterogeneous catalysts and applications. *Chemosphere*, 189, pp. 224–238.
62. Saqib, N.U., Adnan, R. and Shah, I., 2016. A mini-review on rare earth metal-doped TiO_2 for photocatalytic remediation of wastewater. *Environmental Science and Pollution Research*, 23, pp. 15941–15951.
63. Beltran, F.J. and Rey, A., 2017. Solar or UVA-visible photocatalytic ozonation of water contaminants. *Molecules*, 22(7), p. 1177.
64. Tsydenova, O., Batoev, V. and Batoeva, A., 2015. Solar-enhanced advanced oxidation processes for water treatment: Simultaneous removal of pathogens and chemical pollutants. *International Journal of Environmental Research and Public Health*, 12(8), pp. 9542–9561.
65. Duan, X., Sun, H. and Wang, S., 2018. Metal-free carbocatalysis in advanced oxidation reactions. *Accounts of Chemical Research*, 51(3), pp. 678–687.
66. Wols, B.A. and Hofman-Caris, C.H.M., 2012. Review of photochemical reaction constants of organic micropollutants required for UV advanced oxidation processes in water. *Water Research*, 46(9), pp. 2815–2827.

67. Novoselov, K.S., Geim, A.K., Morozov, S.V., Jiang, D.E., Zhang, Y., Dubonos, S.V., Grigorieva, I.V. and Firsov, A.A., 2004. Electric field effect in atomically thin carbon films. *Science*, 306(5696), pp. 666–669.
68. Xia, F., Wang, H., Xiao, D., Dubey, M. and Ramasubramaniam, A., 2014. Two-dimensional material nanophotonics. *Nature Photonics*, 8(12), pp. 899–907.
69. Min, S. and Lu, G., 2012. Enhanced electron transfer from the excited eosin Y to mpg-C_3N_4 for highly efficient hydrogen evolution under 550 nm irradiation. *The Journal of Physical Chemistry C*, 116(37), pp. 19644–19652.
70. Hossain, F., Perales-Perez, O.J., Hwang, S. and Román, F., 2014. Antimicrobial nanomaterials as water disinfectant: Applications, limitations and future perspectives. *Science of the Total Environment*, 466, pp. 1047–1059.
71. Beyth, N., Houri-Haddad, Y., Domb, A., Khan, W. and Hazan, R., 2015. Alternative antimicrobial approach: Nano-antimicrobial materials. *Evidence-Based Complementary and Alternative Medicine*, 2015, p. 246012.
72. Horiuchi, S., Gotou, T., Fujiwara, M., Asaka, T., Yokosawa, T. and Matsui, Y., 2004. Single graphene sheet detected in a carbon nanofilm. *Applied Physics Letters*, 84(13), pp. 2403–2405.
73. Huang, X., Qi, X., Boey, F. and Zhang, H., 2012. Graphene-based composites. *Chemical Society Reviews*, 41(2), pp. 666–686.
74. Hummers Jr., W.S. and Offeman, R.E., 1958. Preparation of graphitic oxide. *Journal of the American Chemical Society*, 80(6), pp. 1339–1339.
75. Yuan, Z., Xiao, X., Li, J., Zhao, Z., Yu, D. and Li, Q., 2018. Self-assembled graphene-based architectures and their applications. *Advanced Science*, 5(2), p. 1700626.
76. Li, R., Guiney, L.M., Chang, C.H., Mansukhani, N.D., Ji, Z., Wang, X., Liao, Y.P., Jiang, W., Sun, B., Hersam, M.C. and Nel, A.E., 2018. Surface oxidation of graphene oxide determines membrane damage, lipid peroxidation, and cytotoxicity in macrophages in a pulmonary toxicity model. *ACS Nano*, 12(2), pp. 1390–1402.
77. Eldesoky, G.E., Baraneedharan, P., Reddy, B.M., Yahya, A.E., Alfakeer, M., Thangadurai, T.D., Nehru, K. and Sephra, P.J., 2021. Fe_2O_3/graphene nanocomposites as heterogeneous Fenton catalysts: Correlation studies on catalyst dosage, graphene loading and adsorption kinetics on methylene blue degradation. *MRS Communications*, 11, pp. 943–949.
78. Karahan, H.E., Wiraja, C., Xu, C., Wei, J., Wang, Y., Wang, L., Liu, F. and Chen, Y., 2018. Graphene materials in antimicrobial nanomedicine: Current status and future perspectives. *Advanced Healthcare Materials*, 7(13), p. 1701406.
79. Roy, E., Patra, S., Kumar, D., Madhuri, R. and Sharma, P.K., 2015. Multifunctional magnetic reduced graphene oxide dendrites: Synthesis, characterization and their applications. *Biosensors and Bioelectronics*, 122, p. 300.
80. Sanchez, V.C., Jachak, A., Hurt, R.H. and Kane, A.B., 2012. Biological interactions of graphene-family nanomaterials: An interdisciplinary review. *Chemical Research in Toxicology*, 25(1), pp. 15–34.
81. Dreyer, D.R., Park, S., Bielawski, C.W. and Ruoff, R.S., 2010. Graphite oxide. *Chemical Society Reviews*, 39(1), pp. 228–240.
82. Liu, S., Zeng, T.H., Hofmann, M., Burcombe, E., Wei, J., Jiang, R., Kong, J. and Chen, Y., 2010. Antibacterial activity of graphite, graphite oxide, graphene oxide, and reduced graphene oxide: Membrane and oxidative stress. *ACS Nano*, 5, pp. 6971–6980.
83. Moon, I.K., Lee, J., Ruoff, R.S. and Lee, H., 2010. Reduced graphene oxide by chemical graphitization. *Nature Communications*, 1 (73), pp. 1–6.
84. Zhao, K., Wang, Y., Wang, W. and Yu, D., 2018. Moisture absorption, perspiration and thermal conductive polyester fabric prepared by thiol–ene click chemistry with reduced graphene oxide finishing agent. *Journal of Materials Science*, 53, pp.14262–14273.
85. Shin, H.J., Kim, K.K., Benayad, A., Yoon, S.M., Park, H.K., Jung, I.S., Jin, M.H., Jeong, H.K., Kim, J.M., Choi, J.Y. and Lee, Y.H., 2009. Efficient reduction of graphite oxide by sodium borohydride and its effect on electrical conductance. *Advanced Functional Materials*, 19(12), pp. 1987–1992.
86. Qi, Z., Bharate, P., Lai, C.H., Ziem, B., Böttcher, C., Schulz, A., Beckert, F., Hatting, B., Mülhaupt, R., Seeberger, P.H. and Haag, R., 2015. Multivalency at interfaces: Supramolecular carbohydrate-functionalized graphene derivatives for bacterial capture, release, and disinfection. *Nano Letters*, 15(9), pp. 6051–6057.
87. Zou, F., Zhou, H., Jeong, D.Y., Kwon, J., Eom, S.U., Park, T.J., Hong, S.W. and Lee, J., 2017. Wrinkled surface-mediated antibacterial activity of graphene oxide nanosheets. *ACS Applied Materials & Interfaces*, 9(2), pp. 1343–1351.

88. Kim, T.I., Kwon, B., Yoon, J., Park, I.J., Bang, G.S., Park, Y., Seo, Y.S. and Choi, S.Y., 2017. Antibacterial activities of graphene oxide–molybdenum disulfide nanocomposite films. *ACS Applied Materials & Interfaces*, 9(9), pp. 7908–7917.
89. Liu, L., Bai, H., Liu, J. and Sun, D.D., 2013. Multifunctional graphene oxide-TiO_2-Ag nanocomposites for high performance water disinfection and decontamination under solar irradiation. *Journal of Hazardous Materials*, 261, pp. 214–223.
90. Deng, C.H., Gong, J.L., Zeng, G.M., Jiang, Y., Zhang, C., Liu, H.Y. and Huan, S.Y., 2016. Graphene–CdS nanocomposite inactivation performance toward Escherichia coli in the presence of humic acid under visible light irradiation. *Chemical Engineering Journal*, 284, pp. 41–53.
91. Gao, P., Liu, J., Sun, D.D. and Ng, W., 2013. Graphene oxide–CdS composite with high photocatalytic degradation and disinfection activities under visible light irradiation. *Journal of Hazardous Materials*, 250, pp. 412–420.
92. Ma, S., Zhan, S., Jia, Y. and Zhou, Q., 2015. Highly efficient antibacterial and Pb (II) removal effects of Ag-$CoFe_2O_4$-GO nanocomposite. *ACS Applied Materials & Interfaces*, 7(19), pp. 10576–10586.
93. Bai, J., Liu, Y. and Jiang, X., 2014. Multifunctional PEG-GO/CuS nanocomposites for near-infrared chemo-photothermal therapy. *Biomaterials*, 35(22), pp. 5805–5813.
94. Zeng, X., Wang, Z., Wang, G., Gengenbach, T.R., McCarthy, D.T., Deletic, A., Yu, J. and Zhang, X., 2017. Highly dispersed TiO_2 nanocrystals and WO_3 nanorods on reduced graphene oxide: Z-scheme photocatalysis system for accelerated photocatalytic water disinfection. *Applied Catalysis B: Environmental*, 218, pp. 163–173.
95. Yu, H., Shi, R., Zhao, Y., Bian, T., Zhao, Y., Zhou, C., Waterhouse, G.I., Wu, L.Z., Tung, C.H. and Zhang, T., 2017. Alkali-assisted synthesis of nitrogen deficient graphitic carbon nitride with tunable band structures for efficient visible-light-driven hydrogen evolution. *Advanced Materials*, 29(16), p. 1605148.
96. Mohammad, I., Jeshurun, A., Ponnusamy, P. and Reddy, B.M., 2022. Mesoporous graphitic carbon nitride/hydroxyapatite (g-C_3N_4/HAp) nanocomposites for highly efficient photocatalytic degradation of rhodamine B dye. *Materials Today Communications*, 33, p. 104788.
97. Zheng, Y., Lin, L., Wang, B. and Wang, X., 2015. Graphitic carbon nitride polymers toward sustainable photoredox catalysis. *Angewandte Chemie International Edition*, 54(44), pp. 12868–12884.
98. Zhang, J., Guo, F. and Wang, X., 2013. An optimized and general synthetic strategy for fabrication of polymeric carbon nitride nanoarchitectures. *Advanced Functional Materials*, 23(23), pp. 3008–3014.
99. Liao, G., He, F., Li, Q., Zhong, L., Zhao, R., Che, H., Gao, H. and Fang, B., 2020. Emerging graphitic carbon nitride-based materials for biomedical applications. *Progress in Materials Science*, 112, p. 100666.
100. Zhao, H., Chen, S., Quan, X., Yu, H. and Zhao, H., 2016. Integration of microfiltration and visible-light-driven photocatalysis on g-C_3N_4 nanosheet/reduced graphene oxide membrane for enhanced water treatment. *Applied Catalysis B: Environmental*, 194, pp. 134–140.
101. Wu, Y., Zhou, Y., Xu, H., Liu, Q., Li, Y., Zhang, L., Liu, H., Tu, Z., Cheng, X. and Yang, J., 2018. Highly active, superstable, and biocompatible Ag/polydopamine/g-C_3N_4 bactericidal photocatalyst: synthesis, characterization, and mechanism. *ACS Sustainable Chemistry & Engineering*, 6(11), pp. 14082–14094.
102. Wang, W., Yu, J.C., Xia, D., Wong, P.K. and Li, Y., 2013. Graphene and g-C_3N_4 nanosheets cowrapped elemental α-sulfur as a novel metal-free heterojunction photocatalyst for bacterial inactivation under visible-light. *Environmental Science & Technology*, 47(15), pp. 8724–8732.
103. Li, R., Ren, Y., Zhao, P., Wang, J., Liu, J. and Zhang, Y., 2019. Graphitic carbon nitride (g-C_3N_4) nanosheets functionalized composite membrane with self-cleaning and antibacterial performance. *Journal of Hazardous Materials*, 365, pp. 606–614.
104. Wang, R., Kong, X., Zhang, W., Zhu, W., Huang, L., Wang, J., Zhang, X., Liu, X., Hu, N., Suo, Y. and Wang, J., 2018. Mechanism insight into rapid photocatalytic disinfection of Salmonella based on vanadate QDs-interspersed g-C_3N_4 heterostructures. *Applied Catalysis B: Environmental*, 225, pp. 228–237.
105. Wang, L., Zhang, X., Yu, X., Gao, F., Shen, Z., Zhang, X., Ge, S., Liu, J., Gu, Z. and Chen, C., 2019. An all-organic semiconductor C_3N_4/PDINH heterostructure with advanced antibacterial photocatalytic therapy activity. *Advanced Materials*, 31(33), p. 1901965.
106. Ma, S., Zhan, S., Jia, Y., Shi, Q. and Zhou, Q., 2016. Enhanced disinfection application of Ag-modified g-C_3N_4 composite under visible light. *Applied Catalysis B: Environmental*, 186, pp. 77–87.
107. Jiang, J., Ou-yang, L., Zhu, L., Zheng, A., Zou, J., Yi, X. and Tang, H., 2014. Dependence of electronic structure of g-C_3N_4 on the layer number of its nanosheets: a study by Raman spectroscopy coupled with first-principles calculations. *Carbon*, 80, pp. 213–221.

108. Lin, Q., Li, L., Liang, S., Liu, M., Bi, J. and Wu, L., 2015. Efficient synthesis of monolayer carbon nitride 2D nanosheet with tunable concentration and enhanced visible-light photocatalytic activities. *Applied Catalysis B: Environmental*, 163, pp. 135–142.
109. Li, K., Zeng, Z., Yan, L., Luo, S., Luo, X., Huo, M. and Guo, Y., 2015. Fabrication of platinum-deposited carbon nitride nanotubes by a one-step solvothermal treatment strategy and their efficient visible-light photocatalytic activity. *Applied Catalysis B: Environmental*, 165, pp. 428–437.
110. Zhao, H., Yu, H., Quan, X., Chen, S., Zhang, Y., Zhao, H. and Wang, H., 2014. Fabrication of atomic single layer graphitic-C_3N_4 and its high performance of photocatalytic disinfection under visible light irradiation. *Applied Catalysis B: Environmental*, 152, pp. 46–50.
111. Wang, Z., Dong, K., Liu, Z., Zhang, Y., Chen, Z., Sun, H., Ren, J. and Qu, X., 2017. Activation of biologically relevant levels of reactive oxygen species by Au/g-C_3N_4 hybrid nanozyme for bacteria killing and wound disinfection. *Biomaterials*, 113, pp. 145–157.
112. Ma, S., Zhan, S., Xia, Y., Wang, P., Hou, Q. and Zhou, Q., 2019. Enhanced photocatalytic bactericidal performance and mechanism with novel Ag/ZnO/g-C_3N_4 composite under visible light. *Catalysis Today*, 330, pp. 179–188.
113. Li, G., Nie, X., Chen, J., Jiang, Q., An, T., Wong, P.K., Zhang, H., Zhao, H. and Yamashita, H., 2015. Enhanced visible-light-driven photocatalytic inactivation of Escherichia coli using g-C_3N_4/TiO_2 hybrid photocatalyst synthesized using a hydrothermal-calcination approach. *Water Research*, 86, pp. 17–24.
114. Li, J., Yin, Y., Liu, E., Ma, Y., Wan, J., Fan, J. and Hu, X., 2017. In situ growing Bi_2MoO_6 on g-C_3N_4 nanosheets with enhanced photocatalytic hydrogen evolution and disinfection of bacteria under visible light irradiation. *Journal of Hazardous Materials*, 321, pp. 183–192.
115. Khan, M.E., Han, T.H., Khan, M.M., Karim, M.R. and Cho, M.H., 2018. Environmentally sustainable fabrication of Ag@ g-C_3N_4 nanostructures and their multifunctional efficacy as antibacterial agents and photocatalysts. *ACS Applied Nano Materials*, 1(6), pp. 2912–2922.
116. Li, Y., Zhang, C., Shuai, D., Naraginti, S., Wang, D. and Zhang, W., 2016. Visible-light-driven photocatalytic inactivation of MS2 by metal-free g-C_3N_4: virucidal performance and mechanism. *Water Research*, 106, pp. 249–258.
117. Song, X., Hu, J. and Zeng, H., 2013. Two-dimensional semiconductors: Recent progress and future perspectives. *Journal of Materials Chemistry C*, 1(17), pp. 2952–2969.
118. Valodkar, M., Rathore, P.S., Jadeja, R.N., Thounaojam, M., Devkar, R.V. and Thakore, S., 2012. Cytotoxicity evaluation and antimicrobial studies of starch capped water soluble copper nanoparticles. *Journal of Hazardous Materials*, 201, pp. 244–249.
119. Mo, S., Chen, X., Chen, M., He, C., Lu, Y. and Zheng, N., 2015. Two-dimensional antibacterial Pd@ Ag nanosheets with a synergetic effect of plasmonic heating and Ag^+ release. *Journal of Materials Chemistry B*, 3(30), pp. 6255–6260.
120. Wang, H., Liu, J., Wu, X., Tong, Z. and Deng, Z., 2013. Tailor-made Au@Ag core–shell nanoparticle 2D arrays on protein-coated graphene oxide with assembly enhanced antibacterial activity. *Nanotechnology*, 24(20), p. 205102.
121. Chen, F., Fang, P., Gao, Y., Liu, Z., Liu, Y. and Dai, Y., 2012. Effective removal of high-chroma crystal violet over TiO_2-based nanosheet by adsorption–photocatalytic degradation. *Chemical Engineering Journal*, 204, pp. 107–113.
122. Wang, G., Feng, W., Zeng, X., Wang, Z., Feng, C., McCarthy, D.T., Deletic, A. and Zhang, X., 2016. Highly recoverable TiO_2–GO nanocomposites for stormwater disinfection. *Water Research*, 94, pp. 363–370.
123. Bhave, R.C. and Lee, B.I., 2007. Experimental variables in the synthesis of brookite phase TiO_2 nanoparticles. *Materials Science and Engineering: A*, 467(1–2), pp. 146–149.
124. Bogala, M.R. and Reddy, R.G., 2016. Thermodynamic assessment of TiO_2 reduction to Ti metal in molten $CaCl_2$. *Journal for Manufacturing Science and Production*, 16(1), pp. 1–11.
125. Yan, M., Chen, F., Zhang, J. and Anpo, M., 2005. Preparation of controllable crystalline titania and study on the photocatalytic properties. *The Journal of Physical Chemistry B*, 109(18), pp. 8673–8678.
126. Straňák, V., Quaas, M., Wulff, H., Hubička, Z., Wrehde, S., Tichý, M. and Hippler, R., 2008. Formation of TiO_x films produced by high-power pulsed magnetron sputtering. *Journal of Physics D: Applied Physics*, 41(5), p. 055202.
127. Watson, S., Beydoun, D., Scott, J. and Amal, R., 2004. Preparation of nanosized crystalline TiO_2 particles at low temperature for photocatalysis. *Journal of Nanoparticle Research*, 6, pp. 193–207.
128. Kobayashi, M., Tomita, K., Petrykin, V., Yoshimura, M. and Kakihana, M., 2008. Direct synthesis of brookite-type titanium oxide by hydrothermal method using water-soluble titanium complexes. *Journal of Materials Science*, 43, pp. 2158–2162.

129. Wang, G., Xing, Z., Zeng, X., Feng, C., McCarthy, D.T., Deletic, A. and Zhang, X., 2016. Ultrathin titanium oxide nanosheets film with memory bactericidal activity. *Nanoscale*, 8(42), pp. 18050–18056.
130. Batzill, M., 2011. Fundamental aspects of surface engineering of transition metal oxide photocatalysts. *Energy & Environmental Science*, 4(9), pp. 3275–3286.
131. Venieri, D., Gounaki, I., Binas, V., Zachopoulos, A., Kiriakidis, G. and Mantzavinos, D., 2015. Inactivation of MS2 coliphage in sewage by solar photocatalysis using metal-doped TiO_2. *Applied Catalysis B: Environmental*, 178, pp. 54–64.
132. Ma, S.L., Zhan, S.L., Jia, Y.N. and Zhou, Q.X., 2015. Superior antibacterial activity of Fe_3O_4-TiO_2 nanosheets under solar light. *ACS Applied Materials & Interfaces*, 7, pp. 21875–21883.
133. Wang, B., Leung, M.K., Lu, X.Y. and Chen, S.Y., 2013. Synthesis and photocatalytic activity of boron and fluorine codoped TiO_2 nanosheets with reactive facets. *Applied Energy*, 112, pp. 1190–1197.
134. Zhang, J., Suo, X., Liu, X., Liu, B., Xue, Y., Mu, L. and Shi, H., 2016. PdO loaded WO_3 composite with $Na_2W_4O_{13}$ flake: A 2-D heterostructure composite material. *Materials Letters*, 184, pp. 25–28.
135. Pal, A., Jana, T.K., Roy, T., Pradhan, A., Maiti, R., Choudhury, S.M. and Chatterjee, K., 2018. MoS_2-TiO_2 nanocomposite with excellent adsorption performance and high antibacterial activity. *ChemistrySelect*, 3(1), pp. 81–90.
136. Wang, X., Lv, F., Li, T., Han, Y., Yi, Z., Liu, M., Chang, J. and Wu, C., 2017. Electrospun micropatterned nanocomposites incorporated with Cu_2S nanoflowers for skin tumor therapy and wound healing. *ACS Nano*, 11(11), pp. 11337–11349.
137. Huang, K., Li, Z., Lin, J., Han, G. and Huang, P., 2018. Two-dimensional transition metal carbides and nitrides (MXenes) for biomedical applications. *Chemical Society Reviews*, 47(14), pp. 5109–5124.
138. Lu, Z., Wei, Y., Deng, J., Ding, L., Li, Z.K. and Wang, H., 2019. Self-crosslinked MXene ($Ti_3C_2T_x$) membranes with good antiswelling property for monovalent metal ion exclusion. *ACS Nano*, 13(9), pp. 10535–10544.
139. Rafieerad, A., Amiri, A., Yan, W., Eshghi, H. and Dhingra, S., 2022. Conversion of 2D MXene to multi-low-dimensional GerMXene superlattice heterostructure. *Advanced Functional Materials*, 32(10), p. 2108495.
140. Wang, W., Feng, H., Liu, J., Zhang, M., Liu, S., Feng, C. and Chen, S., 2020. A photo catalyst of cuprous oxide anchored MXene nanosheet for dramatic enhancement of synergistic antibacterial ability. *Chemical Engineering Journal*, 386, p. 124116.
141. Rasool, K., Mahmoud, K.A, Johnson, D.J., Helal, M., Berdiyorov, G.R., Gogotsi, Y., 2017. Efficient antibacterial membrane based on two-dimensional $Ti_3C_2T_x$ (MXene) nanosheets. *Scientific Reports*, 7, pp. 1598–609.
142. Jastrzębska, A.M., Karwowska, E., Wojciechowski, T., Ziemkowska, W., Rozmysłowska, A., Chlubny, L. and Olszyna, A., 2019. The atomic structure of Ti_2C and Ti_3C_2 MXenes is responsible for their antibacterial activity toward E. coli bacteria. *Journal of Materials Engineering and Performance*, 28, pp. 1272–1277.
143. Pandey, R.P., Rasool, K., Madhavan, V.E., Aïssa, B., Gogotsi, Y. and Mahmoud, K.A., 2018. Ultrahigh-flux and fouling-resistant membranes based on layered silver/MXene ($Ti_3C_2T_x$) nanosheets. *Journal of Materials Chemistry A*, 6(8), pp. 3522–3533.
144. Liu, G., Shen, J., Ji, Y., Liu, Q., Liu, G., Yang, J. and Jin, W., 2019. Two-dimensional Ti_2CT_x MXene membranes with integrated and ordered nanochannels for efficient solvent dehydration. *Journal of Materials Chemistry A*, 7(19), pp. 12095–12104.
145. Anasori, B., Lukatskaya, M.R. and Gogotsi, Y., 2017. 2D metal carbides and nitrides (MXenes) for energy storage. *Nature Reviews Materials*, 2(2), pp. 1–17.
146. Arabi Shamsabadi, A., Sharifian Gh, M., Anasori, B. and Soroush, M., 2018. Antimicrobial mode-of-action of colloidal $Ti_3C_2T_x$ MXene nanosheets. *ACS Sustainable Chemistry & Engineering*, 6(12), pp. 16586–16596.
147. Zhang, D., Liu, H.M., Shu, X., Feng, J., Yang, P., Dong, P., Xie, X. and Shi, Q., 2020. Nanocopper-loaded Black phosphorus nanocomposites for efficient synergistic antibacterial application. *Journal of Hazardous Materials*, 393, p. 122317.
148. Mao, C., Xiang, Y., Liu, X., Cui, Z., Yang, X., Li, Z., Zhu, S., Zheng, Y., Yeung, K.W.K. and Wu, S., 2018. Repeatable photodynamic therapy with triggered signaling pathways of fibroblast cell proliferation and differentiation to promote bacteria-accompanied wound healing. *ACS Nano*, 12(2), pp. 1747–1759.
149. Huang, X.W., Wei, J.J., Zhang, M.Y., Zhang, X.L., Yin, X.F., Lu, C.H., Song, J.B., Bai, S.M. and Yang, H.H., 2018. Water-based black phosphorus hybrid nanosheets as a moldable platform for wound healing applications. *ACS Applied Materials & Interfaces*, 10(41), pp. 35495–35502.

150. Sofer, Z., Sedmidubský, D., Luxa, J., Bouša, D., Huber, Š., Lazar, P., Veselý, M. and Pumera, M., 2017. Universal method for large-scale synthesis of layered transition metal dichalcogenides. *Chemistry–A European Journal*, 23(42), pp. 10177–10186.
151. Bang, G.S., Cho, S., Son, N., Shim, G.W., Cho, B.K. and Choi, S.Y., 2016. DNA-assisted exfoliation of tungsten dichalcogenides and their antibacterial effect. *ACS Applied Materials & Interfaces*, 8(3), pp. 1943–1950.
152. Yang, X., Li, J., Liang, T., Ma, C., Zhang, Y., Chen, H., Hanagata, N., Su, H. and Xu, M., 2014. Antibacterial activity of two-dimensional MoS_2 sheets. *Nanoscale*, 6(17), pp. 10126–10133.
153. Pandit, S., Karunakaran, S., Boda, S.K., Basu, B. and De, M., 2016. High antibacterial activity of functionalized chemically exfoliated MoS_2. *ACS Applied Materials & Interfaces*, 8(46), pp. 31567–31573.
154. Yin, W., Yu, J., Lv, F., Yan, L., Zheng, L.R., Gu, Z. and Zhao, Y., 2016. Functionalized nano-MoS_2 with peroxidase catalytic and near-infrared photothermal activities for safe and synergetic wound antibacterial applications. *ACS Nano*, 10(12), pp. 11000–11011.
155. Alimohammadi, F., Sharifian Gh, M., Attanayake, N.H., Thenuwara, A.C., Gogotsi, Y., Anasori, B. and Strongin, D.R., 2018. Antimicrobial properties of 2D MnO_2 and MoS_2 nanomaterials vertically aligned on graphene materials and Ti_3C_2 MXene. *Langmuir*, 34(24), pp. 7192–7200.
156. Kong, D., Wang, H., Cha, J.J., Pasta, M., Koski, K.J., Yao, J. and Cui, Y., 2013. Synthesis of MoS_2 and $MoSe_2$ films with vertically aligned layers. *Nano Letters*, 13(3), pp. 1341–1347.
157. Li, X., Shan, J., Zhang, W., Su, S., Yuwen, L. and Wang, L., 2017. Recent advances in synthesis and biomedical applications of two-dimensional transition metal dichalcogenide nanosheets. *Small*, 13(5), p. 1602660.
158. Yang, L., Wang, J., Yang, S., Lu, Q., Li, P. and Li, N., 2019. Rod-shape MSN@MoS_2 nanoplatform for FL/MSOT/CT imaging-guided photothermal and photodynamic therapy. *Theranostics*, 9(14), p. 3992.
159. Liu, C., Kong, D., Hsu, P.C., Yuan, H., Lee, H.W., Liu, Y., Wang, H., Wang, S., Yan, K., Lin, D. and Maraccini, P.A., 2016. Rapid water disinfection using vertically aligned MoS_2 nanofilms and visible light. *Nature Nanotechnology*, 11(12), pp. 1098–1104.
160. Cheng, P., Zhou, Q., Hu, X., Su, S., Wang, X., Jin, M., Shui, L., Gao, X., Guan, Y., Nözel, R. and Zhou, G., 2018. Transparent glass with the growth of pyramid-type MoS_2 for highly efficient water disinfection under visible-light irradiation. *ACS Applied Materials & Interfaces*, 10(28), pp. 23444–23450.
161. Di, J., Xia, J., Ji, M., Li, H., Xu, H., Li, H. and Chen, R., 2015. The synergistic role of carbon quantum dots for the improved photocatalytic performance of Bi_2MoO_6. *Nanoscale*, 7(26), pp. 11433–11443.
162. Xiong, J., Song, P., Di, J., Li, H. and Liu, Z., 2019. Freestanding ultrathin bismuth-based materials for diversified photocatalytic applications. *Journal of Materials Chemistry A*, 7(44), pp. 25203–25226.
163. Zhang, X., Wang, X.B., Wang, L.W., Wang, W.K., Long, L.L., Li, W.W. and Yu, H.Q., 2014. Synthesis of a highly efficient BiOCl single-crystalnanodisk photocatalyst with exposing {001} facets. *ACS Applied Materials & Interfaces*, 6(10), pp. 7766–7772.
164. Zhu, W., Li, Z., Zhou, Y. and Yan, X., 2016. Deposition of silver nanoparticles onto two dimensional BiOCl nanodiscs for enhanced visible light photocatalytic and biocidal activities. *RSC Advances*, 6(69), pp. 64911–64920.
165. Khan, M.H., Liu, H.K., Sun, X., Yamauchi, Y., Bando, Y., Golberg, D. and Huang, Z., 2017. Few-atomic-layered hexagonal boron nitride: CVD growth, characterization, and applications. *Materials Today*, 20(10), pp. 611–628.
166. Lee, K.H., Shin, H.J., Lee, J., Lee, I.Y., Kim, G.H., Choi, J.Y. and Kim, S.W., 2012. Large-scale synthesis of high-quality hexagonal boron nitride nanosheets for large-area graphene electronics. *Nano Letters*, 12(2), pp. 714–718.
167. Merenkov, I.S., Myshenkov, M.S., Zhukov, Y.M., Sato, Y., Frolova, T.S., Danilov, D.V., Kasatkin, I.A., Medvedev, O.S., Pushkarev, R.V., Sinitsyna, O.I. and Terauchi, M., 2019. Orientation-controlled, low-temperature plasma growth and applications of h-BN nanosheets. *Nano Research*, 12, pp. 91–99.
168. Shi, Y., Hamsen, C., Jia, X., Kim, K.K., Reina, A., Hofmann, M., Hsu, A.L., Zhang, K., Li, H., Juang, Z.Y. and Dresselhaus, M.S., 2010. Synthesis of few-layer hexagonal boron nitride thin film by chemical vapor deposition. *Nano Letters*, 10(10), pp. 4134–4139.
169. Song, L., Ci, L., Lu, H., Sorokin, P.B., Jin, C., Ni, J., Kvashnin, A.G., Kvashnin, D.G., Lou, J., Yakobson, B.I. and Ajayan, P.M., 2010. Large scale growth and characterization of atomic hexagonal boron nitride layers. *Nano Letters*, 10(8), pp. 3209–3215.
170. Gao, G., Mathkar, A., Martins, E.P., Galvao, D.S., Gao, D., da Silva Autreto, P.A., Sun, C., Cai, L. and Ajayan, P.M., 2014. Designing nanoscaled hybrids from atomic layered boron nitride with silver nanoparticle deposition. *Journal of Materials Chemistry A*, 2(9), pp. 3148–3154.

171. Roy, A.K., Park, B., Lee, K.S., Park, S.Y. and In, I., 2014. Boron nitride nanosheets decorated with silver nanoparticles through mussel-inspired chemistry of dopamine. *Nanotechnology*, 25(44), p. 445603.
172. Wang, Z., Yu, H., Ma, K., Chen, Y., Zhang, X., Wang, T., Li, S., Zhu, X. and Wang, X., 2018. Flower-like surface of three-metal-component layered double hydroxide composites for improved antibacterial activity of lysozyme. *Bioconjugate Chemistry*, 29(6), pp. 2090–2099.
173. Ghadiri, M., Chrzanowski, W. and Rohanizadeh, R., 2014. Antibiotic eluting clay mineral (Laponite®) for wound healing application: an in vitro study. *Journal of Materials Science: Materials in Medicine*, 25, pp. 2513–2526.
174. Zhu, C., Shen, H., Liu, H., Lv, X., Li, Z. and Yuan, Q., 2018. Solution-processable two-dimensional In_2Se_3 nanosheets as efficient photothermal agents for elimination of bacteria. *Chemistry–A European Journal*, 24(71), pp. 19060–19065.
175. Miao, Z., Fan, L., Xie, X., Ma, Y., Xue, J., He, T. and Zha, Z., 2019. Liquid exfoliation of atomically thin antimony selenide as an efficient two-dimensional antibacterial nanoagent. *ACS Applied Materials & Interfaces*, 11(30), pp. 26664–26673.
176. Han, W., Wu, Z., Li, Y. and Wang, Y., 2019. Graphene family nanomaterials (GFNs)—promising materials for antimicrobial coating and film: A review. *Chemical Engineering Journal*, 358, pp. 1022–1037.
177. Kurapati, R., Kostarelos, K., Prato, M. and Bianco, A., 2016. Biomedical uses for 2D materials beyond graphene: Current advances and challenges ahead. *Advanced Materials*, 28(29), pp. 6052–6074.
178. Lu, X., Feng, X., Werber, J.R., Chu, C., Zucker, I., Kim, J.H., Osuji, C.O. and Elimelech, M., 2017. Enhanced antibacterial activity through the controlled alignment of graphene oxide nanosheets. *Proceedings of the National Academy of Sciences*, 114(46), pp. E9793–E9801.
179. Wei, W., Li, J., Liu, Z., Deng, Y., Chen, D., Gu, P., Wang, G. and Fan, X., 2020. Distinct antibacterial activity of a vertically aligned graphene coating against Gram-positive and Gram-negative bacteria. *Journal of Materials Chemistry B*, 8(28), pp. 6069–6079.
180. Cui, H., Gu, Z., Chen, X., Lin, L., Wang, Z., Dai, X., Yang, Z., Liu, L., Zhou, R. and Dong, M., 2019. Stimulating antibacterial activities of graphitic carbon nitride nanosheets with plasma treatment. *Nanoscale*, 11(39), pp. 18416–18425.
181. Tu, Y., Lv, M., Xiu, P., Huynh, T., Zhang, M., Castelli, M., Liu, Z., Huang, Q., Fan, C., Fang, H. and Zhou, R., 2013. Destructive extraction of phospholipids from *Escherichia coli* membranes by graphene nanosheets. *Nature Nanotechnology*, 8(8), pp. 594–601.
182. Karunakaran, S., Pandit, S., Basu, B. and De, M., 2018. Simultaneous exfoliation and functionalization of 2H-MoS_2 by thiolated surfactants: applications in enhanced antibacterial activity. *Journal of the American Chemical Society*, 140(39), pp. 12634–12644.
183. Gurunathan, S., Han, J.W., Dayem, A.A., Eppakayala, V. and Kim, J.H., 2012. Oxidative stress-mediated antibacterial activity of graphene oxide and reduced graphene oxide in *Pseudomonas aeruginosa*. *International Journal of Nanomedicine*, 7, pp. 5901–5914.
184. Xiong, Z., Zhang, X., Zhang, S., Lei, L., Ma, W., Li, D., Wang, W., Zhao, Q. and Xing, B., 2018. Bacterial toxicity of exfoliated black phosphorus nanosheets. *Ecotoxicology and Environmental Safety*, 161, pp. 507–514.
185. Rasool, K., Helal, M., Ali, A., Ren, C.E., Gogotsi, Y. and Mahmoud, K.A., 2016. Antibacterial activity of $Ti_3C_2T_x$ MXene. *ACS Nano*, 10(3), pp. 3674–3684.
186. Tavares, A., Carvalho, C.M., Faustino, M.A., Neves, M.G., Tomé, J.P., Tomé, A.C., Cavaleiro, J.A., Cunha, Â., Gomes, N.C., Alves, E. and Almeida, A., 2010. Antimicrobial photodynamic therapy: study of bacterial recovery viability and potential development of resistance after treatment. *Marine Drugs*, 8(1), pp. 91–105.
187. Chong, Y., Ge, C., Fang, G., Wu, R., Zhang, H., Chai, Z., Chen, C. and Yin, J.J., 2017. Light-enhanced antibacterial activity of graphene oxide, mainly via accelerated electron transfer. *Environmental Science & Technology*, 51(17), pp. 10154–10161.
188. Huang, J., Ho, W. and Wang, X., 2014. Metal-free disinfection effects induced by graphitic carbon nitride polymers under visible light illumination. *Chemical Communications*, 50(33), pp. 4338–4340.
189. Ramasamy, M., Lee, S.S., Yi, D.K. and Kim, K., 2014. Magnetic, optical gold nanorods for recyclable photothermal ablation of bacteria. *Journal of Materials Chemistry B*, 2(8), pp. 981–988.
190. Spuhler, D., Rengifo-Herrera, J.A. and Pulgarin, C., 2010. The effect of Fe^{2+}, Fe^{3+}, H_2O_2 and the photo-Fenton reagent at near neutral pH on the solar disinfection (SODIS) at low temperatures of water containing *Escherichia coli* K12. *Applied Catalysis B: Environmental*, 96(1–2), pp. 126–141.
191. Zepp, R.G., Faust, B.C. and Hoigne, J., 1992. Hydroxyl radical formation in aqueous reactions (pH 3–8) of iron (II) with hydrogen peroxide: the photo-Fenton reaction. *Environmental Science & Technology*, 26(2), pp. 313–319.

192. Rincón, A.G. and Pulgarin, C., 2006. Comparative evaluation of Fe^{3+} and TiO_2 photoassisted processes in solar photocatalytic disinfection of water. *Applied Catalysis B: Environmental*, 63(3–4), pp. 222–231.
193. Ruales-Lonfat, C., Benítez, N., Sienkiewicz, A. and Pulgarín, C., 2014. Deleterious effect of homogeneous and heterogeneous near-neutral photo-Fenton system on Escherichia coli. Comparison with photo-catalytic action of TiO_2 during cell envelope disruption. *Applied Catalysis B: Environmental*, 160, pp. 286–297.
194. Javid, A., Kumar, M., Yoon, S., Lee, J.H. and Han, J.G., 2017. Size-controlled growth and antibacterial mechanism for Cu: C nanocomposite thin films. *Physical Chemistry Chemical Physics*, 19(1), pp. 237–244.
195. Hong, X., Wen, J., Xiong, X. and Hu, Y., 2016. Silver nanowire-carbon fiber cloth nanocomposites synthesized by UV curing adhesive for electrochemical point-of-use water disinfection. *Chemosphere*, 154, pp. 537–545.
196. Singh, S.P., Li, Y., Be'er, A., Oren, Y., Tour, J.M. and Arnusch, C.J., 2017. Laser-induced graphene layers and electrodes prevents microbial fouling and exerts antimicrobial action. *ACS Applied Materials & Interfaces*, 9(21), pp. 18238–18247.
197. Rodriguez-Lazaro, D., Cook, N., Ruggeri, F.M., Sellwood, J., Nasser, A., Nascimento, M.S.J., D'Agostino, M., Santos, R., Saiz, J.C., Rzeżutka, A. and Bosch, A., 2012. Virus hazards from food, water and other contaminated environments. *FEMS Microbiology Reviews*, 36(4), pp. 786–814.
198. Akhil, K., Jayakumar, J., Gayathri, G. and Khan, S.S., 2016. Effect of various capping agents on photocatalytic, antibacterial and antibiofilm activities of ZnO nanoparticles. *Journal of Photochemistry and Photobiology B: Biology*, 160, pp. 32–42.
199. Raghupathi, K.R., Koodali, R.T. and Manna, A.C., 2011. Size-dependent bacterial growth inhibition and mechanism of antibacterial activity of zinc oxide nanoparticles. *Langmuir*, 27(7), pp. 4020–4028.
200. Chen, D., Li, X., Soule, T., Yorio, F. and Orr, L., 2016. Effects of solution chemistry on antimicrobial activities of silver nanoparticles against Gordonia sp. *Science of the Total Environment*, 566, pp. 360–367.
201. Dror-Ehre, A., Mamane, H., Belenkova, T., Markovich, G. and Adin, A., 2009. Silver nanoparticle–*E. coli* colloidal interaction in water and effect on *E. coli* survival. *Journal of Colloid and Interface Science*, 339(2), pp. 521–526.
202. Suri, R.P., Thornton, H.M. and Muruganandham, M., 2012. Disinfection of water using Pt-and Ag-doped TiO_2 photocatalysts. *Environmental Technology*, 33(14), pp. 1651–1659.
203. Venieri, D., Fraggedaki, A., Kostadima, M., Chatzisymeon, E., Binas, V., Zachopoulos, A., Kiriakidis, G. and Mantzavinos, D., 2014. Solar light and metal-doped TiO_2 to eliminate water-transmitted bacterial pathogens: Photocatalyst characterization and disinfection performance. *Applied Catalysis B: Environmental*, 154, pp. 93–101.
204. Fakhri, A., Azad, M., Fatolahi, L. and Tahami, S., 2018. Microwave-assisted photocatalysis of neurotoxin compounds using metal oxides quantum dots/nanosheets composites: Photocorrosion inhibition, reusability and antibacterial activity studies. *Journal of Photochemistry and Photobiology B: Biology*, 178, pp. 108–114.
205. Hemdan, B.A., El Nahrawy, A.M., Mansour, A.F.M. and Hammad, A.B.A., 2019. Green sol–gel synthesis of novel nanoporous copper aluminosilicate for the eradication of pathogenic microbes in drinking water and wastewater treatment. *Environmental Science and Pollution Research*, 26, pp. 9508–9523.
206. Zhan, S., Yang, Y., Shen, Z., Shan, J., Li, Y., Yang, S. and Zhu, D., 2014. Efficient removal of pathogenic bacteria and viruses by multifunctional amine-modified magnetic nanoparticles. *Journal of Hazardous Materials*, 274, pp. 115–123.
207. Jin, Y., Deng, J., Liang, J., Shan, C. and Tong, M., 2015. Efficient bacteria capture and inactivation by cetyltrimethylammonium bromide modified magnetic nanoparticles. *Colloids and Surfaces B: Biointerfaces*, 136, pp. 659–665.
208. Khaydarov, R.A., Khaydarov, R.R. and Gapurova, O., 2013. Nano-photocatalysts for the destruction of chloro-organic compounds and bacteria in water. *Journal of Colloid and Interface Science*, 406, pp. 105–110.
209. Pandiyan, R., Mahalingam, S. and Ahn, Y.H., 2019. Antibacterial and photocatalytic activity of hydrothermally synthesized SnO_2 doped GO and CNT under visible light irradiation. *Journal of Photochemistry and Photobiology B: Biology*, 191, pp. 18–25.
210. Hassouna, M.E.M., ElBably, M.A., Mohammed, A.N. and Nasser, M.A.G., 2017. Assessment of carbon nanotubes and silver nanoparticles loaded clays as adsorbents for removal of bacterial contaminants from water sources. *Journal of Water and Health*, 15(1), pp. 133–144.

211. Sun, L., Du, T., Hu, C., Chen, J., Lu, J., Lu, Z. and Han, H., 2017. Antibacterial activity of graphene oxide/g-C_3N_4 composite through photocatalytic disinfection under visible light. *ACS Sustainable Chemistry & Engineering*, 5(10), pp. 8693–8701.
212. Cheng, R., Xue, X.Y., Liu, L., Wang, J.L., Liu, Y.P., Shen, Z.P., Zhang, T. and Zheng, X., 2018. Removal of waterborne pathogen by nanomaterial-membrane coupling system. *Journal of Nanoscience and Nanotechnology*, 18(2), pp. 1027–1033.
213. Zheng, X., Shen, Z.P., Cheng, C., Shi, L., Cheng, R. and Yuan, D.H., 2018. Photocatalytic disinfection performance in virus and virus/bacteria system by Cu-TiO_2 nanofibers under visible light. *Environmental Pollution*, 237, pp. 452–459.
214. Liga, M.V., Bryant, E.L., Colvin, V.L. and Li, Q., 2011. Virus inactivation by silver doped titanium dioxide nanoparticles for drinking water treatment. *Water Research*, 45(2), pp. 535–544.
215. Abebe, L.S., Su, Y.H., Guerrant, R.L., Swami, N.S. and Smith, J.A., 2015. Point-of-use removal of cryptosporidium parvum from water: independent effects of disinfection by silver nanoparticles and silver ions and by physical filtration in ceramic porous media. *Environmental Science & Technology*, 49(21), pp. 12958–12967.
216. Darwish, A.S., Bayaumy, F.E. and Ismail, H.M., 2018. Photoactivated water-disinfecting, and biological properties of Ag NPs@Sm-doped ZnO nanorods/cuttlefish bone composite: In-vitro bactericidal, cercaricidal and schistosomicidal studies. *Materials Science and Engineering: C*, 93, pp. 996–1011.
217. Hussein, E.M., Ahmed, S.A., Mokhtar, A.B., Elzagawy, S.M., Yahi, S.H., Hussein, A.M. and El-Tantawey, F., 2018. Antiprotozoal activity of magnesium oxide (MgO) nanoparticles against *Cyclospora cayetanensis* oocysts. *Parasitology International*, 67(6), pp. 666–674.
218. Sunnotel, O., Verdoold, R., Dunlop, P.S.M., Snelling, W.J., Lowery, C.J., Dooley, J.S.G., Moore, J.E. and Byrne, J.A., 2010. Photocatalytic inactivation of *Cryptosporidium parvum* on nanostructured titanium dioxide films. *Journal of Water and Health*, 8(1), pp. 83–91.
219. Deng, C.H., Gong, J.L., Zhang, P., Zeng, G.M., Song, B. and Liu, H.Y., 2017. Preparation of melamine sponge decorated with silver nanoparticles-modified graphene for water disinfection. *Journal of Colloid and Interface Science*, 488, pp. 26–38.
220. Biswas, P. and Bandyopadhyaya, R., 2016. Water disinfection using silver nanoparticle impregnated activated carbon: *Escherichia coli* cell-killing in batch and continuous packed column operation over a long duration. *Water Research*, 100, pp. 105–115.
221. Tang, C., Hu, D., Cao, Q., Yan, W. and Xing, B., 2017. Silver nanoparticles-loaded activated carbon fibers using chitosan as binding agent: Preparation, mechanism, and their antibacterial activity. *Applied Surface Science*, 394, pp. 457–465.
222. Behin, J., Shahryarifar, A. and Kazemian, H., 2016. Ultrasound-assisted synthesis of Cu and Cu/Ni nanoparticles on NaP zeolite support as antibacterial agents. *Chemical Engineering & Technology*, 39(12), pp. 2389–2403.
223. Garvey, M., Panaitescu, E., Menon, L., Byrne, C., Dervin, S., Hinder, S.J. and Pillai, S.C., 2016. Titania nanotube photocatalysts for effectively treating waterborne microbial pathogens. *Journal of Catalysis*, 344, pp. 631–639.
224. Li, Q., Li, Y.W., Wu, P., Xie, R. and Shang, J.K., 2008. Palladium oxide nanoparticles on nitrogen-doped titanium oxide: Accelerated photocatalytic disinfection and post-illumination catalytic "memory". *Advanced Materials*, 20(19), pp. 3717–3723.
225. Kang, Z., Wu, C., Dong, L., Liu, W., Mou, J., Zhang, J., Chang, Z., Jiang, B., Wang, G., Kang, F. and Xu, C., 2019. 3D porous copper skeleton supported zinc anode toward high capacity and long cycle life zinc ion batteries. *ACS Sustainable Chemistry & Engineering*, 7(3), pp. 3364–3371.
226. Zhu, X., Wu, D., Wang, W., Tan, F., Wong, P.K., Wang, X., Qiu, X. and Qiao, X., 2016. Highly effective antibacterial activity and synergistic effect of Ag-MgO nanocomposite against Escherichia coli. *Journal of Alloys and Compounds*, 684, pp. 282–290.
227. Wang, W., Chen, X., Liu, G., Shen, Z., Xia, D., Wong, P.K. and Jimmy, C.Y., 2015. Monoclinic dibismuthtetraoxide: A new visible-light-driven photocatalyst for environmental remediation. *Applied Catalysis B: Environmental*, 176, pp. 444–453.
228. Gao, P., Liu, J., Zhang, T., Sun, D.D. and Ng, W., 2012. Hierarchical TiO_2/CdS "spindle-like" composite with high photodegradation and antibacterial capability under visible light irradiation. *Journal of Hazardous Materials*, 229, pp. 209–216.
229. Hu, G., Xu, T., Chen, X., James, T.D. and Xu, S., 2016. Solar-driven broad spectrum fungicides based on monodispersed Cu_7S_4 nanorods with strong near-infrared photothermal efficiency. *RSC Advances*, 6(106), pp. 103930–103937.

230. Xiang, Z., Wang, Y., Ju, P. and Zhang, D., 2017. Controlled synthesis and photocatalytic antifouling properties of $BiVO_4$ with tunable morphologies. *Journal of Electronic Materials*, 46, pp. 758–765.
231. Zhang, Y., Zhu, Y., Yu, J., Yang, D., Ng, T.W., Wong, P.K. and Jimmy, C.Y., 2013. Enhanced photocatalytic water disinfection properties of Bi_2MoO_6–RGO nanocomposites under visible light irradiation. *Nanoscale*, 5(14), pp. 6307–6310.
232. Chen, Y.J., Tseng, C.S., Tseng, P.J., Huang, C.W., Wu, T. and Lin, Y.W., 2017. Synthesis and characterization of Ag/Ag_3PO_4 nanomaterial modified $BiPO_4$ photocatalyst by sonochemical method and its photocatalytic application. *Journal of Materials Science: Materials in Electronics*, 28, pp. 11886–11899.
233. Deng, C.H., Gong, J.L., Ma, L.L., Zeng, G.M., Song, B., Zhang, P. and Huan, S.Y., 2017. Synthesis, characterization and antibacterial performance of visible light-responsive Ag_3PO_4 particles deposited on graphene nanosheets. *Process Safety and Environmental Protection*, 106, pp. 246–255.
234. Chen, X., Hu, B., Xiang, Q., Yong, C., Liu, Z. and Xing, X., 2016. Magnetic nanoparticles modified with quaternarized N-halamine based polymer and their antibacterial properties. *Journal of Biomaterials Science, Polymer Edition*, 27(11), pp. 1187–1199.
235. Chen, L., He, F., Huang, Y., Meng, Y. and Guo, R., 2016. Hydrogenated nanoporous TiO_2 film on $Ti_{25}Nb_3Mo_2Sn_3Zr$ alloy with enhanced photocatalytic and sterilization activities driven by visible light. *Journal of Alloys and Compounds*, 678, pp. 5–11.
236. Zeng, X., Wang, Z., Meng, N., McCarthy, D.T., Deletic, A., Pan, J.H. and Zhang, X., 2017. Highly dispersed TiO_2 nanocrystals and carbon dots on reduced graphene oxide: ternary nanocomposites for accelerated photocatalytic water disinfection. *Applied Catalysis B: Environmental*, 202, pp. 33–41.
237. Niu, P., Zhang, L., Liu, G. and Cheng, H.M., 2012. Graphene-like carbon nitride nanosheets for improved photocatalytic activities. *Advanced Functional Materials*, 22(22), pp. 4763–4770.
238. Zhou, Y., Zhang, Y., Lin, M., Long, J., Zhang, Z., Lin, H., Wu, J.C.S. and Wang, X., 2015. Monolayered Bi_2WO_6 nanosheets mimicking heterojunction interface with open surfaces for photocatalysis. *Nature Communications*, 6(1), p. 8340.
239. Kang, S., Huang, W., Zhang, L., He, M., Xu, S., Sun, D. and Jiang, X., 2018. Moderate bacterial etching allows scalable and clean delamination of g-C_3N_4 with enriched unpaired electrons for highly improved photocatalytic water disinfection. *ACS Applied Materials & Interfaces*, 10(16), pp. 13796–13804.
240. Samak, N.A., Selim, M.S., Hao, Z. and Xing, J., 2022. Immobilized arginine/tryptophan-rich cyclic dodecapeptide on reduced graphene oxide anchored with manganese dioxide for microbial biofilm eradication. *Journal of Hazardous Materials*, 426, p. 128035.
241. Shi, T., Wen, Z., Ding, L., Liu, Q., Guo, Y., Ding, C. and Wang, K., 2019. Visible/near-infrared light response VOPc/carbon nitride nanocomposites: VOPc sensitizing carbon nitride to improve photo-to-current conversion efficiency for fabricating photoelectrochemical diclofenac aptasensor. *Sensors and Actuators B: Chemical*, 299, p. 126834.
242. Li, M., Li, L., Su, K., Liu, X., Zhang, T., Liang, Y., Jing, D., Yang, X., Zheng, D., Cui, Z. and Li, Z., 2019. Highly effective and noninvasive near-infrared eradication of a *Staphylococcus aureus* biofilm on implants by a photoresponsive coating within 20 min. *Advanced Science*, 6(17), p. 1900599.
243. Zhang, L., Chen, P., Xu, Y., Nie, W. and Zhou, Y., 2020. Enhanced photo-induced antibacterial application of graphene oxide modified by sodium anthraquinone-2-sulfonate under visible light. *Applied Catalysis B: Environmental*, 265, p. 118572.
244. Guirguis, A., Polaki, S.R., Sahoo, G., Ghosh, S., Kamruddin, M., Merenda, A., Chen, X., Maina, J.W., Szekely, G. and Dumee, L., 2020. Engineering high-defect densities across vertically-aligned graphene nanosheets to induce photocatalytic reactivity. *Carbon*, 168, pp. 32–41.
245. Zhang, G., Hu, Z., Sun, M., Liu, Y., Liu, L., Liu, H., Huang, C.P., Qu, J. and Li, J., 2015. Formation of Bi_2WO_6 bipyramids with vacancy pairs for enhanced solar-driven photoactivity. *Advanced Functional Materials*, 25(24), pp. 3726–3734.
246. Guan, M., Xiao, C., Zhang, J., Fan, S., An, R., Cheng, Q., Xie, J., Zhou, M., Ye, B. and Xie, Y., 2013. Vacancy associates promoting solar-driven photocatalytic activity of ultrathin bismuth oxychloride nanosheets. *Journal of the American Chemical Society*, 135(28), pp. 10411–10417.
247. Dai, Z., Qin, F., Zhao, H., Ding, J., Liu, Y. and Chen, R., 2016. Crystal defect engineering of aurivillius Bi_2MoO_6 by Ce doping for increased reactive species production in photocatalysis. *ACS Catalysis*, 6(5), pp. 3180–3192.
248. Xiong, T., Huang, H., Sun, Y. and Dong, F., 2015. In situ synthesis of a C-doped $(BiO)_2CO_3$ hierarchical self-assembly effectively promoting visible light photocatalysis. *Journal of Materials Chemistry A*, 3, pp. 6118–6127.

249. Zhang, J.W., Gong, S., Mahmood, N., Pan, L., Zhang, X. and Zou, J.J., 2018. Oxygen-doped nanoporous carbon nitride via water-based homogeneous supramolecular assembly for photocatalytic hydrogen evolution. *Applied Catalysis B: Environmental*, 221, pp. 9–16.
250. Feng, C., Tang, L., Deng, Y., Zeng, G., Wang, J., Liu, Y., Chen, Z., Yu, J. and Wang, J., 2019. Enhancing optical absorption and charge transfer: Synthesis of S-doped h-BN with tunable band structures for metal-free visible-light-driven photocatalysis. *Applied Catalysis B: Environmental*, 256, p. 117827.
251. Cai, T., Wang, L., Liu, Y., Zhang, S., Dong, W., Chen, H., Yi, X., Yuan, J., Xia, X., Liu, C. and Luo, S., 2018. Ag_3PO_4/Ti_3C_2 MXene interface materials as a Schottky catalyst with enhanced photocatalytic activities and anti-photocorrosion performance. *Applied Catalysis B: Environmental*, 239, pp. 545–554.
252. Zhang, Z., Huang, J., Fang, Y., Zhang, M., Liu, K. and Dong, B., 2017. A nonmetal plasmonic Z-scheme photocatalyst with UV-to NIR-driven photocatalytic protons reduction. *Advanced Materials*, 29(18), p. 1606688.
253. Guerra, F.D., Attia, M.F., Whitehead, D.C. and Alexis, F., 2018. Nanotechnology for environmental remediation: Materials and applications. *Mole0cules*, 23(7), p. 1760.
254. Lee, J., Mackeyev, Y., Cho, M., Wilson, L.J., Kim, J.H. and Alvarez, P.J., 2010. C60 aminofullerene immobilized on silica as a visible-light-activated photocatalyst. *Environmental Science & Technology*, 44(24), pp. 9488–9495.
255. Qu, X., Alvarez, P.J. and Li, Q., 2013. Applications of nanotechnology in water and wastewater treatment. *Water Research*, 47(12), pp. 3931–3946.
256. Zhang, L., Lin, C.Y., Zhang, D., Gong, L., Zhu, Y., Zhao, Z., Xu, Q., Li, H. and Xia, Z., 2019. Guiding principles for designing highly efficient metal-free carbon catalysts. *Advanced Materials*, 31(13), p. 1805252.
257. Hlongwane, G.N., Sekoai, P.T., Meyyappan, M. and Moothi, K., 2019. Simultaneous removal of pollutants from water using nanoparticles: A shift from single pollutant control to multiple pollutant control. *Science of The Total Environment*, 656, pp. 808–833.

13 2D Metal Oxides and Their Heterostructures for Photodegradation of Organic Water Pollutants

Lan Wang
School of Environmental Science and Engineering, Shaanxi University of Science and Technology, Xi'an 710021, PR China

Chen Hou
School of Environmental Science and Engineering, Shaanxi University of Science and Technology, Xi'an 710021, PR China

Wei Zhang
School of Environmental Science and Engineering, Shaanxi University of Science and Technology, Xi'an 710021, PR China

Cong Wang
School of Environmental Science and Engineering, Shaanxi University of Science and Technology, Xi'an 710021, PR China

13.1 INTRODUCTION

With the booming economy and increasing requirements for high quality of life, various environment-related contaminants are inevitably released into water media through industry and human activities. The major contaminants in water include organic pollutants, inorganic pollutants, and bio-organisms. Some contaminants with high toxicities, especially the residual toxic organic pollutants, become concentrated in aquatic organisms and even move into the soil and accumulate in plants, which can seriously endanger the ecosystem and human health. These organic pollutants such as organic dyes, pharmaceuticals and personal care products (PPCPs), pesticides, and other common industrial organics have adverse effects on biota, thus, they are of primary concern [1–3]. Conventional water treatment techniques include adsorption, membrane separation, and biological and chemical methods [4–6]. Nevertheless, these methods have some demerits and may cause secondary pollution while treating a diverse class of organic pollutants from industrial/domestic effluents. For example, the adsorption method can only transfer organic compounds from water to the solid phase but cannot decompose them. Thus, it is of pivotal significance to develop highly efficient, eco-friendly, and sustainable degradation technologies to remove these organic pollutants quickly and completely for water remediation.

Advanced oxidation processes (AOPs) have been widely studied and developed for the total oxidation of organic pollutants or their conversion into less toxic compounds, which are extremely efficient and ideal approaches for the treatment of many types of wastewater. The efficiency of AOPs is the result of their nonselective pathway, higher reaction rates, and oxidation potential that

DOI: 10.1201/9781003436942-13

led to reduced toxicity and completely decomposing organic pollutants rather than simply adsorbing or concentrating them for further treatment [7,8]. During AOPs, highly reactive oxygen species (ROS) such as hydroxyl radicals (OH) are generated. Hydroxyl radicals have high standard reduction potential (E (•OH/H_2O)=2.80 V vs. SHE), which can act unselectively on target organic pollutants up to their complete mineralization (H_2O, CO_2, and inorganic ions) [9].

The photocatalytic process as an important and emerging AOP technology can generate strong oxidative OH for the degradation of organic pollutants, in which solar energy is converted into chemical energy through a series of reaction steps, converting complex pollutants into simple and harmless molecules using mild conditions and consuming abundantly available sunlight [6,10]. Since the discovery of photocatalytic water splitting with TiO_2 in 1972 [11], great effort has been made in the development of photocatalysts for efficient photocatalytic processes. In general, the photocatalytic processes are exerted by the redox reactions caused by photoinduced electrons (e^-) and holes (h^+) generated on the solid surfaces of photocatalysts under light irradiation. Several ROS are generated through these reactions with holes and electrons, which are considered to be involved in the actual oxidative and reductive reactions in photocatalysis. However, photogenerated charge carriers (electron and hole) in excited states are not stable and can recombine easily, resulting in low conversion efficiency of photocatalysis. Considering the requirements for efficient light adsorption and photogenerated carrier separation and transfer, preferably high-efficiency photocatalysts should have a high surface area, good crystallinity and stability, and a suitable band structure. Until now, a wide variety of photocatalytic materials have been exploited and used for photocatalytic water purification.

Among various photocatalysts, two-dimensional (2D) materials have attracted increasing attention due to their unique thickness-dependent physicochemical properties. In the past decade, a large array of new 2D photocatalysts, including metal oxide or chalcogenide nanosheets, and carbon nitride, have been developed for photocatalytic applications [12]. In particular, 2D metal oxides and their heterostructures comprise a popular category, as a result of their tunable structure, high stability, Earth abundance, low cost, and appealing catalytic activity, along with a large number of different compounds in this category. With the unique structure of high surface-to-volume ratios, 2D photocatalysts have several advantages, such as high specific surface areas, an abundance of surface-active sites, relatively short migration distance of photogenerated carriers, and a low recombination rate of electrons and holes. To date, enormous types of 2D metal oxides-based photocatalysts have been widely developed and implemented for photocatalytic reactions, including metal oxides (TiO_2, ZnO, WO_3, Fe_2O_3, etc.) and multicomponent metal oxides (Bi_2WO_6, perovskite oxides, etc.) [13,14]. Compared to their bulk counterparts, the increased active sites and the reduced migration distance for 2D materials are beneficial to the potential application in the environment. However, the surface recombination of the photogenerated electrons and holes in the 2D photocatalysts can still lead to relatively low activity in photocatalytic reactions. Thus, 2D metal oxide-based heterostructures have been widely studied by many researchers to tackle the challenges associated with 2D photocatalysts [15]. The charge transfer rate, the separation of the photoinduced electrons and holes, the light absorption, and the redox power can be enhanced by the formation of the heterostructures.

The publications regarding the 2D metal oxides and their heterostructures for photocatalysis rose continuously over the last years, which serves to prove the general trend of an increasing interest in the research community. This chapter begins with a brief description of developed 2D metal oxides-based materials used for photocatalytic degradation of organic pollutants from water, and then focuses on the main photocatalytic principles and the preparation of 2D metal oxides and their heterostructures. A particular focus of the discussion is on different families of 2D materials based on metal oxides, which can be employed for the photocatalytic degradation of organic pollutants. Finally, we conclude with the challenges and perspectives of the development of 2D metal oxides and their heterostructure photocatalysts in the field of water remediation.

13.2 OVERVIEW OF PHOTOCATALYSIS AND MECHANISMS OF PHOTODEGRADATION ON 2D METAL OXIDES

Heterogeneous photocatalysis is a discipline that includes a large variety of reactions. Among them, heterogeneous photocatalytic oxidation using metal oxides as photocatalysts has received more attention as an alternative promising method for eliminating organic compounds from water. The basic photochemical and photophysical principles underlying photocatalysis have been already established and reported in the literature. In general, photocatalysis is defined as a photoinduced redox process based on the band structures and the ability to transport photogenerated charge carriers of photocatalysts. Briefly, the photocatalytic process involves four steps [16,17]: (i) light absorption; (ii) generation of electron–hole (e^-–h^+) pairs; (iii) separation and transfer of photoexcited charge carriers; and (iv) occurrence of surface oxidation (h^+)/reduction (e^-) reactions. Thus, it can be seen that effective incident light absorption and strong reduction/oxidation ability are the fundamental prerequisites for designing desired photocatalysts. Specifically, the photocatalytic reaction is initiated when a photoexcited electron is promoted from the filled valence band (VB) of semiconductor photocatalyst to the empty conduction band (CB) as the absorbed photon energy, $h\nu$, equals or exceeds the band gap (E_g) of the semiconductor photocatalyst leaving behind a hole in the VB. Thus, in concert, an electron and hole pair is formed. The photogenerated electrons and holes in the CB and VB, respectively, transfer toward the surface of the photocatalyst. The volume recombination of electrons and holes then takes place, producing photons or heat. And surface recombination may also occur due to the number of active surface states in the photocatalyst. The electrons at the surface initiate reduction reactions, producing superoxide radicals ($\bullet O_2^-$) with a reductive potential of between +0.5 and –1.5 V vs. NHE, while the holes initiate oxidation reactions and form •OH with an oxidative potential between +1.0 and +3.5 V vs. NHE [17]. The band structures of typical 2D metal oxide semiconductors are presented in Figure 13.1.

Photocatalytic degradation of organic pollutants usually involves multiple steps: (i) 2D metal oxide absorbs photons with energy greater than or equal to E_g to produce e^-–h^+ pairs, (ii) interfacial charge transfer, (iii) formation of ROS via reduction and oxidation processes, (iv) decomposition of organic pollutants, and (v) desorption of intermediates from the 2D metal oxide surface [18,19]. Interfacial charge transfer and ROS generation are represented as major reaction mechanisms. The photogenerated electrons would either recombine with h^+ or randomly transfer to the surface of photocatalysts, which can be further trapped by O_2 to form $\bullet O_2^-$. The photogenerated holes can directly be involved in the photocatalytic oxidation of organic or molecules or trapped by H_2O to form •OH (Figure 13.2). Prolonging the lifetime of photogenerated charge carriers and lowering recombination rate of e^-/h^+ is crucial during the photocatalytic water purification process. The generated h^+, $\bullet O_2^-$ and •OH radicals are capable of degrading organic pollutants to convert them to nontoxic forms or mineralizing them into CO_2 and H_2O. The main reaction mechanisms during the photocatalytic degradation of organic pollutants are highlighted in the following equations:

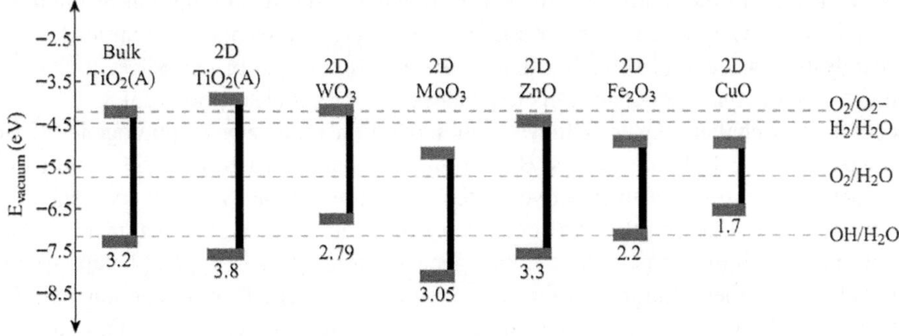

FIGURE 13.1 The band structures of typical 2D metal oxide photocatalysts. (Reproduced with permission from Ref. [17], Copyright © 2019 Royal Society of Chemistry.)

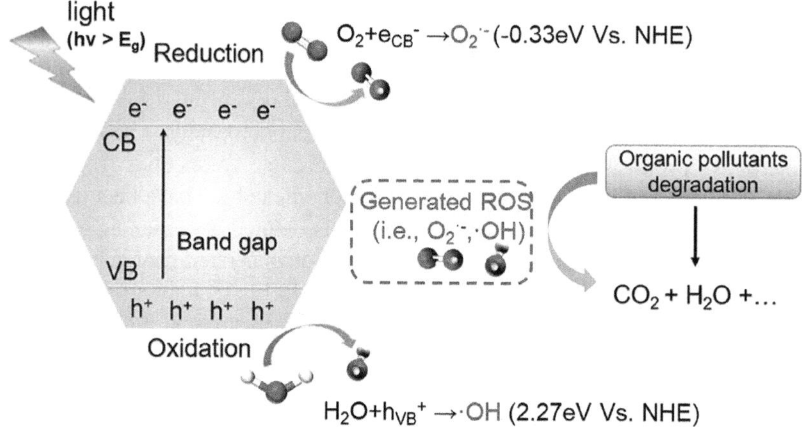

FIGURE 13.2 Mechanisms in the photodegradation of organic water pollutants using 2D metal oxide.

$$\text{Photocatalysts} + \text{light} \rightarrow e^-_{cb} + h^+_{vb} \tag{11.1}$$

$$e^-_{cb} + O_2 \rightarrow \bullet O_2^- \tag{11.2}$$

$$h^+_{vb} + H_2O \rightarrow \bullet OH \tag{11.3}$$

$$\bullet O_2^- / \bullet OH + \text{Pollutants} \rightarrow \text{Intermediates}, CO_2, H_2O \tag{11.4}$$

$$h^+_{vb} + \text{Pollutants} \rightarrow \text{Intermediates}, CO_2, H_2O \tag{11.5}$$

13.3 KEY FEATURES OF 2D NANOSHEETS AND THEIR HETEROSTRUCTURES AS PHOTOCATALYSTS

2D materials are a class of nanomaterials that are single- or few-atom thick crystals and exhibit a sheet-like structure with lateral dimensions of tens to hundreds of nanometers to even micrometers, referred to as "nanosheets" [20,21]. In 2D metal oxides, the in-plane atoms are connected to each other by strong covalent bonds, while the layers are held together by weak van der Waals interactions [22]. Compared with their bulk counterparts, 2D metal oxides exhibit higher specific surface areas, more abundant active sites, shorter migration distances for charge carriers, and more readily tunable band gaps and electronic structures for light harvesting and charge transfer, all of which are crucial for superior photocatalytic performance [13]. However, when the thickness of 2D metal oxides is downsized to the nanometer (ultrathin thickness), the recombination of photogenerated charges can be effectively inhibited by shortening the migration distance of charges to the surface of photocatalysts, but it can also further induce the quantum effect of nanosheets, which enlarges the band gap of semiconductors and leads to a blue-shift of light absorption [23].

In general, it is challenging to achieve high separation efficiency of photogenerated charges on single-component 2D photocatalysts. By building the heterostructure of 2D metal oxides with other conductive matrixes, it is an effective strategy to figure out the two conflicts mentioned above and to create a synergy among the two components that complement the advantages of each other by overcoming their limitations. Over the past decades, four types of typical heterostructures are categorized based on their charge transfer mechanism: type I, type II, p–n junction, and Z-scheme [24]. Besides the charge transfer mechanism, heterostructures can also be divided based on their respective dimensionality. In general, heterostructures based on 2D materials are termed van der Waals heterostructures. To this end, four types of van der Waals heterostructures formed, including 0D/2D, 1D/2D, 2D/2D, and 3D/2D heterostructures (Figure 13.3). With the formation of the 2D

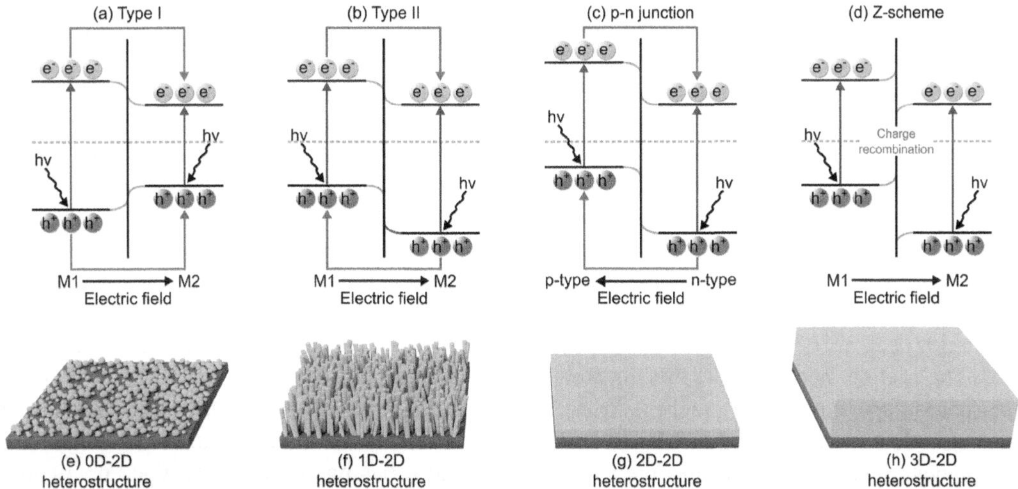

FIGURE 13.3 Possible (a–d) charge transfer mechanisms throughout the heterostructure interface areas and (e–h) different types of heterostructures with various mixed dimensionality. In panels (a–d), M1 and M2 are material 1 and material 2, respectively, while the blue and red lines represent band gap excitation and charge transfer at the interface, respectively. (Reproduced with permission from Ref. [24], Copyright © 2022 Royal Society of Chemistry.)

metal oxides-based heterojunction, the interfacial charge transfer and the separation of photogenerated electron and hole pairs could be dramatically enhanced.

13.4 SYNTHESIS OF 2D METAL OXIDE AND THEIR HETEROSTRUCTURES

13.4.1 Synthesis of 2D Metal Oxide Nanosheets

To date, a variety of methods have been developed to prepare 2D metal oxide nanosheets, which can be either produced by exfoliation of a parent layered material via a top-down approach or synthesized from small molecules with a bottom-up self-assembly method. The schematic diagrams of two methods for forming 2D materials are shown in Figure. 13.4.

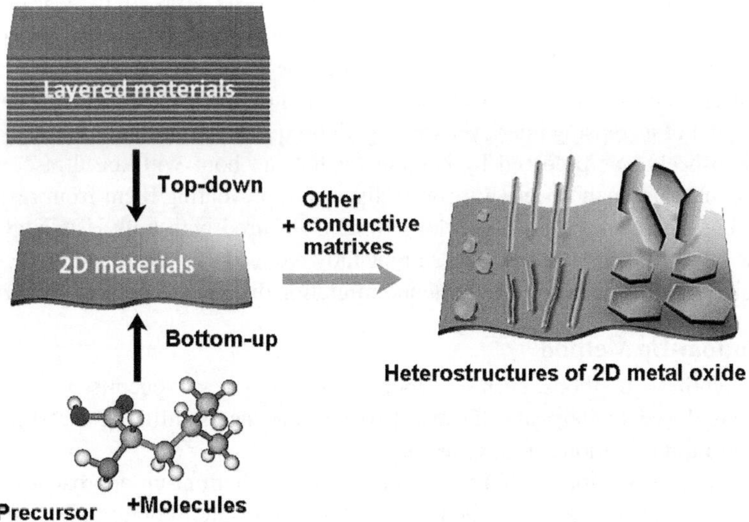

FIGURE 13.4 Schematic illustrations of synthesis of 2D metal oxides and their heterostructures. (Reproduced with permission from Ref. [12]. Copyright © 2016 Royal Society of Chemistry.)

13.4.1.1 Top-Down Method

A top-down multistep approach based on the intercalation and exfoliation of layered metal oxides has been well-documented to prepare 2D metal oxide nanosheets. The exfoliation of bulk-layered materials refers to the processes that use various methods to weaken the electrostatic interaction between the layers and disassemble the ordered layered structure into single individual sheets. The top-down methods can be divided into physical or chemical approaches. Physical methods include the separation of single- or multilayer nanoscale 2D metal oxide materials from large, layered van der Waals solids under mechanical or ultrasonic drive. In chemical methods, it is formed through chemical reactions involving ion exchange or thermal utilization.

The publication of micromechanical cleavage of HOPG (highly ordered pyrolytic graphite) by Novoselov et al. in 2004 provided a strong foundation for the successful preparation of graphene sheets [25]. The peeling mechanism of this method is to apply a normal force to the surface of HOPG with transparent tape, cutting the graphene layer from the surface of the block HOPG. Graphene undergoes countless normal forces as it becomes thinner and thinner, ultimately becoming a single layer of graphene. The graphene sample prepared by this method has many excellent properties. In addition, the van der Waals interaction between layers can be weakened by lateral forces, which can form relative motion between adjacent graphite layers to achieve delamination. Other 2D materials such as metal oxides can also be obtained through mechanical peeling. It is noteworthy that compared to graphene, 2D metal oxide nanosheets produced by mechanical exfoliation have not been extensively studied because the preparation of ultra-pure 2D metal oxides is remarkably difficult. Wang and Sasaki [26] synthesized 2D layered metal oxide nanosheets using a top-down approach (Figure 13.5a). Layered titanates were first prepared via a high-temperature solid-state reaction of a mixture of TiO_2 and alkali metal carbonates and then processed with an acidic solution to produce protonated intermediate via an ion-exchange process. The interlayer of the protonated titanate could be further expanded by replacing the protons with a certain amount of bulky organic ions, e.g., tetra-butylammonium cations (TBA^+). Under appropriate conditions, the layered structure can be exfoliated and induced by a weak shear force such as mechanical shaking in an aqueous solution. The obtained TiO_2 nanosheets exhibited extremely high 2D anisotropy, with very small thicknesses ranging from sub nanometers to a few nanometers, surface charge properties of colloids, clear crystallinity, and composition, as well as a wide band gap.

Ultrasonication, a well-established means in the exfoliation of layered carbon materials, is still the most used technique to exfoliate bulk metal oxides into nanosheets with proven scalability. Hu et al. prepared ultrathin perovskite lead niobate nanosheets (TBA/H) $Pb_2Nb_3O_{10}$ by using a novel high-power ultrasound to peel off layered $HPb_2Nb_3O_{10}$ [28]. First, $HPb_2Nb_3O_{10}$ was prepared through a conventional solid-state reaction and dispersed in (tetra(n-butyl)ammonium hydroxide) TBAOH aqueous solution with intense stirring. At this time, the larger volume of TBA^+ could partially replace interlayer protons through ion exchange to form intermediate products of $TBA/HPb_2Nb_3O_{10}$, achieving the goal of increasing interlayer spacing. Subsequently, a colloidal suspension of (TBA/H) $Pb_2Nb_3O_{10}$ nanosheets was prepared by peeling for half an hour with an ultrasound cell rupture processor. The obtained nanosheets were well dispersed, preventing them from restacking at high pH and allowing them to be stored for a long time. Compared with bulk $HPb_2Nb_3O_{10}$, nanosheets have the same tetragonal crystal structure, significantly reduced thickness, and higher photocatalytic hydrogen evolution activity due to the shortened migration distance of photoexcited electrons.

13.4.1.2 Bottom-Up Method

"Bottom-up" synthesis methods, such as vapor deposition and wet-chemistry self-assembly, have been widely employed to fabricate 2D metal oxide nanosheets utilizing metallic salts or other sources in their liquid solutions or gaseous form.

Chemical vapor deposition (CVD) is one of the typical bottom-up approaches, where volatile precursors are decomposed and deposited on a suitable substrate to obtain 2D metal oxide sheets. For instance, Lee et al. successfully synthesized high-density anatase TiO_2 nanosheets via a CVD process using $TiCl_4$ as a precursor along with oxygen and argon as reaction gases [29]. This method

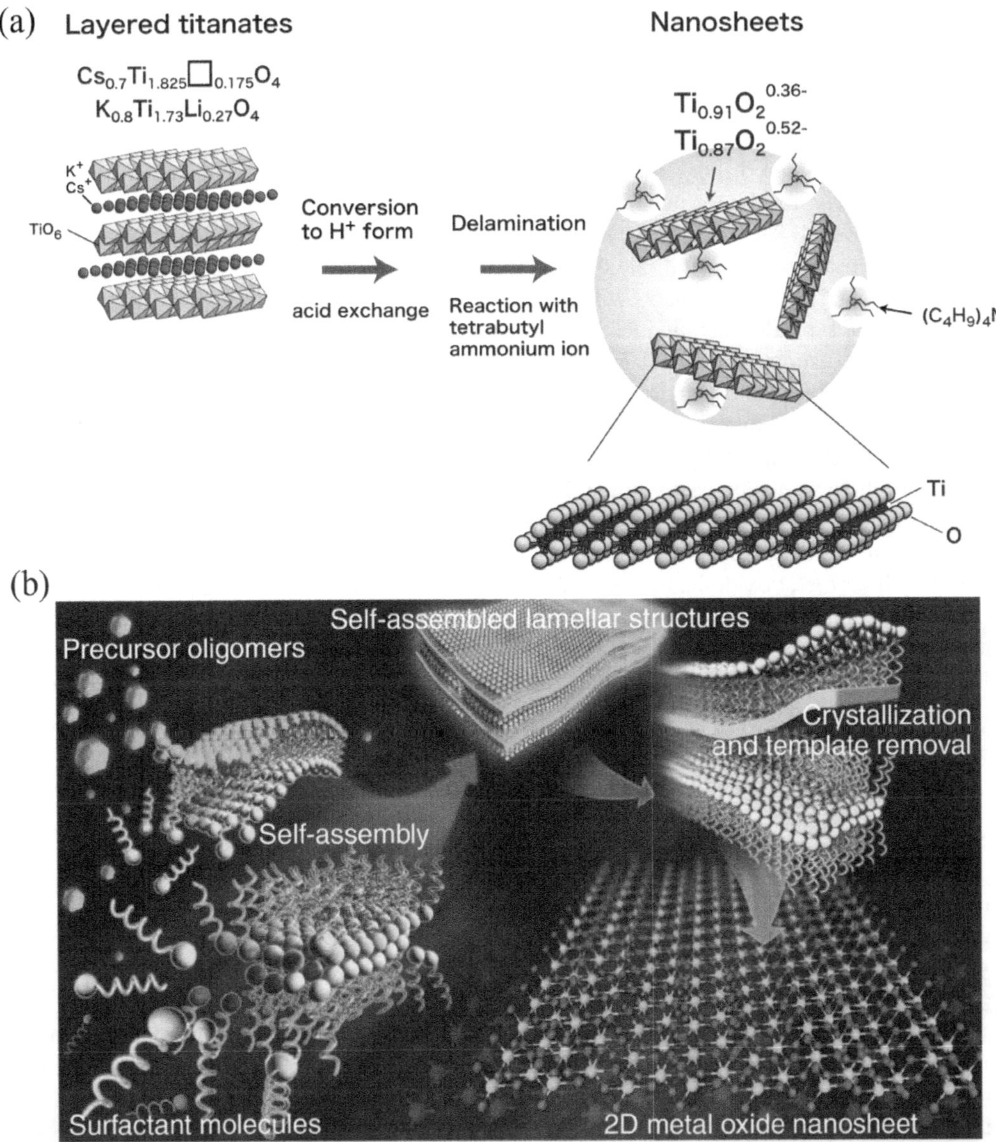

FIGURE 13.5 (a) Schematic illustration of the crystal structure of a typical titanate and its exfoliation into 2D nanosheets. (Reproduced with permission from Ref. [26], Copyright © 2014 American Chemical Society.) (b) The molecular assembly of metal oxide oligomers into lamellar structures with a polymeric surfactant, the condensation of which induces crystallization and generates 2D metal oxide nanosheets. (Reproduced with permission from Ref. [27], Copyright © 2014 Nature Publishing Group.)

yielded vertically grown TiO_2 nanosheets on a silica substrate with exposed high energetic (001) facets, thermodynamically more active and suitable for various applications. Although CVD methods produce high-quality and well-controlled growth of 2D metal oxides, its low productivity and the requirements for harsh conditions handicap their applications. In this regard, wet-chemical synthetic methods have been explored due to their high yields, low cost, and the possibility of scaleup.

Among different wet chemical approaches, solvothermal/hydrothermal synthesis, template-directed methods, and self-assembly processes have been extensively studied for the preparation of 2D nanostructures. Sun et al. [27] provided a generalized synthesis strategy for surfactant self-assembly or bottom-up growth of various ultrathin 2D transition metal oxide nanostructures from molecular precursors, in which amphiphilic block copolymers and short-chain alcohol co-surfactants were intelligently employed as structure-directing agents to confine the stacking and growth of the metal oxide along the chosen direction. The authors formed a reverse layer micelle of polyethylene oxide polypropylene oxide polyethylene oxide (PEO_{20}-PPO_{70}-$PEEO_{20}$, Pluronic P123) surfactant and ethylene glycol (EG) co-surfactant in ethanol solvent. Hydrated inorganic oligomers were limited to forming layered inorganic oligomer clusters within reverse-layered micelles. The addition of EG is crucial for the formation of ultrathin 2D metal oxide nanosheets, as it can form more stable layered phases. Then, perform hydrothermal or solvent heat treatment to improve and induce complete condensation and crystallization. Finally, after removing the surfactant template, ultrathin 2D transition metal oxide nanosheets with good crystallization were collected. The author successfully synthesized typical transition metal oxide 2D nanostructures such as TiO_2, ZnO, Co_3O_4, WO_3, Fe_3O_4, and MnO_2 using this method (Figure 13.5b).

Compared to self-assembly and solvothermal processes, the template-based methods are highly controllable and can be extensively implemented to exfoliate 2D nanomaterials. Based on the types of templates and experimental parameters, the properties of the 2D nanostructures, including their size and morphology, can be easily tuned. Peng et al. [30] reported a general two-step strategy for controlled synthesis of holey 2D transition metal oxide nanosheets with tunable pore sizes using graphene oxides as a sacrificial template. This method can not only synthesize various simple 2D porous transition metal oxide nanosheets but also synthesize mixed oxides such as $ZnMn_2O_4$ and $ZnCo_2O_4$. The obtained 2D porous nanosheets exhibited excellent electrochemical properties such as high reversible capacity, excellent rate performance, and cyclic stability. In addition to the above methods, methods such as epitaxial growth and thermal injection can also be widely used to manufacture 2D materials with different morphologies and sizes [31].

13.4.2 Synthesis of 2D Heterostructures of Metal Oxides

The synthesis method of 2D heterostructures of metal oxides is similar to that of metal oxide nanosheets, mainly involving hydrothermal, solvothermal, self-assembly, sol–gel, and other methods [32,33]. To date, the reported synthetic strategies for heterostructures of 2D metal oxide could be divided into two categories, i.e., in situ growth and ex situ assembly methods. The in situ approach mainly includes the direct growth of low-dimensional nanomaterials on the surface of the 2D metal oxide or the formation of 2D metal oxide in the presence of other nanomaterials, while ex situ assembly methods refer to the pre-synthesis of nanomaterials with defined composition and structure, which are subsequently combined with 2D metal oxide via covalent or non-covalent interactions [12]. Many in situ methods, such as hydrothermal or solvothermal methods, as well as in situ hydrolysis, have been developed to construct 2D-based composite photocatalysts via one-step growth or multi-step conversion approaches. For instance, Wang et al. prepared 2D/2D TiO_2/g-C_3N_4 nano-heterostructures by using a one-step solvothermal synthesis [34]. The unique 2D/2D structure effectively increased the interface between TiO_2 and g-C_3N_4, which made the migration distance of photogenerated carriers to the surface shorter and accelerated the speed of participation in the photocatalytic reaction. Ahmed et al. synthesized 2D sheets of /WO_3 and also reduced graphene oxide/ WO_3 (rGO/WO_3) composite via a well-controlled one-pot hydrothermal method [35]. When rGO was mixed with WO_3 precursor, some parts of WO_3 rectangular nanosheets were wrapped with rGO at different points. This rGO/WO_3 composite was further used for the photocatalytic degradation of methylene blue (MB) and Rhodamine B (RhB) under sunlight irradiation. Wang et al. synthesized ultrathin TiO_2 nanosheet UTNs using a hydrothermal method, which were then used to support the in situ growth of Bi_2MoO_6 nanosheets, producing TiO_2/Bi_2MoO_6 heterojunction [36]. TiO_2 coupled with Bi_2MoO_6 to produce heterojunction with close interface and strong adhesion, which is beneficial for the transfer and migration of charge carriers.

13.5 2D METAL OXIDE AND THEIR HETEROSTRUCTURES FOR PHOTOCATALYTIC DEGRADATION

The types of organic water pollutants are abundant and diverse, mainly including dyes, surfactants, pesticides, personal care products (PCPs), endocrine-disrupting chemicals (EDCs), and some other common industrial chemicals. 2D metal oxide and their heterostructures' photocatalysts have been widely investigated for the photocatalytic elimination of organic pollutants due to their good catalytic performances. Tables 13.1 and 13.2 summarize the representative findings on the photocatalytic performances of 2D metal oxide and their heterostructures for the degradation of organic water pollutants.

TABLE 13.1
Summary of Organic Pollutant Degradation Performances of 2D Metal Oxide Photocatalysts

Photocatalyst	Thickness (nm)	Synthetic Method	Application	Ref.
TiO_2 NS	14	Hydrothermal	Rhodamine B degradation	[37]
Ag-TiO_2 NS	50	Solvothermal	Rhodamine B degradation	[38]
WO_3 NS	4.9	Surfactant-induced self-assembly method	Methyl orange degradation	[39]
WO_3 NS	4	Solvothermal/surface hydrogenation reduction strategy	Metribuzin degradation	[40]
ZnO NS	10	Solvothermal	Methyl orange degradation	[41]
ZnO NS	5–30	Solvothermal-annealing	Phenol degradation	[42]
CuO NS	20	Hydrothermal	Methylene blue degradation	[43]
CuO NS	5–10	Solution-phase method	Rhodamine B degradation	[44]
α-Fe_2O_3 NS	40	Solvothermal	Salicylic acid degradation	[45]
α-Fe_2O_3 NS	16	Facile solvothermal method	Rhodamine B degradation	[46]
Bi_2MoO_6 NS	99	Electrophoretic deposition	Metronidazole degradation	[47]
Bi_2MoO_6 NS	36	Ionic liquid-assisted hydrothermal	Phenol degradation	[48]
$RbBiNb_2O_7$ NS	5–7	Conventional solid-state	Rhodamine B degradation	[49]

TABLE 13.2
Summary of Organic Pollutant Degradation Performances of 2D Metal Oxide-Based Heterostructures

Photocatalyst	Thickness (nm)	Synthetic Method	Application	Ref.
Co_3O_4/TiO_2	1.5	Solvothermal	Enrofloxacin degradation	[50]
TiO_2(B)-BiOBr	10–20	In-situ crystallized	Bisphenol A degradation	[51]
$WO_3/SnNb_2O_6$	3.2	Hydrothermal	Rhodamine B degradation	[52]
WO_3/g-C_3N_4	5	Alcohol-thermal	Rhodamine B degradation	[53]
ZnTPP/ZnO	60	Dip-coating method	Rhodamine B degradation	[54]
ZnO/Au	30–40	Co-precipitation	2-Chlorophenol degradation	[55]
CuO/g-C_3N_4	6	Thermal polymerization	Oxytetracycline degradation	[56]
C/Fe_2O_3	3–5	Hydrothermal	Methylene blue degradation	[57]
α-Fe_2O_3/Ti_3C_2	5–40	Facile ultrasonic assisted self-assembly method	Rhodamine B degradation	[58]
Bi_2WO_6/Ti_3C_2	5–40	In situ growth route	Rhodamine B degradation	[59]
Ti_3C_2–OH/Bi_2WO_6:Yb^{3+}, Tm^{3+}	40–60	Hydrothermal	Rhodamine B degradation	[60]
$MoS_2/PbTiO_3$	110	Hydrothermal	Methylene blue degradation	[61]

13.5.1 TiO_2-Based Photocatalysts

The widespread interest in titanium dioxide (TiO_2) has been shown to be the prominent photocatalyst candidate due to advantages such as non-toxicity, stability, and low cost [62]. However, the wide band gap of 3.2 eV of TiO_2 limits its catalytic activity under visible light irradiation. To solve this problem, the morphology, type, and structure of TiO_2 have been modified to improve the photocatalytic performance [63,64]. Thus, the high surface area properties of 2D nanosheet structures have attracted the major attention of scholars. The 2D structure of TiO_2 nanosheets can promote the rapid transfer of carriers and facilitate the effective separation of carriers, resulting in efficient photocatalytic activity. Dai et al. prepared ultrathin 2D TiO_2 nanosheets with large specific surface area and fast electron transfer by conventional hydrothermal method using tetrabutyl titanate as the titanium source, in which the TiO_2 nanosheets have a mixed crystalline phase of rutile and anatase. Notably, the lifetimes of photogenerated electron and hole pairs are significantly improved attributed to this 2D structure. Moreover, the calcined TiO_2 exhibited increased catalytic activity and could completely degrade MB under visible light irradiation within 10 minutes, and also showed great photocatalytic activity in the degradation of RhB [37]. Moreover, it is well-known that the oxidation state of metal dopants affects the electron density of metal oxide nanocrystals via improving the oxygen vacancies and refining the crystallinity in metal oxide nanocrystals by donating and receiving electrons, as well as reducing the recombination of electron vacancies or facilitating the separation of electron vacancies, which is useful to promote the active electron transfer process for the highly efficient photocatalytic processes. For example, Khatijah et al. synthesized 2D metal-doped TiO_2 nanosheets (TNS) using ammonium hexafluorotitanate hydrolysis in the presence of hexamethylenetetramine and metal precursors, and their typical morphology is a hierarchical nanosheet consisting of random nanotubes with edge length and thickness of 5 μm and 200 nm, respectively. In a typical photocatalytic reaction system, AgTNS converted about 91% of RhB in only 120 minutes, corresponding to a rate constant of 0.018 m^{-1}, which is about two times higher than the pristine TNS and several orders of magnitude higher than the Zn- and Al-doped TNS. The enhancement of crystallinity, the increase of oxygen vacancies, the decrease of defect density, and the photogenerated electron–hole pair complexation rate are the key factors to improve the photocatalytic performance of AgTNS [38].

2D TiO_2 nanosheets are considered to be an efficient photocatalyst due to their larger surface area and more active sites [65]. However, the rapid compounding of photogenerated carriers leads to undesirable photocatalytic efficiency. Compared with single-component photocatalysts, heterostructure photocatalysts combine the advantages of each and exhibit superior photocatalytic performance. Wang et al. developed the sheet-on-sheet (2D/2D) TiO_2/Bi_2MoO_6 heterostructure with a built-in electric field (BIEF) [36], which showed efficient photocatalytic performance for amoxicillin (AMX) degradation (Figure 13.6). The introduction of ultrathin TiO_2 nanosheet (UTNs) not only improved the specific surface area, active site, and pore volume of the system, but also achieved 2D/2D microstructure, which was beneficial for the interfacial charge migration of heterojunction. The difference in Φ and EF between the two materials, derived from DFT calculation, contributed to the formation of (−)TiO_2/(+) Bi_2MoO_6 BIEF that can promote the separation of photogenerated charges in the system. Consequentially, the prepared TiO_2/Bi_2MoO_6 heterostructure photocatalyst showed high efficiency and stable performance for photocatalytic AMX degradation, which was 18.2 and 5.7 times higher than TiO_2 and Bi_2MoO_6, respectively. Wang et al. successfully prepared Z-scheme heterojunction Co_3O_4/titanium dioxide photocatalysts for the photocatalytic degradation of enrofloxacin by growing zero-dimensional Co_3O_4 nanodots in situ on 2D TiO_2 nanosheets. The degradation rate of enrofloxacin was 95.6% within 100 minutes (0.0269 $minutes^{-1}$), and the excellent photocatalytic performance, crystal structure, and microstructure were maintained after four cycles (400 minutes) of photodegradation, suggesting that the system has favorable stability. The Z-scheme heterojunction can simultaneously inhibit photoinduced electron–hole pairs complexation and facilitate photoinduced charge-carrier transfer, thus producing more active species for final photocatalytic degradation [50].

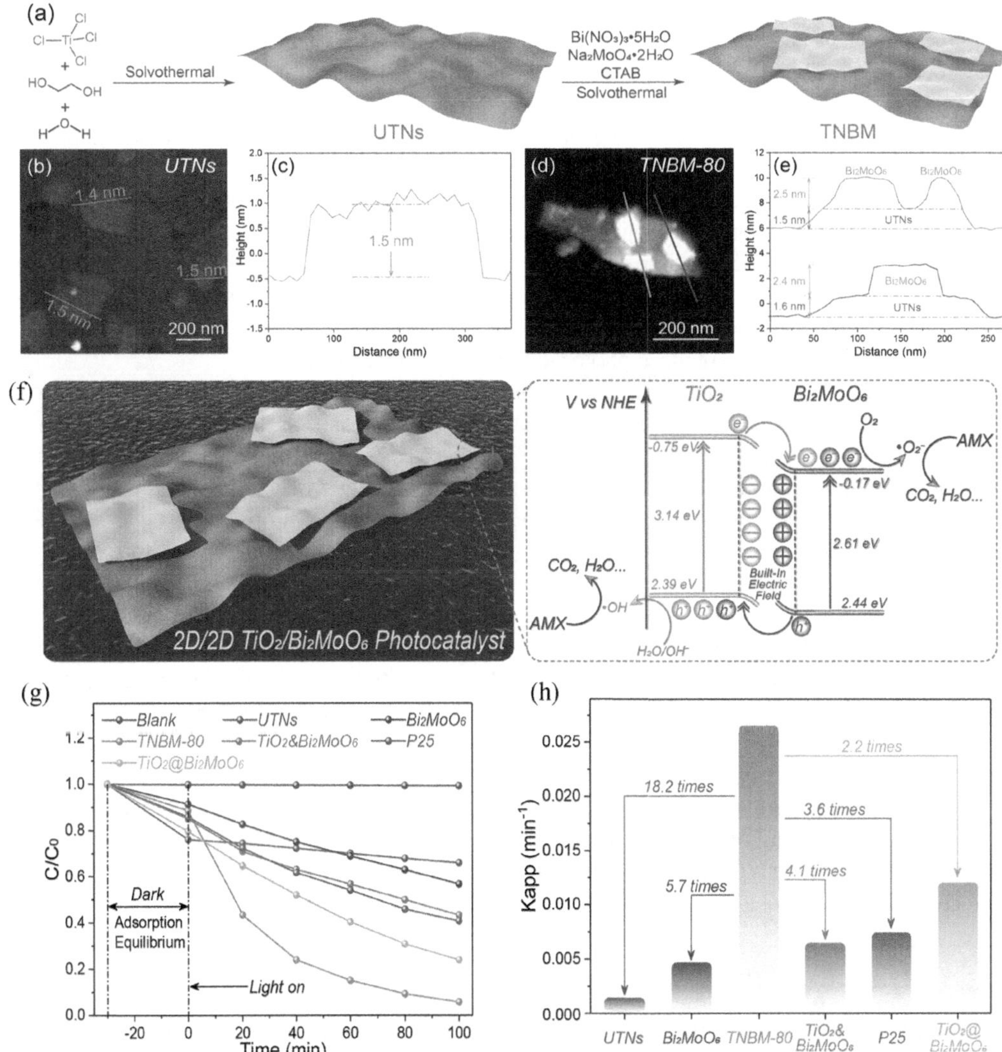

FIGURE 13.6 (a) Schematic illustration for the synthesis of UTNs and TiO_2/Bi_2MoO_6 nanohybrid. (b) AFM image and (c) height profile of UTNs. (d) AFM image and (e) height profile of TiO_2/Bi_2MoO_6 nanohybrid. (f) Structure illustration and proposed photocatalytic mechanism of TiO_2/Bi_2MoO_6. (g) Time-concentration plots and (h) k_{app} of AMX photocatalytic degradation for the prepared TiO_2/Bi_2MoO_6. (Reprinted with permission from Ref. [36], Copyright © 2022 Elsevier.)

13.5.2 ZnO-Based Photocatalysts

Due to the advantages of low cost, non-toxicity, high redox potential, and strong adsorption capacity, zinc oxide (ZnO) has been widely investigated for photocatalytic applications. However, its practical application is severely limited due to the wide theoretical band gap of 3.37 eV and the fast recombination of photogenerated carriers [66,67]. It has been found that the preparation of catalysts with nanosheet structures can effectively improve photocatalytic performance because of the availability of more adsorption and reaction sites, and the large pore volume, narrow pore size distribution, regular channel structure, and significant nano-size effect in the pore structure of nanosheets are favorable for catalytic reactions [68–70]. With the combination of structural advantages from 2D and porous materials, 2D porous materials often show great performance as photocatalysis. Lee et al. prepared 2D porous zinc oxide (ZnO) by ammonia-assisted phase transformation and calcination using coordination polymers containing zinc ions and fumaric acid ligands as raw materials [71]. Under 365 nm light irradiation, the 2D porous ZnO exhibited higher photocatalytic degradation of dyes. It was indicated that the pore morphology of ligand polymer-derived 2D ZnO has significant effects on the photocatalytic performance than its crystal structure, surface area, and chemical state. Lu et al. employed a solvothermal method to prepare a novel ZnO micro/nanoarchitecture in aqueous ethylenediamine (EDA) solution, consisting of a dense network of nanosheets constructed on hexagonal pyramidal microcrystals (core part). Notably, this structure of ZnO exhibited a strong structure-induced enhancement of photocatalytic performance and showed superior photocatalytic performance and durability than other nanostructured ZnO (e.g., nanoparticles and nanoneedle) in photodegradation of methyl orange (MO) ascribed to its higher surface-to-volume ratio and stability against aggregation. The present study provides an effective route for the enhancement of structure-induced photocatalytic performance [41].

However, the practical application of single-component ZnO materials remains challenging because of the rapid clustering of their electron–hole pairs and the confined spectral response range. Therefore, it is essential to investigate the improvement of ZnO-based photocatalysts to address environmental requirements. Wang et al. designed and fabricated a novel 2D/2D Z-scheme heterojunction between $SnNb_2O_6$ nanosheets and flower-shaped ZnO nanosheets through a simple two-step hydrothermal method [72]. Compared to the bare $SnNb_2O_6$ and ZnO, the as-prepared heterostructure displayed a remarkably improved photocatalytic performance. The reaction rate constant over 20%-$SnNb_2O_6$/ZnO heterostructure in RhB degradation was almost 15.03 and 3.95 times higher than pristine $SnNb_2O_6$ and ZnO, respectively. The enhanced photocatalytic activity can be attributed to the broad optical absorption, facilitated separation of photogenerated electron–hole pairs, and improved redox ability. Specifically, as the CB of pure $SnNb_2O_6$ is located at a higher position than the CB of ZnO, when $SnNb_2O_6$ and ZnO contact and form the heterojunction, the electron will migrate from $SnNb_2O_6$ to ZnO until the heterojunction reaches an equilibrium state. Moreover, an internal electric field from the $SnNb_2O_6$ to ZnO was built at the $SnNb_2O_6$/ZnO interface as a result of the charge accumulation, which might induce the bending of energy bands. When irradiated under light, the photogenerated material can shift from ZnO to $SnNb_2O_6$ on account of this electric field. Consequently, the obtained composite photocatalyst exhibited outstanding photodegradation efficiencies for RhB and phenol.

13.5.3 Fe_2O_3-Based Photocatalysts

As an n-type semiconductor with a 2.1 eV band gap, hematite (α-Fe_2O_3) is the most thermodynamically stable, non-toxic, corrosion-resistant iron oxide under ambient conditions [73]. Owing to its natural abundance, low cost, chemical stability, and environmental safety, hematite has been intensively studied for potential applications in various areas including photocatalysis. Due to the obvious shape dependence of Fe_2O_3 nanomaterials, the shape-controlled synthesis of them has been a hot topic of research. In particular, the controlled synthesis of interesting 2D hematite nanostructures (e.g., α-Fe_2O_3 nanoplates or nanosheets) was investigated. Sun et al. fabricated α-Fe_2O_3 nanosheet-assembled hierarchical hollow mesoporous microspheres (HHMSs) by a simple microwave-assisted solvothermal method combined with heat treatment. The specific surface areas of α-Fe_2O_3 HHMSs calcined at 400°C and 500°C were 31.50 and 12.63 $m^2 \cdot g^{-1}$, respectively, which were ferromagnetic and had superior photocatalytic activity toward salicylic acid with promising applications in wastewater treatment [45]. Liu et al. prepared ultrathin hematite (α-Fe_2O_3) nanosheets with highly exposed (110) facets by hydrothermal method. The prepared ultrathin α-Fe_2O_3 nanosheets with expanded surface area and increased exposed (110) surface can effectively harvest visible light, inhibit the recombination of electron and hole pairs, and provide excellent photocatalytic activity for the degradation of p-nitrophenol (p-NP) [74].

Frustratingly, the photocatalytic performance of Fe_2O_3 is still much lower than expected because of the sluggish charge transfer kinetics, poor electron–hole pairs separation, and limited specific surface area [75,76], which is a very urgent problem to be solved. Pan et al. successfully prepared a novel 0D/2D z-type heterojunction of nanodiamonds (NDs)/hematite (Fe_2O_3) (NDs/Fe_2O_3) photocatalyst using a solvothermal method with ND nanoparticles tightly anchored on the surface of hexagonal Fe_2O_3 nanosheets. Beneficial to the close contact between 0D NDs nanoparticles and 2D hexagonal Fe_2O_3 nanosheets, the NDs/Fe_2O_3 composites exhibited stronger photocatalytic performance compared to pristine Fe_2O_3. The optimal 15-NDs/Fe_2O_3 composite sample (containing 15 wt% NDs) showed a photocatalytic performance of 85% for 180 minutes, which was about 4.13 times that of pure Fe_2O_3 [77]. Yuan et al. designed γ-Fe_2O_3/$BiVO_4$ (FB) and Ag-γ- Fe_2O_3/$BiVO_4$ (AFB) with significant exciton effects based on the metal effect of Bi and the dielectric shielding effect of 2D photocatalysts. Due to the small exciton binding energy, the thin sheet structure facilitates exciton formation and further promotes the exciton-based energy transfer mechanism. Correspondingly, Ag-γ-Fe_2O_3/$BiVO_4$ degraded 99.4% of sulfadiazine in 90 minutes due to the addition of noble metal Ag as an electron mediator in the Z-scheme heterojunction, which facilitates the generation of more free carriers and generates a large number of excitons on the surface of the photocatalyst, thus promoting the generation of $^1O_2$ [78].

13.5.4 CuO-Based Photocatalysts

Copper oxide (CuO) is a p-type semiconductor with a narrow band gap of 1.2–1.4 eV with promising applications for degrading organic pollutants [44,79]. The unique shape of nanostructures of metal oxides significantly affects their properties and overall functionality for use in particular applications like photocatalytic activity and energy storage devices. Among them, 2D CuO consistently exhibits better performance than bulk CuO during catalysis, which is mainly due to the surface atomic mismatch caused by changes in the surface environment. The mismatch of surface atoms leads to defects on the surface of 2D CuO, which provide a great quantity of active sites for photocatalytic reaction and improve the catalytic rate. Phutanon et al. successfully synthesized CuO catalysts with excellent homogeneity using a self-assembly method, which exhibited a flower-like structure composed of copper oxide nanosheets. These semiconductor particles enhanced the degradation of azo dyes within 240 minutes, and the adsorption intensity decreased gradually with increasing time [80]. Nazim et al. synthesized a porous CuO fragmented nanosheet and used it as a catalyst for the degradation of organic pollutants. The obtained CuO nanosheets have a high optical energy band gap of up to 1.92 eV due to the strong quantum size confinement effect and exhibited an efficient and ultrafast degradation of the Allura Red AC (AR) dye by 96.99% within 6 minutes. According to the Langmuir–Hinshelwood model, the dye degradation kinetics followed a quasi-first order reaction with a rate constant $k = 0.524$ minutes^{-1}, and the half-life ($t_{1/2}$) of the AR dye degradation reaction was 2.5 minutes [81]. Bhattacharjee et al. used microwave heating to synthesize 2D CuO nanoleaves (NLs) in a facile and green manner using glutamic acid as a raw material. The construction of native CuO nanoparticles into 2D CuO nanoparticles was determined by the self-orientation and attachment growth mechanism, where the amino acids adsorbed on the surface of CuO nanoparticles control the self-assembly of CuO nanostructures. With the decrease in grain size, the energy band energy of the synthesized CuO NLs was blue-shifted to 2.1 eV. The reduction times of hazardous nitro compounds such as p-nitrophenol, p-nitroaniline, and 2,4,6-trinitrophenol were 9, 6, and 6 minutes, respectively. In addition, the CuO NLs were effective in degrading the Rose of Bengal and methyl violet 6B dyes [82].

To date, many studies have indicated that CuO-based heterojunctions exhibit excellent performance in the photocatalytic degradation of organic pollutants. Dai et al. introduced layered MXenes as charge transfer bridges into the system of O-doped g-C_3N_4 nanosheets and CuO nanosheets to construct 2D/2D/2D CuO-MXene-OCN S-scheme heterojunctions, which showed much superior photocatalytic performance than CuO-OCN heterojunctions. The degradation efficiency of the catalysts for tetracycline hydrochloride (TC) and diclofenac (DCF) under visible light irradiation during 60 minutes was 98.2% and 92.6%, respectively. This efficient degradation performance was attributed to the MXene nanosheets as electron transport mediators, which can effectively accelerate the transport between CuO nanosheets and OCN nanosheets, facilitating the separation and transfer of carriers. Meanwhile, the 2D/2D/2D structure provided a unique high specific surface area and abundant reaction sites for CuO-MXene-OCN, which significantly improved photodegradation performance [83].

13.5.5 WO_3-Based Photocatalysts

To date, tungsten trioxide (WO_3) as an n-type semiconductor with a band gap of about 2.5–2.8 eV has received much attention for its ability to absorb light in the visible region of the solar spectrum, non-toxicity, and superior catalytic properties, making it one of the promising candidates for photocatalytic degradation of organic pollutants [84,85]. Furthermore, WO_3 can be used as a photocatalyst due to its excellent optical and structural properties, as well as its superior photocorrosion performance in acid solutions [86]. Various shapes of WO_3 photocatalysts have been developed to improve its catalytic performance, among which nanosheets are considered as an ideal surface engineering material because a large number of exposed surfaces can provide sufficient surface-active sites and the specific structure can increase the electron surface migration rate [87,88]. 2D ultrathin WO_3 nanosheets with a thickness of ~4.9 nm were synthesized via a surfactant-induced self-assembly method by Liang et al. Under visible light irradiation, the prepared WO_3 nanosheets could completely degrade 25 mg L^{-1} of methyl orange within 90 minutes, and such excellent photocatalytic activity was attributed to the high specific surface area (121 m^2g^{-1}) as well as the percentage of reactive crystalline surfaces (> 90%) [39]. In addition, the large-scale use of pesticides and antibiotics is causing serious damage to the ecosystem, and addressing this issue is crucial for human drinking water safety. Wu et al. prepared defective WO_3 ultrathin nanosheets with a band gap of about 2.48 eV by the solvent thermal method and low-temperature surface hydrogenation reduction method, which have significantly enhanced visible light absorption properties. The defective WO_3 ultrathin nanosheets could completely degrade highly toxic herbicides and the degradation rate was three times higher than that of the original WO_3 ultrathin nanosheets. The enhanced photocatalytic performance is, on the one hand, due to surface oxygen vacancy defects that facilitated light harvesting and space charge separation. On the other hand, a 2D nanosheet structure with a large specific surface area could provide sufficient surface-active sites. In addition, the photocatalytic activity of the defective WO_3 ultrathin nanosheets remained almost unchanged after 10 cycles, indicating high stability and favorable for practical applications. Thus, this work provides new insights into the preparation of high-performance ultrathin nanosheet oxide photocatalysts for other environmental applications [40].

However, the high recombination rate of photoinduced charges of WO_3 photocatalysts remains a major obstacle to their practical application in photocatalysis [89]. To suppress electron–hole complexation, coupling a suitable co-catalyst with WO_3 to construct a heterostructure is an effective method. Wang et al. reported the construction of 2D/1D heterojunction photocatalysts by hydrothermal growth of WO_3 nanosheets on TiO_2 nanoribbons. The constructed WO_3/TiO_2 heterojunction broadened the photoresponse range and significantly reduced the electron–hole pair complex, which promoted the photogenerated carrier separation. Additionally, the WO_3/TiO_2 heterostructures exhibited higher photocatalytic activity under visible light irradiation compared with the pure WO_3 nanosheets and the TiO_2 nanoribbons, removing 92.8% of RhB pollutants within 120 minutes and maintaining high removal rates after five consecutive cycles [90]. Li et al. constructed $WO_3/BiVO_4/BiOCl$ ternary heterojunctions by a one-step solvothermal method, in which the synergistic catalytic effect of BiOCl, WO_3, and $BiVO_4$ effectively inhibited the photogenerated carrier binding and enhanced the photocatalytic degradation performance. Compared with $BiVO_4$ and binary $BiVO_4/BiOCl$, the $WO_3/BiVO_4/BiOCl$ ternary heterojunction significantly improved the degradation efficiency of RhB under both visible and simulated sunlight irradiation, with the best combination $WO_3/BiVO_4/BiOCl$ showing a photocatalytic degradation rate of RhB (12.5 wt%) than $BiVO_4$ and binary $BiVO_4/BiOCl$ by about 2-fold and 1.5-fold, respectively [91].

Antibiotics are widely used in medical and livestock industries to treat diseases and promote livestock growth [92,93]. However, the massive use and reckless release have caused serious pollution to the ecological environment and posed a major threat to human health. Zhang et al. rationally constructed oxygen-rich vacancy WO_3-modified monolayer Bi_2WO_6 nanosheet composites as atomic-level heterojunctions with highly reactive substances and ultrafast carrier transfer. Compared with the prepared ultrathin WO_3 and Bi_2WO_6 nanosheets, the composites can degrade ciprofloxacin by about 79.5% within 120 minutes under visible light irradiation. The large specific surface area and abundant oxygen vacancies of the composites provided more active sites for the reaction, promoting the generation with more electrons. In addition, atomic-scale heterojunctions improved the interfacial contact area, shortened the carrier transport distance, and achieved the effective separation of photoexcited carriers, all of which were the main reasons for the increased photoactivity [94]. Liu et al. prepared 2D/2D WO_3/BiOBr S-scheme heterojunctions by hydrothermal method, which achieved 98% and 87% removal of tetracycline (TC) and enrofloxacin (ENR), respectively, within 60 minutes. The superior photocatalytic activity was attributed to the effective separation of light-induced photoelectron–hole pairs by the heterojunction, which reduced the recombination of photogenerated electrons and holes [95].

13.5.6 Bi_2WO_6-Based Photocatalysts

As a typical layered material, bismuth tungstate (Bi_2WO_6) is widely used for numerous visible-light-driven photocatalytic degradations due to its stability, non-toxicity, and moderate band gap [96–98]. Due to the unique layer-like structure, Bi_2WO_6 is very favorable for nanosheet forming, which can be tuned from tens of layers to one atomic layer in thickness. Ultrathin 2D nanosheet materials can provide a large number of active sites for reactions due to their huge interlayer space, ultra-high surface area, and tunable electronic properties. 2D Bi_2WO_6 ultrathin nanosheets were prepared and their photocatalytic properties were evaluated by Pang et al. [99]. The narrow band gap and open interlayer structure of the nanosheets facilitated the trapping of light and the rapid separation and migration of carriers, respectively, thus providing better photocatalytic activity for RhB, which was nearly 3.5 times higher than that of Bi_2WO_6 nanoparticles. Cheng et al. synthesized a visible-light-responsive 2D Bi_2WO_6 photocatalyst and used it for the degradation of nitenpyram (NTP) with excellent results. The enhanced photocatalytic activity can be attributed to its unique 2D structure, which leads to a higher carrier separation efficiency. Meanwhile, the inhibitory effect of tap water containing more anions and negative ions was more pronounced compared to rainwater. This work provides a theoretical basis for the application of photocatalysts in environmental water [100]. Usman Ali et al. synthesized cabbagelike floating Bi-Bi_2S_3-Bi_2WO_6@EP (expanded perlite) photocatalysts via a one-pot modified hydrothermal process, which can completely degrade organic pollutants such as 4-chlorophenol, RhB, and methyl orange under simulated sunlight (Figure 13.7). The enhanced photocatalytic effect is due to the surface plasmon resonance (SPR) effect of Bi [101].

Indeed, the Bi_2WO_6 band gap structure allows the construction of effective heterojunctions with other semiconductor materials and retains its strong redox capability [102]. Hence, many studies are focusing on the construction of Bi_2WO_6-based heterojunctions to facilitate the effective separation of photogenerated charges and achieve the efficient catalytic activity. Ma et al. constructed 2D/2D GO-Ag/P/BWO (GO: graphene oxide; Ag: Ag, Ag_2O; P: $BiPO_4$; and BWO: Bi_2WO_6) heterojunction photocatalysts with the advantages of small crystal size, strong light absorption, and fast charge transfer, which are beneficial for the improvement of photocatalytic activity. The experimental results showed that the heterojunctions exhibited excellent degradation of Rh B under visible light irradiation. And the GO(II)-Ag/P/BWO showed superior photocatalytic degradation performance compared to another reported GO-modified Bi-based photocatalysts [103]. Li et al. synthesized efficient 2D/2D rGO-Bi_2WO_6 composites with face-to-face tight contact interfaces and full visible light absorption by a simple hydrothermal method. The 2 wt%-rGO-Bi_2WO_6 with high contact interface quality exhibited the highest degradation kinetic rate of $(5.53 \pm 0.75) \times 10^{-2}$ L·mg^{-1}·minutes^{-1} for tetracycline (TC), which was 2.4 times higher than that of the original Bi_2WO_6 and 2.1 times higher than that of the 2 wt%-rGO-Bi_2WO_6 with poorer interface quality [104]. Qian et al. prepared 2D/2D $Bi_4O_5Br_2$/Bi_2WO_6 Z-scheme heterojunctions. Notably, the heterojunction showed a photocatalytic efficiency of 91.0% for the degradation of ciprofloxacin (CIP) in visible light and could still effectively degrade CIP after five consecutive cycles. The tight heterogeneous interface between

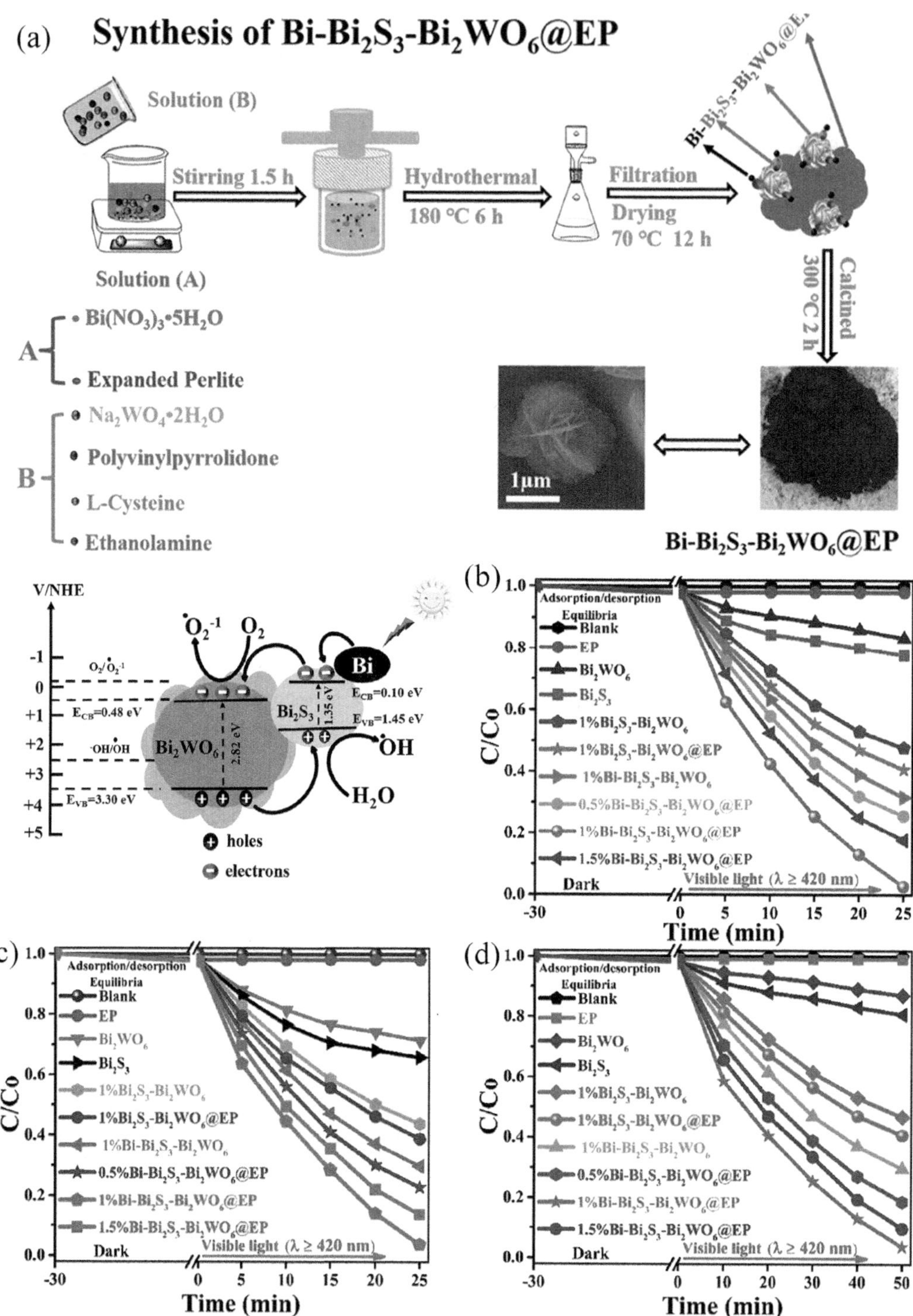

FIGURE 13.7 (a) Schematic presentation of the fabrication process for Bi-Bi_2S_3-Bi_2WO_6@EP nanocomposite and diagrammatic illustration of energy band levels for i-Bi_2S_3-Bi_2WO_6 and Bi-Bi_2S_3-Bi_2WO_6@EP heterojunctions. (b–d) Photocatalytic degradation of MO, RhB, and 4-CP. (Reproduced with permission from Ref. [101], Copyright © 2019 American Chemical Society.)

$Bi_4O_5Br_2$ and 2D Bi_2WO_6 promoted rapid migration and separation of photoinduced carriers, which enhanced the photocatalytic activity. This work provides a rational understanding of the developing 2D/2D Z-scheme heterogeneous junction environment remediation [105].

13.5.7 Perovskite Oxides-Based Photocatalysts

The promising application of perovskite oxides in the field of photocatalysis is attributed to their stability in water and relatively high hydrogen precipitation activity [106]. The present study reports that almost all layered perovskites show remarkably efficient UV-driven photocatalysis. To date, 2D and quasi-2D nanostructures have been a hot topic in the research of nanotechnology due to their unique physical–chemical properties that potentially benefit building blocks for environmental remediation and other fields [107,108]. Therefore, the development of 2D structured perovskite oxides for the treatment of organic pollutants in the field of photocatalysis has become a hot topic. Deng et al. prepared ferroelectric chalcogenide $KNbO_3$ (KNO) nanosheets in cubic and orthorhombic phases, and the results showed that the photocatalytic degradation activity of cubic $KNbO_3$ nanosheets was significantly superior to that of orthorhombic $KNbO_3$ for RhB dye. This might be attributed to its special 2D morphology, unique electronic structure, and higher symmetry, which are favorable for the excitation and transport of electrons [109]. Gao et al. prepared $BiFeO_3$ chalcogenides with a 2D layered columnar structure by a simple one-pot hydrothermal method, which could completely degrade RhB under visible light for 180 minutes. The robust photocatalytic activity of the $BiFeO_3$ catalyst was attributed to a large number of oxygen vacancies and the unique 2D layered columnar nanostructure promoting effective interfacial charge transfer [110].

Modifying the energy band structure of perovskite photocatalysts to enhance photocatalytic activity is one of the effective methods commonly used. Li et al. prepared a novel 2D/2D layered P-$La_2Ti_2O_7$/Bi_2WO_6 contact heterojunction by a simple two-step method to enhance the photocatalytic performance of $La_2Ti_2O_7$ by energy level engineering. The conversion of RhB was close to 99.02% after visible light irradiation for 80 minutes. The presence of large interfaces between the heterostructures resulted in efficient interfacial charge transfer, which yielded long-lived charge separation [111]. Sun et al. prepared a layered photocatalyst $Ni_xSr_{1-x}TiO_3$ by doping nickel chalcogenide and then loaded it on reduced graphene oxide with a large specific surface area, which successfully prepared a high-performance heterojunction photocatalyst. The addition of Ni formed a new energy level to change the energy band structure, effectively reducing the band gap width and expanding the light absorption range to achieve the effective degradation of MB [112]. In addition, Qin et al. prepared MoS_2/$PbTiO_3$ composites with S-type heterojunctions by in situ hydrothermal method, which exhibited extended optical response and enhanced charge separation with excellent photocatalytic removal of organic pollutants. The removal rate of tetracycline (TC) hydrochloride reached 90% within 120 minutes under simulated daylight. It is worth mentioning that the composite also has excellent photocatalytic removal of other organic pollutants such as dyes [61]. The Z-scheme 2D/2D BiOCl/$K^+Ca_2Nb_3O_{10}^-$ heterostructure synthesized by Liang et al. exhibited superior photocatalytic performance in simulating the degradation of TC by sunlight irradiation, which was 6.41 and 1.75 times higher than that of the pristine BiOCl and K^+CNO^- nanosheets, respectively. The improved photocatalytic performance is attributed to the synergistic effect of the heterostructure, which can promote the effective separation and transfer of photoinduced carriers [113].

13.5.8 Others

Transition metal oxides have wide applications in the photocatalytic field due to their excellent properties of low cost, pollution-free, and high stabilization. Besides the abovementioned typical metal oxides that have been widely used for photocatalytic degradation of organic pollutants, other 2D binary metal oxide nanosheets have also been investigated for photocatalytic materials for organic pollutants, such as Co_3O_4, V_2O_5, and MnO_2. Among the many transition metal oxides, spinel tricobalt tetraoxide (specifically, Co_3O_4) is the favorable material owing to its distinct features [114]. Furthermore, as a p-type semiconductor, Co_3O_4 is commonly used to enhance visible light capture and improve photonic efficiency. Yang et al. prepared 2D Co_3O_4 nanosheets with porous structure and special morphology. The 2D nanosheet structure and porous properties of the samples not only increased specific surface area and active sites for photodegradation reactions, but also facilitated mass transfer, light scattering, and harvesting as well as inhibited photogenerated carrier recombination. In addition, the prepared Co_3O_4 nanosheets exhibited superior photocatalytic activity due to their 2D porous nanosheet structure and high specific surface area ($51.36\,m^2 \cdot g^{-1}$) compared to commercial Co_3O_4 [115].

13.6 CONCLUSION AND OUTLOOK

In the past decades, a huge development has been made in 2D metal oxides and their heterostructures, and it is still advancing rapidly. This chapter elaborated on the latest progress and advancements in 2D metal oxide photocatalysis for water environmental remediation, with an emphasis on 2D metal oxides-based heterostructures. 2D metal oxides can be prepared by different processes; however, control over thickness, lateral dimensions, stacking, electronic and optical properties, and atomic arrangements are key issues. However, despite the fact that a wide variety of 2D materials is applicable for photocatalytic degradation of organic water pollutants, the photocatalytic performances are far from ideal due to several drawbacks including wide band gaps, which limit the effective utilization of solar light, and dissatisfied separation efficiency of electron–hole pairs. The incorporation of other conductive matrixes to form 2D metal oxides-based heterojunction nanocomposites can overcome these challenges and maximize the photocatalytic efficiency. There is no doubt that by constructing correct designs of 2D metal oxide-based heterostructures nanoscale interface induced by the combination of mixed oxides semiconductor deliver exceptional synergistic structure reactivity that provides considerable superior performance of photodegradation of organic pollutants. However, there are still several existing issues and drawbacks that need to be addressed. Novel techniques to prevent the re-stacking of individual ultrathin sheets, while retaining their high availability of active sites, must be explored. Deeper fundamental studies are still needed to fully unravel the expected theoretical advantages of such heterostructures. Additionally, the toxicity of these photocatalysts, as well as the by-products from photodegradation reactions of organic compounds, must be better investigated and understood.

ACKNOWLEDGMENTS

Financial support by the National Natural Science Foundation of China (Grant No. 22076114), the Key Research and Development Program of Shaanxi (2021SF-452), and the "Thousand Talents Program" of Shaanxi province of China is gratefully acknowledged.

REFERENCES

1. Y. Yang, Y.S. Ok, K.H. Kim, E.E. Kwon, Y.F. Tsang, Occurrences and removal of pharmaceuticals and personal care products (PPCPs) in drinking water and water/sewage treatment plants: A review, *Sci. Total Environ.*, 596 (2017) 303–320.
2. D.A. Yaseen, M. Scholz, Textile dye wastewater characteristics and constituents of synthetic effluents: A critical review, *Inter. J. Environ. Sci. Technol.*, 16 (2019) 1193–1226.
3. Y. Yang, X.R. Zhang, J.Y. Jiang, J.R. Han, W.X. Li, X.Y. Li, K.M.Y. Leung, S.A. Snyder, P.J.J. Alvarez, Which micropollutants in water environments deserve more dttention globally? *Environ. Sci. Technol.*, 56 (2022) 13–29.

4. A. Saravanan, P.S. Kumar, S. Jeevanantham, S. Karishma, B. Tajsabreen, P.R. Yaashikaa, B. Reshma, Effective water/wastewater treatment methodologies for toxic pollutants removal: Processes and applications towards sustainable development, *Chemosphere*, 280 (2021) 130595.
5. P.V. Nidheesh, C. Couras, A.V. Karim, H. Nadais, A review of integrated advanced oxidation processes and biological processes for organic pollutant removal, *Chem. Eng. Commun.*, 209 (2022) 390–432.
6. V.K. Sharma, M.B. Feng, Water depollution using metal-organic frameworks-catalyzed advanced oxidation processes: A review, *J. Hazard. Mater.*, 372 (2019) 3–16.
7. Y. Deng, R.Z. Zhao, Advanced oxidation processes (AOPs) in wastewater treatment, *Curr. Pollut. Rep.*, 1 (2015) 167–176.
8. D.S. Ma, H. Yi, C. Lai, X.G. Liu, X.Q. Huo, Z.W. An, L. Li, Y.K. Fu, B.S. Li, M.M. Zhang, L. Qin, S.Y. Liu, L. Yang, Critical review of advanced oxidation processes in organic wastewater treatment, *Chemosphere*, 275 (2021) 130104.
9. P. Canizares, R. Paz, C. Saez, M.A. Rodrigo, Costs of the electrochemical oxidation of wastewaters: A comparison with ozonation and Fenton oxidation processes, *J. Environ. Manage.*, 90 (2009) 410–420.
10. A. Pattnaik, J.N. Sahu, A.K. Poonia, P. Ghosh, Current perspective of nano-engineered metal oxide based photocatalysts in advanced oxidation processes for degradation of organic pollutants in wastewater, *Chem. Eng. Res. Des.*, 190 (2023) 667–686.
11. A. Fujishima, K. Honda, Electrochemical photolysis of water at a semiconductor electrode, *Nature*, 238 (1972) 37–38.
12. B. Luo, G. Liu, L. Wang, Recent advances in 2D materials for photocatalysis, *Nanoscale*, 8 (2016) 6904–6920.
13. G. Zhuang, J. Yan, Y. Wen, Z. Zhuang, Y. Yu, Two-dimensional transition metal oxides and chalcogenides for advanced photocatalysis: Progress, challenges, and opportunities, *Solar RRL*, 5 (2020) 2000403.
14. F. Haque, T. Daeneke, K. Kalantar-Zadeh, J.Z. Ou, Two-dimensional transition metal oxide and chalcogenide-based photocatalysts, *Nanomicro Lett.*, 10 (2018) 23.
15. H. Hou, X. Zhang, Rational design of 1D/2D heterostructured photocatalyst for energy and environmental applications, *Chem. Eng. J.*, 395 (2020) 125030.
16. U.I. Gaya, A.H. Abdullah, Heterogeneous photocatalytic degradation of organic contaminants over titanium dioxide: A review of fundamentals, progress and problems, *J. Photochem. Photobiol. C Photochem. Rev.*, 9 (2008) 1–12.
17. S.J. Phang, L.-L. Tan, Recent advances in carbon quantum dot (CQD)-based two dimensional materials for photocatalytic applications, *Catal. Sci. Technol.*, 9 (2019) 5882–5905.
18. K. Wei, Y. Faraj, G. Yao, R. Xie, B. Lai, Strategies for improving perovskite photocatalysts reactivity for organic pollutants degradation: A review on recent progress, *Chem. Eng. J.*, 414 (2021) 128783.
19. Y. Nosaka, A.Y. Nosaka, Generation and detection of reactive oxygen species in photocatalysis, *Chem. Rev.*, 117 (2017) 11302–11336.
20. C. Tan, X. Cao, X.J. Wu, Q. He, J. Yang, X. Zhang, J. Chen, W. Zhao, S. Han, G.H. Nam, M. Sindoro, H. Zhang, Recent advances in ultrathin two-dimensional nanomaterials, *Chem. Rev.*, 117 (2017) 6225–6331.
21. M.A. Timmerman, R. Xia, P.T.P. Le, Y. Wang, J.E. Ten Elshof, Metal oxide nanosheets as 2D building blocks for the design of novel materials, *Chemistry*, 26 (2020) 9084–9098.
22. D. Akinwande, N. Petrone, J. Hone, Two-dimensional flexible nanoelectronics, *Nat. Commun.*, 5 (2014) 5678.
23. M.Z. Qin, W.X. Fu, H. Guo, C.G. Niu, D.W. Huang, C. Liang, Y.Y. Yang, H.Y. Liu, N. Tang, Q.Q. Fan, 2D/2D Heterojunction systems for the removal of organic pollutants: A review, *Adv. Colloid Interface Sci.*, 297 (2021) 102540.
24. A.R. Fareza, F.A.A. Nugroho, F.F. Abdi, V. Fauzia, Nanoscale metal oxides–2D materials heterostructures for photoelectrochemical water splitting-a review, *J. Mater. Chem. A*, 10 (2022) 8656–8686.
25. K.S. Novoselov, A.K. Geim, S.V. Morozov, D. Jiang, Y. Zhang, S.V. Dubonos, I.V. Grigorieva, A.A. Firsov, Electric field effect in atomically thin carbon films, *Science*, 306 (2004) 666–669.
26. L. Wang, T. Sasaki, Titanium oxide nanosheets: Graphene analogues with versatile functionalities, *Chem. Rev.*, 114 (2014) 9455–9486.
27. Z. Sun, T. Liao, Y. Dou, S.M. Hwang, M.S. Park, L. Jiang, J.H. Kim, S.X. Dou, Generalized self-assembly of scalable two-dimensional transition metal oxide nanosheets, *Nat. Commun.*, 5 (2014) 3813.

28. Y. Hu, L. Guo, Rapid preparation of perovskite lead niobate nanosheets by ultrasonic-assisted exfoliation for enhanced visible-light-driven photocatalytic hydrogen production, *ChemCatChem*, 7 (2015) 584–587.
29. W.-J. Lee, Y.-M. Sung, Synthesis of anatase nanosheets with exposed (001) facets via chemical vapor deposition, *Cryst. Growth & Design*, 12 (2012) 5792–5795.
30. L. Peng, P. Xiong, L. Ma, Y. Yuan, Y. Zhu, D. Chen, X. Luo, J. Lu, K. Amine, G. Yu, Holey two-dimensional transition metal oxide nanosheets for efficient energy storage, *Nat. Commun.*, 8 (2017) 15139.
31. J. Yang, K. Kim, Y. Lee, K. Kim, W.C. Lee, J. Park, Self-organized growth and self-assembly of nanostructures on 2D materials, *FlatChem*, 5 (2017) 50–68.
32. J. Theerthagiri, S. Chandrasekaran, S. Salla, V. Elakkiya, R.A. Senthil, P. Nithyadharseni, T. Maiyalagan, K. Micheal, A. Ayeshamariam, M.V. Arasu, N.A. Al-Dhabi, H.-S. Kim, Recent developments of metal oxide based heterostructures for photocatalytic applications towards environmental remediation, *J. Solid State Chem.*, 267 (2018) 35–52.
33. N. Mahmood, I.A. De Castro, K. Pramoda, K. Khoshmanesh, S.K. Bhargava, K. Kalantar-Zadeh, Atomically thin two-dimensional metal oxide nanosheets and their heterostructures for energy storage, *Energy Storage Mater.*, 16 (2019) 455–480.
34. P. Wang, J. Wang, Y. Zhu, R. Shi, D. Wang, P. Yang, Interface nanoarchitectonics of TiO_2/g-C_3N_4 2D/2D heterostructures for enhanced antibiotic degradation and Cr(VI) reduction, *Langmuir*, 38 (2022) 11068–11079.
35. B. Ahmed, A.K. Ojha, A. Singh, F. Hirsch, I. Fischer, D. Patrice, A. Materny, Well-controlled in-situ growth of 2D WO_3 rectangular sheets on reduced graphene oxide with strong photocatalytic and antibacterial properties, *J. Hazard. Mater.*, 347 (2018) 266–278.
36. Y. Wang, G. Zuo, J. Kong, Y. Guo, Z. Xian, Y. Dai, J. Wang, T. Gong, C. Sun, Q. Xian, Sheet-on-sheet TiO_2/Bi_2MoO_6 heterostructure for enhanced photocatalytic amoxicillin degradation, *J. Hazard. Mater.*, 421 (2022) 126634.
37. Z. Dai, X.Z. Song, F. Tang, X. Kang, S. Liu, H. Abe, S. Ohara, Z. Tan, Preparation of 2D ultrathin titanium dioxide nanosheets with enhanced visible-light photocatalytic activity, *Micro Nano Lett.*, 16 (2021) 313–318.
38. S.K. Md Saad, A. Ali Umar, M.I. Ali Umar, M. Tomitori, M.Y. Abd Rahman, M. Mat Salleh, M. Oyama, Two-dimensional, hierarchical Ag-doped TiO_2 nanocatalysts: Effect of the metal oxidation state on the photocatalytic properties, *ACS Omega*, 3 (2018) 2579–2587.
39. Y. Liang, Y. Yang, C. Zou, K. Xu, X. Luo, T. Luo, J. Li, Q. Yang, P. Shi, C. Yuan, 2D ultra-thin WO_3 nanosheets with dominant {002} crystal facets for high-performance xylene sensing and methyl orange photocatalytic degradation, *J. Alloys Compd.*, 783 (2019) 848–854.
40. J. Wu, P. Qiao, H. Li, L. Ren, Y. Xu, G. Tian, M. Li, K. Pan, W. Zhou, Surface-oxygen vacancy defect-promoted electron-hole separation of defective tungsten trioxide ultrathin nanosheets and their enhanced solar-driven photocatalytic performance, *J. Colloid Interface Sci.*, 557 (2019) 18–27.
41. F. Lu, W. Cai, Y. Zhang, ZnO Hierarchical micro/nanoarchitectures: Solvothermal synthesis and structurally enhanced photocatalytic performance, *Adv. Funct. Mater.*, 18 (2008) 1047–1056.
42. D. Liu, Y. Lv, M. Zhang, Y. Liu, Y. Zhu, R. Zong, Y. Zhu, Defect-related photoluminescence and photocatalytic properties of porous ZnO nanosheets, *J. Mater. Chem. A*, 2 (2014) 15377.
43. S.P. Meshram, P.V. Adhyapak, U.P. Mulik, D.P. Amalnerkar, Facile synthesis of CuO nanomorphs and their morphology dependent sunlight driven photocatalytic properties, *Chem. Eng. J.*, 204–206 (2012) 158–168.
44. M.P. Rao, P. Sathishkumar, R.V. Mangalaraja, A.M. Asiri, P. Sivashanmugam, S. Anandan, Simple and low-cost synthesis of CuO nanosheets for visible-light-driven photocatalytic degradation of textile dyes, *J. Environ. Chem. Eng.*, 6 (2018) 2003–2010.
45. T.W. Sun, Y.J. Zhu, C. Qi, G.J. Ding, F. Chen, J. Wu, alpha-Fe_2O_3 nanosheet-assembled hierarchical hollow mesoporous microspheres: Microwave-assisted solvothermal synthesis and application in photocatalysis, *J. Colloid Interface Sci.*, 463 (2016) 107–117.
46. J. Cai, S. Chen, M. Ji, J. Hu, Y. Ma, L. Qi, Organic additive-free synthesis of mesocrystalline hematite nanoplates via two-dimensional oriented attachment, *CrystEngComm*, 16 (2014) 1553–1559.
47. M. Chahkandi, M. Zargazi, Water EPD based of 2D-Bi_2WO_6 ultrathin film on innovative designed substrates: Efficient photocatalytic degradation of binary antibiotics, *J. Mol. Liq.*, 335 (2021) 116153.
48. A. Pancielejko, J. Łuczak, W. Lisowski, G. Trykowski, D. Venieri, A. Zaleska-Medynska, P. Mazierski, Ionic liquid as morphology-directing agent of two-dimensional Bi_2WO_6: New insight into photocatalytic and antibacterial activity, *Appl. Surf. Sci.*, 599 (2022) 153971.

49. W. Xiong, H. Porwal, H. Luo, V. Araullo-Peters, J. Feng, M.-M. Titirici, M.J. Reece, J. Briscoe, Photocatalytic activity of 2D nanosheets of ferroelectric Dion–Jacobson compounds, *J. Mater. Chem. A*, 8 (2020) 6564–6568.
50. Y. Wang, C. Zhu, G. Zuo, Y. Guo, W. Xiao, Y. Dai, J. Kong, X. Xu, Y. Zhou, A. Xie, C. Sun, Q. Xian, 0D/2D Co_3O_4/TiO_2 Z-Scheme heterojunction for boosted photocatalytic degradation and mechanism investigation, *Appl. Catal. B*, 278 (2020) 119298.
51. L. Han, B. Li, H. Wen, Y. Guo, Z. Lin, Photocatalytic degradation of mixed pollutants in aqueous wastewater using mesoporous 2D/2D TiO_2(B)-BiOBr heterojunction, *J. Mater. Sci. Technol.*, 70 (2021) 176–184.
52. X. Ma, W. Ma, D. Jiang, D. Li, S. Meng, M. Chen, Construction of novel $WO_3/SnNb_2O_6$ hybrid nanosheet heterojunctions as efficient Z-scheme photocatalysts for pollutant degradation, *J. Colloid Interface Sci.*, 506 (2017) 93–101.
53. Y. Li, Y. Lu, X. Jia, Z. Ma, J. Zhang, 2D/1D Z-scheme $WO_3/g\text{-}C_3N_4$ photocatalytic heterojunction with enhanced photo-induced charge-carriers separation, *J. Phys. D Appl. Phys.*, 55 (2022) 434005.
54. Y. Xu, C. Li, H. Huang, L. Huang, L. Peng, H. Pan, Hydrophobic functionalization of ZnO nanosheets by in situ center-substituted synthesis for selective photocatalysis under visible irradiation, *Part. Part. Syst. Char.*, 36 (2019) 1800403.
55. S. Chidambaram, M.K. Ganesan, M. Sivakumar, S. Pandiaraj, M. Muthuramamoorthy, S. Basavarajappa, A. Abdullah Al-Kheraif, M.L. Aruna Kumari, A.N. Grace, Au integrated 2D ZnO heterostructures as robust visible light photocatalysts, *Chemosphere*, 280 (2021) 130594.
56. M. Wang, C. Jin, J. Kang, J. Liu, Y. Tang, Z. Li, S. Li, $CuO/g\text{-}C_3N_4$ 2D/2D heterojunction photocatalysts as efficient peroxymonosulfate activators under visible light for oxytetracycline degradation: Characterization, efficiency and mechanism, *Chem. Eng. J.*, 416 (2021) 128118.
57. P. Zhang, X. Yang, Z. Zhao, B. Li, J. Gui, D. Liu, J. Qiu, One-step synthesis of flowerlike C/Fe_2O_3 nanosheet assembly with superior adsorption capacity and visible light photocatalytic performance for dye removal, *Carbon*, 116 (2017) 59–67.
58. H. Zhang, M. Li, J. Cao, Q. Tang, P. Kang, C. Zhu, M. Ma, 2D a-Fe_2O_3 doped Ti_3C_2 MXene composite with enhanced visible light photocatalytic activity for degradation of Rhodamine B, *Ceram. Int.*, 44 (2018) 19958–19962.
59. D. Zhao, C. Cai, Layered Ti_3C_2 MXene modified two-dimensional Bi_2WO_6 composites with enhanced visible light photocatalytic performance, *Mater. Chem. Front.*, 3 (2019) 2521–2528.
60. H. Fang, Y. Pan, M. Yin, C. Pan, Enhanced photocatalytic activity and mechanism of $Ti_3C_{2-}OH/Bi_2WO_6$:Yb^{3+}, Tm^{3+} towards degradation of RhB under visible and near infrared light irradiation, *Mater. Res. Bull.*, 121 (2020) 110618.
61. Y. Qin, K. Xiao, S. Sun, Y. Wang, C. Kang, Fabrication of a novel pyramidal 3D MoS_2/2D $PbTiO_3$ nanocomposites and the efficient photocatalytic removal of organic pollutants: Effects of the $PbTiO_3$ internal electric field and S-scheme heterojunction formation, *Appl. Surf. Sci.*, 616 (2023) 156431.
62. W. Guo, F. Zhang, C. Lin, Z.L. Wang, Direct growth of TiO_2 nanosheet arrays on carbon fibers for highly efficient photocatalytic degradation of methyl orange, *Adv. Mater.*, 24 (2012) 4761–4764.
63. A. Kumar, Y. Singla, M. Sharma, A. Bhardwaj, V. Krishnan, Two dimensional S-scheme Bi_2WO_6-TiO_2-Ti_3C_2 nanocomposites for efficient degradation of organic pollutants under natural sunlight, *Chemosphere*, 308 (2022) 136212.
64. R.K. Sonker, G. Hitkari, S.R. Sabhajeet, S. Sikarwar, Rahul, S. Singh, Green synthesis of TiO_2 nanosheet by chemical method for the removal of Rhodamin B from industrial waste, *Mater. Sci. Eng. B*, 258 (2020) 114577.
65. J. Li, X. Li, G. Wu, J. Guo, X. Yin, M. Mu, Construction of 2D Co-TCPP MOF decorated on B-TiO_{2-x} nanosheets: Oxygen vacancy and 2D-2D heterojunctions for enhancing visible light-driven photocatalytic degradation of bisphenol A, *J. Environ. Chem. Eng.*, 9 (2021) 106723.
66. C.B. Ong, L.Y. Ng, A.W. Mohammad, A review of ZnO nanoparticles as solar photocatalysts: Synthesis, mechanisms and applications, *Renew. Sust. Energ. Rev.*, 81 (2018) 536–551.
67. Y.V. Kaneti, J. Yue, X. Jiang, A. Yu, Controllable synthesis of ZnO nanoflakes with exposed $(10\bar{1}0)$ for enhanced gas sensing performance, *J. Phys. Chem. C*, 117 (2013) 13153–13162.
68. R. Guo, Mesoporous nitrogen doped zinc oxide nanosheets with enhanced photocatalytic activity, *Int. J. Electrochem. Sci.*, 15 (2020) 8459–8470.
69. S. Kumar, N.L. Reddy, A. Kumar, M.V. Shankar, V. Krishnan, Two dimensional N-doped ZnO-graphitic carbon nitride nanosheets heterojunctions with enhanced photocatalytic hydrogen evolution, *Int. J. Hydrogen Energy*, 43 (2018) 3988–4002.

70. Y. Chen, J. Li, B. Zhai, Y. Liang, Enhanced photocatalytic degradation of RhB by two-dimensional composite photocatalyst, *Colloid Surf. A*, 568 (2019) 429–435.
71. J. Lee, J. Sung Lee, Two-simensional (2D) porous zinc oxides derived from the coordination polymer via chemical-assisted Phase transition, *Eur. J. Inorg. Chem.*, 26 (2022) e202200603.
72. H. Wang, J. Yu, X. Zhan, L. Chen, Y. Sun, H. Shi, Direct 2D/2D Z-scheme $SnNb_2O_6$/ZnO hybrid photocatalyst with enhanced interfacial charge separation and high efficiency for pollutants degradation, *Appl. Surf. Sci.*, 528 (2020) 146938.
73. C. Wang, X. Cheng, X. Zhou, P. Sun, X. Hu, K. Shimanoe, G. Lu, N. Yamazoe, Hierarchical alpha-Fe_2O_3/NiO composites with a hollow structure for a gas sensor, *ACS Appl. Mater. Interfaces*, 6 (2014) 12031–12037.
74. H. Liu, Y. Guo, N. Wang, B. Liu, Y. Zhang, H. Liu, R. Chen, Controllable synthesis and photocatalytic activity of ultrathin hematite nanosheets, *J. Alloys Compd.*, 771 (2019) 343–349.
75. S. Paulose, R. Raghavan, B.K. George, Graphite oxide-iron oxide nanocomposites as a new class of catalyst for the thermal decomposition of ammonium perchlorate, *RSC Adv.*, 6 (2016) 45977–45985.
76. N.A. Arzaee, M.F. Mohamad Noh, A.A. Halim, M.A. Faizal Abdul Rahim, N.S. Haziqah Mohd Ita, N.A. Mohamed, S.N. Farhana Mohd Nasir, A.F. Ismail, M.A. Mat Teridi, Cyclic voltammetry-A promising approach towards improving photoelectrochemical activity of hematite, *J. Alloys Compd.*, 852 (2021) 156757.
77. J. Pan, F. Guo, H. Sun, M. Li, X. Zhu, L. Gao, W. Shi, Nanodiamond decorated 2D hexagonal Fe_2O_3 nanosheets with a Z-scheme photogenerated electron transfer path for enhanced photocatalytic activity, *J. Mater. Sci.*, 56 (2021) 6663–6675.
78. Y. Yuan, Y. Liu, X. Xie, Y. Wen, M. Song, J. He, Z. Wang, 2D defect-engineered Ag-doped gamma-Fe_2O_3/$BiVO_4$: The effect of noble metal doping and oxygen vacancies on exciton-triggering photocatalysis production of singlet oxygen, *Chemosphere*, 322 (2023) 138176.
79. K. Omri, A. Bettaibi, K. Khirouni, L. El Mir, The optoelectronic properties and role of Cu concentration on the structural and electrical properties of Cu doped ZnO nanoparticles, *Physica B Condens. Matter*, 537 (2018) 167–175.
80. N. Phutanon, P. Pisitsak, H. Manuspiya, S. Ummartyotin, Synthesis of three-dimensional hierarchical CuO flower-like architecture and its photocatalytic activity for rhodamine b degradation, *J. Sci. Adv. Mater. Devices*, 3 (2018) 310–316.
81. M. Nazim, A.A.P. Khan, A.M. Asiri, J.H. Kim, Exploring rapid photocatalytic degradation of organic pollutants with porous CuO nanosheets: Synthesis, dye removal, and kinetic studies at room temperature, *ACS Omega*, 6 (2021) 2601–2612.
82. A. Bhattacharjee, M. Ahmaruzzaman, Microwave assisted facile and green route for synthesis of CuO nanoleaves and their efficacy as a catalyst for reduction and degradation of hazardous organic compounds, *J. Photoch. Photobiol. A*, 353 (2018) 215–228.
83. S. Dai, Y. Wang, L. Xiao, G. Hao, Y. Hu, G. Zhang, W. Jiang, 2D/2D/2D CuO-MXene-OCN heterojunction with enhanced photocatalytic removal of pharmaceuticals and personal care products: Characterization, efficiency and mechanism, *J. Alloys Compd.*, 919 (2022) 165873.
84. Y.M. Hunge, A.A. Yadav, M.A. Mahadik, V.L. Mathe, C.H. Bhosale, A highly efficient visible-light responsive sprayed WO_3/FTO photoanode for photoelectrocatalytic degradation of brilliant blue, *J. Taiwan Inst. Chem. E*, 85 (2018) 273–281.
85. Z. Sun, R. Huo, C. Choi, S. Hong, T.-S. Wu, J. Qiu, C. Yan, Z. Han, Y. Liu, Y.-L. Soo, Y. Jung, Oxygen vacancy enables electrochemical N_2 fixation over WO_3 with tailored structure, *Nano Energy*, 62 (2019) 869–875.
86. K.D. McDonald, B.M. Bartlett, Photocatalytic primary alcohol oxidation on WO_3 nanoplatelets, *RSC Adv.*, 9 (2019) 28688–28694.
87. Y. Wang, W. Jiang, W. Luo, X. Chen, Y. Zhu, Ultrathin nanosheets g-C_3N_4@Bi_2WO_6 core-shell structure via low temperature reassembled strategy to promote photocatalytic activity, *Appl. Catal. B*, 237 (2018) 633–640.
88. W. Dang, W. Wang, Y. Yang, Y. Wang, J. Huang, X. Fang, L. Wu, Z. Rong, X. Chen, X. Li, L. Huang, X. Tang, One-step hydrothermal synthesis of 2D WO_3 nanoplates@ graphene nanocomposite with superior anode performance for lithium ion battery, *Electrochim. Acta*, 313 (2019) 99–108.
89. L. Zhou, Y. Li, S. Yang, M. Zhang, Z. Wu, R. Jin, Y. Xing, Preparation of novel 0D/2D Ag_2WO_4/WO_3 Step-scheme heterojunction with effective interfacial charges transfer for photocatalytic contaminants degradation and mechanism insight, *Chem. Eng. J.*, 420 (2021) 130361.

90. L. Wang, K. Xu, H. Tang, L. Zhu, Vertical growth of WO_3 nanosheets on TiO_2 nnanoribbons as 2D/1D heterojunction photocatalysts with improved photocatalytic performance under visible light, *Catalysts*, 13 (2023) 556.
91. H. Li, Y. Chen, W. Zhou, H. Jiang, H. Liu, X. Chen, T. Guohui, $WO_3/BiVO_4/BiOCl$ porous nanosheet composites from a biomass template for photocatalytic organic pollutant degradation, *J. Alloys Compd.*, 802 (2019) 76–85.
92. L. Qin, Z. Zeng, G. Zeng, C. Lai, A. Duan, R. Xiao, D. Huang, Y. Fu, H. Yi, B. Li, X. Liu, S. Liu, M. Zhang, D. Jiang, Cooperative catalytic performance of bimetallic Ni-Au nanocatalyst for highly efficient hydrogenation of nitroaromatics and corresponding mechanism insight, *Appl. Catal. B*, 259 (2019) 118035.
93. X. Zhou, C. Lai, D. Huang, G. Zeng, L. Chen, L. Qin, P. Xu, M. Cheng, C. Huang, C. Zhang, C. Zhou, Preparation of water-compatible molecularly imprinted thiol-functionalized activated titanium dioxide: Selective adsorption and efficient photodegradation of 2, 4-dinitrophenol in aqueous solution, *J. Hazard. Mater.*, 346 (2018) 113–123.
94. M. Zhang, C. Lai, B. Li, D. Huang, S. Liu, L. Qin, H. Yi, Y. Fu, F. Xu, M. Li, L. Li, Ultrathin oxygen-vacancy abundant WO_3 decorated monolayer Bi_2WO_6 nanosheet: A 2D/2D heterojunction for the degradation of ciprofloxacin under visible and NIR light irradiation, *J. Colloid Interface Sci.*, 556 (2019) 557–567.
95. C. Liu, S. Mao, H. Wang, Y. Wu, F. Wang, M. Xia, Q. Chen, Peroxymonosulfate-assisted for facilitating photocatalytic degradation performance of 2D/2D $WO_3/BiOBr$ S-scheme heterojunction, *Chem. Eng. J.*, 430 (2022) 132806.
96. H. Huang, K. Liu, K. Chen, Y. Zhang, Y. Zhang, S. Wang, Ce and F Comodification on the crystal structure and enhanced photocatalytic activity of Bi_2WO_6 photocatalyst under visible light irradiation, *J. Phys. Chem. C*, 118 (2014) 14379–14387.
97. T. Hu, H. Li, R. Zhang, N. Du, W. Hou, Thickness-determined photocatalytic performance of bismuth tungstate nanosheets, *RSC Adv.*, 6 (2016) 31744–31750.
98. G. Zhang, Z. Hu, M. Sun, Y. Liu, L. Liu, H. Liu, C.-P. Huang, J. Qu, J. Li, Formation of Bi_2WO_6 bipyramids with vacancy pairs for enhanced solar-Driven photoactivity, *Adv. Funct. Mater.*, 25 (2015) 3726–3734.
99. B. Pang, S. Liu, Y. Tu, X. Wang, Controllable synthesis and enhanced photoactivity of two-dimensional Bi_2WO_6 ultra-thin nanosheets, *ChemistrySelect*, 6 (2021) 5381–5386.
100. Y.-H. Cheng, J. Chen, H.-N. Che, Y.-H. Ao, B. Liu, Ultrafast photocatalytic degradation of nitenpyram by 2D ultrathin Bi_2WO_6: Mechanism, pathways and environmental factors, *Rare Metals*, 41 (2022) 2439–2452.
101. U.A. Khan, J. Liu, J. Pan, H. Ma, S. Zuo, Y. Yu, A. Ahmad, M. Iqbal, S. Ullah, B. Li, One-Pot fabrication of hierarchical floating Bi-Bi_2S_3-Bi_2WO_6/expanded perlite [hotocatalysts for efficient photocatalysis of organic contaminants utilized sunlike illumination, *Ind. Eng. Chem. Res.*, 58 (2019) 9286–9299.
102. H. Yin, C. Yuan, H. Lv, K. Zhang, X. Chen, Y. Zhang, Y. Zhang, Fabrication of 2D/1D Bi_2WO_6/C_3N_5 heterojunctions for efficient antibiotics removal, *Powder Technol.*, 413 (2023) 118083.
103. Q. Ma, J. Ming, X. Sun, N. Liu, G. Chen, Y. Yang, Visible light active graphene oxide modified Ag/$Ag_2O/BiPO_4/Bi_2WO_6$ for photocatalytic removal of organic pollutants and bacteria in wastewater, *Chemosphere*, 306 (2022) 135512.
104. X. Li, H. Zhang, X. Du, S. Wang, Q. Zhang, H. Li, F. Ye, Efficient visible-light-driven degradation of tetracycline by a 2D/2D rGO-Bi_2WO_6 heterostructure, *Environ. Res.*, 212 (2022) 113326.
105. X. Qian, Y. Ma, M. Arif, J. Xia, G. He, H. Chen, Construction of 2D/2D $Bi_4O_5Br_2/Bi_2WO_6$ Z-scheme heterojunction for highly efficient photodegradation of ciprofloxacin under visible light, *Sep. Purif. Technol.*, 316 (2023) 123794.
106. S. Pan, J. Shi, M. Zhang, M. Wu, Y. Cen, W. Guo, Y. Zhu, Photocatalytic performance enhancement of two-dimensional Ruddlesden-Popper type perovskite $K_2La_2Ti_3O_{10}$ by nitrogen-doping, *Mater Res. Express*, 6 (2019) 075047.
107. W. Gao, C. Wu, M. Cao, J. Huang, L. Wang, Y. Shen, Thickness tunable SnS nanosheets for photoelectrochemical water splitting, *J. Alloys Compd.*, 688 (2016) 668–674.
108. J. Annett, G.L. Cross, Self-assembly of graphene ribbons by spontaneous self-tearing and peeling from a substrate, *Nature*, 535 (2016) 271–275.
109. M. Deng, M. Ye, T. Li, H. Huang, W.-X. Yuan, H. Jiang, P. Lin, X. Zeng, S. Ke, Synthesis of ferroelectric $KNbO_3$ nanosheets by liquid exfoliation of layered perovskite K_2NbO_3 F, *J. Alloys Compd.*, 698 (2017) 357–363.

110. X. Gao, Y. Dai, F. Fu, X. Hua, 2D laminated cylinder-like $BiFeO_3$ composites: Hydrothermal preparation, formation mechanism, and photocatalytic properties, *Solid State Sci.*, 62 (2016) 6–12.
111. J. Li, Y. Zhao, M. Xia, H. An, H. Bai, J. Wei, B. Yang, G. Yang, Highly efficient charge transfer at 2D/2D layered P-$La_2Ti_2O_7$/Bi_2WO_6 contact heterojunctions for upgraded visible-light-driven photocatalysis, *Appl. Catal. B*, 261 (2020) 118244.
112. Z. Sun, H. Zhang, X. Shen, Y. Su, H. Liu, H. Yu, Preparation of highly efficient Ni-doped layered perovskite grafted graphene oxide $Ni_xSr_{1-x}TiO_3$-RGO heterojunction photocatalyst with enhanced visible-light photocatalytic activity for MB degradation, *Mater. Chem. Phys.*, 273 (2021) 125119.
113. X. Liang, Y. Zhang, D. Li, B. Wen, D. Jiang, M. Chen, 2D/2D BiOCl/$K^+Ca_2Nb_3O_{10}^-$ heterostructure with Z-scheme charge carrier transfer pathways for tetracycline degradation under simulated solar light, *Appl. Surf. Sci.*, 466 (2019) 863–873.
114. S. Jamil, M.R.S.A. Janjua, T. Ahmad, The synthesis of flower shaped microstructures of Co_3O_4 by solvothermal approach and investigation of its catalytic activity, *Solid State Sci.*, 36 (2014) 73–79.
115. J. Yang, M. Wang, S. Zhao, Y. Liu, W. Zhang, B. Wu, Q. Liu, Petal-biotemplated synthesis of two-dimensional Co_3O_4 nanosheets as photocatalyst with enhanced photocatalytic activity, *Int. J. Hydrogen Energy*, 44 (2019) 870–879.

14 2D Graphene Nanocomposites for Removal of Organic Pollutants

Elakkiya Venugopal, Suresh Babu Naidu Krishna, and Sharangouda J. Patil

14.1 INTRODUCTION

The carbon-based nanomaterials, which have desirable structural and physiochemical characteristics, include 0D fullerenes, 1D nanotubes, and 2D graphene. Chemical bonding, structure, and dimensionality all have a role in how these materials behave. Four valence electrons, two each in the 2s and 2p subshells, make up carbon's ground state. Bonds with other carbon atoms are created through sp hybrid orbitals when one of the 2s electrons moves to an open 2p orbital. Sp, sp^2, and sp^3 hybrid orbitals are among various varieties, and carbon atoms possessing those orbitals form bonds with other carbon atoms nearby. The current chapter focuses on carbon atoms with sp^2 bonds, i.e., 2D graphene materials [1].

The grid of classification utilized in Figure 14.1 divides various graphene forms into three categories: the average lateral dimension, the number of graphene layers, and the atomic oxygen/carbon ratio. Based on the number of layers and lateral dimensions provided in the literature, various materials drawn at the box's six corners indicate the best-case scenarios. Although it is possible to stretch the values to the microscale, the three axis values are related to the GBMs at the nanoscale [2]. Graphene, the most well-known form of carbon, is composed of carbon atoms arranged in two-dimensional planar hexagonal units. It is recognized for having excellent physical, thermal, and electrical properties. In other words, graphene is only one sheet of graphite, a different macroscopic form of carbon. Graphene's structure is similar to that of a honeycomb made of polycyclic aromatic hydrocarbons (Figure 14.2). By sp2 hybridization with nearby carbon atoms, carbon atoms in single layer graphene form a ring of benzene in which each atom contributes an unpaired electron. Graphene has an incredibly stable structure and is only 0.35 nm thick, or 1/200,000th of the diameter of a human hair. The lattice plane twists, applying an external force that the bond between carbon atoms is strong enough to resist, preventing atom reconfiguration [3,4].

2D graphene materials have many attractive physical, chemical, and mechanical properties, and studies on the interrelation between these properties make them useful for various novel applications [5] (Figure 14.3). Graphene has the highest tensile strength and modulus. Several studies have been done to evaluate the mechanical properties of graphene. Using atomic force microscopy, the force and elastic modulus of graphene were initially investigated. The results showed that the elastic stiffness and Young's modulus were, respectively, 2.0 TPa and 1.0 TPa [6]. Many remarkable properties of graphene include optical transparency, mechanical strength, electric conductivity, and thermal conductivity. Graphene is a very light substance with a planar density of 0.77 mg m^{-2}. There are just two carbon atoms in the graphene ring since every atom at the vertices is shared through three-unit rings. Graphene's extremely thin and lightweight nature is further enhanced by the fact that it is composed of just one carbon atomic layer. Graphene has an extremely high permeability of 97.7%, which means merely 2.3% of visible light is absorbed by it due to its one atomic layer [7].

DOI: 10.1201/9781003436942-14

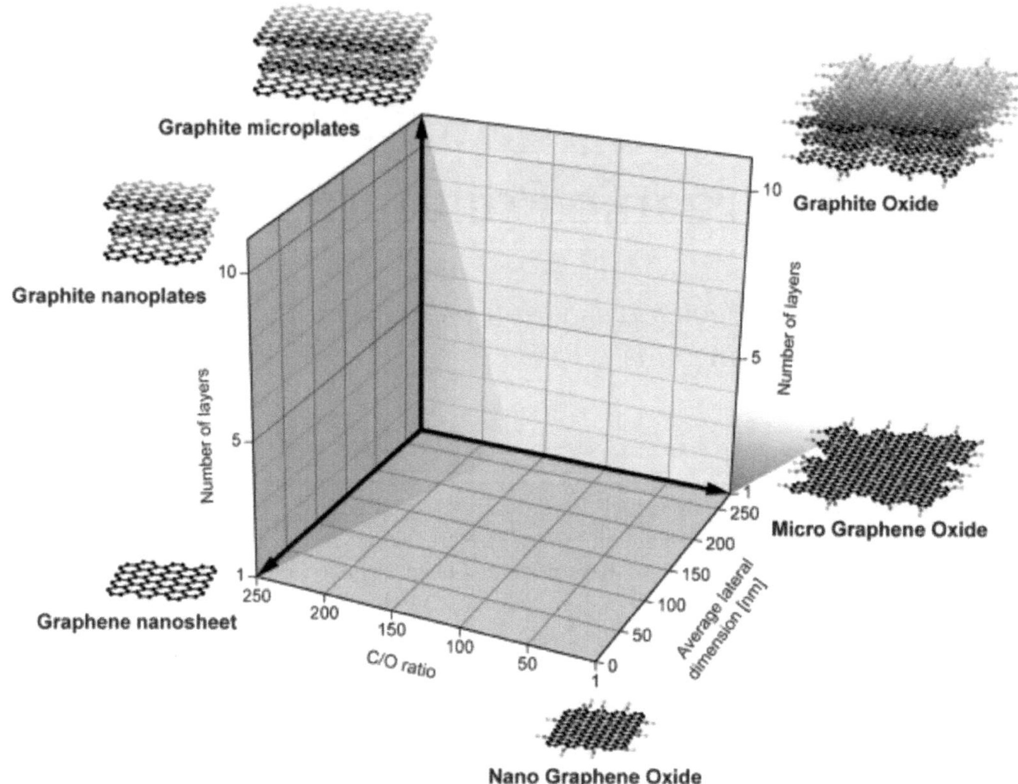

FIGURE 14.1 Classification grid for graphene nanomaterial. (Reproduced with permission from Ref. [2], Copyright © 2014 Wiley.)

14.2 OVERVIEW OF SYNTHESIS STRATEGIES OF 2D GRAPHENE

Many techniques for producing graphene have been developed recently. Figure 14.3 provided the structure and classification of several 2D materials. Liquid-phase exfoliation, molecular beam epitaxy, mechanical exfoliation from graphite, chemical vapor deposition (CVD), reduction of graphene oxide, and surface segregation are the most often used techniques for producing appropriate graphene layers. Every tactic has benefits and drawbacks. Precise control over the quantity and flawless structure of graphene layers throughout a substrate is still a major and crucial issue. As a result, efforts to refine the production process to realize different features of graphene layers are underway [8]. Table 14.1 outlines the benefits and drawbacks of the top-down and bottom-up synthesis strategies [9].

a. **Mechanical exfoliation**

Mechanical exfoliation is commonly referred to as the Scotch tape method or the peel-off procedure. To create graphene, layers of graphene are separated by applying an adhesive tape. During this process, the graphene breaks off in layers, but finally it breaks into multiple layers of graphene flakes. Under a light microscope, the tape's binding to a particular substrate—acetone—breaks, causing particles on SiO_2/Si surface that vary in size and thickness. New tape is used for a final peeling after that. Due to the tedious and inaccurate nature of this process, the material that is produced is often used to study the properties of graphene prior to its commercial application. This approach can also make use of other agents, such as an electric field, epoxy glue, and the transfer printing approach [11,12].

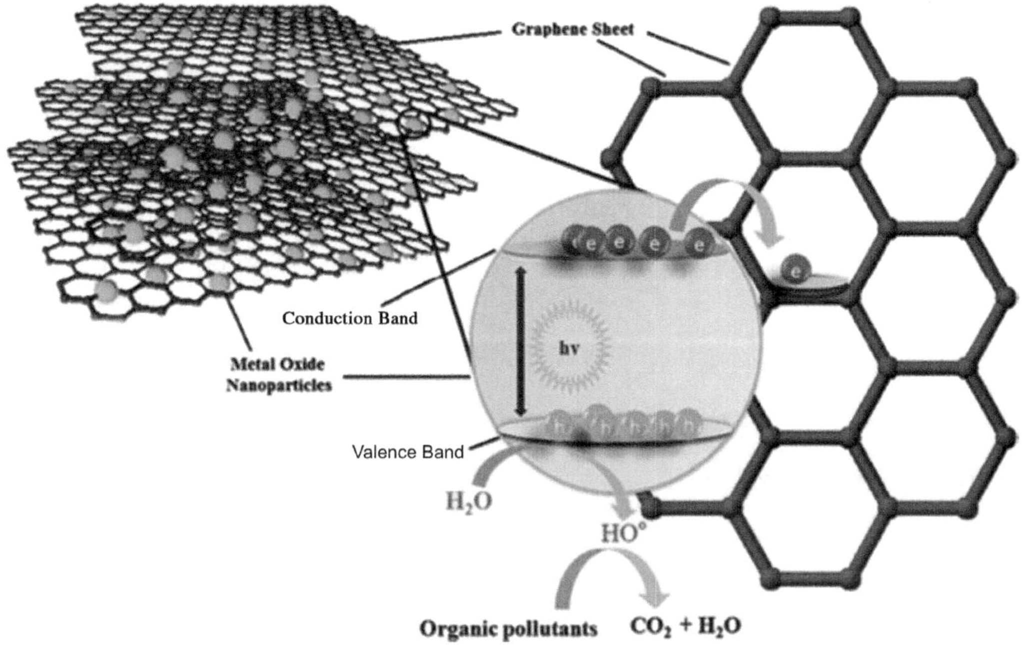

FIGURE 14.2 Hexagonal lattice of graphene. (Reproduced with permission from Ref. [3], Copyright © 2016 Intech.)

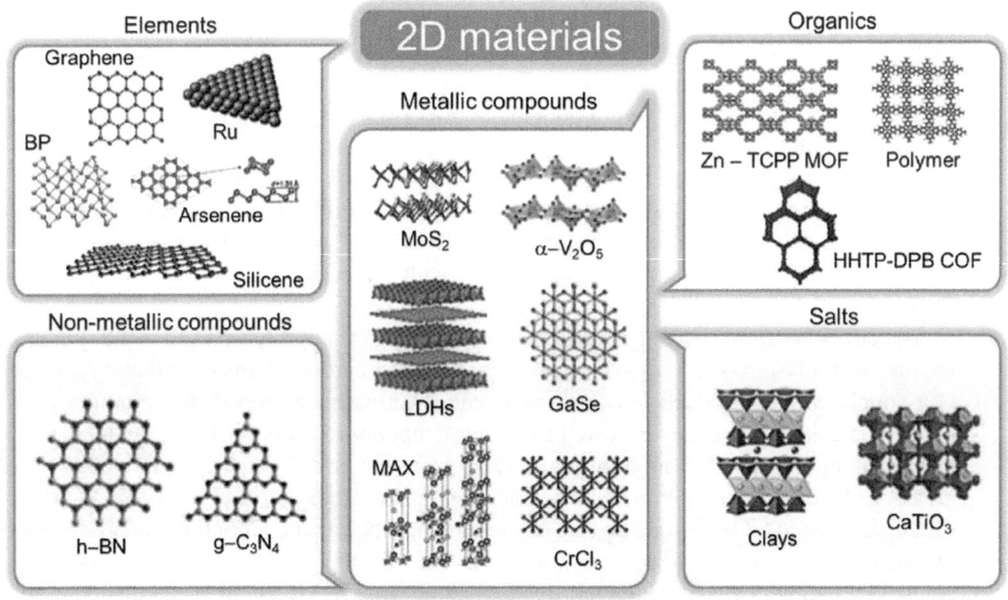

FIGURE 14.3 Structure and classification of different 2D materials. (Reproduced with permission from Ref. [18], Copyright © 2022 Wiley.)

b. **Electrochemical exfoliation**

Electrochemical methods of graphite exfoliation have emerged as a straightforward yet highly productive method for producing graphene in large quantities in recent years. In an aqueous or non-aqueous electrolyte, graphite foils, rods, plates, and powders are employed

TABLE 14.1
Summary of Top-down and Bottom-up Methods of Graphene's Growth Mechanism

Method	Thickness	Lateral	Advantage	Disadvantage
Micromechanical exfoliation	Few layers	μm to cm	Unmodified and large size graphene sheets	Very small-scale production
Electrochemical exfoliation	Single to few layers	500–700 nm	High electrical conductivity of the functionalized graphene	High cost of ionic liquids
Direct sonication of graphene	Both single and multiple layers	μm	Inexpensive and unmodified graphene	Low yield
Reduction of carbon monoxide (CO)	Multiple layers	Sub-μm	Unoxidized sheets	Contamination with α-Al_2S and α-Al_2O_3
Epitaxial growth on SiC	Few layers	Up to cm size	Very large area of pure graphene	Very small scale
Unzipping of carbon nanotubes	Multiple layers	Few μm long nanoribbons	Size depends on the starting nanotubes	Expensive and oxidized graphene
CVD	Few layer	Very large (cm)	Large size and high quality	Small production scale

Reproduced with permission from [10], Copyright @ 2016 Springer.

as electrodes, and electrode expansion is promoted by the application of an electric current. The electrodes can be classified as anodic (positive) or cathodic (negative) depending on how much power is applied. Luheng et al.'s study [13] employed deionized water as the electrolyte and pure graphite electrodes containing PSS (Polysodium-4-styrenesulfonate). The rods of graphite were stored in an electrochemical cell containing electrolyte. A 5V continuous current was employed. Following a brief period of electrolysis, a dark material gathered near the anode. For 4 hours, the exfoliation process was repeated in order to extract the substance from the cell. The product was spun at 1000 rpm and then slowly poured out. It was discovered that the resulting dispersion was remarkably stable. To make powdered dry graphene, the dispersion was vacuum-dried after being cleaned with alcohol and deionized water. Weighing the dry powder and sediment allowed us to calculate the yield [14,15].

c. **Pyrolysis**

The Greek terms pyro and lysis are the source of the term pyrolysis. Pyro is the symbol for fire, while lysis is the symbol for division. Creating carbon atoms on a metal surface is a simple way to produce few-layer graphene. One popular process for creating graphene is heating silicon carbide (SiC) to a high temperature. Si desorbs at high temperatures, leaving C atoms behind to partially form graphene layers. This procedure has been greatly enhanced by the continuous mm-scale graphene sheet production at 750°C on a thin nickel-plated SiC substrate layer. This method has the benefit of continually producing graphene layers over the whole SiC-coated surface. On the other hand, large-scale graphene fabrication is not feasible with this technique. At 1000 K, the thermal degradation of ethylene is approached in a similar way. This synthesis method's ability to produce very pure graphene monolayers is one of its benefits [16,17].

A bottom-up manufacturing method called CVD is widely used to create high-grade graphene. A surface substrate and a gas molecule are combined in a reaction chamber at a certain pressure, temperature, and gas flow rate. Quartz reaction chambers, mass flow controllers, pumps, thermocouples for determining temperatures, gas delivery systems, vacuum systems, energy systems, and computers

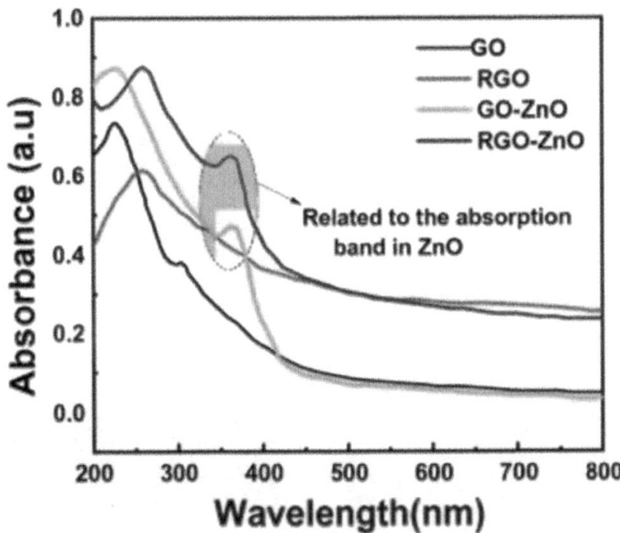

FIGURE 14.4 Absorbance spectra of GO, RGO, GO-ZnO, and RGO-ZnO nanocomposites. (Reproduced with permission from Ref. [19], Copyright © 2023 Springer Nature.)

for auto-control are typical parts of a CVD equipment. Several substrates, including stainless steel, copper (Cu), nickel (Ni), and iron (Fe), are used in the CVD process to produce graphene films. The carbon is typically supplied by methane (CH_4) and acetylene (C_2H_2). Thermal CVD and plasma-enhanced CVD are the two CVD techniques utilized to activate the carbon source (PECVD).

The GO UV–Vis spectra, as shown in Figure 14.4, show two absorption peaks: a shoulder at about 300 nm, linked to the $n \rightarrow \pi^*$ transition of the C=O bonds, and one at about 230 nm, most likely induced by the $\pi \rightarrow \pi^*$ transition of the C–C bonds. On the other hand, with RGO, the π^* transition peak shifts to 260 nm, signifying the regaining of the connected structure and the removal of certain groups from the GO surface. This indicates an increase in π-electron concentration and structural ordering, which is consistent with the restoration of the sp^2 carbon and maybe with atom rearrangement. The GO absorption peak and the major ZnO absorption peak are the two absorption peaks that were found at 230 and 366 nm, accordingly, in the GO-ZnO spectra. Band edge absorption is observed in RGO–ZnO hybrids at 361 nm, approximately 9 nm less shifted to the blue than the band gap absorption of bulk ZnO at 370 nm. The lower feature size of ZnO may have a quantum confinement effect, explaining this discrepancy. Furthermore, in comparison to pure RGO, there is a little blue-shift (from 260 to 258) in the absorption peak of RGO-ZnO [19].

High-quality graphene flakes can be produced using a straightforward hydrodynamics-based method, according to Zhang et al. [20]. One simple needle valve was used as an exfoliating instrument. The produced graphene flakes had an average density and length of 2.3 nm, and about 71% of them had lesser than five layers, according to the data. Figure 14.5a shows a Raman spectrum of produced graphene with a reference of raw graphite. For these two graphitic materials, three different peaks were observed: the D band (~1350 cm^{-1}), the 2D band (~2700 cm^{-1}), and the G band (~1580 cm^{-1}). The produced graphene's D/G (ID/IG) intensity ratio was 0.10, which was less than the ultrasonication-exfoliated graphene's value of 0.29 [21]. Figure 14.5b displays the XRD results for C60/rGO and graphite. Graphite's XRD pattern displays a noticeable peak at $2\theta = 27°$. The graphitic peak moves to 14.6° after oxidation, indicating an increase in the interfacial gap. After graphite oxide was chemically reduced, the peak disappeared. The exfoliation of graphite oxide's layered structures was proposed as the cause of the steep peak. Additionally, it is evident from the XRD pattern that fullerenes (C60) were successfully integrated into the graphene surface. High-quality SEM pictures of graphene nanosheets are shown in Figure 14.5c. The present investigation employed a

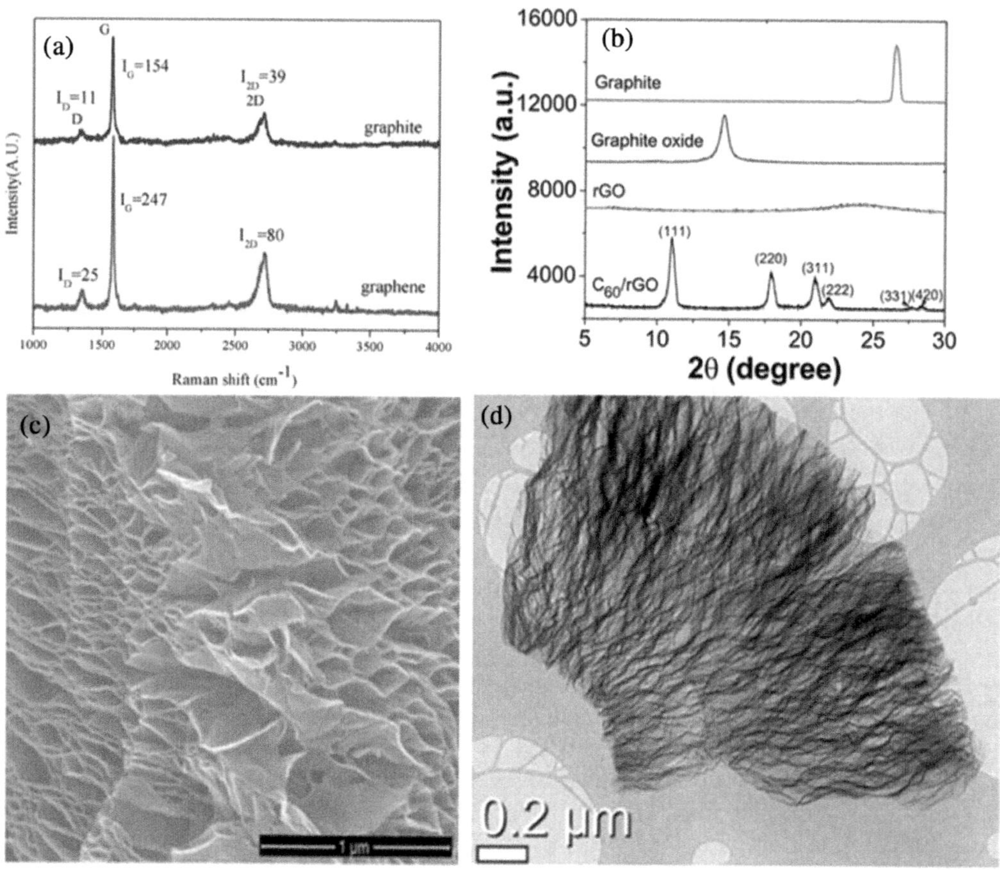

FIGURE 14.5 (a) Bulk graphite and graphene Raman spectroscopy, (b) XRD diffraction patterns of graphite and its derivatives, (c) SEM images of reduced graphene oxide, and (d) HRTEM images of the rGO stack. (Reproduced with permission from Ref. [22], Copyright © 2021 PMC.)

modified Hummers' method to produce graphene oxide and a solid-state microwave irradiation approach to decrease it. A crinkled stack of extraordinarily thin, porous graphene oxide nanosheets was visible in the SEM pictures. Figure 14.5d shows a HRTEM image of a reduced graphene oxide. The rGO morphology has an interesting doughnut-like appearance [22].

14.3 ADSORPTION OF ORGANIC WATER POLLUTANTS WITH 2D GRAPHENE NANOCOMPOSITES

Over the last few decades, there has been an increase in interest in the development of resilient and effective wastewater treatment procedures. Numerous techniques have been employed for this, including advanced oxidation, coagulation, membrane processing, chemical precipitation, extraction, photocatalysis, and adsorption. One of these strategies that have been proven to be effective in removing OPs from wastewater is adsorption. The process of eliminating particular contaminants from a liquid solution by inducing the solute, or adsorbate, to stick to a solid object, or adsorbent, is known as adsorption, which is a fundamental surface phenomenon. Adsorption is more efficient than other alternative processes at getting rid of even minute levels of dangerous pollutants. It also has a simpler design and costs less. Because they have more oxygen-containing groups with functions (carboxyl groups, epoxy groups, and hydroxyl groups) and a higher adsorption capacity

TABLE 14.2
Summary of Adsorption Capacities by Graphene, Graphene Analogs, and Composite Graphene

Graphene	Pollutants	Adsorbent Capacity (mg g^{-1})
Graphene-like layered molybdenum disulfide	Antibiotic doxycycline	310
(Fe_3O_4)-modified graphene nanoplatelets	Antibiotic amoxicillin	14.10
GO	Au (III)	108.34
	Pd (II)	80.78
	Pt (IV)	71.38
EDTA-magnetic GO	Cu (II)	301.2
	Hg (II)	268.4
	Pb (II)	508.4
Polyamide–graphene	Sb (III)	158.2
GO/cellulose membranes	Co (II)	15.5
	Ni (II)	14.3
	Cu (II)	26.6
	Zn (II)	16.7
	Cd (II)	26.8
	Pb (II)	107.9
Chitosan/GO	Au (III)	1076.65
	Pd (II)	216.92
GO/$NiFe_2O_4$	Cr (III)	25.0
	Pb (II)	45.4
Graphene	Bisphenol A	182
GO	Cu (II)	294
	Zn (II)	345
	Cd (II)	530
	Pb (II)	1119
Fe_3O_4/rGO nanocomposite	Ametryn	57.64
GO	Co (II)	21.28
GO	Acid Orange 8	29.0
	Direct Red 23	15.3
GO	Basic Yellow 28	68.5
	Basic Red 46	76.9
GO/iron oxide	MB	39
GO	Pb (II)	250
GO	Ni (II)	38.61
GO	Th (IV)	58.59

than pristine-oxidized graphene nanomaterials, graphene nanomaterials (rGO and GO) have drawn growing attention for their adsorption uses. The latter produces reactive sites and a surface with a negative charge for the grafting of various functional groups, thereby increasing surface area and capacity for adsorption. Depending on the characteristics, the relative adsorption capabilities of various OPs upon oxidized graphene range from 20 to >7000 mg g^{-1} (Table 14.2).

14.3.1 Adsorption with Non-modified 2D Graphene Nanocomposites

Graphene-based components and GNS have been studied as potential next-generation adsorbents for treating water and wastewater due to their extremely hydrophilic surface, reveal layer shape, and strong adsorption attraction to organic pollutants, metallic ions, and other organic pollutants. Graphene oxide (GO), an oxidized version of graphene, has different adsorption processes and a lower surface hydrophobicity than GNS. A wide range of organic chemicals, including halogenated aliphatic compounds, polycyclic aromatic hydrocarbons (PAHs), medications, plasticizers, dyes, insecticides, and polychlorinated biphenyls (PCBs), were adsorbed using GNS. The background chemistry of the water as well as a number of factors related to the physicochemical characteristics of GNS and OCs influence OC adsorption by GNS.

Over the past 10 years, scientists have examined how GNS affects the fate, toxicity, and changes of organic compounds in aquatic environments. Studies show that in natural systems, the cellular uptake and mobility of OCs were modified upon their adsorption onto GNS. According to a recent analysis, GNS negatively affected the absorption and mobility of OC cells in the environment. Therefore, it is essential to comprehend OC adsorption by GNS in order to appropriately evaluate these environmental effects. Carbonaceous adsorbents' ability to absorb organic compounds (OCs) is influenced by their surface chemistry and shape. Characterizing adsorption reactions among adsorbent–adsorbate, adsorbent–solvent, and adsorbate–solvent are frequently done by physical, chemical, and electrostatic interactions. Various types of GNS, including magnetic, colloidal, and sulfonated GNS, were additionally used in certain studies [23,24].

The material is composed of a thick sheet structure with pores and wrinkled edges, which suggests complex flake-like stacking based on the observed morphology of GO (Figure 14.6a). The surface layer of GO may become more distorted and rougher than that of rGO due to the occurrence of functional groups contains oxygen such as carboxyl in the periphery, hydroxyl, and epoxide groups present in the core. These groups might interfere by disrupting the interconnected microstructure of graphene oxide (GO), leading to wrinkles and an increase in the intermediate gap in graphite. As can be seen in Figure 14.6b, there is a significant change in the surface shape of rGO after the solvothermal reduction of GO. The rGO featured a thin, layered structure that was clearly capable of folding into a fiber. The flakes are arranged unevenly because surface distortion prevents face-to-face stacking. As shown in Figure 14.6c–f, the amount of MB in the water-based solution was determined both prior to and after different GO and rGO dose applications. 350 mg L^{-1} was the initial concentration of MB. However, after injecting 5 mg of GO and rGO, the initial concentration of MB decreased to 4.57 and 25.36 mg L^{-1}, respectively. The amount of MB adsorbed fell from 2887.30 to 1381.71 mg g^{-1} when the dosage of GO was raised from 2 to 5 mg. In a similar vein, the total amount of MB adsorbed dropped by 2658.32 to 898.58 mg g^{-1} when the quantity of rGO increased from 2 to 5 mg. Figure 14.6 illustrates how equilibrium adsorption capacity decreased, while total removal efficiency increased as adsorbent dose increased.

The removal of MB increased from 82.49% to 98.69% at 2 and 5 mg of GO, accordingly. At rGO dosages of 2 and 5 mg, respectively, the elimination efficiency rose from 78.81% to 92.76% with similar results. It is evident that the MB molecules present in the fluid saturate the readily accessible sorption sites at lower adsorbent dosages, reducing the effectiveness of their removal. Figure 14.6 illustrates how contact time affects removal performance. As contact time increases, removal efficiency increases. The data show that at the same dosage, GO has greater clearance efficiency than rGO. The greatest efficacy of removal is 93.47% for rGO and 98.67% for GO [25].

14.3.2 Adsorption with Metal- and Metal Oxides-Modified 2D Graphene Hybrids

GO and its derivatives are particularly interesting for the treatment of wastewater because of their abundance of surface-active groups, such as carboxyl, hydroxyl, and epoxy groups, in addition to their 2D ultrathin structure, which provides them large wide surface areas. Because carboxylated GO (GO-COOH) has a large between-layer space, a large number of carboxyl groups as solely

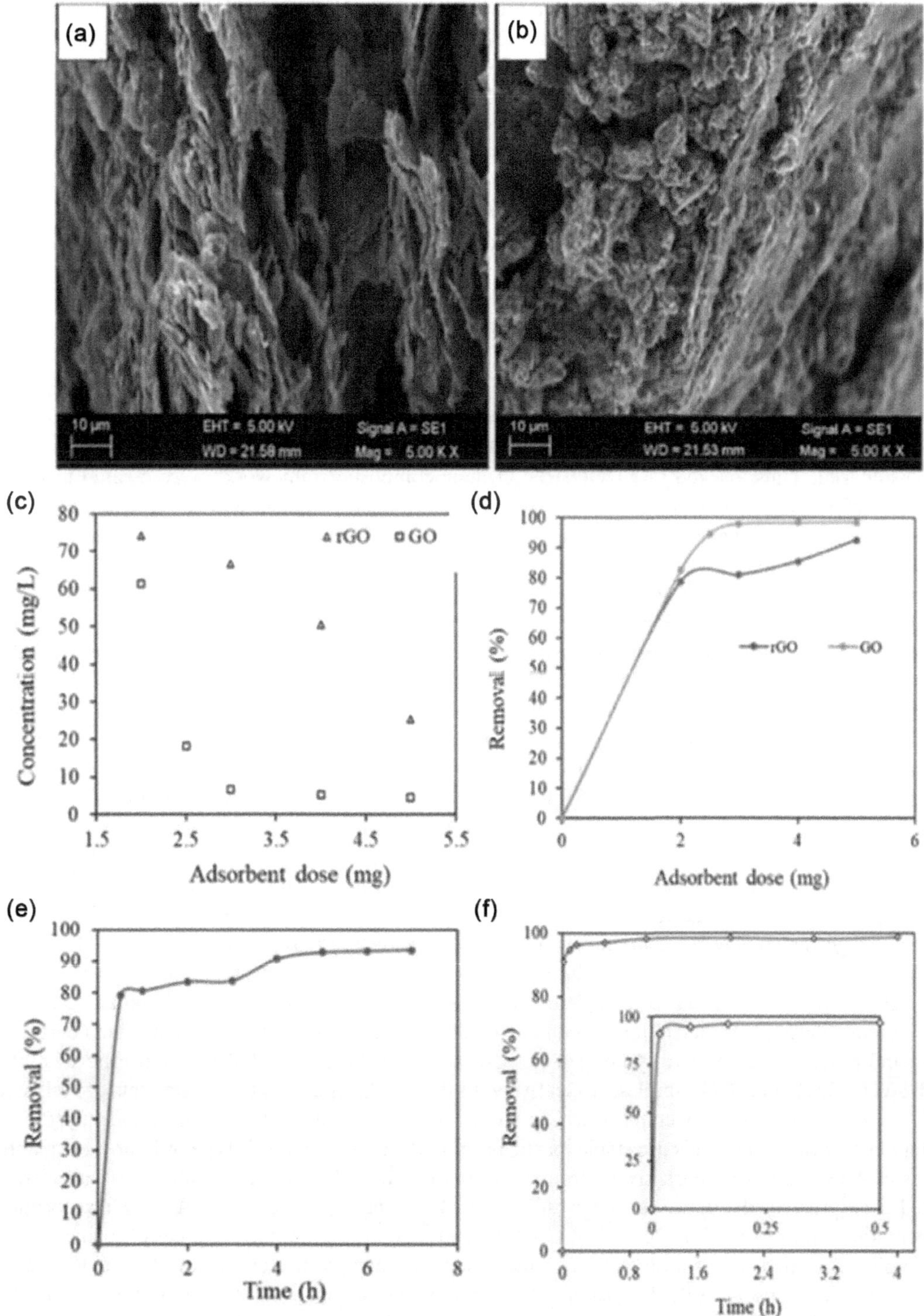

FIGURE 14.6 SEM images of (a) GO and (b) rGO. Change in MB concentration with increasing GO and rGO dose (c) and percent MB removal with increasing GO and rGO dose (d). Percent removal efficiency of (c) rGO and (f) GO with time (h). (Reproduced with permission from Ref. [25], Copyright © 2022 Elsevier.)

active species, and an obvious preference for organic dyes, it is one of the largest and most useful graphene oxides and is easier to use for the elimination of organic dye-related pollutants from wastewater. GO-COOH nanosheets, on the other hand, continue to clump and agglomerate in liquid solution due to their elevated surface strength and particular surface area, which decreases the quantity of contact-active sites that are accessible for absorption and produces a lack of absorption efficiency [21]. It is crucial to lessen GO-COOH nanosheet aggregation in order to preserve their initial active sites during dye absorption.

Dadashi et al. [26] used a hydrothermal system with some hydroxyl groups to generate Fe_3O_4 particles. It didn't appear that the Fe_3O_4 particle–GO connection was robust enough to be applied in a removal of dye regeneration process. Rather, carboxylated GO and Fe_3O_4 with carboxyl groups of chemicals were used to enhance the tiny structures of the Fe_3O_4/GO nanocomposites. The uniform distribution of Fe_2O_3 nanosheets between graphene nanosheets created a distinctive sheet-on-sheet nanostructure. In addition, a significant quantity of carboxyl groups in the Fe_3O_4-COOH NPs may promote the Fe_3O_4-NPs combination with poly(allylamine)hydrochloride (PAH), hence enhancing the NPs' stability [27]. Thanks to electrostatic attraction, eFe_3O_4-COOH NPs coated in PAH molecules may be loaded into pleated GO-COOH sheets simply and steadily as a matrix supporting material. Thus, packed GO-COOH/Fe_3O_4 nanocomposite frameworks were created utilizing a layer-by-layer (LbL) assembly process as part of an effective, quick, and scalable strategy. The resulting sandwiched nanostructure can prevent GO-COOH nanosheet agglomeration and improve specific surface area by introducing as many sites of activity for adsorption as possible [28].

The hydroxyl radicals (OH) were studied in order to quantify the enhanced photocatalytic performance of the Bi_2MoO_6/reduced graphene oxide aerogel (BMO/GA) hybrid in relation to photonic efficiency and clarify the mechanism. The quantity of fluorescence that 2-hydroxyterephthalic acid—which has a noticeable fluorescent characteristic signal—generated during the chemical interactions with terephthalic acid and •OH radicals was used to quantify the amount of •OH radicals. Furthermore, because the prepared materials react better to visible light, more •OH radicals are formed during the breakdown process. Figure 14.7a and b shows a significant bright peak, indicating the production of •OH radicals during the breakdown process in MB. Additionally, Figure 14.7a and b shows how, in each instance of the samples BMO, GA, BMO/GA-5, BMO/GA-15, and BMO/GA-10, the PL intensity rose over time. These findings demonstrated that the addition of GA can enhance the photocatalytic efficiency of BMO. Figure 14.7c depicts the BMO/GA composite's adsorption and photocatalytic reaction scheme based on the aforementioned experiments. Three mechanisms are involved in this mechanism: (i) MB molecules are absorbed into the 3D porous framework of GA; (ii) the BMO/GA composite as a whole absorbs external light sources; and (iii) the charge transfer reactions produce radical species that degrade the simulated pollutant MB [29].

Acetaldehyde (AcH) and chloramphenicol (CAP), colorless organic contaminants found in indoor air and water, were chosen to symbolize the adsorption capability and photocatalytic efficacy of the Bi2WO_6, BiOI, and rGO samples, respectively. In the blank experiments, the amounts of AcH and CAP were not significantly changed when exposed to visible light and without a photocatalyst. This implies that the compounds are stable, particularly in the presence of this type of light. The quantities of AcH and CAP progressively dropped during the degradation with visible light illumination and absorption in the dark, as shown in Figure 14.8 using a photocatalyst. A 1-hour adsorption technique removed 15.33%–25.33% of CAP, depending on the exact surface areas of the generated samples; a 12-hour adsorption technique increased the quantity of AcH removed from 10.89% to 22.22%. The 1 wt% rGO/Bi_2WO_6 sample showed the highest capacity for adsorption for both CAP (25.33%) and AcH (22.22%) across the samples due to its largest specific surface area. Interestingly, a rise in rGO has no discernible effect on the adsorption capacity. For the 1 wt% rGO/Bi_2WO_6/BiOI sample, the photodegradation efficiencies of AcH and CAP approach 100% as soon as 2 and 4 hours, respectively. The concentrations of CAP and AcH decrease in the presence of visible light [30].

In order to understand why adding rGO to the Bi_2WO_6/BiOI composite enhanced its photocatalytic performance for the photodegradation of AcH and CAP, a few possible processes are further

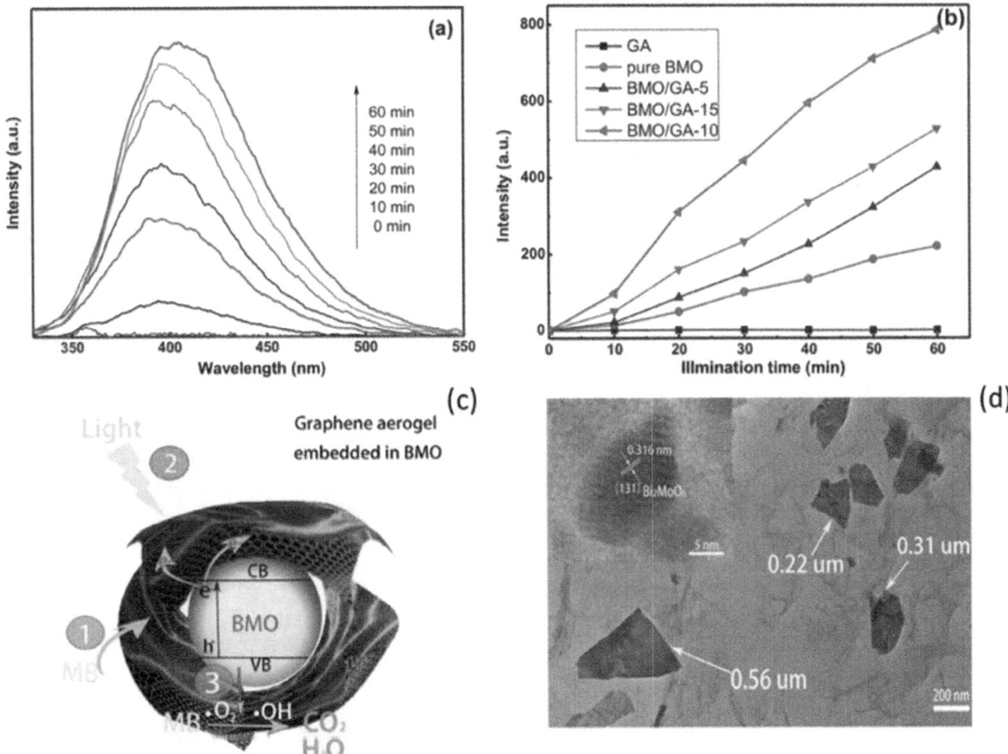

FIGURE 14.7 (a) PL spectral changes observed during illumination for the BMO/GA-10 composite. (b) Comparison of PL intensity at 405 nm against irradiation time for GA, BMO, BMO/GA-5, BMO/GA-10, and BMO/GA-15 samples. (c) Schematic illustration of the photocatalytic enhancement mechanism of BMO/GA composite. (d) TEM and HRTEM (insert) images of BMO/GA-10 composite. (Reproduced with permission from Ref. [29], Copyright © 2018 MDPI.)

discussed here. Here, both the effects of the rGO and the generated p–n heterojunction are considered. Electrons can move from n-type Bi_2WO_6 to p-type BiOI through a p–n heterojunction that is created when they interact till their Fermi levels match. Consequently, the valence band (VB) edge of Bi_2WO_6 is situated lower than that of BiOI, while the conduction band (CB) edge of BiOI rises higher than that of Bi_2WO_6. This creates an inner electrical field within the p–n heterojunction connection at equilibrium. While electrons from the CB of BiOI go to the more positively situated CB of Bi_2WO_6, vacancies from the VB of BiOI shift to the VB of Bi_2WO_6 during the photocatalytic event. OH and other active compounds involved in the photodegradation process develop when vacancies in the VB respond directly with AcH and CAP molecules or with H_2O on the surface of the composite. •Oxygen and electrons in the CB react to form O_2^-, which targets the molecules of AcH and CAP. Consequently, a quicker process of separating the electrons and holes generated by photolysis lowers recombination and boosts the efficiency of photocatalysis. While NO_3^-, Cl^-, NH_4^+, CO_2, and H_2O can be produced at different phases of the photodegradation of CAP, it is probable that the photodegradation of AcH will yield CO_2 and H_2O.

14.3.3 Adsorption with Carbon-Modified 2D Graphene Nanocomposites

Because of its large surface area and plentiful active sites, graphene oxide (GO) was employed widely as a very effective adsorbent over the last 10 years. It was shown that GO has a thickness of one C atom and functions as a two-dimensional single layer adsorbent. The basal plane was

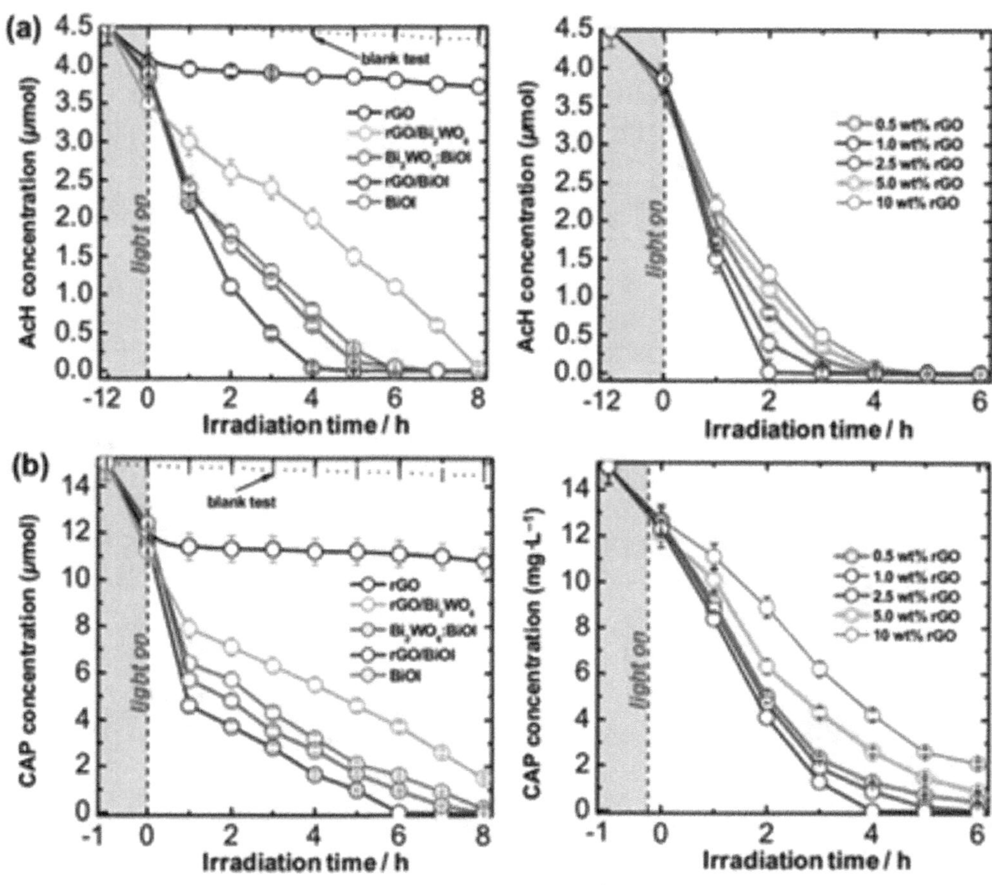

FIGURE 14.8 In an aqueous solution of the produced materials, adsorption and photocatalytic activity were used to remove (a) gaseous acetaldehyde and (b) chloramphenicol under the influence of visible light. (Reproduced with permission from Ref. [30], Copyright © 2018 Elsevier.)

embellished with multiple oxygenated active groups (–OH and C–O–C), whereas the layer edge was adorned with (-COOH). Such activated function groups are thought to have access to accessible negative sites of activity that can form interactions with cationic contaminants. The adsorption efficiency of sodium alginate (SA), graphene oxide (GO), and hydroxyethyl cellulose (HEC; SA-HEC/GO bio-adsorbent hydrogel beads), graphene oxide/gelatin/graphene oxide (SGGO) nanocomposite, graphene oxide/hollow mesoporous silica (GO-HMS) composite, polyaniline/graphene oxide (PANI/GO) nanocomposite, graphene oxide-chitosan-EDTA (GO-EDTA-CS) nanocomposite, and graphene oxide-PEI modified (GO) to form (GO/PEI) sponge, decoration of GO with zinc oxide nanoparticles (ZnO) (GO/ZnO), graphene oxide (GO) modified with isocyanate (MDI), and then (EDTA) (EDTA/MDI/GO) have all been demonstrated to be effective in the removal of organic pollutants. [31]. Figure 14.9 provides a schematic depiction of the procedures involved in the production, adsorption, regeneration, and reuse of GO–C composite.

Two different functional groups that are both attached to the same C-atom—hydroxyl (–OH) and carbonyl (C=O)—combine to form O=C–OH. This peculiar pairing of both groups (–OH and C=O) produces a weak acid group that is polar and extremely electronegative. Resonance between the two O-atoms sustained the carboxylate (COO) anion during deprotonation. This makes it easier for the carboxyl groups and the cationic contaminants to form effective complexes [33]. A number of GO carboxylation techniques were examined, including NaOH/chloroacetic acid and hydrobromic acid/

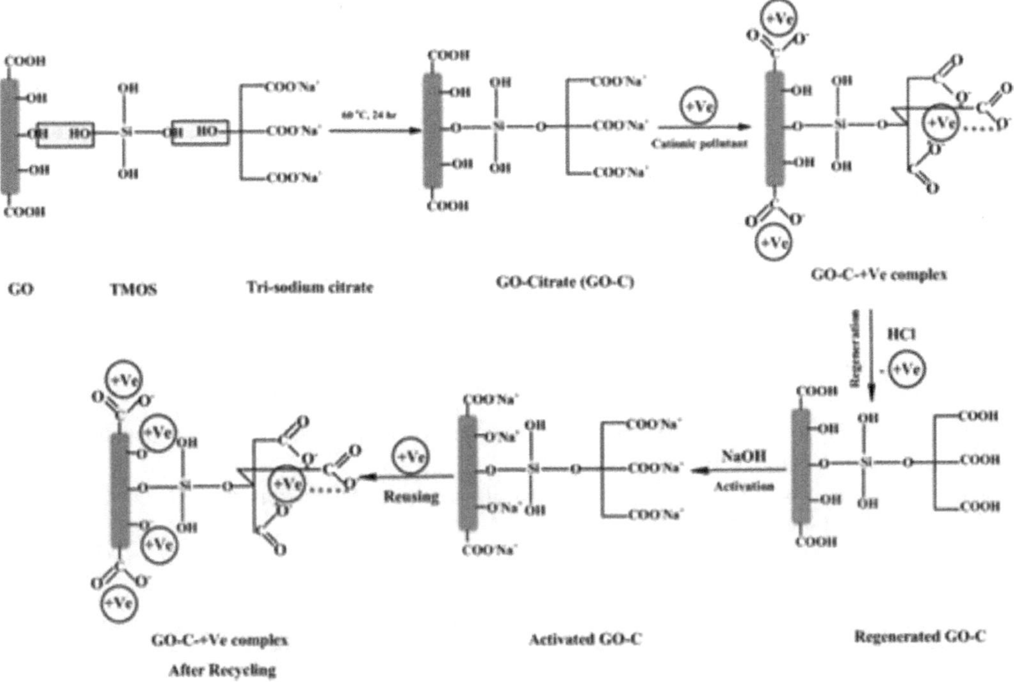

FIGURE 14.9 GO-C composite steps for regeneration/reuse, adsorption, and preparation. (Reproduced with permission from Ref. [32], Copyright © 2022 Springer Nature.)

oxalic acid. Citric acid has one (–COOH) group and one (–OH) group,, as well as two (–COOH) groups. It is an easy-to-find, visually appealing complexing agent. It is also environmentally safe. Seven (O-donor) sites are available from these groupings to chelate cationic species. As a linking and bridging spacer, these donor centers may cluster in different arrangements surrounding the cation ion [34].

14.3.4 Adsorption with Polymers-Modified 2D Graphene Nanocomposites

Several polymer/GO composites have been produced and used to remove organic contaminants from liquid. Large loading capacities of polymeric materials result in a synergistic adsorption effect when paired with GO. For the production of polymer nanocomposites, merging, in situ polymerization, and solution mixing are common techniques. Under extreme conditions, they might alter the molecular structure and produce a change in properties. These multi-step procedures also run the risk of introducing additional pollutants. Dielectric barrier discharge (DBD) is an alternate method that activates the surface of typical substrates enabling functionalization without changing the properties of the support material [35].

Dopamine may have applications as an adsorbent for both organic and inorganic pollutants due to its functional groups related to catechol and amine. Its exceptional adhesive properties also make it the perfect material for coating for a range of supports. Dopamine's active groups and GO's large surface area can be able to work together to produce synergistic adsorption properties. Quinones, an oxidized by-product of catechol, are very susceptible to forming covalent bonds with organic molecules through Schiff base formation or the Michael addition process. A benzene ring's amine and hydroxy groups allow metals to interact with one another. By using catechol chemistry to induce regulated self-polymerization of dopamine, a thin layer of polydopamine can be produced on the GO surface. The polydopamine coating gets rid of the GO-related frequent aggregation problem.

For at least 2 months, polydopamine/GO did not precipitate and was completely soluble in water. The polydopamine coating's thickness is based on the weight fraction of dopamine. Up to 70% total fraction of dopamine, no precipitation was found [36]. Table 14.2 provides a summary of adsorption capacities by graphene, graphene analogs, and composite graphene.

14.4 PHOTODEGRADATION OF ORGANIC POLLUTANTS WITH 2D GRAPHENE NANOCOMPOSITES

With no secondary pollution, photocatalysis provides a long-term, environmentally safe way to remove organic pollutants from wastewater. Heterogeneous photocatalysis is based on complex processes that, in the presence of light, result in the production of electron–hole pairs. UV and/or visible light irradiation from sunlight is momentarily absorbed by a photocatalyst, which is generally accessible as natural irradiation energy or an illuminated light source as shown in Figure 14.10a [37]. To help with the photodegradation of organic pollutants in wastewater, bulk or powder-like photocatalysts are added. Organic molecules are adsorbed before the photocatalyst activity is started. Visible or ultraviolet (UV) light can excite electrons from a semiconductor's VB to its CB, as long as the absorbed light energy is equal to or greater than the semiconductor's band gap and leaves the corresponding holes in the VB. Hydroxyl radicals (OH) are produced when the superoxide anion (O_2) produced by the reaction of the photoexcited electrons with the oxygen adsorbed on the surface of the photocatalyst mixes with H^+ [38]. Meanwhile, the holes create hydroxyl radicals (OH) when they combine with H_2O. Due to their great oxidation

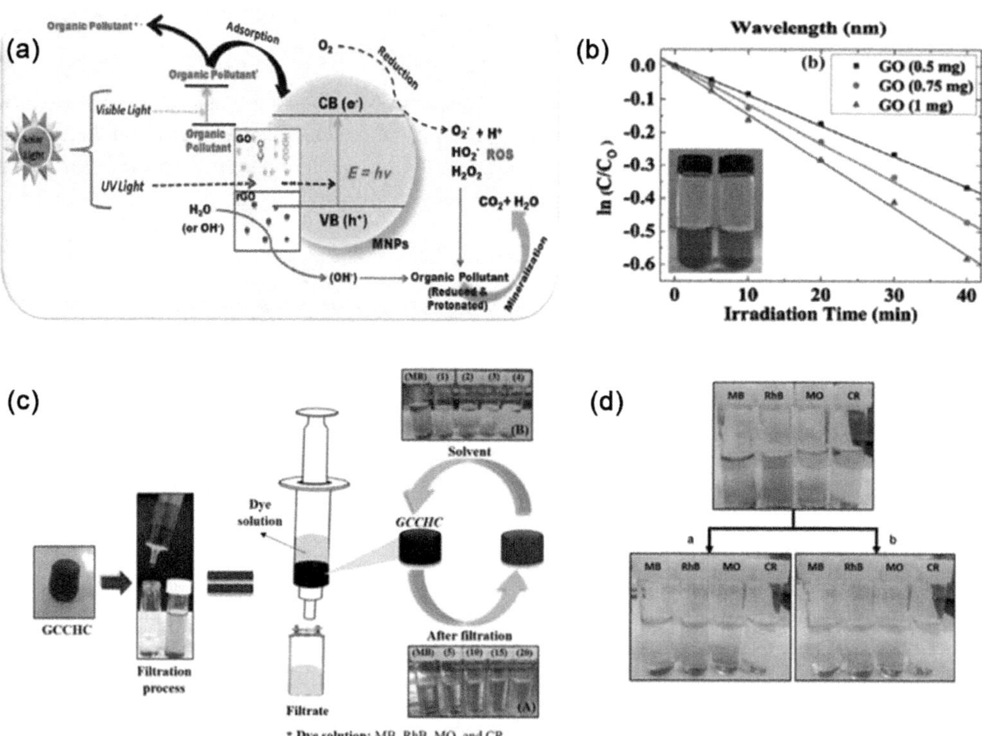

FIGURE 14.10 (a) Mechanism for the photocatalytic degradation of organic pollutants. (b) Absorption graph for MB removal by GO. (c) The filtration–adsorption process using the GCCHC. (d) The color of the treated dye solutions using GC_15 composite hydrogel column via (i) direct filtration and (ii) filtration after 24 hours of immersion. (Reproduced with permission from Ref. [40], Copyright © 2016 InTech.)

capacity, hydroxyl radicals (OH) can completely oxidize organic molecules into CO_2 and H_2O without causing any other pollutants. First, they can oxidize organic molecules into molecular fragments or intermediates. A photochemical reaction requires photoexcitation, in accordance with the previously outlined photocatalytic process. Furthermore, by quickly reuniting the electrons produced by light and vacancies on the semiconductor surface, the reaction time between the oxygen and the active radicals can be decreased. Therefore, the primary technique for developing high-performance semiconductor photocatalysts is to modify the band gap to enhance light absorption [39].

Krishnamoorthy provided a description of GO nanostructures' photocatalytic properties. Resazurin (RZ) reduction was used to quantify the photocatalytic capacity under 350 nm UV light as an estimate of irradiation time. The band gap of GO was determined to be 3.26 eV. Various dosages of GO (0.5, 0.75, and 1 mg) and 10 mL of resazurin mixture at a concentration of 1.5 mg L^{-1} were utilized for the photocatalytic tests. The change in color from blue to pink and the change in the bands adsorption of the adsorption spectra confirmed that RZ had been converted into resorufin (RF) in the results. The pseudo-order approach provided a well-fitting solution for reducing the amount of RZ into RF [40]. Fast photogenerated electron transfer from graphene nanosheets that are conductive reduces photogenerated electron and hole recombination and increases metal oxides' photocatalytic activity. Thus, effective graphene transport and metal oxide photocatalysis can be combined to generate the photocatalytic process of graphene/metal oxides. While exposed to UV or visible light, graphene acts as an electron storage tank to decrease electron–hole pair recombination. In graphene, electrons from metal oxides naturally migrate to the elevated Fermi level of 4.6 eV. Moreover, the specific surface area of the composite is increased by the efficient decrease of nanostructure agglomeration and re-stacking through metal oxide nanostructure decorating on graphene nanosheets. The adsorption of organic molecules is facilitated by the functional groups on the graphene surface, which function as transient active sites. As a result, adding graphene significantly raises metal oxides' photocatalytic efficacy [41]. When creating high-performance graphene/metal oxide photocatalysts, the type of graphene utilized is a major factor in improving composites' photodegradation performance, in addition to the metal oxide selection and microstructure design. In the creation of graphene/metal oxide composite photocatalysts, conductive graphene, hydrophilic graphene oxide (GO), and modified graphene—which includes graphene quantum dots, heteroatom doping, and a three-dimensional network or framework—are frequently employed as conductive or active substrates [42].

14.5 PHOTODEGRADATION WITH MODIFIED 2D GRAPHENE NANOCOMPOSITES

14.5.1 Photodegradation with Doped and Metal Oxide Hybrids of 2D Graphene Nanocomposites

The combined effect of metal oxides and graphene sheets results in a higher photocatalytic activity in metal oxide/graphene hybrids than in separate metal oxides. Thus, a lot of effort has been put into creating new synthetic methods, creating a range of microstructures, adding metal or nonmetal atoms to them, and creating films or frameworks that can stand alone. To eliminate organic pollution, several graphene/metal oxide binary composites are being investigated, including graphene/TiO_2 [43], graphene/ZnO [44], graphene/SnO_2 [45], graphene/WO_3 [46], graphene/Fe_2O_3 [47], and graphene/Cu_2O [48]. Other metal oxides under investigation include V_2O_5, CeO_2, Nb_2O_5, NiO, CO_3O_4, MoO_3, and In_2O_3 [49].

a. **Graphene/TiO_2 composites**

The combined effect of metal oxides and graphene sheets results in a higher photocatalytic activity in metal oxide. Owing to its chemical stability, low cost, and ease of

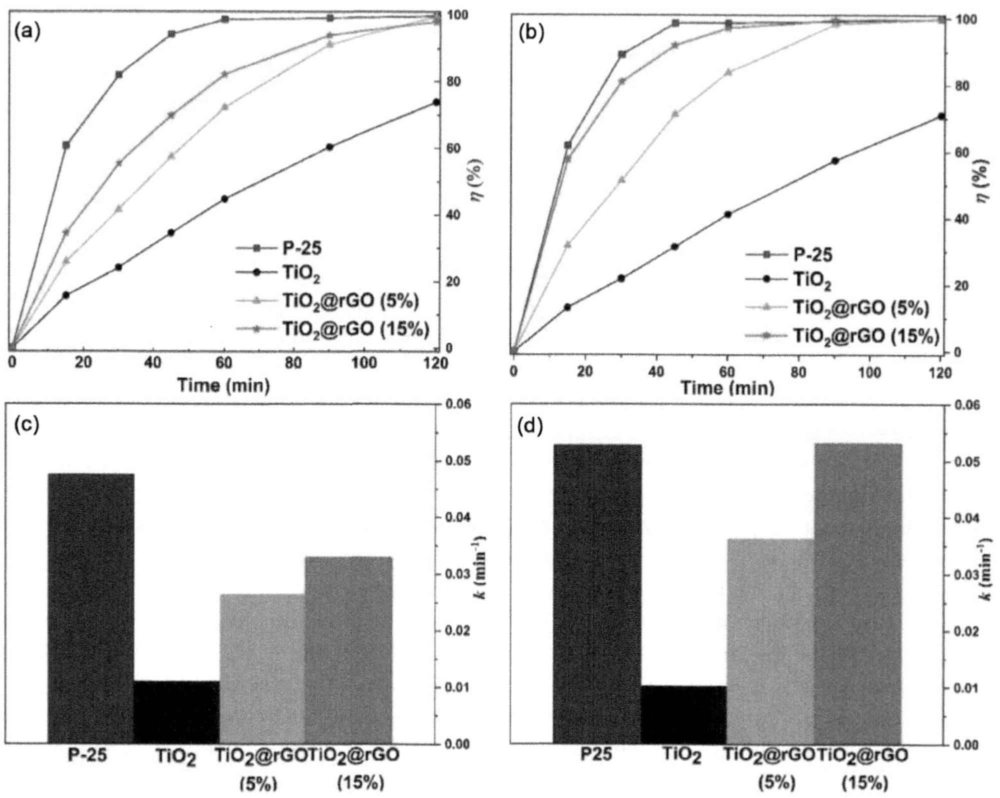

FIGURE 14.11 The percentage of (A) MB and (B) RhB degradation efficiency by the photocatalytic activity of the prepared catalysts under the radiation; rate constant k of (C) MB and (D) RhB degradation for the as-prepared photocatalysts. (Reproduced with permission from Ref. [51] Copyright © 2021 MDPI.)

production, TiO_2, a common n-type semiconductor, finds extensive application in photocatalysis. However, the large band gap of TiO_2 limits the absorption of light in the UV light spectrum. Graphene nanosheets are combined with TiO_2 to increase photocatalytic activity by decreasing electron–hole pair recombination and expanding the light absorption range [50]. Compared to the produced TiO_2 photocatalyst, the TiO_2@rGO nanocomposites significantly accelerated the degradation of RhB and MB, as shown in Figure 14.11. After 120 minutes, the MB and RhB decolorization of both produced nanocomposites was remarkably effective and resembled that of the reference P-25 material. It was demonstrated that TiO_2@rGO (5%) exhibited 99.9% RhB photodecomposition photoactivity and 99.6% MB degradation. The nanocomposite TiO_2@rGO (15%), which included a higher percentage of rGO (98.1% for MB and 99.8% for RhB), showed comparable photodegradation results. On the other hand, the generated TiO_2 particles' photoactivity evaluation was significantly lower than that of the nanocomposites and reference P-25 material. The combined effect of metal oxides and graphene sheets results in a higher photocatalytic activity in metal oxide. For RhB and MB, the TiO_2 degradation rates were 70.9% and 73.9%, respectively. The dyes MB and RhB underwent photodegradation at rates that corresponded to pseudo-first-order kinetics. It was shown that in the degradation of both dyes, the pseudo-first-order rate constants of TiO_2@rGO nanocomposites were substantially higher than those of pure TiO_2 nanoparticles. The structural composition of the organic pollutants had a major impact on the photocatalytic process. MB, a cationic dye, and RhB, a zwitterionic dye, showed notable levels of photodegradation under nanocomposites' irradiation that were linked to the presence of rGO based on the outcomes of photocatalytic investigations. The combined effect of metal oxides and graphene sheets results in a higher photocatalytic activity in metal oxide. When compared to P-25 TiO_2, the photocatalytic activity of all composite materials for MB photodegradation was lower, but for RhB, it was about the same as that of P-25 TiO_2. For both colors, pure synthetic TiO_2 exhibited the least amount of photoactivity [51].

b. **Graphene/SnO_2 composites**

SnO_2 is an n-type semiconductor with a band gap width of 3.189 eV. Since it's non-toxic, inexpensive, and simple to make, photocatalysis has made extensive use of it. However, it mostly absorbs UV light from sunshine because of its wide band gap. To increase visible light absorption and hence enhance photocatalytic activity, several SnO_2/GO combinations have been developed [52]. Tin oxide (SnO_2) hierarchical nanorods are used in the sol–gel method to create graphene oxide (GO) in weight percentages of 1%, 2%, and 3%. The produced hierarchical nanorods (HNRs SnO_2@GO) feature a tetragonal rutile structure with notable crystallinity, according to X-ray diffraction investigation. The produced SnO_2@GO nanocomposites exhibit the distinctive bands of both SnO_2 and GO, as determined by an FTIR study. SEM examination of the morphology of the produced nanocomposite materials showed graphene sheets encircled by hierarchical SnO_2 nanoparticles. EDX analysis indicates that all the samples are composed of Sn and O elements (for SnO_2) and Sn, O, and C elements (for SnO_2 and GO composite). Under visible light irradiation, the photocatalytic performance of the produced GO coated SnO_2 samples was investigated for the degradation of MB dye. Consequently, the high degradation efficiency of 96% for MB dye is demonstrated by the 2% weight of GO with SnO_2 [53] (Figure 14.12).

c. **Graphene/WO_3 composites**

WO_3, a notable n-type semiconductor with a 2.8 eV band gap, has found extensive application in photocatalysis due to its notable absorption in the visible light spectrum and physicochemical stability. Many WO_3/graphene composites were developed recently that are extremely efficient visible light photocatalysts. WO_3 nanorods were produced on rGO sheets using a solvothermal technique by Paik et al. [54]. The well-defined WO_3–rGO interface reduced the recombination of photogenerated charges, enabling the improved composite to reach a degradation efficiency of over 94.1% with visible light in 150 minutes [54,55]. The degradation rate of MB dye during visible light irradiation was used to

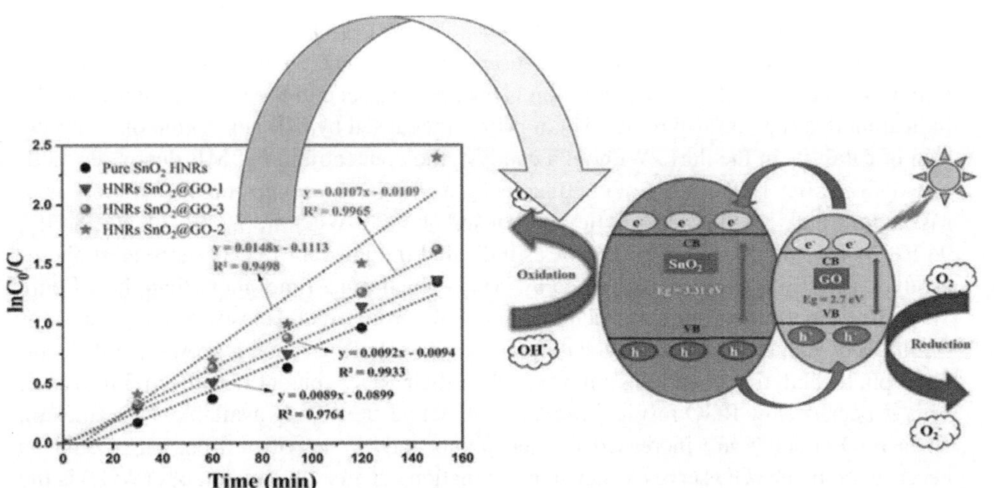

FIGURE 14.12 SnO_2 with GO for the removal of methylene blue. (Reproduced with permission from Ref. [53], Copyright © 2022 Elsevier.)

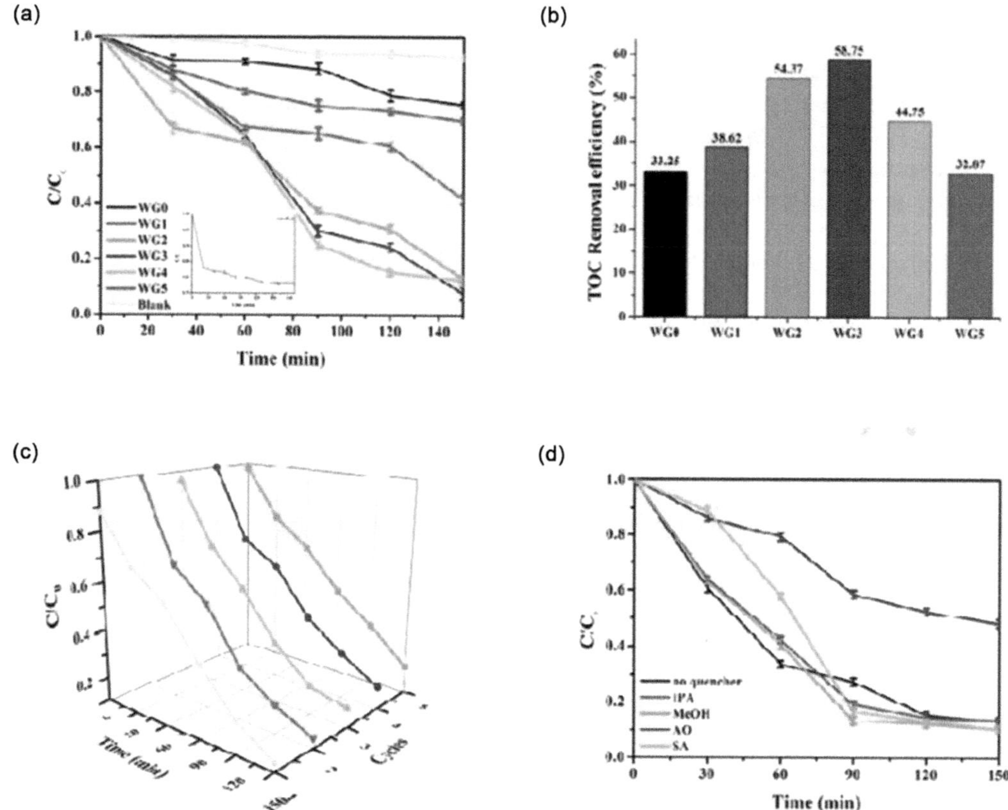

FIGURE 14.13 The degradation efficiency (a) and the TOC removal efficiencies (b) of 15mg/L MB by the as-prepared catalysts (WG0-WG5), the 5 recycling runs of WG3 for MB oxidation degradation (c), and photocatalytic degradation of MB solution over WG3 with and without the quenchers (d). The inset (a) is the pure absorption rate of WG3 for MB (15 mg L^{-1}) in dark. (Reproduced with permission from Ref. [54], Copyright © 2018 Elsevier.)

calculate the photocatalytic efficiency of the as-obtained samples. To reach an equilibrium adsorption state, a 40-minute dark adsorption experiment was conducted prior to illumination. Less than 20% of the concentration is changed, as seen in the inset of Figure 14.13a, indicating that the removal of MB is not primarily caused by MB adsorption on a succession of catalysts in the dark. Without a catalyst, the concentration of MB decreases modestly (7.8%) after 150 minutes of continuous light irradiation, suggesting that photolysis of MB is minimal. The photocatalytic efficiencies of WG0-WG5 are 22.3%, 29.5%, 85.1%, 94.1%, 85.6%, and 57.0%, respectively, as indicated in Figure 14.13a. The strongest photocatalytic activity is evidently exhibited by WG3, which is 4.2 times more than that of pure WO_3. The experimental findings demonstrate that adding rGO nanosheets is essential for significantly enhancing WO_3 photocatalytic activities. Furthermore, it shows that the catalysts' photocatalytic capabilities improve when their RGO content rises from 3 to 6 wt%. This is because low RGO results in inadequate active sites being available. Nevertheless, when rGO contents are increased further, photocatalytic activities drop. This could be because too much rGO serves as new recombination centers. As a result, 6% (WG3) is the ideal weight ratio for rGO and WO_3. As everyone is aware, the TOC result may indicate how much organic matter has been mineralized. The associated TOC removal efficiencies of WG0–WG5 are 33.25%, 38.62%, 54.37%, 58.75%, 44.75%, and 32.87%, respectively, as shown in Figure 14.9b. This finding suggests that catalysis not only breaks down the MB chromophore but also partially mineralizes MB into smaller molecules like CO_2 and H_2O. The goal of the recycling experiment was to determine whether synthesized WG3 was stable and photocatalytically repeatable when exposed to visible light. Figure 14.13c displayed MB's cycle-running degradation performances. The fact that WG3's photocatalytic capacity did not significantly decrease after five cycles of MB photodegradation indicates that the material is stable and reusable when it comes to MB photodegradation.

Isopropanol (IPA), ammonium oxalate (AO), methanol (MeOH), and sodium azide (SA) scavenged the key active species of O_2^-, h^+, •OH, and $^1O_2$ during the photocatalytic process, in that order, to clarify the photodegradation mechanism of MB. As Figure 14.13d illustrates that adding IPA, MeOH, or SA will not reduce WG3's photocatalytic activity. However, when AO is added to the solution, the speed of MB degradation above the WG3 sample is significantly decreased by around 50%. This suggests that h^+ is the main oxidative species and that other radicals are less important in the MB process breakdown [54].

d. **Graphene/other metal oxide composites**

To boost photocatalytic activity, graphene has been combined with several n-type semiconductors, such as CeO_2, V_2O_5, NiO, Nb_2O_5, MoO_3, Co_3O_4, and In_2O_3. The rGO/V_2O_5 nanocomposite was created using a simple solution mixing method, according to Aawani et al. [56]. Compared to pure V_2O_5 rods, graphene nanosheets coated with V_2O_5 rods exhibited superior MB dye degrading activity due to the existence of a conductive graphene medium. Shanavas et al. [57] hydrothermally treated V_2O_5 nanorods via rGO suspension to create a rGO-V_2O_5 nanocomposite. The band gap of the nanocomposite decreased from 2.26eV (V_2O_5) to 1.60eV upon the addition of graphene, resulting in an increase in both charge separation and photocatalytic efficiency. Compared to visible and UV light, the photodegradation efficacy of the mercury lamp is 85% for the discoloration of a solution of MB in 255 minutes. Because of the Z-scheme photoexcited charge transfer process in the nanocomposite, the photocatalytic degradation capacity of 3D/3D/2D-Co_3O_4/Bi_2O_3/rGO ternary nanocomposites toward tetracycline and ibuprofen was greatly boosted. The photocatalytic degradation efficiency of the nanocomposite was enhanced by the 2D rGO nanosheets, which decreased photoexcited charge carrier recombination [58]. The synthesized Ag-Bi_2O_3-rGO composite's photocatalytic H_2 evolution activity (432.7 mol $g^{-1}h^{-1}$) was 5.5 times higher than that of Ag-Bi_2O_3 (101.2 mol $g^{-1}h^{-1}$) NPs and bare-Bi_2O_3 (78.6 mol $g^{-1}h^{-1}$). Additionally, the MB dye was eliminated by Bi_2O_3, Ag-Bi_2O_3, and Ag-Bi_2O_3-rGO composite in 90 minutes by 63.96%, 74.9%, and 100%, respectively. Ag-Bi_2O_3-rGO exhibited enhanced electrochemical nitrite detection properties (LOD of 19.792 M) [59]. In situ precipitation of Cu_2O and Bi_2O_3 nanocrystals on the surface of rGO was used to create RGO-Cu_2O/Bi_2O_3 composite photocatalysts, which degrade tetracycline when exposed to visible light. This Z-scheme photocatalytic system used RGO as a solid-state medium for the transfer of photogenerated electrons from the CB of Bi_2O_3 to the VB of Cu_2O [60]. Acetaminophen is degraded by an all-solid-state ternary Z-scheme composed of the components Bi_2O_3, rGO, and Mo_nO_{3n-1} up to 77% in groundwater samples. It was discovered that rGO is crucial to the material's performance because it helps to delay the recombination period of photogenerated electron/hole (e^-/h^+) pairs, significantly enhancing Bi_2O_3's photocatalytic activity [61]. Some of the graphene composites studied for organic pollutant removal are listed in Table 14.3.

Rhodamine B (RhB) dye degradation was used to measure the photocatalytic efficiency of the samples as they were prepared under UV light. Figure 14.14 displays the colorless RhB dye solution after 180 minutes, as well as the 20-minute decline in Bi_2O_3-WO_3/rGO NCs' RhB absorption at 553 nm. The produced samples' UV photocatalytic ability was measured using the intensity of the RhB solution peak at 553 nm. The blue RhB dye mixture turns colorless after 180 minutes, and the amount of RhB absorbed at 553 nm decreases

TABLE 14.3
Graphene Oxide Composite for Removal of Organic Pollutants

S. No	Catalyst	Synthesis Method	Dye/Antibiotic Degradation	Condition	Removal Amount	Time	Ref.
1	Cu_2O/MoS_2/ rGO	Microwave	Tetracycline and ciprofloxacin	Visible light and ultrasonic irradiation	Tetracycline (20 mg L^{-1}) and Ciprofloxacin (10 mg L^{-1})	10 minutes	[63]
2	MoS_2/RGO	One-pot hydrothermal	Methylene blue	Sonocatalytic	98%	10 minutes	[64]
3	MoS_2/RGO	Hydrothermal method	Ranitidine	Visible light	74%	60 minutes	[65]
4	MoS_2/rGO	Hydrothermal method	Rhodamine B	Visible light	80%	4 hours	[66]
5	RGO/ZnO/ MoS2	Hydrothermal method	Aniline	Visible light	100%	210 minutes	[67]
6	MoS_2/rGO	Hydrothermal methods	Perfluorooctanoic acid	Fluorescent irradiation	77%	1.5 hours	[68]
7	Mn-MoS_2/rGO	Hydrothermal method	Rhodamine B	Visible light	90%	240 minutes	[69]
8	rGO/ZnS-MoS_2	Solvent thermal method	2-Nitrophenol	Visible light	74.05%	–	[70]
9	RGO-MoS_2-$NiCo_2O_4$	Hydrothermal	Rhodamine B	Visible light	95%	90 minutes	[71]
10	Cu/rGO/MoS_2	Solvothermal method	Rhodamine B	Visible light	100%	5 minutes	[72]
11	SnS_2/RGO	One-pot hydrothermal synthesis	Remazol brilliant red Remazol brilliant blue	Visible light	97% 99.7%	–	[73]
12	SnS_2/RGO	Precipitation	Methyl orange	Visible light	91.7%	10 minutes	[74]
13	SnS_2/RGO/ BiOBr	One-pot hydrothermal	Methyl orange norfloxacin	Visible light	99.15% 81.6%	15 minutes	[75]
14	TiO_2/SnS_2/ RGO	Liquid-phase exfoliation and solvothermal method	Rhodamine B	Visible light	–	–	[76]
15	$AgInS_2/SnS_2$/ RGO	Hydrothermal	Norfloxacin	Visible light	95%	120 minutes	[77]
16	CdS-SnS-SnS_2/ rGO	One-pot solvothermal method	Ibuprofen	Visible light	84.46%	60 minutes	[78]
17	SnS_2/RGO	Hydrothermal	Acid orange 7	Visible light	81%	180 minutes	[79]
18	SnS_2-Fe_3O_4/ rGO	-	Diazinon	Visible light	99%	40 minutes	[80]
19	Bi_2S_3/TiO_2/ RGO	Facile solvothermal method	Methylene blue	Visible light	95%	90 minutes	[81]
20	Z-scheme Bi_2S_3/RGO/ $BiVO_4$	One-pot hydrothermal method	Glyphosate	Visible light	96%	–	[82]

(*Continued*)

TABLE 14.3 (*Continued*)
Graphene Oxide Composite for Removal of Organic Pollutants

S. No	Catalyst	Synthesis Method	Dye/Antibiotic Degradation	Condition	Removal Amount	Time	Ref.
21	Bi_2S_3/RGO	Facile single-step process using D-penicillamine	Crystal violet	Visible light	97%	100 minutes	[83]
22	$LaTiO_2N/Bi_2S_3$	Solvothermal	Tetracycline	Visible light Sunlight	96.4% 80.2%	90 minutes	[84]
23	MoS_2/rGO/ Bi_2S_3	Fabrication	Berberine chloride Tetracycline hydrochloride	Carbon felt—anode BiOBr—cathode	90%	–	[85]
24	rGO-$NiWO_4$/ Bi_2S_3	Green sol–gel	Methylene blue Methyl orange dyes	Visible light	57.71%	40 minutes	[86]
25	Bi_2S_3-TiO_2-RGO	Facile one-step hydrothermal	Rhodamine B	Visible light	99.5%	50 minutes	[87]
26	Chitosan-based Fe_2O_3/rGO/ Bi_2S_3	Hydrothermal	Rhodamine B Methylene blue	Visible light	100%	120 minutes	[88]
27	(rGO)/ Bi_2S_3–BiOBr	One-step solvothermal	2-Nitrophenol Cr(VI)	Visible light	95% 67%	60 minutes	[89]
28	CdS/rGO/Fe2+	Photo-Fenton	Phenol	Visible light	100%	1 hour	[90]
29	$BiFeO_3$/CdS/ rGO	Facile hydrothermal	Methylene blue Methyl orange	Visible light	92% 89%	90 minutes	[91]
30	CdS/ RGO/g-C_3N4	Solvothermal method	Atrazine	Visible light	90.50%	5 hours	[92]
31	RGO/CdS	One-step solvothermal	Methylene blue	Visible light	94%	–	[13]
32	Fe_3O_4@β-CD/ rGO	Solvothermal method	Bisphenol A	Catalytic activity	78.2%	5 cycles	[93]
33	CdS/AgBr-rGO	Thermal method	Rhodamine B	Visible light	99%	60 minutes	[77]
34	UiO-67/CdS/ rGO-x	Thermal method	Ofloxacin	Visible light	93.4%	30 minutes	[77]
35	TiO_2-CdS/rGO	Solvothermal	Methylene blue Rhodamine B	Visible light	97.5% 93.5%	20 minutes	[94]
36	CdS-Bi_2MoO_6/ RGO	Solvothermal	Ciprofloxacin	Visible light	90%	60 minutes	[95]
37	RGO–CdS/ZnS	Hydrothermal	Tetracycline	Visible light	90%	60 minutes	[96]
38	rGO/ ZnS-MoS2	Solvent thermal	2-Nitrophenol	Visible light	74%	60 minutes	[70]
39	ZnS–TiO2/ RGO	Solvent thermal	Methylene blue	Visible light	90%	60 minutes	[97]
40	ZnS/RGO	One-step solvothermal method	Methyl orange	Visible light	90%	60 minutes	[98]
41	ZnS–Ag_2S-RGO	Hydrothermal	Rhodamine B	Simulated sunlight	54.08%	60 minutes	[99]
42	S-rGO/ZnS	Facile one-pot	2-Chlorophenol	Sunlight	99.3%	4 hours	[100]
43	(RGO)–CdS/ ZnS	Hydrothermal	Tetracycline	Visible light	90%	60 minutes	[96]

(*Continued*)

TABLE 14.3 (*Continued*)
Graphene Oxide Composite for Removal of Organic Pollutants

S. No	Catalyst	Synthesis Method	Dye/Antibiotic Degradation	Condition	Removal Amount	Time	Ref.
44	ZnS–RGO	Sol–gel synthesis	Brilliant Blue Brilliant Yellow	Visible light	90%	180 minutes	[101]
45	RGO-ZnS	Facile ultrasound-assisted hydrothermal	Reactive black 5	Sunlight	90%	300 minutes	[102]
46	ZnS-rGO	Simple precipitation	Congo Red	Sunlight	98%	60 minutes	[103]
47	MoS2/rGO/ Cu2O	Simple hydrothermal	Methylene blue Rhodamine B	Sunlight	98%	45 minutes	[104]
48	MoS_2/RGO	One-step hydrothermal	Methylene blue	Sonication	95%	28 minutes	[105]
49	SnO_2/MoS_2/ rGO	Hydrothermal	Methylene blue	Sunlight	90%	75 minutes	[106]
50	rGO/$ZnIn_2S_4$	Facile one-pot hydrothermal	Naproxen	Visible light	99%	60 minutes	[107]

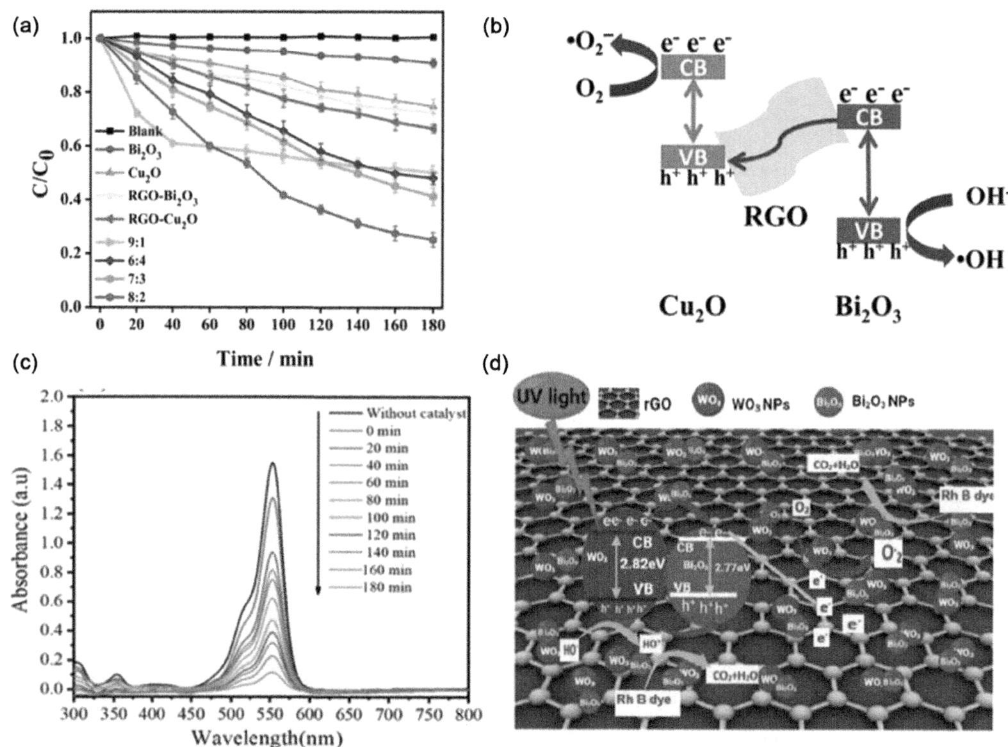

FIGURE 14.14 (a) The degradation of TC over the sample with different Cu_2O/Bi_2O_3 ratios and the same graphene amount (50 mg). (b) Separation process of photoexcited electron–hole on RGO. (c) UV light absorbance of Rh B solutions as a function of irradiation time: Bi_2O_3-WO3/rGO NCs. (d) Possible photocatalytic mechanism scheme of the Bi_2O_3-WO_3/rGO NCs under UV irradiation. (Reproduced with permission from Ref. [62], Copyright © 2023 ACS and [60], Copyright (2016), Elsevier.)

every 20 minutes. The degradation of the RhB dye involves a number of chemical reactions that result in the generation of free radicals, including superoxide and hydroxyl. RhB dye and Bi_2O_3-WO_3/rGO NCs were exposed to UV light. Furthermore, irradiating WO_3 NPs resulted in the production of electron–hole pairs in the VB and CB. The electrons generated by the VB are moved to the CV. Because of their proximity, electrons are said to move from the CB of WO_3 to the CB of Bi_2O_3. As a result, the electrons that travel to Bi_2O_3 CB are transported right away to the rGO. Superoxide radicals (O^2), which mix with water molecules to destroy RhB dye, can be produced by these electrons reacting with oxygen molecules (O_2) in water. Hydroxyl radicals (HO^-) and H_2O are simultaneously absorbed by the remaining holes in WO_3, leading to the generation of hydroxyl radicals ($HO^•$). Furthermore, these free oxygen radicals have the ability to break them down into smaller molecules. Furthermore, it was these free oxygen radicals that broke down the RhB dye into smaller molecules. The findings show that rGO increased degrading efficiency in Bi_2O_3-WO_3/rGO NCs by decreasing electron–hole pair recombination [62].

14.6 PHOTODEGRADATION WITH METAL SULFIDES INTERFACED 2D GRAPHENE NANOHYBRIDS

Apart from metallic oxides, metal sulfides are also combined with graphene to produce photocatalysts that degrade organic contaminants in water more effectively. Zou et al. [108] revealed the production of rGO-ZnS composites over optoelectronic devices. When the composite was exposed to UV light for 4 hours, it was shown to have photocatalytic activity on RhB. This resulted in a degradation efficiency of 84, which was 34% greater than that of ZnS nanorods. Four cycles of the same composite were used, and the RhB degrading efficiency remained constant. Zhou et al. [108] synthesized CdS and rGO composites.

Because of rGO's effectiveness as an acceptor electron, the photocatalytic ability of the rGO–CdS hybrids on Congo red dye degradation increased the efficiency attained with CdS to 90%. Ternary systems, such as rGO–TiO_2–ZnO and rGO–TiO_2–Cu_2O, have been synthesized and employed as effective photocatalysts to degrade organic pollutants in water. Additional graphene-containing substances that have been used are rGO-$BiPO_4$, GO-Ag_2CO_3, and GO-Ag_3PO_4 [2]. Cu_2O/MoS_2/rGO, a ternary compound produced by Zou et al. [63] using a microwave technique, reduces the rate of fast electron–hole recombination and is effective against ciprofloxacin and tetracycline degradation. MoS_2/rGO synthesized via one-pot hydrothermal method studied for the removal of methylene blue and it achieved 98% degradation under sonocatalysis [64]. The same composite was investigated for ranitidine degradation and NDMA generation potential elimination in visible light, and it was found to degrade 74% of ranitidine in 60 minutes [65]. Under visible light irradiation, the ternary compound rGO–MoS_2–$NiCo_2O_4$ demonstrated 95% degradation of Rhodamine B in 90 minutes. The dual charge transfer pathway between the interfacial layer of $NiCo_2O_4$ and rGO–MoS_2 to rGO enables the ternary composite, rGO–MoS_2–$NiCo_2O_4$, to perform better toward photocatalytic activity than bare $NiCo_2O_4$ and rGO–MoS_2 [71]. This process produces more photoinduced charge carriers and suppresses the electron–hole recombination process.

Superior photoelectrochemical efficiency and photocatalytic breakdown of norfloxacin were demonstrated by 2D $AgInS_2$/SnS_2/rGO photocatalysts with numerous connections, with a high degradation rate of up to 95% [77]. Over the course of 60 minutes, the CdS-SnS-SnS_2/rGO composite effectively removed 84.46% of the ibuprofen. In order to increase the separation efficiency of photogenerated carriers, rGO worked as a transfer mediator, moving electrons from the SnS CB to the SnS_2 CB at the heterointerface. These electrons then flowed to the CdS CB via a different heterojunction [78]. Higher photocatalytic activity is demonstrated by Bi_2S_3/rGO produced via an easy single-step procedure with d-penicillamine serving as a sulfur source than by pure Bi_2S_3 and marketed TiO_2. In 100 minutes, around 97% of the crystal violet dye in Bi_2S_3/rGO was destroyed, compared to only 50% in pure Bi_2S_3 [83].

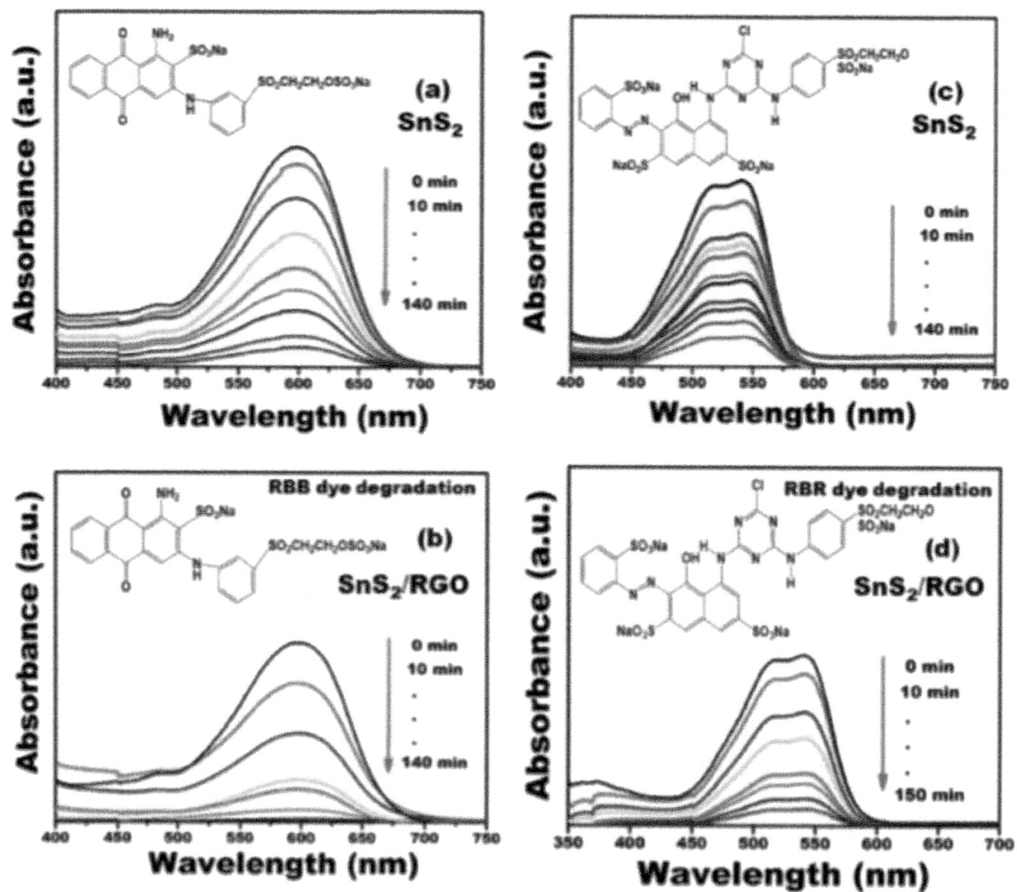

FIGURE 14.15 UV–visible spectra for the photocatalytic degradation of (a, b) RBB dye and (c, d) RBR dye by SnS_2 and SnS_2/rGO under visible light photodegradation. (Reproduced with permission from Ref. [73], Copyright © 2019 Elsevier.)

Remazol Brilliant Red (RBR) and Remazol Brilliant Blue (RBB) were hydrothermally produced and degraded by SnS_2 quantum dots and nanodiscs supported on rGO sheets under visible light radiation. The BET surface area increased to approximately 103.55 m^2g^{-1} from approximately 71.09 m^2g^{-1} upon the addition of rGO, and the photodegradation of the RBB and RBR dyes improved to approximately 99.7% and 97% correspondingly. The comparison of the data of the photocatalytic breakdown of RBB and RBR dye by SnS_2 and SnS_2/rGO is shown in Figure 14.15 [73].

14.7 CONCLUSION AND FUTURE OUTLOOK

Generally speaking, graphene species occur as a honeycomb-type or hexagonal network with one or more layers. Graphene species are mainly made of sp^2 hybridized C atoms. Their unique delocalized-conjugated aromatic system, large surface area ratio, strong light adsorption, easily accessible reactive functional groups, and higher electrical conductivity make them widely used in a range of fields. The advancement of graphene-based composites is a result of the continuous quest for novel strategies to support distinct 2D/3D designs. Consequently, this study covers the production processes with graphene and its various forms as photocatalysts/adsorbents for a range of uses. A number of 2D/3D graphene-based composite components have been researched in an effort to develop a more cost-effective, environmentally friendly, and toxic-free catalyst that will address the growing issue of water pollution. In summary, as the demand for environmentally friendly and efficient water treatment technology rises, 2D/3D GR-based composites will become more and more significant. The composite's catalytic efficiency is increased by the combined action of graphene as well as metal oxide semiconductors upon the breakdown of various water dye contaminants, including methylene blue and phenol. Composites with far more functionalities, such as enhanced surface area, a broad pH reaction range, wide sunlight absorption, enhanced adsorption capacity, large electron transportation speed, and minimal carrier (electrons and holes) recombination, are the result of combining exciting graphene properties with size-related nanomaterial properties.

REFERENCES

1. Ng, L.W., Hu, G., Howe, R.C., Zhu, X., Yang, Z., Jones, C.G., Hasan, T., Ng, L.W., Hu, G., Howe, R.C., & Zhu, X., (2019). Structures, properties, and applications of 2D materials. In: *Printing of Graphene and Related 2D Materials: Technology, Formulation and Applications*, pp. 19–51. https://doi.org/10.1007/978-3-319-91572-2_2.
2. Wick, P., Anna, E.L., Melanie, K., Harald F.K., Kostas K., Bengt, F., Kenneth A.D., et al. (2014). Classification framework for graphene-based materials. *Angewandte Chemie International Edition*, 53(30): 7714. https://doi.org/10.1002/anie.201403335.
3. Son, Y.W., Cohen, M.L., & Louie, S.G. (2006). Energy gaps in graphene nanoribbons. *Physics Reviews Letters*, 97(21): 216803. https://doi.org/10.1103/PhysRevLett.97.216803
4. Zhen, Z., & Zhu, H. (2018). Structure and properties of graphene. In: *Graphene* (pp. 1–12). Academic Press. https://doi.org/10.1016/B978-0-12-812651-6.00001-X.
5. Pérez-Ramírez, E.F, Luz-Asunción M de la, Martínez-Hernández A.L., & Velasco-Santos C. (2016). Graphene materials to remove organic pollutants and heavy metals from water: Photocatalysis and adsorption. *Semiconductor Photocatalysis - Materials, Mechanisms and Applications*. IntechOpen. https://doi.org/10.5772/62777.
6. Li, X., Shen, R., Ma, S., Chen, X., & Xie, J. (2018). Graphene-based heterojunction photocatalysts. *Applied Surface Science*, 430: 53–107. https://doi.org/10.1016/j.apsusc.2017.08.194.
7. Nair, R.R., Blake, P., Grigorenko, A.N., et al., (2008). Fine structure constant defines visual transparency of graphene. *Science*, 320(5881): 1308. https://doi.org/10.1126/science.1156965.
8. Kumar, R., Singh, R., & Singh, D. (2016). Synthesis strategies for graphene chapter: 12. In: *Graphene Science Handbook: Fabrication Methods*, pp. 71–110, https://www.routledgehandbooks.com/doi/10.1201/b19606-5.
9. Bhuyan, M.S.A., Uddin, M.U., Islam, M.M., Bipasha, F.A., & Hossain, S.S. (2016). Synthesis of graphene. *International Nano Letters*, 6: 65–83. https://doi.org/10.1007/s40089-015-0176-1.
10. Van Noorden, R., (2012). Production: Beyond sticky tape. *Nature*, 483(7389): S32–S33. https://doi.org/10.1038/483S32a.
11. Edwards, R.S., & Coleman, K.S. (2013). Graphene synthesis: Relationship to applications. *Nanoscale*, 5(1): 38–51. https://doi.org/10.1039/C2NR32629A.
12. Wang, X., Hongwei, T., Yan, Y., Huan, W., Shumin, W., Weitao, Z., & Yichun, L. (2012). Reduced graphene oxide/CdS for efficiently photocatalystic degradation of methylene blue. *Journal of Alloys and Compounds*, 524: 5–12. https://doi.org/10.1016/j.jallcom.2012.02.058.
13. Luheng, W., Ding T., & Wang, P. (2009). Influence of carbon black concentration on piezoresistivity for carbon-black-filled silicone rubber composite. *Carbon*, 47(14): 3151–3157. https://doi.org/10.1016/j.carbon.2009.06.050.
14. Shams, S.S., Zhang, R., & Zhu, J., (2015). Graphene synthesis: A review. *Material Science and Polymers*, 33(3): 566–578. https://doi.org/10.1515/msp-2015–0079.
15. Hibino, H., Hiroyuki, K., & Masao, N., (2010). Graphene growth on silicon carbide. *NTT Technical Review*, 8: 8.
16. Juang, Z.Y., Wu, C.Y., Lo, C.W., Chen, W.Y., Huang, C.F., Hwang, J.C., Chen, F.R., Leou, K.C., & Tsai, C.H., (2009). Synthesis of graphene on silicon carbide substrates at low temperature. *Carbon*, 47(8): 2026–2031. https://doi.org/10.1016/j.carbon.2009.03.051.
17. Shanmugam, V., Rhoda, A.M., Karthik, B., Sidique, G., Avishek, C., Yongming, T., Rasoul E.N., Michael, F., Gabriel, S., & Oisik, D. (2002). A review of the synthesis, properties, and applications of 2D materials. *Particle & Particle Systems Characterization*, 39(6): 2200031. https://doi.org/10.1016/j.matchemphys.2023.127332.

18. Ebrahimi, N., Neghabi, M., Zadsar, M. et al. (2023). Synthesis and characterization of linear/nonlinear optical properties of graphene oxide and reduced graphene oxide-based zinc oxide nanocomposite. *Scientific Reports*, 13: 1496. https://doi.org/10.1038/s41598-023-28307-7.
19. Zhiliang Z., Jin H., Wu C., & Ji, J. (2018). Efficient production of high-quality few-layer graphene using a simple hydrodynamic-assisted exfoliation method. *Nanoscale Research Letters*, 13:1–8. https://doi.org/10.1186/s11671-018-2830-9.
20. Zhang, K., Zhang, Y., & Wang, S. (2013). Enhancing thermoelectric properties of organic composites through hierarchical nanostructures. *Scientific Reports*, 3: 3448. https://doi.org/10.1038/srep03448.
21. Ibrahim, A., Klopocinska, A., Horvat, K., & Abdel Hamid, Z. (2021). Graphene-based nanocomposites: Synthesis, mechanical properties, and characterizations. *Polymers*, 2613(17): 2869. https://doi.org/10.3390/polym13172869.
22. Zhenfeng, Z., Wang, X., Qiu, J., Lin, J., Xu, D., Zhang, C., Lv, M., & Yang, X. (2014). Three-dimensional graphene-based hydrogel/aerogel materials. *Reviews in Advanced Material Science*, 36: 137–151.
23. Gamze, E., Onur, G., Apul, F., & Perreault, T. (2017). Adsorption of organic contaminants by graphene nanosheets: A review. *Water Research*, 126: 385–398. https://doi.org/10.1016/j.watres.2017.08.010.
24. Nusrat, Z.J., Roy, H., Reaz, A.H., Arshi, S., Rahman, E., Firoz, S.H., & Islam, M.S. (2022). A comparative study on sorption behavior of graphene oxide and reduced graphene oxide towards methylene blue. *Case Studies in Chemical and Environmental Engineering*, 6: 100239. https://doi.org/10.1016/j.cscee.2022.100239.
25. Guo, S., Gaoke, Z., Yadan, G., & Jimmy, C.U. (2013). Graphene oxide–Fe_2O_3 hybrid material as highly efficient heterogeneous catalyst for degradation of organic contaminants. *Carbon*, 60: 437–444. https://doi.org/10.1016/j.carbon.2013.04.058.
26. Dadashi, F., Akbari Ahami, M., Rabbani Esfahani, F., Turner, M., & Nejati, C.H.S. (2020). Experimental and molecular dynamics study on dye removal from water by a graphene oxide-copper-metal organic framework nanocomposite. *Journal of Water Process Engineering*, 34: 101180. https://doi.org/10.1016/j.jwpe.2020.101180.
27. Zaied, B.K., Nasrullah, M.N., Abdullah, W., Zularisam, S., & Lakhveer, K.S. (2020). Application of 2D Graphene-Based nanomaterials for pollutant removal from advanced water and wastewater treatment processes. In: *Adapting 2D Nanomaterials for Advanced Applications*. Chapter: 9. ACS Symposium Series. https://doi.org/10.1021/bk-2020-1353.ch009.
28. Liu, X., Jue, W., Yuming, D., Hexing, L., Yongmei, X., & Haijun, W., (2018). One-step synthesis of Bi2MoO6/reduced graphene oxide aerogel composite with enhanced adsorption and photocatalytic degradation performance for methylene blue. *Materials Science in Semiconductor Processing*, 88: 214–223. https://doi.org/10.1016/j.mssp.2018.08.012.
29. Hojamberdiev, M., Zukhra, C.K., Renato, V.G., Kunio, Y., Nobuhiro, M., Katsuya, T., Masashi, H., & Kiyoshi, O. (2018). Reduced graphene oxide-modified Bi_2WO_6/BiOI composite for the effective photocatalytic removal of organic pollutants and molecular modeling of adsorption. *Journal of Molecular Liquids*, 268: 715–727. https://doi.org/10.1016/j.molliq.2018.07.110.
30. Abd-Elhamid, A.I., Elgoud, E.M.A., Emam, S.S., et al. (2022). Superior adsorption performance of citrate modified graphene oxide as nano material for removal organic and inorganic pollutants from aqueous solution. *Scientific Reports*, 12: 9204. https://doi.org/10.1038/s41598-022-13111-6.
31. Abu Elgoud, E.M., Abd-Elhamid, A.I., Emam, S.S., & Aly, H.F., (2022). Selective removal of some heavy metals from Lanthanide solution by graphene oxide functionalized with sodium citrate. *Scientific Reports*, 12(1): 13755. https://doi.org/10.1038/s41598-022-17949-8.
32. Liu, Y., Deng, R., Wang, Z. & Liu, H. (2012). Carboxyl-functionalized graphene oxide–polyaniline composite as a promising supercapacitor material. *Journal of Material Chemistry*, 22: 13619–13624. https://doi.org/10.1039/C2JM32479B.
33. Zabiszak, M. et al. (2018). Carboxyl groups of citric acid in the process of complex formation with bivalent and trivalent metal ions in biological systems. *Journal of Inorganic Biochemistry*, 182: 37–47. https://doi.org/10.1016/j.jinorgbio.2018.01.017.
34. Hu, R., Songyuan, D., Dadong, S., Ahmed, A., Bashir, A., & Xiangke, W. (2015). Efficient removal of phenol and aniline from aqueous solutions using graphene oxide/polypyrrole composites. *Journal of Molecular Liquids*, 203: 80–89. https://doi.org/10.1016/j.molliq.2014.12.046.
35. Dong, Z., Wang, D., Liu, X., Pei, X., Chen, L., & Jin, J. (2014). Bio-inspired surface-functionalization of graphene oxide for the adsorption of organic dyes and heavy metal ions with a superhigh capacity. *Journal of Material Chemistry*, 2: 5034–5040. https://doi.org/10.1039/C3TA14751G.
36. Kumar, B. (2020). Graphene-and graphene oxide-bounded metal nanocomposite for remediation of organic pollutants. https://doi.org/10.5772/intechopen.92992.

37. Djurišić, A.B., Leung, Y.H., & Ching Ng, A.M. (2014). Strategies for improving the efficiency of semiconductor metaloxide photocatalysis. *Materials Horizons*, 1: 400–410. https://doi.org/10.1039/C4MH00031E.
38. Haque, F., Daeneke, T., Kalantar-zadeh, K., & Ou, J.Z. (2017). Two-dimensional transition metal oxide and chalcogenide-based photocatalysts. *Nano-Micro Letters*, 10: 23. https://doi.org/10.1007/s40820-017-0176-y.
39. Pérez-Ramírez E.F., Luz-Asunción, M., de la, Martínez-Hernández, A.L., & Velasco-Santos, C. (2016). Graphene materials to remove organic pollutants and heavy metals from water: photocatalysis and adsorption. In: *Semiconductor Photocatalysis - Materials, Mechanisms and Applications*. IntechOpen. https://doi.org/10.5772/62777.
40. Singh, N., Prakash, J., & Gupta, R.K. (2017). Design and engineering of high-performance photocatalytic systems based on metal oxide–graphene–noble metal nanocomposites. *Molecular Systematics and Design Engineering*, 2: 422–439. https://doi.org/10.1039/C7ME00038C.
41. Li, X., Shen, R., Ma, S., Chen, X., & Xie, J. (2018). Graphene-based heterojunction photocatalysts. *Applied Surface Science*, 430: 53–61. https://doi.org/10.1016/j.apsusc.2017.08.194.
42. Liao, G., Dongyun, Z., Jinxian, Z., Jing., Y., Bingyan., L., & Laisheng, L. (2016). Efficient mineralization of bisphenol A by photocatalytic ozonation with TiO_2-graphene hybrid. *Journal of the Taiwan Institute of Chemical Engineers*, 67: 300–305. https://doi.org/10.1016/j.jtice.2016.07.035.
43. Pang, Y.L., Su, F.T., Steven, L., Ahmad, Z.A., Hwai, C.C.H.W., Woon, C.C., Abdul, W.M., & Ebrahim, M., (2018). Enhancement of photocatalytic degradation of organic dyes using ZnO decorated on reduced graphene oxide (rGO). *Desalination Water Treatments*, 108: 311–321. https://doi.org/10.5004/dwt.2018.21947.
44. Li, Y., Xiaofang, W., Wingkei, H., Kangle, L., Qin, L., Mei, L., & Shun, C.L. (2018). Graphene-induced formation of visible-light-responsive SnO_2-Zn_2SnO_4 Z-scheme photocatalyst with surface vacancy for the enhanced photoreactivity towards NO and acetone oxidation. *Chemical Engineering Journal*, 336: 200–210. https://doi.org/10.1016/j.cej.2017.11.045.
45. Zhou, M., Jianhui, Y., & Peng, C. (2012) Synthesis and enhanced photocatalytic performance of WO_3 nanorods@ graphene nanocomposites. *Materials Letters*, 89: 258–261. https://doi.org/10.1016/j.matlet.2012.08.081.
46. Yang, X., Chen, C., Li, J., Zhao, G., Ren, X., & Wang, X., 2012. Graphene oxide-iron oxide and reduced graphene oxide-iron oxide hybrid materials for the removal of organic and inorganic pollutants. *RSC Advances*, 2(23): 8821–8826. https://doi.org/10.1039/C2RA20885G.
47. Li, B., Huaqiang, C., Gui, Y., Yuexiang, L., & Jiefu, Y. (2011). Cu_2O reduced graphene oxide composite for removal of contaminants from water and supercapacitors. *Journal of Materials Chemistry*, 21(29): 10645–10648. https://doi.org/10.1039/C1JM12135A.
48. Hong, X., Wang, X., Li, Y., Fu, J., & Liang, B., (2020). Progress in graphene/metal oxide composite photocatalysts for degradation of organic pollutants. *Catalysts*, 10: 921. https://doi.org/10.3390/catal10080921.
49. Guo, Q., Zhou, C., Ma, Z., & Yang, X. (2019) Fundamentals of TiO_2 photocatalysis: Concepts, mechanisms, and challenges. *Advances in Materials*, 31: 1901997. https://doi.org/10.1002/adma.201901997.
50. Kocijan, M., Lidija, Ć., Davor, L., Katarina, M., Ivana, B., Tina, R., Matejka, P. et al. (2021). Graphene-based TiO_2 nanocomposite for photocatalytic degradation of dyes in aqueous solution under solar-like radiation. *Applied Sciences*, 11(9): 3966. https://doi.org/10.3390/app11093966.
51. Jiang, H., Wang, R., Wang, D., Hong, X. & Yang, S. (2019). SnO_2/diatomite composite prepared by solvothermal reaction for low-cost photocatalysts. *Catalysts*, 9: 1060. https://doi.org/10.3390/catal9121060.
52. Perumal, V., Sabarinathan, A., Chandrasekar, M., Subash, M., Inmozhi, C., Uthrakumar, R., Abdulgalim, B.I., et al. (2022). Hierarchical nanorods of graphene oxide decorated SnO_2 with high photocatalytic performance for energy conversion applications. *Fuel*, 324: 124599. https://doi.org/10.1016/j.fuel.2022.124599.
53. Tie, L., Yu, C., Zhao, Y., Chen, H., Yang, S., Sun, J., Dong, S., & Sun, J. (2018). Fabrication of WO3 nanorods on reduced graphene oxide sheets with augmented visible light photocatalytic activity for efficient mineralization of dye. *Journal of Alloys Compounds*, 769: 83–91. https://doi.org/10.1016/j.jallcom.2018.07.176.
54. Paik, T., Cargnello, M., Gordon, T.R., Zhang, S., Yun, H., Lee, J.D., Woo, H.Y., Oh, S.J., Kagan, C.R., Fornasiero, P., et al. (2018). Photocatalytic hydrogen evolution from substoichiometric colloidal WO_{3-x} nanowires. *ACS Energy Letters*, 3: 1904–1910. https://doi.org/10.1021/acsenergylett.8b00925.
55. Shanmugam, M., Alsalme, A., Alghamdi, A., & Jayavel, R. (2015). Enhanced photocatalytic performance of the graphene-V_2O_5 nanocomposite in the degradation of methylene blue dye under direct sunlight. *ACS Applied Materials and Interfaces*, 7: 14905–14911. https://doi.org/10.1021/acsami.5b02715.

56. Aawani, E., Memarian, N., & Dizaji, H.R. (2019). Synthesis and characterization of reduced graphene oxide–V_2O_5 nanocomposite for enhanced photocatalytic activity under different types of irradiations. *Journal of Physics and Chemistry Solids*, 125: 8–15. https://doi.org/10.1016/j.jpcs.2018.09.028.
57. Shanavas, S., Haija, M.A., Singh, D.P., Ahamad, T., Roopan, S.M., Le, A.V., Acevedo, R., & Anbarasan, P.M. (2022). Development of high efficient Co_3O_4/Bi_2O_3/rGO nanocomposite for an effective photocatalytic degradation of pharmaceutical molecules with improved interfacial charge transfer. *Journal of Environmental Chemical Engineering*, 10(2): 107243. https://doi.org/10.1016/j.jece.2022.107243.
58. Nethravathi, P.C., Manjula, M.V.S., Sakar, D.M., & Suresh, D. (2023). Eco-friendly preparation of $Bi2O_3$, Ag-Bi_2O_3 and Ag-Bi_2O_3-rGO nanomaterials and their photocatalytic H_2 evolution, dye degradation, nitrite sensing and biological applications. *Journal of Photochemistry and Photobiology A: Chemistry*, 435: 114295. https://doi.org/10.1016/j.jphotochem.2022.114295.
59. Shen, H., Wang, J., Jiang, J., Luo, B., Mao, B., & Shi, W. (2017). All-solid-state Z-scheme system of RGO-Cu_2O/Bi_2O_3 for tetracycline degradation under visible-light irradiation. *Chemical Engineering Journal*, 313: 508–517. https://doi.org/10.1016/j.cej.2016.11.161.
60. Rubio-Govea, R., Carolina, O.N., Sergio, F.L., Netzahualpille, H., Jürgen, M., Alejandra, G., & Nancy O. (2020). Bi_2O_3/rGO/$MonO_3$n-1 all-solid-state ternary Z-scheme for visible-light driven photocatalytic degradation of bisphenol A and acetaminophen in groundwater. *Journal of Environmental Chemical Engineering* 8(5): 104170. https://doi.org/10.1016/j.jece.2020.104170.
61. Alaizeri, Z.M., Hisham A.A., Saad A., Akhtar, M.J., & Ahamed, M. (2023). Bi_2O_3-doped WO3 nanoparticles decorated on rGO sheets: simple synthesis, characterization, photocatalytic performance, and selective cytotoxicity toward human cancer cells. *ACS Omega* 8(28): 25020–25033. https://doi.org/10.1021/acsomega.3c01644.
62. Selvamani, P.S., Vijaya, J.J., Kennedy, L.J., Mustafa, A., Bououdina, M., Joice, S., & Jothi, R. (2021). Synergic effect of Cu_2O/MoS_2/rGO for the sonophotocatalytic degradation of tetracycline and ciprofloxacin antibiotics. *Ceramics International*, 47(3): 4226–4237. https://doi.org/10.1016/j.ceramint.2020.09.301.
63. Zou, L., Qu, R., Gao, H., Guan, X., Qi, X., Liu, C., & Lei, X. (2019). MoS_2/RGO hybrids prepared by a hydrothermal route as a highly efficient catalytic for sonocatalytic degradation of methylene blue. *Results in Physics*, 14: 102458. https://doi.org/10.1016/j.rinp.2019.102458.
64. Zou, X., Zhang, J., Zhao, X., & Zhang, Z. (2020). MoS_2/RGO composites for photocatalytic degradation of ranitidine and elimination of NDMA formation potential under visible light. *Chemical Engineering Journal*, 383: 123084.
65. Phan, T., Thuy, T., Thi, T.H.N., Ha, T.H., Thanh, T.T., Le, T.N., Van, T.N., Vy, A.T., Thi, L.N., Hong, L.N., & Vien V. (2021). Hydrothermal synthesis of MoS_2/rGO heterostructures for photocatalytic degradation of rhodamine B under visible light. *Journal of Nanomaterials*, 2021: 1–11. https://doi.org/10.1155/2021/9941202.
66. Parisa, G., Fattahi, M., Rasekh, B., & Fatemeh, Y. (2020). Developing the ternary ZnO doped MoS_2 nanostructures grafted on CNT and reduced graphene oxide (RGO) for photocatalytic degradation of aniline. *Scientific Reports*, 10(1): 4414. https://doi.org/10.1038/s41598-020-61367-7.
67. Norsham, I., Najwa, M., Kavirajaa, P., Sambasevam, S., Shahabuddin, A.H.J., & Baharin, S.R.A. (2022). Photocatalytic degradation of perfluorooctanoic acid (PFOA) via MoS_2/rGO for water purification using indoor fluorescent irradiation. *Journal of Environmental Chemical Engineering*, 10(5): 108466. https://doi.org/10.1016/j.jece.2022.108466.
68. Phan, T., Thuy, T., Thi, T.H.N., Ha, T.H., Thanh, T.T., Le, T.N., Van, T.N., Vy, A.T., Thi, L.N., Hong, L.N., & Vien V. (2021). Hydrothermal synthesis of MoS_2/rGO heterostructures for photocatalytic degradation of rhodamine B under visible light. *Journal of Nanomaterials*, 2021: 1–11. https://doi.org/10.1155/2021/9941202.
69. Hu, X., Deng, F., Huang, W., Zeng, G., Luo, X., & Dionysios, D.D. (2018). The band structure control of visible-light-driven rGO/ZnS-MoS_2 for excellent photocatalytic degradation performance and long-term stability. *Chemical Engineering Journal*, 350: 248–256. https://doi.org/10.1016/j.cej.2018.05.182.
70. Chakrabarty, S., Mukherjee, A., & Basu, S. (2018). RGO-MoS_2 supported $NiCo_2O_4$ catalyst toward solar water splitting and dye degradation. *ACS Sustainable Chemistry & Engineering*, 6(4): 5238–5247. https://doi.org/10.1021/acssuschemeng.7b04757.
71. Liang, H., Peng, H., Yufei, Z., Zhimin, F., Jiaoning, T., & Junfeng N. (2019). Fabrication of Cu/rGO/MoS_2 nanohybrid with energetic visible-light response for degradation of rhodamine B. *Chinese Chemical Letters*, 30(12): 2245–2248. https://doi.org/10.1016/j.cclet.2019.05.046.

72. Dashairya, L., Sharma, M., Basu, S., & Saha, P. (2019). SnS_2/RGO based nanocomposite for efficient photocatalytic degradation of toxic industrial dyes under visible-light irradiation. *Journal of Alloys and Compounds*, 774: 625–636. https://doi.org/10.1016/j.jallcom.2018.10.008.
73. Babu, S.G., Athira, S.V., Neppolian, B., & Ashokkumar, M. (2015). SnS_2/rGO: An efficient photocatalyst for the complete degradation of organic contaminants. *Materials Focus*, 4(4): 272–276. https://doi.org/10.1166/mat.2015.1247.
74. Zhang, R., Lu, C., Youfeng, C., Qiong, H., Yu, L., Tongqing, Z., Yi, L., et al. (2021). Lamellar insert SnS_2 anchored on BiOBr for enhanced photocatalytic degradation of organic pollutant under visible-light. *Colloids and Surfaces A: Physicochemical and Engineering Aspects*, 618: 126444. https://doi.org/10.1016/j.colsurfa.2021.126444.
75. Weiping, Z., Xiao, X., Zeng, X., Li, Y., Zheng, L., & Caixia, W. (2016). Enhanced photocatalytic activity of TiO_2 nanoparticles using SnS_2/RGO hybrid as co-catalyst: DFT study and photocatalytic mechanism. *Journal of Alloys and Compounds*, 685: 774–783. https://doi.org/10.1016/j.jallcom.2016.06.199.
76. Zhang, S., Yimeng, W., Zan, C., Jian, X., Jun, H., Yuan, H., Changzheng, C., Honglai, L., & Hualin, W. (2020). Simultaneous enhancements of light-harvesting and charge transfer in UiO-67/CdS/rGO composites toward ofloxacin photo-degradation. *Chemical Engineering Journal*, 381: 122771. https://doi.org/10.1016/j.cej.2019.122771.
77. Liang, M., Yajing, Y., Ying, W., & Yan, Y. (2020). Remarkably efficient charge transfer through a double heterojunction mechanism by a CdS-SnS-SnS_2/rGO composite with excellent photocatalytic performance under visible light. *Journal of Hazardous Materials*, 391: 121016. https://doi.org/10.1016/j.jhazmat.2019.121016.
78. Rahimi, A., Iraj, K., & Seyed, E.M.G., (2018). Synthesis and investigation of SnS_2/RGO nanocomposites with different GO concentrations: Structure and optical properties, photocatalytic performance. *Journal of Materials Science: Materials in Electronics*, 29: 4449–4456. https://doi.org/10.1007/s10854-017-8392-2.
79. Pirsaheb, M., Hiwa, H., Anvar, A., & Zeinab, J. (2022). Persulfate activation by magnetic SnS_2-Fe_3O_4/rGO nanocomposite under visible light for detoxification of organophosphorus pesticide. *Journal of Molecular Liquids*, 364: 119975. https://doi.org/10.1016/j.molliq.2022.119975.
80. Liu, Y., Yidan, S., Xiang, L., & Hexing, L. (2017). A facile solvothermal approach of novel Bi_2S_3/TiO_2/RGO composites with excellent visible light degradation activity for methylene blue. *Applied Surface Science*, 396: 58–66. https://doi.org/10.1016/j.apsusc.2016.11.028.
81. Luo, X.-L., Yang, S.-Y., Wang, Z.-L., & Xu, Y.-H. (2023). Synthesis of Z–scheme Bi_2S_3/RGO/$BiVO_4$ photocatalysts with superior visible light photocatalytic effectiveness for pollutant degradation. *Separation and Purification Technology*, 318: 123966. https://doi.org/10.1016/j.seppur.2023.123966.
82. Vadivel, S., Kamalakannan, V.P., & Balasubramanian, N. (2014). d-Penicillamine assisted microwave synthesis of Bi_2S_3 microflowers/RGO composites for photocatalytic degradation—A facile green approach. *Ceramics International*, 40(9): 14051–14060. https://doi.org/10.1016/j.ceramint.2014.05.133.
83. Sharma, S.K., Kumar, M.A., Sharma, G., Stadler, F.J., Naushad, M., Ghfar, A.A., & Tansir, A. (2020). $LaTiO_2N$/Bi_2S_3 Z-scheme nano heterostructures modified by rGO with high interfacial contact for rapid photocatalytic degradation of tetracycline. *Journal of Molecular Liquids*, 311: 113300. https://doi.org/10.1016/j.molliq.2020.113300.
84. Yu, T., Qingsong, L., Zhiyuan Z., Wenwei, W., Lifen, L., Jinlong, Z., Changfei, G., & Tao, Y. (2021). Construction of a photocatalytic fuel cell using a novel Z-scheme MoS_2/rGO/Bi_2S_3 as electrode degraded antibiotic wastewater. *Separation and Purification Technology*, 277: 119276. https://doi.org/10.1016/j.seppur.2021.119276.
85. Liaquat, H, Muhammad, I., Shoomaila, L., Sajid, I., Nazim, H., & Bilal, M. (2022). Citric acid-capped $NiWO_4$/Bi_2S_3 and rGO-doped $NiWO_4$/Bi_2S_3 nanoarchitectures for photocatalytic decontamination of emerging pollutants from the aqueous environment. *Environmental Research*, 212: 113276. https://doi.org/10.1016/j.envres.2022.113276.
86. Li, Y. Zhongmin, L., Yaru, L., Yongchuan, W., Jitao C., Yanjun, L., & Ping, N. (2018). One-step hydrothermal synthesis of Bi_2S_3-TiO_2-RGO composites with enhanced visible light photocatalytic activities. *Nano*, 13(05): 1850051. https://doi.org/10.1142/S1793292018500510.
87. Ramasamy, N., Kavitha, N.P., Mathew, A., & Balasubramanian, N. (2023). Development of chitosan@Fe_2O_3/rGO/Bi_2S_3 as a new eco-friendly photocatalyst for enhancing the catalytic stability and superior degradation of organic pollutants. *Research on Chemical Intermediates*, 49(6): 2603–2624. https://doi.org/10.1007/s11164-023-05001-x.

88. Li, H., Fang, D., Yang, Z., Li, H., Chunhua, Q., & Xubiao, L. (2019). Visible-light-driven Z-scheme rGO/ Bi_2S_3–BiOBr heterojunctions with tunable exposed BiOBr (102) facets for efficient synchronous photocatalytic degradation of 2-nitrophenol and Cr (VI) reduction. *Environmental Science: Nano*, 6(12): 3670–3683. https://doi.org/10.1039/C9EN00957D.
89. Jiang, Z., Lingzhi, W., Juying, L., Yongdi, L., & Jinlong, Z. (2019). Photo-Fenton degradation of phenol by CdS/rGO/Fe^{2+} at natural pH with in situ-generated H_2O_2. *Applied Catalysis B: Environmental*, 241: 367–374. https://doi.org/10.1016/j.apcatb.2018.09.049.
90. Alijani, H., Majid, A., & Hannaneh, K. (2022). Efficient photocatalytic degradation of toxic dyes over $BiFeO_3$/CdS/rGO nanocomposite under visible light irradiation. *Diamond and Related Materials*, 122: 108817. https://doi.org/10.1016/j.diamond.2021.108817.
91. Jo, W., & and Selvam, N.C.S. (2017). Z-scheme CdS/g-C_3N_4 composites with RGO as an electron mediator for efficient photocatalytic H_2 production and pollutant degradation. *Chemical Engineering Journal*, 317: 913–924. https://doi.org/10.1016/j.cej.2017.02.129.
92. Zhang, Y., Zhuang, C., Lincheng, Z., Panpan, W., Yalong, Z., Yuxian, L., & Fei, W. (2019). Heterogeneous fenton degradation of bisphenol a using Fe_3O_4 β-CD/rGO composite: Synergistic effect, principle and way of degradation. *Environmental Pollution*, 244: 93–101. https://doi.org/10.1016/j.envpol.2018.10.028.
93. Wang, L., Ming, W., Wenya, W., Nchare, M., Zhenwei, W., & Shuzhen, L. (2016). Photocatalytic degradation of organic pollutants using rGO supported TiO_2-CdS composite under visible light irradiation. *Journal of Alloys and Compounds*, 68(3): 318–328. https://doi.org/10.1016/j.jallcom.2016.05.111.
94. Chen, Z., Jianxing, L., Xiangyang, X., Guangyu, H., & Haiqun, C. (2020). CdS–Bi_2MoO_6/RGO nanocomposites for efficient degradation of ciprofloxacin under visible light. *Journal of Materials Science*, 55(14): 6065–6077. https://doi.org/10.1007/s10853-020-04413-z.
95. Tang, Y., Xinlin, L., Changchang, M., Mingjun, Z., Pengwei, H., Longbao, Y., Jianming P., Weidong, S., & Yongsheng, Y. (2015). Enhanced photocatalytic degradation of tetracycline antibiotics by reduced graphene oxide–CdS/ZnS heterostructure photocatalysts. *New Journal of Chemistry*, 39(7): 5150–5160. https://doi.org/10.1039/C5NJ00681C.
96. Qin, Y.L., Zhao, W.W., Sun, Z., Liu, X.Y., Shi, G.L., Liu, Z.Y., Ni, D R. & Ma, Z.Y. (2019). Photocatalytic and adsorption property of ZnS–TiO_2/RGO ternary composites for methylene blue degradation. *Adsorption Science & Technology*, 37(9–10): 764–776. https://doi.org/10.1177/0263617418810932.
97. Qin, Y., Zheng, S., Wenwen, Z., Zhenyu, L., Dingrui, N., & Zongyi, M. (2017). Improved photocatalytic properties of ZnS/RGO nanocomposites prepared with GO solution in degrading methyl orange. *Nano-Structures & Nano-Objects*, 10: 176–181. https://doi.org/10.1016/j.nanoso.2017.05.005.
98. Reddy, D., Ma, A.R., Choi, M.Y., & Kim, T.K (2015). Reduced graphene oxide wrapped ZnS–Ag_2S ternary composites synthesized via hydrothermal method: Applications in photocatalyst degradation of organic pollutants. *Applied Surface Science*, 324: 725–735. https://doi.org/10.1016/j.apsusc.2014.11.026.
99. Alafif, Z.O., Muzammil, A., Ansari, M.O., Kumar, R., Rashid, R., Madkour, M., & Barakat, M.A. (2019). Synthesis and characterization of S-doped-rGO/ZnS nanocomposite for the photocatalytic degradation of 2-chlorophenol and disinfection of real dairy wastewater. *Journal of Photochemistry and Photobiology A: Chemistry*, 377: 190–197. https://doi.org/10.1016/j.jphotochem.2019.04.004.
100. Kashinath, L., Keerthiraj, N., Shivanna, S., Ajayan, V., & Kullaiah, B. (2017). Microwave treated sol–gel synthesis and characterization of hybrid ZnS–RGO composites for efficient photodegradation of dyes. *New Journal of Chemistry*, 41(4): 1723–1735. https://doi.org/10.1039/C6NJ03716J.
101. Mahvelati, S., Tahreh, E.K., Goharshadi, M.S., & Zohreh, N. (2018). ZnS@ reduced graphene oxide nanocomposite as an effective sunlight driven photocatalyst for degradation of reactive black 5: A mechanistic approach. *Separation and Purification Technology*, 202: 326–334. https://doi.org/10.1016/j.seppur.2018.04.001.
102. Dalal, C., Anjali, K.G., Nimisha J., Abbas, R.N., Rajneesh, K.P., Shyam, K.C., & Sumit, K.S. (2023). Sunlight-assisted photocatalytic degradation of azo-dye using zinc-sulfide embedded reduced graphene oxide. *Solar Energy*, 251: 315–324. https://doi.org/10.1016/j.solener.2023.01.017.
103. Gopalakrishnan, A., Satyam, P.S., & Sushmee, B. (2020). Reusable, free-standing MoS_2/rGO/Cu_2O ternary composite films for fast and highly efficient sunlight driven photocatalytic degradation. *Chemistry Select*, 5(6): 1997–2007. https://doi.org/10.1002/slct.201904932.
104. Ahani, F., Maisam, J., Javad, M., & Mohammad, H.R. (2023). Enhancing sonocatalytic dye pollutant degradation using MoS_2/RGO nanocomposites: An optimization study. *Water Resources and Industry*, 30, 100223. https://doi.org/10.1016/j.wri.2023.100223.
105. Agboola, P.O., & Imran, S. (2022). Facile fabrication of SnO_2/MoS_2/rGO ternary composite for solar light-mediated photocatalysis for water remediation. *Journal of Materials Research and Technology*, 18: 4303–4313. https://doi.org/10.1016/j.jmrt.2022.04.109.

106. Fu, K., Yishuai, P., Chao, D., Jun, S., & Huiping, D. (2020). Reduced graphene oxide/$ZnIn_2S_4$ nanocomposite photocatalyst with enhanced photocatalytic performance for the degradation of naproxen under visible light irradiation. *Catalysts*, 10(6): 710. https://doi.org/10.3390/catal10060710.
107. Chakraborty, K., Chakrabarty, S., Das, P., Ghosh, S., & Pal, T. (2016). UV-assisted synthesis of reduced graphene oxide zinc sulfide composite with enhanced photocatalytic activity. *Materials Science and Engineering B*, 204: 8–14. https://doi.org/10.1016/j.mseb.2015.11.001.
108. Zou, L., Wang, X., Xu, X., & Wang, H. (2016). Reduced graphene oxide wrapped CdS composites with enhanced photocatalytic performance and high stability. *Ceramics International*, 42: 372–378. https://doi.org/10.1016/j.ceramint.2015.08.119.

15 2D Layered Double Hydroxides Nanomaterials for the Removal of Water Pollutants

Jijoe Prabagar Samuel and Harikaranahalli Puttaiah Shivaraju

15.1 INTRODUCTION

Since the advent of industrialization and urbanization in the 1920s, the proliferation of wastewater, residual matter, and waste gas has culminated in an unprecedented degree of water pollution. Unlike organic pollutants, heavy metal ions and radionuclides possess elevated solubility in aqueous environments and can accumulate within organisms and traverse the food chain [1]. Over the past several decades, researchers have focused on developing efficient methods to eliminate these pollutants from wastewater. Currently, traditional water treatment technologies encompass ion exchange, chemical precipitation, oxidation–reduction procedures, and adsorption [2–5]. Of these techniques, adsorption and advanced oxidation processes (AOPs) have garnered substantial research attention and are deemed one of the most effective methods due to their simplicity, low startup cost, ease of implementation, and heightened sensitivity to pollutants [6].

In recent times, the focus of research in this field has shifted towards the development of adsorbents with augmented affinity, capacity, and selectivity for target pollutants [7–9]. Clay mineral materials, carbon-based nanomaterials, metal–organic framework (MOF) materials, and zero-valent iron materials are widely utilized for the elimination of a range of pollutants from aqueous solutions [10–13]. Despite their high removal efficiency, these materials have associated disadvantages, such as high preparation costs, unstable properties, and the potential for secondary pollution [14,15]. In contrast, layered double hydroxides (LDHs) exhibit moderate chemical stability, low cost, and non-toxicity, which have garnered considerable attention from both academic and popular communities [16]. Over the past several decades, substantial progress has been made in the synthesis and characterization of LDH composite materials.

In this context, 2D LDHs have emerged as promising alternatives for water treatment due to their unique physical and chemical properties [17–19]. LDHs are composed of positively charged metal cations and anionic layers of hydroxides, which form a sandwich-like structure [20]. The interlayer space of LDHs can adsorb a wide range of pollutants, making them an effective solution for water treatment. Recent advances in the synthesis of 2D LDHs have led to the development of materials with enhanced adsorption capacity and selectivity [21,22]. Synthesis methods include co-precipitation, hydrothermal synthesis, and sol–gel methods. In addition, the incorporation of various functional groups into the LDH structure has been shown to improve the removal of specific contaminants, such as heavy metals and organic compounds [23,24]. Characterization techniques, such as X-ray diffraction (XRD), transmission electron microscopy (TEM), and Fourier transform infrared (FTIR) spectroscopy, have been used to study the structure and surface properties of 2D LDHs [16]. These techniques have provided valuable information on the mechanism of adsorption,

DOI: 10.1201/9781003436942-15

which involves the interaction between the pollutants and the LDH surface. LDHs have been shown to effectively remove various water pollutants, including heavy metals, organic compounds, and inorganic ions, due to their high adsorption capacity and favourable surface chemistry.

Further, 2D LDHs are widely employed as a promising photocatalyst for the removal of organic pollutants from water, due to their high surface area, tunable composition, and catalytic properties. Common metals used in the construction of LDHs-based composites for photocatalytic applications include Mg, Al, Zn, and Fe. These metals have been chosen because they can be easily incorporated into the layered structure of LDHs, leading to the formation of stable and efficient photocatalysts. The photocatalytic activity of LDH materials can be attributed to their unique properties, such as the presence of basic/acidic sites, redox centres, possible oxygen vacancies, and electronic structures that determine their band gap and optical properties. The basic and acidic sites present in the LDH materials play a crucial role in the adsorption and activation of organic pollutants, while the redox centres facilitate the transfer of electrons during the photocatalytic process. Additionally, the possible oxygen vacancies present in the LDHs structure can act as trapping sites for photogenerated charge carriers, leading to enhanced photocatalytic activity. The electronic structure of LDH materials also plays a crucial role in their photocatalytic properties. The band gap of the materials determines their ability to absorb light of different wavelengths, while their optical properties are determined by the presence of transition metal ions, which can exhibit intense absorption in the visible region of the electromagnetic spectrum. Overall, the unique combination of properties exhibited by LDH materials makes them promising candidates for the photocatalytic degradation of organic pollutants in water.

In this chapter, the focus will be on recent advances in 2D LDHs for the removal of water pollutants, including the synthesis methods, characterization techniques, and mechanism of action. The chapter will provide a comprehensive overview of the current state of the field and highlight the potential of 2D LDHs as a promising solution for water treatment as both sorbents and photocatalytic active materials for adsorption and degradation of water pollutants including metal ions and organics.

15.2 OVERVIEW OF STRUCTURE PROPERTIES OF LAYERED DOUBLE HYDROXIDES

LDHs are clay minerals with a two-dimensional layered structure. LDHs are composed of laminates with a positive charge and interlayer anions with a negative charge, bound together through non-covalent bonds [16]. This composition gives rise to a large surface area and a high ion exchange capacity [25]. LDHs are known for their versatile properties, including stability, reactivity, and biocompatibility, making them suitable for a wide range of applications [26,27]. The unit layer of LDHs has a positive charge and is composed of a shared edge of MgO_6 octahedrons [28]. The positive charge of the laminates in LDHs is compensated by the exchangeable anions located in the interlayer region, resulting in the neutral charge of the LDHs as a whole. The interlayer space is occupied by water molecules. The chemical formula for LDHs can be written as $[M_{1-x}{}^{2+}M_x{}^{3+}(OH)_2]^{x+}[(A^{n-})_{x/n}\ mH_2O]^{x-}$, where M^{2+} and M^{3+} denote divalent and trivalent cations, respectively [29]. X represents the molar ratio of M^{3+} to the sum of M^{2+} and M^{3+}. A^{n-}, with -n charges, represents the interlayer anion and m is the number of crystal water molecules present in the LDHs. The schematic illustration of a typical LDH is displayed in Figure 15.1.

LDHs have gained widespread use in the treatment of environmental pollution because of their advantageous characteristics, including the memory effect, enhanced thermal stability, and the capacity for interlayer anion exchange [30]. LDH materials can be classified based on their composition and structure. Compositionally, LDHs can be categorized as either hydrotalcite-like or hydrocalumite-like, depending on the type of metal ions used to construct the layered structure. Structurally, LDHs can be categorized as either pure or mixed, depending on the type of interlayer anions present. Mixed LDHs can have a combination of different types of interlayer anions, leading

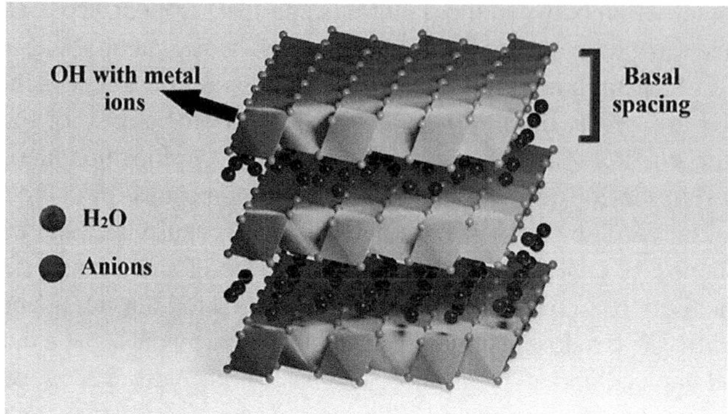

FIGURE 15.1 Schematic representation of LDH.

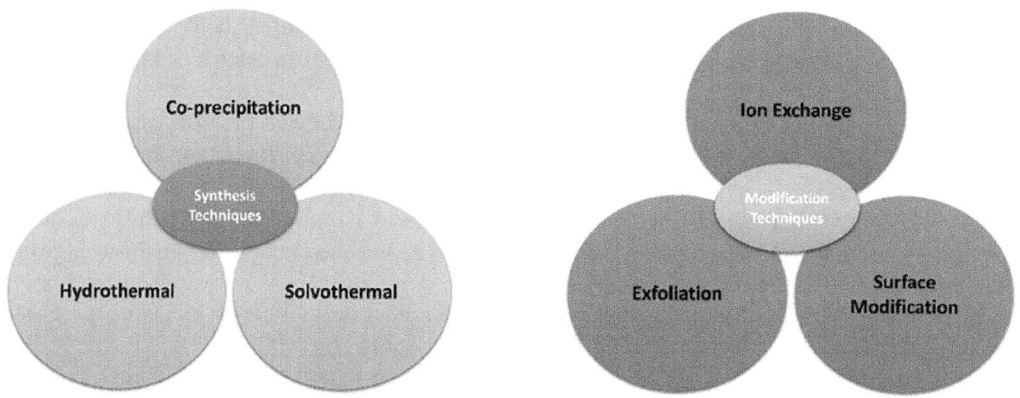

FIGURE 15.2 Synthesis and modification techniques of LDH.

to a wider range of applications. The mentioned unique traits make LDHs a favoured solution in the field. They are also relatively easy to synthesize in the laboratory using methods such as co-precipitation, hydrothermal, ion exchange, and sol–gel techniques. A significant challenge in the synthesis of pristine LDHs is the limited number of functional groups present, which severely hampers their removal ability. Efforts to enhance the removal ability of pristine LDHs have led researchers to investigate a range of modification techniques, including surface modification, calcination, anion intercalation, and the preparation of composite materials. These explorations have resulted in substantial progress in the creation of multi-functional LDHs as highly effective adsorbents.

15.3 SYNTHESIS AND MODIFICATION TECHNIQUES OF LDHS

LDHs are a class of layered materials that have a unique combination of properties such as high thermal stability, hydrophilicity, and chemical reactivity, making them suitable for a wide range of applications. Synthesis and modification techniques of LDHs play a crucial role in tailoring their properties for specific applications (Figure 15.2).

15.3.1 Synthesis Techniques

15.3.1.1 Co-precipitation

This is the most commonly used method for the synthesis of LDHs. It involves the simultaneous precipitation of metal cations and anions in an aqueous solution, followed by the formation of

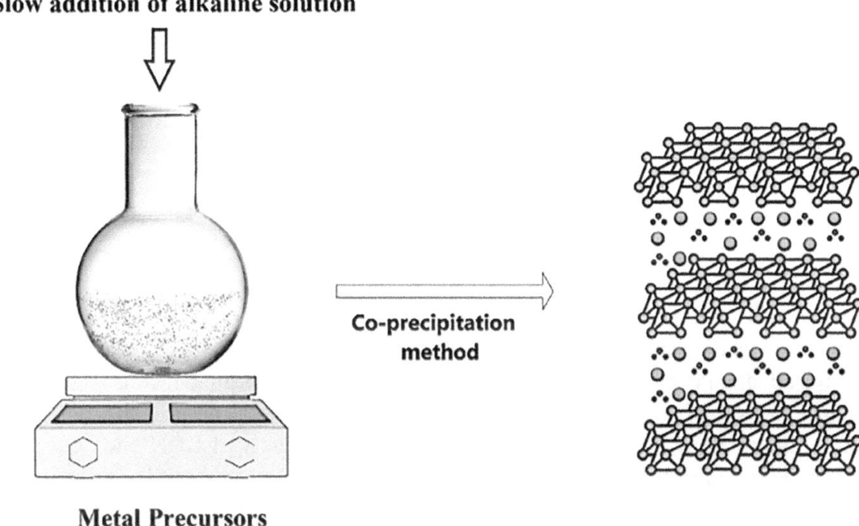

FIGURE 15.3 Co-precipitation synthesis of LDH.

layered structures through ion exchange and ageing processes [31]. Here, metal cations and anions are simultaneously precipitated from an aqueous solution to form the LDH compound. The metal cations and anions are typically precipitated by adding a suitable base, such as ammonia, to the solution to raise the pH. The pH is adjusted until the desired metal hydroxide species form, and the mixture is then cooled to cause precipitation. The precipitate is then washed, filtered, and dried to produce the final LDH product. The co-precipitation method allows for the facile synthesis of LDHs with high purity and control over the interlayer spacing between the metal hydroxide layers and the interlayer anions. This control over the interlayer spacing is particularly important for controlling the band gap of the material, which affects its light absorption capacity and photocatalytic properties. Furthermore, the co-precipitation method allows for the incorporation of other materials into the LDHs structure, such as metal oxides or carbon-based nanomaterials, to form composite or heterostructures. These hybrid structures can exhibit improved photocatalytic activity compared to pure LDHs due to their enhanced surface area, improved electronic properties, and synergistic effects between different materials. Figure 15.3 illustrates the schematic representation of LDH co-precipitation synthesis.

Recently, Yadav and his team performed the synthesis of nitrate intercalated MgAl-LDH nanopowder in the co-precipitation method in the presence of refluxing conditions for the adsorption of methyl orange dye from the aqueous environment [32]. This method involves the precipitation of metal hydroxides and the subsequent formation of the LDH structure. The addition of nitrate to the reaction mixture leads to the intercalation of nitrate anions into the LDH structure, resulting in the formation of nitrate-intercalated MgAl-LDH nanopowder. A new method for synthesizing Fe_2O_3@ SiO_2@MgAlCl-LDH materials was proposed that involves a single step of co-precipitation under atmospheric conditions [33]. The method allows for the insertion of chlorine ions into the Fe_2O_3@ SiO_2@MgAlCl-LDH layers. The development of an efficient halogen-doped (g-C_3N_4/LDHs) hybrid through the use of a controlled co-precipitation method was proposed by Hu and his team for the treatment of organic pollutants in water bodies [34]. The hybrid material has improved properties compared to either g-C_3N_4 or LDH alone, and the co-precipitation method allows for control over the formation of the hybrid, leading to consistent and reproducible results. Koilraj and his team fabricated a novel 2D/2D GO/LDH for the sorption of radionuclides from water medium [35]. The surface functional group of GO played a significant role in eliminating positively charged radioactive particles known as cationic radionuclides through interaction. The increased ability of the GO/LDH composite to absorb is a result of the collaborative effect between the adsorbents (GO and LDH) and the substances being absorbed. Overall, the co-precipitation method is simple and cost-effective, making it a popular method for synthesizing LDHs that can be employed for wastewater treatment.

15.3.1.2 Hydrothermal Synthesis

Hydrothermal synthesis is a method for preparing LDHs through the reaction of metal cations and anions in an aqueous solution under high temperature and pressure conditions. This method involves the use of a high-pressure reactor, such as an autoclave, where the metal cations and anions are dissolved in a solution and subjected to elevated temperatures and pressures for a set period [36]. The conditions within the reactor promote the formation of the LDH compound, which precipitates from the solution as the pressure and temperature are lowered.

The main advantage of the hydrothermal synthesis method is that it allows for the preparation of high-quality LDHs with precise control over their crystal structure, size, and morphology. Additionally, the method can be used to synthesize a wide variety of LDH compounds, including those with novel or rare metal cations [37,38]. However, the hydrothermal synthesis method can also be more complex and time-consuming compared to other methods, such as co-precipitation, and requires specialized equipment and conditions [39]. Furthermore, the reaction conditions can sometimes result in the formation of impurities, which may need to be removed through additional processing steps. The resulting LDHs have a 2D layered structure, similar to those obtained by co-precipitation, and their physicochemical and electronic properties are also influenced by this structure. For pure LDHs synthesized by the hydrothermal method, the 2D configuration can lead to a highly crystalline structure with a high degree of order and uniformity [40]. The interlayer spacing and composition of the interlayer anions can be precisely controlled, which allows for the tuning of the band gap and light absorption capacity of the material. The presence of metal cations in the LDHs structure can also lead to redox reactions that are important for their photocatalytic activity. For LDH-based composite or heterostructures synthesized by the hydrothermal method, the 2D configuration can facilitate the integration of different materials into the LDH structure. The highly crystalline and ordered structure of LDHs obtained by the hydrothermal method also facilitates the formation of heterojunctions between different materials, which can enhance their photocatalytic activity. The construction of a novel 2D NiFe-LDH with ultra-low palladium ion doping through a one-step hydrothermal method was performed by Liu and his team to investigate the purification of water [41]. The method involves the introduction of palladium ions through palladium chloride, which results in a Pd/NiFe-LDH with improved properties compared to traditional nickel iron-LDHs. A novel nanocomposite, montmorillonite@MgAl-LDH, was created by combining montmorillonite with hydrogen (H^+) and chloride (Cl^-) ions for the removal of anionic dyes in water medium which was achieved through a single-step process that involved both an acid-salt treatment and a hydrothermal method [42]. Zhao and his team synthesized a loofah-like nanostructured material by growing cobalt hydroxide fluoride intercalated with nickel manganese LDH on cloth, accomplished through a straightforward two-step hydrothermal process for water pollutant sensing [43]. Therefore, the use of hydrothermal methods for the synthesis of LDH provides the potential for various environmental applications, particularly in the treatment of water.

15.3.1.3 Solvothermal Synthesis

Solvothermal synthesis is a method used to prepare LDHs by utilizing a high-temperature and high-pressure solvent environment. This technique involves the reaction of metal cations and anions in a solvent under specific temperature and pressure conditions to form LDHs [44]. The solvothermal process usually involves the use of a high-boiling solvent, such as ethylene glycol or 1,4-butanediol, to dissolve the metal salts and hydroxide ions, followed by heating at a specific temperature and pressure [45,46]. The reaction conditions are carefully controlled to promote the formation of LDHs with specific compositions, structures, and morphologies. Solvothermal synthesis offers several advantages over other synthesis methods, including the ability to synthesize LDHs with high crystallinity and uniform particle size. It also allows for the synthesis of LDHs with unique compositions and structures that cannot be obtained using other methods. Additionally, solvothermal synthesis is scalable and can be used to produce large quantities of LDHs for industrial applications. Overall, solvothermal synthesis is a valuable technique for the preparation of LDHs, and it provides a versatile platform for the synthesis of high-quality LDH materials with specific compositions and structures.

15.3.2 Modification Techniques

15.3.2.1 Ion Exchange

Ion exchange is a widely used method for the preparation of LDHs by exchanging interlayer anions in a pre-formed LDH material with new anions. This technique involves the immersion of the LDH material in an aqueous solution containing the desired anions, followed by stirring to facilitate the exchange of anions between the LDH material and the solution [47]. The exchanged anions are intercalated into the interlayer space of the LDH structure, leading to the formation of a new LDH material with a different composition. Ion exchange is a versatile and convenient method for the preparation of LDHs with specific compositions and structures. It allows for the easy modification of the interlayer anions in a pre-formed LDH material, providing a simple and effective way to tailor the properties of the LDH material to meet specific requirements. Additionally, ion exchange can be easily scaled up for large-scale production of LDHs, making it a popular choice for industrial applications. For pure LDHs synthesized via ion exchange, the 2D configuration can lead to a high surface area and porosity, which is beneficial for their photocatalytic activity. The interlayer spacing can also be adjusted by the size of the replaced anions, which can affect the material's band gap and light absorption capacity [48]. Furthermore, the presence of metal cations in the structure can facilitate redox reactions that are critical for their photocatalytic activity. In the case of LDH-based composite or heterostructures synthesized via ion exchange, the 2D structure can enable the incorporation of other materials into the LDH structure. For example, metal nanoparticles, metal oxides, or carbon-based nanomaterials can be introduced through ion exchange to create composite structures with enhanced photocatalytic activity. The 2D layered structure of LDHs also promotes the formation of heterojunctions between different materials, which can increase their photocatalytic activity. Additionally, the electronic structure and band gap of LDHs synthesized by ion exchange can be altered by doping with foreign atoms such as transition metals or nonmetals, which can also impact their photocatalytic activity [42]. The high surface area and porosity of LDHs produced through ion exchange can also aid in the formation of oxygen vacancies, which can serve as active sites for the photocatalytic degradation of organic pollutants. The schematic illustration of ion exchange synthesis of LDH is presented in Figure 15.4.

Su and team investigated the combination of MgAl-LDH and coconut shell biochar resulting in the creation of the composite LDH, which was designed for adsorbing and removing lead and copper from water [49]. Fabrication of a well-organized hybrid material by replacing chloride ions with 2,4,5-trichlorophenoxyacetic herbicide in a layered double hydroxide through ion exchange was studied for the removal of pollutants from wastewater sources [50]. Overall, ion exchange is a valuable technique for the preparation of LDHs, and it provides a flexible platform for the synthesis of LDH materials with specific compositions and structures. The ease and versatility of the ion exchange method make it a popular choice for both academic and industrial research in the field of LDHs.

15.3.2.2 Surface Modification

Surface modification of LDHs is a technique that involves altering the properties of the surface of LDHs to enhance their performance in various applications. This can be achieved by a variety of methods, including chemical modification, physical modification, and biological modification. Chemical modification involves the introduction of functional groups onto the surface of the LDHs,

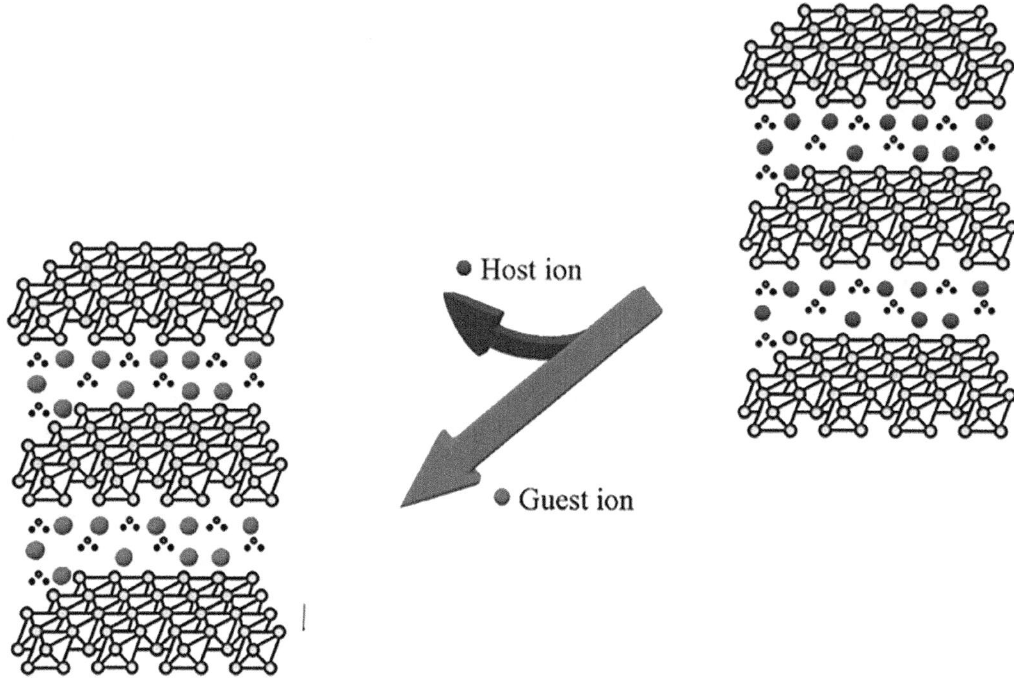

FIGURE 15.4 Ion exchange of LDH.

while physical modification involves changing the physical properties of the LDHs, such as size and shape [51,52]. The biological modification involves the integration of biological molecules onto the surface of the LDHs, allowing for interactions with biological systems [53]. These modifications can result in enhanced stability, improved reactivity, and increased specificity, making LDHs more suitable for use in various applications, including drug delivery, catalysis, and environmental remediation.

One common surface modification technique is the deposition of metal nanoparticles onto the surface of LDHs. This can create composite materials with enhanced photocatalytic activity due to the synergistic effects between the LDHs and metal nanoparticles [54]. The metal nanoparticles can act as electron sinks, facilitating charge separation and reducing recombination, while the LDHs can provide high surface area and redox centres for the photocatalytic reaction. The 2D configuration of the LDHs can further enhance these properties by providing a large surface area for metal nanoparticle deposition. Another surface modification technique is the functionalization of LDHs with organic molecules, which can alter the surface chemistry and introduce new electronic states [55]. For example, the functionalization of LDHs with carboxylic acid groups can increase their surface area and introduce acidic sites that can enhance the photocatalytic degradation of organic pollutants. The 2D configuration of LDHs can also play a role in this process by providing a high surface area for functionalization and ensuring a uniform distribution of the functional groups. In addition to surface modification, the electronic structure and band gap of LDHs can also be tuned by controlling the interlayer spacing and the type of interlayer anions. For example, the interlayer spacing can be controlled by adjusting the synthesis conditions or by intercalating different anions between the layers. This can influence the band gap and light absorption properties of LDHs, ultimately affecting their photocatalytic activity. Dinari and Neamati investigated the surface modification of 2D CaFe-LDH intercalated with citrate anions using a silane coupling agent for the removal of heavy metals from an aqueous solution [19]. The surfaces of low-cost diatomite and bentonite were altered using nickel iron-LDHs to analyze the removal of model organic dye from an aqueous medium [56]. In conclusion, surface modification of 2D LDHs is a versatile approach to improve their performance and versatility for various applications. The choice of modification method depends on the desired properties and the intended application.

15.3.2.3 Exfoliation

Exfoliation of LDHs is a process by which the layered structure of LDHs is separated into individual layers, resulting in a significant increase in the surface area and reactivity of the material [57]. This technique has been widely studied in the field of materials science and is used in various applications, including catalysis, drug delivery, and water treatment. The most common method of exfoliating LDHs is mechanical exfoliation, which involves grinding the material in the presence of a solvent to produce a dispersion of LDH nanosheets [58]. Another method is chemical exfoliation, which uses chemical agents to dissolve the interlayer cations and expand the interlayer space, leading to the separation of the layers. It is important to note that the choice of exfoliation method depends on the specific properties of the LDH material and the desired application. For example, mechanical exfoliation is generally preferred for large-scale production, while chemical exfoliation is used for more precise control over the size and thickness of the exfoliated layers.

Sarkar and his team performed the investigation of the photocatalytic degradation of organic pollutants through synchronized in situ exfoliation of Ti-based ZnCr-LDH utilizing a copolymer [59]. Synthesis of a ZnAl-LDH over silver hybrid material via a facile as well as governable exfoliation process was studied for the evaluation of its plasmonic photodegradation of phenol [60]. Yang and his team performed the construction of a hybrid material through the exfoliation and restacking process, which combines NiAl-LDH with copper phthalocyanine to demonstrate the photocatalytic ability in removing Rhodamine 6G [61]. In summary, the exfoliation of LDHs is a crucial step in unlocking the potential of this versatile material for a wide range of applications. Further research is needed to optimize the exfoliation process and to explore new methods for the efficient production of high-quality LDH nanosheets. These are some of the commonly used synthesis and modification techniques for LDHs. The choice of technique depends on the desired properties of the final material and the application for which it will be used.

15.4 APPLICATIONS IN WATER TREATMENT

LDHs have gained significant attention in the field of water treatment due to their excellent performance in removing various contaminants from water. In this section, we will discuss the applications of LDHs in water treatment which is schematically represented in Figure 15.5. LDHs have been used for the removal of heavy metals, organic pollutants, and other harmful substances from water. Additionally, LDHs can also be used as adsorbents and catalysts in water treatment processes.

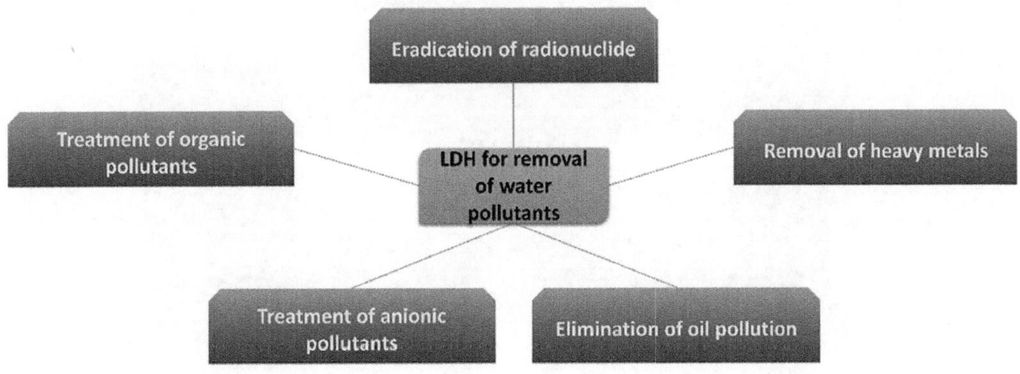

FIGURE 15.5 LDH employed for different water treatment.

The unique properties of LDHs make them promising materials for improving water quality and addressing water scarcity issues.

15.4.1 Water Pollutant Removal through Adsorption

15.4.1.1 Eradication of Radionuclide

Generating nuclear energy is an essential aspect of a cost-effective and sustainable energy generation plan. Nonetheless, advancing nuclear energy generation methods necessitates establishing efficient management procedures for the radioactive waste generated throughout the nuclear fuel cycle. Radioactive elements are among the most hazardous pollutants in the environment due to their extended half-life, elevated radioactivity, and potential harm to living organisms. 2D LDH-based materials have been a subject of significant interest for many years due to their adaptable composition and structure. Additionally, 2D LDHs and their derivatives have found extensive application in water purification due to their unique structure and exceptional ion exchange capacity. The initial report on the removal of radionuclides using 2D LDH materials dates back to the 1990s, when Kang et al. [62] presented a revolutionary study that demonstrated the capture of sodium pertechnetate using 2D LDHs through an ion exchange mechanism. Since then, the enhancement of radionuclide elimination and solidification using 2D LDHs and their derivatives has been a key focus of research in this field, and notable advancements have been made.

2D LDHs are capable of eliminating or immobilizing radioactive pollutants primarily through the mechanisms of surface adsorption or anionic clay intercalation [63]. Surface adsorption is the process in which a nuclide adheres to either the LDH surface or the guest species sites that have been employed to customize the 2D LDH, causing the development of either a firmly established molecular or atomic layer on the surface of the 2D LDH material or a clustering effect on the guest species sectors. During the process of surface adsorption, the properties of either the 2D LDH or the guest species attached to it can instigate the adsorption of the radionuclide. The interlayer anion-exchange mechanism comprises the exchange of anions between the interlayer spaces with the desired nuclide anions. This process is predominantly impacted by the characteristics of the charge-balancing anions present in the interlayer region, the density of layer charge, and the distance between the interlayers [64]. The graphical presentation of radionuclide adsorption on 2D LDH is illustrated in Figure 15.6.

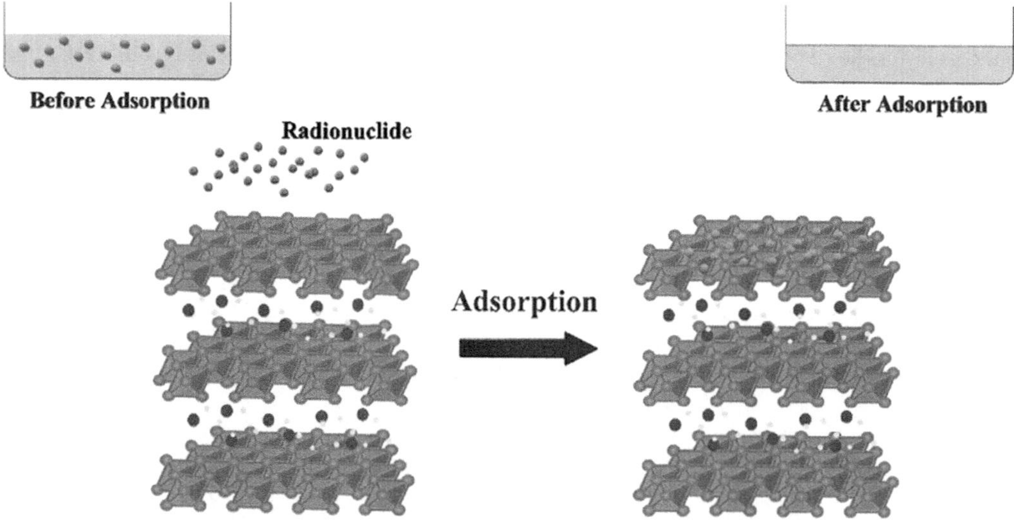

FIGURE 15.6 Adsorption mechanism of radionuclide on 2D LDH.

Gu et al. engineered a unique flower-shaped structure consisting of ultrathin 2D MXene/MgAl-LDH and examined its effectiveness in removing uranium and europium [65]. The technique of in situ anchoring of 2D LDHs nanosheets offered a practical solution to tackle the limitations associated with restricted layer space, inadequate action sites, and low stability of 2D MXene sheets in their original form. Adsorption experiments indicated that the synthesized 2D LDH composite can capture uranium and europium due to its high porosity and various functional groups as well as demonstrated remarkable uranium (241.0 mg g^{-1}) and europium (97.1 mg g^{-1}) trapping abilities. Similarly, a new material called Prussian blue attached to 2D CoCr-LDH was developed to effectively eliminate both cesium and iodine from radioactive solutions at the same time [66]. The primary mechanism for eliminating cesium involved redox-coupled transport of alkali metal ions, which resulted in the formation of cesium tetracobalt (II) triferricyanide species at the surface of LDH. In contrast, the removal of iodine occurred mainly through ion exchange with intercalated chloride ions, along with the redox impact of cobalt ions. Overall, LDH-based materials are extensively employed for the eradication of radionuclides from the water source and have displayed capable results.

15.4.1.2 Removal of Heavy Metals

The existence of heavy metal ions in the environment has become a significant global health concern due to their toxic properties and persistence. These heavy metals can originate from various anthropogenic activities, such as wastewater and effluents from different industrial sectors, including plastics, battery manufacturing, mining, paper, electroplating, fertilizer, and steel [67]. Heavy metals like lead, mercury, cadmium, arsenic, and chromium are the most commonly found heavy metal pollutants in the environment. They are hazardous to both human health and the environment, as they can accumulate in living organisms, including humans, and cause adverse effects. These heavy metals enter the food chain and subsequently affect the entire ecosystem [68]. The process of bioaccumulation of heavy metals in living organisms, such as fish and plants, is concerning, as it can result in serious health problems for humans. Chronic exposure to heavy metals can cause neurological, cardiovascular, renal, and hepatic diseases [69]. These effects can occur due to heavy metal-induced oxidative stress, inflammation, and DNA damage. To mitigate the harmful effects of heavy metals, the development of effective methods for their detection and removal from the environment is necessary. Various scientific techniques such as ion exchange, adsorption, and precipitation have been proposed to remove heavy metal pollutants from water and wastewater [68,69].

Research in the field has indicated that LDHs demonstrate a noteworthy capacity for the adsorption of heavy metals [70]. The specific structural properties of LDHs that promote their high adsorption capacity and selectivity for heavy metals are primarily related to their layered structure and the presence of interlayer anions [71]. LDHs have a layered structure consisting of positively charged metal hydroxide layers and negatively charged interlayer anions, which allows for the formation of a wide range of metal hydroxide layers with varying cationic and anionic compositions [16]. This layered structure provides a large surface area and numerous exchangeable sites for the adsorption of heavy metal ions. The adsorption capacity and selectivity of LDHs are further enhanced by the presence of exchangeable anions in the interlayer region. These exchangeable anions can be readily exchanged with other anions in solution, allowing for the selective adsorption of specific heavy metal ions. The selectivity of LDHs can also be influenced by the size and charge of the metal ions, as well as the chemical composition of the interlayer anions. In addition to their layered structure and exchangeable anions, the high surface area and abundance of hydroxyl groups on the surface of LDHs also contribute to their adsorption performance. The hydroxyl groups can form hydrogen bonds with heavy metal ions, enhancing their adsorption capacity. This property is often measured by assessing the adsorption capacity and selectivity of these materials, indicating that they can effectively remove toxic metals from their environment [72]. This is particularly relevant in light of increasingly stringent environmental regulations, which necessitate the removal of toxic contaminants from various sources. The exceptional adsorption performance of functionalized LDHs towards heavy metals is primarily attributed to their unique layered structure and surface chemistry [42]. These materials possess a high surface area and are capable of undergoing ion exchange reactions, which allow them to selectively and effectively adsorb heavy metals from aqueous solutions [73]. Zhu and his team investigated a novel layered cationic material functionalized with phosphonate that was synthesized through an intercalation method for the efficient removal of perilous metals from an aqueous environment [74]. The modified LDH demonstrated remarkable chelation adsorption capabilities, particularly for zinc (281.36 mg g^{-1}) and iron (206.03 mg g^{-1}). The adsorption process was found to be chemisorption, involving monolayer interaction as suggested by the fitting results of the adsorption isotherm models. The obtained results suggested that the functionalized layered cationic framework material loaded with phosphonate may be a promising candidate for removing hazardous metals from wastewater. Similarly, Guan and his team synthesized FeMg-LDH@bentonite using an easy co-precipitation technique in situ which is capable of eliminating heavy metals from an aqueous environment [75]. The researchers used the consignment adsorption technique to investigate how various factors affect the ability of FeMg-LDH@bentonite to remove heavy metals. The study found that the adsorption of heavy metals onto FeMg-LDH@bentonite was best described by the pseudo-second-order scheme, and the Langmuir model was effective in simulating isotherms. The maximal adsorption capacity of cadmium and lead was found to be 510.2 and 1397.62 mg g^{-1}, respectively, which is higher than that of conventional adsorbents. Overall, the exceptional adsorption properties of functionalized LDHs make them a promising candidate for the development of effective heavy metal adsorbents.

15.4.1.3 Treatment of Organic Pollutants

Organic pollutants are highly diverse and can pose various risks to human health and the environment. They are predominantly released into the environment as a result of industrial processes and ultimately find their way into different bodies of water. Dyes, pharmaceuticals, and persistent organic pollutants are among the well-known types of organic pollutants [76]. Certain organic pollutants are highly soluble in fats and can build up in tissues, causing bioaccumulation. This accumulation can have harmful effects on the health of humans, wildlife, and other organisms [77]. Because of their long half-lives, these pollutants can persist in the environment for prolonged periods, posing a serious threat to environmental safety.

In recent years, there has been growing interest in exploring the use of LDHs and their composites for removing different types of harmful organic contaminants from wastewater. This is because these materials have a large surface area, excellent capability for anion exchange, customizable properties, and low toxicity. LDHs have been utilized for the elimination or management of various organic pollutants, such as dyes, drugs, and natural organic substances [21,78,79]. Ge et al. developed a flexible hybrid material by coating carbon dots on a MgAl-LDH [80]. To achieve this, they modified the surface of MgAl-LDH with sodium dodecylbenzene sulfonate, which served as a carbon precursor. They then dehydrated the hydrotalcite at 450°C under nitrogen flow, causing ligand carbonization to take place. The composite of LDH exhibited excellent adsorption capabilities for three different dyes, with the highest adsorption capacity ranging from 3628.9 to 5174.1 mg g^{-1}. Therefore, the amazing characteristics of LDH can help in the eradication of various classes of organic pollutants present in water bodies.

15.4.1.4 Treatment of Anionic Pollutants

The presence of anionic pollutants in water is a significant problem worldwide. This type of pollution includes substances like phosphate, fluorides, nitrite, sulfides, nitrate, chloride, and arsenate, which can have harmful effects on both humans and the environment [81]. Unlike metal ions, eliminating anionic pollutants from water is difficult due to their unique physicochemical properties. One proposed solution is ion exchange, and LDHs have shown promise as an effective adsorbent for anions, making them a potentially suitable solution for removing anionic contaminants from water [68,82,83].

Vu and his team created beads using MgAl-LDH and sodium alginate [84]. These beads were made through a process that involved combining MgAl-LDHs and sodium alginate in a crosslinking reaction. The team then tested the beads to see how effective they were at removing nitrates from both synthetic solution and groundwater. Finally, they also assessed the beads' ability to slowly release nitrates. The impact of pH on the adsorption of nitrate was not significant across a broad range of pH values from 4 to 10. The maximum capacity for nitrate adsorption using the Langmuir model was approximately 20–22 mg of nitrate per gram. However, the presence of other anions alongside the nitrate may have a negative effect on the removal of nitrate. Besharatlou et al. synthesized LDH adsorbents of manganese aluminium and copper aluminium using co-precipitation, while copper aluminium oxides were prepared by calcination of CuAl-LDH at 500°C for sulphate removal from aqueous media [85]. According to the findings, the adsorption capacity of calcinated LDH towards sulphate is superior to that of pristine LDH. Hybrid nanocomposites were produced by spawning an inorganic system in the occurrence of organic amalgams to enhance the selectivity and capacity of calcinated LDH. By utilizing the co-precipitation method, Khalil and his team synthesized composites of MgAl-LDH doped on the activated carbon surface with different total metal loading of magnesium and aluminium [86]. The impact of the magnesium and aluminium total metal loading on activated carbon was studied to assess its effectiveness in eliminating phosphate ions from the water medium. The results of the adsorption experiments revealed that the adsorption capacity for phosphate considerably increased as the magnesium and aluminium metal loading on the activated carbon surface increased. The composite (15 wt% magnesium and aluminium) exhibited the maximum Langmuir adsorption amount of phosphate, which was 337.2 mg of phosphate per unit mass of the adsorbent, at 1 g L^{-1} of adsorbent, pH approximately 6, and 22°C. Therefore, LDHs have demonstrated the potential to effectively remove negatively charged pollutants from water.

15.4.2 Elimination of Oil Pollution

The worldwide output of discarded oil is approximated to be roughly one million tons annually owing to diverse actions executed by residential, lodging, manufacturing, and vehicular domains [87,88]. Regrettably, an overwhelming majority of such oil wastes, approximately 95%, are either burned, discarded in dumps, or discharged into water sources. As reported by the American Petroleum Institute, a mere litre of waste oil can pollute up to one million litres of potable water, emphasizing the gravity of the disposal and management of waste oil as a critical global issue [89]. To remove oil from water, superhydrophobic and superoleophilic materials are typically utilized. Functionalized LDHs have emerged as a promising option for the remediation of oil pollution. While the surface of LDHs is naturally hydrophilic due to the presence of copious surface hydroxyl clusters, it can be effortlessly altered using various chemical modification approaches [72]. For instance, the introduction of certain organic species onto the surface of LDHs can render them hydrophobic.

Yue and her team successfully created a sturdy and inexpensive adaptable membrane by utilizing the in situ growth method and hydrophobic adjustment, using LDH nanosheets on cellulose assistance [90]. To create the membrane, they first used the in situ growth technique to fabricate a rough surface made of ZnAl-LDH nanosheets. By attaching a silane link agent to the membrane, the authors were able to achieve a superhydrophobic surface. The membrane demonstrated the ability to exhibit both superhydrophobic and superoleophilic properties at the same time. The membrane displayed excellent separation performance with a high separation efficiency of over 94.4%, good recyclability, and chemical durability. It also showed outstanding separation capabilities for emulsified water-in-oil mixtures stabilized by surfactants with a separation efficiency of less than 25 ppm and a good flux of 500 $Lm^{-2}h^{-1}$, without requiring additional energy. The team led by Liu was able to create a textile that exhibits both superhydrophobic and superoleophilic properties, which can be effectively used for separating oil and water as well as selectively adsorbing oil [91]. The team created this textile by modifying a commercially available textile with LDH microscopic crystals and nonpolar molecules. Additional experiments conducted by the team showed that the textile they created is not only suitable for use as a membrane material in effectively separating oil and water mixtures with a high separation efficiency of over 97% but can also be used as a bag to selectively adsorb oil from water. Zhang and his team were able to successfully fabricate stainless steel mesh that has been modified to exhibit both superhydrophobic and superoleophilic properties coated with a novel NiAl-LDHs material, which demonstrated excellent oil–water separation capability [92]. The mesh that was created showed ideal superhydrophobic properties, with a strong water-repellent property and excellent droplet mobility, resulting in a self-cleaning surface. Moreover, the team also investigated the separation efficiency of the coated mesh for a range of oil/water mixtures, and the results demonstrated an efficiency of up to 98% in a large proportion of the tested oil and water mixtures. Thus, based on the aforementioned recent studies, LDH-based materials have the potential to be a highly effective option for removing oil from bodies of water.

15.4.3 LDHs Photocatalysts for Degradation of Water Organic Pollutants

2D LDHs have gained significant interest as photocatalysts due to their unique layered structure and tunable physicochemical properties. One of the key factors that contribute to their enhanced photocatalytic performance is the modulation of their electronic, optical, and band gap properties. In terms of electronic properties, LDHs exhibit high electron mobility, which allows for efficient charge transport and separation, leading to improved photocatalytic activity [93]. The modulation of electronic properties can be achieved through various methods, such as doping with transition metals, heteroatom doping, and surface modification. In addition to electronic properties, the optical properties of LDHs can also be modulated for enhanced photocatalytic activity. The absorption spectrum of LDHs can be tuned by adjusting the composition, morphology, and size of the material [94]. This results in increased light absorption and utilization efficiency, leading to improved photocatalytic performance.

The band gap of LDHs can also be modulated for enhanced photocatalytic activity. The band gap is an important factor that determines the energy required for electron transfer and affects the overall photocatalytic performance [95]. Moreover, 2D LDHs can be interfaced with 0D, 2D, and 3D composites to further enhance their photocatalytic performance. For example, the interface of 2D LDHs with 0D metal particles or carbon quantum dots (QDs) can improve the charge transfer and separation due to the formation of Type II heterojunctions. Figure 15.7 shows the schematic of a Type II heterojunction formed between 2D LDH composites. Finally, the interface of 2D LDHs with

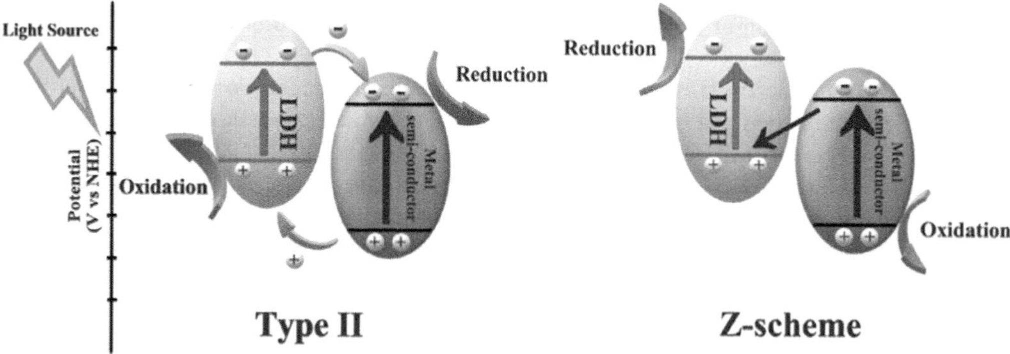

FIGURE 15.7 Type II and Z-scheme illustration of 2D LDH heterojunction photocatalyst.

3D composites such as TiO_2 can also improve their photocatalytic activity through the formation of Z-scheme or S-scheme interfaces. Figure 15.7 shows the schematic of a Z-scheme interface formed between 2D LDH composites.

A Z-scheme heterojunction made of 2D NiFe-LDH and bismuth oxide iodine was developed by Zhu and his team for the photo-Fenton oxidation of tetracycline via the solvothermal method [96]. The Z-scheme heterojunction is a type of photocatalytic system that mimics the natural photosynthesis process in plants, where two different semiconductor materials are used to create a cascade of electron transfer reactions. In this case, 2D NiFe-LDH and bismuth oxide iodine are used to facilitate the photo-Fenton oxidation of tetracycline, which is a widely used antibiotic that can cause environmental pollution. A composite material made up of g-C_3N_4/Ce-MgAl-LDH 2D composite was produced using a simple one-step solvothermal process for photocatalytic eradication of organic dye [97]. The composite material made up of g-C_3N_4/Ce-MgAl-LDH has shown superior photocatalytic activity compared to its individual components. This is due to the synergistic effect between g-C_3N_4 and Ce-MgAl-LDH, which enhances the efficiency of the photocatalytic reaction. Zhu and his team engineered Ag-coated ZnTi-LDH photocatalyst to degrade the model Rhodamine B dye from an aqueous medium [98]. The outcomes acquired from photoluminescence (PL) spectroscopic measurement and transient photocurrent (I–V) analysis both demonstrated the occurrence of Schottky interfaces between LDH and Ag nanoparticles. The electron spin resonance (ESR) demonstrated that hydroxyl radicals were the dominant reactive species in the process of photodegradation. The better photocatalytic capability of the composite is due to the surface plasmon resonance (SPR) effect of Ag nanoparticles when exposed to visible light, as well as the existence of Schottky interfaces between LDH and Ag nanoparticles.

Asif and his team created a CoAl-LDH nanosheet membrane with numerous active sites, which they hoped to use for the catalytic degradation of micropollutants. They combined the LDH nanosheets with peroxymonosulfate, a powerful oxidizing agent that can break down a wide range of organic pollutants. The CoAl-LDH nanosheet membrane would act as a catalyst, activating peroxymonosulfate and providing the active sites necessary for the reaction to occur [99]. The activation efficiency of peroxymonosulfate by the LDH membrane was found to be excellent, with 87.8% removal of the probe chemical (ranitidine) at 0.2 mM peroxymonosulfate. This percentage was higher than that achieved by the conventional LDH (5–20 mg)/ peroxymonosulfate (0.2 mM) system, which typically achieved a removal rate of 37%–44%. The LDH membrane used in this study had a mass of 0.63 mg.

15.5 CONCLUSION AND FUTURE PERSPECTIVES

In conclusion, 2D LDH nanomaterials have shown great potential in the removal of water pollutants. LDHs possess a unique layered structure that allows for the efficient adsorption and removal of various types of contaminants from water, including heavy metals, radionuclides, organic pollutants, anionic pollutants, and oil pollution. Moreover, the synthesis and modification techniques of LDHs have been extensively studied, enabling the customization of LDHs to target specific pollutants in water. Future perspectives for the use of 2D LDH nanomaterials in water treatment are promising. One area of research is the optimization of LDH properties, such as their size, composition, and surface chemistry, to enhance their adsorption capacity and selectivity towards different pollutants. Another direction is the development of new LDHs-based materials with improved stability and reusability, which would reduce the cost and environmental impact of water treatment processes. Additionally, there is a need to study the potential toxicological effects of LDHs on the environment and human health to ensure their safe use. Overall, 2D LDH nanomaterials have the potential to contribute to the development of sustainable and efficient water treatment technologies, which are crucial for addressing the growing challenges of water pollution and scarcity.

REFERENCES

1. Liu, Y., Pang, H., Wang, X., Yu, S., Chen, Z., Zhang, P., Chen, L., Song, G., Saleh, A.N., Omar, R.S., Wang, X. Zeolitic imidazolate framework-based nanomaterials for the capture of heavy metal ions and radionuclides: A review. *Chemical Engineering Journal*, 406 (2021) 127139. https://doi.org/10.1016/J.CEJ.2020.127139.
2. Rathi, B.S., Kumar, P.S. Application of adsorption process for effective removal of emerging contaminants from water and wastewater. *Environmental Pollution*, 280 (2021) 116995. https://doi.org/10.1016/J.ENVPOL.2021.116995
3. Deng, Y., Zhao, R. Advanced oxidation processes (AOPs) in wastewater treatment. *Current Pollution Reports*, 1(3) (2015) 167–176. https://doi.org/10.1007/S40726-015-0015-Z.
4. Rashed, I.G.A.A., Afify, H.A., Ahmed, A.E.M., Ayoub, M.A.E.S. Optimization of chemical precipitation to improve the primary treatment of wastewater. *New Pub: Balaban*, 51(37–39) (2013) 7048–7056. https://doi.org/10.1080/19443994.2013.792147.
5. Levchuk, I., Rueda Márquez, J.J., Sillanpää, M. Removal of natural organic matter (NOM) from water by ion exchange – A review. *Chemosphere*, 192 (2018) 90–104. https://doi.org/10.1016/J.CHEMOSPHERE.2017.10.101.
6. Alam, G., Ihsanullah, I., Naushad, M., Sillanpää, M. Applications of artificial intelligence in water treatment for optimization and automation of adsorption processes: Recent advances and prospects. *Chemical Engineering Journal*, 427 (2022) 130011. https://doi.org/10.1016/J.CEJ.2021.130011.
7. Iwuozor, K.O., Emenike, E.C., Aniagor, C.O., Iwuchukwu, F.U., Ibitogbe, E.M., Okikiola, T.B., Omuku, P.E., Adeniyi, A.G. Removal of pollutants from aqueous media using cow dung-based adsorbents. *Current Research in Green and Sustainable Chemistry*, 5 (2022). 100300. https://doi.org/10.1016/J.CRGSC.2022.100300.
8. Rasheed, T. Covalent organic frameworks as promising adsorbent paradigm for environmental pollutants from aqueous matrices: Perspective and challenges. *Science of the Total Environment*, 833 (2022) 155279. https://doi.org/10.1016/J.SCITOTENV.2022.155279.
9. Okoro, H.K., Pandey, S., Ogunkunle, C.O., Ngila, C.J., Zvinowanda, C., Jimoh, I., Lawal, I.A., Orosun, M.M., Adeniyi, A.G. Nanomaterial-based biosorbents: Adsorbent for efficient removal of selected organic pollutants from industrial wastewater. *Emerging Contaminants*, 8 (2022) 46–58. https://doi.org/10.1016/J.EMCON.2021.12.005.
10. Awang, N.A., Wan Salleh, W.N., Aziz, F., Yusof, N., Ismail, A.F. A review on preparation, surface enhancement and adsorption mechanism of biochar-supported nano zero-valent iron adsorbent for hazardous heavy metals. *Journal of Chemical Technology & Biotechnology*, 98(1) (2023) 22–44. https://doi.org/10.1002/JCTB.7182.
11. Fu, M., Deng, X., Wang, S. Q., Yang, F., Lin, L. C., Zaworotko, M. J., & Dong, Y.. Scalable robust nanoporous Zr-based MOF adsorbent with high-capacity for sustainable water purification. *Separation and Purification Technology*, 288 (2022) 120620. https://doi.org/10.1016/J.SEPPUR.2022.120620.
12. Husien, S., El-taweel, R. M., Salim, A. I., Fahim, I. S., Said, L. A., & Radwan, A. G. Review of activated carbon adsorbent material for textile dyes removal: Preparation, and modelling. *Current Research in Green and Sustainable Chemistry*, 5 (2022) 100325. https://doi.org/10.1016/J.CRGSC.2022.100325.
13. Hacıosmanoğlu, G. G., Mejías, C., Martín, J., Santos, J. L., Aparicio, I., & Alonso, E.. Antibiotic adsorption by natural and modified clay minerals as designer adsorbents for wastewater treatment: A comprehensive review. *Journal of Environmental Management*, 317 (2022) 115397. https://doi.org/10.1016/J.JENVMAN.2022.115397.
14. Anderson, A., Anbarasu, A., Pasupuleti, R. R., Manigandan, S., Praveenkumar, T. R., & Aravind Kumar, J. Treatment of heavy metals containing wastewater using biodegradable adsorbents: A review of mechanism and future trends. *Chemosphere*, 295 (2022) 133724. https://doi.org/10.1016/J.CHEMOSPHERE.2022.133724.
15. Gupta, V. K., & Suhas. Application of low-cost adsorbents for dye removal – A review. *Journal of Environmental Management*, 90(8) (2009) 2313–2342. https://doi.org/10.1016/J.JENVMAN.2008.11.017.
16. Jijoe, P. S., Yashas, S. R., & Shivaraju, H. P. Fundamentals, synthesis, characterization and environmental applications of layered double hydroxides: A review. *Environmental Chemistry Letters*, 19(3) (2021). https://doi.org/10.1007/s10311-021-01200-3.
17. el Rouby, W. M. A., El-Dek, S. I., Goher, M. E., & Noaemy, S. G. Efficient water decontamination using layered double hydroxide beads nanocomposites. *Environmental Science and Pollution Research*, 27(16) (2020) 18985–19003. https://doi.org/10.1007/s11356-018-3257-7.

18. Das, J., Patra, B. S., Baliarsingh, N., & Parida, K. M. Adsorption of phosphate by layered double hydroxides in aqueous solutions. *Applied Clay Science*, 32(3–4) (2006) 252–260. https://doi.org/10.1016/j.clay.2006.02.005.
19. Dinari, M., & Neamati, S. Surface modified layered double hydroxide/polyaniline nanocomposites: Synthesis, characterization and Pb^{2+} removal. *Colloids and Surfaces A: Physicochemical and Engineering Aspects*, 589 (2020) 124438. https://doi.org/10.1016/J.COLSURFA.2020.124438.
20. Prabagar, J. S., Sneha, Y., Tenzin, T., Shahmoradi, B., Rtimi, S., Wantala, K., Jenkins, D., & Shivaraju, H. P. Photocatalytic transfer of aqueous nitrogen into ammonia using nickel-titanium-layered double hydroxide. *Environmental Science and Pollution Research* (2022) 1–11. https://doi.org/10.1007/S11356-022-24726-7.
21. Morimoto, K., Tamura, K., Iyi, N., Ye, J., & Yamada, H. Adsorption and photodegradation properties of anionic dyes by layered double hydroxides. *Journal of Physics and Chemistry of Solids*, 72(9) (2011) 1037–1045. https://doi.org/10.1016/j.jpcs.2011.05.018.
22. Khenifi, A., Derriche, Z., Mousty, C., Prévot, V., & Forano, C. Adsorption of glyphosate and glufosinate by Ni_2AlNO_3 layered double hydroxide. *Applied Clay Science*, 47(3–4) (2010) 362–371. https://doi.org/10.1016/j.clay.2009.11.055.
23. Pang, H., Wu, Y., Wang, X., Hu, B., & Wang, X. Recent advances in composites of graphene and layered double hydroxides for water remediation: A review. *Chemistry – An Asian Journal*, 14(15) (2019) 2542–2552. https://doi.org/10.1002/ASIA.201900493.
24. Tao, Q., He, H., Li, T., Frost, R. L., Zhang, D., & He, Z. Tailoring surface properties and structure of layered double hydroxides using silanes with different number of functional groups. *Journal of Solid State Chemistry*, 213 (2014) 176–181. https://doi.org/10.1016/J.JSSC.2014.02.032.
25. He, J., Wei, M., Li, B., Kang, Y., Evans, D. G., & Duan, X. Preparation of layered double hydroxides. *Structure and Bonding*, 119 (2005) 89–119. https://doi.org/10.1007/430_006.
26. Laipan, M., Yu, J., Zhu, R., Zhu, J., Smith, A. T., He, H., O'Hare, D., & Sun, L.. Functionalized layered double hydroxides for innovative applications. *Materials Horizons*, 7(3) (2020) 715–745. https://doi.org/10.1039/C9MH01494B.
27. Li, F., & Duan, X. Applications of layered double hydroxides. *Structure and Bonding*, 119 (2005) 193–223. https://doi.org/10.1007/430_007.
28. Yang, Y., Zhao, X., Zhu, Y., & Zhang, F. Transformation mechanism of magnesium and aluminum precursor solution into crystallites of layered double hydroxide. *Chemistry of Materials*, 24(1) (2012) 81–87. https://doi.org/10.1021/CM201936B.
29. Bukhtiyarova, M. v. A review on effect of synthesis conditions on the formation of layered double hydroxides. *Journal of Solid State Chemistry*, 269 (2019) 494–506. https://doi.org/10.1016/J.JSSC.2018.10.018.
30. Feng, J., He, Y., Liu, Y., Du, Y., & Li, D. Supported catalysts based on layered double hydroxides for catalytic oxidation and hydrogenation: general functionality and promising application prospects. *Chemical Society Reviews*, 44(15) (2015) 5291–5319. https://doi.org/10.1039/C5CS00268K.
31. Theiss, F. L., Ayoko, G. A., & Frost, R. L. Synthesis of layered double hydroxides containing Mg^{2+}, Zn^{2+}, Ca^{2+} and Al^{3+} layer cations by co-precipitation methods - A review. *Applied Surface Science*, 383 (2016) 200–213. https://doi.org/10.1016/j.apsusc.2016.04.150.
32. Yadav, B. S., & Dasgupta, S. Effect of time, pH, and temperature on kinetics for adsorption of methyl orange dye into the modified nitrate intercalated MgAl LDH adsorbent. *Inorganic Chemistry Communications*, 137 (2022) 109203. https://doi.org/10.1016/J.INOCHE.2022.109203.
33. Chengqian, F., Yimin, D., Ling, C., Zhiheng, W., Qi, L., Yaqi, L., Ling, C., Bo, L., Yue-Fei, Z., Yan, L., & Li, W. One-step coprecipitation synthesis of Cl− intercalated Fe_3O_4@SiO_2 @MgAl LDH nanocomposites with excellent adsorption performance toward three dyes. *Separation and Purification Technology*, 295 (2022) 121227. https://doi.org/10.1016/J.SEPPUR.2022.121227.
34. Hu, J., Sun, C., Wu, L., Zhao, G., Liu, H., & Jiao, F. Halogen doped g-C_3N_4/ZnAl-LDH hybrid as a Z-scheme photocatalyst for efficient degradation for tetracycline in seawater. *Separation and Purification Technology*, 309 (2023) 123047. https://doi.org/10.1016/J.SEPPUR.2022.123047.
35. Koilraj, P., & Sasaki, K. 2D/2D graphene oxide-layered double hydroxide nanocomposite for the immobilization of different radionuclides. In *2D Functional Nanomaterials* (2021) 21–30. https://doi.org/10.1002/9783527823963.ch2.
36. Liu, L., Deng, Q., White, P., Dong, S., Cole, I. S., Dong, J., & Chen, X.-B. Hydrothermally prepared layered double hydroxide coatings for corrosion protection of Mg alloys – a critical review. *Corrosion Communications*, 8 (2022) 40–48. https://doi.org/10.1016/J.CORCOM.2022.07.001.

37. Dang Van, C., Kim, S., Kim, M., & Lee, M. H. (2023). Effect of rare-earth element doping on NiFe-layered double hydroxides for water oxidation at ultrahigh current densities. *ACS Sustainable Chemistry and Engineering*. https://doi.org/10.1021/ACSSUSCHEMENG.2C05060.
38. Wang, L., Li, B., Zhao, X., Chen, C., & Cao, J. Effect of rare earth ions on the properties of composites composed of ethylene vinyl acetate copolymer and layered double hydroxides. *PLOS ONE*, 7(6) (2012) e37781. https://doi.org/10.1371/JOURNAL.PONE.0037781.
39. Duan, M., Liu, S., Jiang, Q., Guo, X., Zhang, J., & Xiong, S. (2022). Recent progress on preparation and applications of layered double hydroxides. *Chinese Chemical Letters*, 33(10), 4428–4436. https://doi.org/10.1016/J.CCLET.2021.12.033.
40. Boppella, R., Choi, C. H., Moon, J., & Ha Kim, D. Spatial charge separation on strongly coupled 2D-hybrid of rGO/La_2Ti_2O7/NiFe-LDH heterostructures for highly efficient noble metal free photocatalytic hydrogen generation. *Applied Catalysis B: Environmental*, 239 (2018) 178–186. https://doi.org/10.1016/j.apcatb.2018.07.063.
41. Liu, D., Liu, J., Xue, B., Zhang, J., Xu, Z., Wang, L., Gao, X., Luo, F., & Li, F. Bifunctional Water Splitting Performance of NiFe LDH Improved by Pd^{2+} Doping. *ChemElectroChem*, 10 (2023) e202201025. https://doi.org/10.1002/CELC.202201025.
42. Dai, X., Jing, C., Li, K., Zhang, X., Song, D., Feng, L., Liu, X., Ding, H., Ran, H., Zhu, K., Dai, N., Yi, S., Rao, J., & Zhang, Y. Enhanced bifunctional adsorption of anionic and cationic pollutants by MgAl LDH nanosheets modified montmorillonite via acid-salt activation. *Applied Clay Science*, 233 (2023) 106815. https://doi.org/10.1016/J.CLAY.2023.106815.
43. Zhao, S., Zhang, Y., Wu, J., Ling, C., Tang, X., Xing, Y., Yu, H., Huang, K., Zou, Z., & Xiong, X. Self-supported loofah-like Co(OH)F@NiMn-LDH hierarchical core–shell nanosheet arrays as efficient electrocatalyst for hydrazine sensing. *Microchemical Journal*, 185 (2023),108255. https://doi.org/10.1016/J.MICROC.2022.108255.
44. Meng, X., Feng, M., Zhang, H., Ma, Z., & Zhang, C. Solvothermal synthesis of cobalt/nickel layered double hydroxides for energy storage devices. *Journal of Alloys and Compounds*, 695 (2017) 3522–3529. https://doi.org/10.1016/J.JALLCOM.2016.11.419.
45. Huang, Q., Chen, Y., Yu, H., Yan, L., Zhang, J., Wang, B., Du, B., & Xing, L. Magnetic graphene oxide/MgAl-layered double hydroxide nanocomposite: One-pot solvothermal synthesis, adsorption performance and mechanisms for Pb^{2+}, Cd^{2+}, and Cu^{2+}. *Chemical Engineering Journal*, 341 (2018) 1–9. https://doi.org/10.1016/J.CEJ.2018.01.156.
46. Whalen, T., Vansaders, B., Vakifahmetoglu, C., Mughal, A., Zlotnikov, E., Cho, S. B., & Riman, R. E. Solvothermal synthesis of acmite conversion coatings on steel. *Journal of the American Ceramic Society*, 96(11) (2013) 3656–3661. https://doi.org/10.1111/JACE.12594.
47. Meyn, M., Beneke, K., & Lagaly, G. Anion-exchange reactions of layered double hydroxides. *Inorganic Chemistry*, 29(26) (1990) 5201–5207. https://doi.org/10.1021/IC00351A013.
48. Hu, H., Wageh, S., Al-Ghamdi, A. A., Yang, S., Tian, Z., Cheng, B., & Ho, W. NiFe-LDH nanosheet/carbon fiber nanocomposite with enhanced anionic dye adsorption performance. *Applied Surface Science*, 511(2020) 145570. https://doi.org/10.1016/j.apsusc.2020.145570.
49. Su, X., Chen, Y., Li, Y., Li, J., Song, W., Li, X., & Yan, L. Enhanced adsorption of aqueous Pb(II) and Cu(II) by biochar loaded with layered double hydroxide: Crucial role of mineral precipitation. *Journal of Molecular Liquids*, 357 (2022) 119083. https://doi.org/10.1016/J.MOLLIQ.2022.119083.
50. Mourid, E. H., Lakraimi, M., & Legrouri, A. Preparation of well-structured hybrid material through ion exchange of chloride by 2,4,5-trichlorophenoxyacetic herbicide in a layered double hydroxide. *Materials Chemistry and Physics*, 278 (2022) 125570. https://doi.org/10.1016/J.MATCHEMPHYS.2021.125570.
51. Awassa, J., Soulé, S., Cornu, D., Ruby, C., & El-Kirat-Chatel, S. Understanding the role of surface interactions in the antibacterial activity of layered double hydroxide nanoparticles by atomic force microscopy. *Nanoscale*, 14(29) (2022) 10335–10348. https://doi.org/10.1039/D2NR02395D.
52. Chen, Y., Wu, L., Yao, W., Chen, Y., Zhong, Z., Ci, W., Wu, J., Xie, Z., Yuan, Y., & Pan, F. A self-healing corrosion protection coating with graphene oxide carrying 8-hydroxyquinoline doped in layered double hydroxide on a micro-arc oxidation coating. *Corrosion Science*, 194 (2022) 109941. https://doi.org/10.1016/J.CORSCI.2021.109941.
53. Yan, W. J., Xu, S., Tian, X. Y., Min, J. J., Liu, S. C., Ding, C. J., Wang, N. L., Hu, Y., Fan, Q. X., Li, J. S., & Zeng, H. Y. Novel bio-based lignosulfonate and $Ni(OH)_2$ nanosheets dual modified layered double hydroxide as an eco-friendly flame retardant for polypropylene. *Colloids and Surfaces A: Physicochemical and Engineering Aspects*, 655 (2022) 130195. https://doi.org/10.1016/J.COLSURFA.2022.130195.
54. Ding, X., Wu, L., Chen, J., Zhang, G., Xie, Z., Sun, D., Jiang, B., Atrens, A., & Pan, F.. Enhanced protective nanoparticle-modified MgAl-LDHs coatings on titanium alloy. *Surface and Coatings Technology*, 404 (2020) 126449. https://doi.org/10.1016/j.surfcoat.2020.126449.

55. Pavlovic, M., Szerlauth, A., Muráth, S., Varga, G., & Szilagyi, I. Surface modification of two-dimensional layered double hydroxide nanoparticles with biopolymers for biomedical applications. *Advanced Drug Delivery Reviews*, 191 (2022) 114590. https://doi.org/10.1016/j.addr.2022.114590.
56. Sriram, G., Bendre, A., Altalhi, T., Jung, H. Y., Hegde, G., & Kurkuri, M. Surface engineering of silica based materials with Ni–Fe layered double hydroxide for the efficient removal of methyl orange: Isotherms, kinetics, mechanism and high selectivity studies. *Chemosphere*, 287 (2022) 131976. https://doi.org/10.1016/J.CHEMOSPHERE.2021.131976.
57. Mao, N., Zhou, C. H., Tong, D. S., Yu, W. H., & Cynthia Lin, C. X. Exfoliation of layered double hydroxide solids into functional nanosheets. *Applied Clay Science*, 144 (2017) 60–78. https://doi.org/10.1016/J.CLAY.2017.04.021.
58. Qin, M., Li, S., Zhao, Y., Lao, C.-Y., Zhang, Z., Liu, L., Fang, F., Wu, H., Jia, B., Liu, Z., Wang, W., Liu, Y., Qu, X., Qin, M., Li, S., Zhao, Y., Zhang, Z., Liu, L., Fang, F., Liu, Y. Unprecedented synthesis of holey 2D layered double hydroxide nanomesh for enhanced oxygen evolution. *Advanced Energy Materials*, 9(1) (2019) 1803060. https://doi.org/10.1002/AENM.201803060.
59. Sarkar, A. N., Kumari, S., Jagadevan, S., Panda, A. B., & Pal, S. Simultaneous *in situ* exfoliation of titanate and Zn-Cr layered double hydroxides with a copolymer for photocatalytic degradation of organic pollutants. *ACS Applied Polymer Materials*, 4(8) (2022) 6132–6147. https://doi.org/10.1021/ACSAPM.2C00934.
60. Lestari, P. R., Takei, T., Yanagida, S., & Kumada, N. Facile and controllable synthesis of Zn-Al layered double hydroxide/silver hybrid by exfoliation process and its plasmonic photocatalytic activity of phenol degradation. *Materials Chemistry and Physics*, 250 (2020) 122988. https://doi.org/10.1016/J.MATCHEMPHYS.2020.122988.
61. Yang, M., Zhang, Z., Ma, J., Liu, L., Wang, M., Pan, B., Li, J., Xu, J., & Tong, Z. Fabrication of intercalation hybrid of Ni–Al layered double hydroxide with Cu(II) phthalocyanine via exfoliation/restacking route and photocatalytic activity on elimination of Rhodamine 6G. *Journal of Inclusion Phenomena and Macrocyclic Chemistry*, 87(1–2) (2017) 117–125. https://doi.org/10.1007/S10847-016-0684-2.
62. Kang, M. J., Rhee, S. W., Moon, H., Neck, V., & Fanghänel, T. Sorption of MO^{4-} (M=Tc, Re) on Mg/Al layered double hydroxide by anion exchange. *Radiochimica Acta*, 75(3) (1996) 169–173. https://doi.org/10.1524/RACT.1996.75.3.169.
63. Gu, P., Zhang, S., Li, X., Wang, X., Wen, T., Jehan, R., Alsaedi, A., Hayat, T., & Wang, X. Recent advances in layered double hydroxide-based nanomaterials for the removal of radionuclides from aqueous solution. *Environmental Pollution*, 240 (2018) 493–505. https://doi.org/10.1016/J.ENVPOL.2018.04.136.
64. Zhang, R., Ai, Y., & Lu, Z. Application of multifunctional layered double hydroxides for removing environmental pollutants: Recent experimental and theoretical progress. *Journal of Environmental Chemical Engineering*, 8(4) (2020) 103908. https://doi.org/10.1016/J.JECE.2020.103908.
65. Gu, P., Zhang, S., Ma, R., Sun, M., Wang, S., Wen, T., & Wang, X. Layered double hydroxides nanosheets in-situ anchored on ultrathin MXenes for enhanced U(VI) and Eu(III) trapping: Excavating from selectivity to mechanism. *Separation and Purification Technology*, 288 (2022) 120641. https://doi.org/10.1016/J.SEPPUR.2022.120641.
66. Kim, J., Kang, J., & Um, W. Simultaneous removal of cesium and iodate using prussian blue functionalized CoCr layered double hydroxide (PB-LDH). *Journal of Environmental Chemical Engineering*, 10(3) (2022) 107477. https://doi.org/10.1016/J.JECE.2022.107477.
67. Almomani, F., Bhosale, R., Khraisheh, M., kumar, A., & Almomani, T. Heavy metal ions removal from industrial wastewater using magnetic nanoparticles (MNP). *Applied Surface Science*, 506 (2020) 144924. https://doi.org/10.1016/j.apsusc.2019.144924.
68. Bashir, A., Malik, L. A., Ahad, S., Manzoor, T., Bhat, M. A., Dar, G. N., & Pandith, A. H. Removal of heavy metal ions from aqueous system by ion-exchange and biosorption methods. *Environmental Chemistry Letters*, 17(2) (2019) 729–754. https://doi.org/10.1007/s10311-018-00828-y.
69. El-Dib, F. I., Mohamed, D. E., El-Shamy, O. A. A., & Mishrif, M. R. Study the adsorption properties of magnetite nanoparticles in the presence of different synthesized surfactants for heavy metal ions removal. *Egyptian Journal of Petroleum*, 29(1) (2020) 1–7. https://doi.org/10.1016/j.ejpe.2019.08.004.
70. Feng, X., Long, R., Wang, L., Liu, C., Bai, Z., & Liu, X. A review on heavy metal ions adsorption from water by layered double hydroxide and its composites. *Separation and Purification Technology*, 284 (2022) 120099. https://doi.org/10.1016/J.SEPPUR.2021.120099.
71. Dong, Y., Kong, X., Luo, X., & Wang, H. Adsorptive removal of heavy metal anions from water by layered double hydroxide: A review. *Chemosphere*, 303 (2022) 134685. https://doi.org/10.1016/j.chemosphere.2022.134685

72. Yang, D., Wang, Y., Zhao, J., Dai, J., Yan, Y., Chen, L., & Ye, J. Strong coupling of super-hydrophilic and vacancy-rich g-C3N4 and LDH heterostructure for wastewater purification: Adsorption-driven oxidation. *Journal of Colloid and Interface Science*, 639 (2023) 355–368. https://doi.org/10.1016/J.JCIS.2023.02.060.
73. Guan, X., Yuan, X., Zhao, Y., Wang, H., Wang, H., Bai, J., & Li, Y. Application of functionalized layered double hydroxides for heavy metal removal: A review. *Science of the Total Environment*, 838 (2022) 155693. https://doi.org/10.1016/J.SCITOTENV.2022.155693.
74. Zhu, S., Chen, Y., Khan, M. A., Xu, H., Wang, F., & Xia, M. In-depth study of heavy metal removal by an etidronic acid-functionalized layered double hydroxide. *ACS Applied Materials and Interfaces*, 14(5) (2022) 7450–7463. https://doi.org/10.1021/ACSAMI.1C22035.
75. Guan, X., Yuan, X., Zhao, Y., Bai, J., Li, Y., Cao, Y., Chen, Y., & Xiong, T. Adsorption behaviors and mechanisms of Fe/Mg layered double hydroxide loaded on bentonite on Cd (II) and Pb (II) removal. *Journal of Colloid and Interface Science*, 612 (2022) 572–583. https://doi.org/10.1016/J.JCIS.2021.12.151.
76. Du, C., Zhang, Y., Zhang, Z., Zhou, L., Yu, G., Wen, X., Chi, T., Wang, G., Su, Y., Deng, F., Lv, Y., & Zhu, H.. Fe-based metal organic frameworks (Fe-MOFs) for organic pollutants removal via photo-Fenton: A review. *Chemical Engineering Journal*, 431 (2022) 133932. https://doi.org/10.1016/j.cej.2021.133932.
77. Yu, G. W., Laseter, J., & Mylander, C. Persistent organic pollutants in serum and several different fat compartments in humans. *Journal of Environmental and Public Health* (2011). https://doi.org/10.1155/2011/417980.
78. Shen, X., Zhu, Z., Zhang, H., Di, G., Chen, T., Qiu, Y., & Yin, D. Carbonaceous composite materials from calcination of azo dye-adsorbed layered double hydroxide with enhanced photocatalytic efficiency for removal of Ibuprofen in water. *Environmental Sciences Europe*, 32(1) (2020). https://doi.org/10.1186/s12302-020-00351-4.
79. Hou, L., Li, X., Yang, Q., Chen, F., Wang, S., Ma, Y., Wu, Y., Zhu, X., Huang, X., & Wang, D. Heterogeneous activation of peroxymonosulfate using Mn-Fe layered double hydroxide: Performance and mechanism for organic pollutant degradation. *Science of the Total Environment*, 663 (2019) 453–464. https://doi.org/10.1016/j.scitotenv.2019.01.190.
80. Ge, J., Lian, L., Wang, X., Cao, X., Gao, W., & Lou, D. Coating layered double hydroxides with carbon dots for highly efficient removal of multiple dyes. *Journal of Hazardous Materials*, 424 (2022) 127613. https://doi.org/10.1016/J.JHAZMAT.2021.127613.
81. Lito, P. F., Aniceto, J. P. S., & Silva, C. M. Removal of anionic pollutants from waters and wastewaters and materials perspective for their selective sorption. *Water, Air, & Soil Pollution*, 223(9) (2012) 6133–6155. https://doi.org/10.1007/S11270-012-1346-7.
82. Fu, D., Kurniawan, T. A., Avtar, R., Xu, P., & Othman, M. H. D. Recovering heavy metals from electroplating wastewater and their conversion into Zn_2Cr-layered double hydroxide (LDH) for pyrophosphate removal from industrial wastewater. *Chemosphere*, 271 (2021) 129861. https://doi.org/10.1016/J.CHEMOSPHERE.2021.129861.
83. He, W., Ai, K., Ren, X., Wang, S., & Lu, L. Inorganic layered ion-exchangers for decontamination of toxic metal ions in aquatic systems. *Journal of Materials Chemistry A*, 5(37) (2017) 19593–19606. https://doi.org/10.1039/C7TA05076C.
84. Vu, C. T., Wu, T. MgAl-layered double hydroxides/sodium alginate beads for nitrate adsorption from groundwater and potential use as a slow-release fertilizer. *Journal of Cleaner Production*, 379 (2022) 134508. https://doi.org/10.1016/J.JCLEPRO.2022.134508.
85. Besharatlou, S., Anbia, M., Salehi, S. Optimization of sulfate removal from aqueous media by surfactant-modified layered double hydroxide using response surface methodology. *Materials Chemistry and Physics*, 262 (2021) 124322. https://doi.org/10.1016/J.MATCHEMPHYS.2021.124322.
86. Khalil, A. K. A., Dweiri, F., Almanassra, I. W., Chatla, A., & Atieh, M. A. Mg-Al layered double hydroxide doped activated carbon composites for phosphate removal from synthetic water: Adsorption and thermodynamics studies. *Sustainability*, 14(12) (2022) 6991. https://doi.org/10.3390/SU14126991.
87. Li, S., Zhang, L., Tian, S., He, Y., & Guo, X. Mineralized cupric phosphate/alginate gel alternately multilayer-wrapped nanofibrous membrane with robust anti-crude oil pollution for oily wastewater purification. *Journal of Membrane Science*, 669 (2023) 121280. https://doi.org/10.1016/J.MEMSCI.2022.121280.
88. Karlapudi, A. P., Venkateswarulu, T. C., Tammineedi, J., Kanumuri, L., Ravuru, B. K., Dirisala, V. ramu, & Kodali, V. P. Role of biosurfactants in bioremediation of oil pollution-a review. *Petroleum*, 4(3) (2018) 241–249. https://doi.org/10.1016/J.PETLM.2018.03.007.
89. Srinivasan, N., Thangavelu, K., & Uthandi, S. Microbial fermentation of waste oils for production of added-value products. In: *Sustainable Microbial Technologies for Valorization of Agro-Industrial Wastes* (2022) 305–327. https://doi.org/10.1201/9781003191247-15.

90. Yue, X., Zhang, T., Yang, D., Qiu, F., Li, Z., Zhu, Y., & Yu, H. Oil removal from oily water by a low-cost and durable flexible membrane made of layered double hydroxide nanosheet on cellulose support. *Journal of Cleaner Production*, 180 (2018) 307–315. https://doi.org/10.1016/J.JCLEPRO.2018.01.160.
91. Liu, X., Ge, L., Li, W., Wang, X., & Li, F. Layered double hydroxide functionalized textile for effective oil/water separation and selective oil adsorption. *ACS Applied Materials and Interfaces*, 7(1) (2015) 791–800. https://doi.org/10.1021/AM507238Y.
92. Zhang, L., Gong, Z., Jiang, B., Sun, Y., Chen, Z., Gao, X., & Yang, N. Ni-Al layered double hydroxides (LDHs) coated superhydrophobic mesh with flower-like hierarchical structure for oil/water separation. *Applied Surface Science*, 490 (2019) 145–156. https://doi.org/10.1016/J.APSUSC.2019.06.064.
93. Prabagar, J. S., Vinod, D., Sneha, Y., Anilkumar, K. M., Rtimi, S., Wantala, K., & Shivaraju, H. P. Novel gC_3N_4/MgZnAl-MMO derived from LDH for solar-based photocatalytic ammonia production using atmospheric nitrogen. *Environmental Science and Pollution Research* (2022) 1–14. https://doi.org/10.1007/S11356-022-24997-0.
94. Jijoe Samuel, P., Yadav, S., Thinley, T., Hosakote Shankara, A., Vikram P R, H., Gurupadayya, B. M., Anil Kumar, K. M., & Shivaraju, H.P. Visible light irradiation driven CO_2 reduction into hydrocarbons on tri-metallic based layered double hydroxide. *Materials Today: Proceedings* (2022). https://doi.org/10.1016/J.MATPR.2022.10.304.
95. Cai P., Ci S., Wu N., Hong Y., Wen Z. Layered structured CoAl/CdS-LDHs nanocomposites as visible light photocatalyst. *Physica Status Solidi (A) Applications and Materials Science*, 214(6) (2017) 1600910. https://doi.org/10.1002/pssa.201600910.
96. Zhu C., Wang Y., Qiu L., Yang W., Yu Y., Li J., Liu Y. Z-scheme NiFe LDH/$Bi_4O_5I_2$ heterojunction for photo-Fenton oxidation of tetracycline. *Journal of Alloys and Compounds*, (2023) 169124. https://doi.org/10.1016/J.JALLCOM.2023.169124.
97. Huang X., Xu X., Yang R., Fu X. Synergetic adsorption and photocatalysis performance of g-C_3N_4/Ce-doped MgAl-LDH in degradation of organic dye under LED visible light. *Colloids and Surfaces A: Physicochemical and Engineering Aspects*, 643 (2022) 128738. https://doi.org/10.1016/J.COLSURFA.2022.128738.
98. Zhu Y., Zhu R., Zhu G., Wang M., Chen Y., Zhu J., Xi Y., He H. Plasmonic Ag coated Zn/Ti-LDH with excellent photocatalytic activity. *Applied Surface Science*, 433 (2018) 458–467. https://doi.org/10.1016/j.apsusc.2017.09.236.
99. Asif M.B., Kang H., Zhang Z. Gravity-driven layered double hydroxide nanosheet membrane activated peroxymonosulfate system for micropollutant degradation. *Journal of Hazardous Materials*, 425 (2022) 127988. https://doi.org/10.1016/J.JHAZMAT.2021.127988.

16 2D Graphitic Carbon Nitrides Hybrids for the Removal of Organic Water Pollutants

Ponnaiah Sathish Kumar, Vellaichamy Balakumar, Govindaraj Sangami, and Keiko Sasaki

16.1 INTRODUCTION

Water resources and the environment have been polluted as a result of rapid population growth, changes in lifestyles, and new industrial developments. Water scarcity and pollution of the water resources are among the most critical environmental problems, since potable water is essential for life and survival depends on it [1,2]. As of 2012, 780 million people around the world lack access to safe, clean drinking water, according to a report released by the World Health Organization (WHO). One-third of the world's population – 2.4 billion people – lives in water-stricken countries, and this percentage is expected to increase by two-thirds by 2025 [3]. Water resources have been inappropriately managed, and industrial pollution has affected access to drinkable water sources. This has posed a critical threat to humans, other living organisms, and even the survival of humans [4]. A number of major water contaminants, including heavy metals, pharmaceuticals, organic dyes, pesticides, personal care products, and herbicides, can pose serious risks to living creatures. These compounds are widely used in industry and can cause water pollution when they enter the environment due to their large quantities [5]. As a result of contaminants in water, the water emits an unpleasant odor, prevents sunlight from reaching it, and reduces microorganism activity. Using contaminated water and contact with it can result in severe health problems. In addition to being toxic, these contaminants can cause allergies, cyanosis, nausea, heart disease, headaches, cancer, and respiratory damage [6]. In addition to their resistance to heat, chemical reactions, and light irradiation, these contaminants can contaminate water sources for a long period of time if released into the environment. In light of this, water and wastewater sources must be cleaned immediately of contaminants [7].

16.2 OVERVIEW OF VISIBLE LIGHT PROPERTIES AND DEGRADATION MECHANISM OF 2D GCN

There is a need to remove pollutants or reduce their concentrations below the allowable level in effluents. There are a number of methods used to remove contaminants from wastewater, advanced oxidation processes (AOPs), including ion exchange, membrane separation, biological processes, coagulation, and adsorption [8]. Because of its simplicity, high efficiency, zero waste generation, low cost, and in situ generation of high oxidant species like hydroxyl radicals, AOPs are one of the most efficient methods for removing various pollutants from aquatic environments [9]. Over the past few decades, numerous AOPs have been developed, including electrochemical oxidation, Fenton-like oxidation, ozonation, microwaves, sonolysis, and photocatalysis. Among all AOPs, photocatalysis has gained increasing attention because of its nontoxic materials, lower cost, efficiency under mild conditions using potential sunlight, and relatively high chemical stability of the

DOI: 10.1201/9781003436942-16

catalyst [10]. Furthermore, photocatalysis has the potential to completely destroy, degrade, or mineralize organic pollutants. Photocatalysis can be divided into two categories, solar/visible-light-active photocatalysis and ultraviolet (UV) active photocatalysis, which is the preferred method [11]. Due to the stronger intensity of UV than solar/visible light, UV photocatalysis generally performs better than solar/visible light photocatalysis; however, solar/visible light photocatalysis is preferred to UV light photocatalysis because of its economics and has been regarded as an environmentally friendly method for removing pollutants [12]. Graphitic carbon nitride ($g\text{-}C_3N_4$=GCN) is a polymeric nonmetallic semiconductor with several advantages in photocatalysis, including its low-cost elements of carbon and nitrogen, eco-friendly nature, chemical stability, nontoxicity, abundance, biocompatibility, tunable band gap (1.8–2.7 eV), nonpollution properties, and stability to safe pH conditions (pH 0–14) [13]. Based on theoretical studies, GCN can have a surface area of up to $2500\,m^2g^{-1}$. The ability to absorb visible light in the spectrum (400–500 nm) is one of the most unique and attractive properties for advanced research on GCNs [14]. This chapter examines the effects of operating parameters on organic water pollutants degradation with GCN photocatalysts. In addition to a brief overview of GCN preparation techniques and modification techniques to enhance its photocatalytic activity, composites based on GCN will be provided. By writing this chapter, we hope to not only continue the development of visible-light-active photocatalysis, but also to address the importance of pilot-scale tests and even real-world applications of visible-light-active photocatalysts.

16.3 PHOTOCATALYSTS FOR VISIBLE LIGHT

The visible-light-active photocatalyst consists of GCN materials, light harvesting antennas, and active species, and it is a type of AOP in which radicals are generated after photoexcitation of GCN materials. Detailed information about the mechanism and materials of photocatalysis is crucial for research on photocatalysis. This section presents a review of the mechanisms and materials of photocatalysis [15].

16.3.1 Photocatalytic Destruction Mechanism

In general, the mechanism of visible-light-activated photocatalysis uses visible energy to produce radicals and other active species and destruct pollutants (Figure 18.1). A series of reactions can occur in visible photoactive catalysts due to light absorption for contaminant degradation, which have been extensively reported in many literatures [2] and summarized as follows:

$$CN + h\nu \rightarrow CN\,(h^+ + e^-)$$
$$h^+ + H_2O \rightarrow \bullet OH + H^+$$
$$h^+ + OH^- \rightarrow \bullet OH$$
$$e^- + O_2 \rightarrow \bullet O_2^-$$
$$O_2^- + H^+ \rightarrow \bullet OOH$$
$$\bullet OOH + \bullet O_2^- \rightarrow OOH^- + O_2$$
$$OOH^- + h^+ \rightarrow \bullet OOH$$
$$2\bullet OOH \rightarrow O_2^- + H_2O_2$$
$$H_2O_2 + \bullet O_2^- \rightarrow \bullet OH + OH^- + O_2$$
$$H_2O_2 + hv \rightarrow 2\bullet OH$$
$$\text{Organic water pollutant} + (\bullet OH,\ h^+,\ \bullet OOH \text{ or } O_2^-) \rightarrow \text{degradation products} + CO_2 + H_2O$$
$$h^+ + e^- \rightarrow \text{heat}$$

if visible light energy ($h\nu$, where ν is the light's frequency and h is called Planck's constant and equal to 6.62608×10^{-34} Js) is absorbed by the photocatalyst.

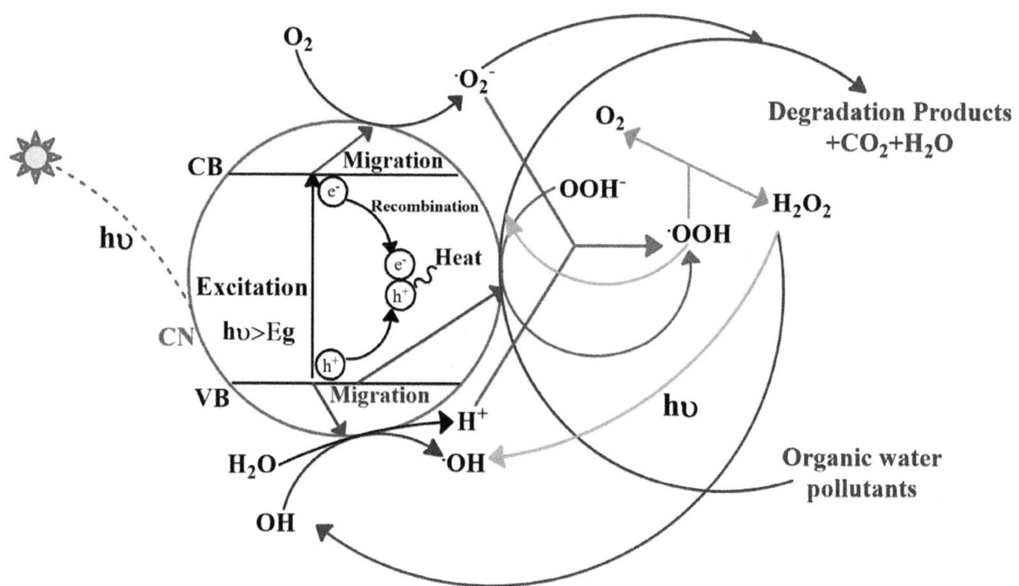

FIGURE 16.1 The possible photocatalytic reaction mechanism of GCN.

16.4 SYNTHESIS STRATEGIES FOR GRAPHITIC CARBON NITRIDE

The polymer GCN is one of the hoariest composites, first discovered by Berzelius in 1834 and named "melon". CN has received more attention since the prediction of β-GCN in 1989 [16]. Other predicted phases of CN are α-CN, cubic-CN, pseudocubic-CN, and graphitic-CN [17]. CN was first discovered by Franklin in 1922 by using $Hg(SCN)_2$ in thermal decomposition [18]. As the most stable CN allotrope under ambient conditions, polymeric CN has enticed the consideration of researchers who are conducting important studies on the synthesis of CN-based compounds. The statistical results of the publication trend on this material.

The People's Republic of India, China, Japan, Germany, Saudi Arabia, United States, Iran, South Korea, England, and Australia are the top ten countries in the study of photocatalytic applications of CN. As can be observed, interest in this topic has grown, as seen by the gradual rise in publications from only 25 in 2000 to 1863 in 2019. The majority of these studies fall into the following categories, according to an analysis of the reports' context: nanoscience and nanotechnology, chemistry, energy fuels, environmental engineering, physics, material science, chemical engineering, applied physics, material science coating films, and condensed matter physics. CN has been successfully produced using low-cost N-rich feedstocks such as dicyandiamide ($C_2H_4N_4$), cyanamide (CH_2N_2), guanidinium chloride (CH_6ClN_3), urea (CH_4N_2O), melamine ($C_3H_6N_6$), guanidinium thiocyanate ($C_2H_6N_4S$), and thiourea (CH_4N_2S) [19]. Chemical vapor deposition, electrochemical deposition, sonochemical exfoliation, microwave-assisted synthesis, and thermal condensation are among the methods of synthesis that are frequently utilized [20].

The physical and chemical characteristics of the produced graphitic-CN, such as structure, surface area, porosity, photoluminescence, absorption capacity, and C/N ratio, may be significantly influenced by a variety of feedstocks and treatment techniques (Table 16.1) according to studies [21]. Instruments like Fourier transform infrared (FTIR) spectroscopy, high-resolution transmission electron microscopy (HRTEM), diffuse reflectance spectra in the UV–visible range (DRS/UV–vis), X-ray photoelectron spectroscopy (XPS), and X-ray powder diffraction (XRD) are frequently used to confirm the presence of CN [22,23]. Density function theory (DFT) is typically used to identify the VB and CB calculations [24].

TABLE 16.1
Structure, Various Precursors and Synthesis Methods Have been Used in Synthesizing 2D CN

Structure	N-rich Feedstocks	Synthesis Methods	
NH, N, H$_2$N, NH$_2$	Cyanamide (CH_2N_2)	Chemical synthesis methods	Hydrothermal
	Urea (CH_4N_2O)		Solvothermal
	Melamine ($C_3H_6N_6$)		Sol–gel
			Pyrolysis
	Thiourea (CH_4N_2S)	Physical synthesis methods	Chemical vapor deposition
	Guanidinium chloride (CH_6ClN_3)		Physical vapor deposition
	Dicyandiamide ($C_2H_4N_4$)		Gas Condensation
	Guanidinium thiocyanate ($C_2H_6N_4S$)		Solid State Reaction
		Mechanical synthesis methods	Mechanical exfoliation

16.5 DIFFERENT MORPHOLOGIES, ADVANTAGES AND DISADVANTAGES OF CN

The synthesis parameters can be changed to change the dimension of CN. For the production of CN in various dimensionalities (Figure 18.2), including zero-dimensional quantum dots, one-dimensional micronanotubes, micronanorods and micronanowires, two-dimensional nanosheets (NS) and films, and three-dimensional bulk forms, various modifications have been made [25]. Specific processes can convert 0D and 2D CN into 1D CN. Its photocatalytic activity is increased by the huge surface area of 2D-CN materials, which facilitates effective interfacial charge transfer through improved surface contact [26]. According to Figure 18.2, the structure of CN is principally made up of the two most noticeable components, carbon and nitrogen, as well as a number of functional moieties, including Lewis basic functions, pi-bonds, H-bonding motifs, and Brönsted basic functions [27]. The optical characteristics and photocatalytic activity of CN are primarily improved by the elemental buildup, in-plane holes, and tiny pore size. The CNs network experiences electron delocalization (π-network of CN), resulting in the merger of electrons and holes, which are then consumed in oxidation and reduction reactions, respectively. By optimizing the electronic structure of CN and functionalizing the -NH-, =N-, and -NH_2 groups that are already present on the surface, better optical characteristics can be produced [28]. However, pristine CN has lower photocatalytic activity due to its stacked two-dimension structure (van der Waals force) and low surface area [29]. The surface condition, which includes defects and doping, functional groups, and the structural condition, which includes porosity, morphology, and thickness, both have a significant impact on the performance of CN [30]. In addition to these preponderances, CN has a low specific surface area, a narrow absorption band at visible light wavelengths (beyond 470 nm), a low quantum yield, and a high recombination rate because of weak van der Waals forces between adjacent CN layers [31]. Pure CNs substantial photocatalytic activity and practical uses are severely constrained by these disadvantages. As a result, several methods have been proposed by researchers to increase the photoactivity of this substance.

16.5.1 Zero-Dimensional CN Nanostructures

Electrons and holes are unable to travel freely in zero-dimensional CN (also known as CNQDs). This substance has received extra attention because it possesses important qualities such as water solubility,

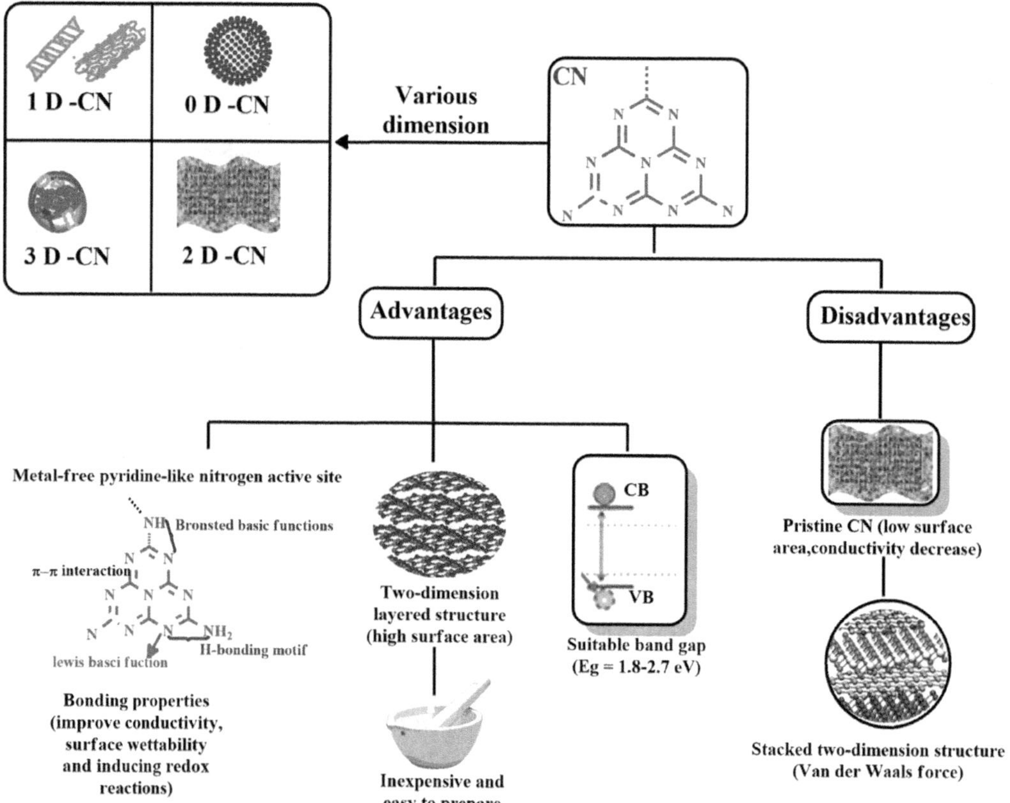

FIGURE 16.2 Various dimensionalities, advantages, and disadvantages of CN.

adequate photoexcited electron transfer, affordability, nontoxicity, effective upconversion features, size efficacy, and photochemical resistance [32]. Due to these outstanding characteristics, there has been a significant increase in research on the use of CNQDs for photocatalysis in recent years. Several techniques, including hydrothermal [33], solvothermal [34], thermal polymerization [35], and ultrasonic-assisted method [36], have been used to create CNQDs with an average size of 10 nm.

16.5.2 One-Dimensional CN Nanostructures

One-dimensional CN (1D CN) is a different morphology that has drawn significant interest and significance in academia because of its exceptional electrochemical and optical capabilities [37]. By altering variables like aspect ratio (length to diameter), diameter, and length, these qualities might be tuned for the right photoexcitation [38]. Micronanorod CN has demonstrated greater photocatalytic efficiency as compared to bulk CN. Bulk CN could be compressed and a 1D structure could be made by using physical methods for their mechanical, thermal, chemical effects, ultrasonic, and high pressure [39]. By using a variety of techniques, including the nanocasting technique [40], thermal polymerization [41], CVD [42], the ionic liquid promoted method [43], and the template-assisted approach [44], one-dimensional CN has been created in which electrons can only move in one direction.

16.5.3 Two-Dimensional CN Nanostructures

Due to nitrogen's strong electron donor properties in CN, which are absent from graphene, interest has grown in 2D-CN materials that resemble graphene [45]. The N-bridged poly(tri-s-triazine) has a lot of defects, which act as active sites for electron conductivity. Two-dimensional CN nanosheets have received a great deal of study attention because of their exceptional qualities, including their extensive surface area, advanced pillared structure, great length to width, large surface anchoring groups for cocatalysts, and ultrathin thickness [46]. As a result of the interactions between 2D CN and other connected 2D semiconductors, heterojunction interfaces are extended, making it easier to extend the lifetime of e^-/h^+ and ultimately enable successful photocatalytic applications [47]. By destroying the hydrogen bonds that connect stacks of polymeric units, the exfoliation of bulk CN will result in the production of one or more layers of structures, or occasionally nanoparticles rather than nanosheets, increasing the specific surface area [48]. In addition to increasing surface area by exfoliation, 2D CN also improved charge carrier separation thanks to a shorter carrier migration distance compared to bulk, which increased photocatalytic activity [49]. CN has been exfoliated to nanosheets using a variety of solvents, including ethanol, isopropanol, methanol, and 1,3-butanediol. Additionally, 2D layers of CN can be created by exfoliating materials in liquid phase in any polar solvent, such as water. After 12 hours of ultrasonic treatment, hydroxyl-rich nanosheets of CN may be produced using water as the solvent [50]. Mesoporous graphitic carbon nitride is produced by Solvothermal in an ethanol–water solution [51]. Additionally, a metal-free 2D/2D CN-based photocatalyst that is affordable and widely available will lower application costs [52]. By using bottom-up methods, CN films can be synthesized without aggregating or developing cracks, which results in a homogeneous structure, a durable coating, and good contact with the substrate [53].

16.5.4 Three-Dimensional CN Nanostructures

When compared to other morphologies of CN, three-dimensional CN (or 3D CN) has a higher Brunauer–Emmett– Teller (BET) surface area, which can significantly increase the amount of active sites needed for surface reactions by light harvesting and, as a result, more appropriately perform in adsorption [54]. To put it another way, 3D CN enables simple interfacial transport, a clear diffusion pathway, and simple dispersion of active sites with various dimensions [55]. Over time, scientists have begun to pay more and more attention to these benefits [56]. To create 3D CN, a variety of methods have been used, such as hydrothermal/solvothermal synthesis [57], heating-cold polymerization [58], photopolymerization [59], template approach [60], and freeze-drying method [61].

16.6 APPLICATION OF 2D CN IN ORGANIC WASTEWATER POLLUTANTS TREATMENT

A variety of processes can be used to create 2D CN. However, the precursors show high conversion yields when they are annealed in a standard laboratory furnace. The ease of this method is its main benefit, and because precursors like urea and melamine are inexpensive, it doesn't come at a significant cost. This strategy is the most popular, best, and offers the chance of wide-scale application.

16.6.1 Adsorption with 2D CN

On the surface of the adsorbent, the adsorption process takes place. As a result, the physicochemical characteristics of the materials' outside parts are extremely important to the process progress. Due to the chemical characteristic of its surface, 2D CN can also be used in the adsorptive removal of organic contaminants from aqueous solutions [62]. However, given the high organic contaminant adsorption performances, the data that have already been published are intriguing. Numerous interactions are related to the adsorption of organic compounds. Organic molecules can be drawn to them via interaction, weak London dispersion forces, van der Waals forces, or dipole–dipole forces, as demonstrated by the instance of commonly used activated carbon in 2D CN. The type of interaction is directly related to the surface's chemical makeup [63]. Figure 16.3 shows a schematic representation of potential adsorbate and adsorbent interactions on the 2D-CN surface.

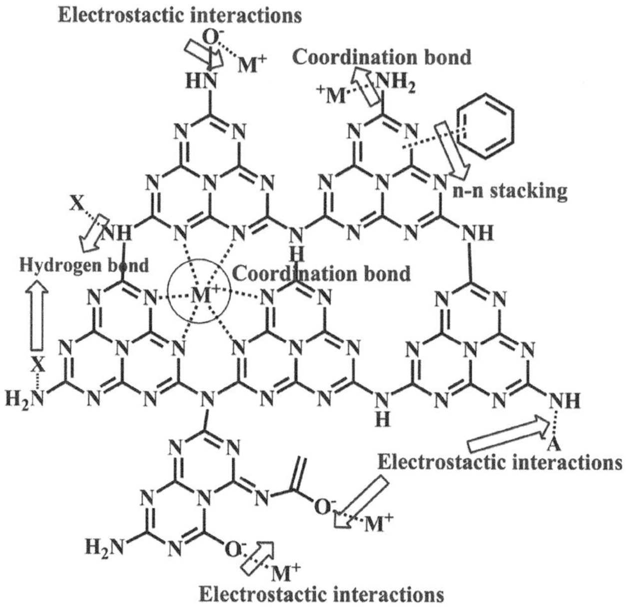

FIGURE 16.3 Potential interactions between adsorbate and adsorbent on the surface of 2D-CN are presented schematically.

The adsorption of water molecules on the tri-s-triazine-based structure has been elucidated by Aspera et al. [64]. The outcomes show that H_2O molecules have adhered to nitrogen atoms. Both of its O-H bonds are simultaneously parallel to the adsorbent surface and face the surface. The adsorption energy was estimated to be 0.82 eV, while the barrier energy for the process was 0.02 eV. They claimed that the defective layers are preferred by water molecules to adsorb on, which has a significant impact on water splitting. Additionally, according to them [64], the distorted layers of the CN have increased activity in the water photodecomposition process. Despite the fact that a different recent study looked into the impact of pH on the methylene blue (MB) dye's ability to adsorption using CN, this study shows that the solution pH increases the dye adsorption rate onto CN. Since H^+ ions are present in relatively high amounts at lower pH levels, they compete with MB as a cationic dye throughout the adsorption process [65]. Another study looked into the MB dye's ability to adsorb to Biochar/CN. According to the findings, at a pH level of 11, this adsorption increases by up to 96.7% [66]. According to Nezampour et al., in the case of nitrobenzene adsorption, the researchers used graphitic carbon nitride immobilized on two mesoporous substrates, MCM-41 and Al-MCM-41. These adsorbents' respective adsorption capabilities were found to be 40 and 85 mg g^{-1}, with a short equilibrium reaching time of just 5 minutes [67]. Malachite green can be adsorbed by boron-doped CN (1 wt% of B) with a capacity of 54 mg g^{-1} for contact duration of 30 minutes, according to research by Azimi et al. [68]. Three adsorbents were tested in the study by Yousefi et al. [69]. CN and CN that had undergone Hummers method oxidation at 22° and 60° using a solution of sulfuric acid and potassium permanganate. They compared the adsorption of tetracycline, an antibiotic, and six dyes (MO, basic fuchsin, RhB, rose Bengal, MG, and MB) onto these materials. Every time, oxidation enhanced adsorption efficiency. As an illustration, the ability of Na-doped CN produced by cyanamide and NaCl to remove methyl blue was investigated. Analyses showed that electrostatic interactions caused MB to adsorb on Na-doped CN. The doped material has a significantly higher maximum absorption of methyl blue than the undoped material. In actuality, the adsorption was 367 mg g^{-1}, and equilibrium was reached under 1 minute. It is important to note that the high porosity of the synthetic adsorbent was responsible for this extraordinary adsorption [70].

16.6.2 Photocatalytic Degradation of Organic Pollutants using 2D-CN

The application of bare CN is constrained by its quick recombination. In order to address these problems and enhance the photocatalytic performance of CN, we create heterogeneous structure-based photocatalysts based on CN. The separation of light-induced electron–hole pairs can be facilitated by using this strategy. When it comes to the treatment of organic wastewater, photocatalytic composites containing CN have been widely used in the photocatalytic destruction of organic pollutants in water, including MB, methyl orange (MO), rhodamine B (RhB), antibiotics, and phenol [71]. The catalyst is activated by light radiation to produce powerful oxidative holes and reducible electrons, which is the major mechanism for deterioration. The holes and electrons generate hydroxyl and superoxide radicals (OH and $^{\bullet}O_2$, respectively). These radical groups can effectively mineralize organic contaminants due to their potent oxidative or reducible characteristics. The two-dimensional CN structure's composite materials have high pore volume and porosity and specific surface area at the same time. This allows for more efficient organic pollutant adsorption and a faster rate of degradation [72]. Textile wastewater frequently contains highly toxic and chemically stable organic compounds, including dyes, surfactants, oils, acid, alkali, solvents, and some metal salts. These compounds are ranked as one of the worst industrial pollutants. According to Perez-Osorio et al. (2020), 280,000 tons of textile dyes alone are generated annually around the world. It could seriously contaminate aquatic environments that are receiving it if proper treatment is not provided before release [73]. Remazol Turquoise Blue (RTB) was removed by Das and Mahalingam (2019) by immobilizing TiO_2, rGO, and CN in a polystyrene film and exposing it to sunshine. After 90 minutes, 60% of the color was decolorization and 51.43% of the substance was degraded [74]. Additionally, J. U. Choi et al. (2019) created a Cu-loaded CN/1D hydrogenated black TiO_2 nanofiber (CuCNBTNF) for the removal of aqueous dye pollutants, which reached 86.2% for RhB in 5 hours [75]. The sulfuric acid-treated graphitic carbon nitride (SA-CN) embedded porous cellulose network (CN/CA film) that reduced 95% of Cr(VI) in 100 minutes and removed 99% of RhB in 150 minutes. The resulting films had a thickness of around 10 m, were incredibly flexible and strong, and had SA-CN evenly dispersed throughout the cellulose acetate network [76]. A mix of CN nanosheets and Zr-based metal–organic framework (UiO-66) photocatalysts (UiO-66/CN sheets) was able to completely degrade MB in 240 minutes [77]. Chegeni et al. (2019) had a comparable outcome from V_2O_5/S-CN for MB and phenol treatment, achieving a 99% and 89% elimination rate after 60 minutes, respectively [78]. In order to improve the photocatalytic activity of CN/V_2O_5, immobilized Ag on the surface, which may completely decrease p-nitrophenol (NP) in 8 minutes when exposed to sunlight. The photocatalyst only requires 60 minutes to completely eliminate 4-NP and 4-AP in the dark, according to their findings [79]. *N*-methyl-2-pyrrolidone consuming liquid exfoliation approach was used to create metal-free black phosphorus (BP)/polymeric CN photocatalysts, which demonstrated enough photoactivity in the activation of oxygen processes due to efficient charge transfer because of the preferable energy configuration [80]. In other words, within 15 minutes of irradiation, 98% of RhB was degraded. In comparison to pure carbon nitride, the rate constant for the degradation of RhB contaminants utilizing the 10% BP/CN composite was four times higher. Melamine and phosphonitrilic chloride trimer ($Cl_6N_3P_3$) were utilized as precursors in a two-step procedure to create phosphorus-doped CN (P(x%)-CN [81], and a solvent-thermal technique was used to coat the obtained P(x%)-CN on TiO_2 nanoparticles in order to create a P(x%)-CN/TiO_2 photocatalyst. In the degradation of MB under visible light, the results of measuring the photocatalytic activity of manufactured composites showed that P(0.1%)-CN/TiO_2 composite gave the highest activity, which is around 5 times higher than that of pure CN. For the photodegradation of RhB, SiO_2/CN (Si-CN) composite photocatalyst modified SiO_2/CN (Si-CN-HO) with surface hydroxylation were created [82]. In order to create two blended slurries, SiO_2 and melamine are first ground separately with ethanol. Second, to create a Si-CN composite photocatalyst, the two combined slurries were ground and calcined in a muffle furnace. In order to accomplish the surface hydroxylation modification of Si-CN-HO, the produced Si-CN composites were heated in an H_2O_2 solution. RhB was photocatalyzed by hydroxylated Si-CN with a 20.9 times greater efficiency than with CN when exposed to sunlight. Sol–gel technique was used to create Mo, N codoped $ZnIn_2S_4$/CN material (M, N-ZIS/CN) for the photodegradation of MB. First, CN was produced through calcination. Then, 150 mL of distilled water was added, along with zinc acetate and indium chloride, and agitated for 30 minutes before thioacetamide was added. Then, drop by drop, the CN solution dissolved in distilled water was added. To create a gel, the fluid was heated and dried. To create $ZnIn_2S_4$/CN material, the dried gel was washed and vacuum dried. By adding ammonium molybdate and N-N-dimethylformamide to the solution for making $ZnIn_2S_4$/CN, M, N-ZIS/CN nanosheets were produced, and the remaining steps were the same as for making $ZnIn_2S_4$/CN. SEM and TEM investigations revealed that M, N-ZIS/CN was a small and loose 2D nanoflower, whereas $ZnIn_2S_4$ was a thick nanoflower structure and CN was a thin nanosheet. These findings suggest that codoping boosted the composite material's specific surface area, shortened the charge transfer path and its active site, and decreased $ZnIn_2S_4$ particle aggregation [83]. The most development in the degradation of organic contaminants employing composite photocatalysts based on CN is presented in Table 16.2.

TABLE 16.2
Summary of the Latest Progress in the Degradation of Organic Pollutants by Using CN-Based Composite Photocatalysts

Composite	Synthesis Method	Pollutants	Illumination	Degradation (min)/(%)	Reaction Rate Constant (min^{-1})	Ref.
Ag_3PO_4/ polyaniline@CN	Calcination	MCP	150 W Xe arc lamp	50/99.6	–	[14]
TiO_2/rGO/CN	Staudenmaier's/ calcination	RTB	Sunlight	90/51.43	–	[74]
SA-CN/CA film	Thermal polymerization/ chemical exfoliation	RhB; Cr (VI)	Sunlight	150 ;100 /99; 95	0.03761 –	[76]
UiO-66/CN	Calcination/ thermal	MB	350 W Xe lamp	240/100	–	[77]
V_2O_5/S-CN	Thermal	MB; phenol	Sunlight	60; 60/99; 89	–	[78]
Ternary Ag/CN/ V_2O_5	Photodeposition	4-NP	Sunlight	8/100	0.041	[79]
PTCDA and CN	Calcination	BPA	300 W Xe lamp	60/96	0.0501	[84]
MIL-125/Ag/CN	Calcination/ ultrasonication	4-NP	500W Xe lamp	30/82	0.025	[85]
Ag/CN/CNF	Electrospinning / Calcination	4-NP	500W Xe lamp	50/100	0.094	[86]
Oxygen-rich EGCN	Calcination/ sonication	BPA	300 W Xe lamp	90/99.8	0.045	[87]
Pd/mpg-CN	Calcination/ sonication	BPA	350 W Xe lamp	180/93.9	–	[88]
Au/CN	Calcination/stirring	BPA	Natural sunlight	90/98	–	[89]
Ag/mpg-CN/PMS	Calcination/stirring	BPA	300 W Xe lamp	60/98	–	[90]

(Continued)

TABLE 16.2 (*Continued*)
Summary of the Latest Progress in the Degradation of Organic Pollutants by Using CN-Based Composite Photocatalysts

Composite	Synthesis Method	Pollutants	Illumination	Degradation (min)/(%)	Reaction Rate Constant (min^{-1})	Ref.
N–CN@In_2S_3 nanocomposites	Reflux-polycondensation	MB	200 W W lamp	10/98	0.053	[91]
N–ZnO-CN	Calcination/hydrothermal	RhB	500W Xe lamp	180/97	0.0154	[92]
O-doped CN	Calcination/microwave digestion	BPA	1000W Xe lamp	180/93	–	[93]
Mpg-CN/Ag/ZnO nanowires	Thermal polycondensation/dip-coating	Direct orange 26	250 W Hg lamp	120/94	0.0225	[94]
Thiophene-doped-CN	Calcination	BPA	300 W Xe lamp	100/99	0.0249	[95]
O-doped CN	Calcination/Hydrothermal	RhB	300W Xe lamp	3/99	–	[96]
Ag-CN/$LaFeO_3$	Calcination/ultrasonication	MB/Tetracycline	300W Xe lamp	90/98 120/94	0.0414	[97]
Multi-element doped CN	Calcination	Naproxen	500W Xe lamp	90/93	0.00851	[98]
Sm_2O_3/S–CN	Ultrasonication/Calcination	MB	500 W Xe lamp	150/93	0.0146	[99]
Mn/P codoped CN	Hydrothermal/Calcination	RhB	500 W Xe lamp	150/99	0.0192	[100]
N-CQDs/CN/PDS	Calcination/Hydrothermal	Tetracycline hydrochloride	500 W Xe lamp	60/90	–	[101]
N-doped CN	Hydrothermal/Calcination	RhB	350 W Xe lamp	60/94	–	[102]
$NiCo_2O_4$/CN	Calcination/Ultrasonication/Stirring	Tetracycline hydrochloride	500 W Xe lamp	30/98	0.1168	[103]
Cu–O–CN	Calcination	BPA	500 W Xe lamp	100/99	–	[104]
C–CN/Ag_2O/α-Fe_2O_3	Stirring/Calcination	Acid red 14	SMD lamps	90/90	–	[105]
KF, KCl, KBr doped CN	Calcination	MB	250 W Xe lamp	100/99	0.04808, 0.03315, 0.01747	[106]
Ag–CN	Calcination	RhB	300 W Xe lamp	30/93	0.12695	[107]
Boron-doped-CN	Calcination	RhB	300 W Xe lamp	40/100	0.1075	[108]

16.6.3 Electrochemical Destruction of Organic Wastewater Treatment

Due to its quick removal rate, straightforward operation method, and lack of secondary contamination, electrochemical catalytic oxidation technology has recently gained considerable attention for the removal of various contaminants [109]. The MoS_2/CN bifunctional electrocatalysts were created by Monga and coauthors [110] using a straightforward microwave technique, and the electrochemical removal of MB was studied using the MoS_2/CN modified GCE. The highest peak current was obtained with an electrode immersed in a 5:1 MoS_2/CN catalyst solution at pH=7. Guo and coauthors used a hydrothermal approach to create the CN/CNT composites, and the finished materials were used to create functional electrodes for the electro-peroxone process [111]. Evidently, only 27% of the oxalate was degraded by pure CNT's catalytic ozonation process in 60 minutes. The n-CN/CNT composites' removal performances were improved after being combined with CN. Additionally, several catalysts' adsorption abilities were improved. The outcomes further showed that the extraordinary degradation efficiencies were due to the catalysts' catalytic ozonation throughout the preparation phase. In a similar manner, CN-wrapped TiO_2 nanotube arrays were designed and created by Wang and coauthors using two-step procedures that included electrochemical anodization methods and the impregnation under vacuum conditions. The as-prepared samples displayed excellent photoelectrocatalytic activity for the removal of phenol when exposed to visible light [112]. Sun and coauthors used a straightforward electrospinning technique to create the CN/$BiVO_4$ photoanode, and they then used the photoanode in its prepared form to photoelectrocatalytically remove diclofenac sodium model pollution [113]. The ordered CN/ZnO NRAs were designed and created by Kuang and coauthors, and the as-prepared samples displayed impressive photoelectrocatalytic activity for the destruction of MB when exposed to visible light [114].

16.6.4 Catalytic Removal

With very quick kinetics and exceptional stability, the Au-CNx combination demonstrated outstanding catalytic activity toward the reduction of 4-nitrophenol in an aqueous medium in the presence of sodium borohydride ($NaBH_4$). However, the Au-CNx catalyst completes the conversion of 4-NP in under 15 seconds. In addition, the RhB, MB, and methyl red (MR) organic dyes can be removed from water using the Au-CNx catalyst as an absorbent. Despite using high concentrations of dyes for degradation, excellent photodegradation rate constants for MR (0.02 minute), MB (0.024 minute), and RhB (0.024 minutes) were observed [115]. Only 96% of the reduction of 4-NP with AuNPs doped mesoporous carbon nitride (m-CN-Au) reported by M. Antonietti and his coworker occurred with $NaBH_4$ in 300 seconds [116].

16.7 CONCLUSION AND FUTURE OUTLOOK

The elimination of organic water pollutants using various 2D-CN-based composites used for their abatement on a global scale was covered in this chapter. An extensive overview of the properties of CN and its various morphologies, as well as a variety of synthetic techniques and the noteworthy successes of the created composites in terms of treatment objectives, has been provided. Developing heterojunction composites, morphological control, metal and nonmetal ion doping, and the use of sensitizers are some of the techniques for minimizing the number of defects present in pure CN and improving the removal efficiency. It is now a time- and money-saving, cost-effective material for in situ cleanup of polluted organic water contaminants thanks to advancements in fabrication techniques. This can be attributed to an increase in surface area, a longer charge carrier lifetime, and a wider wavelength range of absorbance in the produced composites as compared to pure CN. Uncontrolled release of these materials into the natural habitats is documented as a result of the rapid expansion in the synthesis of new CN-based composites and their widespread use in water and/or wastewater treatment, particularly in the degradation of developing contaminants. Therefore, basic knowledge regarding the effective use and production of tailored composites is crucial to enhancing their environmental friendliness.

REFERENCES

1. Li, D., Zuo, Q., Wu, Q., Li, Q., Ma, J. 2021. Achieving the tradeoffs between pollutant discharge and economic benefit of the Henan section of the South-to-North Water Diversion Project through water resources-environment system management under uncertainty, *J. Cleaner Prod.* 321:128857.
2. Sun, Y., O'Connell, D.W. 2022. Application of visible light active photocatalysis for water contaminants: A review, *Water Environ Res.* 94(10):e10781.
3. Abtahi, M., Dobaradaran, S., Torabbeigi, M., Jorfi, S. Gholamnia, R. Koolivand, A. Darabi, H. Kavousi, A. Saeedi, R. 2019. Health risk of phthalates in water environment: Occurrence in water resources, bottled water, and tap water, and burden of disease from exposure through drinking water in tehran, Iran, *Environ. Res.*173:469–479.
4. Iqbal, N., Agrawal, A., Verma, A., Kumar, J. 2021. Encapsulation of water soluble pesticides for extended delivery of pesticides without contaminating water bodies, *J. Environ. Sci. Health.* 56(5):458–466.
5. Palansooriya, K.N., Yang, Y., Tsang, Y.F., Sarkar, B., Houd, D., Cao, X., Meers, E., Rinklebe, J., Kim, K.-H., Ok, Y.S. 2020. Occurrence of contaminants in drinking water sources and the potential of biochar for water quality improvement: A review, *Crit. Rev. Environ. Sci. Technol.* 56(6):549–611.
6. Shaniba, C., Akbar, M., Ramseena, K., Raveendran, P., Narayanan, B.N., Ramakrishnan, R.M. 2020. Sunlight-assisted oxidative degradation of cefixime antibiotic from aqueous medium using TiO_2/nitrogen doped holey graphene nanocomposite as a high performance photocatalyst. *J. Environ. Chem. Eng.* 8:102204.
7. Ponnaiah, S.K., Periakaruppan, P., Arumuganathana, T., Jeyaprabha B. 2019. Effectual light-harvesting and electron-hole separation for enhanced photocatalytic decontamination of endocrine disruptor using Cu_2O/BiOI nanocomposite, *J. Photochem. Photobiol.*, A 380:111860.
8. Kumar, K.S., Vellaichamy, B., Paulmony, T. 2019. Visible light active metal-free photocatalysis: N-doped graphene covalently grafted with g-C_3N_4 for highly robust degradation of methyl orange, *Solid State Sci.* 94:99–105.
9. Zhang, H.X., Nengzi, L.C., Wang, Z.J., Zhang, X.Y., Li, B., Cheng, X.W. 2020. Construction of Bi_2O_3/CuNiFe LDHs composite and its enhanced photocatalytic degradation of lomefloxacin with persulfate under simulated sunlight. *J. Hazard. Mater.* 383:121236.
10. Wang, J.L., Xu, L.J. 2012. Advanced oxidation processes for wastewater treatment: Formation of hydroxyl radical and application, *Crit. Rev. Environ. Sci. Technol.* 42:251–325.
11. Hassaan, M.A., Nemr, A.E., El-Zahhar, A.A., Idris, A.M., Alghamdi, M.M., Sahlabji, T., Said, T.O. 2022. Degradation mechanism of Direct Red 23 dye by advanced oxidation processes: A comparative study, *Toxin Rev.* 41(1):38–47.
12. Ponnaiah, S.K., Periakaruppan, P., Vellaichamy, B., Nagulan, B. 2018. Efficacious separation of electron–hole pairs in CeO_2-Al_2O_3 nanoparticles embedded GO heterojunction for robust visible-light driven dye degradation, *J. Colloid Interface Sci.* 512:219–230.
13. Sierra, M., Borges, E., Esparza, P., Méndez-Ramos, J., Martín-Gil, J., Martín-Ramos, P. 2016. Photocatalytic activities of coke carbon/g-C_3N_4 and Bi metal/Bi mixed oxides/g-C_3N_4 nanohybrids for the degradation of pollutants in wastewater. *Sci. Technol. Adv. Mater.* 17(1):659–668.
14. Balasubramanian, J., Ponnaiah, S.K., Periakaruppan, P., Kamaraj, D. 2020. Accelerated photodeterioration of class I toxic monocrotophos in the presence of one-pot constructed Ag_3PO_4/polyaniline@g-C_3N_4 nanocomposite: Efficacy in light harvesting, *Environ. Sci. Pollut. Res.* 27:2328–2339.
15. Divakaran, K., Baishnisha, A., Balakumar, V., Perumal, K.N., Meenakshi, C., Kannan, R.S. 2021. Photocatalytic degradation of tetracycline under visible light using TiO_2@sulfur doped carbon nitride nanocomposite synthesized via in-situ method, *J. Environ. Chem. Eng.* 9(4):105560.
16. Hao, Q., Jia, G., Wei, W., Vinu, A., Wang, Y., Arandiyan, H., Ni, B.J. 2020. Graphitic carbon nitride with different dimensionalities for energy and environmental applications, *Nano Res.* 13:18–37.
17. Patnaik, S., Swain, G., Parida, K.M. 2020. Photo-/electro-catalytic applications of visible light-responsive porous graphitic carbon nitride toward environmental remediation and solar energy conversion, *Green Photocatalysts for Energy and Environ. Process*, 36:211–246.
18. Ong, W.J., Tan, L.L., Ng, Y.H., Yong, S.T., Chai, S.P. 2016. Graphitic carbon nitride (g-C_3N_4)-based photocatalysts for artificial photosynthesis and environmental remediation: Are we a step closer to achieving sustainability? *Chem. Rev.* 116:7159–7329.
19. Kalyani, A.K.M., Rajeev, R., Benny, L., Cherian, A.R., Varghese A. 2023. Surface tuning of nanostructured graphitic carbon nitrides for enhanced electrocatalytic applications: A review, *Mater. Today Chem.* 30:101523.

20. Obregón, S. 2023. Exploring nanoengineering strategies for the preparation of graphitic carbon nitride nanostructures, *Flat Chem.* 38:100473.
21. Wang, A., Wang, C., Fu, L., Wong-Ng, W., Lan, Y. 2017. Recent advances of graphitic carbon nitride-based structures and applications in catalyst, sensing, imaging, and LEDs, *Nano-Micro Lett.* 9:47.
22. Praus, P., Smýkalová, A., Foniok, K., Velíšek, P., Cvejn, D., Žádný, J., Storch, J. 2020. Post-synthetic derivatization of graphitic carbon nitride with methane sulfonyl chloride: Synthesis, characterization and photocatalysis, *Nanomater.* 10(2):193.
23. Suter, T., Brázdová, V., McColl, K., Miller, T.S., Nagashima, H., Salvadori, E., Sella, A., Howard, C.A., Kay, C.W.M., Corà, F., McMillan, P.F. 2018. Synthesis, structure and electronic properties of graphitic carbon nitride films, *J. Phys. Chem. C* 122:25183–25194.
24. Liu, J. 2015. Origin of high photocatalytic efficiency in monolayer g-C_3N_4/CdS heterostructure: A hybrid DFT study, *J. Phys. Chem. C* 119:28417–28423.
25. Taghipour, S., Ataie-Ashtiani, B., Hosseini, S.M., Yeung, K.L. 2022. Graphitic carbon nitride-based composites for photocatalytic abatement of emerging pollutants, in *Micro and Nano Technologies* pp. 175–214. https://doi.org/10.1016/B978-0-12-823961-2.00001-X.
26. Tan, L., Nie, C., Ao, Z., Sun, H., An, T., Wang, S. 2021. Novel two-dimensional crystalline carbon nitrides beyond g-C_3N_4: Structure and applications, *J. Mater. Chem. A* 9:17–33.
27. Sun, Y.-P., Ha, W., Chen, J., Qi, H.-Y., Shi, Y.-P. 2016. Advances and applications of graphitic carbon nitride as sorbent in analytical chemistry for sample pretreatment: A review, *TrAC Trends Anal. Chem.* 84(A):12–21.
28. Zhu, J., Xiao, P., Li, H., Carabineiro, S.A.C. 2014. Graphitic carbon nitride: Synthesis, properties, and applications in catalysis, *ACS Appl. Mater. Interfaces* 6(19):16449–16465.
29. Alwin, E., Zieliński, M., Suchora, A., Gulaczyk, I., Piskuła, Z., Pietrowski, M. 2022. High surface area, spongy graphitic carbon nitride derived by selective etching by Pt and Ru nanoparticles in hydrogen, *J. Mater. Sci.* 57:15705–15721.
30. Sakakibara, N., Shizuno, M., Kanazawa, T., Kato, K., Yamakata, A., Nozawa, S., Ito, T., Terashima, K., Maeda, K., Tamaki, Y., Ishitani O. 2023. Surface-specific modification of graphitic carbon nitride by plasma for enhanced durability and selectivity of photocatalytic CO_2 reduction with a supramolecular photocatalyst, *ACS Appl. Mater. Interfaces* 15(10):13205–13218.
31. Liu, X., Ma, R., Zhuang, L., Hu, B., Chen, J., Liu, X., Wang, X. 2020. Recent developments of doped g-C_3N_4 photocatalysts for the degradation of organic pollutants, *Crit. Rev. Environ. Sci. Technol.* 51(8):751–790.
32. Zheng, C., Qiu, X., Han, J., Wu, Y., Liu, S. 2019. Zero-dimensional-g-CNQD-coordinated two-dimensional porphyrin MOF hybrids for boosting photocatalytic CO_2 reduction, *ACS Appl. Mater. Interfaces* 11(45):42243–42249.
33. Liu, Q., Chen, T., Guo, Y., Zhang, Z., Fang, X. 2016. Ultrathin g-C_3N_4 nanosheets coupled with carbon nanodots as 2D/0D composites for efficient photocatalytic H_2 evolution, *Appl. Catal. Environ.* 193:248–258.
34. Li, K., Su, F.Y., Zhang, W.D. 2016. Modification of g-C_3N_4 nanosheets by carbon quantum dots for highly efficient photocatalytic generation of hydrogen, *Appl. Surf. Sci.* 375:110–117.
35. Wang, Y., Liu, X., Liu, J., Han, B., Hu, X., Yang, F., Xu, Z., Li, Y., Jia, S., Li, Z., Zhao, Y. 2018. Carbon quantum dot implanted graphite carbon nitride nanotubes: Excellent charge separation and enhanced photocatalytic hydrogen evolution, *Angew. Chem. Int. Ed.* 57:5765–5771.
36. Wang, W., Zeng, Z., Zeng, G., Zhang, C., Xiao, R., Zhou, C., Xiong, W., Yang, Y., Lei, L., Liu, Y., Huang, D., Cheng, M., Yang, Y., Fu, Y., Luo, H., Zhou, Y. 2019. Sulfur doped carbon quantum dots loaded hollow tubular g-C_3N_4 as novel photocatalyst for destruction of Escherichia coli and tetracycline degradation under visible light, *Chem. Eng. J.* 378:122132.
37. Tahir, M., Cao, C., Mahmood, N., Butt, F.K., Mahmood, A., Idrees, F., Hussain, S., Tanveer, M., Ali, Z., Aslam, I. 2014. Multifunctional g-C_3N_4 nanofibers: A template-free fabrication and enhanced optical, electrochemical, and photocatalyst properties, *ACS Appl. Mater. Interfaces* 6:1258–1265.
38. Kumar, S., Karthikeyan, S., Lee, A.F. 2018. g-C_3N_4-based nanomaterials for visible light-driven photocatalysis, *Catalysts* 8(2):74.
39. Wang, Y., Liu, L., Ma, T., Zhang, Y., Huang, H. 2021. 2D graphitic carbon nitride for energy conversion and storage, *Adv. Funct. Mater.* 31:2102540.
40. Zheng, Y. Lin, L. Ye, X. Guo, F. Wang, X. 2014. Helical graphitic carbon nitrides with photocatalytic and optical activities, *Angew. Chem. Int. Ed.* 53:11926–11930.

41. Li, X.H., Zhang, J., Chen, X., Fischer, A., Thomas, A., Antonietti, M., Wang, X. 2011. Condensed graphitic carbon nitride nanorods by nanoconfinement: Promotion of crystallinity on photocatalytic conversion, *Chem. Mater.* 23:4344–4348.
42. Xiao, L., Liu, T., Zhang, M., Li, Q., Yang, J. 2019. Interfacial construction of zero-dimensional/one dimensional g-C_3N_4 nanoparticles/TiO_2 nanotube arrays with Z-scheme heterostructure for improved photoelectrochemical water splitting, *ACS Sustain. Chem. Eng.* 7:2483–2491.
43. Lin, Z., Wang, X. 2014. Ionic liquid promoted synthesis of conjugated carbon nitride photocatalysts from urea, *Chem. Sus. Chem* 7:1547–1550.
44. Xu, Z., Kong, L., Wang, H., Ma, Q., Shen, X., Wang, J., Premlatha, S. 2022. Soft-template assisted preparation of hierarchically porous graphitic carbon nitride layers for high-performance supercapacitors, *J. Appl. Polym. Sci.* 139:e52947.
45. Bu, S., Yao, N., Hunter, M.A., Searles D.J., Yuan, Q. 2020. Design of two-dimensional carbon-nitride structures by tuning the nitrogen concentration, *Comput. Mater.* 6:128.
46. Xie, L., Li, X., Wang, X., Ge, W., Zhang, J., Jiang, J., Zhang, G. 2019. Bimetallic Pd/Co embedded in two-dimensional carbon-nitride for Z-Scheme photocatalytic water splitting, *J. Phys. Chem. C* 123(3):1846–1851.
47. Xu, H., She, X., Fei, T., Song, Y., Liu, D., Li, H., Yang, X., Yang, J., Li, H., Song, L., Ajayan, P.M., Wu, J. 2019. Metal-oxide-mediated subtractive manufacturing of two-dimensional carbon nitride for high-efficiency and high-yield photocatalytic H_2 evolution, *ACS Nano* 13(10):11294–11302.
48. Ji, J., Wen, J., Shen, Y., Lv, Y., Chen, Y., Liu, S., Ma, H., Zhang, Y. 2017. Simultaneous noncovalent modification and exfoliation of 2D carbon nitride for enhanced electrochemiluminescent biosensing, *J. Am. Chem. Soc.* 139(34):11698–11701.
49. Jia, J., White, E.R., Clancy, A.J., Rubio, N., Suter, T., Miller, T.S., McColl, K., McMillan, P.F., Brázdová, V., Corà, F., Howard, C.A., Law, R.V., Mattevi, C., Shaffer, M.S.P. 2018. Fast exfoliation and functionalisation of two-dimensional crystalline carbon nitride by framework charging, *Angew. Chem. Int. Ed.* 57:12656–12660.
50. Huang, Y., Wang, Y., Bi, Y., Jin, J., Ehsan, M.F., Fu, M., He, T. 2015. Preparation of 2D hydroxyl-rich carbon nitride nanosheets for photocatalytic reduction of CO_2, *RSC Adv.* 5:33254–33261.
51. Nguyen, P.A., Nguyen, T.K.A., Dao, D.Q., Shin, E.W. 2022. Ethanol solvothermal treatment on graphitic carbon nitride materials for enhancing photocatalytic hydrogen evolution performance, *Nanomaterials.* 12(2):179.
52. Kumar, P., Laishram, D., Sharma, R.K., Vinu, A., Hu, J., Kibria, Md.G. 2021. Boosting photocatalytic activity using carbon nitride based 2D/2D van der waals heterojunctions, *Chem. Mater.* 33(23):9012–9092.
53. Wang, L., Tong, Y., Feng, J., Hou, J., Li, J., Hou, X., Liang, J. 2019. G-C_3N_4-based films: A rising star for photoelectrochemical water splitting, *Sustain. Mater. Technol.* 19:e00089.
54. Sheng, Y., Wei, Z., Miao, H., Yao, W., Li, H., Zhu, Y. 2019. Enhanced organic pollutant photodegradation via adsorption/photocatalysis synergy using a 3D g-C_3N_4/TiO_2 free-separation photocatalyst, *Chem. Eng. J.* 370:287–294.
55. Gao, H., Cao, R., Zhang, S., Yang, H., Xu, X. 2019. Three-dimensional hierarchical g-C_3N_4 architectures assembled by ultrathin self-doped nanosheets: Extremely facile hexamethylenetetramine activation and superior photocatalytic hydrogen evolution, *ACS Appl. Mater. Interfaces* 11:2050–2059.
56. Li, X., Xiong, J., Gao, X., Huang, J., Feng, Z., Chen, Z., Zhu, Y. 2019. Recent advances in 3D g-C_3N_4 composite photocatalysts for photocatalytic water splitting, degradation of pollutants and CO_2 reduction, *J. Alloys Compd.*, 802:196–209.
57. Hu, J., Chen, D., Li, N., Xu, Q., Li, H., He, J., Lu, J. 2018. 3D aerogel of graphitic carbon nitride modified with perylene imide and graphene oxide for highly efficient nitric oxide removal under visible light, *Small* 14:1800416.
58. Tan, L., Yu, C., Wang, M., Zhang, S., Sun, J., Dong, S., Sun, J. 2019. Synergistic effect of adsorption and photocatalysis of 3D g-C_3N_4-agar hybrid aerogels, *Appl. Surf. Sci.*, 467(468):286–292.
59. Kumru, B., Shalom, M., Antonietti, M., Schmidt, B. 2017. Reinforced hydrogels via carbon nitride initiated polymerization, *Macromolecules*, 50:1862–1869.
60. Zhang, R., Ma, M., Zhang, Q., Dong, F., Zhou, Y. 2018. Multifunctional g-C_3N_4/graphene oxide wrapped sponge monoliths as highly efficient adsorbent and photocatalyst, *Appl. Catal. B Environ.* 235:17–25.
61. Wan, W., Yu, S., Dong, F., Zhang, Q., Zhou Y. 2016. Efficient CN/graphene oxide macroscopic aerogel visible-light photocatalyst, *J. Mater. Chem. A*, 4:7823–7829.

62. Mao, H., Zhang, Q., Cheng, F., Feng, Z., Hua, Y., Zuo, S., Cui, A., Yao, C.2022. Magnetically separable mesoporous Fe_3O_4@g-C_3N_4 as a multifunctional material for metallic ion adsorption, oil removal from the aqueous phase, photocatalysis, and efficient synergistic photoactivated fenton reaction, *Ind. Eng. Chem. Res.* 61(25):8895–8907.
63. Fulazzaky, M.A. 2019. Study of the dispersion and specific interactions affected by chemical functions of the granular activated carbons, *Environ. Nanotechnology, Monit. Manag.*, 12:100230.
64. Aspera, S.M., David, M., Kasai, H. 2010. First-principles study of the adsorption of water on tri-s-triazine-based graphitic carbon nitride, *Jpn. J. Appl. Phys.*, 49(11):115703.
65. Mohanty, L., Dash, S.K. 2020. Adsorptive removal of MB Dye by graphitic-C_3N_4 from industrial effluents, *Int. J. Scientific Technol. Res.*, 9(3):2029–2034.
66. Dutta, D.P., Dagar, D. 2020. Efficient selective sorption of cationic organic pollutant from water and its photocatalytic degradation by $AlVO_4$/g-C_3N_4 nanocomposite, *J. Nanosci. Nanotechnol.*, 20(4):2179–2194.
67. Nezampour, F., Ghiaci, M., Masoomi, K. 2018. Activated carbon and graphitic carbon nitride immobilized on mesoporous silica for adsorption of nitrobenzene, *J. Chem. Eng. Data*, 63:1977–1986.
68. Azimi, E.B., Badiei, A., Ghasemi, J.B. 2019. Efficient removal of malachite green from wastewater by using boron-doped mesoporous carbon nitride, *Appl. Surf. Sci.*, 469:236–245.
69. Yousefi, M., Villar-Rodil, S., Paredes, J.I., Moshfegh, A.Z. 2019. Oxidized graphitic carbon nitride nanosheets as an effective adsorbent for organic dyes and tetracycline for water remediation, *J. Alloys Compd.*, 809:151783.
70. Fronczak, M., Krajewska, M., Demby, K., Bystrzejewski, M. 2017. Extraordinary adsorption of methyl blue onto sodium-doped graphitic carbon nitride. *J. Phys. Chem. C* 121:15756.
71. Wang, J., Zhang, T., Mei, Y., Pan, B. 2018. Treatment of reverse-osmosis concentrate of printing and dyeing wastewater by electro-oxidation process with controlled oxidation-reduction potential (ORP). *Chemosphere* 201:621–626.
72. Ozturk, E., Karaboyacı, M., Yetis, U., Yigit, N.O., Kitis, M. 2015. Evaluation of integrated pollution prevention control in a textile fiber production and dyeing mill, *J. Clean. Prod.* 88:116–124.
73. Solís, M., Solís, A., Inés Pérez, H., Manjarrez, N., Flores, M. 2012. Microbial decolouration of azo dyes: A review, *Process Biochem.* 47(12):1723–1748.
74. Das, S., Mahalingam, H. 2019. Dye degradation studies using immobilized pristine and waste polystyrene TiO_2/rGO/g-C_3N_4 nanocomposite photocatalytic film in a novel airlift reactor under solar light. *J. Environ. Chem. Eng.* 7(5):103289.
75. Choi, J.U., Kim, Y.G., Jo, W.K. 2019. Multiple photocatalytic applications of non-precious Cu-loaded g-C_3N_4/hydrogenated black TiO_2 nanofiber heterostructure. *Appl. Surf. Sci.* 473:761–769.
76. Wang, S.Y., Li, F., Dai, X.H., Wang, C.J., Lv, X.T., Waterhouse, G.I.N., Fan, H., Ai, S.Y. 2020. Highly flexible and stable carbon nitride/cellulose acetate porous films with enhanced photocatalytic activity for contaminants removal from wastewater. *J. Hazard. Mater.* 384:121417.
77. Zhang, Y., Zhou, J., Feng, Q., Chen, X., Hu, Z. 2018. Visible light photocatalytic degradation of MB using UiO-66/g-C_3N_4 heterojunction nanocatalyst. *Chemosphere* 212:523–532.
78. Chegeni, M., Goudarzi, F., Soleymani, M. 2019. Synthesis, characterization and application of V_2O_5/S-doped graphitic carbon nitride nanocomposite for removing of organic pollutants. *Chem. Select* 4:13736–13745.
79. El-Sheshtawy, H.S., El-Hosainy, H.M., Shoueir, K.R., El- Mehasseb, I.M., El-Kemary, M. 2019. Facile immobilization of Ag nanoparticles on g-C_3N_4/V_2O_5 surface for enhancement of post-illumination, catalytic, and photocatalytic activity removal of organic and inorganic pollutants. *Appl. Surf. Sci.* 467:268–276.
80. Zheng, Y., Yu, Z., Ou, H., Asiri, A.M., Chen, Y., Wang, X. 2018. Black phosphorus and polymeric carbon nitride heterostructure for photoinduced molecular oxygen activation. *Adv. Funct. Mater.* 28:1705407.
81. Li, Z., Jiang, G., Zhang, Z., Wu, Y., Han, Y. 2016. Phosphorus-doped g-C_3N_4 nanosheets coated with square flake-like TiO_2: Synthesis, characterization and photocatalytic performance in visible light. *J. Mol. Catal. A: Chem.* 425:340.
82. Sun, S., Li, C., Sun, Z., Wang, J., Wang, X., Ding, H. 2021. In-situ design of efficient hydroxylated SiO_2/g-C_3N_4 composite photocatalyst: Synergistic effect of compounding and surface hydroxylation. *Chem. Eng. J* 416:129107.
83. Ma, M., Lin, Y., Maheskumar, V., Li, P., Li, J., Wang, Z., Zhang, M., Jiang, Z., Zhang, R. 2022. A highly efficient (Mo, N) codoped $ZnIn_2S_4$/g-C_3N_4 Z-scheme photocatalyst for the degradation of methylene blue. *Appl. Surf. Sci.* 585:152607.

84. Zhang, J., Zhao, X., Wang, Y., Gong, Y., Cao, D., Qiao, M. 2018. Peroxymonosulfate-enhanced visible light photocatalytic degradation of bisphenol A by perylene imide-modified g-C_3N_4, *Appl. Catal. B: Environ.* 237:976–985.
85. Yang, Z., Xu, X., Liang, X., Lei, C., Cui, Y., Wu, W., Yang, Y., Zhang, Z., Lei, Z. 2017. Construction of heterostructured MIL-125/Ag/g-C_3N_4 nanocomposite as an efficient bifunctional visible light photocatalyst for the organic oxidation and reduction reactions, *Appl. Catal. B: Environ.* 205:42.
86. Bo, Y., Yongkun, L., Guohua, J., Depeng, L., Weijiang, Y., Hua, C., Lei, L., Qin, H. 2017. Preparation of electrospun Ag/g-C_3N_4 loaded composite carbon nanofibers for catalytic applications, *Mater. Res. Exp.* 4:015603.
87. Sahu, R.S., Shih, Y.-H., Chen, W.-L. 2021. New insights of metal free 2D graphitic carbon nitride for photocatalytic degradation of bisphenol A, *J. Hazard. Mater.* 402:123509.
88. Chang, C., Fu, Y., Hu, M., Wang, C., Shan, G., Zhu, L. 2013. Photodegradation of bisphenol A by highly stable palladium-doped mesoporous graphite carbon nitride (Pd/mpg-C_3N_4) under simulated solar light irradiation, *Appl. Catal. B* 142–143:553–560.
89. Hak, C.H., Sim, L.C., Leong, K.H., Chin, Y.H., Saravanan, P. 2018. Sunlight photodeposition of Gold nanoparticles onto graphitic carbon nitride (g-C_3N_4) and application towards the degradation of bisphenol A, *IOP Conf. Ser. Mater. Sci. Eng.*, 409:012008.
90. Wang, Y., Zhao, X., Cao, D., Wang, Y., Zhu, Y. 2017. Peroxymonosulfate enhanced visible light photocatalytic degradation bisphenol A by single-atom dispersed Ag mesoporous g-C_3N_4 hybrid, *Appl. Catal. B*, 211:79–88.
91. Vuggili, S.B., Gaur U.K., Sharma, M. 2022. 2D/2D facial interaction of nitrogen-doped g-C_3N_4/In_2S_3 nanosheets for high performance by visible-light-induced photocatalysis, *J. Alloys Compd.* 2022:163757.
92. Kumar, S., Kumar, A., Kumar, A., Balaji, R., Krishnan, V. 2018. Highly efficient visible light active 2D-2D nanocomposites of N-ZnO-g-C_3N_4 for photocatalytic degradation of diverse industrial pollutants, *Chem. Select* 3(6):1919–1932.
93. Bai, X., Wang, X., Lu, X., Hou, S., Sun, B., Wang, C., Jia T., Yang, S. 2021. High crystallinity and conjugation promote the polarization degree in O-doped g-C_3N_4 for removing organic pollutants, *Cryst. Eng. Comm.* 2021:1366–1376.
94. Hassani, A., Faraji M., Eghbali, P. 2020. Facile fabrication of mpg-C_3N_4/Ag/ZnO nanowires/Zn photocatalyst plates for photodegradation of dye pollutant, *J. Photochem. Photobiol. A* 400:112665.
95. Ge, F., Huang, S., Yan, J., Jing, L., Chen, F., Xie, M., Xu, Y., Xu H., Li, H. 2021. Sulfur promoted n-π^* electron transitions in thiophene-doped g-C_3N_4 for enhanced photocatalytic activity, *Chin. J. Catal.* 42(3):450–459.
96. Huang, Y., Ning, L., Feng, Z., Ma, G., Yang, S., Su, Y., Hong, Y., Wang, H., Peng, L., Li, J. 2021. Graphitic carbon nitride nanosheets with low O N1-doping content as efficient photocatalysts for organic pollutant degradation, *Environ. Sci. Nano* 8(2):460–469.
97. Zhang, W., Ma, Y., Zhu, X., Liu, S., An, T., Bao, J., Hu, X., Tian, H. 2021. Fabrication of Ag decorated g-C_3N_4/$LaFeO_3$ Z-scheme heterojunction as highly efficient visible-light photocatalyst for degradation of methylene blue and tetracycline hydrochloride, *J. Alloys Compd.*, 864:158914.
98. Mafa, P.J., Malefane, M.E., Idris, A.O., Liu, D., Gui, J., Mamba, B.B., Kuvarega, A.T. 2022. Multielemental doped g-C_3N_4 with enhanced visible light photocatalytic activity: Insight into naproxen degradation, kinetics, effect of electrolytes, and mechanism, *Sep. Purif. Technol.* 282:120089.
99. Jourshabani, M., Shariatinia, Z., Badiei, A. 2018. Synthesis and characterization of novel Sm_2O_3/S-doped g-C_3N_4 nanocomposites with enhanced photocatalytic activities under visible light irradiation, *Appl. Surf. Sci.*, 427:375–387.
100. Wang, D., Dong, X., Lei, Y., Lin, C., Huang, D., Yu, X., Zhang, X. 2022. Fabrication of Mn/P co-doped hollow tubular carbon nitride by a one-step hydrothermal–calcination method for the photocatalytic degradation of organic pollutants, *Catal. Sci. Technol.* 12(18):5709–5722.
101. Chen, H., Zhang, X., Jiang, L., Yuan, X., Liang, J., Zhang, J., Yu, H., Chu, W., Wu, Z., Li, H., Li, Y. 2021. Strategic combination of nitrogen-doped carbon quantum dots and g-C_3N_4: Efficient photocatalytic peroxydisulfate for the degradation of tetracycline hydrochloride and mechanism insight, *Sep. Purif. Technol.* 272:118947.
102. Wang, G., Zhao, Y., Ma, H., Zhang, C., Dong, X., Zhang, X. 2021. Enhanced peroxymonosulfate activation on dual active sites of N vacancy modified g-C_3N_4 under visible-light assistance and its selective removal of organic pollutants, *Sci. Total Environ.* 756:144139.
103. Jiang, J., Wang, X., Yue, C., Liu, S., Lin, Y., Xie, T., Dong, S. 2021. Efficient photoactivation of peroxymonosulfate by Z-scheme nitrogen-defect-rich $NiCo_2O_4$/g-C_3N_4 for rapid emerging pollutants degradation, *J. Hazard. Mater.* 414:125528.

104. Song, H., Guan, Z., Xia, D., Xu, H., Yang, F., Li, D., Li, X. 2021. Copper-oxygen synergistic electronic reconstruction on g-C_3N_4 for efficient non-radical catalysis for peroxydisulfate and peroxymonosulfate, *Sep. Purif. Technol.* 257:117957.
105. Amani-Ghadim, A.R., Tarighati Sareshkeh, A., Nozad Ashan, N., Mohseni-Zonouz, H., Seyed Ahmadian, S.M., Seyed Dorraji, M.S., Gholinejad, M., Bayat, F. 2021. Photocatalytic activity enhancement of carbon-doped g-C_3N_4 by synthesis of nanocomposite with Ag_2O and α-Fe_2O_3, *J. Chin. Chem. Soc.* 68(11):2118–2131.
106. Prabakaran, E., Velempini, T., Molefe, M., Pillay, K. 2021. Comparative study of KF, KCl and KBr doped with graphitic carbon nitride for superior photocatalytic degradation of methylene blue under visible light, *J. Mater. Res. Technol.* 15:6340–6355.
107. Tian, C., Tao, X., Luo, S., Qing, Y., Lu, X., She J., Wu, Y. 2018. Cellulose nanofibrils anchored Ag on graphitic carbon nitride for efficient photocatalysis under visible light, *Environ. Sci. Nano* 5(9):2129–2143.
108. Su, L.X., Liu, Z.Y., Ye, Y.L., Shen, C.L., Lou, Q., Shan, C.X. 2019. Heterostructured boron doped nanodiamonds @ g-C_3N_4 nanocomposites with enhanced photocatalytic capability under visible light irradiation, *Int. J. Hydrogen Energy*, 944(36):19805–19815.
109. Hu, S.G., Hu, J.P., Sun, Y.F., Zhu, Q., Wu, L.S., Lu, B.C., Xiao, K.K., Liang, S., Yang, J.K., Hou, H.J. 2021. Simultaneous heavy metal removal and sludge deep dewatering with Fe(II) assisted electrooxidation technology, *J. Hazard. Mater.* 405:124072.
110. Monga, D., Ilager, D., Shetti, N.P., Basu, S., Aminabhavi, T.M. 2020. 2D/2D heterojunction of MoS_2/g-C_3N_4 nanoflowers for enhanced visible-light-driven photocatalytic and electrochemical degradation of organic pollutions, *J. Environ. Manag.* 274:111208
111. Guo, Z., Cao, H.B., Wang, Y.X., Xie, Y.B., Xiao, J.D., Yang, J., Zhang, Y. 2018. Highly activity of g-C_3N_4/multiwall carbon nanotube in catalytic ozonation promotes electro- peroxone process, *Chemosphere*, 201:206–213.
112. Wang, H., Liang, Y.H., Liu, L., Hu, J.S., Cui, W.Q. 2018. Highly ordered TiO_2 nanotube arrays wrapped with g-C_3N_4 nanoparticles for efficient charge separation and increased photoelectrocatalytic degradation of phenol, *J. Hazard. Mater.* 344:369–380.
113. Sun, J.Y., Guo, Y.P., Wang, Y., Cao, D., Tian, S.C., Xiao, K., Mao, R., Zhao, X. 2018. H_2O_2 assisted photoelectrocatalytic degradation of diclofenac sodium at g-C_3N_4/$BiVO_4$ photoanode under visible light irradiation, *Chem. Eng. J.* 332:312–320.
114. Kuang, P.Y., Su, Y.Z., Chen, G.F., Luo, Z., Xing, S.Y., Li, N., Liu, Z.Q. 2018. g-C_3N_4 decorated ZnO nanorod arrays for enhanced photoelectrocatalytic performance, *Appl. Surf. Sci.* 358:296–303.
115. Bhowmik, T. Kundu, M.K. Barman, S. 2015. Ultra small gold nanoparticles–graphitic carbon nitride composite: An efficient catalyst for ultrafast reduction of 4-nitrophenol and removal of organic dyes from water, *RSC Adv.* 5:38760–38773.
116. Li, X.-H., Wang, X., Antonietti, M. 2012. Mesoporous g-C_3N_4 nanorods as multifunctional supports of ultrafine metal nanoparticles: Hydrogen generation from water and reduction of nitrophenol with tandem catalysis in one step, *Chem. Sci.* 3:2170–2174.

17 2D MXenes Nanomaterials for Removal of Organic Wastewater Contaminants

Lindani Mdlalose, Lerato Hlekelele, and Vongani Chauke

17.1 INTRODUCTION

The research and development of two-dimensional (2D) materials was prompted and advanced after the discovery of the remarkable physical properties of single/multiple layered graphene. This hastily encouraged more research on 2D materials in the form of manipulating the structure of graphene through exfoliation, altering the starting material with readily available layered precursors such as graphite-like hexagonal boron nitride or dichalcogenides or even layered oxides [1]. This then stimulated the development of more 2D materials including the birth of MXenes. MXenes are a group 2D transition metal carbides, carbonitrides, and nitrides discovered in 2011 [2]. Their single flakes are denoted by a chemical formula $M_{n+1}X_nT_x$ (n = 1 to 4), which designates transition metals alternating layers (M) enclosed by carbon/nitrogen (X) layers with attached terminations Tx ($-O_2$, $-F_2$, $-OH_2$, $-Cl_2$) on the external transition metal surfaces [3]. Due to their intriguing electrical and optical properties, they play numerous roles in photodetectors. Additionally, their distinctive mechanical, chemical, and physical properties allow MXenes to be altered by various surface terminations and transition metals.

The atomically narrow structure of 2D MXenes makes it an appropriate alternative material for water purification technologies. Additionally, its large surface area, excellent mechanical strength, and numerous functional groups on their surfaces make it a suitable candidate for the uptake of contaminants from aqueous medium [4]. MXenes water purification interest is facilitated by its unique adsorptive, antibacterial, and reductive properties, which are further augmented by high electrical conductivity.

With the intensive industrialization and vast agricultural systems, the release of toxic contaminants into ground and surface water continues to be a strain on the environment. In this chapter, the potential use of 2D MXenes derivatives for organic contaminants (such as dyes, antibiotics, and pharmaceuticals) removal is addressed. This entails mechanistic pathways of using MXene-based materials as adsorbents, water purification membranes, and photocatalysts. The ability of MXene-based composites in showing catalytic activity toward diverse pollutants and superior selectivity toward specific pollutants will be discussed.

17.2 SYNTHESIS AND STRUCTURE ATTRIBUTES OF TI_3C_2 MXENE MATERIALS

MXenes construction techniques involve selective acid etching of either MAX/non-MAX parents and the chemical transformation method or bottom-up, as shown in Figure 17.1. The schematic representation in Figure 17.1a indicates three steps, where the first stage involves the exfoliation activity of $TiAlC_2$, followed by how the Al atoms are replaced after the reaction with hydrofluoric acid to form OH, and lastly, the breaking of hydrogen bonds and parting of nanosheets after being sonicated in methanol [5]. Figure 17.1b simply represents the chemical vapor deposition to produce

DOI: 10.1201/9781003436942-17

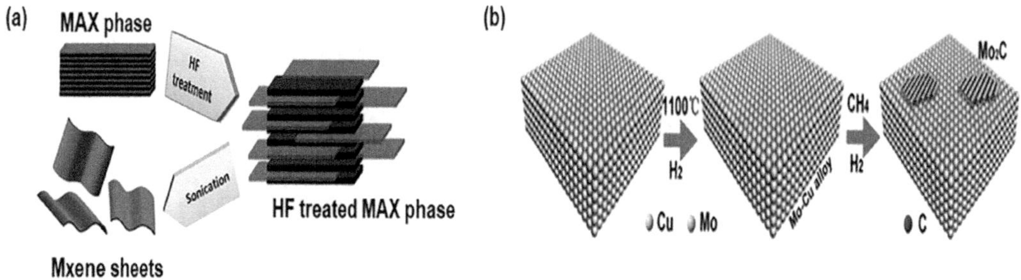

FIGURE 17.1 Schematic of MXene synthesis methods. (a) Selective etching approach and (b) bottom-up construction. (Reproduced with permission from Ref. [5], Copyright @ 2012 American Chemical Society.)

2D carbide crystals [6]. Here, the molten copper dissolves molybdenum (Mo), permitting the growth of α-Mo_2C crystals after the reaction of Mo atoms with carbon atoms resulting from the decomposition of methane at the melt surface.

The selective acid etching method is most preferred due to its feasibility aspects, the production yields, cost-effectiveness, and controllability of the process [7]. In comparison to covalent/ionic/metallic combined M-X bond, the weaker M-A metallic bond permits the viability of selective extraction of the "A" atomic layers from the MAX phase ternary. Mostly, hydrofluoric acid (HF) is used for MXenes production, where it reacts with low bonding energy A-elements in the exfoliation process resulting in gaseous hydrogen and fluorides and subsequently an accordion-like structure of $M_{n+1}X_n$. With over ten types of A-elements in MAX phase reported, for MXenes production, only aluminum (Al) and silicon (Si) have been etched. Generally, in Al-contained MAXs, there are limited valence electrons that are simple to be etched in comparison to the Si; additionally, moderate conditions are required to etch M_2AX phase in comparison to M_3AX_2 and M_4AX_3 phases. When it comes to the wet acid etching procedure, Carbides are much simpler to prepare than nitrides and carbon nitrides [7]. As a result, there are relatively few nitrides that have been successfully prepared through wet acid etching.

As mentioned earlier, MXenes can also be produced using non-MAX precursors via a selective acid etching approach [8]. In this process, weak bonded elements are utilized instead of single A-elements. Non-MAX precursor of Mo_2Ga_2C was used to produce Mo_2C MXene with the extraction of binary layers of "A" element [8]. It has been observed that some MXenes are tricky to obtain using the established wet acid etching of MAX, while they can simply be constructed from their non-MAX precursors.

Selective etching method can also be used to produce MXenes without using an acid [9]. This process entails anodic etching in alkaline aqueous electrolyte consisting of a binary solution of ammonium chloride and tetramethylammonium hydroxide, and the Ti_3AlC_2 rods are applied as electrodes. The protons required in the etching process, which are generally from the acid in the acid selective method are replaced by the electric current in this instance.

Apart from the commonly implemented selective etching approaches for MXene production, the chemical transformation is gaining interest. The in situ chemical transformation process for MXenes production is especially relevant to high-performance heterostructures [10]. The enhanced electrical conductivity of MXenes was achieved via ammoniation of parent carbides. The nitrides of molybdenum and vanadium were produced through Mo_2C and V_2C ammoniation at 600°C. However, the obtained structure varies where Mo_2N maintains a hexagonal structure, whereas the V_2C changes into a diverse layered assembly of trigonal V_2N and cubic VN. The top-down selective etching methods are still preferrable over bottom-up construction methods. This might be due to MXenes complex crystal lattice and its multi-elemental components.

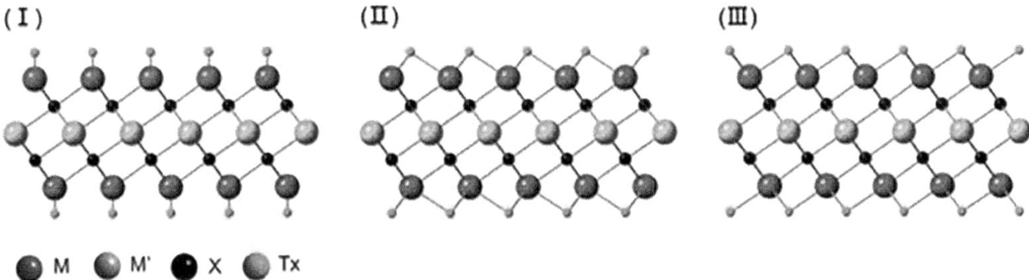

FIGURE 17.2 Three possible positions of surface terminations on MXenes. (Reproduced with permission from Ref. [12], Copyright @ 2016 Royal Chemical Society.)

The basic morphological properties of pristine MXene powder acquired from selective acid etching procedure generally show an accordion-like layered structure [11]. Its crystal structure usually inherits the hexagonal atomic lattice of their MAX parents. In MXenes surface terminations, the chemically active transition metals remain on the surface of the stacked sequence. The MXenes produced via fluoride contained acid etching will essentially show electronegative hydrophilic surface terminations which include -OH, -O, and F. Anionic terminations (-Cl, -S) with large bonding energies can also be obtained. The three potential surface termination sites are indicated in Figure 17.2: (i) exactly on the surface of metal atoms, (ii) in a hollow location in between the top metal atoms, and (iii) resonating site between the following packing layer of X atoms [12]. The evaluated bond lengths are 2.1 Å, 1.9 Å, and 0.97 Å for Ti-F, Ti-O, and O-H, respectively. With all the functionalized terminations, the oxygen terminated characteristics show the substantial adsorption energy, which is followed by -F, -OH, and -H being the least.

MXenes in their pristine form occupy metallic conductivities and electron densities close to the Fermi level because of the d-electrons it contains from the transition metals ("M") [13]. The influence of electronic properties is greater in well-ordered double-M structures with outer transition metal layers in comparison to metal inner layer. This is however different when the MXenes are functionalized as each functional group will contribute its own electronic properties. Through functionalization, the density of states (DOS) substantially gets reduced by terminations leading to the lifting of d-band above the Fermi level and subsequently band gap creation. This is enabled by the hybridization between d-electrons of the transition metal and p-orbitals functional groups. The electronic structures exhibited by -F and -OH groups are similar because they can only accept a single electron, whereas oxygen terminated shows a different behavior as it can allow two electrons [14]. Therefore, the electronic structure of MXenes is determined by the type of "M" and "X" atoms, together with the terminations. There are investigations that have proven that band gap is also allied to material doping and modulation strain [15]. When Sc_2CO_2 was doped with niobium atoms, its band gap decreases from 2.0 to 1.8 eV and it further reduces to 1.4 eV with the uniaxial strain of 3% [15]. The high Young modulus, electric conductivities, and thermal properties with amendable band gap are some of the remarkable factors that contribute to MXenes uniqueness. Among other 2D materials like graphene, MXenes hydrophilic surfaces and high metallic conductivities present them distinguishable.

Heterostructures consisting of intriguing electronic properties can be fabricated on MXenes base and on other 2D materials, as indicated in Figure 17.3a. The designed heterostructures facilitate the manipulation and alteration of electronic properties by proximity impact and the interface induced strain [16]. The formation of heterostructures and electronic behavior for MXenes were projected using theoretical calculations. Due to MXenes diversity and dichalcogenides transition metal, the appropriate heterostructures can be designed with minimal lattice mismatch to form and examine

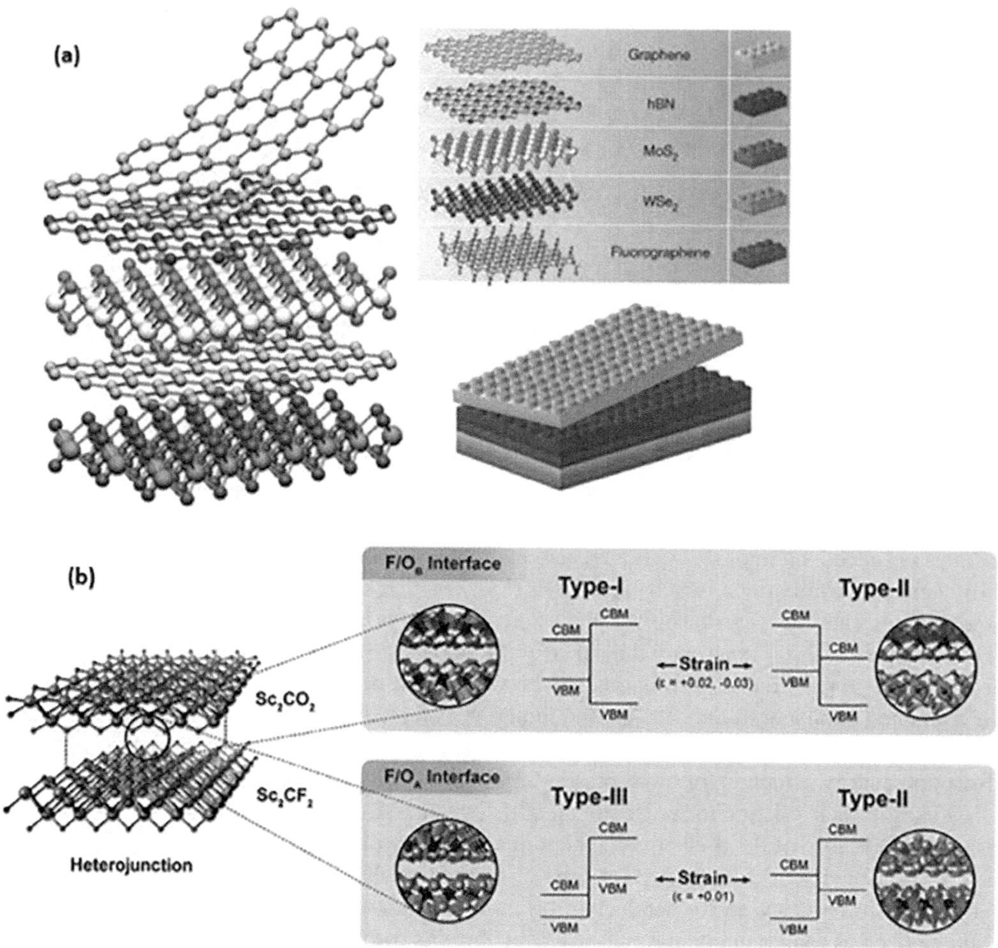

FIGURE 17.3 (a) Demonstration of heterostructure formation for different layers and (b) illustration of MXene heterojunctions and impacts of strain and band properties functionalization. (Reproduced with permission Ref. from [19], Copyright @ 2015 American Chemical Society.)

the interfaces [17]. The MoS_2/Ti_2C, for example, will occupy strong chemical bonding due to unsaturated surface states, but a layer-by-layer van der Waals (vdW)-form non-covalent is demonstrated in MoS_2/Ti_2CF_2 and $MoS_2/Ti_2C(OH)_2$ interfaces [18]. The metallic nature of Ti_2C transforms the electronic structure of MoS_2 to the metallic kind, whereas the other two heterostructures maintain their semiconducting properties. As indicated in Figure 17.3b, different heterostructures can be achieved by the combination of different MXene layers [19]. In this case, 2D Sc_2C (Sc_2CF_2, $Sc_2C(OH)_2$, and Sc_2CO_2) formed type I, II, and III heterostructures. Type I heterostructure describes the complete overlap between the band gaps creating a straddling gap. In type II, the valence and the conduction band borders of one semiconductor are lower than the band edges of the other semiconductor. While in type III heterostructure, there is an overlap in conduction and valence bands from separate semiconductors, creating a broken gap. This indicates that Schottky barriers in either a semiconductor or metal heterostructure can be manipulated through an anticipated functionalization and applied electric fields. When using the first principle measurement, it was discovered that O-terminated MXenes are capable of forming a hole barrier-free interaction with WSe_2, whereas the OH-terminated ones can form an electron barrier-free interaction with 2D semiconductors [10].

17.3 2D MXENE-BASED COMPOSITES FOR DEGRADATION OF ORGANIC WATER CONTAMINANTS

17.3.1 PHOTOCATALYTIC DEGRADATION WITH 2D MXENE COMPOSITES

The distinctive properties exhibited by the MXenes structure which include high metallic conductivity, different functional groups, layered assembly, and large specific surface area make it a potential cocatalyst in photocatalysis. The core structure of Ti_3C_2 is set to be the photogenerated electrons acceptor that enables photogenerated charges partitioning in the Schottky barriers as it has metallicity and a lower Fermi level compared to most semiconductors in the photocatalysis process. Additionally, for enhanced photocatalytic activity, the morphology and the size of MXenes cocatalyst are crucial. Therefore, a well-arranged multilayered accordion MXenes is likely to exhibit a robust matrix when used to support even growth and fine dispersion of photocatalysts (Figure 17.4) [20]. It is observed that single-layer Ti_3C_2 has intimate contact with TiO_2 hence it's likely to show improved photogenerated charge carrier separation in comparison to multilayered Ti_3C_2. As a result, more prudence is required when tuning MXene thickness for photocatalyst preparation to ensure that photocatalytic activity is not compromised. It is however mentioned that in multilayered Ti_3C_2-based photocatalysts, Ti_3C_2 has a great potential to assemble semiconductors to enhance the photogenerated charge transfer and separation while acting as electron acceptor.

For a multilayered stacked Ti_3C, it is simpler to fabricate via HF etching [21]. Several semiconductors like graphitic carbon nitride (g-C_3N_4), metal oxides, metal sulfide, and metal–organic frameworks (MOFs) have been constructed onto multilayered MXenes for application in photocatalysis [21]. Improved photocatalytic activity was observed when g-C_3N_4 semiconductor was used [22]. In the investigation, a chosen precursor (melamine) was combined with Ti_3C_2 MXene followed by calcination under inert environment. In this one-step synthesis procedure, a uniform 2D lamellar g-C_3N_4 was formed and the even distribution on the multilayered surface was verified by scanning electron microscopy (SEM) analysis as shown in Figure 17.5(b). The findings represented in Figure 17.5a showed that the Schottky junction between Ti_3C_2 and g-C_3N_4 effectively restrained the recombination of photoinduced carriers. The electron immigration was well facilitated by exceptional Ti_3C_2 conductivity and components interface. This demonstrated that g-C_3N_4/Ti_3C_2 composite could be a potential approach for photoconversion applications.

Because of the chemical stability and other physicochemical properties, g-C_3N_4 in photocatalysis has been widely investigated and is attractive to technologies related to environmental and energy issues. However, g-C_3N_4 pristine photocatalyst still has challenges as its catalytic performance is insufficient due to the recombination of high charge, low specific surface area, and limited use of

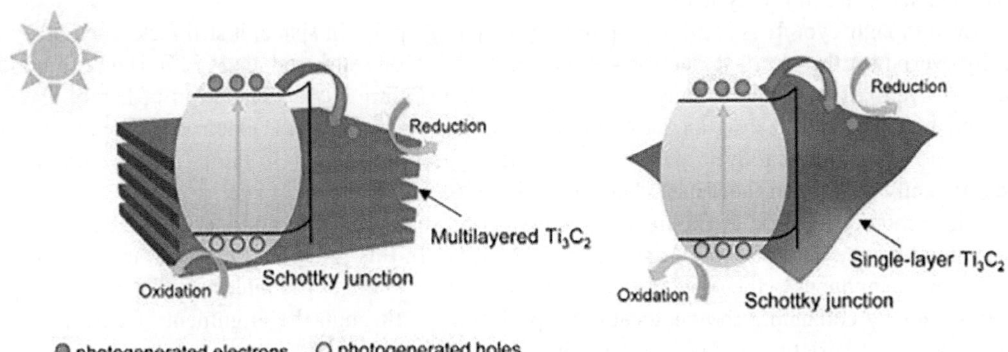

FIGURE 17.4 Typical mechanism of Ti_3C_2 MXenes on supported semiconductor-based photocatalysis. (Reproduced with permission from Ref. [20], Copyright @ 2020 American Chemical Society.)

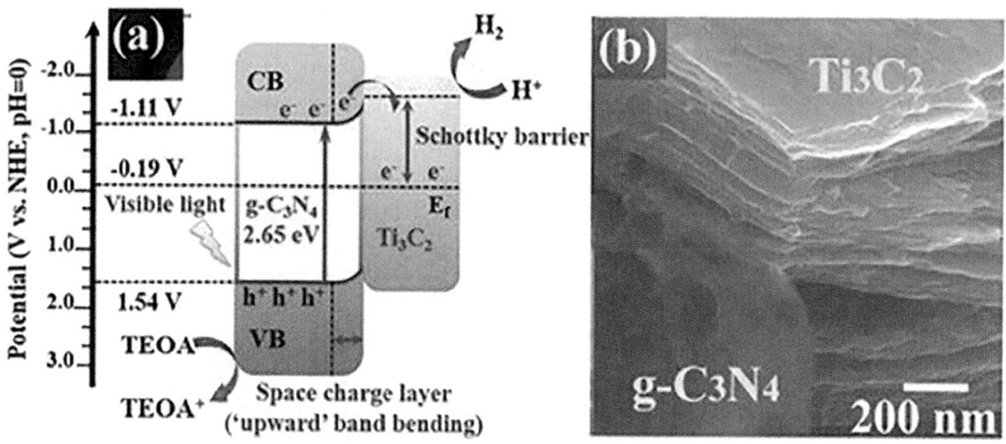

FIGURE 17.5 (a) Schematic presentation of photocatalytic activity and (b) SEM image of g-C_3N_4/Ti_3C_2 composite. (Reproduced with permission from Ref. [22], Copyright @ 2020 Elsevier.)

visible light. As a result, its catalytic performance is further enhanced by tailoring with transition metal sulfide [20]. The blended heterojunction between g-C_3N_4 and metal sulfide serves as the best approach to boost the spatial separation of photogenerated electron pairs leading to improved photocatalytic activity. Based on the charge separation mechanism for several heterostructures among g-C_3N_4 and joined metal sulfides, the resulting composite photocatalyst is classified into five heterojunction categories [23]. The five different heterojunction systems are shown in Figure 17.6, namely: (1) Type I, Type II, (3) p–n heterojunction, (4) Schottky junction, and (5) Z-scheme heterojunction. A closer look at Figure 17.6 shows that for the Type I heterojunction with straddling alignment, of the two semiconductors, the VB and CB are quite lower for semiconductor 1. These are found to be relatively higher than those of semiconductor 2; as a result, once the electrons or energy are gained from light irradiation, the photogenerated holes will move from semiconductor 1 VB to the 2nd semiconductor. While this occurs, the photoinduced electron will be able to migrate from the CB of semiconductor 1 to the CB of the second semiconductor.

Type II heterojunction is highly required for photocatalysis performance. Here, both semiconductors with similar band potentials are strongly bonded to create a stable heterostructure, as shown in Figure 17.6b. The optimal band position presented by Type II heterojunctions assists in gaining enough separation of photogenerated electron–hole pairs; hence, the most reported heterojunction photocatalysts are of this type [24].

Even though Type II is ideal for separating the charge pairs in space, it still lacks the ability to fully overpower the ultrafast electron–hole recombination on semiconductors [25]. Hence, a notion of p–n heterojunction photocatalysts is recommended (Figure 17.6c). This kind of heterojunction complements by introducing an additional electric field to speed up the electron–hole pairs movement to advance the catalytic performance. The separation efficacy in the p–n heterojunction is significantly faster than the Type II heterojunction [26].

The Schottky junction in Figure 17.6d is made of a semiconductor and metallic material that is beneficial for the assembly of space-charge separation. In this junction, the electrons can transfer from one component to the other at a very fast rate. This occurs at the interface of the two components, thereby enhancing the photocatalytic performance through the alignment of Fermi energy levels and the restriction of charge recombination [27].

The Z-scheme heterojunction in Figure 17.6e and f is addressing issues associated with not only charge transfer for efficient separation of photogenerated charge carries but also the oxidation and reduction on the semiconductors. The other mentioned heterojunctions are mainly focusing

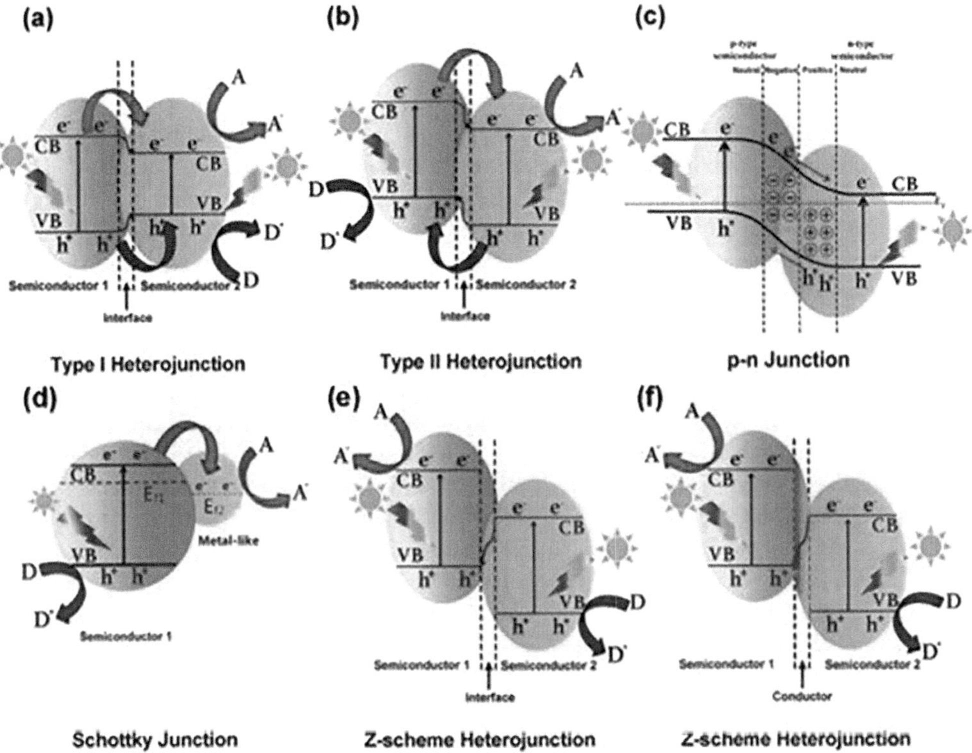

FIGURE 17.6 Heterojunctions for photocatalysis of nanocomposites with different band structures: (a) Type I heterojunction, (b) Type II heterojunction, (c) p–n junction, (d) Schottky junction, (e) Z-scheme heterojunction (no electron mediator), and (f) indirect Z-scheme (electron mediator). A, D, and EF denote the electron acceptor, electron donor, and fermi level, respectively. (Reproduced with permission from Ref. [23], Copyright @ 2019 Elsevier.)

on advancing the electron–hole separation efficacy. There are similarities that are noted between Z-scheme and Type II in their band structure with varying charge transfer paths (Figure 17.6). To differentiate the two, when there is a mediator in the heterojunction, it is projected that the photocatalyst was driven by the Z-scheme, while the absence of the mediator indicates a synergistic formation of direct Z-scheme and Type II heterojunction [28,29].

The plasticity properties and visible light activity of spinel ferrite, MFe_2O_4 (M = Ni, Zn, Co, Cu, etc.), drew much attention on its use in photocatalysis. The combination of $CuFe_2O_4$ with a p-type semiconductor has a potential to form a cocatalyst for visible light response [30]. Even though $CuFe_2O_4$ is recommendable, these nanoparticles are susceptible to agglomeration resulting from intense magnetic interaction during the synthesis process. The introduction of 2D MXenes comes into play to solve these issues by providing heterostructure-assisted catalyst. MXenes in this case were able to increase the dispersity of $CuFe_2O_4$, preventing agglomeration [31]. The altered terminal functional groups of MXenes managed to control the band gap and collected significant number of electrons near its Fermi level. The MXene therefore served as an electron reservoir, which resulted in effective carrier movement [31]. This proved that incorporating or fixing particles on MXenes does not only provide suitable dispersity but also advances the materials reactive sites and space-charge regions with improved carrier recombination. The produced $CuFe_2O_4/Ti_3C_2$ nanocomposite was applied in the photodegradation of sulfamethazine. The band gap measured by UV–Vis DRS was 0.63 eV, while the valence band analyzed using X-ray photoelectron spectroscopy (XPS) VB spectrum was 0.33 eV of which $CuFe_2O_4$ alone had 0.31 eV. The composite further demonstrated that a significant number of photogenerated electrons in $CuFe_2O_4$ were at the Fermi level under light illumination, suggesting the metallic behavior of Ti_3C_2 MXene [31].

The characteristics of MXenes, including rich electron density, good metal conductivity, and adjustable band gap, are crucial for implementation as photocatalysts. Additionally, MXenes can simply be combined with other photocatalysts to create a heterojunction. A photocatalyst is required to produce numerous reactive species to successfully degrade organic pollutants. The catalytic activity of MXene-based materials has been highly explored for environmental pollution applications. With changing societal modernization, environmental contamination from existing/emerging organic contaminants highlights photocatalysis as a viable process for environmental remediation. MXene is a 2D titanium carbide with an adjustable band gap of 0.92–1.75eV [32]. The MXenes in its pristine form suffer from charge carrier recombination resulting in reduction of quantum efficacy and photocatalytic performance. As a result, photocatalytic properties are improved by alteration to its heterostructure and Schottky junction. In comparison to traditional carbonaceous materials, the Ti in Ti_3C_2 MXenes has a high redox capability, which makes it simple to capture photogenerated electrons leading to accelerated separation and migration of photoexcited charge carriers [32]. This provides more advantage for MXenes to degrade organic compounds under light irradiation.

$Ti_3C_2T_x$ MXene showed excellent performance in the degradation of methylene blue dye under UV irradiation [33]. Methylene blue removal efficiency ranged between 62% and 81% when exposed to UV irradiation, while in the dark, methylene blue removal was facilitated by electrostatic attraction between the negatively charged surface of $Ti_3C_2T_x$ and cationic dye. However, it was noticed that over time, the oxidation of $Ti_3C_2T_x$ in the presence of dispersed O_2 to form TiO_2 was a challenge in this aspect [33]. When comparing the performance of $Ti_3C_2T_x$ material on a positively charged dye, methylene blue with the anionic acid blue, the former dye degraded favorably by 81%, where the latter degraded only 62% under UV irradiation over a 5-hour period [34]. In the absence of UV irradiation (in the dark), 18% degradation was attained for positively charged methylene blue, while no degradation was achieved for anionic acid blue. The surface functional group of $Ti_3C_2T_x$ MXene consists of -OH and -F, making it advantageous for adsorption interaction with cationic methylene blue over anionic acid blue. The great degradation performance under UV irradiation is facilitated by the formation of titanium tetrahydroxide and titanium dioxide generating the photocatalytic outcome. In this investigation, the challenge associated with MXene oxidation to dissolved O_2 over time to produce TiO_2 was cautiously controlled by improving methylene blue degradation in the dark and accelerating the photodegradation under UV irradiation [34].

To enhance the visible-light-driven degradation, cadmium sulfide (CdS) was incorporated onto Ti_3C_2/TiO_2 MXene to form a $CdS/Ti_3C_2/TiO_2$ hybrid via a facile calcination and hydrothermal methods [35]. The synthesized nanohybrid outperformed the unmodified MXene where the optimized $CdS/Ti_3C_2/TiO_2$ material showed effective degradation for methylene blue, rhodamine B, sulfachloropyridazine, and phenol under visible light illumination ($\lambda \geq 420$ nm). The enhanced photocatalytic activity of the CdS-modified nanohybrid material is caused by the increased electron–hole separation efficacy resulting in free radicals that play a crucial role in oxidizing the studied organic contaminants [35]. Cerium dioxide (CeO_2) was also incorporated on Ti_3C_2 MXene to produce nanosheets cube-like composite for the degradation of tetracycline, a pharmaceutical contaminant, under sunlight illumination [36]. The CeO_2/Ti_3C_2-MXene composite transmission electron microscopy (TEM) image in Figure 17.7a demonstrates the cube-like morphology of CeO_2 grown on Ti_3C_2-MXene nanosheets. The improved photocatalytic performance was confirmed by characterization with UV–Vis in Figure 17.7b, as the metallic character of Ti_3C_2 MXene showed no absorption band, whereas the composite presented improved light absorption capability in the visible region compared to pure CeO_2. The CeO_2 absorption edge was centered at 410 nm, corresponding to the band gap of 3.16 eV, as indicated in Figure 17.7c.

The optimized composite also exhibited a strong interfacial effect in the CeO_2/Ti_3C_2-MXene heterojunction when compared to CeO_2, as indicated by the lower interfacial resistance of the

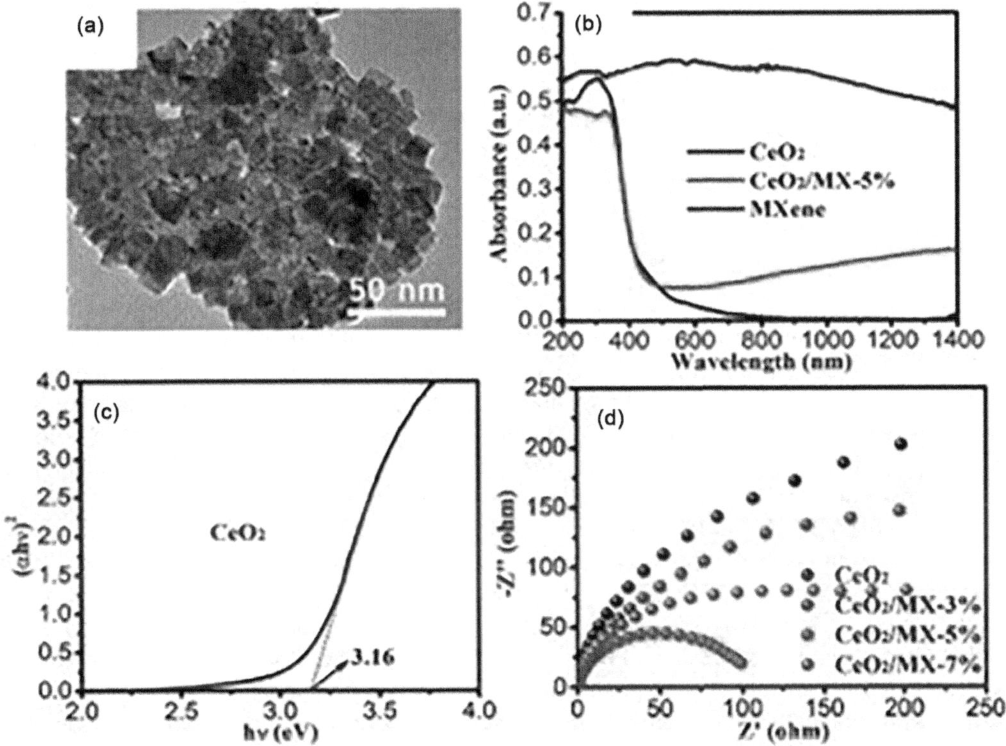

FIGURE 17.7 (a) TEM image of CeO_2/Ti_3C_2-MXene; (b) UV–vis spectra of the CeO_2, Ti_3C_2-MXene, and CeO_2/Ti_3C_2-MXene; (c) band gap energies of the CeO_2 and CeO_2/Ti_3C_2-MXene; and (d) EIS curves of CeO_2 and different compositions of CeO_2/Ti_3C_2-MXene. (Reproduced with permission from Re. [36], Copyright @ 2019 Elsevier.)

electrochemical impendence spectroscopy (EIS) curve in Figure 17.7d. The presence of Ti_3C_2 on CeO_2 significantly improved the degradation due to an improved Schottky junction between CeO_2 and Ti_3C_2 in comparison to bare CeO_2.

17.3.2 Peroxide-Assisted Degradation with 2D Ti_3C_2 MXene Composites

Hydrogen peroxide (H_2O_2) is an environmentally friendly multifunctional oxidant used in various applications including environmental remediation. However, H_2O_2 production procedure is complicated, expensive, and releases toxic by-products, which then restricts its application. Alternative safe ways of H_2O_2 production that are simple and require low energy consumption are desirable. The activation of H_2O_2 using solid catalysts is also preferred for organic pollutant degradation. Various valence states of transition metals in MXenes enable its fabrication to make capable catalysts. For high reactivity toward H_2O_2, iron and copper are commonly used in heterogeneous catalysis [37]. However, because iron suffers from a rapid dissipation and a narrow pH, copper is frequently used. In addition, there is a higher rate constant for cupric ions on H_2O_2 over a wide pH range [37]. Therefore, using cupric ions is advantageous in improving the catalytic performance of MXenes.

A solvothermal method was used to make cobalt (II,III) oxide (Co_3O_4)-modified MXene nanocomposite [38]. The proper distribution of Co_3O_4 on the interior and exterior wall of MXene inhibited the restacking of Ti_3C_2 layers. As a result, excellent performance was obtained for the degradation of methylene blue and Rhodamine B, and the process was repeatable. The degradation process was immensely influenced by superoxide and hydroxyl radicals.

17.3.2.1 H_2O_2-Fenton Degradation

There are various AOPs used in wastewater treatment works due to their reactive oxygen species and good performance in the degradation of contaminants. One of the recommended AOPs includes the Fenton process as a potential technology for wastewater treatment. This process entails degradation of organic contaminants via the activation of H_2O_2 with ferrous ions (Fe^{2+}) to generate ·OH, as indicated in the following equation [39]:

$$H_2O_2 + Fe^{2+} \rightarrow \cdot OH + OH^- \quad (17.1)$$

However, this process suffers from excessive use of ferrous salt, and for the reaction to proceed, it must be carried out in an acidic medium [40]. It is therefore vital to create an appropriate method to encourage the redox cycling of F^{3+}/Fe^{2+}. Heterogeneous catalysis came in hand for the improvement of Fenton reaction, resolving the acidic conditions constraints in the traditional Fenton process [40]. To trigger Fe^{3+}/Fe^{2+} redox cycling, Ti_3C_2 MXene was used as a reducing agent where it assisted in inhibiting the hydrolysis of iron. It further decreased the effects of secondary pollution [41]. To determine the effects of Fenton reaction using MXene, the Fe^{3+} pathway was investigated in dark conditions [39]. This was done to determine the impact of MXenes in the Fenton reaction for degradation of bisphenol A as a contaminant. The study showed that there was no degradation effect for MXene/Fe^{3+} and MXene/H_2O_2, respectively, toward bisphenol A, but when MXene was combined with Fe^{3+} and H_2O_2, degradation efficacy significantly improved [39]. This was attributed to the initial reaction of MXene with Fe^{3+} to form Fe^{2+}, which was able to activate H_2O_2 and form ·OH, subsequently leading to the degradation of bisphenol A. The generation of Fe^{2+} was verified by XPS analysis in Figure 17.8a where different bands associated with Fe^{2+} generation were observed. Further, the complex mechanism was determined using ethylenediamine tetraacetic acid (EDTA) and oxalic acid (OA) as complexing agents, respectively, as indicated in Figure 17.8b. EDTA entirely prevented the generation of Fe^{2+} followed by OA, while Fe^{3+}/MXene complex upheld the significant production of Fe^{2+} ions. It was however highlighted that the 35% degradation of bisphenol A obtained in dark conditions improved six times when exposed to visible light.

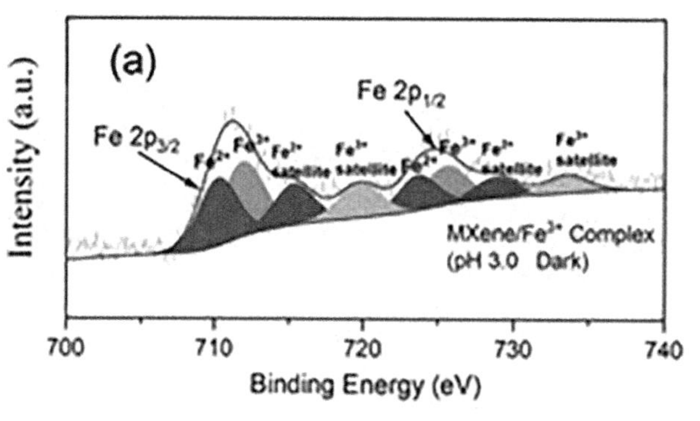

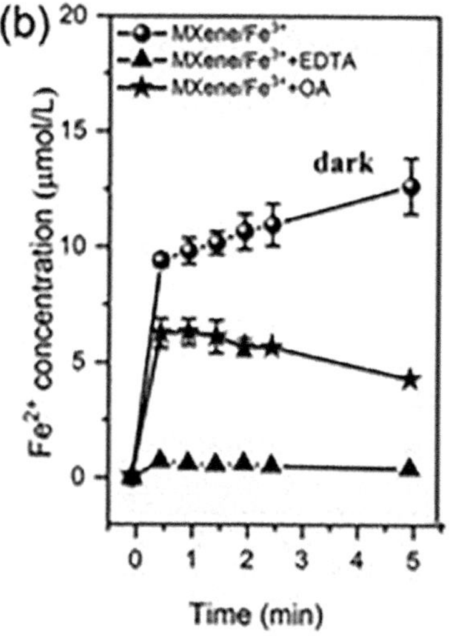

FIGURE 17.8 (a) Fe 2p XPS spectra of the MXene/Fe^{3+} complex under dark conditions and (b) generation of Fe^{2+} ions in the MXene/Fe^{3+} system in the existence or absence of different chelating agents. (Reproduced with permission from Ref. [39], Copyright @ 2023 Elsevier.)

17.3.2.2 Photocatalytic Generation of H_2O_2 Combined with Fenton Reaction

The synergistic effect of photocatalysis and Fenton oxidation has been considered as an effective way to improve organic contaminants degradation by triggering the redox cycling of Fe^{3+}/Fe^{2+}. The performance of Ti_3C_2 MXene modified with Fe^{3+}/H_2O_2 system was investigated for degradation of bisphenol A under visible light [39]. This process was advantageous in a way that the amount of the catalyst used was significantly reduced for application when compared to just using the Fenton process alone. The analyses of Fe^{3+}/Ti_3C_2 complex implied that MXene interacts with Fe^{3+} by breaking the Ti-C bond to speed up the Fe^{3+}/Fe^{2+} cycle. This activity is enabled by photogenerated electrons under visible light. The reaction mechanism for bisphenol A degradation by implementing photocatalysis together with Fenton technology was shown in Figure 17.9. The schematic diagram (Figure 17.9) shows how MXene/Fe^{3+} complex behaves under visible light. MXene oxidizes and Fe^{3+} gets reduced to Fe^{2+}, which is responsible for the degradation of bisphenol A via a Fenton reaction with H_2O_2. Notably, the complex demonstrated lower redox potential and charge transfer resistance, resulting in significant reduction of Fe^{3+}. Further, there was a rapid transfer of photogenerated electrons on MXene surface; as a result, holes recombination was inhibited [39].

17.3.3 Persulfate-Activated Degradation with 2D Ti_3C_2 Composites

Persulfate activation in heterogeneous catalysis normally consists of peroxydisulfate or peroxymonosulfate with other metals to produce sulfate radical, hydroxyl, or the respective active species

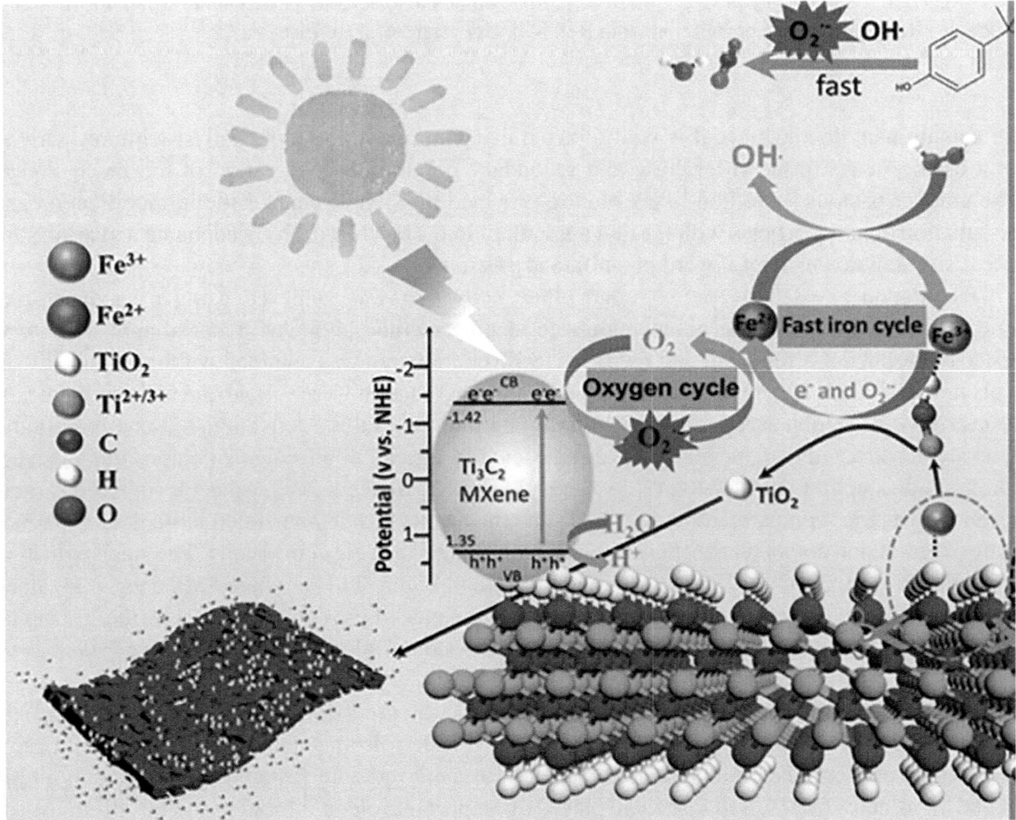

FIGURE 17.9 Schematic illustration of the proposed photodegradation mechanism of bisphenol A on MXene/Fe^{3+}/H_2O_2 system under visible light. (Reproduced with permission from Ref. [39], Copyright @ 2023 Elsevier.)

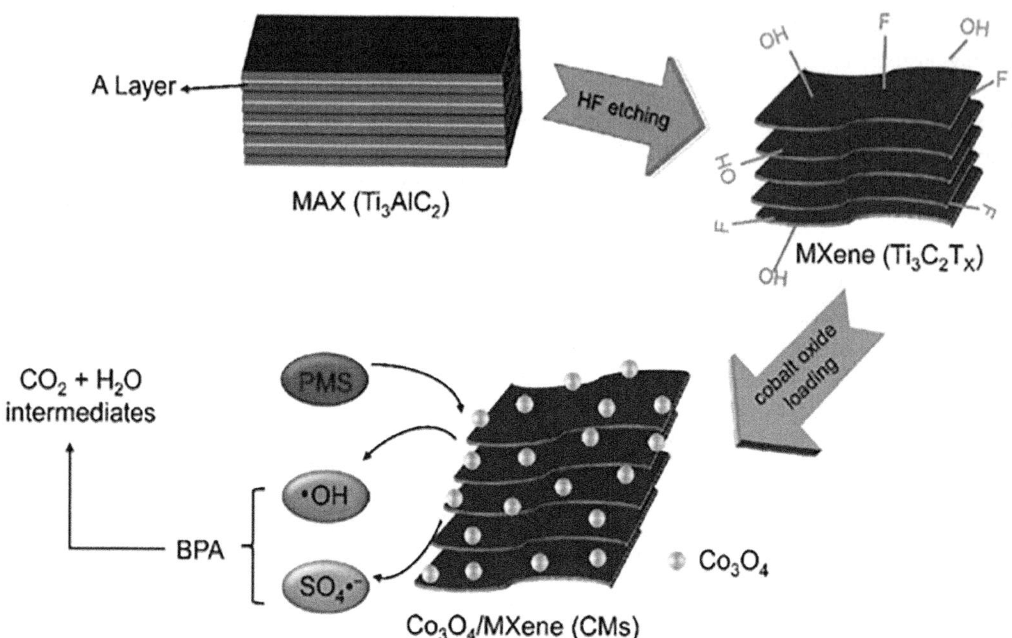

FIGURE 17.10 Sandwich-like Co_3O_4/MXene (CMs) composite to activate peroxymonosulfate for BPA degradation. (Reproduced with permission from Ref. [42], Copyright @ 2018 Elsevier.)

for contaminant degradation. It is vital to avoid using a large amount of catalyst with low activity as it usually leads to metal leaching and secondary pollution by generation of too much sludge. The unique spacious accordion-like structure provided by MXene enables the immobilization and distribution of nanoparticles with limited agglomeration. This then delivers adequate active sites for effective catalytic behavior toward persulfate activation.

Degradation assisted by peroxymonosulfate activation was achieved using a sandwich-like Co_3O_4/MXene catalyst made via a single-step heating method [42]. The material application was tested on bisphenol A removal, and optimally, 95% degradation was obtained within 7 minutes over a pH range of 4–10. The major oxidizing species for degradation were sulfate and hydroxyl radicals, respectively. To exhibit exceptional reactivity for peroxymonosulfate activation, MXene-based catalysts are constructed by controlling the dimensionality instead of altering its composition or structure, as indicated in Figure 17.10 [42]. As a result, $Ti_3C_2T_x$ that is in a nanosheet form has desirable reaction dynamics compared to bulk $Ti_3C_2T_x$ and the former has demonstrated more than 300 times better degradation competencies than the latter for various organic compounds. The steep reduction of bisphenol A by Co_3O_4/MXene composite is shown in Figure 17.11a, where PMS and Co_3O_4 alone showed limited to no activity. The outstanding performance of the composite was further shown by its recyclability study in Figure 17.11b. The bisphenol A removal rates remained above 90% for five cycling tests, confirming stability of the material.

Reduction of Ferric iron (Fe^{3+}) with Ti_3C_2 MXene was used to investigate peroxymonosulfate activation in the removal of sulfamethoxazole over a wide pH range [43]. Under optimized conditions, complete degradation of sulfamethoxazole was achieved, which was within 5 minutes (at pH 7) and 10 minutes (pH 9). Ti_3C_2 MXene played an important role in inhibiting the Fe^{3+} from hydrolyzing through its confinement impacts. Both sulfate and hydroxyl radicals were the influential reactive species in the overall catalytic system.

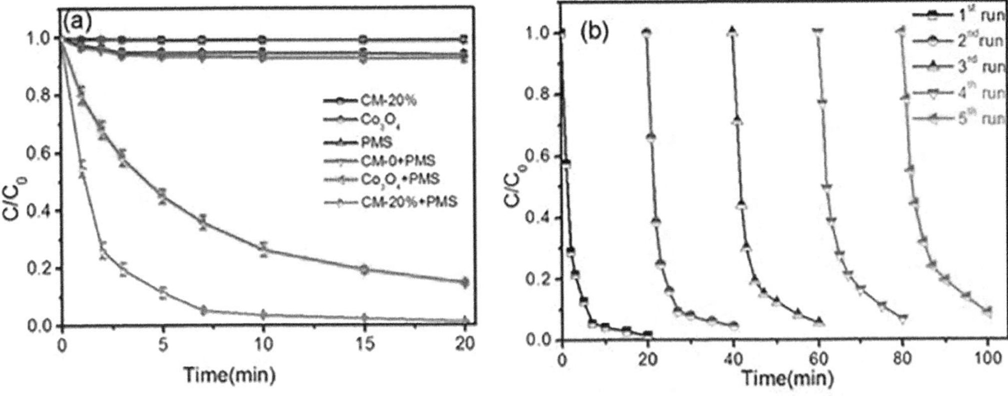

FIGURE 17.11 (a) Bisphenol A removal efficiency for different materials and (b) recyclability of Co_3O_4/MXene optimized composite. (Reproduced with permission from Ref. [42], Copyright @ 2018 Elsevier.)

17.4 2D MXENES-BASED ADSORBENTS FOR THE REMOVAL OF ORGANIC CONTAMINANTS

Among different types of wastewater treatment technologies, adsorption is regarded the most feasible due to fast equilibrium efficacy, ease in operation, high adsorption capacity, and low adsorbent cost. Nonetheless, the drawbacks of traditional adsorbents, including poor selectivity, sluggish rates of adsorption, and low-capacity removal, have forced science research toward the evolution and development nanomaterials that possess many potential applications in water remediation, including the adsorption technology [44]. The MXene structure consists of recommendable features that are fundamental in adsorption processes. These include the large surface area and abundant active sites that can capture different pollutants in aqueous systems. For a comprehensive adsorption behavior of 2D MXenes, different adsorption parameters are implemented to determine the removal mechanism for different organic pollutants.

17.4.1 Pure Ti_3C_2 Sorbent

In various organic contaminants present in water, organic dyes are mostly used in large quantity in their respective industries because they are mostly reported in water pollution. They are detected in high and even low concentrations and adversely affect the ecosystem as they are quite difficult to biodegrade [45]. So, depending on the organic pollutant form, whether cationic or anionic, MXene-based adsorbents are tuned to improve the removal capacity of the targeted compounds. Ti_3C_2 MXenes have a negatively charged surface and have shown to have better adsorption capacities with cationic dyes due to attractive interaction [46]. Even though the pure Ti_3C_2Tx MXene could remove cationic dyes, there was an outstanding performance when the material was treated with an alkaline solution. More improvements were obtained for the removal of cationic dyes (methyl blue and safranine) and neutral red dye when carboxyl functionalized MXenes were used instead of pure MXenes [47].

Methylene blue dye was effectively removed through adsorption in aqueous environments comprising of Ti_3C_2Tx MXene [48]. The material possessed remarkable adsorption toward cationic dyes, which was best characterized by a Freundlich isotherm, as shown in Figure 17.12.

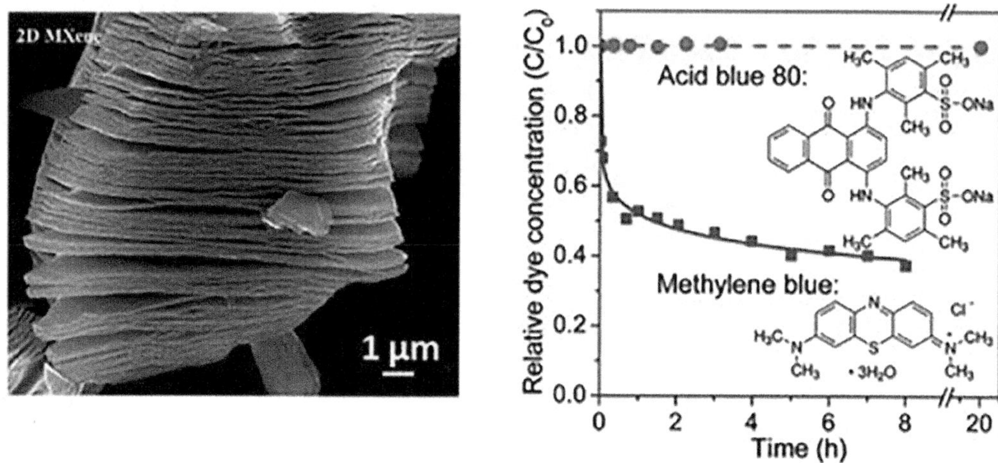

FIGURE 17.12 Illustration of the 2D MXene and Freundlich isotherm adsorption. (Reproduced with permission from Ref. [48], Copyright @ 2014 Royal Chemistry Society.)

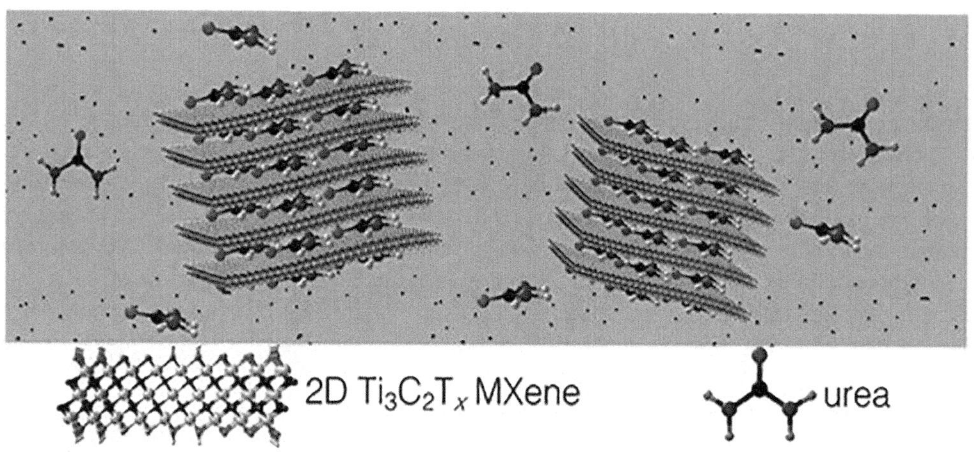

FIGURE 17.13 Adsorption mechanism of urea on 2D MXene adsorbent. (Reproduced with permission from Ref. [49], Copyright @ 2018 American Chemical Society.)

17.4.2 Ti_3C_2 Interfaced with Metal Oxide and Metal Sulfides as Sorbents

Upon modification, the MXene adsorbent exterior intensified, and active groups were readily available for contaminants uptake. With the growing concerns of eutrophication in wastewater works, the removal of urea in wastewater can be practiced using MXene-based adsorbents. Figure 17.13 shows how urea is selectively adsorbed on dialysate using different 2D MXenes-based materials ($Ti_3C_2T_x$, Ti_2CT_x, and $Mo_2TiC_2T_x$) [49]. From the adsorption mechanism, it was discovered that the removal was driven by hydrogen bonding formation between the MXene oxygen-containing interlayer groups and urea molecules.

Pharmaceutical pollutants are removed from wastewater mostly by the assistance biological and sedimentation processes. These, however, could lack some sensitivity particularly at lower contaminant concentrations as they are resistant in nature. The 2D MXene structure in adsorption technology could rather be efficient to solve those challenges since it has accordion-like assembly supported by hexagonally closed packed structure. To accelerate the spacing and surface termination between the layers, MXene was intercalated with sodium ions [50]. The produced fabricated MXene nanocomposite had improved removal of ciprofloxacin of 99.7%, which was two times higher than the unmodified MXene. Additionally, the adsorption kinetics of the analyte was significantly faster for the sodium intercalated MXene due to the accessibility of the active surface sites allowing faster diffusion. To enhance the adsorption of tetracycline in aqueous environment, MXene was alkaline intercalated and further incorporated with nickel metal [51]. The adsorption mechanism and significant improvement of tetracycline removal were presumed to be facilitated by the hydroxyl site and complexation with nickel metal.

The pollution of air by volatile organic compounds (VOCs) has been on the rise since industrialization. These VOCs are a threat to the human health and animals. Developing adsorbents to help in minimizing some of these VOCs is essential and important. Ti_3C_2 MXene nanoparticles were electrostatically adsorbed on Bi_2WO_6 nanoplates for efficient electron/hole separation [52]. However, for adsorption purposes after density functional theory (DFT) simulations, Ti_3C_2 showed string chemical adsorption for HCHO and CH_3COCH_3, while Bi_2WO_6 demonstrated weak physical adsorption. These strong adsorption "bonds" were attributed to the charge transfer between the Ti_3C_2 MXene and the VOC molecules, as described by differential charge density and Bader charge analysis (Huang, 2022).

Multilayered (ML) MXene-based nanocomposites that were composed of Ti_3C_2 MXene nanosheets and transition metal oxides nanomaterials including zinc oxide (ZnO), bismuth molybdate (Bi_2MoO_6), and tin dioxide (SnO_2) were successfully prepared to adsorb water-soluble organic compounds [53]. The ML metal oxide-modified Ti_3C_2 MXene were modified with oil to improve their hydrophilic nature of the composites for effective adsorption of methylene blue. The adsorption capacities of ZnO/ML-Ti_3C_2, SnO_2/ML-Ti_3C_2, and Bi_2MoO_6/ML-Ti_3C_2 toward the removal of methylene blue were 155, 163 and 152 mg g^{-1}, respectively. The Langmuir model was assumed for the adsorption that the ML-Ti_3C_2-based nanocomposites, insinuating that the monolayer covering of the nanocomposite is the main adsorption mechanism. In terms of data fitting of adsorption kinetics, the pseudo-first-order and second-order models were assumed. The pseudo-second-order model accurately simulates the adsorption process of methylene blue over the composites of more than 150 mg L^{-1} (R^2>0.95). In the initial 40 minutes of adsorption, a rapid adsorption stage by each material is observed (Figure 17.14), where MXene showed the highest adsorption capacity, followed by Bi_2MoO_6/ML-Ti_3C_2, then SnO_2/ ML-Ti_3C_2, and lastly ZnO/ML-Ti_3C_2.

The higher adsorption capacity of MXene was attributed to the abundant peripheral and low hanging chemical bonds, which may make the composite material to have physical and chemical adsorption capabilities, thus ensuring increased removal of pollutants. The synthesis and characterization of ternary Ti_3C_2-OH/In_2S_3/CdS were documented for applications in the visible-light-driven photocatalytic degradation and adsorption of Rhodamine B [54]. During the adsorption

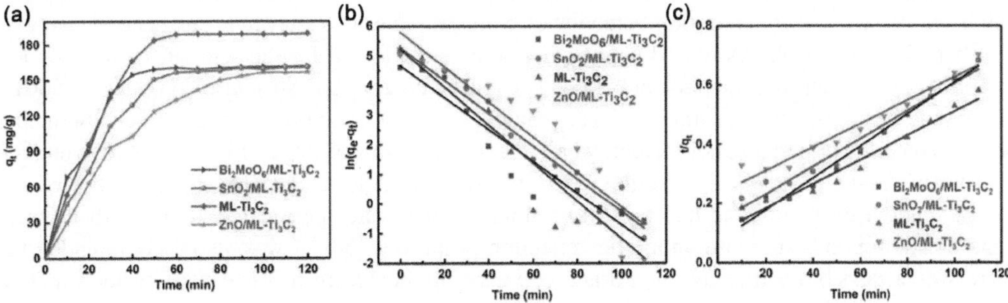

FIGURE 17.14 (a) Adsorption kinetics curves, (b) pseudo-first-order models, and (c) pseudo-second-order models of methylene blue. (Reproduced with permission from Ref. [53], Copyright @ 2022 Elsevier.)

experiments, Ti_3C_2-OH/In_2S_3/CdS showed remarkable adsorption properties for Rhodamine B molecules when compared to pure CdS and In_2S. This was attributed to the distinguishing later spacing of Ti_3C_2-OH/In_2S_3/CdS, which encourages the adsorption of organic pollutants. The adsorption efficiency for different derivatives of Ti_3C_2-OH/In_2S_3/CdS was ~81%.

17.4.3 Factors Influencing 2D MXenes-Based Nanocomposite Adsorption

The robustness of the adsorbent material is an important factor for regulating the adsorption behavior. In the MXene structure, hydroxylated carbonitrides with diverse carbon and nitrogen atom proportions are the outmost thermodynamically preferred and stable phase [55]. Adsorption parameters, including initial concentration of the analyte, the solution pH, contact time, temperature, and the ionic strength, are crucial factors that propel the adsorption of the MXene surface. When viewing on the solution pH, electrostatic interaction between the adsorbent and the adsorbate is crucial despite the nature of the contaminant. In the case of MXene-based adsorbents, there is a strong negative charge present, and therefore, adsorption is mainly preferred in alkaline conditions for cationic pollutants removal [56]. The increase adsorption efficiency at higher pH is attributed to pH point of zero charge (PZC) of $Ti_3C_2T_x$ and cationic pollutant (like methylene blue) solution pH allowing the electrostatic attraction. At lower solution pH, the hydrogen cation (H^+) and cationic methylene blue pollutant compete, resulting in minimal adsorption on the MXene adsorbent surface.

The adsorption behavior of MXene has also been investigated by varying the effect of temperature during the adsorption process. Generally, the viscosity of water is broken down by heat allowing greater diffusion and better potential for adsorption on a material surface. Changes in Gibbs free energy and entropy variations predict whether adsorption process will occur spontaneously while variations in enthalpy determine whether the reaction is endothermic or exothermic. For the adsorption of methylene blue using $Ti_3C_2T_x$ investigation, adsorption capacity increased with the increase in temperature; this was supported by the obtained negative Gibbs free energy as well as positive entropy and enthalpy changes [56].

17.5 MXENE-BASED MEMBRANES

Membrane technology is well investigated for water purification applications, particularly emerging organic pollutants. What makes it advantageous is its manufacturing consistency and provision of low carbon print. Membranes fabrication using MXenes involves different approaches, as indicated in Figure 17.3 [57]. For production of a lamellar structure (Figure 17.15a), MXenes are used as a matrix; for (b), they are combined with other nanomaterials to produce a hybrid membrane, whereas in (c), they are used as coating agents to modify the membrane surface. The manufacturing of these membranes is commonly produced via vacuum-assisted filtration, hot press, and drop casting developments [57].

On MXene assembled membranes, the regular interlayer space on nanosheets is seemingly contribute to achieve great separation behavior of contaminants in wastewater purification. An even slit-shaped designed MXene-based membranes were investigated for the separation of antibiotics in polluted water and organics solvents [58]. The stacked membrane structure consisted of 500 nm membrane thickness and 1.35 nm interlayer spacing. The best antibiotic separation was obtained in aqueous medium and in ethanol solvent, where the mechanism was driven by size-selective molecular filtering and Coulombic interaction. The comparison made in this investigation showed that the MXene-based membrane had up to two times higher water permeance than another typical nanofiltration membrane with antibiotics rejection properties [58]. It was also recommended that there was a possibility that the membrane could be applicable in medical wastewater water for drug separation. The incorporation of TiO_2 nanoparticles onto the MXene structure extends water transportation passages that are limited in pristine membranes, thereby cultivating antifouling and permeance behavior [57]. Additionally, the MXene distinct features including hydrophilicity and

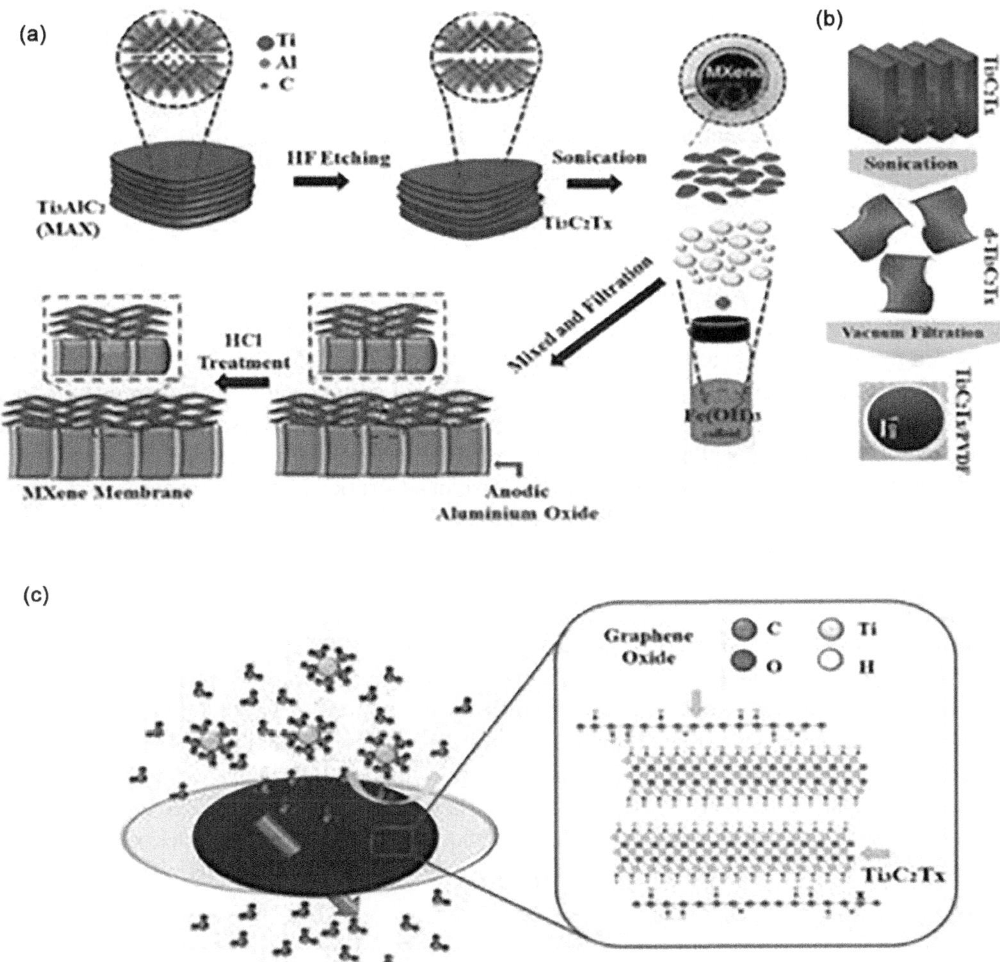

FIGURE 17.15 Manufacturing diagram of MXene supported membranes on (a) anodic aluminum oxide, (b) polyvinylidene fluoride, and (c) polycarbonate. (Reproduced with permission from Ref. [57], Copyright @ 2022 Elsevier.)

integral negative charge enable the combination with other functionalities to enhance membrane fouling resistance and attraction and repulsion interactions. A 2D MXene membrane was designed and tested using a forward osmosis system [59]. This was a size- and charge-selective rejection-controlled process. The investigation was performed on alkali, alkaline, and methylthioninium (MB^+) dye cation. An extremely low permeation rate of 10^{-4}mol h.m^2 was obtained by the thick membrane for cations with large radius, including cation dye. For dead-end nanofiltration process, Ti_3C_2Tx membranes exhibited a rejection rate of 92.3% for Congo red dye with a water flux of 115 l m^{-2}hour^{-1} [60].

Membrane thickness significantly affects the pollutants removal and flux performance [61]. A 2D MXene ($Ti_3C_2T_x$) membrane modified with Ag nanoparticles was investigated for the removal of Rhodamine B, methylene green, and bovine serum albumin, with removal rates of 79.9%, 92.3%, and>99%, respectively. For all these removals, the minimal membrane thickness was optimized. When increasing the thickness, water flux reduced drastically. The silver-modified membrane demonstrated flux recover of 91%–97%, whereas pristine MXene only showed 81%–86%. The presence of silver nanoparticles enhanced the composite hydrophilicity and supplementary nanopores [61].

17.6 REGENERATION OF MXENE-BASED NANOCOMPOSITES FOR REUSABILITY

For practical application of MXene-based nanomaterials on a larger scale, it is essential to investigate the reusability of the material to minimize cost and to limit the production of secondary waste from the used catalyst/adsorbent. Numerous studies have indicated that MXene nanocomposites are stable and can be reused for multiple cycles without losing integrity.

2D MXene functionalized with magnetic iron oxide nanoparticles via in situ growth methods showed intriguing properties for methylene blue adsorption [62]. The magnetic composite demonstrated stability and reusability for up to five cycles, with an initial removal efficiency of around 85% and a removal efficiency of 77% on the fifth cycle. The MXene iron oxide nanocomposite was recovered by collecting it with an external magnet, then systematically washing with ethanol and drying it at 70°C. Carboxyl-modified MXene incorporated with polyethylene polyimide and poly (acrylic acid) formed in a core shape showed effective removal of three dyes and could be reused for eight cycles [47]. The removal rate was maintained at 64.4% after eight cycles compared to 85.6%, which was the first adsorption removal. The decrease in adsorption after several cycles was associated with minor loss of exterior by-products and recurring rinsing. The commonly used agents for regeneration of MXene-based nanomaterials for adsorption process include NaOH, HNO_3, HCl, Na_2EDTA, $SC(NH_2)_2$, and $Ca(NO_3)_2$ [50]. Oxidation of TiO_2 on MXene structure is still a challenge in its regeneration process as it tends to affect its stability by reducing active adsorption sites.

To further improve the stability of MXene, polypyrrole was introduced to assist the exfoliation of MXene via in situ polymerization [63]. The produced composite was used for selective removal of methylene blue in aqueous solution. The stability of the composite was maintained as there was nearly no oxidation achieved. For catalytic performance, catalytic stability and reusability were evaluated on a copper oxide modified MXene with peroxymonosulfate activation [64]. Degradation of carbamazepine was over 91.1% for five catalytic runs. While still on peroxymonosulfate activation, cobalt oxide nanoparticle MXene composite showed stability over 30 cycles for 1,4-dioxane degradation [65]. However, there are still concerns on metal leaching for long-term recyclability and pertinence of MXene-based catalysts.

17.7 CONCLUSION

This book chapter attempted to encapsulate the remarked and exceptional 2D MXene nanomaterials. The curiosity for further research or carbon material after the discovery of single/multilayered graphene afforded many products including 2D)materials. The astonishing properties that afford them applications in different fields including the removal of organic pollutants were discussed. Their distinctive mechanical, chemical, and physical properties allow MXenes to be altered by various surface terminations and transition metals, which were also included in the chapter. The synthesis, modification, and characterization of MXenes are discussed. The application of MXenes and their modified derivates in water remediation as photocatalysts as well as adsorbents was discussed.

REFERENCES

1. Dadashi Firouzjaei, Mostafa, Mohammadsepehr Karimiziarani, Hamid Moradkhani, Mark Elliott, and Babak Anasori. MXenes: The two-dimensional influencers. *Materials Today Advances* 13 (2022) 10–13. https://doi.org/10.1016/j.mtadv.2021.100202.
2. Xu, Hao, Aobo Ren, Jiang Wu, and Zhiming Wang. Recent advances in 2D MXenes for photodetection. *Advanced Functional Materials* 30 (2020) 1–16. https://doi.org/10.1002/adfm.202000907.
3. Wyatt, Brian C., Andreas Rosenkranz, and Babak Anasori. 2D MXenes: Tunable mechanical and tribological properties. *Advanced Materials* 33 (2021) 1–15. https://doi.org/10.1002/adma.202007973.
4. Khalid, Zaied Bin, Mohd Nasrullah, Abdullah Nayeem, Zularisam Abd Wahid, Lakhveer Singh, and Santhana Krishnan. Application of 2D graphene-based nanomaterials for pollutant removal from advanced water and wastewater treatment processes. *ACS Symposium Series* 1353 (2020) 191–217. https://doi.org/10.1021/bk-2020-1353.ch009.

5. Naguib, Michael, Olha Mashtalir, Joshua Carle, Volker Presser, Jun Lu, Lars Hultman, Yury Gogotsi, and Michel W. Barsoum. Two-dimensional transition metal carbides. *ACS Nano* 6 (2012) 1322–31. https://doi.org/10.1021/nn204153h.
6. Geng, Dechao, Xiaoxu Zhao, Linjun Li, Peng Song, Bingbing Tian, Wei Liu, Jianyi Chen, Controlled growth of ultrathin Mo_2C superconducting crystals on liquid Cu surface. *2D Materials* 4 (2017). https://doi.org/10.1088/2053-1583/aa51b7.
7. Jiang, Xiantao, Artem V. Kuklin, Alexander Baev, Yanqi Ge, Hans Ågren, Han Zhang, and Paras N. Prasad. Two-dimensional MXenes: From morphological to optical, electric, and magnetic properties and applications. *Physics Reports* 848 (2020) 1–58. https://doi.org/10.1016/j.physrep.2019.12.006.
8. Halim, Joseph, Sankalp Kota, Maria R. Lukatskaya, Michael Naguib, Meng Qiang Zhao, Eun Ju Moon, Jeremy Pitock, et al. Synthesis and characterization of 2D molybdenum carbide (MXene). *Advanced Functional Materials* 26 (2016) 3118–27. https://doi.org/10.1002/adfm.201505328.
9. Yang, Peizhen, Songrong Li, Liang Xiaofu, An Xiaojing, Dongfang Liu, and Wenli Huang. Singlet oxygen-dominated activation of peroxymonosulfate by CuO/MXene nanocomposites for efficient decontamination of carbamazepine under high salinity conditions: Performance and singlet oxygen evolution mechanism. *Separation and Purification Technology* 285 (2022) 120288. https://doi.org/10.1016/j.seppur.2021.120288.
10. Jeon, Jaeho, Yereum Park, Seunghyuk Choi, Jinhee Lee, Sung Soo Lim, Byoung Hun Lee, Young Jae Song, Jeong Ho Cho, Yun Hee Jang, and Sungjoo Lee. Epitaxial synthesis of molybdenum carbide and formation of a Mo_2C/MoS_2 hybrid structure via chemical conversion of molybdenum disulfide. *ACS Nano* 12 (2018) 338–46. https://doi.org/10.1021/acsnano.7b06417.
11. Naguib, Michael, Murat Kurtoglu, Volker Presser, Jun Lu, Junjie Niu, Min Heon, Lars Hultman, Yury Gogotsi, and Michel W. Barsoum. Two-dimensional nanocrystals produced by exfoliation of Ti_3AlC_2. *Advanced Materials* 23 (2011) 4248–53. https://doi.org/10.1002/adma.201102306.
12. Hope, Michael A., Alexander C. Forse, Kent J. Griffith, Maria R. Lukatskaya, Michael Ghidiu, Yury Gogotsi, and Clare P. Grey. NMR reveals the surface functionalisation of Ti_3C_2 MXene. *Physical Chemistry Chemical Physics* 18 (2016) 5099–102. https://doi.org/10.1039/c6cp00330c.
13. Ronchi, Rodrigo Mantovani, Jeverson Teodoro Arantes, and Sydney Ferreira Santos. Synthesis, structure, properties and applications of MXenes: Current status and perspectives. *Ceramics International* 45 (2019) 18167–88. https://doi.org/10.1016/j.ceramint.2019.06.114.
14. Pang, Jinbo, Rafael G. Mendes, Alicja Bachmatiuk, Liang Zhao, Huy Q. Ta, Thomas Gemming, Hong Liu, Zhongfan Liu, and Mark H. Rummeli. Applications of 2D MXenes in energy conversion and storage systems. *Chemical Society Reviews* 48 (2019) 72–133. https://doi.org/10.1039/c8cs00324f.
15. Guo, Jianxin, Yong Sun, Baozhong Liu, Qingrui Zhang, and Qiuming Peng. Two-dimensional scandium-based carbides (MXene): Band gap modulation and optical properties. *Journal of Alloys and Compounds* 712 (2017) 752–59. https://doi.org/10.1016/j.jallcom.2017.04.149.
16. Geim, Andre K. and Grigorieva, Irina V. Van der waals heterostructures. *Nature* 499 (2013) 419–25. https://doi.org/10.1038/nature12385.
17. Ma, Zhinan, Zhenpeng Hu, Xudong Zhao, Qing Tang, Dihua Wu, Zhen Zhou, and Lixin Zhang. Tunable band structures of heterostructured bilayers with transition-metal dichalcogenide and mxene monolayer. *Journal of Physical Chemistry C* 118 (2014) 5593–99. https://doi.org/10.1021/jp500861n.
18. Li, Xinru, Ying Dai, Yandong Ma, Qunqun Liu, and Baibiao Huang. Intriguing electronic properties of two-dimensional MoS_2/TM_2CO_2 (TM=Ti, Zr, or Hf) hetero-bilayers: Type-II Semiconductors with tunable band gaps. *Nanotechnology* 26 (2015). https://doi.org/10.1088/0957-4484/26/13/135703.
19. Lee, Youngbin, Yubin Hwang, and Yong Chae Chung. Achieving type I, II, and III heterojunctions using functionalized Mxene. *ACS Applied Materials and Interfaces* 7 (2015) 7163–69. https://doi.org/10.1021/acsami.5b00063.
20. Huang, Kelei, Chunhu Li, Haozhi Li, Guangmin Ren, Liang Wang, Wentai Wang, and Xiangchao Meng. Photocatalytic applications of two-dimensional Ti_3C_2MXenes: A review. *ACS Applied Nano Materials* 3 (2020) 9581–603. https://doi.org/10.1021/acsanm.0c02481.
21. Nguyen, Thang Phan, Dinh Minh Tuan Nguyen, Dai Lam Tran, Hai Khoa Le, Dai Viet N. Vo, Su Shiung Lam, Rajender S. Varma, Mohammadreza Shokouhimehr, Chinh Chien Nguyen, and Quyet Van Le. MXenes: Applications in electrocatalytic, photocatalytic hydrogen evolution reaction and CO_2 reduction. *Molecular Catalysis* 486 (2020) 110850. https://doi.org/10.1016/j.mcat.2020.110850.
22. Li, Jinmao, Li Zhao, Shimin Wang, Jin Li, Guohong Wang, and Juan Wang. In situ fabrication of 2D/3D $g\text{-}C_3N_4/Ti_3C_2$ (MXene) heterojunction for efficient visible-light photocatalytic hydrogen evolution. *Applied Surface Science* 515 (2020) 145922. https://doi.org/10.1016/j.apsusc.2020.145922.
23. Ren, Yijie, Deqian Zeng, and Wee Jun Ong. Interfacial engineering of graphitic carbon nitride ($g\text{-}C_3N_4$)-based metal sulfide heterojunction photocatalysts for energy conversion: A review. *Cuihua Xuebao/Chinese Journal of Catalysis* 40 (2019) 289–319. https://doi.org/10.1016/s1872-2067(19)63293-6.

24. Chen, Wei, Yu Xiang Hua, Ying Wang, Ting Huang, Tian Yu Liu, and Xiao Heng Liu. Two-dimensional mesoporous g-C_3N_4 nanosheet-supported mgin2s4 nanoplates as visible-light-active heterostructures for enhanced photocatalytic activity. *Journal of Catalysis* 349 (2017) 8–18. https://doi.org/10.1016/j.jcat.2017.01.005.
25. Chu, Jiayu, Xijiang Han, Zhen Yu, Yunchen Du, Bo Song, and Ping Xu. Highly efficient visible-light-driven photocatalytic hydrogen production on CdS/Cu_7S_4/g-C_3N_4 ternary heterostructures. *ACS Applied Materials and Interfaces* 10 (2018) 20404–11. https://doi.org/10.1021/acsami.8b02984.
26. Guo, Feng, Weilong Shi, Huibo Wang, Mumei Han, Hao Li, Hui Huang, Yang Liu, and Zhenhui Kang. Facile fabrication of a CoO/g-C_3N_4 p-n heterojunction with enhanced photocatalytic activity and stability for tetracycline degradation under visible light. *Catalysis Science and Technology* 7 (2017) 3325–31. https://doi.org/10.1039/c7cy00960g.
27. He, Kelin, Jun Xie, Mingli Li, and Xin Li. In situ one-pot fabrication of g-C_3 N_4 nanosheets/NiS cocatalyst heterojunction with intimate interfaces for efficient visible light photocatalytic H_2 generation. *Applied Surface Science* 430 (2018) 208–17. https://doi.org/10.1016/j.apsusc.2017.08.191.
28. Qin, Hao, Rui Tang Guo, Xing Yu Liu, Wei Guo Pan, Zhong Yi Wang, Xu Shi, Jun Ying Tang, and Chun Ying Huang. Z-Scheme MoS_2/g-C_3N_4 heterojunction for efficient visible light photocatalytic CO_2 reduction. *Dalton Transactions* 47 (2018) 15155–63. https://doi.org/10.1039/c8dt02901f.
29. Liu, Yazi, Huayang Zhang, Jun Ke, Jinqiang Zhang, Wenjie Tian, Xinyuan Xu, Xiaoguang Duan, Hongqi Sun, Moses O Tade, and Shaobin Wang. 0D (MoS_2)/2D (g-C_3N_4) heterojunctions in Z-scheme for enhanced photocatalytic and electrochemical hydrogen evolution. *Applied Catalysis B: Environmental* 228 (2018) 64–74. https://doi.org/10.1016/j.apcatb.2018.01.067.
30. Zhang, Huoli, Man Li, Changxin Zhu, Qingjie Tang, Peng Kang, and Jianliang Cao. Preparation of magnetic α-$Fe_2O_3/ZnFe_2O_4@Ti_3C_2$ MXene with excellent photocatalytic performance. *Ceramics International* 46 (2020) 81–8. https://doi.org/10.1016/j.ceramint.2019.08.236.
31. Cao, Yang, Yu Fang, Xianyu Lei, Bihui Tan, Xia Hu, Baojun Liu, and Qianlin Chen. Fabrication of novel $CuFe_2O_4$/MXene hierarchical heterostructures for enhanced photocatalytic degradation of sulfonamides under visible light. *Journal of Hazardous Materials* 387 (2020) 122021. https://doi.org/10.1016/j.jhazmat.2020.122021.
32. Feng, Xiaofang, Zongxue Yu, Yuxi Sun, Runxuan Long, Mengyuan Shan, Xiuhui Li, Yuchuan Liu, and Jianghai Liu. Review MXenes as a new type of nanomaterial for environmental applications in the photocatalytic degradation of water pollutants. *Ceramics International* 47 (2021) 7321–43. https://doi.org/10.1016/j.ceramint.2020.11.151.
33. Peng, Jiahe, Xingzhu Chen, Wee Jun Ong, Xiujian Zhao, and Neng Li. Surface and heterointerface engineering of 2D MXenes and their nanocomposites: Insights into electro- and photocatalysis. *Chem* 5 (2019) 18–50. https://doi.org/10.1016/j.chempr.2018.08.037.
34. Tanweer, Mohd Saquib, and Masood Alam. Novel 2D nanomaterial composites photocatalysts: Application in degradation of water contaminants. In *2D Nanomaterials for Energy and Environmental Sustainability*. Materials Horizons: From Nature to Nanomaterials (2022) 75–96. https://doi.org/10.1007/978-981-16-8538-5_4.
35. Liu, Qiaoran, Xiaoyao Tan, Shaobin Wang, Fang Ma, Hussin Znad, Zhangfeng Shen, Lihong Liu, and Shaomin Liu. MXene as a non-metal charge mediator in 2D layered $CdS@Ti_3C_2@TiO_2$ composites with superior Z-scheme visible light-driven photocatalytic activity. *Environmental Science: Nano* 6 (2019) 3158–69. https://doi.org/10.1039/c9en00567f.
36. Shen, Jiyou, Jun Shen, Wenjing Zhang, Xiaohui Yu, Hua Tang, Mingyi Zhang, Zulfiqar, and Qinqin Liu. Built-in electric field induced CeO_2/Ti_3C_2-MXene schottky-junction for coupled photocatalytic tetracycline degradation and CO_2 reduction. *Ceramics International* 45 (2019) 24146–53. https://doi.org/10.1016/j.ceramint.2019.08.123.
37. Wang, Chongqing, Yijun Cao, and Hui Wang. Copper-based catalyst from waste printed circuit boards for effective fenton-like discoloration of rhodamine B at neutral PH. *Chemosphere* 230 (2019) 278–85. https://doi.org/10.1016/j.chemosphere.2019.05.068.
38. Luo, Shanshan, Ran Wang, Juanjuan Yin, Tifeng Jiao, Kaiyue Chen, Guodong Zou, Lun Zhang, Jingxin Zhou, Lexin Zhang, and Qiuming Peng. Preparation and dye degradation performances of self-assembled MXene-Co_3O_4 nanocomposites synthesized via solvothermal approach. *ACS Omega* 4 (2019) 3946–53. https://doi.org/10.1021/acsomega.9b00231.
39. Xu, Suzhou, Can Liu, Xuping Jiang, Xinyu Wang, Shijin Zhang, Yanting Zhang, Qingguo Wang, Weiling Xiong, and Jing Zhang.Ti_3C_2 MXene promoted Fe^{3+}/H_2O_2 fenton oxidation: Comparison of mechanisms under dark and visible light conditions. *Journal of Hazardous Materials* 444 (2023) 130450. https://doi.org/10.1016/j.jhazmat.2022.130450.

40. Wang, Zhongjuan, Ye Du, Peng Zhou, Zhaokun Xiong, Chuanshu He, Yang Liu, Heng Zhang, Gang Yao, and Bo Lai. Strategies based on electron donors to accelerate Fe(III)/Fe(II) cycle in fenton or fenton-like processes. *Chemical Engineering Journal* 454 (2023). https://doi.org/10.1016/j.cej.2022.140096.
41. Song, Haoran, Daoyuan Zu, Changping Li, Rui Zhou, Yuwei Wang, Wei Zhang, Shiting Pan, et al. Ultrafast activation of peroxymonosulfate by reduction of trace Fe^{3+} with Ti_3C_2 MXene under neutral and alkaline conditions: Reducibility and confinement effect. *Chemical Engineering Journal* 423 (2021) 130012. https://doi.org/10.1016/j.cej.2021.130012.
42. Liu, Yuxin, Rui Luo, Yang Li, Junwen Qi, Chaohai Wang, Jiansheng Li, Xiuyun Sun, and Lianjun Wang. Sandwich-like Co_3O_4/MXene composite with enhanced catalytic performance for bisphenol A degradation. *Chemical Engineering Journal* 347 (2018) 731–40. https://doi.org/10.1016/j.cej.2018.04.155.
43. Ding, Mingmei, Wei Chen, Hang Xu, Chunhui Lu, Tao Lin, Zhen Shen, Hui Tao, and Kai Zhang. Synergistic features of superoxide molecule anchoring and charge transfer on two-dimensional $Ti_3C_2T_x$ MXene for efficient peroxymonosulfate activation. *ACS Applied Materials and Interfaces* 12 (2020) 9209–18. https://doi.org/10.1021/acsami.9b20530.
44. Sun, Yubing, and Ying Li. Potential environmental applications of MXenes: A critical review. *Chemosphere* 271 (2021) 129578. https://doi.org/10.1016/j.chemosphere.2021.129578.
45. Jena, Rajesh K., Himadri Tanaya Das, Braja N. Patra, and Nigamananda Das. MXene-based nanomaterials as adsorbents for wastewater treatment: A review on recent trends. *Frontiers of Materials Science* 16 (2022) 1–16. https://doi.org/10.1007/s11706-022-0592-x.
46. Chen, Junyu, Qiang Huang, Hongye Huang, Liucheng Mao, Meiying Liu, Xiaoyong Zhang, and Yen Wei. Recent progress and advances in the environmental applications of MXene related materials. *Nanoscale* 12 (2020) 3574–92. https://doi.org/10.1039/c9nr08542d.
47. Li, Kaikai, Guodong Zou, Tifeng Jiao, Ruirui Xing, Lexin Zhang, Jingxin Zhou, Qingrui Zhang, and Qiuming Peng. Self-assembled MXene-based nanocomposites via layer-by-layer strategy for elevated adsorption capacities. *Colloids and Surfaces A: Physicochemical and Engineering Aspects* 553 (2018) 105–13. https://doi.org/10.1016/j.colsurfa.2018.05.044.
48. Mashtalir, Olha, Cook, Kevin, Mochalin, Vadym, Crowe, Martin, Barsoum, Michel and Gogotsi Yury Dye adsorption and decomposition on two-dimensional titanium carbide in aqueous media. *Journal of Materials Chemistry A* 2 (2014) 14334–38. https://doi.org/10.1039/c4ta02638a.
49. Meng, Fayan, Mykola Seredych, Chi Chen, Victor Gura, Sergey Mikhalovsky, Susan Sandeman, Ganesh Ingavle, et al. MXene sorbents for removal of urea from dialysate: A step toward the wearable artificial kidney. *ACS Nano* 12 (2018) 10518–28. https://doi.org/10.1021/acsnano.8b06494.
50. Ahmaruzzaman, Md. MXene-based novel nanomaterials for remediation of aqueous environmental pollutants. *Inorganic Chemistry Communications* 143 (2022) 109705. https://doi.org/10.1016/j.inoche.2022.109705.
51. Dao, Xuan, Hongxun Hao, Jingtao Bi, Shiyu Sun, and Xin Huang. Surface complexation enhanced adsorption of tetracycline by ALK-MXene. *Industrial and Engineering Chemistry Research* 61 (2022) 6028–36. https://doi.org/10.1021/acs.iecr.2c00037.
52. Huang, Guimei, Shuangzhi Li, Lijun Liu, Leifan Zhu, and Qiang Wang. Ti_3C_2 MXene-modified Bi_2WO_6 nanoplates for efficient photodegradation of volatile organic compounds. *Applied Surface Science* 503 (2020) 1–8. https://doi.org/10.1016/j.apsusc.2019.144183.
53. Zhang, Li, Pingping Ma, Li Dai, Zhen Bu, Xueying Li, Wei Yu, Yiran Cao, and Jie Guan. Removal of pollutants via synergy of adsorption and photocatalysis over MXene-based nanocomposites. *Chemical Engineering Journal Advances* 10 (2022) 100285. https://doi.org/10.1016/j.ceja.2022.100285.
54. Fang, Hongjun, Yusong Pan, Minyun Yin, Linfeng Xu, Yuan Zhu, and Chengling Pan. Facile synthesis of ternary Ti_3C_2–OH/Ln_2S_3/CdS Composite with efficient adsorption and photocatalytic performance towards organic dyes. *Journal of Solid State Chemistry* 280 (2019) 120981. https://doi.org/10.1016/j.jssc.2019.120981.
55. Enyashin, Andrey and Ivanovskii, Alexander. Two-dimensional titanium carbonitrides and their hydroxylated derivatives: Structural, electronic properties and stability of MXenes Ti_3C_2-$XN_x(OH)_2$ from DFTB calculations. *Journal of Solid State Chemistry* 207 (2013) 42–48. https://doi.org/10.1016/j.jssc.2013.09.010.
56. Jun, Byung Moon, Jiyong Heo, Nader Taheri-Qazvini, Chang Min Park, and Yeomin Yoon. Adsorption of selected dyes on $Ti_3C_2T_x$ MXene and Al-based metal-organic framework. *Ceramics International* 46 (2020) 2960–68. https://doi.org/10.1016/j.ceramint.2019.09.293.
57. Lim, Gim Pao, Chin Fhong Soon, A. A. Al-Gheethi, Marlia Morsin, and Kian Sek Tee. Recent progress and new perspective of MXene-based membranes for water purification: A review. *Ceramics International* 48 (2022) 16477–91. https://doi.org/10.1016/j.ceramint.2022.03.165.

58. Li, Zhong Kun, Yanying Wei, Xue Gao, Li Ding, Zong Lu, Junjie Deng, Xianfeng Yang, Jürgen Caro, and Haihui Wang. Antibiotics separation with MXene membranes based on regularly stacked high-aspect-ratio nanosheets. *Angewandte Chemie - International Edition* 59 (2020) 9751–56. https://doi.org/10.1002/anie.202002935.
59. Ren, Chang E., Kelsey B. Hatzell, Mohamed Alhabeb, Zheng Ling, Khaled A. Mahmoud, and Yury Gogotsi. Charge- and size-selective ion sieving through $Ti_3C_2T_x$ MXene membranes. *Journal of Physical Chemistry Letters* 6 (2015) 4026–31. https://doi.org/10.1021/acs.jpclett.5b01895.
60. Han, Runlin, Xufeng Ma, Yongli Xie, Da Teng, and Shouhai Zhang. Preparation of a new 2D MXene/PES composite membrane with excellent hydrophilicity and high flux. *RSC Advances* 7 (2017) 56204–10. https://doi.org/10.1039/c7ra10318b.
61. Pandey, Ravi P., Kashif Rasool, Vinod E. Madhavan, Brahim Aïssa, Yury Gogotsi, and Khaled A. Mahmoud. Ultrahigh-flux and fouling-resistant membranes based on layered silver/MXene ($Ti_3C_2T_X$) nanosheets. *Journal of Materials Chemistry A* 6 (2018) 3522–33. https://doi.org/10.1039/c7ta10888e.
62. Zhang, Ping, Mingxue Xiang, Huiling Liu, Chenkai Yang, and Shuguang Deng. Novel two-dimensional magnetic titanium carbide for methylene blue removal over a wide PH Range: Insight into removal performance and mechanism. *ACS Applied Materials and Interfaces* 11 (2019) 24027–36. https://doi.org/10.1021/acsami.9b04222.
63. Shi, Xian ying, Meng hang Gao, Wen wen Hu, Dan Luo, Shao zhong Hu, Ting Huang, Nan Zhang, and Yong Wang. Largely enhanced adsorption performance and stability of MXene through in-situ depositing polypyrrole nanoparticles. *Separation and Purification Technology* 287 (2022) 120596. https://doi.org/10.1016/j.seppur.2022.120596.
64. Yang, Peizhen, Songrong Li, Liang Xiaofu, An Xiaojing, Dongfang Liu, and Wenli Huang. Singlet oxygen-dominated activation of peroxymonosulfate by CuO/MXene nanocomposites for efficient decontamination of carbamazepine under high salinity conditions: Performance and singlet oxygen evolution mechanism. *Separation and Purification Technology* 285 (2022) 120288. https://doi.org/10.1016/j.seppur.2021.120288.
65. Li, Wei, Kuanchang He, Longxiang Tang, Qian Liu, Kui Yang, Yi Di Chen, Xin Zhao, Kai Wang, Hui Lin, and Sihao Lv. Peroxymonosulfate activation by oxygen vacancies-enriched MXene nano-Co_3O_4 Co-catalyst for efficient degradation of refractory organic matter: Efficiency, mechanism, and stability. *Journal of Hazardous Materials* 432 (2022) 128719. https://doi.org/10.1016/j.jhazmat.2022.128719.

Index

For Product Safety Concerns and Information please contact our EU representative GPSR@taylorandfrancis.com Taylor & Francis Verlag GmbH, Kaufingerstraße 24, 80331 München, Germany

Batch number: 10392095

Printed by Printforce, the Netherlands